Advances in Intelligent Systems and Computing

Volume 917

Series editor

Janusz Kacprzyk, Systems Research Institute, Polish Academy of Sciences, Warsaw, Poland
e-mail: kacprzyk@ibspan.waw.pl

The series "Advances in Intelligent Systems and Computing" contains publications on theory, applications, and design methods of Intelligent Systems and Intelligent Computing. Virtually all disciplines such as engineering, natural sciences, computer and information science, ICT, economics, business, e-commerce, environment, healthcare, life science are covered. The list of topics spans all the areas of modern intelligent systems and computing such as: computational intelligence, soft computing including neural networks, fuzzy systems, evolutionary computing and the fusion of these paradigms, social intelligence, ambient intelligence, computational neuroscience, artificial life, virtual worlds and society, cognitive science and systems, Perception and Vision, DNA and immune based systems, self-organizing and adaptive systems, e-Learning and teaching, human-centered and human-centric computing, recommender systems, intelligent control, robotics and mechatronics including human-machine teaming, knowledge-based paradigms, learning paradigms, machine ethics, intelligent data analysis, knowledge management, intelligent agents, intelligent decision making and support, intelligent network security, trust management, interactive entertainment, Web intelligence and multimedia.

The publications within "Advances in Intelligent Systems and Computing" are primarily proceedings of important conferences, symposia and congresses. They cover significant recent developments in the field, both of a foundational and applicable character. An important characteristic feature of the series is the short publication time and world-wide distribution. This permits a rapid and broad dissemination of research results.

More information about this series at http://www.springer.com/series/11156

Michael E. Auer · Thrasyvoulos Tsiatsos
Editors

The Challenges of the Digital Transformation in Education

Proceedings of the 21st International Conference on Interactive Collaborative Learning (ICL2018) - Volume 2

 Springer

Editors
Michael E. Auer
Carinthia University of Applied Sciences
Villach, Kärnten, Austria

Thrasyvoulos Tsiatsos
Department of Informatics
Aristotle University of Thessaloniki
Thessaloniki, Greece

ISSN 2194-5357   ISSN 2194-5365   (electronic)
Advances in Intelligent Systems and Computing
ISBN 978-3-030-11934-8   ISBN 978-3-030-11935-5   (eBook)
https://doi.org/10.1007/978-3-030-11935-5

Library of Congress Control Number: 2018968529

This Springer imprint is published by the registered company Springer Nature Switzerland AG
The registered company address is: Gewerbestrasse 11, 6330 Cham, Switzerland

Preface

ICL2018 was the 21st edition of the International Conference on Interactive Collaborative Learning and the 47th edition of the IGIP International Conference on Engineering Pedagogy.

This interdisciplinary conference aims to focus on the exchange of relevant trends and research results as well as the presentation of practical experiences in Interactive Collaborative Learning and Engineering Pedagogy.

ICL2018 has been organized by Aristotle University of Thessaloniki, Greece, from 25 September to 28 September 2018 in Kos Island.

This year's theme of the conference was "Teaching and Learning in a Digital World".

Again outstanding scientists from around the world accepted the invitation for keynote speeches:

- Stephanie Farrell, Professor and Founding Chair of Experiential Engineering Education at Rowan University (USA)—2018–19 President of the American Society for Engineering Education. Speech title: Strategies for Building Inclusive Classrooms in Engineering.
- Demetrios Sampson. Ph.D. (ElectEng) (Essex), PgDip (Essex), B.Eng./M.Eng. (Elec) (DUTH), CEng—Golden Core Member, IEEE Computer Society— Professor, Digital Systems for Learning and Education, University of Piraeus, Greece. Speech title: Educational Data Analytics for Personalized Learning in Online Education.
- Rovani Sigamoney, UNESCO Engineering Programme. Speech title: UNESCO —Engineering the Sustainable Development Goals.

In addition, three invited speeches have been given by

- Hans J. Hoyer, IFEES, United States of America. Speech title: The work of IFEES and GEDC towards a new Quality in Engineering Education.
- David Guralnick, Kaleidoscope Learning, United States of America. Speech title: Creative Approaches to Online Learning Design.

Furthermore, five very interesting workshops and one tutorial have been organized:

- Tutorial titled "Improving Practical Communication Skills Through Participation in Collaborative English Workshops" by Edward Pearse Sarich (Shizuoka University of Art and Culture, Japan); Mark Daniel Sheehan (Hannan University, Japan), and Jack Ryan (Shizuoka University of Art and Culture, Japan).
- Workshop titled "Evaluation of Experimental Activities by Diana Urbano and Maria Teresa Restivo" (University of Porto, Portugal).
- Workshop titled "Machine Learning and Interactive Collaborative Learning" by Panayotis Tzinis and Irene Tsakiridou (Google Developer Experts).
- Workshop titled "Teaching and Learning Electrical Engineering and Computer Science in High School with a STEM Approach" by Arturo Javier Miguel-de-Priego (Academia de Ingeniería y Ciencia Escolar, Perú).
- Workshop titled "Introduction to BLE System Design Using PSoC® 6 MCUs" by Patrick Kane (Cypress).

Since its beginning, this conference is devoted to new approaches in learning with a focus to collaborative learning and engineering education. We are currently witnessing a significant transformation in the development of education. There are two essential and challenging elements of this transformation process that have to be tackled in education:

- the impact of globalization on all areas of human life and
- the exponential acceleration of the developments in technology as well as of the global markets and the necessity of flexibility and agility in education.

Therefore, the following main themes have been discussed in detail:

- Collaborative learning
- Lifelong learning
- Adaptive and intuitive environments
- Ubiquitous learning environments
- Semantic metadata for e-learning
- Mobile learning environments applications
- Computer-aided language learning (CALL)
- Platforms and authoring tools
- Educational Mashups
- Knowledge management and learning
- Educational Virtual Environments
- Standards and style guides
- Remote and virtual laboratories
- Evaluation and outcomes assessment
- New learning models and applications
- Research in Engineering Pedagogy
- Engineering Pedagogy Education
- Learning culture and diversity

- Ethics and Engineering Education
- Technical Teacher Training
- Academic–industry partnerships
- Impact of globalization
- K-12 and pre-college programmes
- Role of public policy in engineering education
- Women in engineering careers
- Flipped classrooms
- Project-based learning
- New trends in graduate education
- Cost-effectiveness
- Real-world experiences
- Pilot projects/Products/Applications.

The following special sessions have been organized:

- Entrepreneurship in Engineering Education (EiEE 2018)
- Digital Technology in Sports (DiTeS)
- Talking about Teaching 2018 (TaT'18)
- Multicultural Diversity in Education and Science
- Tangible and Intangible Cultural Heritage digitization and preservation in modern era (TICHE-DiPre)
- Advancements in Engineering Education and Technology Research (AEETR).

Also, the "1st ICL International Student Competition on Learning Technologies" has been organized in the context of ICL2018.

The following submission types have been accepted:

- Full paper, short paper
- Work in progress, poster
- Special sessions
- Round-table discussions, workshops, tutorials, doctoral consortium, students' competition.

All contributions were subject to a double-blind review. The review process was very competitive. We had to review near 526 submissions. A team of about 375 reviewers did this terrific job. Our special thanks go to all of them.

Due to the time and conference schedule restrictions, we could finally accept only the best 186 submissions for presentation.

Our conference had again more than 225 participants from 46 countries from all continents.

ICL2019 will be held in Bangkok, Thailand.

Michael E. Auer
ICL Chair
Thrasyvoulos Tsiatsos
ICL2018 Chair

Committees

General Chair

Michael E. Auer   CTI, Villach, Austria

ICL2018 Conference Chair

Thrasyvoulos Tsiatsos   Aristotle University of Thessaloniki, Greece

International Chairs

Samir A. El-Seoud	The British University in Egypt, Africa
Neelakshi Chandrasena Premawardhena	University of Kelaniya, Sri Lanka, Asia
Alexander Kist	University of Southern Queensland, Australia/Oceania
Arthur Edwards	Universidad de Colima, Mexico, Latin America
Alaa Ashmawy	American University Dubai, Middle East
David Guralnick	Kaleidoscope Learning New York, USA, North America

Programme Co-chairs

Michael E. Auer	CTI, Villach, Austria
David Guralnick	Kaleidoscope Learning New York, USA
Hanno Hortsch	TU Dresden, Germany

Technical Programme Chairs

Stavros Demetriadis   Aristotle University of Thessaloniki, Greece
Sebastian Schreiter   IAOE, France
Ioannis Stamelos   Aristotle University of Thessaloniki, Greece

IEEE Liaison

Russ Meier   Milwaukee School of Engineering, USA

Workshop and Tutorial Chair

Barbara Kerr   Ottawa University, Canada

Special Session Chair

Andreas Pester   Carinthia University of Applied Sciences, Austria

Demonstration and Poster Chair

Teresa Restivo   University of Porto, Portugal

Awards Chair

Andreas Pester   Carinthia University of Applied Sciences, Austria

Publication Chair

Sebastian Schreiter   IAOE, France

Senior PC Members

Andreas Pester   Carinthia University of Applied Sciences, Austria
Axel Zafoschnig   Ministry of Education, Austria
Doru Ursutiu   University of Brasov, Romania
Eleonore Lickl   College for Chemical Industry, Vienna, Austria
George Ioannidis   University of Patras, Greece
Samir Abou El-Seoud   The British University in Egypt
Tatiana Polyakova   Moscow State Technical University, Russia

Program Committee

Agnes Toth	Hungary
Alexander Soloviev	Russia
Anastasios Mikropoulos	Greece
Armin Weinberger	Germany
Athanassios Jimoyiannis	Greece
Charalambos Christou	Cyprus
Charalampos Karagiannidis	Greece
Chris Panagiotakopoulos	Greece
Christian Guetl	Austria
Christos Bouras	Greece
Christos Douligeris	Greece
Chronis Kynigos	Greece
Cornel Samoila	Romania
Costas Tsolakis	Greece
Demetrios Sampson	Greece
Despo Ktoridou	Cyprus
Dimitrios Kalles	Greece
Elli Doukanari	Cyprus
Hanno Hortsch	Germany
Hants Kipper	Estonia
Herwig Rehatschek	Austria
Igor Verner	Israel
Imre Rudas	Hungary
Ioannis Kompatsiaris	Greece
Istvan Simonics	Hungary
Ivana Simonova	Czech Republic
James Uhomoibhi	UK
Jürgen Mottok	Germany
Martin Bilek	Czech Republic
Matthias Utesch	Germany
Michalis Xenos	Greece
Monica Divitini	Norway
Nael Barakat	USA
Pavel Andres	Czech Republic
Rauno Pirinen	Finland
Roman Hrmo	Slovakia
Santi Caballé	Spain
Stavros Demetriadis	Greece
Teresa Restivo	Portugal
Tiia Rüütmann	Estonia
Vassilis Komis	Greece
Viacheslav Prikhodko	Russia

Victor K. Schutz   USA
Yiannis Dimitriadis   Spain
Yu-Mei Wang   USA

Local Organization Chair

Stella Douka   Aristotle University of Thessaloniki, Greece

Local Organization Committee Members

Hippokratis Apostolidis   Aristotle University of Thessaloniki, Greece
Agisilaos Chaldogeridis   Aristotle University of Thessaloniki, Greece
Olympia Lilou   Aristotle University of Thessaloniki, Greece
Andreas Loukovitis   Aristotle University of Thessaloniki, Greece
Angeliki Mavropoulou   Aristotle University of Thessaloniki, Greece
Nikolaos Politopoulos   Aristotle University of Thessaloniki, Greece
Panagiotis Stylianidis   Aristotle University of Thessaloniki, Greece
Christos Temertzoglou   Aristotle University of Thessaloniki, Greece
Efthymios Ziagkas   Aristotle University of Thessaloniki, Greece
Vasiliki Zilidou   Aristotle University of Thessaloniki, Greece

Contents

Remote and Virtual Laboratories

New Learning Models and Applications

Introducing Augmented Reality and Internet of Things at Austrian Secondary Colleges of Engineering

Andreas Probst[1]([⊠]), Martin Ebner[2], and Jordan Cox[3]

[1] HTL Ried, Ried im Innkreis, Austria
Andreas.Probst@eduhi.at
[2] Graz University of Technology, Graz, Austria
[3] PTC Inc, Needham, USA

Abstract. In Austria technical education is taught at federal secondary colleges of engineering (HTL) at a quite high level of ISCED 5. Despite mechanical engineering, design with industrial standard 3D programs being state of the art in industry and education, technologies using Internet of Things (IoT) and Augmented Reality (AR) are still at an early stage. This publication describes the introduction of the IoT platform Thingworx at Austrian HTL for mechanical engineering, with focus on AR and IoT from an educational perspective as well as the training aspects from the platform developers' perspective. In addition, an assessment using a system usability scale (SUS) test among students and teachers has been undertaken to obtain knowledge of how students and teachers see the usability of the IoT and AR platform Thingworx. The assessment results are presented in this paper.

Keywords: Augmented reality (AR) · Internet of things (IoT)
Engineering education

1 Introduction

Through the development of technology, the products of today which are connected to the internet are quite complex and require a constant exchange of operational data via the Internet. For instance, passenger cars are a good example of how rapidly mechatronic products have been integrated with computing power, something that was unthinkable not so long ago. The computing power and connectivity possibilities that such devices contain provide brand new avenues for product development. Particularly when the customers have received the products and are first using it, we can obtain data which allows us to improve the next product generation's technology. In addition, AR can be used in product development and to support technicians and companies' staff in several service cases.

Porter and Heppelmann [1] describe the five different stages of industry boundaries, where product and smart product are state of the art. Currently smart connected product is on the focus list of every technology leader and should be from the authors perspective on the list for short-term educational implementation too.

© Springer Nature Switzerland AG 2019
M. E. Auer and T. Tsiatsos (Eds.): ICL 2018, AISC 917, pp. 3–12, 2019.
https://doi.org/10.1007/978-3-030-11935-5_1

Furthermore, interesting research work about IoT and AR has been conducted. For example AR for usability testing was done by Choi and Mittal [2]. They did not only do simple augmentation, they also undertook a project and conducted a survey about AR in a very early stage of product development, and used different items like a play card or a 3d printed handheld to use the AR. Concerning IoT Abraham [3] did a project and survey in which they established IoT connections with the sensor data of the participating students' cellphones. He et al. [4] describe in their research work the process of introducing IoT into STEM undergraduate education, whereas Cin and Callaghan [5] investigated IoT technologies by using the college building as an IoT lab.

The basis for IoT technologies is the increasing number of connected objects via the internet [6] (see Fig. 1 left) as well as decreasing costs of sensors [7] (see Fig. 1 right) and the possibility to connect them via the Internet.

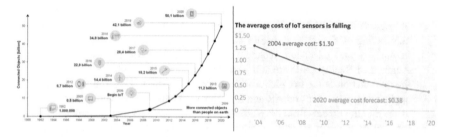

Fig. 1. Development of Internet of Things [6] and average costs of IoT sensors [7]

However, the disadvantage of IoT applications is the security aspect. Sensor and user data and their connection via cloud services to an IoT platform can be manipulated or be the target of cyber-attacks [8]. Bradley et al. [9, p. 67] describes some incidents of this.

2 Training IoT and AR Developers—A Company's Perspective

2.1 Overview of Current Challenges

The new technology waves of IoT and AR have introduced significant problems for companies in terms of training their own workforce but also in terms of recruiting new employees. It is difficult to find university graduates who are trained or even familiar with these technologies. Most universities are not teaching IoT and AR in the context of industrial usage because these technologies are so new and difficult to maintain.

Technology companies who create these IoT and AR solutions usually maintain "education" departments whose responsibility it is to work with universities to provide access to the company's technology solutions as well as some limited curriculum. These technology provider companies usually build networks of reseller partners who provide educational services to universities. A company like PTC works with hundreds of partners to reach thousands of universities throughout the world. IoT and AR offerings are however, sparse because of how complex these technologies are and how recently they have been on the market. Even if technology providers offer their solutions for free, universities are not able to utilize these services because there is not any curriculum or courses or because the faculty is in the process of learning it.

This complexity has prevented many universities from offering any courses on IoT and AR and especially in the context of industrial use cases. It is difficult to create and maintain an IoT server with the capability to connect to diverse data sources and sensors while providing easy access to students. AR also presents its own challenges since it is an integration of 3D CAD data, animation and IoT data. It takes time for faculty to learn the technologies and then prepare curriculum and digital platforms so that students have access.

2.2 New Changes in Education

Significant changes are also happening in the education world at the same time that IoT and AR are changing industry. Today one out of every four university students are actively engaged in online courses. Unlike the online education courses of the past 20 years, today's online courses are of much higher quality and are available to be consumed anywhere, at any time and at any speed. A group of online education digital platforms have been launched in the past 5 years to deliver these courses. Platforms like UDEMY, LYNDA, UDACITY, EDX, HBS, iMooX etc. provide students 24/7 access and provide mentoring services, communities, and credentials.

Technology companies are beginning to align with these approaches to education. Companies will either partner with these digital platform providers to build courses and credentialed learning paths or they will build their own platforms. An example of this is the ANDROID developer's nanodegree offered by Google and UDACITY where students can learn to be an android developer and receive a credential certifying their completion of the nanodegree.

Other companies like PTC and IBM have chosen to build their own digital learning platforms. PTC offers courses on becoming an IoT developer on their IOTU.com platform where students can learn about IoT and then build apps using PTC's IoT solution ThingWorx. All of the courses are free with the exception of the capstone exam which when a student passes they are credentialed as an IoT developer. IBM launched their Watson IOT Academy online education platform in November of 2016 and offers over seventy 2.5 h courses on IoT.

2.3 Opportunities for Partnerships

This new revolution in education can be a real opportunity for faculty at universities. Integrating online courses into the curriculum of a university course, "flips" the classroom. Students can learn outside the classroom the way industry professionals do using these online platforms and then come into the classroom where the faculty can bring context and strategy to the technologies the students have learned. This partnership between universities and technology companies provides students with the latest in technology learning and helps universities stay on top of these rapidly changing technologies.

PTC's online education platform was launched in June of 2017 and since then over 11,000 students have registered and taken courses. University students who register for courses find themselves learning alongside industry professionals and developers. They engage in online communities with them. They also get hands-on experience with PTC's IoT solution ThingWorx in a highly scaffolded environment. The courses are story based so that student finds themselves in a fictitious settlement on Mars having to develop IoT solutions.

Augmented Reality poses the same challenges as IoT for universities. Again, education platforms like IoTU.com can help by providing courses such as "The Fundamentals of AR using Studio." Where students learn about AR and then build two industrial AR use cases like the one shown here where students build an AR service work experience for replacing the servo motor on an industrial robot (Fig. 2).

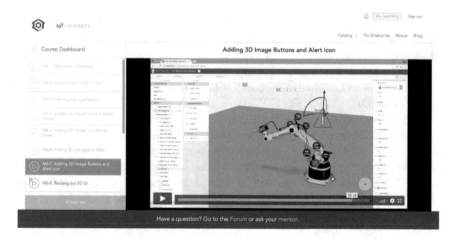

Fig. 2. IoT University online training working with Thingworx

Coupling these types of online education experiences with strategic initiatives like HTL provides students with state-of-the-art education and prepares them for the new workforce. Consortiums like HTL are the way of the future where universities and industries partner to provide students with the best education possible and prepare them for their future careers.

Companies are looking for IoT developers and AR specialists who can work in these new technologies. Partnerships between universities, technology providers, and online education providers is the most effective method of providing this new workforce.

3 Introducing IoT and AR to Secondary Colleges—An Educational Perspective

Since the future workforce in engineering will have to work with AR und IoT technologies and tools [6], there seems to be a chance to introduce it to students within their education. Fernandez-Miranda et al. [10] mention that "students will have to master the combination of mechanical engineering and IT". They also find that universities will educate the future workforce with the skills needed, but see a "close relationship between the competencies that the students acquire at the universities with the needed professional profile". Education at Austrian HTL with their compulsory internships seems to address exactly these topics.

Schuh et al. [6, p. 1390] mention several challenges in product development for Industry 4.0:

- "How should products and the product development be orientated?"
- "Which data is available and which role does it play?"
- "How is an Industry 4.0-specific communication and collaboration defined?"
- "Which resources, methods and tools are needed for Industry 4.0?"

Some might believe that classic approaches and methodologies could be substituted for by IoT technologies. Like the situation in the 90s as CAD technologies were introduced in industry and education, they did adopt the way we are designing machines. For example, hand sketching is still part of engineers' education, but creating technical drawings and documents is done with CAx technologies. Moreover, new jobs like IT technicians or CAD administrators were created. In connection with this Robertson and Radcliffe [11] investigated the impact of CAD tools on creative problem solving in engineering design. They found amongst other issues "... the strengths of the current, most widely used 3D mechanical CAD programs lie more at the detailed stage of design than the conceptual stage" The new IoT technologies will adopt and support engineering daily practice, and present new possibilities to engineers but of variable strength in the different stages of the development process

3.1 Use Cases and Benefits of Using AR and IoT in Engineering Lessons

Especially for education of mechanical engineers there seem to be the opportunities to use AR and IoT technologies into the following daily engineering lessons:

- mechanical engineering design education
- laboratories
- general engineering education to explain the mechanical structures of a machine
- workshops to create work or assembly instructions

Additionally, AR could for instance be used in general education to explain the sun system in a 3d model to students

From the authors perspective there seem to be some benefits in using AR and IoT technologies in engineering lessons:

- Good visualization of machines, equipment and maybe factories
- Good supplement for currently-used technical drawings, which are sometimes hard to understand
- Easy transfer of knowledge about how things are working

  Concerning the IoT platform Thingworx the benefits are:

- Usage of existing 3d CAD data of engineering design lessons
- No programming skills are needed to create an AR experience
- Creation of an AR experience is easy to do and very quick
- Sensor data can be accessed within the Thingworx platform

3.2   Challenges Using AR and IoT Technologies

Whereas introducing AR to engineering design lessons is comparatively easy because of having a lot of 3D CAD data available, setting up and lecturing in an IoT lab is a challenge. Figure 3 shows workflow regarding how 3d CAD data for AR and sensor data are collected and processed to generate an IoT experience available to users' mobile devices or hololens.

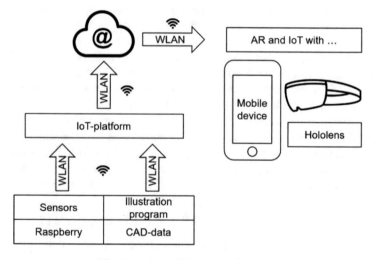

Fig. 3. AR and IoT with Thingworx

Unlike CAD programs, which are installed once a year by the user, the program Thingworx and the technology behind it is changing constantly, the web-application user-interface can even change overnight. Additionally, there is a lot of information available via the web, but few people with AR and IoT knowledge available—like every technology in its early stages, this is especially a problem if someone is seeking support or instructors. Unlike CAD software no company has years of experience with this kind of software and technology, this sometimes leads to a bottleneck related with instructors and support technicians.

3.3   Training Courses for AR and IoT Technologies at Austrian HTL

The aim is to introduce IoT and AR technologies in all HTL for mechanical engineering and industrial engineering in order to integrate current topics into teaching on the one hand and to increase the attractiveness of these branches of education on the other. With regard to the introduction, a task force of the Austrian Federal Ministry of Education was first formed with the aim of familiarizing themselves with the IoT and AR topics. For this purpose, a basic training course and an advanced training course were conducted. The IoT topics were trained using a Raspberry Pi with sensors connected with cables and an IoT training server (see Fig. 4 left).

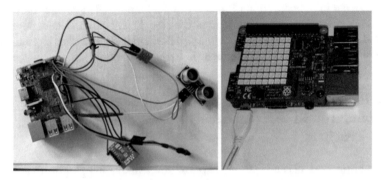

Fig. 4. Raspberry with sensors and cables (left), Raspberry Pi with sensor board (right)

At the same time, a server was installed at the HTL Mödling which has been operated since then. For wider use in the approx. 35 HTL, colleagues from this task force will hold further internal training sessions, but with modified content, which is adapted to the HTL needs. The reason for using HTL teachers as instructors is that the number of instructors available for these technologies is also low in companies. 50 teachers are planned to participate in a basic training course in autumn 2018 and an advanced training course in spring 2019. The aim is to integrate the IoT and AR topics with the academic year 2018/19 into the HTL in the classroom.

With regard to the training environment, some changes will be made for the training courses in the academic year 2018/19 compared to the first training courses. The IoT server operated by HTL Mödling is intended as a platform for the IoT application and user administration. In addition to Raspberry Pi, an Arduino is also offered as a possible platform. Both the Raspberry Pi and the Arduino use a sensor board with approx. 5 sensors (see Fig. 4 right) instead of individual sensors with cables (see Fig. 4 left) which eliminates the wiring. The sensors on the boards are already configured and programmed for use in training, and the focus of the training itself is on integrating the sensors and their measured data. This is important for the authors because the training participants are mechanical and industrial engineers with limited programming knowledge. In addition, the authors expect these factors to increase acceptance of the IoT platform. Further training courses are planned for the next few years with the aim, among other things, of controlling actuators.

3.4   Integration of AR and IoT Technologies into Curriculum

There is currently an agreement with the Austrian Federal Ministry of Education to introduce AR and IoT in education at different HTL without changing the current curricula. The reason for this is the fact that a curriculum change sometimes takes several years and with the added risk that the changes made are no longer up-to-date. This is to be feared especially with such dynamic technologies as AR and IoT are. Additionally, the task force has worked out how AR and IoT technologies can be integrated into daily engineering lessons of all 5 years of HTL education. Besides using AR and IoT technologies in mechanical engineering design lessons, the focus is also on the areas of engineering workshops and laboratories. At least one aim is to establish an AR and IoT lab in the education of mechanical and industrial engineers. Additionally, AR and IoT present the possibility to establish remote laboratories. Andujar et al. [12] conducted some research work about remote laboratories for electrical engineering and reported their findings which are referenced here.

4   Evaluation of the Thingworx AR Platform Through a Standardized SUS Assessment Among Students and Teachers

To find out how students and teachers rate working with the IoT platform Thingworx, a field survey was conducted using a standardized System Usability Scale (SUS) assessment. The SUS was originally created by Brooke [13] and permits analyzation of a diverse array of products and services. Referring to Martin-Gutierrez et al. [14] the SUS "… comprises 10 questions covering the different aspects of a system's usability, such as the need of support, training, and complexity so it is highly valuable as a tool for measuring the usability of a certain system." SUS responses are based on a Likert scale of 1 (totally disagreed) to 5 (completely agreed). The online questionnaire was given to all HTL persons who had had experience with Thingworx, teachers as well as their students.

While a total of 46 people from six Austrian secondary colleges of engineering participated in the survey, 3 of them did not complete the whole test. The remaining 43 responses were from 33 students including two female students, in their 3rd, 4th and 5th year of college education. All of the attending 10 teachers were male. The students had an average of 38.33 h and the teachers had an average of 44.56 h experience with AR and the IoT platform. Included in this average of hours of experience are one student with 500 h and one teacher with 200 h experience. Without these two high single values, the mean value for students is 13 h and for teachers 25 h.

The survey was done with an SUS usability questionnaire originally created by John Brooke in 1986. Since the test is well known and documented, only the questions and the results are displayed in Table 1.

Despite the minimal amount of programming required to generate an AR the overall students' rate for the SUS-score was 56.52%, whereas teachers rate it 63.5%.

Table 1. SUS test

	Question	Students n = 33	Teachers n = 10
1	I think that I would like to use this system frequently	3.39	3.70
2	I found the system unnecessarily complex	2.61	2.40
3	I thought the system was easy to use	3.42	3.70
4	I think that I would need the support of a technical person to be able to use this system	3.24	2.80
5	I found the various functions in this system were well integrated	3.58	3.50
6	I thought there was too much inconsistency in this system	2.24	2.20
7	I would imagine that most people would learn to use this system very quickly	3.15	3.60
8	I found the system very cumbersome to use	2.91	2.20
9	I felt very confident using the system	3.24	3.10
10	I needed to learn a lot of things before I could get going with this system	3.18	2.60
	SUS score	56.52	63.50

Concerning Martin-Gutierrez et al. [14, p. 759] "A product's usability is considered acceptable for values higher than 55%." Improving the usability of the Thingworx platform appears to be a necessity, as does increasing the number of practice hours for students and teachers.

5 Conclusion and Outlook

Augmented Reality and Internet of Things are currently in their introduction phase at Austrian secondary colleges of engineering (HTL). Though the first steps are (like every new technology and software) quite interesting and labor-intensive, these technologies might have a positive impact on engineering education, from the authors perception especially in mechanical engineering. The opportunities to introduce the technologies into daily engineering lessons have been determined, attempts to use them without any changes in existing mechanical engineering curriculum at HTL are being made. Although, students and teachers gave the usability only average ratings of the utilized IoT platform Thingworx, there seems to be a good opportunity to increase future ratings with more practice. In a former publication [15] particularly the enjoyment of working with AR was rated highly. Therefore, the authors hope that the use of these technologies will make mechanical engineering more attractive for potential students, since mechanical engineering is presented in a new and modern way without old fashioned prejudices, new and diverse groups of people, for example female students, are being addressed.

References

1. Porter, M.E., Heppelmann, J.E.: How smart, connected products are transforming competition. Harvard Business Review, no. November, https://hbr.org/2014/11/how-smart-connected-products-are-transforming-competition. (2014)
2. Choi, Y.M., Mittal, S.: Exploring benefits of using augmented reality for usability testing. In: DS 80-4 Proceedings of the 20th International Conference on Engineering Design (ICED 15), vol. 4, Design for X, Design to X, Milan, Italy, 27–30 July 2015
3. Abraham, S.: Using internet of things (IoT) as a platform to enhance interest in electrical and computer engineering. In: ASEE Annual Conference & Exposition, New Orleans, Louisiana (2016)
4. He, J., Lo Chia-Tien, D., Xie, Y., Lartigue, J.: Integrating internet of things (IoT) into STEM undergraduate education: case study of a modern technology infused courseware for embedded system course. In: The Crossroads of Engineering and Business: Frontiers in Education, pp. 1–9. Bayfront Convention Center, Erie, PA, Erie, PA, USA, 2016, 12–15 Oct 2016
5. Chin, J., Callaghan, V.: Educational living labs: a novel internet-of-things based approach to teaching and research. In: 2013 9th International Conference on Intelligent Environments (IE), pp. 92–99, Athens, Greece, 16–17 [i.e. 18–19] July 2013
6. Schuh, G., Rudolf, S., Riesener, M.: Design for industrie 4.0. In: Proceedings of the DESIGN 2016, Dubrovnik, pp. 1387–1396 (2016)
7. Goldman Sachs: BI intelligence estimates, The average cost of IoT sensors is falling. [Online] Available: https://www.theatlas.com/charts/BJsmCFAl. Accessed on 27 May 2018
8. Covington, M.J., Carskadden, R.: Threat implications of the internet of things. In: Podins, K. (ed.) 2013 5th International Conference on Cyber Conflict (CyCon), Tallinn, Estonia, Piscataway, NJ: IEEE, 4–7 June 2013
9. Bradley, D., et al.: The internet of things—the future or the end of mechatronics. Mechatronics **27**, 57–74 (2015)
10. Fernández-Miranda, S.S., Marcos, M., Peralta, M.E., Aguayo, F.: The challenge of integrating Industry 4.0 in the degree of Mechanical Engineering. Procedia Manufacturing **13**, 1229–1236 (2017)
11. Robertson, B.F., Radcliffe, D.F.: Impact of CAD tools on creative problem solving in engineering design. Comput. Aided Des. **41**(3), 136–146 (2009)
12. Andujar, J.M., Mejias, A., Marquez, M.A.: Augmented reality for the improvement of remote laboratories: an augmented remote laboratory. IEEE Trans. Educ. **54**(3), 492–500 (2011)
13. Brooke, J.: System Usability Scale (SUS): A Quick-And-Dirty Method Of System Evaluation User Information. Reading, UK: Digital Equipment Co Ltd., p. 43 (1986)
14. Martín-Gutiérrez, J., Fabiani, P., Benesova, W., Meneses, M.D., Mora, C.E.: Augmented reality to promote collaborative and autonomous learning in higher education. Comput. Hum. Behav. **51**, 752–761 (2015)
15. Probst, A., Ebner, M.: Introducing augmented reality at secondary colleges of engineering. In: Proceedings of E&PDE 2018, 20th International Conference on Engineering and Product Design Education, London, (2018)

Model of Integration of Distance Education in a Traditional University: Migration of Cross-Cutting Courses to Distance Learning

Amadou Dahirou Gueye$^{(\boxtimes)}$, Marie Hélène Wassa Mballo,
Omar Kasse, Bounama Gueye, and Mouhamadou Lamine Ba

IFoAD, Alioune Diop University, Bambey, Senegal
{dahirou.gueye,mariehelene.mballo,omar.kasse,bounama.
gueye,mouhamadoulamine.ba}@uadb.edu.sn

Abstract. Traditional higher education institutions face new challenges, including a new generation of highly connected students who focus on digital tools to learn, learn and stay in touch with the outside world. Indeed, the advent of the internet and the increasing use of digital tools (computers, smartphones, tablets, etc.) require adapting learning models and teaching methods in higher education institutions in Africa and elsewhere in the world. Senegal, which in recent years has begun implementing distance education in public universities, finds in these measures something to take a new boost to improve educational content while optimizing limited resources. However, the distance learning methods offered in these universities are generally geared towards certifying and degree-based distance learning. This does not solve the problems encountered in classical pedagogical training. Based on these limitations, we propose a new model of distance education at the service of a classical university for the optimization of human, financial, material and educational resources. The model is implemented at Alioune Diop University in Bambey, a classical public university in Senegal; and the results obtained have a positive impact on the pedagogy and resources of the university.

Keywords: Distance learning · Classical university · Optimization
of resources · Crosscutting courses · Pooling of resources · Mutualisation

1 Introduction

To meet the development challenge, emerging countries such as Senegal rely heavily on the education of their youth. Thus, in recent years a considerable effort has been made in this direction, as evidenced by the increasing gross enrolment rate in elementary education in Senegal, which stood at between 80 and 93% in 2016 [1].

This growth is also noted in higher education. The number of baccalaureate holders continues to grow from year to year. This growth raises the problem of reception structures in higher education and also of human and financial resources. Indeed, the state of Senegal, despite a clear desire to increase its power in terms of universities and the recruitment of higher education teachers through a new policy of higher education

© Springer Nature Switzerland AG 2019
M. E. Auer and T. Tsiatsos (Eds.): ICL 2018, AISC 917, pp. 13–24, 2019.
https://doi.org/10.1007/978-3-030-11935-5_2

reform [2], is facing the harsh reality of the lack of financial resources inherent to an underdeveloped country.

This raises the question of how to optimize the material and human resources (classrooms, practical work materials, teaching staff, etc.) of Senegal's public universities, without sacrificing the quality of the education provided. The use of digital technologies in education can undoubtedly help to solve the problem of massification of enrolments but also and above all the problem of optimising material and human resources in higher education institutions. Some distance learning models based on digital tools have already been tested in Senegal; see Sect. 2.2 for more details. However, many of these models offer a fully-fledged education system and therefore do not contribute to solving the difficulties that a traditional university may face.

On this basis, we propose in this paper an effective model for integrating distance education in a so-called classical university. The main idea is to share all the cross-cutting courses, i.e. given in training of the same levels but of different specialities and to roll them out online. The rest of this paper is organized as follows.

In Sect. 2, we review the literature on the integration of ICT in education at the global level and particularly in Senegal. In this same section, we highlight the challenges of more efficient ICT integration in Senegal. On the basis of these challenges, we propose and detail, in Sect. 3, our model of distance education in the service of a traditional university. This model will allow a traditional university to optimize its human and pedagogical resources through the opportunities offered by distance education. Finally, we present the results of the implementation of our solution in the concrete case of the Alioune Diop University of Bambey, a classical university present in Senegal.

2   State of the Art

In this part, we present a review of the literature on the integration of ICTs in education at the global level and in Senegal.

2.1   Integration of ICT in Education at the Global Level

The importance of information technology integration in education is well established. Indeed, a number of studies [3–5] have demonstrated the positive contribution of ICT to learner behavior. In other words, the successful integration of ICT into teaching increases the performance and success rate of learners with a real motivation that is noted. The authors of the book "Integration of ICT in Higher Education" [6, 7] identify several advantages that could result from a good integration of ICT in university pedagogy, namely:

- better quality and greater efficiency in teaching;
- a higher graduation rate than in previous years;
- easier access to information resources; and
- an increased presence on global training markets.

The advent of the Internet [8] has greatly revolutionized the integration of ICT into teaching, which ensures real-time learning regardless of the geographical location of learners. Indeed, we are currently witnessing, on the one hand, totally online and distance education [9], and, on the other hand, hybrid forms of teaching [10]. This leads us to talk about the concept of ICT-assisted e-learning. This area is of great interest to researchers working on computer environments for learning. The main concern of ICT-assisted e-learning actors is how to successfully integrate ICT into teaching.

Thus this community around computer environments for learning accompany and support universities, institutes and companies that wish to offer distance and online training. In this paper we focus on the integration of ICT in higher education. Thus, universities that only provide online training are called virtual universities. They are found all over the world, for example the Virtual University of Senegal [11], the Canadian Virtual University [12], the TELUQ in Quebec [13], the Western Governors University [14] of the United States, the Open University [15] in England, the Virtual University [16] of the Université Libre de Bruxelles (ULB) in Belgium, etc. In Asia we are also witnessing the establishment of virtual universities such as the Open University Institute of Hanoi (IUOH) in Vietnam [17].

In sum, we can retain that the success of ICT in the university system depends on a good planning strategy which in turn depends on the following elements [3, 18]: the deployment of physical and technological infrastructures, didactic and pedagogical resources and teachers' skills.

This integration of ICT has enabled higher education institutions to embark on the digital path which provides concrete solutions to the current international problems facing higher education. Among these issues, we can note the accessibility to training, adaptability and enrichment of teaching methods. Furthermore, universities are also integrating distance education to find ways of strengthening their role and place in the democratization of access to higher education.

2.2   Integration of ICT in Senegalese Education

Senegal is one of the countries in sub-Saharan Africa that are taking initiatives to promote the development and use of new technologies for and in education. Such momentum is supported from the outset by major partners such as Microsoft through the "ltl research" program [19] for innovations in teaching and learning through ICTs. Senegal was the only African country selected in this Itl Research program when it was launched.

According to the ranking of the International Telecommunication Union (ITU), which measures the ICT Development Index (IDI) in the world, Senegal is pursuing significant development, first among WAEMU countries and was ranked 2nd behind Côte d'Ivoire, with an IDI of 2.66 [20] in 2017. Moreover, the project to interconnect universities [21], currently being finalized between the National Telecommunications Company [22], the State Computer Agency [23] and the Ministry of Higher Education, Research and Innovation of Senegal [24], demonstrates its willingness to effectively support ICTE. Hence the importance of promoting Education and Research Networks for access to digital training content and for research. In addition, many distance education projects in Senegal are supported by partners such as the Support Project for

Higher Education (Programme d'Appui à l'Enseignement Supérieur PAES) [25] and the Support Project for the Development of Information and Communication for Education (Projet d'Appui au Développement des Technologies de l'Information et de la Communication pour l'éducation PADTICE) [26] as well as other partners such as the Francophone University Agency (Agence Universitaire de la Francophonie AUF) [27].

Senegal is currently at the forefront of the use of information and communication technologies materialized by results, findings, and actions at the level of connectivity and ICT strategy, network infrastructure and equipment. The ICTE actions are carried on the one hand by the Virtual University of Senegal UVS [11] and on the other hand by the Distance Training Institutes created in the classical higher education institutions. However, these institutions that are active in distance education are confronted on the one hand with the challenges of learners' isolation and dropping out, although some authors are trying to find solutions by improving the functionalities of training platforms [28, 29]. On the other hand, distance education practices in these institutions are heterogeneous and depend on the policies in need of distance education. Therefore, we consider that the integration of ICT in higher education in Senegal, seen here from the point of view of the supply of ODL in southern universities and higher education institutions, is confronted less with the availability of technological equipment and infrastructure than with social, human and purely organizational factors.

An effective solution to its challenges will probably have to take into account the current hyper-media dimension in the process of integrating ICT in education [21]. In this context, hyper media refers to the disproportionate resources allocated to equipment at the expense of the training of trainers and users.

The model of integration of distance education in a traditional university that we propose in this work, reserves, thanks to digital, a significant part to the training of trainers and learners to the tools of digital.

3   Proposal of Our Model

In this paper, we develop and present a new model of distance education that is adapted to conventional education. The originality and innovative character of this model come from the pedagogical approach used, the targeted teaching, the technological support, and the practical application framework. Indeed, the model defines an educational approach of mutualisation of so-called transversal courses.

Mutualisation consists in identifying crosscutting courses (i.e. courses that are of the same nature but are held separately in specialities at a given level) and in getting teachers to agree on a single course at the same level. For example, the Expression and Communication Skills (ECS) course, which was taught at the second, third and fourth year levels with 9 separate cohorts, is reduced to one course per level, i.e. a total of 3 courses common to all cohorts at the same level.

After pooling, these courses are scripted and transposed online and remotely with a learner monitoring system based on synchronous and asynchronous tutorials, provided by specialty teachers. This tutoring system is reinforced by the integration of knowledge tests, allowing the learner to gauge his level throughout the learning process.

As far as the teaching taken into account is concerned, the model is intended to be transversal to all the training courses of a university and is thus oriented towards common or so-called transversal courses to all the departments of a classical university, e.g. expression and communication techniques and English. This gives our model a some flexibility of use in any classical university and an adaptability with the easy inclusion of new crosscutting teachings.

In the following we detail the different outlines of our model for integrating distance education in a traditional university.

3.1 Process for Identifying Cross-Cutting Components

The process of migration of the so-called transversal online courses started with the identification of the transversal constituent elements (CE) of the different teaching units of a classical university. This is the first phase of our model of integrating distance education into a conventional university.

Outlined in Fig. 1 with a sequence diagram, our process of identifying transversal CE is carried out using a 3-step methodology, based on the models and syllabi of the different formations, as follows:

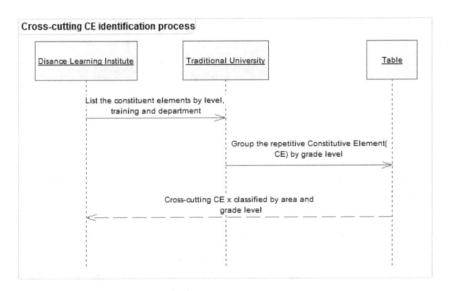

Fig. 1. Process for identifying cross-cutting components

- Identification of candidate CE: crossing of training models of the same level, but of different specialities;
- Selection of so-called transversal CE: exploitation of the syllabi of the candidate CE, and those with the same content are considered crosscutting.

- Final validation by specialty teachers: the list of components resulting from the previous step is presented to a panel of teachers in the field. In this list, each CE is then examined by the experts for final validation.

This three-stage methodology was tested in the training courses of the Alioune Diop University of Bambey, Senegal. The result of the candidate CE identification phase, by crossing the training models, is given in Fig. 2 for the "Expression and Communication Techniques" CE for License levels 1, 2 and 3.

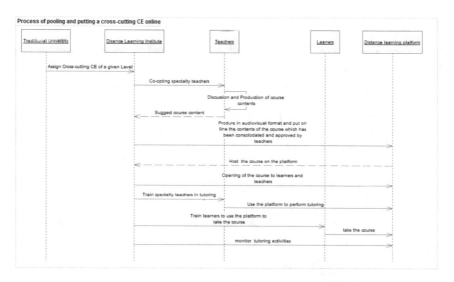

Fig. 2. Process for pooling and putting a cross-cutting component online

After identifying the transversal CE, we proceed to the next two stages of our model, which consist in sharing and putting the lessons online. We detail these two phases in the following.

3.2   Process of Mutualisation and Implementation of Transversal Lessons

The second stage of our model is the pooling of cross-cutting teaching of a purely pedagogical nature, our pooling strategy is based on the use of the expertise of teachers in the targeted speciality. In the concrete case of Alioune Diop University of Bambey, the ECT course was the target and teachers of the said subject were asked to work on mutualisation. The sequence diagram in Fig. 3 details the different stages of mutualisation; the work is carried out by a group of specialist teachers, under the supervision of a coordinator, who proposes a common content at the end.

Once the pooling is done, we move on to the stage of putting the course online. The shared course is first scripted by the techno-pedagogical team of the Open Training Institute (IFoAD) of Bambey according to the pedagogical objectives. The course is also produced in audio-visual format for the interactive aspect. For online and distance

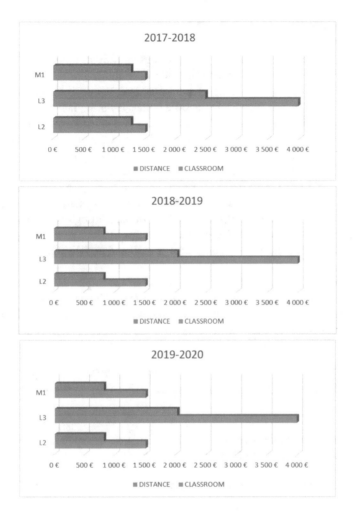

Fig. 3. Financial impact of the shift from face-to-face to distance education for the cross-cutting component in "Expression and communication techniques"

learning, we have opted for the MOODLE learning management system (LMS) [30], which is the world leader in open distance and online learning platforms. The process of putting course content online, stabilized and validated by a committee of specialty teachers, is schematized in Fig. 2.

Once the course is online (see Sect. 4 for the results of the online version of the course), tutors and learners are trained in the use of the distance learning platform and tutoring. At the start of the lessons, the stages of learner follow-up and tutorials are launched. These two steps are the last two steps in our model.

4   Achievements

In this section, we present the achievements resulting from the application of our model for the transversal component "expression and communication techniques", taught at Alioune Diop University in training of the same level but of different specialties.

The specificity of this transversal module comes from the fact that it is a language course. For the validation of our model in a practical framework, we proposed to totally migrate and teach this module remotely for the benefit of the departments in charge of pedagogy in a classical university. Such a decision was motivated by a preliminary financial study on the budgetary impact of transversal teaching in Expression and Communication Techniques (ECT) over a period of 3 years. This study concerns only the nine (9) ECT component that are currently being conducted by IFoAD. Figure 3 (more precisely the summary table at the bottom of the said figure) shows a significant amortization of the cost of the ECT component from the first year and a gain of 8620 EUR over the 3 years, in the scenario where it is shared and dispensed entirely online and remotely.

Our preliminary study of the financial impact of the implementation of our model was based on the training models of the Alioune Diop University of Bambey departments and uses the following formulas (Eqs. 1 and 2) to make the comparison between the cost of a face-to-face transversal course and the cost at a distance education over a term, i.e. 12 weeks:

– **Classroom teaching cost** $= ((hourly\_rate * tutoring\_grps\_Nbr * 11.44EUR) +$
$(restaurant\_price * 12) + (average\_housing\_price * 12) + (transport\_price * 12))$

$$(1)$$

where restaurant_price, average_housing_price and transport_price are equals to 3.81 euros, 15.25 EUR and 3.81 EUR respectively.

– **Distance learning cost** $= (uploading\_cost + (hourly\_rate * 2 * 8\,weeks))$ (2)

where **uploading_cost** and **hourly_rate** are respectively worth 457.5 EUR et 8.39 EUR.

In addition, the pooling of teaching (Sect. 2.2) has enabled us to have a single course for a given level of study, based on the compilation of several course experiences. This has allowed us to have the best courses in our distance learning platform [31]. The shared course is divided, in its scripted and online version, into two main parts: the entry system and the learning system. In the entry system, we present to the student the prerequisites of the course, the general objective, the specific objectives (cf Fig. 4). As shown in Fig. 5, there is also an electronic file in pdf format presenting the course, as well as the video presenting the course and the presentation of the course (pdf file).

LICENCE 2

TECHNIQUE D'EXPRESSION ET DE COMMUNICATION (TEC)

 *MUTUALISATION*

Pré-requis

Avoir fait le cours de TEC Licence 1

Objectif général

Le cours va permettre aux étudiants de maîtriser les caractéristiques des différents types de texte

Objectifs spécifiques

Identifier les types de texte

Produire différents types de texte

Fig. 4. Presentation of entry system

PRESENTATION GENERALE

Fig. 5. Course presentation

Communication tools in our learning platform are made available to learners and tutors. Among others, we have for example the Virtual Classroom resource which is a synchronous chat space and the Exchange Space resource which is a discussion forum. The Class resource also allows synchronous communications with more advanced features such as voice and video communication.

The second part of the course corresponds to the learning system. In our scenario we have opted for a button display, where each number corresponds to a chapter.

As part of our monitoring system, a tutoring session is scheduled once a week between the tutor and the students. This allows students to ask questions about their concerns about the course and possibly for tutors to correct tutorials posted on the

Fig. 6. List of participants in synchronous tutoring

platform. Figure 6 shows the list of students connected during synchronous tutoring. This functionality of our platform, allowing to know in real time the learners' connection rate, is part of the other monitoring levers during learning.

The results of the first experiments of our model in a classical university in Senegal showed that the transition from traditional to distance education produces results that strongly contribute to the optimization of the university's resources.

5   Conclusion

In this paper, we have proposed a new model for the mutualisation of transversal teaching in order to optimize the human, financial, material and pedagogical resources of a traditional university. To achieve this, we have adopted a participatory and inclusive approach towards the authorities and departments of the different entities of the university in charge of classical pedagogy. The mutualisation approach was first to identify certain basic courses of the departments taught in the human sciences (languages), economics and legal and political sciences, and then to mutualise them by level and speciality in order to produce the best courses in these disciplines which will be totally at a distance. The results are conclusive and are being conducted at Alioune Diop University in Bambey, a public Senegalese university through its Distance Learning Institute. The same model can be applied to any conventional university, at the national and international level, wishing to optimize its resources and give itself the means to strengthen its role and its place in the democratization of access to higher education. As a research perspective, we plan to use and possibly extend our model for sharing English courses. Unlike the TEC, the English course has its own specificities that may require a much more sophisticated strategy of course writing, tutoring, and learner follow-up.

References

1. Jacquemin, M., Dia, H.: Les enfants hors ou en marge du système scolaire classique au Sénégal: Etude «ORLECOL»: synthèse analytique. IRD: fdi:010069024, Dakar, Unicef, 2016, p. 92 (2016)
2. Conseil présidentiel sur l'enseignement supérieur et la recherche, 2013, [Online] Available: http://www.cres-sn.org/sites/default/files/cnaes_decisions19aout-l.pdf
3. Mastafi, M.: Intégrer les TIC dans l'enseignement: Quelles compétences pour les enseignants?. Formation et Profession. **23**(2) 2015
4. Balanskat, A., Blamire, R., Kefala, S.: The ICT Impact Report: A Review of Studies of ICT Impact on Schools in Europe. (2006)
5. Machin, S., McNally, S., Silva, O.: New Technology in Schools: Is There a Payoff?. (2006)
6. Karsenti, T., Larose, F.: Les TIC au coeur des pédagogies universitaires. Québec: Presses de l'Université du Québec. Google Scholar
7. Mequanint, D., Lemma, D.: L'intégration des TIC en pédagogie dans les pays en voie de développement. Revue internationale d'éducation de Sèvres, **67**, 75–84 (2014)
8. Pelgrum et Law: Les TIC et l'éducation dans le monde. UNESCO, Institut international de planification de l'éducation, Paris (2004)
9. Taskin, E.: Distance education: a flexible teaching and learning delivery method. In: IEEE International Conference on Application of Information and Communication Technologies (AICT 2009), Baku, Azerbaijan (2009). ISBN: 978-1-4244-4739-8. https://doi.org/10.1109/icaict.2009.5372601
10. Jimenez, M., Bartolomei-Suarez, S., Ochoa, Y., Santiago, W.: A synchronous distance education hybrid model of college-level credits for high-school students. In: Frontiers in Education Conference (FIE), 2016 IEEE, pp. 1–5. Eire, PA, USA, 2016. https://doi.org/10.1109/fie.2016.7757723
11. Université Virtuelle du Sénégal. [Online]. Available: http://www.uvs.sn/. (2015)
12. CVU.: UNIVERSITE VIRTUELLE CANADIENNE (2009). [Online]. Available: http://www.cvu-uvc.ca/englishFR.html. Accès le 16 Septembre 2013
13. TELUQ.: [Online] Available: https://www.teluq.ca/
14. wgu.: Western Governors University (2013). [Online]. Available: http://www.wgu.edu/. [Accès le 16 septembre 2013]
15. OPEN.: The pen University. [En ligne]. (2013). Available: http://www.open.ac.uk/. Accès le 16 septembre 2013
16. ULB.: Université Libre de Bruxelles. (2013) [Online]. Available: http://www.ulb.ac.be/. Accès le 16 septembre 2013
17. Cursus, T.: THOT CURSUS formation et culture numérique. (2009) [Online]. Available: http://cursus.edu/article/1752/une-universite-virtuelle-pour-rejoindre-population/. Accès le 16 septembre 2013
18. Hernandez, R.M.: Impact of ICT on education: challenges and perspectives. J. Educ. Psychol. 337–347 (2017)
19. Shear, L., Gallagher, L., Patel, D.: Innovative Teaching and Learning Research: 2011 Findings and Implications. SRI International, Nov 2011. Available: https://www.sri.com/work/publications/innovative-teaching-and-learning-research-2011-findings-and-implications
20. ITU.: ICT Development Index. (2017). [Online]. Available: http://www.itu.int/net4/ITU-D/idi/2017/

21. snRER.: Réseau pour l'Enseignement Supérieur et la Recherche du Sénégal. [Online]. Available:http://snrer.edu.sn/index.php?option=com_content&view=article&id=50&Itemid= 115
22. SONATEL.: Société National de Télécommunication. [Online]. Available: http://sonatel.sn/
23. ADIE.: Agence de l'Informatique de l'État. [Online]. Available: https://www.adie.sn/
24. MESRI.: Ministère de l'Enseignement supérieur, de la Recherche et de l'Innovation. [Online] Available: http://www.mesr.gouv.sn/
25. PAES.: Projet d'Appui à l'Enseignement Supérieur. [Online]. Available: https://www.afdb. org/fileadmin/uploads/afdb/Documents/Project-and-Operations/Multinational_-_Support_ for_Higher_Education_in_WAEMU_Countries_-_Appraisal_Report.pdf
26. PADTICE.: Projet d'Appui au Développement des Technologies de l'Information et de la Communication pour l'éducation. [Online]. Available: http://www.unesco.org/new/fr/dakar/ about-this-office/single-view/news/padtice_a_project_to_enhance_african_students_daily_life/
27. AUF.: Agence universitaire de la Francophonie. [Online]. Available: https://www.auf.org/
28. Faye, P.M.D., Gueye, A.D., Lishou, C.: Proposal of a virtual classroom solution with WebRTC integrated on a distance learning platform. In: Proceedings of the 19th ICL, 1st ed., vol. 544. Springer International Publishing, Berlin (2017). eBook ISBN: 978-3-319-50337-0, Series ISSN: 2194-5357, Copyright 2017
29. Gueye A.D., Faye P.M.D., Lishou C.: Optimization of practical work for programming courses in the context of distance education. In: Auer, M., Zutin, D. (eds.) Online Engineering & Internet of Things. Lecture Notes in Networks and Systems, Publisher Name: vol. 22. Springer, Cham, 2018. https://doi.org/10.1007/978-3-319-64352-6_72, Print ISBN : 978-3-319-64351-9, Online ISBN:978-3-319-64352-6
30. MOODLE.: [Online]. Available: https://moodle.org/?lang=fr_ca
31. Plateforme de mise en ligne de cours transversaux de l'IFoAD. [Online]. Available: https:// ifoad.uadb.sn/mutualisation/

Academic Determination of Technical Information Optimization Due to Information and Communication Technologies

Denys Kovalenko, Nataliia Briukhanova, Oleksandr Kupriyanov$^{(\boxtimes)}$,
and Tetiana Kalinichenko

Ukrainian Engineering Pedagogics Academy, Kharkiv, Ukraine
a_kupriyanov@uipa.edu.ua

Abstract. The article substantiates the existence of the problem in the optimization of technical information in the process of designing the content of training for the future personnel by means of information and communication technologies in the system of professional education. It was determined requirements for training and scientific sources of technical information, types of defects, as well as the algorithm of the transformation of such text structures as logical, syntactic, semantic, aspectual, communicative, functional and notional, categorizing and graphical, informative. It was established that the following didactic materials are received as a result of constructive activities: structure chart, topic presentation, text (text description), summary. One stressed on the fact that these structures contribute to the optimization of the content of the training of personnel.

Keywords: Information and communication technologies · Technical texts structure · Learning content optimization

1 Problem Statement

With the spread of globalization, the need arose to determine, harmonize and adopt the regulations in the field of education taking into account the requirements of the international and European systems of standards and certification [1, 2].

Peculiarities of professional educational activity of teachers of technical disciplines lie in the fact that, under the massive reorganization of industrial enterprises, the range of professional duties of modern employees is constantly evolving. As a result—the acquisition of another value of certain activities of teachers, in particular, activities to improve the training content.

The dynamic nature of the content of professional training complicates the preparation of the necessary training and methodical support. The most common mistakes of technical texts are as found: the presence of multi-meaning terms that mislead a student; replacement of a definition by condition; incorrect interpretation of the physical meaning of phenomena, processes of reality; the change in the discipline terminology of the value of the general technical term; violation of the definition rules; the presence of unnecessary information or the lack of relevant information in the texts; violation of the text coherence; discrepancy of content fragments with the paragraph title, etc.

M. E. Auer and T. Tsiatsos (Eds.): ICL 2018, AISC 917, pp. 25–34, 2019.
https://doi.org/10.1007/978-3-030-11935-5_3

Any of these disadvantages lead to misunderstanding of the new material by students complicating its assimilation.

Improvement of the training content requires the teacher of technical disciplines to have the skills to determine the necessary and sufficient information for the skilled execution of professional tasks by the future employee, search for such information and choice of ways for its transformation into the curriculum discipline, choice of means for submitting information, use of modern information and communication technologies at the stage of designing the training content and at the stage of transferring experience when interacting with students.

Therefore, the actual development of the algorithm for the design of the training content by optimizing the technical information, implemented by the teacher of technical disciplines in preparation for sessions and necessary to be put in the basis of the development of the methodology for the preparation of such teacher to the constructive activity.

2   Analysis of Recent Research Works and Publications

The training content has long been studied, and there has been a lot of attempts to make it efficient Today, there are formulated regulations, principles and rules for the development of the training content, characterized sources of training information, identified components of the training content, as well as requirements for their development and submission [3–6]. Transformation of information into the training con-tent is performed by filtration, compression of information, editing information, structuring the training material. Usually, the key tool in transforming the information is the rules for work with concepts investigated by the formal logics (correlation of concepts, division of their volumes, and disclosure of contents) [7–12]. Though, the use of knowledge of the formal logics to optimize the technical information requires additional research. In addition, the training content is a complex information system, which requires a comprehensive solution. Not less than the logical activity, the constructive activity of the teacher involves other text structures [6, 7, 10, 11, 13]. There are also available new possibilities of computer technology at the stages of the search, processing of sources of technical information, presentation of results and transfer of new knowledge to students [14].

Further research has been conducted on the structuring of training materials [13] and the combination of different types of multimedia content [15]. The authors [17] have developed the practice of working with the training content as with a product has its life cycle and requires periodic updates. We revealed significant differences of good content for engineering disciplines [15, 16]. One found that the design of content of engineering education should involve predominantly graphical types of materials, and the text should not be based solely on the transfer of information content [18]. The automation of designing the content is a rather complex task, and we know the attempt of its implementation in the training of pilots [19].

The aim of the work is to substantiate and develop the algorithm for designing the content of training the future personnel by optimizing the technical information at the level of such text structures as logical, syntactic, semantic, aspectual, communicative, functional and notional, categorizing and graphical, informative according to pedagogical principles of commitment, availability, consistency, truth, etc., by means of information and communication technologies.

3 Statement of Basic Material and the Substantiation of the Obtained Results

Updating the content of training of future personnel by optimizing the technical information for each training topic is determined by:

- characteristics of the specialty and the goal of the professional training;
- place of discipline and topics in the professional training in a particular specialty;
- degree of readiness of students to accept new knowledge.

Sources of the necessary technical information are textbooks, workshops, instruction booklets, instructions, guidelines and recommendations, tutorials, articles, monographs, etc.

As a result of the conversion activity, we received such documents as block diagram (chart), topic presentation, text (text description), summary. It is necessary to call them didactic materials since they reflect the training content at the level of the training topic.

The fundamental base for the optimization of the technical information and designing didactic materials is made of theories with the subject matter in the form of information search resulting in one or another didactic material. Designing the didactic material requires the knowledge of such text structures as logical, syntactic, semantic, aspectual, communicative, functional and notional, categorizing and graphical, informative. These structures are based on original technical texts and didactic materials on the topic, but with a different method and quality of organization and connection of structural elements. In didactic materials, the method of organization of all structures is determined by the purpose and conditions of specific professional training.

The transformation of existing technical texts into didactic materials makes the constructive activity of the teacher. The first step is to find sources of technical information. The search for sources is a rather daunting task and resources and search techniques are different for each field of knowledge.

The second stage requires the analysis of technical texts on the subject of truth, the accuracy of information, its demonstrability, completeness of doses, structure, logic, optimal sequence in consideration of objects of reality, clarity of wording, achievements of the required degree of specification or generalization of meaningful elements, accessibility, etc. As a result of this analysis, one chooses for the further work those sources that best reflect didactic requirements.

Nowadays, there are more than 300 methods of such analysis. The main are as follows: experimental, social, structural and functional, organometric, etc.

For the professional educational institution, the most appropriate one is the organometric method based on the use of personal experience of the teacher. The essence of this method is that the teacher by selecting the required quality indicators assesses each of the information sources with a certain amount of points. Using the results of the qualitative and quantitative analysis, he chooses the best of the sources. The accuracy of this method depends on the skill of the teacher but when comparing several sources this provision does not affect the results as when evaluating different sources one usually uses the same criteria.

The total amount of points N_i used to assess the quality of textbooks is calculated using the following formula:

$$N_i = \sum_{i=1}^{n} K_i \cdot P_{ij}$$

where
i number of quality indicators;
n number of indicators;
K_i coefficient of the indicator value;
P_{ij} assessment of the degree of implementation of the quality indicator in the source i

Quality indicators are divided into groups: those that characterize the information source based on general requirements to textbooks, purposes; those that characterize actual technical texts based on the content and way of presenting the information.

Indicators of the first group include: conformity of the source to the level of professional education and specialty, presence of target guidance, completeness of the indicative basis of activity, visibility, system of goals and tasks for the practice of mental actions and test of the level of training, availability of instructions that guide students in their didactic activity, etc. Indicators of the second group include: conformity of the training content to conditions of organizing the training process; ensuring the validity of the training information; conformity of the volume of the training information to abilities of students; ensuring the availability of the training material; ensuring a certain semantic completeness of information doses; providing a clear sequence of material presentation; ensuring the logic of material design; ensuring the persuasiveness of the training material, etc.

Automation of the calculation of the total amount of points can be performed by means of electronic tables.

Then, the teacher begins to transform technical texts into the didactic material, correcting them if necessary. With those requirements that the teacher uses to approach the quality of technical texts, he/she controls the didactic material. Moreover, the transition from technical texts to the didactic material must necessarily take place when using the appropriate rules of text-formation. These rules come from the requirements for texts and are classified according to a specific group of homogeneous elements that constitute a particular text structure.

For the stage of structuring the topic content, practical forms of implementing the principles of structuring and visualizing during the work of the teacher over the content of the topic play an important role. Such forms of visualizing the content and its structure include: matrix of relations, chart of the training information and its structural and logical scheme.

We would like to consider the determining the sequence of presentation of the training material on the example of the topic "Individualization Training Tools". Based on the analysis of knowledge, we formed the table of key concepts (Table 1, the middle part of the table is deleted to save space). This table lists the knowledge prior to the study of the topic (provided knowledge) and the knowledge formed as a result of studying the topic. The provided knowledge includes subclauses 1–28 (in Table 1, they are highlighted in grey).

Table 1. Key concepts of the training material

No.	Concept	No.	Concept
1	Training	2	Training tools
3	Training technical tools	4	Classification of training technical tools
5	Training methods	6	Knowledge control methods
7	Training individualization	8	Individual characteristics of teachers
...	...	...	...
25	Classification of computer training tools	26	Multimedia tools
27	Electronic textbook	28	Distance training system
29	Training system	30	Classification of training systems
...	...	...	...
93	Modernization of controlling training systems	94	Readiness
95	Formation of readiness	96	Components of readiness
97	Diagnostics of readiness	98	Formation of knowledge, skills
99	Level of readiness		

To build the structural and content model and form the optimal sequence for presenting the training material, we will use the results presented in the research [20].

In accordance with these provisions, the structure of the training material is presented in the form of a chart. First, we will consider the initial chart of the structural and content model. To account the mutual relations of concepts, we designed the adjacency matrix $M = |m_{ij}|$, where $m_{ij} = 1$, if the study of the concept i should be preceded by the study of the concept j, and $m_{ij} = 0$, if concepts are not related.

At first, we designed the original chart of concepts. We clarified the load of the chart with informational communications that once again confirms the didactic complexity of the training material, which is "at the intersection" of many disciplines. Also, using software tools [20], we designed the chart of concepts in the multilevel and parallel form shown in Fig. 1. It reflects the sequence of presentation of the training material.

According to the results of the program that performs the analysis of the chart of concepts, we received the sequence of presentation of concepts in the study of theoretical material regarding the presentation of the selected topic. Thus, the sequence of presentation that provides the best logic of presentation must correspond to chains by section in Table 2.

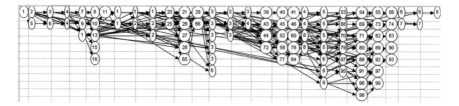

Fig. 1. Chart of key concepts in the multilevel and parallel form

Table 2. Sequence of particulars of educational material

1	2	3	4	5	6	7	8	9	10	11	12	13	14	15	16	17	18	19	20
Sections																			
1	1	2	3	4	7	8	13	9	10	11	12	14	15	16	17	18	19	20	21
21	22	23	24	25	26	27	28	29	30	31	32	33	34	35	36	37	38	39	
65	24	25	27	66	67	26	22	28	29	34	35	36	30	31	32	38	39	40	
41	42	43	44	45	46	47	48	49	50	51	52	53	54	55	56	57	58	59	60
Sequence of particulars																			
37	42	43	59	60	44	45	46	47	50	52	72	77	84	85	86	87	88	89	90
61	62	63	64	65	66	67	68	69	70	71	72	73	74	75	76	77	78	79	80
91	92	93	94	95	96	97	98	99	33	78	81	23	5	6	48	49	51	53	54
81	82	83	84	85	86	87	88	89	90	91	92	93	94	95	96	97	98	99	100
55	56	62	57	63	64	58	68	69	61	70	71	73	74	75	76	79	80	82	83

Finding a particular inaccuracy, error, invalid way of presenting the information, the teacher sets their nature (logical, semantic, syntactic, etc.) by identification with an appropriate structure and chooses methods to overcome these obstacles.

At the same time, the same concept can be dealt with by any other text structure—semantic formed by the meaning of words and expressions. Based on the definition of the semantic structure, it is clear that the adequacy of the essence of processes, objects, phenomena of reality is provided, firstly, by the accurate selection of sign systems that express the exact meaning and, secondly, by the correct installation of all possible links between these values (whole–part, set–item, class–subclass, object–options, process–property, phenomenon–characteristic, reason–consequence, essence–phenomenon, law–manifestation, goal–means of achieving, condition–action).

We developed and used templates of plans of topics on parameters of processes, specifics of labour processes, as well as laws, equations, balances, etc. Templates are created with the help of MSWord, file type *.dotm. After opening the file template, a document for the filling with didactic materials will be created on its basis.

Further, it remains to determine the degree of plan specificity guided by:

- peculiarities of perception and understanding of the students of the new information (technical text with a number of unities that do not exceed seven shall refer to one of the categories; otherwise, it is necessary to divide the text with subordinate headers);

– level of preparation of students (the poorer the preparation is and the more difficult the text is, the more preferable one- or two-stage categorization is; grounds for the multi-stage categorization, in addition to good training of students, include a diverse range of issues covered by the teacher, and a large text volume).

On the basis of the plan, in the sequences specified by the plan, the following didactic materials, like text and summary, are designed. They differ from each other by the level of closeness (openness) of information and capacity of means of information expression. Moreover, a text is primary and a summary is secondary

The further consideration of the topic, result and process of constructive activity will be conducted at the level of each of the text structures.

Designing the text starts by defining the type of its communicative structure, elements of which are topic and rheme. We can consider the name of every plan clause as the topic (T) and meaningful fragments marked by them as the rheme (R). In this case, making a transition "plan-text", it is important that the information selected and arranged by the teacher is fully and at the desired level opened the essence incorporated in the header. This requirement is implemented by systematic comparison of the data to be selected with the topic and already processed information. One or another type of relationship occurs between T and R. The most convenient for closing the information when designing notes is the ratio of the serial correlation of T and R with a constant topic. In this case, each repeated mention of the topic is excluded. The ratio of simple linear correlation of T and R is definitely not subject to changes.

The next structure that determines the quality of the didactic material in documents is the informative one. Such components as relevant and non-relevant information make up the excess information with the only difference—the relevant information is the information that provides the understanding and perception of the new information and the non-relevant information is the information that prevents the perception of new information (unnecessary repetitions, extra details, etc.). The boundary between them is pretty mobile and dependent on the degree of readiness of students to study a new topic.

The design of notes, depending on the degree of its completeness, is that one takes only main, basic provisions (they include definition, classification, graphical symbols, list of parameters, modes of operation, algorithm of activity, etc.) from the new information and hides and uses if necessary the relevant information (definitions, descriptions, illustrations, examples, etc.). To do this, it is advisable to use hyperlinks, outline MSWord, tags <details> and <summary> in HTML5 (Fig. 2), and other technologies.

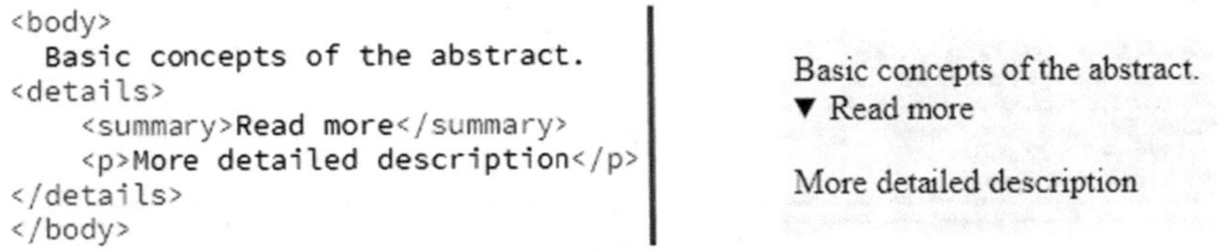

```
<body>
  Basic concepts of the abstract.
<details>
    <summary>Read more</summary>
    <p>More detailed description</p>
</details>
</body>
```

Basic concepts of the abstract.
▼ Read more

More detailed description

Fig. 2. Design of hidden parts of the document using HTML5

The recommended algorithm for designing the training content by optimizing the technical information is shown in Fig. 3.

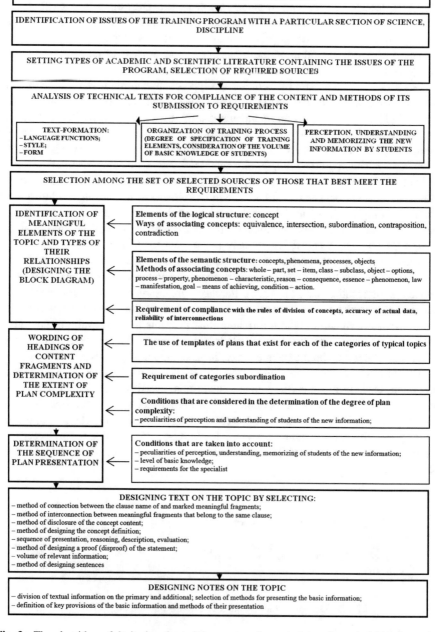

Fig. 3. The algorithm of designing the training content by optimizing the technical information

4 Conclusions

The orientation of professional education on the market of goods and services, which is developed dynamically, leads to several contradictions, including contradictions between the increasing volume of training material and insufficient time allotted for its assimilation in training plans and programs, as well as between the level of representation of training information and its actual condition. Optimization of the technical information by means of information and communication technologies is one of the ways to solve existing disputes and provide answers to the following questions: what kind of scientific information to choose, how to obtain and to include it in the content of training of specialists; how much training information is considered necessary and sufficient for the future specialist; how to present the technical information (sequence, method of presentation) to make it accurate, available.

Thus, it is established that the basics of technical text-formation are logical, syntactic, semantic, aspectual, communicative, functional and notional, categorizing and graphical, informative. These structures are formed by semantic, linguistic or graphical elements with an extensive and multilevel network of connections built over pedagogical principles of commitment, availability, consistency, truth, etc., by means of information and communication technologies facilitate the optimization of the training content of personnel.

It is still necessary to develop methods of preparing the teacher of technical disciplines to the constructive activity, methods of forming skills of synthesis of the internal logic of the technical information with psychological processes and mechanisms of training; ways of presenting the training material using new opportunities of computer engineering and technology, definition of metrics for the evaluation of its volume.

References

1. A Tuning Guide to Formulating Degree Programme Profiles, Including Programme Competences and Programme Learning Outcomes. University of Deusto Press, 93p (2010)
2. Standards and Guidelines for Quality Assurance in the European Higher Education Area (ESG), Brussels, Belgium (2015)
3. Bespalko, V.P.: Components of Pedagogical Technology, 192p. Pedagogy, Moscow (1989)
4. Dietl, P.J.: Teaching, learning and knowing: educational philosophy and theory. 5, 1–25 (1973)
5. Doblaev, L.P.: The Semantic Structure of the Text and the Problems of Understanding It, 176p. Pedagogy, Moscow (1982)
6. Martin, J.R.: Explaining, Understanding and Teaching. McGraw-Hill, New York (1970)
7. Blumenau, D.I.: The Problems of Curtailing Scientific Information, 166p. Science, Leningrad (1982)
8. Hirst, P.H.: The Logical and psychological aspects of teaching a subject. In: Peters, R.S. (ed.) The Concept of Education. Routledge and Kegan Paul, London (1979)
9. Hintikka, J., Hintikka, M.: Sherloc Holmes Confronts modern logic: towards a theory of information—seeking through questioning. In: Barth, E.J., Martens, J. (eds.) Theory of Argumentation. Benjamins, Amsterdam (1982)

10. Hintikka, J.: The semantics of questions and the questions of semantics. Acta Philosophica Fennica **28**(4) (1976)
11. van Merrienboer, J.J.G.: Training Complex Cognitive Skills: A Four-Component Instructional Design Model for Technical Training, 345p. Educational Technology Publications, Inc., Englewood Cliffs (1997)
12. Philippi, J.: How to design instruction: from literacy task analysis to curriculum. **91**, 237
13. Clark, R.C.: Developing Technical Training: A Structured Approach for Developing Classroom and Computer-Based Instruction Materials, 3rd edn, 276p. Wiley, New York (2008)
14. Kovalenko, D., Bondarenko, T.: Cloud monitoring of students' educational outcomes on basis of use of BYOD concept. In: Auer, M., Guralnick, D., Simonics, I. (eds.) Teaching and Learning in a Digital World. ICL 2017. Advances in Intelligent Systems and Computing, vol. 715, pp. 766–773. Springer, Cham (2018)
15. Baukal, C.E., Jr., Ausburn, L.J.: Working engineers' multimedia type preferences. Austr. J. Eng. Educ. 1–10 (2017)
16. Kovalenko, O., Kupriyanov, O., Zelenin H.: Content elements of training teachers of engineering disciplines. In: Proceedings of 15th International Conference on Interactive Collaborative Learning and 41st International Conference on Engineering Pedagogy, Villach, Austria, 26–28 September 2012
17. Gafford, W., Thropp, S., Lucas, L., Aplanalp, J., Archibald, T., Clem, J.: advancing integration of technical data and learning content management (2009)
18. Swanson, R.A., Torraco, R.J.: Technical training's challenges and goals. Tech. Skills Train. **5**(7), 18–22 (1994)
19. Vaughan, D.S., Mitchell, J.L., Yadrick, R.M., Perrin, B.M., Knight, J.R.: Research and development of the training decisions system (1989)
20. Yaschun, T.V.: Evaluation of the quality of educational and cognitive activity in the system "student-computer" (on the example of the disciplines of the cycle "Informatics"), 190p. Ph.D. Dissertation, Kharkov (1999)

Promoting Students Engagement in Pre-class Activities with SoftChalk in Flip Learning

R. D. Senthilkumar$^{(\boxtimes)}$

Department of Math and Applied Sciences, Middle East College, PB 79,
KOM, PC. 124 Al Rusayl, Muscat, Oman
senthil@mec.edu.om

Abstract. This study aims to find the effectiveness of SoftChalk based lessons in engaging students in their pre-class activities in flip learning by determining the amount of learning attained using exploratory method among the first year engineering degree students at the Middle East College (MEC), Oman. Students pursuing Engineering Physics and Engineering Science Modules during the spring 2017 semester are formed into two groups, namely Experimental Group (EG) and Control Group (CG). Pre-class lessons/activities are prepared using Softchalk and converted as SCORM packages and used for the students of EG. The normal lessons which contain narrow study materials and short videos are used for the students of CG. SoftChalk based SCORM packages are uploaded in Moodle for students and monitored their performance. Face-to-face classes are redesigned for the EG based on their performance in the pre-class activities while for the CG predesigned face-to-face class activities are used. The same tutor taught both groups of students of both modules for 15 weeks. Both the coursework assessments and end semester exam are used as a measuring instrument and their scores are statistically analysed. Results of the statistical analysis revealed that the EG has attained higher learning (14.94% in coursework and 29.83% in end semester) in comparison with the CG in Engineering Physics module. Similarly, in the Engineering Science module, EG has attained 16.34 and 15.77% higher learning in coursework and end semester, respectively, compared to that of CG.

Keywords: Flipped learning · SoftChalk · Scorm · Physics education
Enginnring education

1 Introduction

In this digital world, the innovative use of educational technologies provides educational institutions great opportunities for their educators to design media enhanced, interactive, more inclusive and engaging learning environments even beyond the classroom walls. Therefore, education at all levels can no longer be assimilated to a group of students in a classroom following a teacher with a textbook.

Flipped learning which is a part of active learning strategy, has also been recognized as an innovative and effective instructional strategy [1]. One of the key

© Springer Nature Switzerland AG 2019
M. E. Auer and T. Tsiatsos (Eds.): ICL 2018, AISC 917, pp. 35–43, 2019.
https://doi.org/10.1007/978-3-030-11935-5_4

dimensions of flipped learning strategy is pre-class activities where students need to acquire some essential knowledge for the deeper investigation of flipped topics during face-to-face classes. Thus, engaging students in pre-class activities is essential and a challenging task for teachers [2].

In addition, the success of flipped learning completely depends on the planning of lessons, designing the study materials for student's active involvement in completing pre-class, face-to-face class and post-class activities [3].

Generally, educators use videos and narrow study materials for pre-class activities where they couldn't effectively engage the students. This can be overcome by designing lessons using SoftChalk which is an information and communication tool (ICT) that help educators to create their own interactive, personalized and engaging lessons. SoftChalk also supports the educators to monitor the effectiveness of students learning in pre-class [4].

In this study, an effort has been made by designing lessons in (Sharable Content Object Reference Model) SCORM format using SoftChalk for the active participation of students in completing their pre class activities and to monitor their progress.

The SCORM is basically a comprehensive suite of eLearning standards to enable durability, portability, accessibility, interoperability and reusability of the eLearning content. SCORM was an initiative by US government in 1997 and developed by Advance Distributed Learning [5] with the main objective to produce course content (SCO's or shareable content Object), which is to work on any Learning Management System (LMS).

2   Goal

This study aims to find the effectiveness of SoftChalk based lessons in engaging students in their pre-class activities under flip learning by determining the amount of learning attained using exploratory method among the first year engineering degree students at the Middle East College (MEC), Oman.

3   Research Questions and Hypotheses

Research question of this study is "What is the difference in learning between the experimental and control group?" and to test the question, the following hypotheses are set:

Null Hypothesis (H_0): There is no difference in learning attained by experimental group compared to that of control group.
Alternative Hypothesis (H_a): There is a difference in learning attained by experimental group compared to that of control group.

4 Methodology

4.1 Experimental Design

A quasi-experimental design is used in this research to investigate the effectiveness of SoftChalk based pre-class activities on students' academic progression.

First year engineering students pursuing Engineering Physics and Engineering Science Modules under the various programmes in the Middle East College during the spring 2017 semester are formed into two groups, namely Experimental Group (EG) and Control Group (CG). There were two sessions, namely A and B, in both the modules during spring 2017. The Session A of both Engineering Physics (no. of students = 11) and Engineering Science modules (no. of students = 16) are assigned as EG while the Session B of Engineering Physics (no. of students = 13) and Engineering Science (no. of students = 17) modules are assigned as CG.

Pre-class lessons/activities are prepared using Softchalk and converted as SCORM packages and used for the students of EG. The normal lessons which contain narrow study materials and short videos are used for the students of CG. SoftChalk based SCORM packages are uploaded to Moodle for students' access. The screenshot of a lesson used in the experimental group is shown in Fig. 1. The advantage of this package is sending alerts to educators through emails upon completion of lessons and learning activities, such as quizzes, problems solving, etc. by students. Students' performance in the SCORM based pre-class activities are monitored from Moodle and face-to-face classes are designed for the EG while for the CG predesigned face-to-face class activities are used. In order to minimize the effect of the tutor's teaching skills as

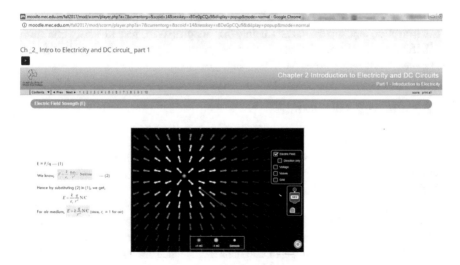

Fig. 1. The screenshot showing narrow texts and simulation from PhET which is embedded in a SoftChalk based lesson, designed by tutor and uploaded in Moodle for the experimental group in engineering physics module

a variable, the same tutor taught both group of students of Engineering Physics and Engineering Science modules for a period of 15 weeks. The coursework assessments (closed book test and lab test) and the end semester exam are used as a measuring instrument to quantify the students learning. The implementation of experimental design used in this study is shown in Fig. 2.

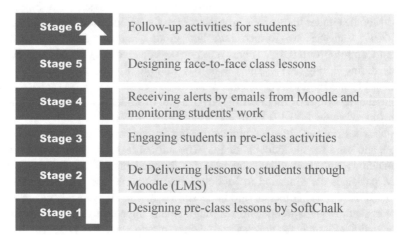

Fig. 2. SoftChalk in flipped learning strategy used in this research study

4.2 Statistical Analysis

The coursework assessments and end semester scores of both the modules are subjected to a descriptive statistical analysis using the Shapiro-Wilk test to verify the normal distribution of the sample. Research hypotheses are tested by analysing the scores by the Levene's test using the IBM SPSS package.

5 Results and Discussions

The descriptive statistics for the coursework assessment and end semester scores obtained by students in this study along with the difference in mean scores (ΔL) between the EG and CG of Engineering Physics and Engineering Science modules, are provided in Tables 1 and 2, respectively.

Table 2 shows the mean scores obtained in Engineering Physics by the working groups: EG 81.41 and CG 66.47 in coursework ($\Delta L = 14.94\%$); EG 79.23 and CG 49.40 in end semester assessments ($\Delta L = 29.83\%$). Table 2 provides the means scores obtained by the working groups in Engineering Science module: EG 83.79 and CG 67.46 in coursework ($\Delta L = 16.34\%$); and in end semester, EG 77.15 and CG 61.38 with the difference in learning $\Delta L = 15.77\%$.

Table 1. Descriptive statistics of coursework assessments and end semester scores of engineering physics module

Parameters	Coursework		End Semester	
	EG	CG	EG	CG
Mean	81.41	66.47	79.23	49.40
Median	87.02	67.70	78.00	52.00
Std. deviation	10.69	13.56	11.45	19.00
Minimum	63.10	43.96	55.50	74.50
Maximum	95.30	91.20	95.00	96.67
$\Delta Learning$ (ΔL)	81.41–66.47 = 14.94%		79.23–49.40 = 29.83%	

Table 2. Descriptive statistics of coursework assessments and end semester scores of engineering science module example table

Parameters	Coursework		End semester	
	EG	CG	EG	CG
Mean	83.79	67.46	77.15	61.38
Median	83.65	73.00	78.50	66.50
Std. deviation	6.66	20.96	16.28	26.21
Minimum	72.00	15.00	42.00	06.00
Maximum	94.10	96.30	99.50	100.00
$\Delta Learning$ (ΔL)	83.79 − 67.46 = 16.34%		77.15 − 61.38 = 15.77%	

Table 3 presents the results of Shapiro-Wilk test which is to verify the normal distribution of the sample collected from the coursework and end semester scores. The p-values of working groups (EG and CG) for the coursework and end semester scores in both engineering physics and engineering science modules are greater than 0.05. This reveals that the sample scores collected in this study are normally distributed.

Table 3. Normality test results

Module	Assessment	Group	Shapiro-Wilk test		
			Statistic	df	Sig.
Engineering physics	Coursework	EG	0.904	11	0.206
		CG	0.987	13	0.998
	End semester	EG	0.973	11	0.919
		CG	0.951	13	0.616
Engineering science	Coursework	EG	0.937	16	0.318
		CG	0.945	17	0.378
	End semester	EG	0.937	16	0.318
		CG	0.945	17	0.378

Table 1 demonstrates that, in both coursework and end semester assessments, the mean score of EG where the pre-class lessons, designed by SoftChalk, are used significantly greater than that of CG scores in Engineering Physics module. Similarly, it is noted from Table 2 that mean scores of EG in coursework and end semester are higher than that of CG scores in Engineering Science module. These observations support the Alternative Hypothesis, that there is a difference in learning attained by experimental group compared to that of control group. Further, the data is analysed by Levene's Test and Independent Student's t-test, at a 5% significance level, for the equality of means of independent samples due to the normal distribution of the scores. The purpose of these tests is to verify the learning between EG and CG scores obtained in engineering physics and engineering science modules.

Table 4 represents the Levene's test result which is in the expected direction and significant as $p < 0.05$. This reveals that there is a significant change in the mean scores of experimental groups with respect to their respective control groups in both engineering physics and engineering science modules. Further, the two-tailed p-values from the t-test are also less than 0.05 ($p < 0.05$). This indicates that the acceptance of Alternative Hypothesis and the difference in mean scores of the experimental groups and control groups are statistically significant. That is, the higher scores achieved by experimental groups is not because of a chance but due to the pre-class lessons designed by SoftChalk.

According to the researchers, theory should become a useful and meaningful tool for understanding practical, rather than an end in itself [6]. The knowledge acquired by the students from SoftChalk based pre-class lessons also developed the skills for performing well in the laboratory classes as EG achieved higher scores in laboratory tests which are the part of coursework assessment.

Researchers [7, 8] have said that the learners should be helped to arrange their learning process flexibly according to individual needs. Further, learning with digital media offers a high level interactive access to the learning activities and enhances a learner's ability for the acquisition of knowledge [9, 10]. This investigative study proves that the SoftChalk based pre-class activities help tutors to monitor students' progression and it also provides enhanced learning experience, as students can learn based on their learning style and own pace.

Students' perceptions on the effectiveness of SoftChalk based lessons in pre-class activities is determined through a questionnaire analysis. Students are asked to indicate their opinion on a 5-point Likert scale: Strongly disagree = 1; Disagree = 2; Neutral = 3; Agree = 4; Strongly Agree = 5. This helped to determine the degree of agreement with the statements related to the use of SoftChalk. The statements included in the questionnaire are given below:

Q1. I find easy in learning through SoftChalk based lessons.
Q2. SoftChalk allows me to interact with lessons and learn at my own pace.
Q3. SoftChalk supports learnings based on my learning styles.
Q4. Study materials and learning activities used in Softchalk based lessons supported in developing my reasoning and problem solving skills.
Q5. Through SoftChalk based lessons I have acquired knowledge and deep understanding of topics flipped.

Table 4. Summary of Levene's test and student's t-test of engineering physics and engineering science modules for a 5% significance level

			Levene's test for equality of variances		Student's t-test for equality of means						95% confidence interval of the difference	
			F	Sig.	t	df	Sig. (2-tailed)	Mean difference	Std. error difference		Lower	Upper
Engineering physics module	Course work	Equal variances assumed	0.238	0.031	2.955	22	0.007	14.940	5.056		4.454	25.427
		Equal variances not assumed			3.015	21.915	0.006	14.940	4.955		4.662	25.218
	End Semester	Equal variances assumed	4.582	0.044	4.728	22	0.000	29.823	6.307		16.742	42.903
		Equal variances not assumed			4.907	20.560	0.000	29.823	6.078		17.166	42.479
Engineering science module	Course work	Equal variances assumed	11.966	0.002	2.977	31	0.006	16.338	5.488		5.144	27.531
		Equal variances not assumed			3.054	19.378	0.006	16.338	5.349		5.155	27.520
	End semester	Equal variances assumed	2.414	0.030	2.061	31	0.048	15.774	7.655		0.161	31.387
		Equal variances not assumed			2.089	6.970	0.046	15.774	7.549		0.282	31.266

The questionnaire analysis report is given in Fig. 3 which shows that Likert scale values of all statements is almost closer to 5 (Strongly Agree). This indicates that students have a positive attitude towards the use of SoftChalk based lessons. Further questionnaire analysis reveals that SoftChalk help students for interacting with lessons and learn at their own pace.

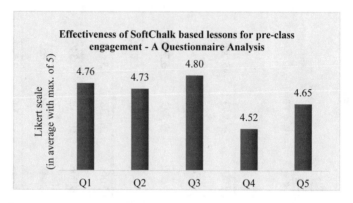

Fig. 3. Report on effectiveness of SoftChalk in flipped learning strategy

6 Conclusion

An experimental study on the use of SoftChalk for engaging students in pre-class activities in flipped learning, by determining the amount of learning attained by the first year engineering students studying engineering physics and engineering science modules at the Middle East College, Oman is studied. The results obtained from this investigative study confirm the Alternative Hypothesis that there is a difference in learning attained by the students in the experimental groups, in both the modules, who used SoftChalk based pre-class lessons compared to those who do not. It is found that the use of SoftChalk in flipped learning has increased the higher percentage of learning: 14.94 and 29.83% (in engineering physics); 16.34 and 15.77% (in engineering science) in coursework and end semester, respectively. This may be due to the face-to-face class activities which are designed based on students' performance in the pre-class activities. In addition, the knowledge acquired by the students from SoftChalk based pre-class lessons also developed the skills for performing well in the laboratory classes. This study also reveals the students' perception on the effectiveness of SoftChalk based lessons in pre-class activities. Students indicated that pre-class activities allows them to learn at their own pace and acquire thorough knowledge and deep understanding through interactive simulations, quizzes, etc. This research study demonstrated that it is viable to track student's performance on their pre-class activities and to promote the student-led learning by which students' academic progression can be enhanced. Different learning styles and cultures also can be accommodated easily by designing the lessons with SoftChalk. Hence, this learning strategy has the capability to engage each student and make us more effective tutors. Further, an elaborate research will be continued on engaging students in pre-class activities in mathematical modules at the Middle East College.

Acknowledgements. Author would like to express gratitude to the students who have participated in this research study.

References

1. Hwang, G.J., Lai, C.-L., Wang, S.-Y.: Seamless flipped learning: a mobile technology enhanced flipped classroom with effective learning strategies. J. Comput. Educ. **2**(4), 449–473 (2015)
2. Nouri, Jalal: The flipped classroom: for active, effective and increased learning—especially for low achievers. Int. J. Educ. Technol. Higher Educ. **13**(33), 1–10 (2016). https://doi.org/10.1186/s41239-016-0032-z
3. Raymond, S.: The Effectiveness of the flipped classroom. Honors Projects, Paper 127 (2014)
4. Brown, J., Ashley, W., Wang, L.: Flipping your classroom with SofChalk [online], 10 Nov 2017. Available from https://softchalk.com/webinar/innovators-in-online-learning-webinar-flipping-your-classroom/
5. Fletcher, J.D., Tobias, S., Wisher, R.A.: Learning anytime, anywhere: advanced distributed learning and the changing face of education. Educ. Res. 96–102 (2007)
6. Korthagen, F.A.J., Kessels, J., Koster, B., Lagerwerf, B., Wubbels, T.: Linking Practice and Theory: The Pedagogy of Realistic Teacher Education. Lawrence Erlbaum Associates, Mahwah, NJ (2001)
7. Oliver, R., Harper, B., Wills, S., Agostinho, S., Hedberg, J.: Describing ICT-based learning designs that promote quality learning outcomes. In: Beetham, H., Sharpe, R. (eds.), Rethinking Pedagogy for a Digital Age, pp. 64–80. Routledge, New York (2007)
8. Wrenn, Jan, Wrenn, Bruce: Enhancing learning by integrating theory and practice. Int. J. Teach. Learn. Higher Educ. **21**(2), 258–265 (2009)
9. Lee, Hee-Sun, Linn, Marcia C., Varma, Keisha, Liu, Ou Lydia: How do technology-enhanced inquiry science units impact classroom learning? J. Res. Sci. Teach. **47**(1), 71–90 (2010)
10. McKnight, K., O'Malley, K., Ruzic, R., Horsley, M. K., Franey, J. J., Bassett, K.: Teaching in a digital age: how educators use technology to improve student learning. J. Res. Technol. Educ. **48**(3) (2016)

Significance of Psychological and Pedagogical Training in Developing Professional Competence of Engineers

Tatiana A. Baranova, Elena B. Gulk$^{(\boxtimes)}$, Anastasia V. Tabolina, and Konstantin P. Zakharov

Department of Engineering Education and Psychology, Institute of Humanities, Peter the Great St. Petersburg Polytechnic University, Saint Petersburg, Russia
baranova.ta@flspbgpu.ru, {super.pedagog2012, stasy335k}
@yandex.ru, sladogor@gmail.com

Abstract. The paper considers the peculiarities of studying psychological and pedagogical disciplines in today's university of engineering, defines the role of these disciplines in developing the professional competence of would-be engineers. The paper focuses on the significance of using contemporary humanitarian technologies, active and interactive forms and methods of learning and cooperative learning. The authors present a model of organizing psychological and pedagogical training for engineers, providing the results of experimental work carried out at the St. Petersburg Polytechnic University over a period of three years.

Keywords: Professional competence · Psychological and pedagogical disciplines · Humanitarian technologies · Active and interactive learning methods · Universal competencies

1 Context

The significance of humanities, including psychological and pedagogical training, for preparing an engineer is confirmed in papers of A. G. Andreev, A. A. Verbitsky, A. I. Subetto, Y. G. Fokin, G. P. Schedrovitsky, and others [1–5].

The authors emphasize the necessity to select and to implement techniques which would take into account the specificity of engineering activity and achievements of humanities, seek new interactive dialogue forms of education targeted at the improvement of studying humanities, including psychological and pedagogical disciplines [6].

A. G. Andreev points out that the problem of humanitarization of technical education is typical for the entire global educational space. That said, there is a strong correlation between a country's position in the modern world and the attention it dedicates to the "integration of social, humanitarian, and technical education" [1, p. 96].

Russian researchers (A. P. Tryapitsina, S. A. Pisareva, etc.) emphasize the importance of creating a humanitarian educational environment that would combine education and culture in universities. In turn, universities, including engineering and

© Springer Nature Switzerland AG 2019
M. E. Auer and T. Tsiatsos (Eds.): ICL 2018, AISC 917, pp. 44–53, 2019.
https://doi.org/10.1007/978-3-030-11935-5_5

technical universities, should create conditions for the development of capacity and readiness for cross-cultural interdisciplinary dialog and innovation and direct students toward universal human values and concepts [7, p. 13].

The modern world is a world of humanitarian (soft) technologies of influencing people. These technologies modify the understanding of the role of humanitarian knowledge in the professional training of engineers. G. P. Schedrovitsky made a great contribution to the development of the new approach in the 1970s. He emphasized the need for a synthesis of technical, naturalistic, and socio-humanitarian knowledge in modern engineering, pointing out that such synthesis was hindered by the unreadiness of humanities due to their inadequacy to the challenges of the time. Humanities should therefore move away from abstract knowledge and traditional methodology toward technological knowledge and creative project-oriented methodology [5, p. 251].

G. P. Schedrovitsky's ideas are manifested in the development of the technological capabilities of humanitarian knowledge and its increasing focus on the transformation of reality through activity instead of its mere explanation. Thus, humanitarian technologies transfer humanities into the field of human activity based on humanitarian knowledge.

A. A. Verbitsky points out that the educational process should provide conditions for students to gain professional skills through situations that simulate actual professional activity. At the same time, there is an ongoing transition from educational activity to quasi-professional, educational professional, and, finally, real activity [2].

2 Purpose

The authors focus on the problems of humanities at the university of engineering. Such problems include fewer academic hours that are assigned to humanities, extremely large academic groups, inability to adapt humanities to students majoring in engineering, insufficient readiness of lecturers and students to use humanitarian technologies.

The above-stated contradictions allowed forming the problem of the research, i.e. to identify conditions, under which psychological and pedagogical disciplines would contribute to increase the professional competence of students majoring in engineering.

Objectives:

1. To develop a model of organizing psychological and pedagogical training of students majoring in engineering.
2. To study the development of universal competencies of students majoring in engineering as a component of their professional competence.
3. To identify the peculiarities of psychological and pedagogical disciplines at the university of engineering and the resourcefulness of these disciplines for developing the professional competence of engineers.
4. To determine and to verify conditions of introduction, content, methodological support of the model of organizing psychological and pedagogical training of students majoring in engineering.
5. To generalize the results of the research.

3 Approach

The contemporary engineering activity is complex and diverse. It requires that a specialist should have not only professional, but also universal competencies, which are related to the specialist's ability to communicate, to interact effectively at the interpersonal and intercultural level, to work in a team, to have skills of self-study, self-organization, personal development, to have established values, responsibility for the professional choice. These abilities become more and more important in the situation when today's engineers deal with high-tech equipment and get involved in multi-level organizational management. An engineer of the future should have skills of collaborative work, of taking into account personality of subordinates and colleagues, skills of teaching and training staff.

Consequently, development of the above-mentioned abilities and skills becomes an urgent didactic problem in higher education.

Disciplines of the psychological and pedagogical cycle as a component of humanitarian education for students majoring in engineering have a special resource to solve this problem. The content, methodology, techniques of psychological and pedagogical disciplines are aimed at the modern knowledge of humanities, contemporary humanitarian technologies, personal development; they have humanitarian emphasis.

Besides, psychological and pedagogical disciplines matter in educating students. The disciplines have a high motivating potential, as they enable young people to improve psychological and pedagogical literacy, develop skills of self-control and self-organization. All this leads to harmonization of interpersonal relations in the society as a whole.

Research methods are the following: theoretical—overview of psychological and pedagogical literature; empirical—questionnaires, performance analysis, expert assessment, testing; methods of data processing—mathematical statistics methods.

Applied techniques:

1. Final testing of students' knowledge.
2. Oral and written tasks for the assessment of communication skills, capacity for teamwork, efficient cooperation with other subjects of professional activity with due regard for social and cultural differences.
3. Stephenson's Q methodology.
4. V. F. Ryakhovsky's test for communication skills.
5. A. A. Karelin's test for communication skills.

4 Result

The result of the research will be development and introduction of the model of organizing psychological and pedagogical training of students majoring in engineering based on using humanitarian technologies, active and interactive forms and methods of learning and cooperative learning; working out the educational content on the basis of solving professional tasks of organizational management necessary for engineers.

The model is a system of interrelated stages: (1) model development; (2) a motivational stage aimed at building students' readiness to develop a new type of educational and professional activity through the activation of their personal experience, joint planning of instruction, creating an atmosphere of empathy and dialogue; (3) the stage of solving educational and professional problems, aimed at acquiring students the experience of communication in the future professional activity; (4) stage of reflection, evaluation of the results obtained, focused on checking the identified pedagogical conditions for the development of the communicative competence of engineering students. Model structure is shown in Fig. 1.

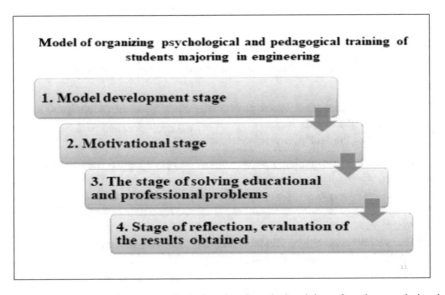

Fig. 1. Model of organizing psychological and pedagogical training of students majoring in engineering

Each stage is revealed through specification of individual components of the process. These components are shown in Fig. 2.

The conceptual core of psychological and pedagogical disciplines comprises educational and professional tasks of the communicative element of an engineer's organizational and managerial activity.

The following types of tasks are identified:

(1) to cooperate with other subjects of engineering activity, taking into account social and cultural differences;
(2) to organize efficient collaboration during meetings, negotiations, consultancy, workflow management, development and implementation of projects;
(3) to take part in seminars, conferences, meetings of the scientific and technical community with public presentations;

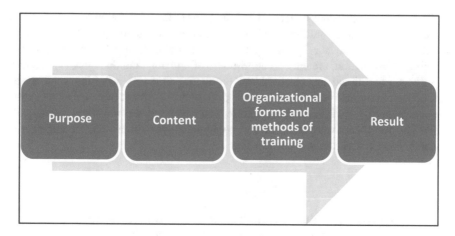

Fig. 2. Structure of stages of psychological and pedagogical training

(4) to produce written texts based on the collection, processing, analysis, and systematization of information;
(5) to design and implement professional self-education.

The basic form of organizing psychological and pedagogical training is collective organization, which involves interaction between all participants of the training process divided into variable pairs. At any specific time, 50% of students assume the role of trainers, while the other 50% are trainees. This ensures maximum student activity and their engagement in the training process while also providing an opportunity for interpersonal communication. Students divided into variable pairs interact through dialog with changing roles (trainer–trainee), processing of new information, its consolidation by mutual training, and immediate testing of the results of training. At the same time, new methods are implemented, such as the associative dialog method (technique–opinion poll, tandem work training, method of addition into text and image), method of monitors, method of paragraph-by-paragraph examination of text, flow chart method, mutual training method) developed by Russian researchers based on the didactic concept of Dyachenko [8].

Implementation of new forms and methods into the training process requires special preparation, since it involves adopting a new type of educational activity. The implementation process is divided into several stages:

(1) motivational stage;
(2) preliminary familiarization with the principles of working in variable pairs;
(3) unsupervised work in variable pairs combined with familiar activities;
(4) "external speech" stage—work in variable pairs while verbalizing the sequence of actions;
(5) "internal speech" stage—work in variable pairs without relying on algorithms;
(6) creative application stage, which allows for free use of the associative dialog method in the training process.

In accordance with these stages, we have organized the training process and selected forms, methods, and techniques for each class.

We have also developed criteria and selected a diagnostic apparatus for assessing the efficiency of the developed model. Along with test methods, it includes oral and written tasks for the assessment of communication skills, capacity for teamwork, efficient cooperation with other subjects of professional activity with due regard for social and cultural differences. We have identified the following levels of capacity development: adaptive, productive, and creative. Respondents were examined before and after psychological and pedagogical training.

We have been carrying out experimental work for three years at Peter the Great St. Petersburg Polytechnic University. The study involved 120 third- and fourth-year students majoring in engineering. There were 77 people in experimental groups and 43 in the control group. Students were taught one of the psychological and pedagogical disciplines (Pedagogy and Psychology, Psychology, Psychology of Communication) over the course of a semester. Teaching in control groups was based on a lecture-seminar model with the occasional use (25%) of active and interactive techniques. Efficiency criteria of this model were development indices for particular components of the communicative competence. The indices were measured before and after the experiment in the experimental and control groups.

The examined parameters show positive dynamics (at the level of statistically significant differences).

The main results are presented below. The results of final knowledge tests of respondents in the field of business communication psychology are shown in Fig. 3.

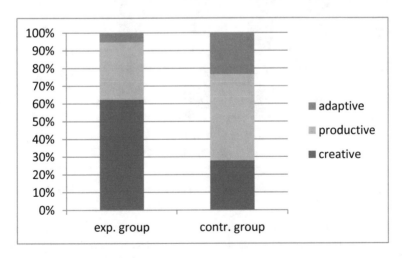

Fig. 3. Results of final testing in the experimental and control groups

The results of control tests in the experimental and control groups are shown in Figs. 4 and 5.

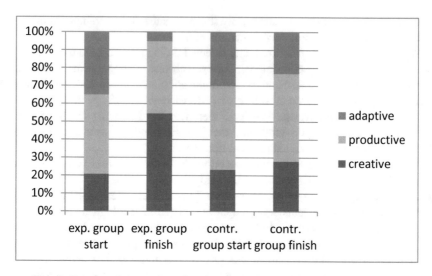

Fig. 4. Results of parameters of oral communication skills development level

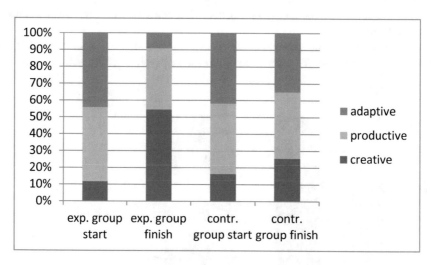

Fig. 5. Results of parameters of written communication skills development level

Capacity for teamwork and efficient cooperation with other subjects of professional activity, taking into account social and cultural differences were analyzed through expert assessment of participation of the students of the experimental and control groups in business games held at the beginning and at the end of the semester. The obtained results are shown in Fig. 6.

Positive changes are also observed in the average level of development of communication skills (A. A. Karelin's test) of the students of the experimental groups.

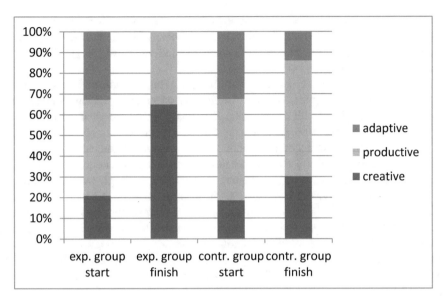

Fig. 6. Results of parameters of capacity for teamwork and efficient cooperation with other subjects of professional activity

The results of V. F. Ryakhovsky's test show that at the beginning of psychological and pedagogical training in experimental and control groups only 22.5% of students have a normal level of sociability. The level of most students (70%) is much higher. For instance, 43% of students are overly sociable, 24% are excessively sociable, and 3% are painfully sociable. Only 7.5% of students tend to be unsociable. Thus, it can be concluded that most students are not afraid of making contact with new people, have a propensity for demonstrative behavior, and want to be the focus of attention. However, they do not correlate their behavior strategies with those of other people and are unwilling to make efforts to maintain contact or hear another person out. The experiment has revealed statistically significant differences in the assessment of sociability level for the experimental groups: $p = 0.002$ ($p < 0.01$). The number of students with a high level of sociability has increased to 40%, while 53% of students qualify as overly sociable. Despite certain flaws inherent to this level, in general there is a positive trend of changes in the nature of student sociability, because prior to the experiment their sociability was excessive. After the formative experiment, the number of excessively sociable students dropped to 13%, and the number of painfully sociable students reduced to 0%.

The results of W. Stephenson's Q methodology show changes in the behavior patterns of the students in the training group during psychological and pedagogical training. At the beginning, about 50% of students (from 50 to 57%) in the experimental and control groups showed a propensity for dependence in their behavior within the group. Combined with the prevailing tendency to avoid conflict (over 50% of respondents), this is indicative of conformist behavior and willingness to remain neutral in disputes and conflicts within the group. A third of all students show a propensity for independent behavior, while only 3% accept conflict. It should also be noted that some students (from 13 to 20%) show ambivalent behavior, which indicates intrapersonal conflict between the desire to adopt the standards and values of the group and their simultaneous rejection.

After the experiment, the number of students with ambivalent behavior in the experimental groups dropped, whereas the number of students willing to take active part in the interactions within the group and ready to defend their point of view increased. On average, 40% of students show a propensity for accepting conflict. However, the general tendency of avoiding conflict remained unchanged. Combined with the reducing propensity for dependent and ambivalent behavior, this indicates a more conscious choice of behavior strategy within the training group. The number of respondents with ambivalent behavior (20–36%) in the control groups has increased. Intrapersonal conflict and uncertainty in choosing behavior strategies in a group may be a consequence of students feeling indecisive during interactions within the group, because they have gained additional knowledge of the psychology of interpersonal communication but do not have enough experience of its practical application. Statistically significant differences for the experimental groups have been obtained ($p < 0.001$).

5   Conclusions

Thus, the introduction of a model of the process of studying psychological and pedagogical disciplines the faculty of engineering based on use of humanitarian technologies, active and interactive forms and methods of education and training in cooperation, development of educational content based on the decision of professional tasks contribute to the development of universal competences as a component of professional engineering competence, improving the quality of education, on the development of the humanitarian environment in the University engineering. Cooperation in solving educational and professional problems and role reversal during this cooperation facilitated the development of the oral and written communication skills of future engineers as well as their capacity for solving the problems of efficient interpersonal interaction and teamwork.

Comparative analysis of the results of experimental work in the control and experimental groups conducted using statistical data processing methods shows statistically significant changes in the level of sociability of the students in the experimental groups, their communication skills, capacity for oral and written communication, for solving the problems of interpersonal interaction under conditions similar to engineering activities, and for teamwork, taking into into account social and cultural differences. There are changes in the behavior patterns of the students in the training group. The students of the experimental group have started to feel more independent; they are ready for interaction within the group, want to take active part in the life of the group, are not afraid of defending their point of view in disputes, and rake a more conscious approach to choosing the perfect behavior model in the training group.

Thus, the conducted experimental work testing the model of psychological and pedagogical training of students majoring in engineering can be considered successful, and its tasks completed.

References

1. Andreev, A.G.: Study of humanities: next crisis? Higher Educ. Russia **7**, 95–103 (2004)
2. Verbitsky, A.A.: Problems of project-context specialist training. Higher Educ. Today **4**, 2–8 (2015)
3. Subetto, A.I. Essays. Noospherism. In: Zelenov, L.A. (ed.) Noospheric or Non-classical Social Science: Seeking Grounds, 628p. N.A. Nekrasov KSU, Kostroma (2007)
4. Fokin, Y.G.: Teaching and Motivation in Higher Education, 224p. Akademia, Moscow (2002)
5. Schedrovitsky, G.P.: Selection of Works, 800p. School, Culture Politics, Moscow (1995)
6. Gulk, E.B.: Educational process at the technical university through the eyes of its participants. In: Gulk, E.B., Kasyanik, P.M., Kruglikov, V.N., Zakharov, K.P., Olennikova, M.V. (eds.) Advances in Intelligent Systems and Computing, vol. 544, pp. 377–388. ICL, Springer, Berlin (2017)
7. Tryapitsina, A.P., Pisareva, S.A.: Guidelines for updating the content of professional training of future teachers. Man Educ. **3**(48), pp. 12–18 (2016)
8. Dyachenko, V.K.: New Didactics, 496p. Public Education, Moscow (2001)

Studying the Effect of Using E-Learning Through Secure Cloud Computing Systems

Hosam F. El-Sofany[1,2], Samir A. El-Seoud[3(✉)],
and Rahma Tallah H. Farouk[4]

[1] King Khalid University, Abha, Kingdom of Saudi Arabia
helsofany@kku.edu.sa
[2] Cairo Higher Institute for Engineering, Computer Science and Management,
Cairo, Egypt
[3] Faculty of Informatics and Computer Science, British University
in Egypt—BUE, Cairo, Egypt
Samir.elseoud@bue.edu.eg
[4] Future University in Egypt—Faculty of Computer Science and Information
Technology—FUE, Cairo, Egypt
rahmahoussam@gmail.com

Abstract. Cloud computing is a set of IT services offered to users over the web on a rented base. Recently, cloud computing is considered as a new internet based paradigm for hosting and providing hardware and software services the consumers. The main services models for cloud computing includes: software, platform and infrastructure (as a service), to satisfy the needs of different kinds of organizations. *Security* in cloud computing is critical when developing services, especially in the education area, cloud computing will be the basic platform of the future E-learning environment. It provides secure data storage, favorable and easy to use internet services and strong computing infrastructure for e-learning environment. This paper will focus on the effectiveness of using e-learning through cloud computing systems. The research study shows that the cloud computing model is extremely valued for both students and instructors to achieve the course objective. The paper presents the advantage, limitation and the challenges of using cloud computing services (IaaS, SaaS, PaaS) as a good platform for e-learning. The descriptive statistics, i.e. of the selected dependent and independent variables that related to the research samples, will be presented by the authors. Arithmetic mean, standard deviation, coefficient of standard variation, etc., will be include in the descriptive statistics. The results of the research study are mainly based on data from questionnaires of survey that have been held by the authors from to February 25, to May 20, 2018, among a sample of adults.

Keywords: Cloud computing · E-learning, cloud security
Cloud attacks, statistical analysis

M. E. Auer and T. Tsiatsos (Eds.): ICL 2018, AISC 917, pp. 54–63, 2019.
https://doi.org/10.1007/978-3-030-11935-5_6

1 Introduction

Cloud computing architecture is considered as a recent generation of web-based network in the IT field. It is a highly successful model of service oriented computing, and has led to the development of the way computing infrastructure used and managed. It is a recent model for hosting resources and provisioning of services to the consumers. It provides an easy, on-demand access to a centralized shared of computing resources that can be published by a minimal management overhead from the organizations and with a great efficiency. Cloud computing providers depend on the Internet as the intermediary communications medium leveraged to deliver their IT resources to their consumers on a pay-as-you-go basis [1].

In the scientific references, there are several definitions of the cloud computing, but the definition given by the NIST *"National Institute of Standards and Technology"* is an authoritative one [2]. According to NIST definition, the cloud model promotes availability and is composed of five essential characteristics that provide (1) high scalability and elasticity, (2) availability and reliability, (3) performance and opti- mization, (4) accessibility and portability, and (5) manageability and interoperability. Three main services are provided via cloud computing, mainly Software as a service (SaaS), Platform as a service (PaaS) and Infrastructure as a service (IaaS). However, five component architectures comprise cloud computing, mainly clients, applications, infrastructure, platforms and servers. In [3–5], the following four deployment models have been presented.

(1) *Private cloud*: The cloud infrastructure is provided for private used by a single organization comprising of multiple users. It is owned and managed by a specific organization.
(2) *Community cloud*: The cloud infrastructure is provided for specific use by a specific community of consumers from organizations that have shared concerns (e.g., mission, security requirements, policy, and compliance considerations). It may be owned, managed, and operated by one or more of the organizations in the community.
(3) *Public cloud*: The cloud infrastructure is provided for open use by the general public. It owned, managed, and operated by service provider.
(4) *Hybrid cloud*: The cloud infrastructure is consists of two or more cloud infrastructures.

Cloud deployment models together with their internal infrastructure are shown in Fig. 1.

The cloud computing provides the above cloud services as pay-per-use. Therefore the utilization of resource, servers, storage, databases, processing, etc., are charged to the customers by the provider. Cloud services are provided by contracts in the form of *service level agreements* (SLA). The SLA presents and defines the level and quality of service required by the organization.

The paper is organized as follows: in section two, we present the advantages and disadvantages of using cloud computing. In section three we present the important objectives of cloud computing security. In section four we introduce a cloud education architecture model. In section five we introduce a cases study for using E-learning through cloud systems in Egypt and KSA, in specific we present our studies in BUE, FUE and KKU. Finally the paper is concluded in section six.

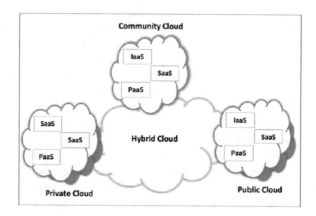

Fig. 1. The distribution of cloud deployment models and their infrastructure

2   Advantages and Disadvantages of Using Cloud Computing

There are many advantages provided by cloud computing environment such as efficiency, integration, scalability and capital reduction [3, 5]:

- **Improve the performance of PCs**: All applications of e-learning will run faster via cloud because of the number of programs and the loading processes into machine memory will radically reduce.
- **Lower the maintenance cost of the software**: The software maintenance cost will be reduced since the software is stored in the cloud. In such case, upgrading of the software will take place automatically on the cloud environment and no need for IT staff to perform this task.
- **Lower the cost of the hardware**: Using the cloud environment to run most cloud computing web-based e-learning applications means no need for powerful PC or hard disk space.
- **Lower the cost of software:** Using the cloud environment to run applications means no need for the organization to install application on each single P and organizations need to pay for only their actual usage of a software package over the cloud.
- **Lower the cost of the IT Infrastructure:** The organization will be able to control its own costs, by using less servers that needed to store the data and the applications.
- **Increase the capacity of storage**: cloud computing offers a very high storage capacity.

- **Automatic software updates**: cloud software updates are done automatically, this means that that the IT department didn't concerned with the software updating and will save the cost of this process.
- **Increase the security of data**: In cloud computing, data will be duplicated automatically, unlike computing using desktop machine. Therefore, if a computer crash took place in the cloud, the data is not destroyed. For this reason, no need for local data backup. Stored data in cloud are secured in the event of viruses as well as any disasters like fires.
- **Accessible through a wide range of devices**: cloud computing facilitates the accessibility of a wide range of devices. Even if you switch devices, the cloud provides access and usage to all of your existing documents and application you were accessing though a different devices such as (PCs, notebooks, smart phones etc.).
- **Easy user interface and group collaboration**: In cloud computing, anyone anywhere with an internet connection can collaborate in real time. It also enables group collaboration from different countries as easily as if they were in the same building.

The disadvantages of Cloud Computing include:

- **Need Internet connection**: Cloud computing needs Internet a connection to access the e-learning applications and documents. Nothing can be accessed on the cloud without Internet connection.
- **Low-speed connections**: Similarly, cloud computing could be very slow in running resources using connection to Internet with low-speed, for example, in in remote regions that only offer dial-up services.
- **Data security issues**: In cloud computing, all data is stored in the cloud. Hence, security is consider to one of the main challenges that hinder the growth of cloud computing. Stored data in the cloud are as the cloud is. Confidential data stored in the cloud could be accessed by any unauthorized users who gain access to the cloud.

3 Cloud Computing Security

Cloud computing Security is critical when developing cloud based e-learning systems and especially when building cloud services. Surveys and scientific studies indicate that "security" is considerably one of the top five concerns. In cloud computing, customer data, information, and programs are stored in the servers of cloud provider. Hence, data security becomes on the main concern. *Security of cloud computing* identifies the following important objectives [4]:

- *Availability*: services should be always available for users at any time, at any location.
- *Authentication*: the identity of individuals involved in the web communication should be assured.
- *Accountability*: in which the cloud computing systems assure no individual can deny its participation in a data transfer between them.

- *Confidentiality*: user's data should be secret in the cloud systems by making it available only to eligible individuals and no unauthorized access to data can be obtained.
- *Integrity*: in which the cloud based systems ensure that data has not been altered in any way while it is stored or while its processing and transport over the cloud networks.

The Security objectives mentioned above required the use of certain security techniques and services to be developed. A *security mechanism* can be defined as a process which aimed to detect or/and prevent a security attacks. A *security service* can be identified as a processing or communication service aimed to enhances the security of data and the information transfers of an entity. These services help in countering security attacks. Security services usually use one or more security mechanism to achieve its objectives [6]. Data security over the cloud could be achieved, for example, via encryption, digital Signature, hashing, identity and access management (IAM), single sign-On (SSO), Public Key infrastructure (PKI), hardened virtual server images, and cloud-based security groups [7]. This research study is an enhancement of research paper presented in [5] that focuses on the application of cloud computing in e-learning environment. The research work presented here shows that the cloud computing platform provides both students and instructors valued mechanisms to achieve the course objective. The paper presents the nature, benefits and cloud computing services, as a platform for e-learning environment.

4   A Cloud Education Architecture Model

The cloud education architecture model is composite of six layers namely: physical hardware layer, virtualization layer, education middleware layer, application program interface layer, management system and security certification system as shown in Fig. 2 [5]:

1. **Physical hardware layer**: contains servers, storage, and network equipments.
2. **Virtualization layer**: consists of three parts: virtual servers, virtual storages, and virtual databases. The goal of this layer is to break completely information islands based on existing regional through the distributed technology and virtualization technology.
3. **Education middleware layer**: consists of three platforms: education platform, teaching administrator, and resources platforms. It is the core and basic business platform layer since all necessary information, including ordinary file and database, attached to it on different computing node.
4. **Application program interface layer**: consists of three modules: user, resource, and course managements. It provides the necessary interface, hosting service and model's scalability.
5. **Management system**: monitors physical condition, virtualization software, hardware and software, open API and enhance the safety of the software platform.
6. **Security system**: monitors identity authentication and authorization, single point login, virtualization soft- ware and hardware access control and audit, the education middleware and open API access control.

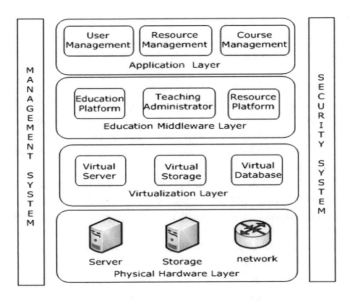

Fig. 2. A cloud education architecture model

5 Using E-Learning Through Cloud Systems in Egypt and KSA

5.1 Method of Using E-Learning Through Cloud in Egypt

A. *The British University In Egypt (BUE)*: *case study*

BUE uses Moodle as cloud based E-learning management system (EMS). Moodle provides the essential tools for the learning process across the university for more than ten years. The E-learning at BUE is hosted on Microsoft Cloud.

BUE Student Record System (SRS) is the corporate student database. Its main purpose is to support all aspects of student—related administration, including admissions, enrolments, fees, registration on modules, scheduling of classes and exams, attendance monitoring, assessment, awards and certificates. The database is available as a web application, so you can connect to it from anywhere using Internet Explorer Browser 8 or above through the address: srs.bue.edu.eg.

The SRS has been integrated with the E-learning. Moodle platform is being used to exam transfer to BUE academic partner London South Bank University (LSBU) in UK and for digital module files.

The academic staff uses Moodle EMS to upload courses materials, assignments, announcements as well as final grades to the students. The population of the research study was random student groups from different departments. The results in this paper are based on modules taught at the BUE. The modules have been conducted at BUE as e-learning modules and traditional classroom modules. At The Interactive assessment, students will be provided with tools to let the educator to know the advantages and disadvantages in terms of their performance and their perceptions of the teacher's performance. Intellectual stimulations in university education can be irreplaceable using web-based education system. It is important for educators to obtain course evaluation to determine how successful and effective a course is taught in the classroom [8].

In the past BUE used paper-based evaluation process at the end of each semester. This process was slow, time consuming, need many qualified staff, questionnaires are often outdated, and the results are often too late to make changes in the classroom. For improving teaching styles, BUE currently using cloud based evaluation process that has the advantage of instant feedback, and appropriate questions could be added, and ease to present statistical analysis results [9].

The main goal of this research study is to show the need for cloud-based e-learning modules and its effectiveness based on the existing system. A web-based survey among the students at the campus had been conducted and reported on their assessments. From 85 to 100 students from different faculties had been engaged at the survey. The students are 18–22 ages, and from different academic years, with online learning experience mainly from the BUE. Figure 3 shows the questionnaire that has been given to the students who participated in the study. The students belong to different colleges and selected randomly over all four academic years. Students were asked to rank ten e-learning system's activities according to their importance and relevance to them. This ranking has five Likert scale ranging from least to very much relevant (i.e., 1 = Least, 2 = Little, 3 = Moderate, 4 = Much, 5 = Very Much).

B. *Future University in Egypt (FUE): case study*

Faculties at the FUE also use Moodle as a cloud based E-learning management system to disseminate course materials, encourage class discussion via forums, and collect student work through assignments and online quizzes. It also facilitate the communication between the students and both lectures and teaching assistants, as both of them can make an announcement for the students if there is something important [10].

Students can use any web-enabled device to access the Moodle. Courses are provided in a password-protected environment, ensuring student privacy within the class [10].

The FUE Moodle provides the essential tools for the learning process. The Lectures and the academic advisors (Teaching Assistants) uses Moodle to upload courses materials, assignments, announcements as well as final grades to the students. On the other hands student can get announcements and notifications from his/her teachers. Students use their own ID's and Username and Password authorization to access the Moodle system. The student's username and password are the same as those he/she will used for the admission and registration of FUE website. For easiest and secure the access, the student username will be "*University ID*", and the password is the "*University ID*" also until the student first login and changes his password if he/she wants. Using Moodle facilitate the communication between the students and the academic staff. Once you are logged into the system as an academies staff or as a student, you can enjoy more and more functions that facilitate the e-learning process.

Student name:	Faculty:
Student ID. No.:	Department: Year:

Questions	**Student Evaluation**
1. Do you access E-learning regularly?	☐ Yes ☐ No ☐ Never
2. How much time do you spend on E-learning per day?	☐ 15 minutes ☐ 30 minutes ☐ 1hour ☐ others
3. Is it well organized/easy to access your content? Explain please	☐ Yes ☐ No ☐ I don't care
4. Does it provide a mean of interaction between you "as a student" and your instructor? Explain please	☐ Yes ☐ No ☐ I don't care
5. Does it provide a mean if interaction among students (collaboration interaction)? Explain please	☐ Yes ☐ No ☐ I don't care
Please rank the following activities according to their importance and relevance to you:	
1. Online quizzes, you will be able to have a quiz online on the e-learning within time limit	☐ Least ☐ little ☐ Moderate ☐ Much ☐Very much
2. Encourage active participation through feedback & comments on various activities and class topics with appropriate monitoring	☐ Least ☐ little ☐ Moderate ☐ Much ☐Very much
3. Collaboration among students and staff through discussion forums to share ideas that would improve research and communication skills collectively	☐ Least ☐ little ☐ Moderate ☐ Much ☐Very much
4. Provide video streamed lectures and labs to your students	☐ Least ☐ little ☐ Moderate ☐ Much ☐Very much
5. share interesting bookmarks of various academic resources and important web sites	☐ Least ☐ little ☐ Moderate ☐ Much ☐Very much
6. Provide interactive chat rooms for students and staff to communicate with each other	☐ Least ☐ little ☐ Moderate ☐ Much ☐Very much
7. Enable voting where students can vote on several topics, activities and ideas through a social network	☐ Least ☐ little ☐ Moderate ☐ Much ☐Very much
8. File sharing system for posting research paper and various research contributions	☐ Least ☐ little ☐ Moderate ☐ Much ☐Very much
9. A hot question activity, where the students could vote on others' question, so the hottest question will be popped up. Teachers will make oral comments on question in classroom	☐ Least ☐ little ☐ Moderate ☐ Much ☐Very much
10. Adding voice narration to PowerPoint slides, where you will be able to read and hear the contents online, in other words the students will be able to hear the instructors' explanation while reading the slides online.	☐ Least ☐ little ☐ Moderate ☐ Much ☐Very much

Fig. 3. E-learning questionnaire for students

5.2 Method of Using E-Learning Through Cloud in Saudi Arabia

A. *King Khalid University (KKU): case study*

Students are the basic of KKU and for them the university was founded in the beginning. KKU provides several services that facilitate student study procedures and keeps pace with the developments in the modern era such as (e-learning services, developmental systems and regular follow-up) that makes it easier for them to obtain educational services from anywhere and at any time [11].

KKU used Blackboard as cloud based E-learning management system. Blackboard can be accessed on-campus or off-campus using only a computer with Internet access. Blackboard provides the essential tools for the learning process across the main campus as well as the university branches. The academic staff uses Blackboard Moodle EMS to upload courses materials, assignments, announcements as well as final grades to the students. The population of the research study was random student groups from the different levels. On the other hands student can get announcements and notifications from his/her teachers and the E-learning Center. Students are used Username and Password authorization to access the Blackboard system. The student's username and password are the same as those he/she will used for the admission and registration of KKU website. For easiest and secure the access, the student username will be "University ID", and the password is the "National ID number" of the student. Using the Blackboard system is very easy to use. Once you are logged into the System, you feel free to explore all the e-learning options [11]. The same web-based survey and questionnaire mentioned above among the students at KKU had been conducted as shown in Fig. 3.

6 Conclusion

This paper focuses on the effectiveness of using e-learning through cloud computing systems. The research study shows that the cloud computing model is extremely valued for both students and instructors to achieve the course objective. The paper presents the advantage, limitation and the challenges of using cloud computing services (IaaS, SaaS, PaaS) as a good platform for e-learning. The descriptive statistics, i.e. of the selected dependent and independent variables that related to the research samples, have been presented. Arithmetic mean, standard deviation, coefficient of standard variation, etc., have been include in the descriptive statistics.

The features of the Cloud Computing platform are quite appropriate for the migration of this learning system, so that we can fully exploit the possibilities offered by the creation of an efficient learning environment that offers personalized contents and easy adaptation to the current education model. Specifically, the benefits considering the integration of an e-Learning system into the cloud can be highlighted as good flexibility and scalability for the resources, including storage, computational.

Accordingly, for answering the question "what are the effects of implementing cloud computing on e-learning system?" we can say that the cloud computing technology, being served as the infrastructure of the service, has had better performance in comparison to the web technology for the e-learning systems. Other studies have proved the above statement as well. They have also stated that the cloud computing system has had lower costs. Therefore, the present study, similar to the previous studies, has introduced cloud computing technology as an appropriate and better technology for the e-learning systems.

References

1. Mell, P., Grance, T.: NIST Definition of Cloud Computing. National Institute of Standards and Technology. October 7, 2009
2. Jansen, W.A.: Cloud Hooks: Security and Privacy Issues in Cloud Computing NIST. In: Proceedings of the 44th Hawaii International Conference on System Sciences—2011
3. Micheal, M.: Cloud Computing: Web-Based Applications That Change the Way You Work and Collaborate Online. Que Publishing, USA (2009). https://books.google.com.eg/books/about/Cloud_Computing.html?id=mzM53Yp9cpUC&redir_esc=y
4. McKendrick, J.: Loud Divide: Senior Executives Want Cloud, Security and IT Managers are Nervous. ZDNet January 2011. http://www.zdnet.com/article/cloud-divide-senior-executives-want-cloud-security-and-it-managers-are-nervous/
5. El-Sofany, H.F., Al Tayeb, A., Alghatani, K., El-Seoud, S.A.: The impact of cloud computing technologies in E-learning. Int. J. Emer. Technol. Learn.–iJET, **8**, 37–43 (2013). http://online-journals.org/i-jet/article/view/2344
6. Chouhan, P., Singh, R.: Security attacks on cloud computing with possible solution. Int. J. Adv. Res. Comput. Sci. Softw. Eng. **6**(1), (2016)
7. Harfoushi, O., Alfawwaz, B., Ghatasheh, N.A., Obiedat, R., Abu-Faraj, M.M., Faris, H.: Data security issues and challenges in cloud computing: a conceptual analysis and review. Commun. Netw. (6), 15–21 (2014). https://www.scirp.org/journal/PaperInformation.aspx?paperID=42813
8. El-Seoud, M.S.A., El-Sofany, H.F., Taj-Eddin, I.A.T.F., Nosseir A., El-Khouly. M.M.: Implementation of web-based education in Egypt through cloud computing technologies and its effect on higher education. High. Educ. Stud. **3**(3) (2013) ISSN 1925-4741, E-ISSN 1925-475X. Published by Canadian Center of Science and Education
9. Theall, M.: Electronic Course Evaluation Is Not Necessarily the Solution. The Technology Source, November/December 2000, Michigan Virtual University. Retrieved August, 2003, from http://technologysource.org/?view=article&id=108
10. https://its.davidson.edu/hc/en-us/articles/115010465907-An-introduction-to-Moodle-for-students. Accessed May 2018
11. www.kku.edu.sa. Accessed May 2018

Better Understanding Fundamental Computer Science Concepts Through Peer Review

Yvonne Sedelmaier$^{(\boxtimes)}$, Dieter Landes, and Martina Kuhn

Coburg University of Applied Sciences and Arts, Coburg, Germany
{yvonne.sedelmaier,dieter.landes}@hs-coburg.de,
martina.kuhn@gmx.com

Abstract. Computer science courses, like many others, encompass subjects that are difficult to understand for students. Reflection on these subjects seems to be important for gaining a better understanding of complex and fundamental concepts in informatics. This paper presents a didactical setting that aims at fostering understanding of such hard subjects. In particular, the evolution of this didactical setting started out from combining ten-minute papers, peer review, and bonus scores. The combination of these didactical elements has been employed in an introductory computer science course that covers, among other things, several topics from theoretical computer science. The paper presents the detailed setup of the learning setting and its underlying goals as well as how it evolved over several years. Furthermore, the paper presents a qualitative and quantitative evaluation of the approach, both from the perspective of students and the perspective of instructors. Evaluation also yields data that expose differences between different versions of the approach. Furthermore, evaluation data also support a critical analysis of potential success and risk factors with respect to the efficacy of the approach, in particular the type of problem statement, the motivation for providing proper feedback, and the role of bonus scores. Overall, the learning setting yields encouraging results, yet offers various options for refinements in future work.

Keywords: Computer science education · Didactics · Peer review
Freshmen students · Higher education

1 Introduction

Similar to other disciplines, computer science builds on some foundations that must be laid before students may advance to more elaborate topics. In computer science, these foundations encompass several topics form theoretical computer science which have an impact on various other issues that are covered later. In addition, most of these concepts are rather abstract since abstraction is one of the key issues in informatics.

More than 15 years of experience in teaching an introductory course in computer science indicate to the authors that freshmen students have difficulties in obtaining a deep understanding of some of these fundamental concepts. For instance, the nature of information as opposed to data is hard to grasp. Similarly, Turing machines as a fundamental model of computation and its relationship to the limits of computability

© Springer Nature Switzerland AG 2019
M. E. Auer and T. Tsiatsos (Eds.): ICL 2018, AISC 917, pp. 64–75, 2019.
https://doi.org/10.1007/978-3-030-11935-5_7

pose difficulties to freshman students. Recursion as opposed to iteration as well as formal languages, automata and their relationship to computability and decidability are also hard issues.

Nevertheless, understanding these topics is vital for computer scientists-to-be. Thus, finding ways to facilitate a deep understanding of such hard issues is an important step in improving computer science education.

This paper presents a didactical setting that combines ten-minute papers, peer review, and bonus scores in order to foster a deeper understanding of abstract and complex material, in particular in an introductory course in computer science. Alongside with technical issues, such a didactical setting needs to address several non-technical competences.

The paper discusses the goals and the didactical underpinning that initiated the development of such a feedback-based didactical setting. Then, the evolution and the details of the didactical setting are described. Furthermore, the paper presents a qualitative and quantitative evaluation of the approach, both from the perspective of students and the perspective of instructors. Evaluation shows that the didactical setting is a promising step forward that may also go with other topics, but also exposes some aspects that might deserve further elaboration. A summary and an outlook on future activities conclude the paper.

2 General Setting

„Foundations of Computer Science" is a compulsory course in the first term of a bachelor program in informatics which aims at introducing freshman students to some of the fundamental concepts of computer science, with an emphasis on software-related issues. In parallel to this course, students attend several other courses that focus on introducing (object-oriented) programming or computer architecture.

Between 100 and 120 students are enrolled in the course which has six contact hours per week. Two contact hours are devoted to exercising while the other four tend to be lecture-like, although they also encompass activating forms of learning such as working in small groups.

Topics covered in the course include:

- Historical evolution of computer science,
- Mission and scope of computer science,
- Models and limits of computation including Turing machines and halting problem,
- Information versus data including an introduction to information theory,
- Requirements and their role in software development,
- Introduction to algorithms including recursion,
- Categories of programming languages and their characteristic features,
- Formal languages, automata and their role in compiling computer programs.

At the end of the term, students need to pass a written exam of 90 min duration.

3 Intended Learning Outcomes and Theoretical Basis

Several years of teaching the course have shown the instructors that some of the covered topics, in particular the more theoretical ones, are hard to understand for a large share of the students. Apart from the inherent complexity of the topics, these difficulties appear to be related to the fact that the students are novices that still have to get accustomed to being university students. At school, pupils need to attend all classes and do the homework that the teacher assigns to them. In contrast, there is no explicit homework at German universities and students are free to decide if they attend a class or not. Furthermore, there are regular examinations across the school year, while it is common at German universities that only a single exam at the end of the term concludes a course. The latter may lead novice students to let things slide due to a perceived lack of pressure and intensify learning efforts only when the final exam is already imminent.

Therefore, we came up with several goals in 2014 that we would want to achieve in the course, alongside to the technical aspects of the covered topics.

3.1 Initial Goals

Goal #1: Students shall reflect on and realize how the various topics in the course relate to each other and what role they play in a broader context.

Goal #2a: Students shall receive appreciation and respect from the instructors as a prerequisite to increase motivation and positive reinforcement.

Goal #2b: Students shall be able to recognize own and others' errors and deficits and accept different perspectives of their peers as a basis to round off their state of knowledge.

Goal #3: Students shall initiate a process of self-reflection on their way of learning and their current state of knowledge.

In particular, goal #3 entails that students should ask themselves whether they

- sufficiently understood the covered topics,
- follow an appropriate learning approach, and
- invested enough effort for the course.

Likewise, students should judge their state of knowledge in relation to their peers in order to obtain some point of reference.

3.2 Didactic Underpinning

Learning settings are frequently devised on an ad-hoc basis without solid theoretical, methodological, and scientific basis. Following such an approach may yield results that are good enough for the particular narrow context, but cannot be generalized.

In order to avoid shortsighted hands-on solutions, more principled approaches are indispensable, and such approaches must be firmly grounded in didactics. Didactics, as the science of teaching and learning, does not mean to apply ready-to-use methods or concepts to a course. Rather didactics establishes a theoretical basis and a solid fundament for making decisions on how students are supposed to learn. This fundament comprises, among other things, an idea of man, theories of learning, and a clear terminology. Didactics as a theoretical foundation supports instructors in making sound and conscious decisions in educational issues. Conscious decisions take into account determining factors such as age of the learners or time and place of learning, etc. Furthermore, didactics determines which learning content, intended learning outcomes, methods and media should be employed, keeping their interplay in mind.

When a learning setting needs to be developed, didactics usually starts out from formulating intended learning outcomes (ILO) explicitly [1]. ILOs are the basis and fundament for all ensuing decisions and specify the knowledge or competences that students are supposed to have after the course.

In the course discussed here, ILOs are cognitive ones rather than competence or performance (see Sect. 3.1) since students are freshmen in informatics and the course covers abstract and complex fundamentals of computer sciences. In a sense, students need to stockpile knowledge for which they do not see immediate use. Rather, this knowledge will be combined and transferred to practice later in advanced studies. Theoretical knowledge is the basis of competence development and must be adapted and transferred to a specific problem in later studies.

Designing a course needs to be based on a sound didactical fundament to substantiate the decisions concerning intended learning outcomes (ILO), content, methods, media, etc. In this case, competence-oriented didactics [2] is the basis for developing the presented didactical setting. Competence-oriented didactics starts from explicitly defining ILOs before analyzing external constraints and deciding on methods and media. Competence-oriented didactics adhere to the primacy of didactics [3]. Furthermore, the presented didactical approach follows the lines of constructivism and constructivist didactics [4, 5].

ILO 1 focusses on recognizing and comprehending complex interrelationships to understand computer science and its theoretical basis in a comprehensive manner. As a didactical consequence, students shall write summaries of relevant topics to identify essential aspects. This method helps students to abstract the learned issues. In addition, students are requested to use complete sentences instead of just key points. Describing an issue in writing helps to point out gaps of understanding, if there are any. The written summaries are supposed to serve as a kind of a mind map at the end of the term and give students an overview of computer sciences. Furthermore, self-written summaries are expected to support students when preparing for the exam.

ILO 2 states that students should recognize their gaps of understanding. As a didactical consequence, students should view the topic from various perspectives, including a change of perspective, and get some impression on how other students describe the interrelationships. This serves as a basis for completing their knowledge. Reading summaries of other students should provoke another view on the covered issues as well as on their own level of knowledge in relation to the others. As a consequence, this should increase motivation in the sense that students get a positive impression and better self-confidence of their level of knowledge in relation to others. Otherwise, they are expected to experience a push from realizing that their peers are more knowledgeable on the topic and recognize own knowledge gaps.

This directly leads to goal #3, namely reflecting on one's own level of competence and drawing behavioral consequences as a result. Students shall get feedback if they understood things correctly, if others have a much deeper understanding of the topic, if they put enough time into their learning process, if they should adapt their way of learning, etc.

4 Peer Review and Peer Feedback as Didactic Setting

Constructivism views learning as establishing links between new information and existing knowledge and experiences [6]. Since the latter are specific and different from any other person's knowledge and experience, each student has to accomplish this individually which presupposes active cognitive involvement of all students. In particular, this also entails that defining "one-size-fits-all ILOs" will not work properly, and that the instructor can only act as a coach and facilitator rather than as a teacher. Each student starts on a different baseline and goes through an individual learning process which consequently leads to individual learning results. Therefore, learning results are not deterministic and cannot be predicted in detail.

Of course, the level of knowledge and experience of the instructors differs significantly from the level from which students start learning. Consequently, it is hard for instructors to oversee which obstacles might prevent students from understanding complex issues and from learning properly, and which knowledge gaps they still have. As a result, chances are that students may adapt to the knowledge level of their peers more effectively than instructors can. Since their level of knowledge tends to be similar, it is easier for students to learn from each other's work and mistakes. It is a matter of trust that students do not feel so stupid if they get feedback from peers instead of instructors. To utilize these facts, a learning setting was developed that trains abstracting the key facts and initiates self-reflection [7]. Core element of the didactical setting was a peer review process since such an approach supports our ILOs perfectly.

Peer review has been shown as beneficial for learning in several disciplines [8, 9]. Yet, peer review in itself is not a didactical setting, but a framework or an idea how a didactical setting might look like. Peer review must be made more specific and adapted to a particular situation in a didactical process. This adaption requires didactical decisions considering conditions of the learning setting such as target group, prior knowledge, ILOs, etc. The result of such didactical decisions is a didactical setting which is grounded on a sound theoretical framework which gives reasons to substantiate each decision.

Peer review is different from peer feedback. Peer feedback typically reports individual behavior observed by peers. It follows strict rules and should be a possibility to further develop personality traits by triggering a self-reflection process. Peer review, in contrast, is a process to improve quality of an artefact through a multi eye principle. Peer review focusses on artefacts while peer feedback focusses on an individual.

5 Evolution of a Didactic Concept

5.1 First Iteration

In 2014, we launched a didactical setting to provide feedback to students on their understanding of specific topics. So far, this didactical setting was deployed three times across the course, in particular in the context of the historical evolution of computer science, Turing machines, and algorithms.

As a first step, the didactical setting encompasses writing a ten-minute paper in class that should summarize a particular section of the course, namely the evolution of computer science, Turing machines, and algorithms. The summary is supposed to be a page in length and shall contain essential issues of the respective topic in a student's own words. Students are asked to write complete sentences on paper.

This method should help students abstract the key issues of the topic and detect gaps of knowledge by writing a brief summary. Furthermore, such a summary establishes a big picture of interrelationships of the topics in computer science. At the end of the term, students should end up with a self-written summary of the complete course to get the big picture of "Foundations of computer science".

Once the ten-minute papers were complete, the second step consisted of reviewing and commenting on the paper of the student in front or in the rear of oneself. Questions that should guide the review were, e.g.,

- Is there something in your peer's paper that you forgot to mention yourself?
- On which aspects would you disagree with your peer?
- Which aspects would you add? Why?

Students had five minutes to write down their comments on the paper they were reviewing and then handed the paper plus comments back to the original author.

This peer review process focusses on goal #2a and #2b. Receiving feedback gives esteem because the summaries are worth reading and commenting. This fosters motivation and encourages engagement in computer science. Goal #2b focusses on recognizing the gaps of knowledge in absolute terms and in relation to other students. Peer review helps students to obtain a different perspective on the topic by reading foreign summaries and to complete their own knowledge.

Instructors cannot handle goals #2a and #2b adequately because there are more than 100 students in class. Furthermore, freshmen students may not participate in an exercise which is corrected by the lecturer which is a person of respect. If they get feedback from peers, they are less afraid of failing or exhibiting some gaps in knowledge. Overall, peer review processes should trigger some type of self-reflection with respect to the learning process and the level of knowledge.

As a third step, students read the comments they received before instructors collected papers and comments. The instructors later looked over papers and comments in order to find out if the intended goals had been achieved. In the first place, the ten-minute papers were intended to stimulate students to reflect on how the various issues of a course section interrelate and how they fit into the big picture. The quality of the papers as such was largely alright although most papers neglected some relevant issues and some contained downright errors. Feedback did not work well—almost all comments were superficial and only few pointed out omissions or errors in the original paper. Apparently, students did not see much benefit in digging into a peer's solution thoroughly. In addition, it seemed hard for students to give comments on a somewhat fuzzy topic that was neither clearly wrong nor clearly correct.

In summary, we largely achieved goals #1 and #2a, but failed on goals #2b and #3 since we could not see indications of meta-reflection on one's own learning approach.

In 2015, we repeated the didactical setting in a modified format. Again, there were two ten-minute papers (historical evolution, Turing machine), but no peer review section. Rather, we offered feedback by the instructors on the collected papers to improve the quality of the feedback. The goals of the exercises were explicitly included in the problem statement to make them clear and unmistakable.

Nevertheless, these modifications did not pay off in terms of achieving goals #2b and #3. Instructors spent considerable effort on commenting the papers. Nonetheless, students were only moderately interested in getting their papers back after instructors had added comments. This situation even did not change near the end of term when students prepared for the final examination. Apparently, students viewed the assignment of writing a paper as some form of examination without prior notice and did not perceive the benefit of the feedback for their own learning.

5.2   Adjustment of Goals

Since feedback in its original format did not work as well as we had hoped we paused the didactical setting in 2016. In 2017, we took a new start with revised goals.

As mentioned above, we observed that students had difficulties in providing meaningful comments on answer that were not clearly correct or clearly wrong. Furthermore, freshmen students are not used to self-responsible and self-initiated learning from school. Rather, they have to be introduced to this kind of learning when they start their studies. Consequently, learning settings are required which support students in this process. In particular, such learning settings need to include activating methods which trigger self-directed learning rather presenting material in an instructive fashion which would leave students largely as passive consumers.

Therefore, goal #3 was seen as less significant at this stage. In contrast, goal #2a gained importance as mechanisms were required to motivate students to really get involved in a solution of one of their peers and provide rich and elaborate comments. As a corollary, a new and more operational goal was added:

Goal #4: Students shall be able to apply their knowledge to solve problems related to specific topics.

5.3   Second Iteration

The revision of the goals led to several modifications of the didactical setting.

First and foremost, the character of the assignment changed. Rather than being directed to the big picture, assignments now became more operative in style, focusing on solving a problem by using the material covered in the respective section of the course. For the students, it is easier to judge if a solution is correct or not.

Secondly, we moved from a purely paper-based approach to a tool-supported one. In particular, we used the learning-management system (LMS) Moodle for submitting a solution electronically, providing comments, and judging the benefits of the feedback. Using a tool-guided process brings in a more mandatory flavor since it is easier, also for one's peers, to track who submitted a deliverable. Due to using an LMS, parts of the assignment moved out of the physical class and turned into offline activities.

As a third modification, we introduced bonus scores for participating in the didactical setting. Bonus scores are only applicable if the final exam has been passed successfully; then, the bonus slightly improves the final grade. In particular, in the final exam, a total of 90 points may be reached and 6 bonus points may be earned on top by taking part in the assignments. More specifically, we grant bonus points for submitting a solution that is not obvious nonsense, but also for writing a comment which is rated useful by its recipient. It is worth noting that writing a comment does not earn a bonus, but only being the author of helpful feedback is awarded with a bonus score.

In 2017, we used the modified didactical setting on two occasions, namely as terminations of the course sections on information/ information theory and on algorithms (iteration/recursion). In the information theory assignment, students had to compute information contents for four different scenarios of drawing cards from a deck of cards based on Shannon's theorem. In the assignment dealing with algorithms, students need to develop an iterative and a recursive algorithm for computing palindromic numbers (i.e. the 196 algorithm) and explain the difference between iteration and recursion based on their solution.

In both cases, students were asked to work on the assignment in class for 30 min (or 45 min, respectively). In class, students were allowed to ask instructors for clarification of the assignment and to discuss ideas with their peers. Students were then requested to submit their solution electronically in Moodle within two days after the class.

Once the submission deadline was reached, Moodle automatically assigned a reviewer for each submission. Only those students who had submitted a solution were eligible as reviewers. Reviewers were asked to provide written comments in Moodle within five days.

Finally, students were given another two days to rate the quality of the review comments they received, if any. Rating was also done in Moodle as a simple helpful/not helpful choice.

6 Evaluation

6.1 Qualitative Evaluation

Lecturers had the impression that the second iteration of the didactical setting worked considerably better than the first one. Specifically, the quality of comments was much better. This may be due to several reasons: First, a major adaptation concerned the assignments. It seems to be easier for students to review an assignment with clear criteria if the answer is either true or false.

Furthermore, using Moodle as a learning management system brought some more engagement compared to the paper-based process. Half of the students participated in the review process even if the number of students decreased in both courses during the term. Since the students are freshmen, a decrease in this dimension is quite normal.

6.2   Quantitative Evaluation

In the first iteration in 2014, we had 91 students enrolled in the course. 67 of them participated in the first assignment concerning the historical evolution. Participation dropped to 40 and 37, respectively, in the assignments on Turing machines and algorithms.

In 2017, 115 students registered for the final exam. 58 handed in a solution for the first assignment on information theory, 48 for the second assignment on algorithms. 23 out of 33 reviews of a solution to the information theory assignment where voted as helpful, while this fraction was rising to 26 valuable comments out of 36 reviews for the assignment on algorithms (Table 1).

Table 1. Participation in the peer review process in 2014 and 2017

2014		
Participants of the course	91	100%
Solutions for assignment #1	67	74%
Solutions for assignment #2	40	44%
Solutions for assignment #3	37	41%
2017		
Participants of the course	115	100%
Solutions for assignment #1	58	50%
Reviews for assignment #1	33	29%
Helpful review comments for assignment #1	23	20%
Solutions for assignment #2	48	42%
Reviews for assignment #2	36	31%
Helpful review comments for assignment #2	26	23%

A systematic evaluation of the didactical approach was accomplished through a survey in 2017. Students were asked if they agree to several statements. The level of (dis)agreement is measured on a 4-point Likert scale. The right side of the scale is the highest score while the left side means no agreement to the statement. 58 students participated in the survey, although not all responses were useable.

Overall, a majority of students agreed to the statement that reviewing someone else's solution was helpful to understand the topic better. This holds true for both assignments in 2017 (see Figs. 1 and 2).

Through both assignments students better know if they understood the topic (Figs. 3 and 4) which was one of the goals of the course (goal #1 and #2b).

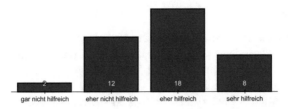

Einem anderen Feedback zu gegeben war …

Fig. 1. Reviewing a foreign assignment was helpful to better understand the topic (assignment #1 2017)

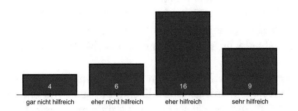

Einem anderen Feedback zu gegeben war …

Fig. 2. Reviewing a foreign assignment was helpful to better understand the topic (assignment #2 2017)

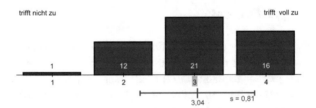

Ich kann jetzt besser einschätzen, ob ich das Thema "Informationstheorie" verstanden habe.

Fig. 3. Better understanding of information theory (assignment #1 2017)

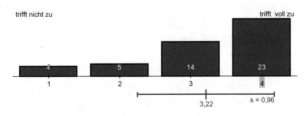

Ich kann jetzt besser einschätzen, ob ich das Thema "Iteration & Rekursion" verstanden habe.

Fig. 4. Better understanding of algorithms (assignment #2 2017)

6.3   Discussion

Overall, peer review for assignments with clearly defined solutions seems to be a suitable way to make students better understand abstract topics of computer science and their interrelationships. As a major benefit, the effort for instructors is moderate while students still perceive individual benefit.

However, instructors initially underestimated the role of assignments. Obviously, the character of the assignment needs to be carefully designed—specifically for freshman students, a more operational character of assignments seems to be favourable over assignments that aim at establishing a big picture. Apparently, assignments that deserve a broader overview tend to overstrain students.

It is helpful to guide students through the review process by learning management systems such as Moodle. Especially freshmen students are not used to didactical methods which require a large degree of self-organization and motivation. Thus, didactical settings should be more guiding and facilitate students developing these skills.

Motivating students to write thorough review comments seems to be a crucial aspect [10]. If the feedback process is too liberal, the quality of comments tends to be unsatisfactory. Even though bonus scores have only a minor effect on the final grade, they still seem to have a major impact on motivation. Approximately half of the respondents of the 2017 survey stated that they would not have participated if there were no bonus scores.

7   Summary and Outlook

Computer science, like many other disciplines, encompasses subjects that are difficult to understand for students, in particular if they are just starting their university studies. Supporting them in getting acquainted to such abstract and complex material and, at the same time, providing them with an evaluation of their performance in dealing with such abstract topics seems to be a useful undertaking.

To that end, this paper presents a didactical setting which uses ten-minute papers on selected topics of an introductory course in computer science. Peer students then review these ten-minute papers in order to provide feedback that students can better accept and understand than feedback from instructors.

Apparently, the strictness of the overall process and the perceived benefit heavily determine the quality of comments in the review process. Learning management systems like Moodle may ensure the former; immediate rewards like bonus scores may warrant the latter.

Overall, the peer review approach in its present shape yields promising rewards. Further work will focus on substantiating the respective evidence and will investigate if the didactical setting may also be used in courses at a later stage in a study program, e.g. in the second year of study. Future work will also focus on better understanding which types of topics and assignments are suitable for reasonable peer review.

Acknowledgements. This work is part of the EVELIN project and funded by the German Ministry of Education and Research (Bundesministerium für Bildung und Forschung) under grants 01PL12022A and 01PL17022A.

References

1. Mager, R.F.: Preparing Instructional Objectives, 2nd edn. Kogan Page, London (1992)
2. Sedelmaier, Y., Landes, D.: A competence-oriented approach to subject-matter didactics for software engineering. Int. J. Eng. Pedagogy (iJEP) **5**(3), 34–44 (2015)
3. Klafki, W.: Didactic analysis as the core of preparation of instruction (Didaktische Analyse als Kern der Unterrichtsvorbereitung). J. Curric. Stud. **27**(1), 13–30 (1995)
4. Siebert, H.: Pädagogischer Konstruktivismus: Lernzentrierte Pädagogik in Schule und Erwachsenenbildung, 3rd edn. Beltz, Weinheim (2005)
5. Terhart, E.: Constructivism and teaching: a new paradigm in general didactics? J. Curric. Stud. **35**(1), 25–44 (2003)
6. Siebert, H.: Didaktisches Handeln in der Erwachsenenbildung: Didaktik aus konstruktivistischer Sicht, 2nd edn. Luchterhand, Neuwied (1997)
7. Sondergaard, H.: Learning from and with peers. In: Proceedings of the 14th Annual ACM SIGCSE Conference on Innovation and Technology in Computer Science Education, pp. 31–35. ACM, New York (2009)
8. Cho, K., MacArthur, C.: Learning by reviewing. J. Educ. Psychol. **103**(1), 73–84 (2011)
9. Cho, Y.H., Cho, K.: Peer reviewers learn from giving comments. Instr. Sci. **39**(5), 629–643 (2011)
10. Turner, S., Pérez-Quiñones, M. A., Edwards, S., Chase, J.: Student attitudes and motivation for peer review in CS2. In: Proceedings of the 42nd ACM Technical Symposium on Computer Science Education, pp. 347–352 (2011)

Students Perceptions in a Flipped Computer Programming Course

Osama Halabi[1](✉), Saleh Alhazbi[1], and Samir Abou El-Seoud[2]

[1] Qatar University, Doha, Qatar
{ohalabi, salhazbi}@qu.edu.qa
[2] The British University in Egypt, Cairo, Egypt
samir.elseoud@bue.edu.eg

Abstract. Flipped classroom pedagogy have gained attention in recent years. This paper reports our study on the attitudes of students towards flipped classroom in introductory programming. Surveys were used to measure the perception using many factors. The study explores the effect of flipped classroom on increasing student-student and student-instructor interaction, ability of student to self-pace through the course, increase the motivation to learn programming. In addition, overall evaluation of the flipped classroom is presented. Flipped classroom provided greater opportunities for communication, more motivation to study programming, and appreciated more by the students who like self-pace. The overall sentiment is positive but still 29.4% of students prefer the traditional teacher led classes.

Keywords: Flipped classroom · Student perception · Student learning
Teaching pedagogy

1 Introduction

Usually, the main method for teaching programing is lecturing followed by a limited time lab. In the lecture, the focus is on syntactic details rather than training students on programming strategies. Most of the faculty recognize the need to adopt more active learning environment and to incorporate different techniques to certain extents in their classes. Although only 36% of faculty think that traditional lecture is good, it has been found that only 60% still teaching using the traditional way [1], and the students are familiar with traditional classroom and generally comfortable [2]. However, the general perception between educators is that it is not the best approach in terms of learning style and flexibility [3].

Recently a flipped classroom teaching model has been gaining popularity. The new paradigm inverts or flips the usual classroom by allowing the students to learn the basic concepts outside the classroom using online education tools like recorded videos. The basic idea is to move the basic knowledge out of the classroom and utilize the class time in more practical activities and to practice and deepen the learned knowledge [3]. The advantage of the new model is still being researched and the need to explore the approach using different criteria is essential to conclude the benefit in classroom. Generally, lecture method impacts students' motivation negatively, this affects students

© Springer Nature Switzerland AG 2019
M. E. Auer and T. Tsiatsos (Eds.): ICL 2018, AISC 917, pp. 76–85, 2019.
https://doi.org/10.1007/978-3-030-11935-5_8

learning of programming. It was found that performance of high motivated students in programming course is better than their peers with low motivation [4, 5]. One approach to promote students' motivation is to improve level of interaction, which creates positive attitudes and eventually enhances students learning [6, 7]. The high level of interaction between students and instructor and between students themselves creates a comfortable learning environment, where students can get feedback that helps them to develop programming strategies. However, one should keep in mind that class activities could be extremely affected by whether students have come to the class prepared or not. The active learning technique used is to allow students to ask peers or instructors for clarification and feedback. When working on assignments in class, students could also provide peer-feedback and send response to received feedback. To support flipped classroom, students must be aware about the technology used, for example, video/audio recording, multimedia, etc. On the other hand, the technology used should be capable of providing immediate and anonymous feedback both for students and for instructors, for example quiz and in-class test.

By using flipped classroom, teaching programming features, that is traditionally takes place in class time, can be achieved before the class using online videos and online materials. This frees up class time, so it can be utilized to improve students' problem-solving skills. Instructor needs to develop educational materials that help students to apply the knowledge they learnt online to write complete programs. Solving problems in class not only improve students' skills, but it allows for more direct feedback from instructor, which makes them more aware of their weakness in applying the knowledge, therefore they focus more on developing algorithmic thinking and writing programs when they study at their own time. Moreover, training students on analyzing and solving problems in class with frequent feedback from instructor will improve students' patience to be more motivated.

Few studies explored the effect of using flipped classroom in teaching programming languages. The study in [8] reported their experience with flipping software engineering course, based on students' reflection it has been found that the class was more effective and engaging that traditional class approach. The impact of applying flipped classroom model in undergraduate introductory programming course has been studied in [9]. It has been observed that the pass rate in final exam increased compared to the traditional class, moreover the competency acquisition enhanced as well. The evaluation of flipped classroom in first year introductory engineering course concluded that there was a slight improvement in assessment grades in flipped class [10]. Sufan et al. [11] explored teaching C programming language using flipped classroom based on SPOC (Small Private Online Course). The results showed improvements the effects of teaching and learning. In [12] a flipped classroom approach was used to teach computer programming course. The students liked the idea of having more time to practice programming in class, and they appreciated the videos developed for the course as it helped them to learn the material.

The objective of this study was to study the students' perceptions of flipped style teaching method in introductory computer programming course. Surveys were used to assess the student attitudes towards flipped classroom and to answer the following research questions:

1. Will flipped classroom improves communication and peer work?
2. The time spent in watching the videos
3. Will the flipped classroom make the student more motivated to learn programming?
4. Will the student self-pace themselves in flipped approach.
5. The overall attitude about flipped classroom approach compared to traditional style

2   Methodology

The design and implementation of flipped model in this course can positively improve students learning and help them to master programming skills. Instead of traditional form of this course that is usually composed of theory and lab, the course had three parts: online activities (out-class), in-class activities, and lab activities using the same approach mentioned in [13]. In essence it includes two components as can be seen in Fig. 1, the first component is the outside component based on the online video-based instruction which followed by the second component of active learning activities inside the classroom. Careful design and implementation is a key for the success of the flipped classroom.

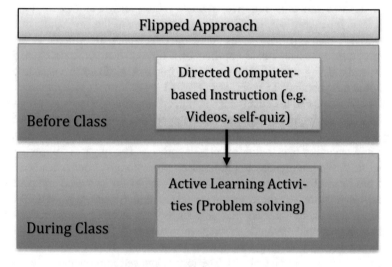

Fig. 1. Flipped classroom activities

Online activities are used to teach students knowledge aspects of the language through online videos and quizzes. It prepares students for training on programming strategies that can be addressed during both in-class and lab activities. The class time devoted to train students to develop multiple necessary skills, which include fixing errors, program comprehension, and writing small programs. The lab activities used to help students analyze problems and practice writing complete programs on computers.

The overall design of the flipped classroom is divided into two steps process. Since programming is an activity that centered around writing code, the activities in classroom focus on problem solving and writing algorithms and programs. The learning activities outside classroom are concentrated on reading programming literature and watching online videos that were carefully prepared in alignment with the course syllabus. The students need Three new activities were introduced to classroom: (1) review session, where the students ask questions about the concepts that they did not understand in the videos to clear any ambiguity in the students' learning. (2) in-class activities, where students were given a problems and asked to think of the algorithm and write a program to solve. (3) video quizzes: a weekly quiz about the weekly videos to check the level of acquiring theoretical knowledge and to motivate the students to watch and come prepared to the class. Engaging the students in pre-class preparation is essential for effective flipped classroom model. Therefore, video quizzes were used as a reward points to motivate the students to more engage with the pre-class work, at the same time it can be used as an indicator of the attainment of the online materials. This will have a positive result on the student participation as reported by [14] where student participation increased to 85 and 76% reported spending more or much time preparing.

2.1 Measures of Affect

The approach was applied on one class of 33 students of first year computer science program. Questionnaire was used to explore the effectiveness and to measure the perception of student about the teaching method. Student attitudes were gathered at the end of the semester. Students reported their response on a 5-points Likert scale (1: Strongly Disagree, 2: Agree, 3: Neutral, 4: Agree, 5: Strongly Agree). The survey asked six questions regarding the course perception, see Table 1 for the list of the questions. At the end of the questionnaire, two free-response questions were asked about what liked most about the class and to comment on recommendation for improvements.

Table 1. The attitudes survey questions

Attitudes Survey
1. The Flipped Classroom gives me greater opportunities to communicate with other students and instructor
2. I like watching the lessons on videos
3. I regularly watch the video assignment
4. I would rather watch a traditional teacher led lesson than a lesson on video
5. I dislike self-pace myself through the course
6. I am more motivated to learn Programming in the Flipped Classroom

3   Results and Discussion

The students were asked if they watch the video on time to understand the percent of pre-class preparation, the result showed that 72.2% of the students always or most of the time watch the video, 22.2% watch some of the time, and only 5.5% rarely watched the videos, see Fig. 2.

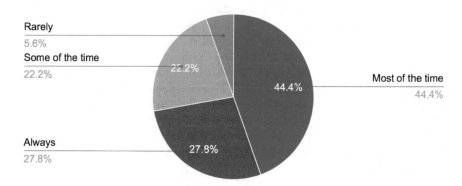

Fig. 2. Do you watch the video on time?

An important result of the flipped approach is related to improve interaction and working with classmate. At the beginning students were hesitant to work in pairs, by the end of the semester the majority agree that the approach helped them to communicate and become socially more comfortable with their classmate. Only 11.8% disagree and generally 88.2% agreed that the flipped class improved their interaction with other students and with the instructor, see Fig. 3.

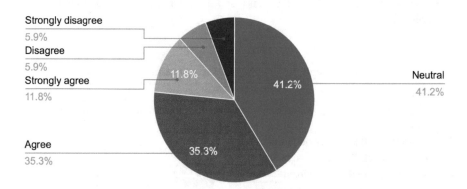

Fig. 3. The Flipped Classroom gives me greater opportunities to communicate with other students and instructor

The result showed that only 17.5% do not like to watch lessons on videos compared to total of 58.8% who strongly like or like to watch online videos, see Fig. 4.

This perception would affect the other results as it seems that 17.6% of the students strongly do not prefer the concept of watching lectures on video and prefer traditional lectures.

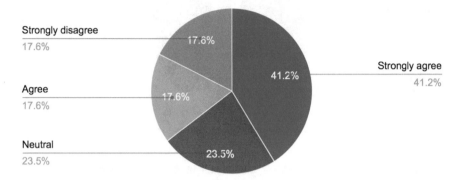

Fig. 4. Do you like watching the lessons on video?

Another indicator of the pre-class preparation is related to watching the video assignment. It has been found that 64.7% who watch regularly the videos assignment compared to 11.8% who do not watch at regular bases as shown in Fig. 5.

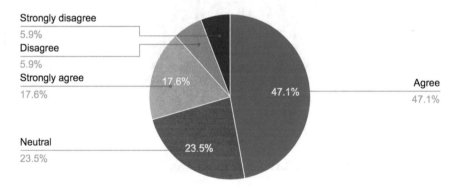

Fig. 5. Do you regularly watch the video assignment?

When asked if student prefer traditional teacher led lesson than a watching the lecture on video. Figure 6 shows that 47.1% of the students like video lesson compared to 29.4% who still like the teacher led lesson which still high percent as one third of the student still prefer the traditional lecturing style.

The flipped classroom pedagogy requires more attention from the students about pre-class work and the ability to pace themselves successfully through the course is an important factor. When students asked about their attitudes toward self-pace studying,

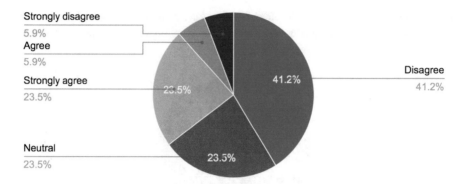

Fig. 6. I would rather watch a traditional teacher led lesson than a lesson video

the result in Fig. 7 shows that 11.8% dislike self-pace study compared to 28.9% who like self-pace study, however, high percent (58.8%) were neutral which may reflect how many students still struggling between traditional and flipped classroom in term of self-pacing themselves through the course.

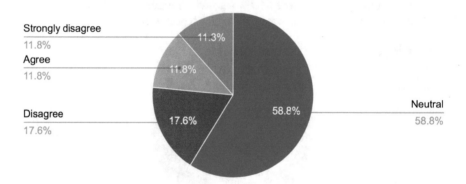

Fig. 7. I dislike self-pace myself through the course

When the students asked if they are more motivated to learn programming in flipped classroom, the results showed that 23.6% disagree and the remain strongly agree, agree, or neutral (Fig. 8).

The overall results showed that the students had a positive attitude about the flipped classroom approach when they asked which approach they prefer if they were given the choice between traditional and flipped classroom, see Fig. 9. 11.8% strongly prefer flipped compared to 5.9% who preferred traditional and 41.2% prefer flipped compared to 23.5% who prefer traditional. In total, 70.6% who prefer or do not mind flipped classroom and 29.4% who still prefer traditional approach.

The student response to the two free-response questions varied with many comments, the most common response that can be conclude was "ability to watch video anywhere and anytime when needed", "ability to review the lectures many times before

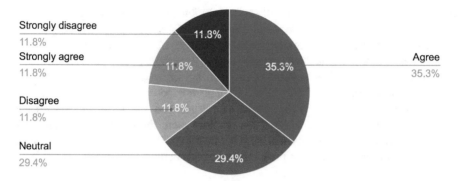

Fig. 8. More motivated to learn programming in flipped classroom

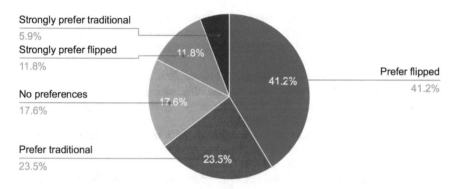

Fig. 9. The overall perception about the flipped classroom

the exams", and "more time for exercises in classroom". As for improvement questions, the most common response is "upload the videos earlier" and "reduce the duration time of each video" which have been noticed by the instructor after few weeks and considered during the course. Nevertheless, students particularly welcomed the fact that they had access to materials like video lectures, they were able to prepare themselves and even learn at their own pace at any time everywhere.

4 Conclusion

This paper presents our research with a flipped classroom. The results indicated that this approach has the potential for successful implementation in introductory computer programming course and this also may support other STEM discipline and improve the outcome in such STEM courses. In our study, we found that students were able to self-paced study and this increased their self-efficacy as well as increase the communication skills and created healthy and motivated environment for student to communicate and work with their classmate. Further analysis of the result shows that a percent between 11.8 and 23.6% of the students tend to negatively affect the result in all attitudes questions and appeared in the overall evaluation. This might be to the fact that there are still students who are resistive to the idea of flipped classes and preferred the traditional teacher led although they admitted the many benefits of the flipped approach as they could not provide explanation or any negative comments on their free-response questions. Also, applying this approach to first year students need to be carefully introduced due to the fact that it is still difficult for these students to self-pace themselves through the course as the results clearly showed this problem. Nevertheless, continues efforts to build a research that explore more the nature, utility, and effectiveness of filliped classroom is necessary before university faculty will be convinced to take such a dramatic change in their instructional practices.

References

1. Lord, S.M., Camacho, M.M.: Effective teaching practices: preliminary analysis of engineering educators. In: 2007 37th Annual Frontiers in Education Conference—Global Engineering: Knowledge Without Borders, Opportunities Without Passports, p. F3C-7–F3C-12 (2007)
2. National Governors Association: Building a Science, Technology, Engineering, and Math Education Agenda: An Update of State Actions. Engineering, 44 (2011)
3. Bergmann, J., Sams, A.: Flip your classroom reach every student in every class every day. Get Abstr. Compress. Knowl. 1–5 (2014)
4. Bergin, S., Reilly, R., Traynor, D.: Examining the role of self-regulated learning on introductory programming performance. First Int. Work Comput. Educ., Res. 81–86 (2005). https://doi.org/10.1145/1089786.1089794
5. Jenkins, T.: The motivation of students of programming. ACM SIGCSE Bull. **33**, 53–56 (2001). https://doi.org/10.1145/507758.377472
6. Lopez-Perez, M.V., Perez-Lopez, M.C., Rodriguez-Ariza, L.: Blended learning in higher education: students' perceptions and their relation to outcomes. Comput. Educ. **56**, 818–826 (2011). https://doi.org/10.1016/j.compedu.2010.10.023
7. Alhazbi S (2015) Active blended learning to improve students' motivation in computer programming courses: A case study
8. Paez, N.M., Martín, N.: A flipped classroom experience teaching software engineering. In: 2017 IEEE/ACM 1st International Workshop on Software Engineering Curricula for Millennials (SECM), pp. 16–20 (2017)
9. Elmaleh, J., Shankararaman, V.: Improving student learning in an introductory programming course using flipped classroom and competency framework. Glob. Eng. Educ. Conf., 49–55 (2017). https://doi.org/10.1109/educon.2017.7942823
10. Salama, G., Scanlon, S., Ahmed, B.: An evaluation of the flipped classroom format in a first year introductory engineering course. In: 2017 IEEE Global Engineering Education Conference (EDUCON). IEEE, pp. 367–374 (2017)
11. An, S., Li, W., Hu, J., et al.: Research on the reform of flipped classroom in computer science of university based on SPOC. ICCSE 2017—12th Int. Conf. Comput. Sci. Educ., 621–625 (2017). https://doi.org/10.1109/iccse.2017.8085567
12. Carlisle, M.C.: Using You Tube to enhance student class preparation in an introductory Java course. In: Proceedings of the 41st ACM Technical Symposium on Computer Science Education—SIGCSE '10, p 470 (2010)

13. Alhazbi, S.: Using flipped classroom approach to teach computer programming. Proc. 2016 IEEE Int. Conf. Teaching Assess Learn. Eng. TALE 2016, 441–444 (2017). https://doi.org/10.1109/tale.2016.7851837
14. Horton, D., Campbell, J.: Impact of reward structures in an inverted course. In: Proceedings of the 2014 Conference on Innovation & Technology in Computer Science Education—ITiCSE '14, pp. 341–341 (2014)

New Approaches in Assistive Technologies Applied to Engineering Education

Andreia Artifice[1], Manuella Kadar[2(✉)], João Sarraipa[1],
and Ricardo Jardim-Goncalves[1]

[1] CTS, UNINOVA, DEE/FCT, Universidade Nova de Lisboa, Lisbon, Portugal
{afva,jfss,rg}@uninova.pt
[2] Computer Science Department, 1 Decembrie 1918 University of Alba Iulia,
Alba Iulia, Romania
mkadar@uab.ro

Abstract. The evolution of technology, new learning theories and universal design made the learning process to become more flexible and adaptable. The 21st century society imposes a need for increased cognitive ability, thus students are required to have a combination of academic knowledge and transferable skills. Assistive technologies have been used not only with students with special needs but in an inclusive environment extending it to other students. Students are becoming more aware and sensitive to their own learning preferences and their own learning styles. However, the task of adapting current educational practices and spread information in student environment is still a challenge. This paper proposes new approaches to assistive technology design imposed by the evolution of technology. Students can now choose how to study, where to study and when to study. Underpinning this change, the paper explores how assistive technologies have evolved into learning technologies by taking into consideration the technological pedagogical content knowledge (TPACK) framework and proposing an extension, based on intelligent systems that combines means of bio signals assessment with emotional state evaluation for engineering students.

Keywords: Assistive technologies · Learning technologies · TPACK

1 Introduction

The evolution of technology, new learning theories and universal design made the learning process to become more flexible and adaptable. The 21st century society imposes a need for increased cognitive ability, thus students are required to have a combination of academic knowledge and transferable skills. Nowadays, assistive technologies have become more widely used in education to support all students, not only students with special needs. Students are becoming more aware and sensitive to their own learning preferences and their own learning styles. However, although the needs and means of flexible and adaptable learning have been uncovered, a huge task remains regarding adapting current educational practices and disseminating information amongst students. Nowadays, Engineering Education must produce technically

M. E. Auer and T. Tsiatsos (Eds.): ICL 2018, AISC 917, pp. 86–96, 2019.
https://doi.org/10.1007/978-3-030-11935-5_9

excellent and innovative graduates, therefore there is a need to enrich and broaden the means and methods to deliver courses, to better adapt those graduates to the global economy.

This paper proposes new approaches to assistive technology design imposed by the evolution of technology. Students can now choose how to study, where to study and when to study. Underpinning this change, the paper explores how assistive technologies have evolved into learning technologies by taking into consideration the technological pedagogical content knowledge (TPACK) framework. The TPACK framework emphasizes how the connections between teachers' understanding of content, pedagogy, and technology are developed to interact with one another to produce effective teaching. The TPACK framework argues that programs which emphasize the development of knowledge and skills in the above mentioned three areas by an isolated manner are doomed to fail. Thus, effective teacher educational and professional development needs to craft systematic, long-term educational experiences where the participants can engage fruitfully in all the three knowledge bases- content, pedagogy, technology- in an integrated manner.

The hereby proposed architecture and solution named **i-TPACK** includes software development knowledge and software design strategies to improve the classic TPACK framework. In i-TPACK both students and teachers are involved in a common effort to design new teaching strategies. Such software design strategies are the outcome of thorough studies on students' emotional state, level of attention during classes, face mimics correlated with students' vital signs such as pulse, hart rate, and brain waves. The investigations have been carried out with techniques such as eye tracking, electroencephalography, electrocardiography. On one hand, teachers are involved in assessing students' behaviour during classes and adapt "*on the fly*" the teaching method to the group's reactions and on the other hand, students can correct their attitude and improve their learning capabilities during classes. Our approach includes the knowledge that teachers acquire when involved in the development of educational software, thus transforming the original concept of TPACK into a smart responsive and corrective system.

This paper is organized as follows: next section is dedicated to literature review. Section 3 entitled "Intelligent Technological Pedagogical Content Knowledge" refers to the design strategy of the smart system. The outcomes are presented in Sect. 4. Finally, the conclusions are drawn in Sect. 5.

2   Literature Review

2.1   Universal Design Theory

Universal Design is the design and composition of an environment that is accessible, comprehendible, and usable by as many people as possible independently of their age, size or having any particular ability or disability [1]. Concerning electronic systems, according to the Disability Act 2005 [2], it is considered "*any electronics-based process of creating products, services or systems so that may be used by any person*".

The application of Universal Design theory has contributed to progress in educational context.

According to Pliner et al. [3], Universal Instructional Design (UID), can be seen as *"an approach for addressing the diverse learning needs of students enrolled in institutions of higher education"*. The same authors argue that it allows to expand institutional teaching methodologies to promote equal access to classroom teaching and learning to all students, despite their learning needs.

Regarding pedagogy, technologies such as text-to-speech software, mind maps, audio recording software and note taking technology are instruments that can be adapted to the learner preferences [4].

2.2 Learning Styles and Theories

According to Dunn, a person's learning style *"is the way that he or she concentrates on, processes, internalizes, an remembers new and difficult academic information or skills"* [5]. Learning styles have been studied from different perspectives. For instance, the Honey-Mumford model refers to four styles [6, 7]: (i) activists—prefer learning by doing, they like group working, consider that repetition is boring, and are characterized by enthusiasm; (ii). reflectors—stand back and observe, they assemble as much information as possible, their strength is data collection and its analysis; (iii) theorists— can adapt their observations into frameworks, and they add learning to existing ones; and (iv) pragmatists—seek and use new ideas, they try to envision the application of new ideas and theories before making a judgment.

The Myers-Briggs model [7, 8] classifies learners in the following categories: (1) extroverts—focus on people, are happy trying things; (2) introverts—focus on ideas, tend to think things; (3) sensors—focus on facts and procedures, usually are practical; (4) intuitors—are focused on meaning, use imagination and are concept-oriented; (5) thinkers—fundament their decisions on logic and rules; (6) feelers— fundament their decisions on personal and humanistic considerations; (7) judgers— follow agendas, aim closure and completeness; (8) perceivers—tend to adapt to circumstances, they will postpone accomplishment until more is known.

Learning styles and preferences have being studied in practical scenarios. For instance, in a study entitled *"Learning Styles an Teaching Styles in College English Teaching"*, Zhou [9] argues that *"an effective means of accommodating these learning styles is for teachers to change their own styles and strategies and provide a variety of activities to meet the needs of different learning styles"*. Consequentially, that increases the probability of a student to be successful. From the practical point of view, the same author, suggests to teachers that want to cover a wide variety of styles to: (1) make liberal use of visuals; (2) assign some repetitive drill exercises; (3) do not spend all time of the class writing on the blackboard; (4) provide explicit instruction in syntax and semantics.

2.3 SETT (Student—Environment—Task—Tools)

The SETT framework (Student—Environment—Task—Tools) provides key questions to help make decisions about which specialized tools and related strategies will make a difference for a student's learning. The SETT Framework is a four-part model that encourages collaborative decision-making in stages of assistive technology service design and delivery [10]. The main features of SETT are:

Shared Knowledge: In SETT framework decisions concerning tools is based on the knowledge about the student, the environment and the task.

Collaboration: The SETT Framework is a tool that both requires and supports the collaboration of the people who will be involved in the decision-making and those who will be impacted by the decisions. Collaboration is not only critical for the SETT Framework, it is also critical to gaining the buy-in necessary for effective implementation of any decisions.

Communication: In SETT Framework communication develops in an active and respectful way.

Multiple Perspectives: Everyone involved brings different knowledge, skills, experience, and ideas to the table. Although multiple perspectives can be challenging at times they are critical to the development of the accurate, complete development of shared knowledge. Not only are the multiple professional perspectives important to include, but also those of the student and the parents. This can make the difference between success and lack-off.

Pertinent information: Although there is much information that is pertinent to decision making, there is information that is not relevant. Knowing where to draw the line is important as a moving target.

Flexibility and Patience: When working through the SETT Framework or using any other means of concerns-identification and solution seeking, there is a tendency to suggest possible solutions before the concerns have been adequately identified. When a solution springs to mind, collaborators are urged NOT to voice it until it is time to talk about the tools because when a solution is mentioned, the conversation shifts immediately from concern-identification to determining the worth or lack off worth of the suggested solution. Even when a team member thinks of the "perfect" solution, silent patience is urged. It might not look quite so perfect when all important factors are discussed.

2.4 TPACK

Technological Pedagogical Content Knowledge (TPACK) was introduced by Koehler and Mishra in 2005 as a conceptual framework that represents the teachers' knowledge needed to effectively teach with technology [11]. TPACK is rooted in Schulman's work on Pedagogical Content Knowledge (PCK) [12, 13]. The PCK concept allows to qualify the teacher's profession. It refers to the integration of teachers' knowledge with content knowledge, in a way that allows students to understand the subject; TPACK is similar since it adds technological knowledge (TK) as necessary part of teacher's profession [14] (Fig. 1).

TPACK represents the required knowledge for teachers who need to integrate technology. Specifically, it refers to the interactions between content, pedagogy and technology in order to teach effectively [15].

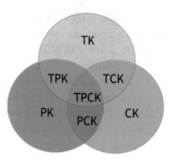

Fig. 1. TPACK [14]

Similar frameworks have been developed both independently and directly out of the TPACK framework. Most of them are based upon Shulman's (1986) model of Pedagogical Content Knowledge. Similar frameworks include (but are not limited to): ICT—Related Pedagogical Content Knowledge (ICTRelated PCK); Knowledge of Educational Technology; Technological Content Knowledge; Electronic Pedagogical Content Knowledge (ePCK); and Technological Pedagogical Content Knowledge—Web (TPCK-W) [16–20].

The TPACK framework has contributed to the teacher's education and professional development. According to the TPACK framework, effective teacher educational and professional development occur in an integrated manner based on the knowledge bases [15]. The same authors argue that a limitation of TPACK framework is a neutral position concerning broader goals of education.

The 21st century requires cognitive skills, necessary for successful learning and achievement, such as critical thinking, problem solving, job and life skills, and synthesis. Additionally, interpersonal skills are required, for instance communication and collaboration [21]. In that context, Mishra et al. [22] proposes seven cognitive tools within TPACK, that are necessary for the new millennium: perceiving, patterning, abstracting, embodied thinking, modelling, playing, and synthesizing.

3   Intelligent Technological Pedagogical Content Knowledge (I-TPACK)

The Intelligent Technological Pedagogical Content Knowledge (i-TPACK) framework promotes the incorporation of software development knowledge and software design strategies as an extension of the TPACK. The system was designed and tested for Engineering Education as the teaching methods focused on engineering processes to define and solve problems using scientific, technical, and professional knowledge bases. Both students and teachers have been involved in the studies (Fig. 2).

The main goal of the research was to study means and to design inclusive technologies to fight student's dropout. Keeping in mind the principles of TPACK framework we propose a system that infers attention analytics based on bio signals of the students. Bio signals are provided by wearable devices for brain waves and heart

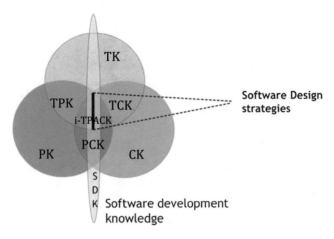

Fig. 2. Architecture of i-TPACK

rates measurements. These are validation methods of the proposed architecture and teaching approach. A case study with scenario is presented for music-based learning. The scope of this analysis is to validate the student's attention level with background music versus no music during a learning process. To achieve this goal, we have evaluated the solution using neuroscience and emotion detection techniques to assess the perception and understanding for personalized learning. In this study the adopted methods explore the hypothesis that a person's physiological state can wield adequate sensorial stimulation to provide diverse levels of information. The solution uses collected data to build a user's musical playlists that tries to match a psychological state with the stimuli evoked by the music that the student is listening to. From that matching, it is possible to improve a person's wellbeing by providing the most adequate music to lift the emotional state of that person.

4   Outcomes

Taking into consideration our approach and the results of the study it has been confirmed that it is possible to improve a person's learning capability with the most suitable music to that person in that moment. The benefits are obvious as the person feels better, will perform better, especially when referring to cognitive functions as studying and learning new subjects. Figure 3 presents the application scenario.

The main application scenario consists of the appropriated environment for the inclusion of a smart system that allows to suggest the appropriated music concerning the person's cognitive state based on bio signals. Such technological environment includes sensor acquisition, specifically electrocardiogram (EKG) and electroencephalogram (EEG); a system engine that analysis and recommends the music and stores an ontology in which appropriated knowledge is represented.

Students attention is an important aspect concerning the learning performance and it is related to the person's physiological activity [23]. A physiological state is

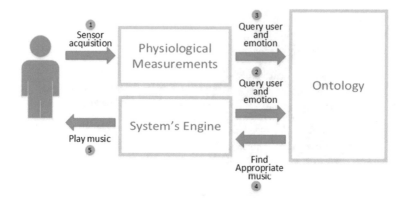

Fig. 3. Application scenario

associated with specific physiological signals that can be captured through technological devices, for instance the ECG and EEG.

Next, will be summarized the results of the experiments that have been performed to develop the scenario. Specifically, the experiments have focused on attention studies in the presence of music as means to improve teaching environments.

4.1 Analysis of Attention with EEG

EEG can perform as an indicator of the level of attention. It can be used to measure the attention spans of students, to help them to improve their learning experience. That indicators are performed from raw signals captured from sensors. EEG is a viable method for determining whether students are attentive.

EEG allows to measure directly the internal state, recording electrical activity of the entire scalp. Furthermore, it can detect changes in real-time.

In the experiment performed, EEG measurement of attention training and assessment system include: functions of data acquisition, signal processing, data analysis, data presentation. Five participants were included in the study. It was used the headset Mindwave connected to the Lucid Scribe software. The computerized Prague test, created by Psycho-technical Institute in Prague, was selected for this study to measure distributed attention of students. The aim of the experiment was to investigate how useful Neurosky Mindwave is in measuring attention levels [24].

The measurements of the MindWave are outlined as follows: raw signal, EEG power spectrum, eSense meters for attention and meditation. The eSense Attention meter indicates the intensity of the user's level of mental 'focus' or 'attention' to determine levels of concentration.

Results achieved by the participants in the test are presented in Table 1. Concerning distributed attention: 3 participants scored excellent, 1 scored good, and 1 sufficient.

Concerning attention levels results are presented in Table 2. The average values are between 7.3 and 67.5.

Table 1. Results achieved by participants in the experiment

No	Students	Age	Gender	First 4 min	Second 4 min	Third 4 min	Forth 4 min	Total
1	M.M.M.	20	F	15	28	25	14	82
2	C. D. I.	20	F	17	16	18	19	70
3	I.T.	20	F	4	4	8	7	23
4	L.S.	21	M	12	26	25	24	87
5	S.I.R.	21	F	18	15	20	29	82

Table 2. Levels of attention achieved by participants in the experiment

No	Students	Age	Gender	First 4 min	Second 4 min	Third 4 min	Forth 4 min	Total
1	M.M.M.	20	F	15	28	25	14	82
2	C. D. I.	20	F	17	16	18	19	70
3	I.T.	20	F	4	4	8	7	23
4	L.S.	21	M	12	26	25	24	87
5	S.I.R.	21	F	18	15	20	29	82

4.2   Analysis of Attention with ECG

To analyse a person's attention during cognitive tasks in the presence of music, a set of experiments has been performed. The main goal of the experiments was to correlate a person's attention with features extracted from ECG signal.

First, a prospective study was developed in the context of eLearning in which participants performed the task of attending a course, including learning and test phases, in the presence versus absence of classical music [25]. At the same time ECG measurements were recorded for further analysis. In the referred analysis it was computed the Heart Rate Variability, i.e. *"the amount of heart rate fluctuations between the mean heart rate"* [26] of the ECG signal and analysed the Low Frequency (LF) and Hight Frequency (HF) features, as can be seen in the Figs. 4 and 5.

The results revealed, according to a study [27], that LF and HF have a decrease when attention decreases. Additionally, the study revealed that the participants were more attentive when they performed the task with classical background music.

The study allowed us to determine the relationship between attention and Low Frequency and High Frequency Hear Rate Variability features extracted from ECG signal in eLearning environment with and without classical music. The outcomes allowed us to conclude that there is a correlation between student's attention and the Heart Rate Variability features (LF and HF) extracted from ECG signal.

The second experiment was developed to test in more detail the applied methodology. The chosen environment was the execution of an immersive cognitive task, namely playing a game. The task was performed with classical music versus annoying music. Results have been corroborated and shown to be aligned with the first performed study.

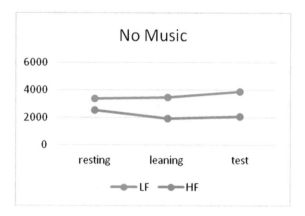

Fig. 4. Average LF(ms^2) and HF (ms^2) for eLearning course with no music [25]

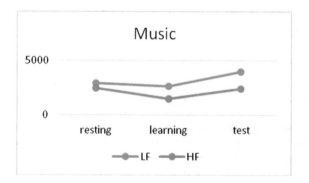

Fig. 5. Average LF(ms^2) and HF (ms^2) for eLearning course music [25]

5   Conclusions

As a general conclusion, technology can be interpreted as the set of theories and techniques allowing the practical use of scientific knowledge in a framework as TPACK. The proposed smart system collects and integrates several bio-signals from the user. It is an innovative solution in the student's dropout scenario since it can suggest automatically the appropriated music to the student to improve his emotional state. It also considers the knowledge that teachers acquire when involved in the development of software for education.

The above-mentioned example can be used by students even beyond the teaching environment so that they can feel better and be extrapolated to other professionals, like the teachers, thus promoting other person's wellbeing either in the teaching environment or in other life circumstances.

Acknowledgements. The authors acknowledge the European Commission for its support and partial funding and the partners of the research project from ERASMUS+: Higher Education—International Capacity Building—ACACIA—Project reference number—561754-EPP-1-2015-1-CO-EPKA2-CBHE-JP, (http://acacia.digital).

References

1. Definition and overview|Centre for Excellence in Universal Design. [Online]. Available: http://universaldesign.ie/What-is-Universal-Design/Definition-and-Overview/. Accessed 17 May 2018
2. O. of the H. of the O. H. S. D. B. Office, "Disability Act 2005." [Online]. Available http://www.achtanna.ie/en.act.2005.0014.7.html. Accessed 17 May 2018
3. Pliner, S.M., Johnson, J.R.: Historical, theoretical, and foundational principles of universal instructional design in higher education. Equity Excell. Educ. **37**(2), 105–113 (2004)
4. Goldrick, M., Stevns, T., Christensen, L.B.: The use of assistive technologies as learning technologies to facilitate flexible learning in higher education. LNCS **8548**, 342–349 (2014)
5. Shaughnessy, M.F.: An interview with Rita Dunn about learning styles. Clear. House **71**(3), 141–145 (1998)
6. Honey, P., Mumford, A.: Manual of Learning Styles, 2nd edn. P. Honey, London (1986)
7. Pritchard, A.: Ways of learning (2009)
8. Myers, I., Myers, P.: Gifts Differing: Understanding Personality. Nicholas Brealey Publishing (2010)
9. Zhou, M.: Learning styles and teaching styles in college english teaching. Int. Educ. Stud. **4**(1), 73–77 (2011)
10. Zabala, J.S.: Using the SETT Framework to Level the Learning Field for Students with Disabilities. Retrieved August, vol. 10, no. Revised, p. 2010 (2005)
11. Herring, M.C., Koehler, M.J., Mishra, P.: Handbook of Technological Pedagogical Content Knowledge (TPACK) for Educators. Routledge, London (2016)
12. Shulman, L.: Knowledge and teaching: foundations of the new reform. Harv. Educ. Rev., 1–23 (1987)
13. Shulman, L.E.: Those who understand: knowledge growth in teaching. Educ. Res. **15**(2), 4–14 (1986)
14. Voogt, J., Fisser, P., Tondeur, J., Johan van Braak, B.: Using Theoretical Perspectives in Developing Understanding of TPACK
15. Koehler, M.J., Mishra, P., Kereluik, K., Shin, T.S., Graham, C.R.: The technological pedagogical content knowledge framework. In: Handbook of Research on Educational Communications and Technology (2014)
16. Angeli, C., Valanides, N.: Pre-service elementary teachers as information and communication technology designers: an instructional systems design model based on an expanded view of pedagogical content knowledge. J. Comput. Assist. Learn. **21**(4), 292–302 (2005)
17. Franklin, C.: Teacher preparation as a critical factor in elementary teachers: use of computers. In: Carlsen, R., Davis, N., Price, J., Weber, R., Dl Willis (eds.) Society for Information Technology and Teacher Education Annual, pp. 4994–4999, Association for the Advancement of Computing in Education, Norfolk, VA (2004)
18. Lee, M.H., Tsai, C.C.: Exploring teachers' perceived self efficacy and technological pedagogical content knowledge with respect to educational use of the World Wide Web. Instr. Sci. **38**(1), 1–21 (2010)

19. Margerum-Lays, J., Marx, R.W.: Teacher knowledge of educational technology: a case study of student/mentor teacher pairs. In: Zhao, Y. (ed.) What Should Teachers Know About Technology? Perspectives and Practices, pp. 123–159. Greenwich, CO (2003)

20. Slough, S., Connell, M.: De fining technology and its natural corollary, technological content knowledge (TCK). In: Crawford, C., et al. (eds.) Proceedings of Society for Information Technology and Teacher Education International Conference, pp. 1053–1059. AACE, Chesapeake, VA (2006)

21. Mishra, P., Kereluik, K.: What 21 century learning? A review and a synthesis. In: Society for Information Technology & Teacher Education International Conference, pp. 2201–3312 (2011)

22. Mishra, P., Koehler, M.J., Henriksen, D.: The 7 trans-disciplinary habits of mind—extending TPACK framework towards 21st century learning. Educ. Technol., 22–28 (2011)

23. Cohen, R.A., Sparling-Cohen, Y.A., O'Donnell, B.F.: The Neuropsychology of Attention. Plenum Press, New York (1993)

24. Kadar, M., Borza, P.N., Romanca, M., Iordăchescu, D., Iordăchescu, T.: Smart testing environment for the evaluation of students' attention. Interact. Des. Architecture(s) J. IxD&A(32), 205–217 (2017)

25. Artífice, A., Ferreira, F., Marcelino-Jesus, E., Sarraipa, J., Jardim-Gonçalves, R.: Student's attention improvement supported by physiological measurements analysis. In: IFIP Advances in Information and Communication Technology (2017)

26. van Ravenswaaij-Arts, C.M., Collee, C.M., Hopman, J.C., Stoelinga, G.B., van Geijn, H.P.: Heart rate variability. Ann. Intern. Med. **118**(6), 436–447 (1993)

27. Tripathi, K., Mukundan, C., Mathew, T.L.: Attentional modulation of heart rate variability (HRV) during execution of PC based cognitive tasks. Ind. J. Aerosp. Med. **47**(1), 1–10 (2003)

The Effect of Educational Environment on Developing Healthy Lifestyle Behavior in University Students

Petr Osipov[1], Julia Ziyatdinova[1(✉)], Liubov Osipova[2], and Elena Klemyashova[3]

[1] Kazan National Research Technological University, Kazan, Russian Federation
posipov@rambler.ru, uliziatd@gmail.com
[2] Kazan State Power Engineering University, Kazan, Russia
lnikosipova@mail.ru
[3] Institute of the Study of Childhood, Family and Education of the Russian Academy of Education, Moscow, Russia
elena.k07@mail.ru

Abstract. The paper aims at defining the positive and negative factors that contribute to developing healthy lifestyle behavior in university students as a major segment of young adult population through creating a motivating educational environment that promotes values to support healthy lifestyle. The authors draw their conclusions on the conditions under which the students are motivated to follow a healthy lifestyle. The hypothesis of the study is that the students can develop healthy lifestyle habits if they realize that they are in charge of their health and the university delegates this responsibility to the students. This idea can be implemented into practice only if the university educational environment provides the necessary conditions. The paper uses a complex approach to the problem of developing healthy lifestyle habits in students. The approach combines theoretical investigations and practical surveys. The general idea is to concentrate on self-directed practices and personal responsibility of the students for their own health.

Keywords: Healthy lifestyle behavior · University students · Physical training

1 Introduction

In recent years, healthy lifestyle behavior development has attracted an increased research interest worldwide. Health is an important aspect in basic human performance. The World Health Organization defines health as "a state of complete physical, mental and social well-being and not merely the absence of disease or infirmity" [1].

In this definition, however, humans are seen as 'knowable systems where imperfections should be fixed' [2]. In fact, health should be seen as a broader and a more positive option, where a person feels capable of doing things and reaching his goals. In this case, we refer to Dietrich Bonhoeffer's definition of health as 'the strength to be' [3].

© Springer Nature Switzerland AG 2019
M. E. Auer and T. Tsiatsos (Eds.): ICL 2018, AISC 917, pp. 97–105, 2019.
https://doi.org/10.1007/978-3-030-11935-5_10

In Dietrich Bonhoeffer's definition, an important value is given to the healthy lifestyle which can be developed at an early age of young adulthood. University students are typical representatives of this age group, and there are some key factors that contribute to the importance of healthy lifestyle behavior development in university students.

Firstly, the universities aim at training competitive professionals who can be successful in their future careers only if they are capable of supporting their health. Healthy lifestyles habits develop at an early age, young people can do it while studying at a university, thus influencing their habits in adulthood and reducing disease, illness or disabilities occurrences. If unhealthy behavior habits are identified and changed at an early age, many health risks can be avoided. These ideas are generally accepted and disseminated. At the same time, however, research data show that student population, in general, experiences many health problems and physical disabilities [4, 5].

Secondly, many scientists, politicians and economists relate the idea of educating a physically and spiritually healthy generation with a resilient society [6, 7]. Resilience can be viewed as an alternative solution to the societal challenges we face today so that a citizen could absorb the shocks that he meets and get over them [8]. A young person who practices healthy lifestyle habits can do this much easier than his peers. Today, the easiest way to reach young people is through the modern information technologies and gadgets.

Numerous efforts have been made to promote new healthy lifestyle technologies to attract the attention of the young people so as to incorporate healthy lifestyle into their everyday habits today and in the future. These technologies contribute to developing the surveillance, self-surveillance and resistance practices while following the step and calories burning targets.

Thus, students, together with their professors can practice self-education skills further important in their lives [9].

The problem of promoting healthy lifestyle in young adult population, university students in particular, is of significance for different countries in different parts of the world. A special attention to this problem is given in engineering universities worldwide as engineers design the future in the digital age, and their personal characteristics, including health status, are very important [10, 11].

Thus, healthy lifestyle is becoming an integral part of the engineering university policy contributing to its internationalization strategy [12, 13] including academic mobility programs [14].

One of the crucial factors that can influence the behavior of students and faculty members in their practice of healthy habits is educational environment in its perception from the student and faculty perspectives [15, 16]. Educational, or learning environment can increase the motivation of students for being healthy, and, in its turn, their practice of healthy lifestyle behavior. The learning environment is of crucial importance for students' extra-curricular activities, including sports and arts as good substituents for unhealthy habits [17].

Therefore, it is necessary to bridge the gap between the demand for healthy lifestyle habits development in students and the insufficiency of key elements in the engineering university educational environment that contribute to developing their healthy life skills.

2 Approach

The hypothesis of the study is that the students can develop healthy lifestyle habits if they realize that they are in charge of their health and the university delegates this responsibility to the students.

This idea can be implemented into practice only if the university educational and learning environment provides the necessary conditions and awards. Among them are regular medical tests and corrections of the students' physical health and fitness, health value education topics and discussions infused into the curriculum of different courses, various contests and awards for healthy lifestyle habits demonstrated by the students in different forms.

Not only the contents, but also the methods of teaching are important in this case. The healthy lifestyle topics should be infused into the teaching and learning materials and implemented through student centered and self-directed teaching and learning practices, finding individual learning pathways for each of the students taking into account the initial status of his health and his attitudes towards healthy lifestyle earlier developed at secondary school and in the family.

All these efforts listed should be taken to overcome the problems that the students face every day, among them there are a number of obstacles to healthy lifestyle behavior.

Apart from obviously unhealthy behaviors, including alcohol, drugs, poor nutrition, low sleep level, most of which are very typical for the young adults opening the world for themselves and getting independence from their parents and families in the university, there are many other everyday factors that can stress students out and block their healthy lifestyle.

These stressing factors include:

- *improper time management* due to the lack of the necessary competencies, poor planning and achievement skills, recurring distractions, poor punctuality, impatience, poorly defined goals, procrastination or perfectionism, and etc. [18];
- *information overload* due to huge volumes of information falling on students, pressure to create and compete, an increased number of channels to receive information, lack of methodologies for quick information processing, and etc. [19];
- *communication problems* due to misunderstanding, conflicts, poor listening skills, failure to self-edit, using wrong tools, oversharing information, culture differences, hierarchy problems, inadequate knowledge, and etc. [20];
- *inflated requirements and expectations* due to skills shortage, student performance challenges, imbalance in support, low motivations, highly competitive environments, and etc. [21].

All the above listed stressing factors should be taken into account when designing an educational environment to motivate students for a healthy lifestyle and success in their studies and future careers.

The paper uses a complex approach to the problem of developing healthy lifestyle habits in students.

The approach combines theoretical investigations and practical surveys. The general idea is to concentrate on self-directed practices and personal responsibility of the students for their own health.

The review and analysis of the existing literature including recent papers and conference reports on the problem gave the following conclusions: negative trends in the students' health status can be prevented on condition the university education contents and technologies change and focus on the positive attitude towards a healthy lifestyle and physical fitness.

An important role in university educational environment for a healthy lifestyle is played by the physical education courses and the attitude towards them.

Therefore, of special importance is the number of contact hours students receive in physical training (or fitness), the equipment of the sports gyms, and the positive feedback given to the students for their achievements in sports in the university by the professors, administrators and peers.

In case the university authorities and peers support and praise the success in sports, the motivation of the students grows, and they can achieve much better results, thus improving their self-esteem and attitude.

A number of previous studies show a positive correlation between the healthy lifestyle and academic achievements [22, 23].

Moreover, the average age of the university students is ideal to develop healthy habits so that they can influence the professional lifestyle and assist in finding new career prospects. Therefore, health care should be one of the priorities in the university education worldwide.

3   Experiment

In order to find out the attitudes of the students towards healthy lifestyle, and the effect of the educational environment, the authors developed and conducted a survey for the engineering university students in B.Sc. degree programs.

The survey consisted of two blocks of questions. The first block included open-ended questions relating to the perception of health and healthy lifestyle, positive factors contributing to healthy lifestyle promotion, and negative factors undermining health.

The second block included questions concerning the influence of different stress factors on healthy lifestyle. The stress factors were divided into four categories, including educational stress factors, knowledge control stress factors, communication stress factors and administrative stress factors. The students were asked to evaluate the importance of each of the factors within the categories using a five point rating scale.

The developed questionnaire was tested for validity and reliability and proved its efficiency.

The survey was conducted anonymously; around 400 freshmen and sophomores participated in it. 34% of the survey participants were partially employed, while 66% did not have jobs, and spent all their time for university studies and other extracurricular activities.

We used simple statistical analysis methods to interpret the collected data. The analysis aimed at finding certain factors and barriers for healthy lifestyle, and correlations between them in order to use them in planning and balancing the university educational environment.

Thus, the approach used was based on theoretical review and analysis of the existing literature on healthy lifestyle behavior in students and results of a survey conducted based on the questionnaire developed by the authors and further tested for reliability and validity.

4 Results

The survey showed a keen and on-growing interest of the university student population towards health and healthy lifestyle issues. Initially, the authors expected no more than 300 survey participants, as this number of printed survey copies was prepared. The survey copies, however, were disseminated and duplicated by the students themselves, thus we received 400 filled out copies, which was more than the initially expected number.

The first open ended question sounded as a request to give a definition to health. The question received a number of different answers. The most popular answers included:

- 'life' (23%),
- 'absence of diseases' (12%),
- 'harmony' (12%),
- 'happiness' (11%).

The other answers included: 'gift', 'future', 'power', 'energy', 'everything we can get', 'well-being', 'joy', and etc.

The analysis of these answers showed that the students gave very diverse definitions of health, they do not consider it as a complex of physical, mental and social well-being as defined by the World Health Organization. Healthy lifestyle is also perceived by the students as a very vague concept.

The typical answers to the second question about a healthy lifestyle included:

- 'life without unhealthy habits' (43%),
- 'sports' (34%),
- 'healthy food' (13%),
- 'keeping regular hours' (7%).

The other answers included: 'reason for existence', 'living according to healthy rules', 'way to success', 'balanced human to nature relations', 'way of thinking', 'views on life', and etc.

We can see that the students do not demonstrate a thorough understanding of a healthy lifestyle as the vital function of developing and keeping human health and fulfilling all the biological and social functions.

Among the positive factors to influence health, the students named:

- 'sport' (85%),
- 'healthy food' (83%),
- 'healthy sleep' (78%),
- 'good ecology' (35%),
- 'hygiene' (9%).

Several students also named 'optimism', 'communications', 'sex', and 'positive attitudes'.

Among the negative factors, the students listed:

- 'alcoholic intoxication' (70%),
- 'tobacco addiction' (67%),
- 'drug addiction' (46%),
- 'unhealthy sleep" (38%),
- 'bad ecology' (34%).

Several students also named 'scandals', 'conflicts', 'nocturnal habits', and 'laziness'.

As we can see, the students gave very detailed answers to the questions concerning the positive and negative factors that influence their health and lifestyle which shows that they are very familiar with their occurrences.

Unfortunately, many students said that they know about the unhealthy habits, but, at the same time, they practice them. Therefore, knowledge is perceived but not implemented in competencies.

The second block of questions included the following four categories of stress factors:

1. Educational stress factors:

   - unclear teaching and learning materials;
   - time pressure;
   - information overload.

2. Achievement stress factors:

   - credits;
   - essays;
   - home tasks;
   - examinations;
   - tests;
   - lack of necessary textbooks.

3. Communication stress factors:

   - rude words;
   - misunderstanding between professors and students;
   - bullying

4. Administration stress factors:

- inflated requirements;
- regular changes of professors and university administrators.

The analysis of the students' answers gave some unexpected answers.

The students said that unclear learning materials and problems in communication with professors often stressed them out, and this influenced their lifestyle. In general the results showed that the strongest stress factors are:

1. credits (71%)
2. time pressure (60%)
3. inflated requirements (56%)
4. tests (51%)
5. misunderstanding between professors and students (48%)
6. unclear teaching and learning materials (46%).

The majority of students admit that the lifestyle is the most influential factor to determine the state of health. Only 23%, however, are absolutely sure that they follow a rational and healthy lifestyle habits.

Almost half of the students (47%) honestly admitted that they do not consider their lifestyle as a healthy one.

In is generally accepted that sport is the basis for a healthy lifestyle. Unfortunately, very few students go in for sports regularly. Only 7% of the survey participants do their morning exercises every day; 48% do it from time to time, while 45% never practice morning exercises.

At the same time, 37% of the survey participants attend sports clubs. The majority of students limit their sports by the physical education classes at university, and 48% of the survey participants consider this university course as an important and necessary one.

In general, the outcomes received from the survey results proved the conclusions made from the review and analysis of the research literature.

5 Conclusions

The study showed that physical, mental and social well-being of the engineering students is very important for their future careers. The university educational environment can influence and develop healthy lifestyle behavior in the following ways:

- providing opportunities for regular health and physical fitness tests with further medical treatment if necessary;
- instilling the value of healthy lifestyle as the necessary prerequisite for personal and professional well-being;
- developing a positive motivation towards an individual path for health promotion;
- mainstreaming self-directed healthy lifestyle behavior;
- delegating the responsibility for the healthy lifestyle to the students.

The health status depends on the lifestyle of the person; therefore we should aim at nurturing the internal desire of the students to practice healthy lifestyle habits. It can be possible through changing and influencing the student mindset and focusing it on the caring attitude towards the surrounding environment and people thus infusing positive ethical values.

These recommendations can be scaled up for developing healthy lifestyle behaviors in students of different countries and universities.

References

1. What is the WHO definition of health? URL: http://www.who.int/suggestions/faq/en/ (accessed on 30 May 2018)
2. Misselbrook, D.: W is for Wellbeing and the WHO definition of health. Br. J. Gen. Pract. Nov; **64**(628), 582 (2014). https://doi.org/10.3399/bjgp14x682381
3. Scott, P.: Beyond Stewardship? Dietrich Bonhoeffer on nature. J. Beliefs Values **18**(2), 193–202 (2006). https://doi.org/10.1080/1361767970180206
4. Wang, D., Xing, X.-H., Wu, X.-B.: Healthy lifestyles of University Students in China and influential factors. Sci. World J. **2013**, Article ID 412950 (2013)
5. Zakaria, A., Abidin, Z.S.Z.: Knowledge and practice of healthy lifestyle among higher institution students. In: 13 April 2014, 9th International Academic Conference, Istanbul (2014)
6. Erin Largo-Wight MS, CHES.: Identity and value development for a healthy lifestyle. Am. J. Health Educ. **38**(2), 104–107 (2013). https://doi.org/10.1080/19325037.2007.10598952
7. Goodyear, V.A., Kerner, C., Quennerstedt, M.: Young people's uses of wearable healthy lifestyle technologies; surveillance, self-surveillance and resistance. Sport Educ. Soc. (2017). https://doi.org/10.1080/13573322.2017.1375907
8. Boin, A., van Eeten, M.J.G.: The resilient organization. Public Manag. Rev. **15**(3), 429–445 (2013). https://doi.org/10.1080/14719037.2013.769856
9. Osipov, P.N., Ziyatdinova, J.N.: Self-education of technical university professors for intercultural communication. In: Joint International IGIP-SEFI Annual Conference 2010; Trnava; Slovakia; 19 Sept 2010–22 Sept 2010; Code 112545 (2010)
10. Ziyatdinova, J., Bezrukov, A., Osipov, P., Sanger, P.A., Ivanov, V.G.: Going globally as a Russian engineering university. In: ASEE Annual Conference and Exposition, Conference Proceedings Volume 122nd ASEE Annual Conference and Exposition: Making Value for Society, Issue 122nd ASEE Annual Conference and Exposition: Making Value for Society, 2015
11. Ziyatdinova, J.N., Osipov, P.N., Bezrukov, A.N.: Global challenges and problems of Russian engineering education modernization. In: Proceedings of 2015 International Conference on Interactive Collaborative Learning, ICL 20154 Nov 2015, Paper# 7318061, pp. 397–400 (2015)
12. Kraysman, N.V., Valeeva, E.E.: Integration of KNRTU into the world community as an example of cooperation with France. In: Proceedings of 2014 International Conference on Interactive Collaborative Learning, ICL 201421 Jan 2015, Paper # 7017886, pp. 862–863 (2014)
13. Ziyatdinova, J., Bezrukov, A., Sanger, P.A., Osipov, P.: Best practices of engineering education internationalization in a Russian Top-20 university. In: 5th Annual ASEE International Forum 2016; New Orleans; United States; 25 June 2016; Code 122760 (2016)

14. Valeeva, R.: Academic mobility is the main tool of the intercultural competence development of engineering students and scholars in China and Russia. In: 2013 International Conference on Interactive Collaborative Learning, ICL 2013, art. no. 6644721, pp. 861–863 (2013). https:// doi.org/10.1109/icl.2013.6644721

15. Osipov, P.N., Ziyatdinova, J.N.: Faculty and students as participants of internationalization. Sotsiologicheskie Issledovaniya **3**(3), 64–69 (Jan 2017)

16. Abualrub, I., Karseth, B., Stensaker, B.: The various understandings of learning environment in higher education and its quality implications. Qual. High. Educ. **19**(1), 90–110 (2013). https://doi.org/10.1080/13538322.2013.772464

17. Vermeulen, L., Schmidt, H.G.: Learning environment, learning process, academic outcomes and career success of university graduates. Stud. High. Educ. **33**(4), 431–451 (2008). https:// doi.org/10.1080/03075070802211810

18. Häfner, A., Oberst, V., Stock, A.: Avoiding procrastination through time management: an experimental intervention study. Educ. Stud. **40**(3), 352–360 (2014). https://doi.org/10.1080/03055698.2014.899487

19. Allen, D., Wilson, T.D.: Information overload: context and causes. New Rev. Inf. Behav. Res. **4**(1), 31–44 (2010). https://doi.org/10.1080/14716310310001631426

20. Chiang, S.-Y.: Dealing with communication problems in the instructional interactions between international teaching assistants and American college students. Lang. Educ. **23**(5), 461–478 (2009). https://doi.org/10.1080/09500780902822959

21. Jackson, D.: Exploring the challenges experienced by international students during work-integrated learning in Australia. Asia Pac. J. Educ. **37**(3), 344–359 (2017). https://doi.org/10.1080/02188791.2017.1298515

22. Lifestyle impact on academic performance. http://www.exercisemed.org/research-blog/lifestyle-impact-on-academi.html (accessed on 30 May 2018)

23. Castro, F., Oliveira, A.: Association between health-related physical fitness and academic performance in adolescents. Revista Brasileira de Cineantropometria & Desempenho Humano **18**(4), 441–449 (2016). https://doi.org/10.5007/1980-0037.2016v18n4p441

24. Shaw, S.R., Gomes, P., Polotskaia, A., Jankowska, A.: The relationship between student health and academic performance: implications for school psychologists. Sch. Psychol. Int. **36**(2), 115–134 (2015)

A Learning Environment for Geography and History Using Mixed Reality, Tangible Interfaces and Educational Robotics

Stefanos Xefteris, George Palaigeorgiou[✉], and Areti Tsorbari

University of Western Macedonia, Florina, Greece
{xefteris,aretitsor}@gmail.com, gpalegeo@uowm.gr

Abstract. Integrating ICT technologies in history and geography teaching may promote critical thinking and bridge the gap between unconstructive information accumulation and an explorative and critical learning approach. The aim of this study was to design, deploy and evaluate a low cost and easy-to-use mixed reality learning environment for interdisciplinary and embodied learning of geography, history and computational thinking. The proposed learning environment is comprised of an augmented 3D-tangible model of southern Europe where students interacted using their fingers, and a second treasure hunt augmented interactive floor depicting historical sites, where students performed tasks with Mindstorms EV3 robots. Students swapped between finger-based and robot-based journeys in Europe in two pairs until the end of the game. In order to evaluate our proposal, six groups of four undergraduate participants played with the environment in 6 sessions and for approximately 45 min. Data were collected with a pre and post knowledge test, an attitudes questionnaire and semi-formal group interviews. Students scored significantly higher in the post-tests and their answers in the questionnaires revealed that the multimodal environment enhanced their engagement and motivation, helped them orient themselves better around Europe's geophysical features, while the robotics treasure hunt consolidated their computational thinking skills, inducing a highly entertaining dimension. This approach better conforms to students' interactive experiences and expectations, gamifies learning and exploits embodied learning opportunities. Students engaged in the two augmented spaces in a perceptually immersive experience which became a more authentic and meaningful educational space.

Keywords: Mixed reality · Tangible interfaces · Educational robotics · Geography learning · History learning

1 Introduction

History provides students with past-to-present anchors and knowledge that can be used to develop a deeper understanding of current society and status quo, but also enables them to make informed decisions about their own social life. While a fascinating subject, learning about history was mostly deemed by students as an uninteresting and boring activity, owed to the fact that up until recently, teaching of history included

© Springer Nature Switzerland AG 2019
M. E. Auer and T. Tsiatsos (Eds.): ICL 2018, AISC 917, pp. 106–117, 2019.
https://doi.org/10.1007/978-3-030-11935-5_11

mainly the endless recitation of dates, facts and events. Students have always had trouble developing their historical understanding, limiting it to just the presented facts, and misinterpreting them out of historical context [1]. Learning history, nowadays, involves actual research and evaluation of historical sources and letting students reach conclusions based on accumulated evidence [2].

Geography on the other hand provides the essential spatial—in counterpoint to history's temporal-dimension, so that neither history nor geography are intelligible without each other [3]. Promoting geographical literacy as early as possible is seen as a highly significant endeavor for early education. The integration of ICT into school curricula has managed to evolve geography teaching by making related activities much more appealing and authentic to students, who are now able to use features such as interactive navigation on 2- or 3D data or access huge databases of geophysical data. ICT offers the opportunity to have more immersive geographical experiences and more time for observation, discussion and analysis [4].

The current research focus is to promote interdisciplinary teaching methods that combine History with Geography (and Maths, Art, Music) which can prove to be instrumental in increasing student motivation and learning effectiveness on all related domains [5]. In this frame of reference, tangible physical maps can play a major role in the development of novel teaching scenarios. Tangible physical maps complement and advance digital cartography (usually represented in screens) and become an invaluable tool for teaching geography in an embodied way. Tangible interfaces promote a sensory engagement of the student and the facilitation of spatial tasks while actively manipulating a digital representation of physical objects [4, 6]. On the other hand, a virtual "fingertrip" over an interactive augmented tangible environment representing historical sites seems to motivate students to engage deeper into the study of historical content [7].

Robotics can also add value to mixed reality tangible interfaces in a "shared reality" concept [8] where the robots act as an extra tangible interface to the mixed reality landscape and function as an agent to the virtual world. Beyond the physicality of the robot, the addition of robotics in a mixed reality scenario enhances the children's computational thinking skills [9]. The integration of robotics in learning environments facilitates the development of high order thinking processes (decomposition, abstraction, pattern recognition, algorithm design) and enables students to improve their problem solving skills [10]. Moreover, educational robotics provide a rich potential for team building and social skills development, enabling students to experiment and create on their own.

In this article, we present a two-layered tangible environment integrating two mixed reality environments that aim to enhance and improve the experience of learning geography, history and practicing computational thinking tasks. The environment offers two interactive surfaces, one table-top and one floor-based. The two tracks depict a journey performed by students via touching a 3d augmented tangible map, coupled with a robotics track where students perform a "Treasure Hunt" with a robotic companion. The goal of the study was to explore the efficacy of this multimodal tangible interface, which was constructed with low-cost and easy-to-find hardware, and which teachers and students can easily reproduce and transform to fit into multiple teaching scenarios.

2 Literature Review

The embodied cognition theory framework, postulates that acting and thinking are intertwined. The way we perceive objects or spaces is affected by the way we engage or explore them tangibly. Our mental representations are directly influenced by the physical world through our body. A variety of theoretical frameworks propose that full-body interaction potentially supports learning by involving users at different levels (affective factors, cognitive aspects, sensorimotor experience). Students create conceptual anchors on which new knowledge is built, by acting out and "physicalizing" processes, relationships etc. [11]. Thus, new interaction technologies provide us with the ability to deploy embodied learning interventions that serve as conceptual leverage. New modalities are constantly being developed, following the precepts of embodied interaction. These environments aim to facilitate embodied experiences of specific concepts, represent abstractions as concrete instances or express specific content via the operalization of actions. The use of educational robotics, mixed reality applications and tangible interfaces offers learning opportunities that need to be explored and exploited, since designing learning activities for such complex environments is an emerging and not yet systematized area of research.

Three research domains constitute the pillars of this study: a. Geography learning and tangible maps b. History learning and ICT and c. Tangible interfaces and educational robotics

Regarding tangible maps, recent studies have indicated that both paper and electronic maps have advantages and limitations in regards to students' spatial thinking skill acquisition [12]. However, the spatial topography of maps is inherently limited since the maps are projected into two dimensions. In that way, tasks about natural limitations or visibility assertions of locations are difficult to accomplish since learners have to reconstruct and reason for them mentally [13]. Tangible 3D physical maps enhanced by new digital forms of interaction, play a major role in contemporary cartography [14] and become an invaluable asset for learning geography in an embodied way [4]. Continuous shape displays, where a digital model is coupled with a physical one through a cycle of sculpting, 3D scanning and computation provide a continuous feedback loop through which the student interacts directly and very naturally with geophysical bodies, are being proposed as tangible interfaces for learning and are of special interest. The FingerTrips approach for teaching geography [4] has been shown to have positive results in altering the learning experience, making it more interactive, facilitating understanding of geographical spatial and geophysical relations.

Examining history learning, several studies have suggested that the use of ICT may motivate students and help them develop historical thinking skills [15] and contribute to the transformation of history learning to an explorative and constructive approach [16]. Apps such as timelines and simulations of historical events allow participants to better understand the concept of time, the successions of historical events and to capture how knowledge was discovered [17]. Museums and public installations are increasingly incorporating digitally-enhanced interactive experiences that provide visitors with a 'multimodal' engagement with the past [18]. There are few examples of embodied learning with tangible interfaces concerning history subjects. Recently, there is a trend to bring closer the classroom with historical installations through affordable and easily reconstructible augmented and embodied learning environments such as the FingerTrips approach, which apart from its geography application has also been used in the context of history teaching [7].

Educational robotics seem to enable educators to implement a wide range of educational approaches to classrooms: discovery learning [19], collaborative learning [20], problem solving [21, 22] competition based learning [23] and compulsory learning [24]. Although educational robotics are usually related to computational thinking [25], the potential for multidisciplinary learning is strong i.e. students can create a catapult as a prop for a tangible re-enactment of a historical battle or a water dam in the area the students inhabit [26]. Robotics have been successfully used in conjunction with drones in out-of-class teaching approaches [27] as well as with wearables and mixed/augmented reality environments in primary education classes [28]. Teaching scenarios integrating robotics with tangibles and mixed reality applications in a gamified context have been described in literature in multiple forms, highlighting the precepts of experiential learning in authentic contexts [29].

Mixed Reality (MR) environments merge the digital with the physical and offer a vivid and immersive audiovisual interface for eliciting body activity. In these environments, authentic and expressive physical activity can be augmented with digital displays that emphasize the metaphor and tools for feedback and reflection [30]. Mixed reality technologies allow students to become part of the system they are trying to familiarize with, and give them the advantage of the insider who can monitor and evaluate the mechanisms and relationships of the domain [30]. Mixed reality environments can function as an umbrella under which multiple technologies can be combined.

3   The Learning Environment

In this study we tried to create a learning scenario for history, geography and computational thinking which combines the following design principles:

1. Exploits embodied interaction with tangible objects.
2. Creates an immersive mixed reality environment in an authentic context where history and geography are intertwined with problem solving activities that also facilitate computational thinking and team work.
3. Creates a differentiated chain of activities which trigger two different modalities (FingerTrips and Robotics) and motivates students to interact with them.

The learning environment is based on two augmented spaces and students have to continuously swap between them. The first space is a 3D augmented interactive map on which students have to perform "FingerTrips", i.e. travel on the map by placing and moving their fingers on an embossed geomorphological path (i.e. the Alps, the Pyrenees, the Apennines etc.). While travelling with their fingers, students have to react to challenges/questions posed by the environment. Some of these challenges prompt them to move to the second augmented space—the interactive floor—in order to program their robots to perform specific tasks for a series of clue-finding missions. The Robotics track changes to a different city every time the students visited a different place with their finger trips. As soon as they perform their programming tasks, they come back to the first augmented space. The whole activity encompasses a "treasure hunt" scenario around Europe. The students—in two groups of two students—have to complete a variety of tasks of increasing difficulty, which ask them to recall and apply prior knowledge but also provide new information in a fun and embodied framework.

For the construction of the FingerTrip [4, 7] model we used a 50 × 75 cm MDF for the base and plasteline for recreating morphological characteristics of the 3D map. A map was then overlaid on the model using a projector as seen in Fig. 1. The game was implemented in Scratch and the interface along with the finger trip with a Makey-Makey board. On the floor, another projector was presenting images on the robotics tracks with dimensions of 1.5 m × 1.13 m as seen in Fig. 2. The robotics track interface was also implemented with Scratch and a second Makey Makey board for the touch-bases. In Fig. 3 the full setup of the environment is presented.

Fig. 1. FingerTrips Board

Fig. 2. Robotics track with EV3 explorers

Fig. 3. Combined view

The game session begins at the augmented 3D map, where all participants play as one team. The journey begins from Corfu and passes through 6 major European cities, exploring the whole routes in between, answering to questions, learning about historical landmarks or geographical information, and finding "clues" that point to the next city. The game prompts questions which are answered (by all participants) via a tangible interface. As soon as the team arrives at a major destination, participants break down into two teams and "turn against each other" hunting for clues with their robots on the floor based track. Their robots take the role of competing "Explorers" who search for clues in historical sites across Europe (Valle dei Tempi in Sicily, the Colosseum and the baths of Caracalla, piazza San Marco in Venice, etc.). Students must follow in each city a projected route and perform specific tasks with their Robot explorers in order to be able to proceed. The interactive floor keeps the time each team needed to complete the robot missions and calculates the score. The programming tasks are evolving from introductory lessons of moving forward/backward, to more advanced uses using sensors, in 4 separate stages. The robotics track is equipped with cardboard "touch-bases" which detect whenever each robot reaches each destination. During all stages, programming instructions are provided to the students in the form of printed cards.

Thus, the intervention included a continuous exchange of activities, from traveling around Europe with Fingertrips to traveling in places with robots.

4   The Study

In order to evaluate the proposed environment, a study was conducted targeting preservice school teachers.

4.1   Participants

Twenty-four (24) students from a Primary Education Department, 13 males and 11 females, participated in the study. The participants played with the FingerTrips and the augmented interactive floor environment in 6 (six) groups of 4 students. Each session lasted about 45 min.

4.2   Procedure

At the beginning of the game brief instructions were given to each group, to help students become familiar with the concept of interacting with the 3D model before starting their FingerTrip game. The researchers offered guidance whenever the participants requested for. Pre and post knowledge tests were given immediately before and after the intervention. At the end of each session, students were also asked to complete an online questionnaire about their experience. All students afterwards, participated in brief group interview.

4.3   Research Instrument

Data collection was based on pre/post tests, an attitudes questionnaire and a semi-formal group interview. Pre and post-tests were identical and were consisted of twelve questions for spatial relations (i.e., *Paris and the Alps are equidistant to the Equator*) and eight questions for information recall questions about geography and history (i.e., *To which mountain range does Mont Blanc belong?*).

The attitudes questionnaire consisted of 25 7-point Likert questions and evaluated the tangible environment in regards to its usability and attractiveness. Some of the questionnaires' items were derived from AttrakDiff [31] and Flow State Scale [32] and composed the following variables:

- *Ease of Use* (3 questions): Measure how easy to use the system is and its learnability;
- *Autotelic experience* (3 questions): Measures the extent to which the system offers internal user satisfaction;
- *Perceived learning* (3 questions): Measures students' perceptions on the educational value of the system;
- *User Focus* (3 questions): Measures the concentration during the use of the system;
- *The learning environment for practicing educational robotics* (3 questions): Measures students' attitudes towards the learning environment as canvas for practicing educational robotics development skills;
- *Pragmatic Quality* (4 questions): Measures the extent to which the system allows a user to achieve his goals;
- *Hedonic Quality-Stimulation* (3 questions): Measures the extent to which the system meets the user's need for innovation and whether it is of interest;
- *Hedonic Quality-Identity* (3 questions): Measures the extent to which the system allows the user to identify with it.

All variables can be considered as consistent since they had satisfactory Cronbach's a as seen in Table 2.

The semi-formal interviews took place immediately after the end of each session and aimed at extracting the qualitative assessments of the students and at allowing them to describe in their own words their experience with the FingerTrips and Robotics virtual space. The questions were focused on what students liked and disliked and their perceptions in regards to the learning effectiveness and efficiency of the environment. All audio-recorded interviews were transcribed and then encoded and compared within and between cases.

5 Findings

5.1 Questionnaires Results

Pre and post scores followed a normal distribution according to Shapiro-Wilk normality test ($p > 0.05$). Paired samples t-test were conducted and the results are presented in Table 1. The students scored significantly higher in the post-test both in spatial relation questions and in information recall questions as seen in Table 1. Hence, we can support that the learning environment had provoked significant learning outcomes.

Table 1. Pre/Post test results

	Pre test mean (SD)	Post test mean (SD)	t	Sig
Information recall	7.09 (1.70)	9.00 (1.98)	−4.757	0.001
Spatial relations	5.57 (1.65)	7.35 (1.07)	−4.229	0.001
Total score	12.65 (2.84)	16.35 (2.56)	−5.334	0.001

Students' answers to the attitudes questionnaire (see Table 2) show that the environment can address the problem of engagement with the historical and geographical content. However, students were also positive in regards to the learning efficiency of the environment and the possibility of exploiting it further for other university courses. They claimed that the environment made learning easier and more intriguing than with traditional teaching methods and that the environment helped them to remain focused on the learning activities.

Table 2. Attitudes questionnaire answers

	Min	Max	Mean	SD	Cronbach's a
Easiness	5.33	7.00	6.28	0.55	0.73
Focus	4.33	7.00	6.19	0.78	0.83
Autotelic experience	5.67	7.00	6.72	0.40	0.75
Learning preference	5.00	7.00	6.57	0.56	0.86
As a platform for learning robotics	4.67	7.00	6.33	0.63	0.76
Pragmatic quality	5.00	7.00	6.13	0.58	0.71
Hedonic identity	5.00	7.00	6.54	0.57	0.82
Hedonic stimulation	4.67	7.00	6.62	0.59	0.72

Students' answers in the mini AttrakDiff questionnaire validated that they considered the functions of the environments as appropriate to achieve the goal of understanding the geographical features and historical information presented (pragmatic quality). Moreover, the variable hedonic quality, which is a measure of pleasure (fun, original, engaging) and avoidance of boredom and discomfort had very high scores. Students' answers indicate that the environment made them identify themselves with it (Hedonic Quality-Identity) and believed that it offered inspiring and novel functions and interactions (Hedonic Quality-Stimulation). Finally, the students stated a strong agreement towards the use of the platform as a canvas for learning robotics ($M = 6.33$, $SD = 63$).

5.2   Interview Results

In accordance to the answers in the questionnaires, students were particularly positive about the intervention and characterized the proposed environment as attractive, fun, playful, pleasant, and creative. Such forms of activity are more suited to their technological expectations and create a more authentic and meaningful learning environment.

> It is a game that we will always be interested in playing.
> It is not a boring thing. Children learn much more easily.

Most students compared the environment with the typical geography and history teaching environments and commented that the new proposal is very different, more interesting and more motivating than typical teaching.

> I think it is a different experience to approach geography in this meaningful way, rather than simply looking at a map, a book or even a computer.
> It is much more different than traditional teaching and I would have preferred it 100%.

The finger-based style of interaction on the map was vivid, real, pleasant or helpful. Students considered the 3D finger trip as an interactive and intriguing experience and claimed that it helped them

(a)   understand better the details of the geomorphology of the map,
(b)   understand better the relative geographical positions of the different sites,
(c)   acquire an overall orientation on the specific map.

FingerTrips gave meaning to the map and the presented narration and brought the students closer to the sites of the scenario.

> It is like going through it [the journey] empirically, it's not just like watching the map, with the finger trip you can feel walking along the mountain ranges.
> I was troubled for example about the geographical location of Rome in relation to Corfu but with the help of the game I understood something I was not sure about
> [Fingertrips are important] because if the map was flat we would not have to touch it with our fingers. We could not understand the morphology. Now, we were in contact with the mountains and the geographical relief.
> It helped us to understand the spatial relations.
> It also helped us in orienting ourselves around Europe.

The students also felt that the mixed reality environment for performing robotic missions was more interesting as a learning canvas than what usually happened in the laboratory course, where robots performed tasks on desks with artificial obstacles. The robots missions were integrated in the overall scenario, while movement seemed to be taking place inside a real physical space. The fact that the interactive model recognized the success of the assigned task for each team, changed the score and the context of use and gave appropriate assistance to the two different groups created a competitive climate that kept the students active and engaged.

> The robotics floor is interesting and more active, like an actual game.
> It certainly helps, in the lab we did everything on a desk, it was not the same in our thinking, the way we perceived it
> The augmented robotics map gives the illusion of a real space
> I loved that it kept score and measuring wins and losses. It was highly competitive and kept us active.

6   Conclusions

Our pilot study indicated that the proposed scenario of integrating a multimodal tangible environment managed to alter the experience of learning European geography and history while it also promoted computational thinking tasks. This approach is highly differentiated from traditional learning approaches, closer to the students' routine environment—which is highly technological and interactive—, gamifies learning and exploits embodied affordances to improve the learning process by making it more effective while keeping it fun and enjoyable. We should underline that students evaluated positively the instructional framework and not only the interaction affordances themselves.

Our intervention consisted of an affordable, reconstructable and easy to make 3D augmented map and an equally affordable and reconstructable robotics track, which gave life to geography and history and offered an enhanced participatory experience to students. Both teachers and students may follow this approach, since they can easily design, develop and build interactive landscapes and tracks for course material of their own. Scratch and Makey Makey board facilitate both teachers and learners to easily deploy such interventions over augmented maps.

The competitive part of our intervention, where students performed robotics tasks on the second augmented environment, added an extra dimension of originality while spurring them to action. Students stated vehemently that the usual laboratory robotics course would be much improved if the learning tasks were also performed on augmented tracks in the form of competition between teams.

We do acknowledge several limitations of our study, first and foremost the small number of participants, or the lack of analysis of the underlying embodied mechanism for learning. More detailed and expanded studies must be done to further explore whether similar multimodal, multi-technologies environment can address students' needs and desires.

References

1. Nokes, J.: Recognizing and addressing the barriers to adolescents "reading like historians". Hist. Teach. **44**, 379–404 (2011)
2. Giannopoulos, D.: Italian presence in the Dodecanese 1912–1943: teaching a history topic in weebly environment. Proc. Comput. Sci. **65**, 176–181 (2015)

3. Baker, A.: Geography and history—bridging the divide. Cambridge university press (2003)
4. Palaigeorgiou, G., Karakostas, A., Skenderidou, K.: FingerTrips: learning geography through tangible finger trips into 3D augmented maps. In: 2017 IEEE 17th International Conference on Advanced Learning Technologies (ICALT), pp. 170–172. IEEE (2017)
5. Bickford, J.H.: Initiating historical thinking in elementary schools. Soc. Stud. Res. Pract. **8**, 60–77 (2013)
6. Mpiladeri, M., Palaigeorgiou, G., Lemonidis, C.: Fractangi: a tangible learning environment for learning about fractions with an interactive number line. In: International Conference on Cognition and Exploratory Learning in the Digital Age (CELDA). International Association for Development of the Information Society(IADIS), pp. 157–164 (2016)
7. Triantafyllidou, I., Chatzitsakiroglou, A.-M., Georgiadou, S., Palaigeorgiou, G.: FingerTrips on tangible augmented 3D maps for learning history. In: Interactive Mobile Communication Technologies and Learning. Springer Cham, pp. 465–476 (2017)
8. Robert, D., Wistorrt, R., Gray, J., Breazeal, C.: Exploring mixed reality robot gaming. In: Proceedings of the Fifth International Conference on Tangible, Embedded, and Embodied Interaction—TEI '11, pp. 125–128 (2011)
9. Eguchi, A.: Computational thinking with educational robotics. In: Proceedings of Society for Information Technology & Teacher Education International Conference. AACE, pp. 79–84 (2015)
10. Atmatzidou, S., Demetriadis, S.: Advancing students' computational thinking skills through educational robotics: a study on age and gender relevant differences. Robot. Auton. Syst. **75**, 661–670 (2016)
11. Lindgren, R., Tscholl, M., Wang, S., Johnson, E.: Enhancing learning and engagement through embodied interaction within a mixed reality simulation. Comput. Educ. **95**, 174–187 (2016)
12. Collins, L.: The impact of paper versus digital map technology on students' spatial thinking skill acquisition. J. Geogr. **117**, 137–152 (2017)
13. Li, N., Willett, W., Sharlin, E., Sousa, M.: Visibility perception and dynamic viewsheds for topographic maps and models. In: Proceedings of the 5th Symposium on Spatial User Interaction. ACM, pp. 39–47 (2017)
14. Petrasova, A., Harmon, B., Petras, V., Mitasova, H.: Tangible modeling with open source GIS (2015)
15. Bogdanovych, A., Ijaz, K., Simoff, S.: The city of Uruk: teaching ancient history in a virtual world. International Conference on Intelligent Virtual Agents, pp. 28–35. Springer, Heidelberg (2012)
16. Blanco-Fernández, Y., Lopez-Nores, M., Pazos-Arias, J., et al.: REENACT: a step forward in immersive learning about Human History by augmented reality, role playing and social networking. Expert Syst. Appl. **41**, 4811–4828 (2014)
17. Galan, J.G.: Learning historical and chronological time: practical applications. Eur. J. Sci. Theol. **12**, 5–16 (2016)
18. Savenije, G.M., de Bruijn, P.: Historical empathy in a museum: uniting contextualisation and emotional engagement. Int. J. Heritage Stud. **23**, 832–845 (2017)
19. Sullivan, F.R., Moriarty, M.A.: Robotics and discovery learning: pedagogical beliefs, teacher practice, and technology integration. J. Technol. Teach. Educ. **17**, 109–142 (2009)
20. Denis, B., Hubert, S.: Collaborative learning in an educational robotics environment. Comput. Hum. Behav. **17**, 465–480 (2001)
21. Alimisis, D., Frangou, S., Papanikolaou, K.: A constructivist methodology for teacher training in educational robotics: the TERECoP course in Greece through trainees' eyes. In: Ninth IEEE International Conference on Advanced Learning Technologies, 2009, ICALT 2009, pp. 24–28 (2009)

22. Ilieva, V.: ROBOTICS in the primary school—how to do it? Auton. Rob. 596–605 (2010)
23. Eguchi, A.: RoboCupJunior for promoting STEM education, 21st century skills, and technological advancement through robotics competition. Rob. Auton. Syst. **75**, 692–699 (2016)
24. Khanlari, A.: Teachers' perceptions of the benefits and the challenges of integrating educational robots into primary/elementary curricula. Eur. J. Eng. Educ. **41**, 320–330 (2016)
25. Afari, E., Khine, M.S.: Robotics as an educational tool: impact of lego mindstorms. Int. J. Inf. Educ. Technol. **7**(6), 437–442 (2017)
26. WeDo|Challenges: The great catapult | Dr. E's WeDo Challenges (2014). https://wedo.dreschallenges.com/the-great-catapult/. Accessed 13 Jan 2018
27. Palaigeorgiou G., Malandrakis, G., Tsolopani, C.: Learning with drones: flying windows for classroom virtual field trips. In: 2017 IEEE 17th International Conference on Advanced Learning Technologies (ICALT). IEEE, pp 338–342 (2017)
28. Honig, W., Milanes, C., Scaria, L., et al.: Mixed reality for robotics. In: IEEE International Conference on Intelligent Robots and Systems, pp. 5382–5387 (Dec 2015)
29. Wang, C.Y., Chi-Hung, C., Chia-Jung, W., et al.: Constructing a digital authentic learning playground by a mixed reality platform and a robot. In: Proceedings of the 18th International Conference on Computers in Education, pp. 121–128 (2010)
30. Lindgren, R., Johnson-Glenberg, M.: Emboldened by embodiment. Educ. Res. **42**, 445–452 (2013)
31. Hassenzahl, M., Monk, A.: The inference of perceived usability from beauty. Hum.-Comput. Interact. **25**, 235–260 (2010)
32. Jackson, S.A., Marsh, H.W.: Development and Validation of a scale to measure optimal experience: the flow state scale. J. Sport Exerc. Psychol. **18**, 17–35 (1996)

Movable, Resizable and Dynamic Number Lines for Fraction Learning in a Mixed Reality Environment

George Palaigeorgiou$^{(\boxtimes)}$, Xristina Tsolopani, Sofia Liakou, and Charalambos Lemonidis

University of Western Macedonia, Florina, Greece
{gpalegeo,xrtsolopani,liakousophia}@gmail.com,
xlemon@uowm.gr

Abstract. Teaching about fractions is a challenging topic for teachers since it includes complex conceptual content. The number line is a visual representation tool that seems to be effective for dealing with fractions. In this study, we aim at transforming the static uncontextualized paper-and-pencil number line into a practical, useful and dynamic measurement tool in an authentic gamified context. We present the mixed reality environment "Marathon of Fractions" which enables learners to position multiple number lines in any place of the augmented space and adjust their characteristics as they wish. The number lines are used for positioning athletes and other props on an augmented stadium. To evaluate our proposal, 28 6th grade students participated in groups of two in 14 sessions which lasted for about 45 min. At the end of each session, students were asked to complete a questionnaire about their experience while sixteen students also participated in brief interviews. Students assessed the environment as effective, enjoyable, innovative, helpful and expressive and claimed that it stole their attention for 45 min even though fraction learning is an unpopular subject matter. They also pinpointed as main advantages of the proposed environment the representational power of the interactive number lines, the authenticity of the environment, the feedback mechanism, the entertaining character of the activity and the collaboration required.

1 Introduction

Teaching about fractions is most often a challenge for teachers since it is an object with complex conceptual content. However, competence with fractions is imperative for solving problems in science, technology, and engineering, and for dealing with everyday activities [1, 2] while their understanding is a prerequisite for learning proportions, ratios, decimals, percentages, and rational numbers, as well as advanced mathematics. Many students consider fractions as "meaningless symbols", are unable to recognize that there are infinite fractions between any two fractions [3], view numerator and denominator as separate numbers and not as a unified whole [4] and usually over-use the part–whole model. The number line is a visual representation tool that has been proposed as a useful tool for students to construct, divide, add, compare, or perform operations with numbers and fractions [5]. There are several reasons for

© Springer Nature Switzerland AG 2019
M. E. Auer and T. Tsiatsos (Eds.): ICL 2018, AISC 917, pp. 118–129, 2019.
https://doi.org/10.1007/978-3-030-11935-5_12

that: the magnitude representation of fractions on a number line helps in perceiving fractions as numbers, the continuity of a number line shows the continuity between numbers and clarifies that there is an infinite number of fractions between any two fractions, and improper and negative fractions are introduced more easily on a number line.

Several researchers suggest that mathematical concepts, such as fractions, can and should be represented in many different ways. Thus, in addition to symbolisms and models, it is also helpful to use and act on real objects that are graspable and tangible. For example, the use of physical objects-manipulatives is quite often in elementary schools. New embodied interaction technologies are being studied for performing physical actions that function as "conceptual leverage" [6] in math. Full-body interaction supports learning by involving users at diverse levels such as sensorimotor experience, cognitive and affective factors etc. [7]. Tangibles seem to have significant value for learning since they limit the input alternatives, reduce the modality on the interface, promote a sensory engagement, and enable the coupling of physical objects with the manipulation of their digital representation.

In this paper, we will introduce a mixed reality environment which offers multiple dynamic, movable and resizable interactive number lines in an authentic and gamified context in order to motivate and engage students, to provide an unprecedented and flexible representation of fractions operations and, thus, help students deepen their understanding of fractions.

2 Literature Review

2.1 Learning with Number Lines

As Siegler et al. [8] suggested, understanding fractions as an abstract concept means recognizing that they are numbers that can be placed on the number line. Wu [9] argues that number line usage has the advantage of making coherent the study of numbers in school math. Fazio and Siegler [4] reported that by placing fractions on a number line, students are offered the ability to compare their sizes, as well as to identify fractions that are equivalent. In addition, integer numbers can also be positioned on a number line, and, hence, students realize that integers may also be written as fractions. Also, two fractions with different denominators can be compared using two different number lines of the same length. Moreover, the use of number lines can be useful for clarifying the concept of fraction density, since an essential characteristic of fractions is that there are infinite fractions between any two of them [10]. It is also possible to extend students' understanding to negative fractions, improper fractions as well as decimals and percentages. By placing the same fraction, decimal and percentage on the same number line, students' advance their flexibility in using different symbolic representations for explicit numbers [11]. Wu [9] argues that the number lines are ideal for fractions representation, since it is easier to divide the whole into equal parts, considering only their length. Thus, the addition, the subtraction and the comparison of fractions become easier to be presented.

Consequently, the numeric line is a geometric representation that gives each number a unique point on the line and an oriented distance from the beginning, both of which reflect its size and direction. It is a useful tool for the development of numerical comprehension, especially when students learn about integers and rational numbers.

2.2   Tangibles, Embodied Learning and Maths

According to theories of tangible and embodied learning, "embodied numerosity" is the idea that the numerical representation of adults is shaped by physical experience, such as finger counting [12]. Therefore, finger counting is not necessary only for a particular stage of development but affects the processing of numbers, even later. Carbonneau et al. [13] found that the use of physical objects in math concepts instruction tends to improve retention, problem-solving, and transfer. Button Matrix [14] used coupled tactile, vibration and visual feedback to highlight features of physical experience with arithmetic concepts and cue reflection on the links between the physical experience and mathematical symbols. Martin and Schwartz [15] designed a TUI that helps students to solve ratio or division problems using TUI objects linked to their digital symbolic representations [16].

Such physical experiences, affect the structure of abstract mental representations of numbers, like that of the mental number line, and this abstract knowledge can have its roots in physical experiences. It seems that mathematical cognition is embodied in two senses: it is based on action and conception and is originated from the natural environment [17]. Original experiences are derived with objects met in nature and not with symbols. So, from the perspective of embodied learning theory, people interactions with physical objects produce the foundation for subsequent "construction" of non-physical entities, contained in formal mathematical definitions [18]. For example, when students talk about the concepts they learn, they often express new knowledge with gestures and bodily expressions proving that gestures are an integral part of communication about mathematical concepts [19]. So, when a student "becomes a thing itself," researchers consider that he has a completely different kind of experience from a student who just watches passively, because the embodied learning promotes links between physical actions and mathematics, in a way that observation cannot do. According to Edwards et al. [18], the motor system is involved in learning in diverse ways such as whole body moves, gestures, gaze, head movement, body posture, object manipulation, rhythm, etc.

In recent years, several studies have tried to reconcile the advantages of number lines with the advantages of embodied learning of math and technology. For example, researchers [20] have found positive results from the use of a digital dance layer, in which students have to compare the size of a number/set of squares displayed on a hypothetical number line, by moving their bodies. Furthermore, researchers [21] have shown that physical movements which are related with the learning objective may have a positive effect on mathematical education, and more specifically on number line estimation and numerical comparison. In this study [21], students managed to connect numerical knowledge with the physical world and body movements, and created mental representations of numerical size. In addition, another study showed that apprentices who moved their entire body to the left or to the right to complete computational estimation activities on a number line presented on an interactive whiteboard, also had positive learning outcomes [22]. Mpiladeri et al. [23] designed an interactive tangible number line, named FRACTANGI, as a conceptual metaphor for helping students to understand and exploit fractions by acting with their hands. Pre-service teachers that used FRACTANGI were enthusiastic with its instructional possibilities and underlined its potential to transform practicing with fractions to an enjoyable and effective learning experience.

3   The Marathon of Fractions

The "Marathon of Fractions" mixed reality environment was designed with the aim of familiarizing students with the use of number lines and of deepening their knowledge of fractions concepts and operations. The basic design premise of the environment was to provide a learning sequence of activities, visual representations, interactions, and physical objects manipulations that would enable even students with limited fractions knowledge to succeed in finishing the game scenario.

The significant differentiation of the environment is that students can use dynamic, movable and resizable number lines to perform fraction tasks in an authentic mixed reality environment. The environment empowers users to put multiple number lines at any time and place in the augmented space and to modify their characteristics as they wish. With this approach, the number lines become a practical visualization, representation and measuring tool in augmented spaces, they promote flexibility in fractions' estimations but also reveal the value of fractions in realistic conditions. This new form of number lines is not a theoretical one but it is placed and used over real objects with specific purposes. Students have to manipulate and "grasp" a fraction with their hands.

The proposed mixed reality environment is based on a miniature wooden stadium (120 × 40 cm) showing two dynamic number lines with the help of a projector (see Fig. 1). Learners can increase or decrease the length of the number lines, move them horizontally, and update the fractional unit (e.g. quarters, fifths, etc.) with the use of two control panels. In addition, there is a third auxiliary number line, which functions as a means of feedback and is triggered by the use of a "help button".

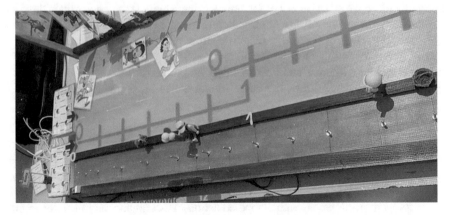

Fig. 1. The Marathon of Fractions

The interaction is based on a game script on which several runners from various countries had begun a race of 2 km, which was interrupted due to heavy rainfall. At that time, each athlete had reached a specific position in the lane which was written down in cards (see Fig. 2). Students had to become the drivers of a car (see Fig. 2) and carry each athlete from the start to the point where they had stopped. The runners' last position is given as a fraction of the main lane or as a fraction in relation to the position of other runners or objects of the wooden miniature. At the bottom of the miniature there were touch points for driving the car. In the event of a mistaken selection, the learner was given the opportunity to click on a help button and watch (on the same surface) a series of steps that simplified the solution. The feedback mechanism was focused on giving instructions on how students had to think, rather than the solutions itself.

Fig. 2. (Left) The runners' cards, (right) a student trying to estimate an athlete's position

Students work in pairs and at each time one has the role of the driver of the car who selects the runners' position while the other has the role of assistant-consultant. The roles swap continuously. Apart from positioning athletes' cards, participants are asked to place aid stations, awards etc. again specified as fractional numbers. The augmented miniature contains several objects which act as fractional references, e.g. trees, animals, police officers, etc. and which provoke students to move, resize and change the number lines several times. For example, at a point, the following instruction is heard: "*Please place the prizes at 8/9 of the distance between the first tree and the basket.*" The player has to initially move the number line, adjust its length, then divide it into nine parts, and finally place the prizes at 8/9 of the number line length. If the player finds the right answer, he is rewarded and his score increases. In the event of a mistake, the game urges him to get help and reduces his score. If he clicks the help button, the road in front of him turns into a screen, which along with the voice instructions, shows him what to press, how to think and how to solve the problem.

To sum up, students during the game among others, they are asked to

- place fractions on the number line,
- adjust the number line length and then place fractions,
- check for fractions equivalence,
- add and subtract fractions with same or different denominators; these operations are expressed as distances and the students have to use two number lines combinatory,
- compare fractions, through embodied actions, gestures and tangible interactions.

Since participants are not stuck in front of a computer, their body is also involved in the learning process by manipulating objects and moving their head, looking right and left, and by counting distances with gestures.

The augmented miniature was created having in mind easiness, affordability, and replicability. The interfaces were created by exploiting two Makey Makey boards attached to a laptop running the web version of Scratch 2, and the viewport of Scratch was projected over the miniature. Such a setting is easy to be reproduced easily by students and teachers.

4   The Study

4.1   Participants

Twenty-eight (28) students from 6th grade of a primary school, 17 boys and 11 girls, participated in a study for evaluating "Marathon of Fractions" game. The participants played with the game in 14 groups of two students. Each session lasted about 45 min.

4.2   Procedure

At the beginning of the game, brief instructions were given to each group of students, to help them familiarize with the manipulation of the dynamic number lines. Afterwards, the researchers offered guidance whenever the participants asked for. At the end of each session, students were asked to complete a questionnaire about their experience while sixteen students also participated in brief interviews. All sessions were also video-recorded.

4.3   Research Instrument

Data collection was based on an attitudes questionnaire and semi-formal group interviews. The attitudes questionnaire consisted of 22.5-point Likert questions and evaluated the tangible environment in regards to its usability and attractiveness. Some of the questionnaires' items were derived from AttrakDiff [24] and Flow State Scale [25] and composed the following variables:

- *Ease of Use* (3 items): Measures how easy to use the system is;
- *Autotelic experience* (3 items): Measures the extent to which the system offers internal user fulfillment;
- *Perceived learning* (3 items): Measures students' perceptions about the educational value of the system;
- *User Focus* (3 items): Measures the students' perceived focus on the learning activities during system usage;
- *Pragmatic Quality* (4 items): Measures the extent to which the system enables a user to achieve his goals;

- *Hedonic Quality-Stimulation* (3 items): Measures the extent to which the system is perceived as innovative and interesting;
- *Hedonic Quality-Identity* (3 items): Measures the extent to which the system lets the user to identify with it.

All questions were 5-point Likert scale questions and all variables can be considered as consistent since they had satisfactory Cronbach's a as seen in Table 1.

Table 1. Students' answers to attitudes questionnaire

	Min	Max	Mean	SD	Cronbach's a
Easiness	3.33	5.00	4.32	0.51	0.75
Focus	2.67	5.00	4.40	0.62	0.82
Autotelic Experience	3.67	5.00	4.67	0.45	0.76
Learning Preference	1	5.00	3.92	1.08	0.83
Pragmatic Quality	2.5	5.00	4.21	0.67	0.75
Hedonic Identity	3.33	5.00	4.62	0.53	0.75
Hedonic Stimulation	3.33	5.00	4.62	0.51	0.80

The semi-formal interviews took place immediately after the end of each session and aimed at identifying the qualitative views of the students about the environment. The questions were focused on what students enjoyed and disliked and their perceptions in regards to the learning success and efficiency of the environment. All audio-recorded interviews were transcribed and then encoded and compared within and between cases.

5 Results

5.1 Answers to Questionnaires

Students overall evaluation of their learning experience was very positive. As seen in Table 1, students assessed the proposed environment as easy to use and indicated that it grabbed their attention for the whole duration of the learning session. Students were delighted with their experience (autotelic experience) and this is of particular importance if we consider that they were interacting with a demanding and difficult to grasp content for 45 min without any interruption. Interestingly, students did not have a converging view on whether the specific approach is a preferable way for learning about fractions in the school environment since the related variable had a moderate positive value and the highest standard deviation. By exploring more the students' answers, it was identified that five students were strongly negative towards exploiting the environment in school (M = 1.99, SD = 0.52) while the rest of the students were positive (M = 4.34, SD = 0.59). It seems that these five students, while they considered positively their experience and the potential of the environment, they believed that the environment is not adequate for learning about fractions in the classroom settings. This finding should be investigated more.

Students' answers in the mini AttrakDiff questionnaire validated that they considered the environment as appropriate to achieve fractions understanding (pragmatic quality). They also assessed the environment as fun, original, and engaging without moments of boredom and discomfort (hedonic quality). Students' answers indicated that they identified themselves with it (Hedonic Quality-Identity) and they believed that it offered inspiring and novel functions and interactions (Hedonic Quality-Stimulation).

5.2 Students' Interviews

In the interviews, the students revealed their initial discomfort when they started the game as the overwhelming majority of them had a negative predisposition towards learning about the fractions. It was a significant achievement that the environment reversed this view and gained their attention. The students felt that the environment was more practical, more participatory and more playful than typical learning in schools. They also argued that the environment helped them to develop their understanding both in relation to number lines and fractional concepts.

> We liked the way you teach math, because we, at school, are only doing theory and practice.
> I liked the whole game, but that was not expected since I did not like the fractions domain, but it was really good.
> We thought that this game was going to be very difficult because it was about fractions, but now that we have played, it was different, easy.
> ...but now I understood better the number lines and I think that the game is useful and the other kids will understand them [number lines] better.
> The environment helped me because of the way the fractions were presented and the nice questions posed, we understood them better.

Students pinpointed the follow 5 elements as the main advantages of the proposed environment:

(A) *the representational power of the interactive number line and its easy handling.* Students used the number lines in different positions, with different sizes and different divisions, thus obtaining a more dynamic and flexible perception of them. The augmentation of the number lines on a space with physical objects rendered them more demonstrative, while their usefulness in the game was instrumental. Hence, the interactive number lines were an easy and responsive tool to use.

> I understood them [number lines] better because, here, they are more expressive and it's like we learnt them by ourselves.
> I confronted many difficulties with school exercises and it seemed much easier here. Here you do not write, you click the buttons and learn if your choice was correct, while at school you are corrected by the teacher and you have to wait, you do not know if you answered correctly [immediately].
> [I liked it] because it makes you think. It does not give you the solutions as the book does. You are thinking. It's even more fun.

(B) *The authenticity of the environment.* The environment represented in a practical way and in real-world contexts, math concepts that students had approached in the past only theoretically. For example, fractions addition, which required the use of both number lines did take a tangible form. Interactive number lines in these examples functioned as computational tools that visualized and helped students make sense of the respective activities.

> For example, when we were talking about the 2/8 which we did not know [they were asked to add this fraction to the position of an athlete], we thought it better in the game and we solved it. The environment helped me. It was easy. It provided more things to put [over a number line] and it was easier to find them while in school, for example, you have one thing and you simply have to put it on a number line.

(C) *The feedback system.* Feedback was structured based on the steps required for handling the interactive number lines. However, feedback guidance also promoted a specific way of thinking. The feedback system always explained to the students how to initially move the beginning of the number line on the adequate position, how to adjust its length, how to change its divisions, and, then, how to identify the fraction on the number line in each activity. That is precisely the way students had to think about completing the tasks.

> It was easier to understand the fractions as explained by the system.
> It also had the help system which helped us a lot... because it explains them [fractions] to us.

(D) *The entertaining character.* The activity was playful and the students underlined that they were learning about fractions while they were having fun. They particularly liked the various details—aids on the augmented model (e.g., small animals, playmobile, athletes' cards), and often compared this type of learning with that of reading a book.

> We do not like the book very much. Here we had a lesson that we enjoyed it.
> That way you understand them [fractions] better, but, we also have a good time.

(E) *Collaboration.* The students noted several times that co-operation and continuous role swapping were very beneficial for the progress of the game and the development of their understanding. The students said they felt more confident within this context of cooperation and stressed that they were able to help each other.

> Yes, we collaborated quite a lot and he helped me a lot.
> We worked together. If someone couldn't continue, the other one helped him.
> Collaboration helped us to win.

5.3 Video Analysis

Students needed some time to familiarize with number line handling and recognize how to apply it to the problems presented. After that, most students were struggling with the way the number line had to be divided. The main problem was that they matched the denominator number with the separation lines and could not discern that the parts of the number line were more by one. This issue is a well identified misconception.

According to video analysis, students avoided using the second number until the end of the first group of activities.

Students also confronted difficulties in activities that required an athlete to be positioned compared to another one (forward or backward). These activities required the simultaneous use of both interactive number lines. Several students couldn't imagine how the two number lines could relate to each other as they haven't done something similar in the past. The difficulty was even higher when the position of the second athlete was specified by a fraction with a different denominator compared to the first. The students had never seen the sum of two such fractions in front of them. The use of the feedback mechanism here was decisive, as also the expressive power of the environment.

Finally, it was interesting that cooperation was more intense at the end of the activities. After each activity, the students became more active: they read the cards, provided explanations and interpretations, suggested solutions, or asked for the opinion of their peers.

6 Discussion

Augmented spaces seem to offer a stimulating alternative to mathematical conceptual tools. From geometry to numeracy and from additions to fractions, mixed reality environments can function as a creative canvas for applying mathematical representations in authentic contexts and in a more dynamic way. Tools such as the number lines can become intriguing interactive toys for the students while playing a game or solving a problem.

In this manuscript, we presented the "Marathon of Fractions" learning environment which had the aim to transform the static uncontextualized paper-and-pencil number line into a practical, useful and dynamic measurement tool that is easily reconfigurable in an authentic context. The environment enables the students to make mistakes, to expose their misconceptions and to drill and practice with a variety of fraction activities. It offers visualizations that address several conceptual issues and it also demands students' embodied actions in order complete their activities. Students assessed the environment as effective, enjoyable, innovative, helpful and expressive and claimed that it stole their attention for 45 min despite the fact that fraction learning is a rather unpopular subject matter.

It is important that the construction of the environment is accessible, affordable and replicable. Such proposals are easier to be diffused in everyday school activities and also offer the potential of exploiting the new trend of the maker culture. Nevertheless, more studies are required in order to identify and evaluate the learning effects of the proposed environment.

References

1. Jordan, N.C., Hansen, N., Fuchs, L.S., Siegler, R.S., Gersten, R., Micklos, D.: Developmental predictors of fraction concepts and procedures. J. Exp. Child Psychol. **116**, 45–58 (2013)
2. Siegler, R.S., Duncan, G.I., Davis-Kean, P.E., Duckworth, K., Claessens, A., Engel, M., Chen, M.: Early predictors of high school mathematics achievement. Psychol. Sci. **23**, 691–697 (2012)
3. Riconscente, M.: Mobile learning game improves 5th graders' fractions knowledge and attitudes. GameDesk Institute, Los Angeles (2011)
4. Fazio, L., Siegler, R.S.: Teaching Fractions. International Academy of Education (2011)
5. Saxe, G.B., Diakow, R., Gearhart, M.: Towards curricular coherence in integers and fractions: a study of the efficacy of a lesson sequence that uses the number line as the principal representational context. ZDM **45**(3), 343–364 (2012)
6. Lindgren, R., Tscholl, M., Wang, S., Johnson, E.: Enhancing learning and engagement through embodied interaction within a mixed reality simulation. Comput. Educ. **95**, 174–187 (2016)
7. Malinverni, L., Pares, N.: Learning of abstract concepts through full-body interaction: a systematic review. J. Educ. Technol. Soc. **17**(4), 100 (2014)
8. Siegler, R.S., Thompson, C.A., Schneider, M.: An integrated theory of whole number and fractions development. Cogn. Psychol. **62**(4), 273–296 (2011)
9. Wu, H.: Teaching fractions according to the Common core standards. Retrieved 24 Feb 2014 (2011)
10. Hannula, M.S.: Locating fraction on a number line. Int. Group Psychol. Math. Educ. **3**, 17–24 (2003)
11. Lemonidis, C.: Mental Computation and Estimation: Implications for Mathematics Education Research, Teaching and Learning. Routledge (2015)
12. Domahs, F., Moeller, K., Huber, S., Willmes, K., Nuerk, H.C.: Embodied numerosity: implicit hand-based representations influence symbolic number processing across cultures. Cognition **116**(2), 251–266 (2010)
13. Carbonneau, K.J., Marley, S.C., Selig, J.P.: A meta-analysis of the efficacy of teaching mathematics with concrete manipulatives. J. Educ. Psychol. **105**(2), 380 (2013)
14. Cramer, E.S., Antle, A.N.: Button matrix: how tangible interfaces can structure physical experiences for learning. In: Proceedings of the Ninth International Conference on Tangible, Embedded, and Embodied Interaction (pp. 301–304). ACM (Jan 2015)
15. Martin, T., Schwartz, D.L.: Physically distributed learning: Adapting and reinterpreting physical environments in the development of fraction concepts. Cogn. Sci. **29**(4), 587–625 (2005)
16. Antle, A.N.: The CTI framework: informing the design of tangible systems for children. In: Proceedings of the 1st International Conference on Tangible and Embedded Interaction, pp. 195–202. ACM (Feb 2007)
17. Alibali, M.W., Nathan, M.J.: Embodiment in mathematics teaching and learning: evidence from learners' and teachers' gestures. J. Learn. Sci. **21**(2), 247–286 (2012)
18. Edwards, L.D., MooreRusso, D., Ferrara, F. (Eds.): Emerging Perspectives on Gesture and Embodiment in Mathematics. IAP (2014)
19. Abrahamson, D., Black, J. B., DeLiema, D., Enyedy, N., Hoyer, D., Fadjo, C.L., Trninic, D.: You're it! Body, action, and object in STEM learning. In: Proceedings of the International Conference of the Learning Sciences: Future of Learning (ICLS 2012), vol. 1, pp. 283–290 (2012)

20. Fischer, U., Moeller, K., Bientzle, M., Cress, U., Nuerk, H.C.: Sensori-motor spatial training of number magnitude representation. Psychon. Bull. Rev. **18**(1), 177–183 (2011)
21. Mavilidi, M.F., Okely, A., Chandler, P., Domazet, S.L., Paas, F.: Immediate and delayed effects of integrating physical activity into preschool children's learning of numeracy skills. J. Exp. Child Psychol. **166**, 502–519 (2018)
22. Fischer, U., Moeller, K., Huber, S., Cress, U., Nuerk, H.C.: Full-body movement in numerical trainings: a pilot study with an interactive whiteboard. Int. J. Serious Games **2**(4), 23–35 (2015)
23. Mpiladeri, M., Palaigeorgiou, G., Lemonidis, C.: Fractangi: A Tangible Learning Environment for Learning about Fractions with an Interactive Number Line. International Association for Development of the Information Society (2016)
24. Hassenzahl, M., Monk, A.: The inference of perceived usability from beauty. Hum.-Comput. Interact. **25**, 235–260 (2010)
25. Jackson, S.A., Marsh, H.W.: Development and validation of a scale to measure optimal experience: the flow state scale. J. Sport Exerc. Psychol. **18**, 17–35 (1996)

Perspectives on the Relevance
of Interdisciplinary Music Therapy

Fulvia Anca Constantin[1]([⊠]) and Stela Drăgulin[2,3]

[1] Transilvania University of Braşov, Braşov, Romania
fulvia.constantin@gmail.com
[2] American-Romanian Academy of Arts and Sciences—ARA, Craiova,
Romania
steladragulin@yahoo.com
[3] Romanian Academy of Scientists—AOSR, Bucuresti, Romania

Abstract. Among healthcare professions, music therapy practice has its own
therapeutic way to address treatment in the areas of psycho-social behaviour,
speech and language, sensorial, motor and cognition of individuals with a
variety of diagnoses: neurological, psychological, physical, or other medical
ones. The practice of music therapy requires specialised music therapists whose
theoretical notions come not only from the fields of Psychology and Music, but
also from Neurology, Psychiatry, Physics, IT or Statistics. The aim of this paper
is on one hand, to explain the various relationships existing among disciplines
and their utility in the education of music therapists, and, on the other hand, to
note the outcome of the interdisciplinary approach on varied categories of students and/or therapy subjects. A broad and complete view and understanding of
the problem leads to a correct approach of the therapy avoiding harm and
offering effective and efficient methods of healing, in order to increase health,
wellbeing and contribution in society. From education to licensing and actual
practice, the therapists acknowledge information about interdisciplinary
approaches including models of interdisciplinary work. Keeping in mind the
distinct, unique features of music therapy and its interdisciplinary characteristics, aspects related to the background education of therapists, to the actual
training as professionals, and also to the communication with specialists from
other fields (working teams) are discussed.

Keywords: Music therapy · Interdisciplinary · Education

1 Introduction

As known, in any kind of therapy practice we focus on three things: on method, on
outcome and on people. If we talk about method we have in mind the type of therapy,
for example music therapy and other expressive art therapies practiced solely or with
music support. The outcome is about changing the person's behaviour or state (e.g.
physical behaviour—less pain, psychological behaviour—less stress, so on). Also,
working with different categories of people implies that a distinction between them
should be made. For example, we could talk about children in special education, about
geriatrics, about adults with mental disabilities, etc. This knowledge helps therapists to

© Springer Nature Switzerland AG 2019
M. E. Auer and T. Tsiatsos (Eds.): ICL 2018, AISC 917, pp. 130–138, 2019.
https://doi.org/10.1007/978-3-030-11935-5_13

do differential work in order to have better results. Thus, the interdisciplinary music therapy is relevant for professionals and the services they provide. The findings highlight the importance of interdisciplinary in education and the results that indicate the value of the music therapy work.

2 Defining Music Therapy

One of the therapy practices with the largest impact on people with severe or less severe problems is music therapy. Simply defining it, music therapy represents the way music addresses the emotional, physical, cognitive and social needs of a person or a group of people. Therefore, through music therapy there are touched aspects of *mind* (a distraction in the time of stress), *brain* (brain waves activated at various stimuli), *body* (a change of rhythm) and *behaviour* (an alteration of the mood). Conducted by specialised music therapists, possible activities include listening to music, playing an instrument or simple drumming, improvising and, not the last, guided imagery. Music therapy sessions are intended to address physical health, communication abilities, cognitive skills, emotional well-being, and diverse interests. Through a variety of methods in accord to the existing health issues that could be treated emotional issues, rehabilitative needs, or chronic conditions. Therefore, by using music we address both, medical and psychological problems. More than being an interdisciplinary object of study, music therapy involves at least three disciplines: music, psychology and medicine (Fig. 1).

Fig. 1. Interdisciplinary music therapy

If music includes from Improvisation to Music Theory and Music Repertory (genres), medicine includes Psychiatrics, Neurology and Alternative Medicine. Although it is said that there is no need to have musical abilities in order to benefit from music therapy, the specialists must have this knowledge so will ensure that the proposed activities satisfy the need of the client.

Music therapy also has a multi-disciplinary and trans-disciplinary approach, through the fact that each music therapy topic is studied holistic through more academic disciplines or professional specializations applying research efforts focused on problems that cross the limits. In the new research studies, more than Music,

Psychology and Medicine, there are included specializations such as Systems of Instrumentation, Acoustics, Ethics, so on (Fig. 2).

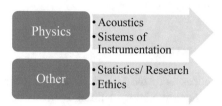

Fig. 2. Interdisciplinarity of music therapy 2

3 Relevance of an Interdisciplinary Specialization in Receptive and Active Music Therapy

According to the need and the treatment aims, music therapist decides which and how to apply the therapy sessions. These have two distinctive elements: one of involvement—the *creative/active* and the other of listening—*receptive/passive process* both of which are applicable to the situation, individually or collectively. [1] In the *creative process*, the music therapist works with clients to actively create or produce music. This may include playing instruments, composing songs, engaging in music/song improvisation, or drumming in individual or group basis. In the *receptive process*, through music listening experiences, the therapist use music to facilitate relaxation. Clients or groups may then discuss thoughts, feelings, or ideas elicited by that music.

Music therapist should be specialised accordingly. If there is a session of active music therapy a minimum knowledge of music and instrument playing is necessary. If there is a session of receptive music therapy basic knowledge of medicine and psychology are, as well, mandatory. Moreover, measurements of the brain activity and of the effect of music on the brain ask for some technical skills.

3.1 Receptive Music Therapy

In receptive music therapy the quality of the session and the actual result are very much influenced by the mood of the receiver. Sometimes it associates with the patient's daily problems at the upper level of the unconscious, but, in depth, the sound experience gains wide dimensions. Similarly, the sound background used in creative art therapy (dance, drawing and others) is important to completing the therapeutic process. In order to achieve this, the music therapist must give proof of his overall specialization. If it is true that any kinds of melodic moving music has therapeutic attributes then why not give the patient to listen to an ideological-emotional piece of music that will stimulate his imagination and creativity, make him affectionate and give him flexibility in thinking? We have the example of the so-called *Mozart effect*, an effect that stimulates the children's entire psychomotor development. [2] The answer implies the fact that the listener's reaction to music is variable, depending on his personality and/or the conditions associated with auditioning. Music therapy raises strong feelings whose appearance is not always the affective response to the melody content. Sometimes experience with a certain piece of music is more decisive than the rhythm, the expressivity or the interval used. But to make a musical analysis there is need for a specialization in music. For example, minor-major alternation builds either tension or relaxation. Here, the knowledge of psychology is used to proper choose the song and to guess as much as possible the reactions of patients.

Analyzing receptive music therapy from a methodological point of view, lead to describing three forms of music therapy:

- the elementary (sound-rhythm) auditioning
- live music together often completed by dance and movement
- audio music auditioning, called *cabinet auditioning* that is therapeutically indicated.

The position adopted during the listening exercise (*the setting*) makes the receptive music therapy effect different, as the perceptions of certain bodily areas or the state of mind give expression to breathing, mimics and gestures. Moreover, the reception of music is accomplished through several components involved in the musical process, which should be taken into account at the time of therapy:

- Sensory component—responsible for selecting sounds/ elements of melody
- Acoustic component—refers to the symmetrical zones of the temporal lobes in which the acoustic sensations are analyzed
- Musical component—discusses the structure of music; its perception is identified in the right brain hemisphere considered to be the center of the music.
- Intellectual-emotional component—at this level there is a somato-visceral impact, translated into a special, real, sometimes physical experience of music [3].

3.2 Active Music Therapy

Active music therapy is a form of therapy that focuses both on the person's (called *patient* or *client*) previous musical experience (mentioning that is of help in drawing the therapeutic scenario), but also on his new musical experiences due to the evaluation and treatment process in which he participates. An unpredictable artistic act is built on the creativity of a person who may have never experienced something similar before.

From a musical point of view, there are four distinct types of experience: improvisation, recreation or interpretation, composition and audition. Each one of these types of musical experiences has unique features, specific applications and its own therapeutic potential. They induce some sensory-motor behavior and require different perceptive and cognitive skills. There is a different interpersonal process, so the emotions and feelings evoked are different. Thus, the role of the music therapist becomes again important as he needs to have the ability to recognize the medical disease, to lead the person in the musical therapeutic process and to conduct and feedback the psychological communication. Improvisation is not judged from the point of view of musical performance, but from the point of view of emotions transmitted, unblocking the necessary emotions and facilitating the appearance of a real, constructive communication. It is taken into account every movement, gesture, look, sound, moments of silence, the position of symmetry/asymmetry between the therapist and patient, etc. Again, the therapist skills and knowledge are of great help in reading and interpreting the body language.

Active (creative) music therapy can be practiced both in group and individually, and, although music produced or interpreted during the session is different and difficult to describe, should be carefully documented by the therapist. And so there are the characteristics of the person's behavior and personality, especially during group sessions. Often using an instrument (or human voice) is an act of courage, especially in improvisation, when shyness must be overcome and makes proof of perseverance and flexibility. Each genre induces a certain attitude, rejection or preference of that genre, reflecting how the person is reporting to that attitude with intense emotional participation. For example, the person who rejects a certain type of music and the person who listens to it can search for the same thing—the need for attention, but tries to get it in opposite ways that do not look alike.

4   More About the Interdisciplinary Music Therapy

Numerous research studies show the benefits of music therapy, from modifying brain waves (creating altered states of consciousness) and heart rate to determining certain reflexes in the body. [4] Through music therapy hidden aggressions, mental or affective disorders, certain mental blockages, and some psychosomatic diseases such as asthma or eating problems can be treated. [5] It is known that music stimulates the release of endorphins (hormones of happiness) which help to relieve fatigue and pain. [6] There are also arguments against music therapy that refer to contraindications in certain psychiatric disorders, epilepsy and post-traumatic stress disorder [7].

Anyway, what music therapy is required to do is to appeal to quality music, to a melodious music with dimensional and formative virtues and to or recommended in accordance with the patient's needs. From this point of view, as we said it before, the therapist must be a trained person in terms of musical education and familiarity with universal musical masterpieces.

No doubt, certain music represents means of mitigating psychic stress, preventing mental illness. The most common question becomes: what music to recommend, how to choose it and what is the *optimal dosage*? The answer is that there is no universal sound panacea, there is no particular composer or music that can be therapeutic for everyone, the choice is made only according to the patient's clinical picture (cultural environment, detested and/or favorite sounds, access to a particular style of music, musical training, etc.) Each individual is being unique and has his own history and sound identity; therefore, accordingly, the therapy must be carefully thought and dosed as a medicine. [8] The therapists need to be prepared to see, to understand and to use all these. Once again we say that they need to have a proper, diverse and solid education.

Music therapy is an efficient process in the therapeutic spectrum. However, a few steps must be taken before selecting songs that correspond to the needs of the patient involved in the therapeutic process and indicate the completion of a psycho-musical testing according to the model developed by J. Verdeau-Pailles [9]. The psycho-musical test includes:

- Completion by the subject of an identity card containing demographic data of the client: name, age, gender, address, profession and fields of interest, civil status (with details if applicable), family relationship, psychological data, medical history and family doctor's name, health status with diagnosis.
- Discussions of the reason and need for music therapy sessions with particularities related to the stages of development: objectives, methods, music choices.
- Studying the receptivity and experience with music: to music in general (the states expressed: relaxation, pleasure, stimulation of imagination or on the contrary: irritability, boredom or anxiety), as well as to a certain kind of music, considering the increase of the person's trust in the effectiveness of music therapy by interviewing the person about personal experiences, tangencies with listening to music in the family or outside it, data about his musical culture and education, particular ways of listening/practicing music, musical genres/composers/favorite singers/bands and creations.

It is advisable to avoid hearing or interpreting parts that are too well known as boredom, nervousness or even rejection may occur, these negative feelings affecting the therapeutic process. Also, the greatest enemy of music therapy is represented by musical bias—the obedient listening of a song to music can lead to the discovery of pleasure.

Music therapy implies recognition as a standing profession and its acceptance by society. And, not the last, it involves openness to therapy in general and music therapy in particular. If music therapy is currently practiced experimentally or as complementary therapy, it is time to be singularized and introduced as regular therapy in hospitals, hospice and palliative facilities, homes care's, and children care's homes. Among music therapists (mainly in Romania) there is a tendency to practice receptive music therapy, partly due to the therapeutic approach: it is seen as psychotherapy combined with musical auditions and partly become of the lack of experience regarding practicing active music therapy. Education—the training of specialized music therapists involves not only devoted, trained musicians, psychologists and medical professionals who become *pupils* and later *masters*, but also physicists, engineers, IT specialists, or neuroscientists. All the measurements and scientific findings have the support of the technicians. For example, the changes in the mood and the activity level when listening to different types of music could be presented through BioRadio recorded signals and Labview software analysis. An example of such analysis appear in Fig. 3.

Also, new entered in the spectrum of measuring devices used in music therapy, Neurosky MindWave is a device used to monitor electrical signals generated by neural activity in the brain. With the meter value eSense there are reported the levels of attention (concentration) and meditation (relaxation). The interface shows the associations between music and the type of frequency transposed in colour. During the music therapy session the user presents an increase in the level of concentration or quite contraire in the level of relaxation, depending on the music genre the person listens to or the preference at the moment [10]. Further analysis is available based on the therapists' education and skills.

Also, for a holistic analysis in music therapy a snapshot of formal methods known in psychology could be used, such as Faces Pain Scale, State—Trait Anxiety Inventory

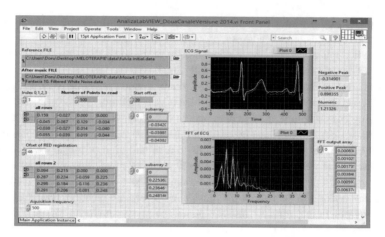

Fig. 3. Example of BioRadio records and labview software analysis

(to assess anxiety), Rating Anxiety in Dementia (effective studying the influence of quiet music), Relax Rating Scale (to assess the relaxation degree), and techniques commonly known as biomedical measurement such as ECG, HR, BP, EEG, EEG or temperature measurement [11].

Intelligent music systems are analyzed using MIDI Toolbox—a compilation of functions for analyzing and visualizing MIDI files in the MATLAB environment. In addition, the toolbox contains "cognitively inspired analytic techniques that are suitable for context-dependent musical analysis that deal with such topics as melodic contour, similarity, key-finding, meterfinding, and segmentation" [12].

5 Transilvania University of Brasov Interdisciplinary Master's Program in Music Therapy

At Transilvania University of Braşov, for example, curricula includes classes on the fields of Music, Psychology, Medicine, Physics, IT, Law, Statistics and Communication. A list of the 2016-2018 Masters' curricula is presented below in Table 1.

As it could be seen at TRansilvania University of Braşov—Music Department the courses are distributed on two years. 80% of the courses are mandatory, 20 % being optional. Among the optional courses are either Music Repertory or Dance Therapy (1st year) and Clinic Music Therapy or Music Therapy in Special Education (2nd year).

All disciplines are taught by professors with different specializations, though most of them are professors of Music or Psychology. The distribution of the fields could be seen in Fig. 4 with Music and Psychology fields representing more than 2/3 of the total disciplines.

Table 1. Music therapy master's curricula 2016–2018

Course	Main field	Year of study
Introduction to music therapy	Music/Psychology	1st year
Active music therapy (instrument)	Music	1st year
Active music therapy (voice)	Music	1st year
Music repertory	Music	1st year
Systems of music education	Music	1st year
Music therapy/psychotherapy	Psychology	1st year
Music therapy and medicine	Medicine	1st year
Alternative methods of healing	Medicine	1st year
Acoustics	Physics	1st year
Emotional development	Psychology	1st year
Instrumentation systems in acoustics and music therapy	Physics/IT	1st year
Evolution in music therapy	Music/Psychology	1st year
Music therapy—Practice I	Music/Psychology	1st year
Musical improvisation	Music	2nd year
Dance therapy	Music/Psychology	2nd year
Music therapy and psychiatric practice	Medicine	2nd year
Mozart effect	Music/Psychology	2nd year
Ethics in music therapy	Law/Medicine/Psychology	2nd year
Habilitation of music therapists	Psychology	2nd year
Stress management	Psychology/Medicine	2nd year
Clinic music therapy	Psychology	2nd year
Music therapy in special education	Psychology	2nd year
Thesis writing	Communication/Statistics/IT	2nd year
Music therapy—Practice II	Music/Psychology	2nd year

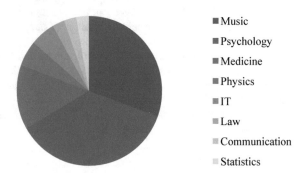

- Music
- Psychology
- Medicine
- Physics
- IT
- Law
- Communication
- Statistics

Fig. 4. Distribution of the disciplines

6 Conclusion

The practice of music therapy requires specialised music therapists mainly in the fields of Psychology, Medicine and Music, but also from Neurology, Psychiatry, Physics, Statistics, IT or Engineering. From education to licensing and actual practice, the therapists acknowledge information about interdisciplinary approaches including models of work. The need for specialists is extended by the need for tools (not only for musical instruments but also for modern, measuring instrumentation) and the need for a network of information, research, and affiliation to international forums. The direction of music therapy is clear: it is necessary, it is beneficial, and it is diverse. The rhythm of music enters the blood, stimulates the movement and orders the time, and some hypnotic states are reached.

References

1. Kraus, W.: The Healing Power of Music and Arts. Limes, Floresti (2014)
2. Iamandescu, I.B.: Receptive Music Therapy. Romanian original. Muzicoterapia receptivă. Premise psihologice și neurofiziologice, aplicații profilactice și terapeutice. InfoMedica, Bucharest (2004)
3. Iamandescu, I.B.: Receptive Music Therapy. Romanian original. Muzicoterapia receptivă. Premise psihologice și neurofiziologice, aplicații profilactice și terapeutice. InfoMedica, Bucharest (2004)
4. Aldridge, D.: The Individual, Health and Integrated Medicine: In Search of A Health Care Aesthetic. Jessica Kingsley Publishers, London (2004)
5. Segerstrom, S.C., Miller, G.E.: Psychological stress and the human immune system: a meta-analytic study of 30 years of inquiry. Psychol. Bull. **130**(4), 601–630 (2004)
6. Leubner, D., Hinterberger, T.: Reviewing the effectiveness of music interventions in treating depression. Frontiers Psychol. **8**, 1109 (2017)
7. Raglio, A., Attardo, L., Gontero, G., Rollino, S., Groppo, E., Granieri, E.: Effects of music and music therapy on mood in neurological patients. World J. Psychiatry **5**(1), 68–78 (2015)
8. Moldovan, A.: About Arttherapy, Interview. Psychologies.ro. pp. 18–23 (2014)
9. Iamandescu, I.B.: Receptive Music Therapy. Romanian original. Muzicoterapia receptivă. Premise psihologice și neurofiziologice, aplicații profilactice și terapeutice. InfoMedica, Bucharest (2004)
10. Girase, P.D., Deshmukh, M.P.: MindWave device wheelchair control. Int. J. Sci. Res. (2016)
11. Skotnicka, M., Mitas, A.W.: About the measurement methods in music therapy. In: Piętka, E., Kawa, J., Wieclawek, W. (eds.) Information Technologies in Biomedicine. Advances in Intelligent Systems and Computing. vol. 4, 284 (2014)
12. Erkkilä, J., Lartillot, O., Luck, G., Riikkilä, K., Toiviainen, P.: Intelligent music systems in music therapy. Music Ther. Today (online) **5**(5) (2004). Available at http://musictherapyworld.net

State-of-the-Art Duolingo Features and Applications

Stamatia Savvani[(⊠)]

University of Nicosia, Nicosia, Cyprus
s.tia.sava@gmail.com

Abstract. Duolingo is a rapidly growing on-line platform for language learning. In this paper learning theories that are embodied in its design are analyzed and certain shortcomings are identified. In the past two years, Duolingo has expanded its platform with the addition of new applications and features. State-of-the-art updates are reviewed in order to uncover whether they address limitations of Duolingo's original design or provide enhancements to the learning platform.

1 Introduction

Duolingo is a web-based language learning platform [10], available as a mobile application. Duolingo's manifesto [2] states that the purpose of the platform is to provide the user with free, fun and personalized education to users globally. How do users learn? The main method that is employed is the translation method. Users are asked to elicit the meaning of sentences in the target language and provide the translation of the sentence in their native language or a language they know well. An extensive and thorough language skill tree guides users through their language learning. Every skill, that consists of two or more lessons, is dedicated to a specific grammar or vocabulary topic. Users are expected to translate sentences, complete matching exercises (match pictures with the corresponding word, or match a word with its definition in the target language) and thus advance in their learning through repetition of vocabulary and grammar structures.

Duolingo's efficiency in language learning has been examined and researchers have concluded that it may be a promising supplementary tool for language learning but does not provide the learner with authentic language [14,28]. When it was introduced in language classrooms, it was observed that it promotes self-directed learning [24] and that it can promote learning two languages at the same time [17]; certain limitations were noticed regarding accuracy of translations and the absence of advanced language use.

© Springer Nature Switzerland AG 2019
M. E. Auer and T. Tsiatsos (Eds.): ICL 2018, AISC 917, pp. 139–148, 2019.
https://doi.org/10.1007/978-3-030-11935-5_14

Since its public release in June 2012, Duolingo has been appended with new languages (currently counting 32 languages available for English Speakers [3]). Most importantly, new features and applications, such as Labs and Tinycards [6,9] (see Sect. 3) are also introduced soliciting innovative ways of learning and interacting. Learning principles other than the translation method are welcomed.

In this paper, we will first identify and comment on learning theories embedded in Duolingo's original design (Sect. 2). In the sections that follow, the new features that are added in Duolingo and have not been previously assessed will be reviewed. Our goal is to critically analyze Duolingo as an example of a successful learning on-line application and draw insights for its future developments as well as for similar digital game language learning applications.

2   Key Learning Theories in a Typical Duolingo Lesson

Duolingo embraces tenets of the Grammar-Translation method [26], Audiolingual method [25] and Digital Game Based Language Learning [13]. In the sections that follow, we analyze how Duolingo's structure is influenced by these language learning theories and identify their strengths and drawbacks.

2.1   Grammar Translation Method

Studying a language through text translation is one of the premature methods in language teaching. Learners in Duolingo are expected to translate sentences in the target language by memorizing vocabulary and grammar structures. Below we revisit learning principles of the Grammar Translation method [26] that are found in the layout of Duolingo lessons.

Learners make constant **transfers from L1** (one's mother, native or first language) **to L2** (one's foreign, second or target language). Duolingo users are expected to understand sentences in L2 while they are given the definition of each word in the sentence in L1. Thus, they are asked to provide a translation for this sentence by employing cognitive skills. Reverse translation (from L2 to L1) is also endorsed in Duolingo.

Similar to the Grammar Translation method, Duolingo encourages **accuracy**. For instance, a sentence will be regarded as incorrect if the article "el" in Spanish is translated as "a or an" instead of the correct "the". However, Duolingo shows leniency for typos and forgives the absence of accents in languages such as Spanish or French, or non capitalization of the nouns in German. In case the user does not use accents or proper capitalization, the translated sentence is not regarded as incorrect but a note is displayed underlining the proper format.

Our main concerns towards employing the Grammar Translation method are that spoken language is ignored and that the user is constantly viewing L2 in regards to L1. In addition to this, users fail to attain the cultural perceptions of L2, as L1 remains their one and only reference point.

2.2 Audiolingual Method

Duolingo embodies practices and theories of the Audiolingual method (ALM), which was developed in the 60s to help learners develop oral proficiency. It is a behaviorist theory in its essence [12], as the learners are led to a self-discovery of rules; then they are drilled on the rules and are rewarded for correct responses. Although Duolingo does not focus on speaking skills, we can find principles of ALM in its lesson structure. Below typical exercises of the ALM are listed that are present in Duolingo lessons [25]:

1. **Replacement:** a word is replaced with another word of the same speech category or topic. For instance, users learning Spanish from English are introduced to sentences like "yo como manzanas" (i.e. "I eat apples") and then a sentence like this "yo como fresas" (i.e. "I eat strawberries") appears. This drills students in a certain grammar structure. In this case, the verb "como" is taught while previously learned vocabulary is revised, e.g. "manzanas", and new words are introduced, e.g. "fresas".
2. **Inflection:** Users are asked to repeat certain structures where the form of a word is changed. For example, first the learner is introduced to singular nouns and then the same nouns are presented in the plural form in the following lesson.
3. **Restoration:** the user is given a set of words and is asked to put them into the correct order so as to form a correct grammatical sentence.

These type of exercises involve repetition and drilling. However, such practices of ALM have received criticism. Linguist Noam Chomsky [12] argued that **creativity** in language is overlooked in ALM and behaviorism, which views language as a habit structure. Chomsky argues that the infinite utterances that a human can develop in a certain language cannot possibly be learned through repetition and reinforcement techniques.

Moreover, **meaningful context** is lacking in Duolingo. Even though the organization of the skill tree in Duolingo is carefully thought through, sentences in the lessons are decontextualized. Hence, the learner fails to relate to the language under instruction [14]. This drawback seems to be resolved in Duolingo's new Stories feature, see Sect. 3.3.

2.3 Digital Game Based Language Learning

Duolingo is an example of Digital Game Based Language Learning (DGBLL) applications. DGBLL concerns "the design and use of a diverse array of digital games for the purpose of learning or teaching a second or foreign language" [13]. Edutainment that is defined as "a hybrid mix of education and entertainment that relies heavily on visual material, on narrative or game-like formats, and on more informal, less didactic styles of address" [15, p. 282] is prevalent in Duolingo. Following an adaptive digital game based learning framework, three elements seem to be core to edutainment design: Multimodal (interaction factors, multimedia elements and narrative), Task (completion of challenges) and

Feedback (assessment of progress and rewards that are meant to stimulate users' motivation) [27]. The way these elements are implemented in a game environment can determine its success. Below we give examples of how these elements manifest themselves in Duolingo:

1. **Multimodal:** sound effects for completing lessons and making correct translations, visuals for new vocabulary words, audio for every word that is present in the sentences in L2
2. **Task:** complete lessons, reach the highest level in each lesson
3. **Feedback:** achievements (badges awarded when users complete predetermined actions), experience points and crown levels (both are awarded upon completing lessons)

Multimedia elements are abundant in Duolingo and language is delivered in different forms (audio, textual, images). Nevertheless, **narrative** (referred to as storytelling), a commonly underlined element of multimodal DGBLL [16,21], is not found in Duolingo. A mystery or drama can keep players engaged in a game [21]. In Duolingo's design this element of interaction is not present as users are not introduced into a world, or characters. In Duolingo, users are asked to complete challenges; however there is not the least flavor text that will invite the learners into their next challenge.

Users are motivated as they are awarded experience points by completing the daily goals they have set a priori. In order to ensure learners' adherence to the goals, their progress in learning should be made clear to them at all times [31]. Duolingo's skill tree is clear and the learner can see how many lessons are remaining until the completion of a skill. However, a recent addition called *Crown levels* does not provide information to the users regarding how far they are from reaching the top crown level. This has caused frustration and discouragement amongst users as it can be observed in the application's forum [4,5].

3   New Features and Applications of Duolingo

Some of Duolingo's newest features are embedded in the platform: e.g. Duolingo Labs (released on December 21, 2017 and in experimental stage) [1], while others are located on different platforms e.g.: Tinycards [9] (released on July 19, 2016) and Duolingo for Schools [7] (released on January 8, 2015). Reviews of Labs and Tinycards are not available in academic literature. Thus, below we provide insight for these applications and suggest further research on their efficiency.

3.1   Duolingo for Schools

A powerful resource for teachers is Duolingo for schools [7]. Teachers can organize classes and have their students sign up in them. For every class the teacher can set specific assignments that will be visible to all students enrolled in the particular classroom. The assignments are regular Duolingo lessons and the student accounts are similar to a regular account of Duolingo. What changes is the teacher's account, through which the teacher can track students' progress. Analytic statistics available to the teacher are gathered from students' logs and include the number of completed lessons, the number of assignments they completed on time or late, as well as the points they earned in each log.

The Duolingo for schools resource could be enhanced further by incorporating its own game elements. For instance, teacher-users could be given the ability to set long term goals for each class and run **competitions** between classes of the same level in order to determine which class meets the goals first, or which class gathers the most points. Badges on language learning platforms have proved to have a positive effect on learner's self-efficacy and language development [29]. Therefore, achievements could also be added in Duolingo for schools, which would be awarded not to an individual but to a class as a whole for completing a series of assignments or goals. Additions as such could spark both competition and collaboration amongst students and can provide extra motivation while working towards a common goal.

The greatest advantage of Duolingo for Schools is that with its immediate feedback to the user and the teacher, it promotes **"visible learning"** [19]. Both sides are allowed to measure learning and detect both weaknesses and strengths in particular linguistic concepts.

3.2   Duolingo Labs

Duolingo Labs include: **Stories**, **Podcasts** and **Events**. In Events, language learners can organize meet ups in their hometown and invite their fellow learners in order to interact in the target language. The potential significance of this feature will not be analyzed in this paper as it is meant for social settings rather than an on-line learning environment.

Duolingo Podcasts are true stories narrated in audio format; podcasts are currently available only in Spanish for English speaking learners. They aim to familiarize the learners with the target language in different meaningful contexts. According to a review of studies on the efficiency of podcasts [18], it has been proven that the incorporation of podcasts in language learning is beneficial both to learners' linguistic performance and engagement.

Duolingo Podcasts seem promising from an educational perspective and can provide learners with authentic L2 material. Moreover, the designers of the platform could consider adding listening comprehension exercises that will check users' understanding of spoken language. In this way, the potential merits of Duolingo Podcasts could be evaluated based on users' responses.

3.3   Duolingo Stories

Duolingo Stories [8] are short humorous stories that help learners practice their language comprehension skills. Currently, they are available in French, Spanish, Portuguese and German. Users can listen to the narration of the story and they are also provided with the textual form of the story.

Stories are meant to be entertaining and they are also somewhat **interactive**. Although users cannot alter the plot or the end of the story, the narration of the story pauses at critical points and waits for the user's input. The platform generates closed-ended questions that test readers' understanding of words and content of the story. Comprehension questions during the narration check in present time whether the users are following and understanding the narration, allowing them time to reread the parts they did not understand and then proceed with the story. Computer-Assisted Language Learning (CALL) applications are argued to be more beneficial to learners' vocabulary acquisition than learning L2 vocabulary with teachers' assistance, as the first promote self-paced and autonomous learning [30]. Frequent pauses in narration can climax reader's interest in the story, as well.

Even though a word's meaning can be deduced from the context that it is found in, it may be distressing to readers if there is a significant number of unknown words in a sentence. For this reason, **glosses** are provided within the stories in Duolingo. Nation defines a gloss as a "brief definition or synonym, either in L1 or L2, which is provided with the text" [22, p. 238]. A gloss can have different formats, e.g. text, picture, audio or a combination of these. In Duolingo Stories, when users hover their mouse over words from the text they are unfamiliar with, the gloss is displayed in textual form and the words are translated in the user's L1 (see Fig. 1).

Cuidando un pájaro

MARCO
Hola Camila.

MARCO
¿Cuándo regresas: mañana o el martes?

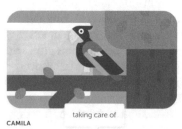

taking care of
CAMILA
Mañana. Gracias por cuidar mi casa y a mi perico.

Fig. 1. A gloss in Duolingo Stories [8]: "cuidar" defined as "taking care of" [9]. Screenshot by author.

In a typical Duolingo lesson, some sentences to be translated by the user sound unnatural e.g. "Yo soy un oso", which translates to "I am a bear" [23]. Apparently, these sentences exist to drill the user on thematic vocabulary, e.g. animals, combining it with previously learned grammar, e.g. the verb "to be". Stories resolve this awkwardness by providing **meaningful** content to the vocabulary expected to be acquired by the user. We support the need to move away from behaviorist CALL and explore communicative CALL applications [20], which will focus on meaning rather than repetition of odd and out of context language chunks.

3.4   Tinycards

Users of Tinycards [9] (available for learners of Spanish, English and Portuguese) can choose to complete or create lessons with the aid of **customizable digital flashcards**, called Tinycards. Users are expected to use their own ideas, images and text to create sets of Tinycards that will review certain vocabulary, grammar or other non-linguistic topics and terms from sciences like architecture, biology etc., always in the target language. Every Tinycard is two-sided (see Fig. 2); thus, on one side the user can have textual input and on the other side a picture that corresponds to the particular text. The platform enables users to upload pictures from their own devices or the web.

Fig. 2. A flashcard in Tinycards [9]. Screen-shot by author.

Users can organize these digital flashcards in decks in order to create a lesson, which can remain private or can be made available to other users in order to study it. In contrast to Duolingo's regular lessons, users who choose to complete a lesson in Tinycards are not asked to translate sentences. They are first shown the digital flashcards, and after they have reviewed them, they are given matching exercises (matching the correct word to the picture) or asked to type in the correct word for the image provided. However, Tinycards do not award experience points or badges to the players or creators of lessons.

From an educational viewpoint, the quality of the lessons, which are produced by users and not by Duolingo, is questionable as the lessons are reviewed by other users who may not be proficient in the target language. Nevertheless,

Tinycards do not only target remembering and comprehension skills but also address higher order thinking skills such as evaluating and creating (according to Bloom's taxonomy [11]), when users attempt to produce their own lessons (see Fig. 3). However, Tinycards lack edutainment elements mentioned in Sect. 2.3, such as experience points and advancing levels. Therefore, users' motivation could be stimulated if such elements were incorporated in Tinycards.

Fig. 3. A lesson in Tinycard [9]. Screen-shot by author.

4   Conclusion

Uncovering the learning methods that Duolingo encompasses, we have identified certain shortcomings. The translation method might be a "safe" method, preferred for teaching beginners, as L1 is used as reference. However, it is not ideal for advanced language levels and does not promote creativity. Rote repetition exercises, encouraged by the ALM method, provide the learner with sufficient practice but not with exposure to meaningful language.

Duolingo's game design could also be enhanced by providing the user with clearer, meaningful language goals and a more personalized experience through narrative. Personalization and meaningfulness could be achieved by asking the personal interests of each user. The platform could then direct users to individualized lessons that would incorporate language related to their interests.

Do the newest Duolingo features address the above summarized limitations? We would argue that some of these limitations are indeed addressed. Duolingo for schools is a great asset that complements the language classroom with immediate feedback from learners. Labs offer a more authentic language experience to the user, as Podcasts provide real-life narratives, and the engaging plot and interactivity of Duolingo Stories offer slightly more advanced language and meaningful context. Tinycards encourage (and rely on) creative skills of users, moving away from rote repetition exercises. Nevertheless, none of the above mentioned features address the absence of practicing oral communication in L2.

Despite their strengths, we have identified several weaknesses regarding recent Duolingo updates. Duolingo for Schools resource only allows the teacher to view student logs. While privacy concerns are valid, collaboration or positive competition could be cultivated if students were aware of their classmates' progress, or if classes were awarded badges as a whole for completing tasks set by the teacher. Podcasts do not give any feedback to learners; if they were implemented as listening exercises, they could test users' understanding of spoken language. The accuracy of language in Tinycards lessons can be questioned, as flashcards can also be created by non-native speakers or beginner users of the target language.

Overall, we encourage the addition of more advanced language content and the release of the newest features, reviewed in this paper, for learners of other languages. As the new Duolingo features seem to support the shift for a more authentic and meaningful language experience, we call for further research towards their efficiency and enhancement.

References

1. Duolingo. https://duolingo.com/. Accessed 14 June 2018
2. Duolingo: About us. https://www.duolingo.com/info. Accessed 26 June 2018
3. Duolingo courses. https://www.duolingo.com/courses. Accessed 26 June 2018
4. Duolingo discussion topic: crown levels: a horrible invention. https://forum.duolingo.com/comment/26840788. Accessed 25 June 2018
5. Duolingo discussion topic: Crown levels not an improvement. https://forum.duolingo.com/comment/26846381. Accessed 25 June 2018
6. Duolingo labs. https://www.duolingo.com/labs. Accessed 14 June 2018
7. Schools, duolingo. https://schools.duolingo.com/. Accessed 14 June 2018
8. Stories. https://stories.duolingo.com/. Accessed 14 June 2018
9. Tinycards. https://tinycards.duolingo.com/. Accessed 14 June 2018
10. What is duolingo? https://support.duolingo.com/hc/en-us/articles/204829090-What-is-Duolingo-. Accessed 26 June 2018
11. Anderson, L.W., Bloom, B.S., Krathwohl, D.R.: A Taxonomy for Learning, Teaching, and Assessing. Longman, New York (2000)
12. Byram, M.: Audiolingual method. In: Routledge Encyclopedia of Language Teaching and Learning, pp. 58–60. Taylor & Francis Ltd/Books (2000)
13. Cornillie, F., Thorne, S.L., Desmet, P.: Digital games for language learning: from hype to insight? ReCALL **24**(3), 243–256 (2012)
14. Cunningham, K.J.: Duolingo. Electr. J. Engl. Second Lang. **19**(1), 1–9 (2015)
15. David Buckingham, M.S.: Parental pedagogies:an analysis of british edutainment magazines for young children. J. Early Child. Literacy **1**(3), 281–299 (1990)
16. Embi, Z.C.: The implementation of framework for edutainment: educational games customization tool. In: Proceedings of the International Symposium on Information Technology, vol. 1, pp. 1–5 (2008)
17. Essa Ahmed, H.B.: Duolingo as a bilingual learning app: a case study. Arab World Engl. J. **7**(2), 255 (2016)
18. Hasan, M.M., Hoon, T.B.: Podcast applications in language learning: a review of recent studies. Engl. Lang. Teach. **6**(2) (2013)
19. Hattie, J.: Visible Learning: A Synthesis of Over 800 Meta-Analyses Relating to Achievement. Routledge, London, New York (2009)

20. Jeffrey Earp, T.G.: Narrative-oriented software for the learning of a foreign language. In: Dettori, G., Tania Giannetti, A.P., Vaz, A. (eds.) Technology-Mediated Narrative Environments for Learning, pp. 27–40. Sense Publishers (2016)
21. Kevin, Y.: Pedagogical gamification. Improve Acad. **32**(1), 335–349 (2013)
22. Nation, I.S.P., Hunston, S.: Learning Vocabulary in Another Language. Cambridge Applied Linguistics, 2nd edn. Cambridge University Press, New York (2013)
23. Nushi, M., Hosein Eqbali, M.: Duolingo: A mobile application to assist second language learning. Teach. Engl. Technol. **17**, 89–98 (2017)
24. Pilar, M.: The case for using duolingo as part of the language classroom experience. Revista Iberoamericana de Educacin a Distancia **19**(1), 83–109 (2016)
25. Richards, J.C., Rodgers, T.S.: The Audiolingual Method, 2nd edn. Cambridge Language Teaching Library, Cambridge University Press, Cambridge (2001). 5070
26. Richards, J.C., Rodgers, T.S.: A Brief History of Language Teaching, 2nd edn. Cambridge Language Teaching Library, Cambridge University Press, Cambridge (2001)
27. Tan, P.H., Ling, S.W., Ting, C.Y.: Adaptive digital game-based learning framework. In: Proceedings of the 2nd International Conference on Digital Interactive Media in Entertainment and Arts, pp. 142–146. ACM (2007)
28. Teske, K.: Duolingo. CALICO Journal **34**(3), 393–401 (2017)
29. Yang, J.C., Quadir, B., Chen, N.S.: Effects of the badge mechanism on self-efficacy and learning performance in a game-based english learning environment. J. Educ. Comput. Res. **54**(3), 371–394 (2016)
30. Chiu, Y.H.: Computerassisted second language vocabulary instruction: a meta-analysis. Brit. J. Educ. Technol. **44**(2), E52–E56 (2013)
31. Zoltan Dornyei, C.M., Ibrahim, Z.: Directed motivational currents. In: Lasagabaster, D., Doiz, A., Sierra, J.M. (eds.) Motivation and Foreign Language Learning: From Theory to Practice, chap. 1, pp. 9–29. John Benjamins Publishing Company (2014)

Pilot Projects/Applications

Experiences from the Introduction of an Automated Lecture Recording System

Herwig Rehatschek(✉)

Medical University of Graz, Virtual Medical Campus, Harrachgasse 21, 8010
Graz, Austria
Herwig.Rehatschek@medunigraz.at

Abstract. Recorded lectures help students to better prepare for exams and offer other important advantages such as time and location flexibility. With respect to existing technology many students consider the recording of lectures as a standard service provided by the university. Hence students expect that all lectures held are recorded and later accessible via a web browser and streaming technology. However, even though there exists a lot of technology supporting the recording of lectures, the introduction of an automated and integrated lecture recording system meeting all requirements of students and teachers is a non-trivial task. It needs a lot of experience and technical know-how since there are no out of the box solutions. Our system is capable to automatically record and synchronize two full HD streams of the lecturer and the slides including audio. All the recorded lessons are accessible via a standard web browser and streaming technology by the students via the university learning management system. In our paper we share lessons learned from a first pilot trial with the system, present evaluation results from the students and feedback from the teachers.

1 Introduction

According to many existing studies [1–3] recorded lectures help students prepare for exams and offer other important advantages such as time and location flexibility. With respect to existing technology, many students consider the recording of lectures as a standard service provided by their university. Hence, students expect that all lectures held are recorded and later accessible via a web browser and streaming technology. Even though there exists a lot of technology supporting the recording of lectures, the introduction of an automated and integrated lecture recording system meeting all requirements of students and teachers is a non-trivial task and needs a lot of experience and technical know-how [4]. There are no out of the box solutions. Furthermore, it has to be considered that the solution must also be accepted by the teachers; possible fears must be covered and important questions such as copyright and data protection must be cleared.

Until now we provided eLectures for students (learning content with the voice of the teacher) [4, 5] combined with complex learning objects such as our virtual microscope [6], animations and simulations. Even though these learning objects are of very high quality they take a lot of time for production and efforts from both sides—technical staff and teachers. A well designed automated lecture recording system shall

© Springer Nature Switzerland AG 2019
M. E. Auer and T. Tsiatsos (Eds.): ICL 2018, AISC 917, pp. 151–162, 2019.
https://doi.org/10.1007/978-3-030-11935-5_15

minimize the efforts on the teacher's side and increase the learning outcome of the students without having much extra work by technical staff.

As a minimum standard for our lecture recording system we defined that our solution must be capable to automatically record and synchronize two full HD video streams. The first video stream shall contain the lecturer and his voice with different pre-defined settings, the second stream shall contain the PC output including audio, i.e. slides or videos shown during the lesson but also the output of a document camera if used. All the recorded lessons shall then be accessible via a standard web browser and streaming technology by the students via the university learning management system.

Next to the technical issues also the acceptance of such a system by the teachers must be taken into consideration. Questions and fears like "nobody will come to my lessons anymore when they are recorded" or copyright issues have to be discussed and solved right ahead.

In our paper we will describe the main issues of the technical specification and the solution we have installed, and the accompanying measures which must be taken into consideration such as teacher acceptance and motivation, copyright questions, the advantages of such a system from teachers and students point of view. We will also state lessons learned from a first pilot trial with the system and present first evaluation results from the students and feedback from the teachers.

2   Existing Solutions

On the market many providers offer hardware components and software for lecture recording systems including Extron, Panopto, Epiphan and StreamAMG. But none of these systems are out of the box solutions. They all require comprehensive technical planning in order to guarantee an optimal adaption to the individual needs. At the time we started the pilot the Extron [7] hardware was according to the technical specification (2016) not capable of recording two full HD streams in parallel nor could they provide a native software solution in order to store and manage the recorded material. With begin of 2018 also Extron now provides a fully integrated solution by adding EMP [8] Entwine to their product portfolio. Entwine is a commercial branch of OpenCast Matterhorn and provides nearly the same functionality than the open source software, however, with full technical support. Panopto [9] and StreamAMG [10] with its product Stream LC offer a software solution for recording and managing self-recorded videos in connection with slides and a streaming portal solution. Both do not natively offer recording hardware for lecture rooms but provide only certified partners. Epiphan [11] provides hardware for live video production and streaming, but no software for management, organisation and broadcast of the recorded material.

3   Implementation

Roughly the following areas can be identified with the introduction of a lecture recording system: selection and mounting of the according hardware elements for the recording environment, definition of the organizational workflow, choosing the software for playout and adapting it to the specific scenario, implementation of software components in order to realize the organizational workflow, storage and archiving.

With respect to the hardware solution we decided to take Epiphan Pearl Rackmount Twin recorders [12], which in 2016 were the only hardware providing the recording of two full HD streams in parallel. For the software play out system of the pilot we have chosen the only existing open source software OpenCast Matterhorn [13]. Based on this framework we defined the organizational and technical workflow.

3.1 Organizational and Technical Workflow

The first step was the definition of an organizational/technical workflow. Since we expected only a number of approximately 50 recordings per term we chose a semi-automated workflow in six steps which is depicted in Fig. 1.

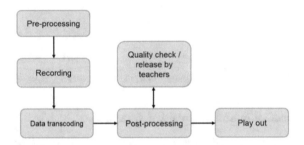

Fig. 1. Organizational-/technical workflow

3.2 Preprocessing

The first step is a pre-processing which contains the scheduling of the recordings. Even though our recording interface is easy to understand by teachers we provide full support also during recording hence we wanted to know when recordings take place. Since we have a central planning of the curriculum at our university we ask teachers to provide recording wishes in advance. The lessons are then scheduled in the appropriate rooms equipped with the recording hardware.

3.3 Data Recording

We decided to work with a partner university in our city who had already eight years of good experience with Epiphan. Since Epiphan also met our technical requirements we opted for Epiphan recording components. All in all we equipped five big lecture rooms, the aula, three seminar rooms and our clinical skills simulation center with recording hardware and with a camera. The recording interface can be easily controlled via a Crestron panel directly placed in the lecture rooms. The camera films the white board and the teacher the PC/Beamer output can be recorded. Both streams are recorded in full HD resolution. Furthermore we provide the teachers with four recording pre-settings. The first setting records PC/Beamer and the teacher's lectern, the second setting records PC/Beamer and the teacher's lectern and the whiteboard, the third setting records PC/Beamer and the whiteboard and the fourth setting records white-board and teacher's lectern but not the PC/Beamer. These easy to understand recording scenarios together with the record, stop and pause button—see Fig. 2—is the entire interface for the teachers to fully automatically record their lessons.

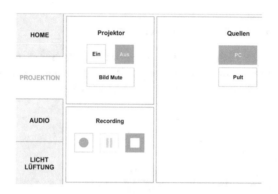

Fig. 2. Teacher Crestron panel interface for recording management

3.4 Data Transmission and Acquisition Interface

After recording is done the data is automatically transmitted as a MPEG-2 transport stream to a storage. Here we programmed an interface which allows us to manage and control the recordings. The interface—see Fig. 3—supports the following functional-ity: notification per E-mail when new recordings are available, download of the streams separately as a ZIP file or side-by-side, adding of metadata, remote control of the recording devices and archiving functionality.

Fig. 3. Data acquisition interface/remote recorder control

Following the defined workflow we use this interface for downloading the recorder material for the post-processing step. Furthermore this interface will manage the archiving of the recorded master material (before post processing), meaning we will permanently store the originally recorded material, the post processed material (within OpenCast server) and the post-processing video editing file.

3.5 Post Production and Quality Check

In the post production step we edit the recorded material. Due to the moderate number of recordings this step is currently done manually. We use Adobe Premiere to add a short introduction and outro sequence containing name of the module, title of the lesson and name of the teacher. Furthermore we cut out sequences of bad quality, enhance the audio level and sometimes add text bubbles when students ask questions without using microphones. Then we publish the material in a secured area of our learning management system (LMS) Moodle [14] where only teachers have access to. The teacher is provided then with a link and performs a quality control of the ready to publish material. All wishes of the teacher will be taken into consideration until he gives his final ok for publishing. Then the video is published on our university LMS to be accessed by the students.

3.6 Playout Software

The challenge for the playout software was to provide a user interface capable of playing two HD streams synchronically and to give students the flexibility for scaling the size of the two streams. So in case the teacher shows something interesting on the whiteboard the video with the teacher can be maximized and in case the information is on the slides this video stream can be enlarged. Furthermore the player should work in all standard web browsers without having to install any plugins and shall work independently of the underlying operating system. Having all these requirements in mind for the pilot trial we chose the only existing open source software in this area which supports these requirements, OpenCast Matterhorn [13].

As depicted in Fig. 4 OpenCast supports with its internal player Theodul the synchronous visualization of two independent video streams. By clicking into the videos the video will be enlarged and the other one minimized, in general different layouts are possible here. The player supports also a zoom function meaning students can zoom into a video stream by using the mouse wheel. This is important when teachers e.g. write on the whiteboard and students want to read it in the video. Also OpenCast provides a slowdown and speed up functionality in case the teachers speaks to fast or to slowly. Last but not least a simple segmentation and optical character recognition (OCR) is offered as well. All in all OpenCast provided all the features needed to start a pilot trial.

However, during the pilot trial this software did not meet our expectations. We encountered difficulties with LDAP (Lightweight Directory Access Protocol) binding and LTI [15] provision. We also discovered severe stability issues which resulted in server failures every few days. So we decided to replace this software by a commercial branch of OpenCast, EMP Entwine [8]. See Chap. 5 Conclusions and lessons learned for further details.

Fig. 4. Playout interface for students utilizing OpenCast Theodul player

3.7 Accompanying Measures

Next to the above described steps another important issue is user acceptance. The best lecture recording solution does not help if not accepted/used by the teachers and students. Primarily for teachers important points include: data protection, launch events, efficient technical support in combination with onsite trainings and demystification of fears. So many teachers are afraid that nobody will come to their lessons anymore once the lesson is recorded. This fear we could demystify by performing a student evaluation where it clearly turned out that students will still come in case they get an added value. See Chap. 4 for more details on this.

In connection with data protection we took two concrete steps: first we introduced a user agreement which each teacher has to accept. It guarantees to make the recorded lessons available to students via our learning management system. Second, we installed a keycard lock mechanism, meaning that only authorized persons (affiliates of the university respectively) are allowed to activate the lecture recording feature in the lecture rooms.

Regarding student issues, the usability of the player, bandwidth and the support of different devices/operating systems are key issues which were met during introduction of the pilot trial.

4 Evaluation Results

With begin of the winter term 2017/18 and having the organizational workflow in place we started a first pilot trial with our brand new lecture recording system. In parallel we initiated a student online evaluation with nine closed questions and one open question. In the evaluated lectures 171 students participated, 85 took part in the anonymous evaluation which results in a return ration of 49.71% which is a very good rate for a pure online evaluation. In this chapter we discuss the results of seven closed questions.

First we asked if recorded lessons help to improve the own learning outcome.

As it can be seen in Fig. 5 93% of the students indicated with a *strong agreement* or *very true* that recorded lessons clearly help them to improve their learning results.

Then we wanted to know whether recorded lessons help to better prepare for exams. As given in Fig. 6 again 93% of the students indicated either with *strong agreement* or with *very true* that recorded lessons help them to better prepare for exams.

Fig. 5. Student evaluation question #1

Further we asked if the user interface for playback provides all functionality nee-ded. Results given in Fig. 7 clearly state that 89% of the students are satisfied with the offered functionality by checking either *strong agreement* or *very true*.

Fig. 6. Student evaluation question #2

In the next step we wanted to know whether the students want an extension of the offered recorded lessons. As it can be seen from the results given in Fig. 8 88% of the students want more recorded lessons, which is also a clear indication that recorded lessons are useful for them.

Fig. 7. Student evaluation question #3

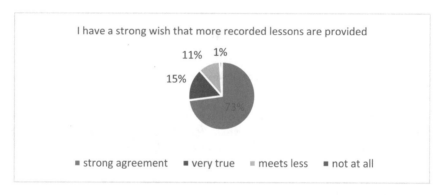

Fig. 8. Student evaluation question #4

In question 5 we asked on which devices students want to view recorded lessons. As depicted in Fig. 9 the vast number of students (62.9%) prefers to view the content on their laptop, only 7.8% on a desktop. This goes in line with our knowledge on student devices where laptops are by far the most. It has to be noted that nearly 30% of the students also want to view recorded lessons on mobile devices such as tablets and smart phones. The latter is too small for visualizing two video streams, but tablets have to be considered. Even though we could not get the player of OpenCast to work with tablets during the pilot phase we will provide it with the new playout software EMP Entwine and hence fulfill the strong wish of the students.

Next we wanted to know whether students watch the lecture recordings in their full length or only parts of it. As visualized in Fig. 10 70% of the students watch the recordings in their full length or at least most of it. Only 27% of the students do watch mostly parts, and only 3% do watch only parts.

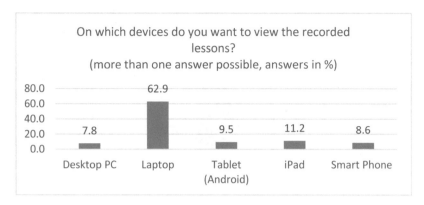

Fig. 9. Student evaluation question #5

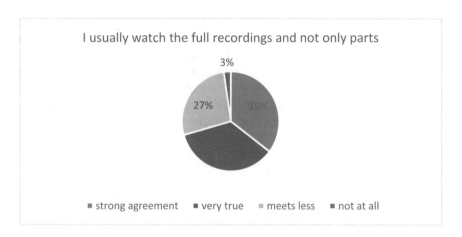

Fig. 10. Student evaluation question #6

Finally we were interested if students still visit the lessons physically even though they know lessons are recorded and they can watch it later on. This of course is also a major question and fear of teachers: "do students still come to my lessons when knowing it is recorded?", "Will I be alone in the room?" As already known from a number of studies like [16] and [17] we knew this is only a myth and according to their results only 16–23% will not visit the lessons anymore. But this is a normal rate also for lessons which are not recorded as well. Since the fear is present at the teachers and we had to prove it we also evaluated this issue. As it can be seen in Fig. 11 only 20% did not visit the lessons anymore physically, which goes completely in line with the known studies. Additionally, it has to be mentioned that due to organizational issues we were not able to plan this lessons overlap free. So many students did not even have the possibility to physically visit the lessons. Many of them replied in the open question that they really would have liked to come there but could not.

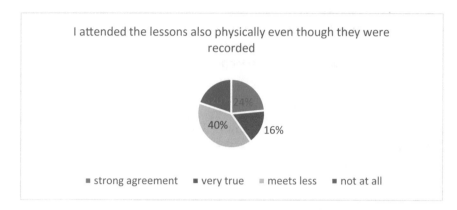

Fig. 11. Student evaluation question #7

Besides the students' feedback we also received feedback from some of the teachers. One teacher was initially very skeptical if the quality is good enough and if even any of the students will ever watch the video. After he had seen the result he turned out to be enthusiastic for the recording system and wanted even to be recorded again in the next term. He performed a self-reflection on his recorded material and looked out for improvements. Two other teachers started to use the recording system spontaneously within their lessons. One teacher wanted to give the students better possibilities to prepare for the exam, the other teacher intended to have the lessons stored for the case of his absence. Once reached a critical mass the system becomes an integral part of the teaching process.

5   Conclusions and Lessons Learned

Advantages of recorded lessons for students include better learning success in terms that they can recap difficult parts of the lesson later on, better preparation for exams, advantage for non-native speakers to recapitulate the lesson, time and location flexibility, additional information next to the slides, individual learning speed and giving students the possibility to study abroad without missing the lectures at the home university.

For teachers advantages include transparency of teaching, possibly sharing of lessons with other universities, increase the quality of teaching by self-reflection, possibility to teach huge groups without having space problems in the rooms and decrease of after-lesson student questions via E-mail due to the fact that students can now recap difficult parts of the lesson for a better understanding.

Furthermore, it can be concluded that in case the university provides high quality recordings of the lessons the bad quality recordings of students via mobile devices will be stopped. This is considered a well desired side effect by many universities since it highly decreases the possible publication of low quality recordings on public platforms such as YouTube.

Epiphan Pearl rackmount twin recorders turned out to be a good choice in terms of programming interface, high quality recording and easy to be controlled. Hardware stability remains an issue—since the beginning of our pilot trial in October already three of the nine twin Pearl recorders failed in functionality and had to be replaced.

For playout we made no good experiences with the open source software OpenCast Matterhorn [13]. As a newcomer to the software we tried to install version 3.2. Even though we had an experienced software engineer we failed to get major functionality such as LDAP and LTI (learning tools interoperability) to function properly. Especially LTI is important for privacy issues since it grants a regulated access policy. LDAP is necessary in order to sync OpenCast with the legal users of the university. Next to this we experienced stability issues resulting in failure of the server every few days. We also could not get the OpenCast internal Theodul player to run on mobile devices such as tablets. This finally brought us to the decision to replace the OpenCast software with the commercial solution EMP Entwine [8].

Evaluation results from students were very encouraging to continue and to foster the recording of lectures. In this connection teachers have to be clearly motivated in order to use the new technology. Many had fears in terms of technical issues, data protection and loss of students in their lessons. Here a very active communication is needed, fears can be met with facts from evaluation results. Additional help is gained when reaching a critical mass of teachers in favor of the lecture recording who finally act as multipliers and motivate their colleagues.

References

1. Mayer, R.: The promise of multimedia learning: using the same instructional design methods across different media. Learn. Instr. **13**(2003), 125–139 (2003). https://doi.org/10.1016/S0959-4752(02)00016-6
2. Krüger, M., Klie, T., Heinrich, A., Jobmann, K.: Interdisziplinärer Erfahrungsbericht zum Lehren und Lernen mit dLectures. In: Breitner, M.H., Hoppe, G. (eds.) E-Learning. Physica-Verlag HD (2005). ISBN 978-3-7908-1588-7
3. Mertens, R., Krüger, A., Vornberger, O.: "Einsatz von Vorlesungsaufzeichnungen", Good Practice Netzbasiertes Lehren und Lernen. Osnabrücker Beiträge zum medienbasierten Lernen **1**, 79–92 (2004)
4. Rehatschek, H.: Design and set-up of an automated lecture recording system in medical education. In: Proceedings of the 20th International Conference on Interactive Collaborative Learning—Volume 715 of the series Advances in Intelligent Systems and Computing, pp 15–20, 27–29 Sep 2017, Budapest, Hungary (2017). ISBN 978-3-319-73209-1, https://doi.org/10.1007/978-3-319-73210-7
5. Rehatschek, H., Aigelsreiter, A., Regitnig, P., Kirnbauer, B.: Introduction of eLectures at the medical university of Graz—results and experiences from a pilot trial. In: Proceedings of the International Conference on Interactive Collaborative Learning (ICL), IEEE Catalog Number: CFP1223R-USB. Villach, Austria 26–28 Sep 2012. ISBN:978-1-4673-2426-7
6. Rehatschek, H., Hye, F.: The introduction of a new virtual microscope into the eLearning platform of the medical university of Graz. In: Proceedings of the 14th conference on interactive collaborative learning (ICL). pp. 10–15. Piešťany, Slovakia 21–23 Sep 2011. ISBN 978-1-4577-1746-8

7. Extron Electronics.: Interfacing switching and control. Mar 2018. URL: http://www.extron.com

8. EMP Entwine and Video Lounge.: Enterprise-class software for management, processing, and playback of media files. Apr 2018. https://www.extron.com/technology/landing/entwine/

9. Panopto. Mar 2018. URL: https://www.panopto.com

10. StreamAMG broadcast quality. Mar 2018. URL: https://www.streamamg.com/

11. Epiphan, capture stream record. Mar 2018. URL: https://www.epiphan.com/

12. Epiphan Pearl 2 Twin. June 2018. URL: https://www.epiphan.com/products/pearl-2/

13. Opencast Matterhorn, open source solution for automated video capture and distribution at scale. Mar 2018. URL: http://www.opencast.org/matterhorn

14. Moodle, Modular Object Oriented Dynamic Learning Environment, Open Source Software. June 2018. URL: https://moodle.org/

15. Learning Tools Interoperability (LTI), specification developed by IMS Global Learning Consortium. April 2019. URL: https://docs.moodle.org/33/en/LTI_and_Moodle

16. Rust, I., Krüger, M.: Der Mehrwert von Vorlesungsaufzeichnungen als Ergänzungsangebot zur Präsenzlehre. In: Wissensgemeinschaften, Köhler, H.T., Neumann, J. Digitale Medien – Öffnung und Offenheit in Forschung und Lehre. pp. 229–239 (2011)

17. Zupanic, B., Horz, H.: Lecture recording and its use in a traditional university course. In ACM SIGCSE Bulletin

18. Rehatschek, H., Hruska, A.: Fully automated virtual lessons in medical education. In: Proceedings of the International Conference on Interactive Collaborative Learning (ICL), IEEE Catalog Number: CFP1323R-ART. pp. 3–8, Kazan, Russian Federation. 25–27 Sep 2013. ISBN 978-1-4799-0153-1

Pilot Project on Designing Competence-Oriented Degree Programs in Kazakhstan

Gulnara Zhetessova[1], Marat Ibatov[2], Galina Smirnova[3],
Damira Jantassova[4(✉)], Valentina Gotting[5], and Olga Shebalina[3]

[1] Strategic Development of Karaganda State Technical University, Karaganda, Kazakhstan
zhetesova@mail.ru
[2] Karaganda State Technical University, Karaganda, Kazakhstan
imaratk@mail.ru
[3] Center of Engineering Pedagogics of Karaganda State Technical University, Karaganda, Kazakhstan
{smirnova_gm, cep.kstu}@mail.ru
[4] Foreign Languages Department of Karaganda State Technical University, Karaganda, Kazakhstan
damira.jantassova@gmail.com
[5] Department « Vocational Education and Pedagogics » of Karaganda State Technical University, Karaganda, Kazakhstan
gottingv@mail.ru

Abstract. The design of methodology of competence-oriented degree programs is considered on the basis of modern labor market requirements deriving from the principle of modularity: the main stages, the principles of design and the mechanism of creating degree programs in higher education has been opened by means of specially developed software. This article establishes a need for competence-based degree programs for modernization of higher education within the context of workforce supply and demand.

Keywords: Competency-based approach · Degree program · Employers

1 Context

In the context of the Message of the President of the Republic of Kazakhstan to the peoples of Kazakhstan, dated from the 31st of January, 2017 "The third modernization of Kazakhstan: global competitiveness" in the frame of the project "Training highly-qualified personnel in the frame of the new model of economy" of the Ministry of education and science of Kazakhstan a pilot project on development of methodology and analytical system of design and assessment of the competence-oriented educational programs, in interaction with industry-specific associations, has been launched. This project has been carried out at Karaganda State Technical University (KSTU) in cooperation with the Center of Bologna Process and Academic Mobility (Astana city). Our University was selected based on its legal status as Innovative and Educational

M. E. Auer and T. Tsiatsos (Eds.): ICL 2018, AISC 917, pp. 163–173, 2019.
https://doi.org/10.1007/978-3-030-11935-5_16

Consortium "Corporate University", and which comprises 80 industrial companies and organizations of the Karaganda.

2 Purpose or Goal

Beginning in 2014, The Republic of Kazakhstan has promoted an increase in academic freedom, and international mobility in its universities in consideration of labor market needs of various defined regions in the country, and for development of appropriate human resources. Recently, the right of universities in Kazakhstan to determine the content of bachelor's degree programs through electives has increased up to 55% for bachelor's degrees, 70% for master's degree programs and 90% for Ph.D. degree programs. As a result, a great number of specialty degree programs offered by various universities now differ in their elective components. The diversity and content variance of degree programs pose difficulties for equivalency. In this context, the urgent goal is to develop common requirements for designing degree programs, based on competence and ranking. Implementation of a methodology for accomplishing this is possible during the process of official registry of degree programs of higher education and its application in practice with the purpose of forming a common digital system of degree program assessment in collaboration with employers.

In the frame of the research work our hypothesis is that design and implementation of competence-based degree programs for engineering universities will enhance higher education content. Optimization and transfer to a practice-oriented approach in training competent specialists, will therefore meet with labor market demand.

This process is focused on reduction to a common denominator of supply and demand on a workforce. This fundamental step brings about a radical interaction between education system and labor market.

Development of learning modular programs on the basis of professional standards can be attributed to reorientation of demand for new abilities and changes of work organization; decline in demand for unqualified manual skills; need for new knowledge and conceptually new content of training; distribution of automated control systems for productions; blurring of lines between professions through decentralization of economic responsibility and development of quality management systems.

3 Approach

In order to confirm our hypothesis and for implementation of the project, we looked at advanced domestic and foreign experiences on the project problem and monitored research on designing bachelor's degree programs in interaction with the labor market, including in the sphere of IT, industrial management, and other fields. In the process of the research a survey of employers was administered on defining the necessary competencies for graduate students in the field; comparative analysis of proposed competencies within activity categories and labor functions of a specialist as indicated by published Professional Standards and other normative documents, which regulate the activity of a specialist in the appropriate branches of the economy.

Validation of modern industry needs in specialists of a new formation has been made. On the basis of the data we have developed a methodology for designing degree programs founded on a competence approach considering the principle of modularity.

On the basis of the methodology there is transfer from knowledge paradigms to a competitive approach. The methodology is based on formation of competences and their projection on the learning outcomes. The result is a description of that which graduate students will know and can do after graduation.

Competence-based approach in this situation is considered as a effective tools for refreshing and optimizing educational content.

The educational program in the system of higher education is developed according to the operating State Classifier of Job Fields and Professional Standards (currently in process of development in Kazakhstan).

Currently, within the fields of specialists, higher education study is independently developed through various degree programs according to the National Qualifications Framework and Professional Standards, and is coordinated with the Dublin Descriptors and the European Qualifications Framework.

In contrast, the proposed degree program has a modular format and its design is carried out in the following sequence:

1. Goal setting of the degree program.
2. Developing a map of the fields of study on the degree program.
3. Development of the qualification profile of a graduate.
4. Development of a competence map of a specialist.
5. Development of a matrix of disciplines forming educational modules.
6. Development of a map of a learning module.
7. Development of content of the degree program.

Main goal setting which is carried out by field experts is the first step of degree program design.

The main goal, which is carried out by field experts is a description of action or the actions demanded for achievement of a result. It is a verb or several verbs in the infinitive, or a verbal noun. The action object is described by means of a noun or several nouns. If there is a need, the action situation will be described.

The main goal, which is carried out by the expert has to be measured and observed and then briefly described within the professional area as a total result of what has to be accomplished in the educational preparation. On the basis of the main goal, which is carried out by the specialist, the purpose of the degree program is formulated.

The goal of degree program is formulated as follows: "Training of specialists for realization ... (according to a main goal which is carried out by the expert).

The map of the fields of study on the degree program is developed according to the State Classifier of Lessons with the higher and postgraduate education of the Republic of Kazakhstan and contains the following components: code and name of field of study, code and name of the educational path, code and name of the educational program.

Qualification profile of a graduate consists of the following components: the degree; list of expert duties; professional category; object of professional activity; types of professional activity; functions of professional activity.

The competence map of a specialist in the educational program of a bachelor degree is submitted by the group of the following competencies: of general education, basic and professional.

Competence of general education (the world outlook competencies of the national idea "Mangilik el" ("Eternal country") and modernization of social conscience in the context of national identity) is the ability of a specialist to carry out personally and socially important productive activities within the social realm.

Competencies of general education of a graduate are the same irrespective of the type of professional activity, i.e. are general for all types of educational programs of a bachelor's degree and are shown through:

- maintaining the culture, own national code, development of factors of the person's competitiveness (computer literacy, knowledge of foreign languages, cultural openness);
- realism and pragmatism of behavior is the ability to live rationally, to lead a healthy life and professional success and also in the field of dialectics of development, socialization in public relations, modern political systems and information and communication technologies in various spheres of activity and also in the field of economics and law, ecology, health and safety, skills of business, leadership, susceptibility to innovations;
- important qualities of graduates: commitment, self-discipline, ambitiousness, responsibility, civic consciousness, communicativeness, tolerance, upgrading of general culture, etc.

Basic competence is the ability of a specialist to solve a set of professional tasks on the basis of integrated knowledge, abilities and experience and also personal qualities allowing carrying out effectively professional activity. This group of competencies are general for specialists within one path of preparation both universal in character and adaptability. They assume maturity of primitive level of ability and readiness for concrete professional activity and function as a basis for formation of professional competencies.

Basic competencies are formed within basic disciplines among which, for example, for engineering specialties: engineering graphics, materials resistance, man-machine study and etc. The cycle of basic disciplines includes a high school component and elective component the ratio of which is independently defined by each university. At the same time disciplines of a cycle are formed according to the departments' offerings, employers and students.

Professional competence is ability of a specialist to solve a set of professional tasks in the chosen field of activity on the basis of concrete knowledge, abilities, and skills.

Professional competencies of a graduate describe a set of the main typical features of any specialty defining concrete orientation, that is a direction of the degree program and are shown in ability of a specialist to solve a set of professional tasks in the chosen field on the basis of concrete knowledge, abilities and skills.

Professional competencies are developed according to each degree program on the basis of Professional Standards factored with labor market demand, expectations of employers, the interests of students and social inquiry of society.

The list of professional competencies is structured according to the main functions or types of professional activity for which the graduate has to be prepared, for example: research, design, production and technological, organizational and administrative competences.

Professional competencies are the most mobile part of a competence map because they are original and determine a profile of preparation and variability of paths by the degree program. The cycle of professional disciplines within which these competencies are formed consists of elective components, which are developed according to departments' offerings, employers and students.

Development of a competence map (professional, basic, of general education) has the general algorithm except where there is active participation in development of professional competencies maps by employers and other stakeholders. The standard starting point is the Professional Standard and also State Classifier of Job Fields.

The projection of Professional Standard (if available) on the degree program is presented in Fig. 1.

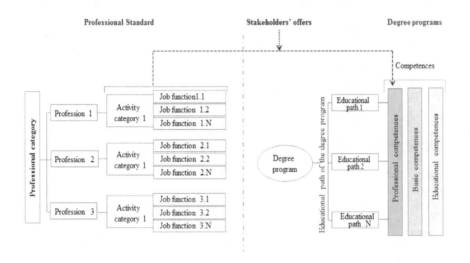

Fig. 1. Projection of the professional standard on the degree program

If the Professional Standard for this profession does not currently exist, it is necessary, in development of the professional competencies map to involve employers and other stakeholders. These may include the branch associations, the Ministry of Labor and Social Protection of the Republic of Kazakhstan, the Ministry of Education and Science of the Republic of Kazakhstan, National Chamber of Entrepreneurs, scientific and other organizations to ensure that the structure of professional competencies will be defined in current trends of the professional activity within in the sphere of development of the graduate of this degree program.

Designing of the competencies map is necessary to consider through the lens of the integrative final outcome of the educational program absorption demanded by the employer.

The format of professional competences map is presented in the Fig. 2. Under educational paths of the degree program can be presented one or several professional competences with the indication of learning outcomes.

Professional competences		Learning outcomes
Professional competence 1	1	
	2	
	...	
Professional competence N	1	
	2	
	...	
Basic competences		**Learning outcomes**
Basic competence 1	1	
	2	
	...	
Basic competence N	1	
	2	
	...	
Educational competences		**Learning outcomes**
Educational competence 1	1	
	2	
	...	
Educational competence N	1	
	2	
	...	

Fig. 2. Professional competences map of a graduate

The algorithm of degree program designing is presented in the Fig. 3.

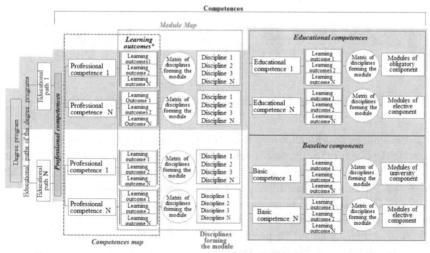

*Learning outcomes have been registered due to active verbs on the purposes of Blum's levels

Fig. 3. Algorithm of designing of degree program's content

The matrix of disciplines forming educational modules is an auxiliary component which allows combining the learning outcomes with disciplines within which these competences will be formed (Fig. 4).

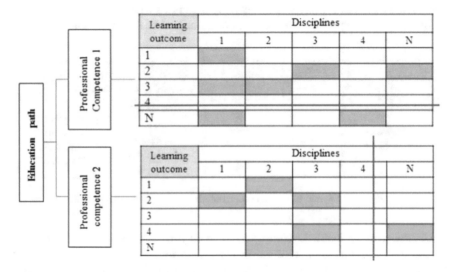

Fig. 4. Matrix of professional competences

Learning outcomes are registered in the left column at the top line of the curriculum discipline on the chosen degree program.

We begin to correlate learning outcomes with disciplines in which they are formed and highlight, within the matrix, the field where they come together.

When we completely fill in a matrix of competencies, it can turn out that some lines or columns won't be highlighted. It means that we have revealed new outcomes which aren't formed in any discipline (in our example it is outcome 4) and we need to bring additional material in the content of some disciplines or to enter new disciplines, which will be responsible for formation of these learning outcomes.

An empty field, not modified in color, within a matrix, means that in the curriculum there is a discipline which "doesn't bear responsibility" for formation of learning outcomes and it is possible to think that it is "unnecessary" (in our example it is Outcome 4). In this case it is necessary to analyze the content of this discipline and to fill it with relevant learning materials to achieve a certain learning outcome, or to exclude it from the curriculum. Thus, revision of contents of the educational program will be carried out with the goal of updating and optimization.

The matrix of competencies can be considered as quality monitoring of determination of the learning outcomes structure, the list and the content of disciplines for the degree module. A full and edited matrix is the basis for designing the structure of this module.

The name of the module is defined according to a competence profile.

Development of a map of a learning module. The map of a learning module on the formation of professional competence is developed "horizontally" in the matrix of competencies (Fig. 5). Learning outcomes are formed in frames of the certain disciplines entering the corresponding module which is responsible for development of professional competence. Maps of the learning modules are developed in the same way on formation of competencies of general education and basic competencies.

Competence name	Name of learning module	Learning outcome	Name of disciplines forming learning outcome	Module discipline
Professional competence	Module 1	1	Discipline 1	Discipline 1
			Discipline 2	Discipline 2
		2	Discipline 3	Discipline 3
			Discipline N	Discipline N
		3	Discipline N	

Fig. 5. The map of the learning module on competence formation

Development of thematic plan of the degree program content in the frame of suitable module. The example of the thematic plan for discipline N (it is developed "vertically" in the matrix of competences) is presented in Fig. 6. Within discipline N learning outcomes are formed (1,2,3, …). The list of topics is shown, which are then built in a logical sequence and the thematic plan of discipline is formed. Developing the thematic plan of disciplines on the basis of competency-based approach will allow for optimization of the disciplines' content and to actualize them.

Professional competences	Learning outcome	Subject name	Workload by types of lessons, hour.				
			Lectures	Laboratories	Practical lessons	Independent learning with teacher's supervision	Independent learning
Professional competence 1	Outcome 1	1					
		2					
	Outcome 2	1					
		2					
		…					
	Outcome 3	1					
		2					
		…					

Fig. 6. Developing the thematic plan of discipline N, forming competence 1

Designing of degree programs is directed to their inclusion in the Registry of educational programs (further—the Registry) representing the information system containing the list of educational programs of higher education divided into subsections and intended for use by the educational organizations and individuals.

The Registry allows automating input, storage, search, maintaining data base of these educational programs that is formed consistent data environment of accounting of all degree programs realized by higher education institutions.

For the goal of conducting monitoring of educational programs, definition of the degree programs which are most demanded by the labor market and an exception of educational programs which aren't corresponding to modern requirements of society; increasing the competitiveness of educational programs it is necessary to introduce the system of their assessment.

Such assessment has to be the most transparent for all interested persons and to stakeholders for definite compliance of educational program to Professional Standards, State Classifier of Lessons and also requirements of employers to specialists of each defined profession.

When developing the Registry of the degree programs, assessment criteria, concept models and architecture of analytical systems for projecting and assessment of competence-oriented bachelor's degree programs have been established. On the results of the assessment in the Registry the rating of degree programs is automatically being formed, which represents a ranked list of active degree programs of specialists training. Competency-based approach in this situation can be considered an effective tool for updating and optimizing of educational content.

The designing system of degree programs (further—the designer of programs) is intended for teachers of universities where they are developers and represents the mechanism of creation of educational programs of higher education on specially developed software.

The designer of programs solves the following problems:

- formation by the system administrator of bases of general education and basic competences;
- formation of professional competences base according to Professional Standards and survey of employers;
- creation of degree programs on a uniform template, at the same time the developer owns full freedom of choice of learning outcomes, tools for their formation and assessment;
- possibility of printing a camera-ready degree program;
- possibility of updating the degree program at any stage of designing;
- the automated process of formation of summary tables: description of degree program, profile map, maps of modules, elements of thematic plans of disciplines;
- creation of innovative degree programs, using bases of professional competences on the existing educational programs;
- access to system 24/7.

The sequence of degree program designing with the help of online designer of programs "The system of degree programs designing" is shown in the Fig. 7.

The distinctive feature of designing systems of educational programs is an opportunity to survey employers online. At the stage of making revisions to the list of selected competencies, the developer has the possibility of sending the list to one or several employers for compilation of the final list of professional competences. In the designer program it is possible to choose an employer from the list, to attach the degree

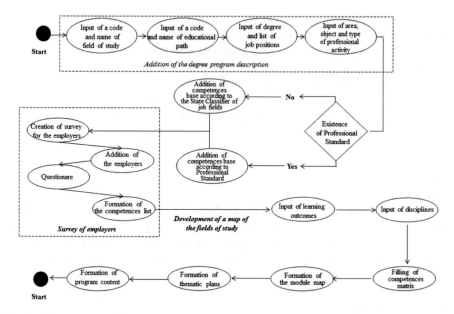

Fig. 7. The sequence of degree program designing with the help of online designer of programs. "The system of degree programs designing"

program and to send the unique generated link e-mail of the employer. The employer can follow the link and choose necessary competences from the list or offer the competences by choosing the action "To offer the competencies".

The result of these operations is a working window, in the program, on the development of the professional competencies list offered by the employer. The developer has an opportunity to add the offered competencies to the general list, and then begin filling in the learning outcomes. Besides, the employer has the possibility of viewing the competencies in the column "Mark of the Employer" which are those most demanded by the employer where numerical expression of outcomes is displayed.

The system automatically carries out structuring the data and allows the developer of educational program editing the forms, adding and downloading the created documents. Thus, use of the program designer allows automating and optimizing the process of degree programs development.

4   Actual or Anticipated Outcomes

Practical result of the research: Development of a methodology for projecting the competence-based degree program in interaction with industry-specific associations; development of guidelines on projecting competence-based degree programs and manual on methodology; design of bachelor degree programs in interaction with the labor market including in the sphere of IT, civil engineering, industrial management and others; development of an online design program for degree programs, the possibility to conduct surveys of employers for identifying the more demanding competences of a specialist and designing the degree programs on developed methodology; development of a Registry for degree programs aimed at ranking and assessment.

In order to examine the efficacy of this methodology we held seminars for representatives of the academic society (42 Universities of Kazakhstan) on the issue of designing competence-based degree programs, and a seminar on assessment of designed programs that included employers, representatives of the National Chamber of entrepreneurs and industry-specific associations.

5   Conclusions/Recommendations/Summary

The developed methodology will allow for revising more than 19% of the degree programs currently offered by universities in Kazakhstan. Moreover, it enables optimization of curricular content and an increase in practice-oriented components in accordance with the requirements of employers.

The proposed system of developing degree programs represents the possibility for automation of degree program design process taking into account the needs of the labor market and to meet the requirements of the National Qualifications Framework, Professional Standards.

The developed system of assessment is maximally transparent for all stakeholders and it concentrates on defining compliances of degree programs with Professional Standards, National Qualifications Framework as well as requirements of employers for the specialists of defined professions.

As a result, the presented methodology offers universities updated tools of projecting and quality assessment of degree programs.

The undertaken research has confirmed the accuracy of the proposed hypothesis and promotes increasing competitiveness of graduates of Kazakhstani universities, integrating into the European zone of the higher education, enhancing attractiveness of the degree programs of the Kazakhstani universities and acceptability of the qualifications in deferent countries as well as determining the transparency between the active systems of higher education.

S.A.M.—A System for Academic Management

Rosa Matias[1,3]([✉]), Angela Pereira[2,3], and Micaela Esteves[1,3]

[1] CIIC-Computer Science and Communication Research—ESTG,
Leiria, Portugal
{rosa.matias,micaela.dinis}@ipleiria.pt
[2] CiTUR Tourism Applied Research Centre—ESTM, Leiria, Portugal
angela.pereira@ipleiria.pt
[3] Polytechnic Institute of Leiria, Leiria, Portugal

Abstract. The Millennial generation was born and grew up in the digital era, so they are natives in using technology, including mobile devices and a whole set of associated applications. The Management Information Systems of Higher Education Institutions must be adapted to the needs of the new technological users. The institutions should integrate in their business processes mechanisms to simplify the exchange of academic data between students and teachers. For instance, grades, schedules, assessments or attendances to classes. Mobile devices can make the difference since students may have access to personalized academic data. In this work, it is proposed a system called System for Academic Management (SAM). It has a desktop application for teachers and a mobile application for students. This system aims to enhance the exchange of personalized academic data between teachers and students, plus accomplishing this generation needs in terms of quick access to information through mobile devices. SAM was developed using two main methodologies. A User-Centered Design methodology for interface development and an Agile Development methodology for software development. In this process low-fidelity and high-fidelity prototypes were developed and tested until acceptance by end-users. The results obtained until now shows that the user specially the teachers appreciate the application and highlighting its usefulness.

Keywords: Mobile application · Management information systems · Higher education

1 Introduction

The usage of mobile devices has been growing rapidly specially for the Millennial generation. They have access to information anywhere, anytime with their mobile devices [1]. Millennials are more likely to use mobile technologies as a means of communication in contrast to the previous generations. Furthermore, the hardware and software evolutions boot the development of mobile applications in many areas, such as, medical, governmental, education [2–4], to satisfy the users' needs in terms of easy access to information.

For a long time, the Management Information Systems (MIS) have been used as a mean to improve the efficiency of school office activities [5]. MIS have a positive

© Springer Nature Switzerland AG 2019
M. E. Auer and T. Tsiatsos (Eds.): ICL 2018, AISC 917, pp. 174–184, 2019.
https://doi.org/10.1007/978-3-030-11935-5_17

impact on data administration including on students' personal data and academic information. Meanwhile, Millennials privilege the usage of mobile devices such as tablets and smartphones. In this sense, MIS should accomplish the tendency by promoting solutions for mobile applications capable of managing personalized academic data. Solutions should simplify the access to educational resources, making it available to the academic community and, at the same time, enhancing the communication between students and teachers. These applications must be simple to use, effective and as automated as possible.

In this paper, researchers present a system called System for Academic Management (S.A.M.) with the aim to facilitate the access to academic information and improve the communication between students and teachers. The system is compounded by a desktop application for teachers and a mobile application for students. This way, students have the guarantee of their privacy. Also, teachers can communicate with students and inform them about lost lessons and attendance rates, among others.

The rest of this article is organized as follows: in Sect. 2, it is presented the overview and motivation; in Sect. 3, the used methodology is described; in Sect. 4, the prototype developed is described. Finally, the conclusion and future work are reported.

2 Overview and Motivation

The Millennial generation, also called Y generation, was born between 1982 and 2002 and is now attending Higher Education Institutions. Mobile devices have significant impact in how information is shared among them. The Millennial generation grew up with digital sources such as social media like Facebook, Instagram, Twitter. Social media is an umbrella term for web-based software and services that allows users to come together online and exchange, discuss, communicate and participate in any form of social interaction [6, 7]. The Millennial generation is now attending Higher Education Institutions which should accomplish the mobile application usage tendency not only for teaching (M-Learning) [4] but also for personal academic data management. MIS of High Education Institutions should integrate business processes dedicated to new communications vehicles.

Mobile devices in education have been largely used as a learning tool, named as M-Leaning [8]. Meanwhile, the usage of mobile devices to facilitate the exchange of academic data and information between teachers and students has not been properly exploited [5]. Information technology in educational management is a field that not only needs in-depth studies, but also effects the school processes.

Many organizations have their data spread across multiple systems and using different technologies which is a problem since there is no data integration. The Higher Education Institutions also suffers from other problems. For instance, academic information such as assessment calendars, grades, tutorial orientation schedules, are spread throughout different systems. Moreover, some of them are not in a digital format. Sometimes, students face the problem of having academic information in different systems, so that difficult them find out the information they need.

Particularity in some Portuguese Higher Education Institutions there are rules regarding mandatory attendance to classes. Consequently, in each lesson, students must sign a presence sheet and later teachers verify the signatures to confirm students' presences. The attendance management is time consuming for both teachers and students since it is executed daily. Moreover, it is hard for students to control the classes absences numbers.

Another issue is related to the way grades are available for students. Generally, teachers release the students' grades by publishing a PDF file in a Learning Management System such as Moodle. However, some students do not appreciate that their grades become visible to all their colleagues. Moreover, teachers have duplicated efforts. For instance, they have their own excel worksheet for grade management, then they released grades in Moodle and finally in the official school private grade system. The same integration problem happens with classes absence registration. It is a hard process specially in the presence of a high number of students.

In this sense, researchers consider the development of a mobile application for students to have a simplified access to academic information. Moreover, a back-office desktop application was considered for teachers to manage subjects which generates the data for students.

3   Methodology

In the following section we will describe the software development methodology and the implementation process. Many of the requirements were obtained by researchers' interactions with the student community and the teaching staff.

3.1   Software Development Methodology

Considering the application target, it is important to have a user-friendly interface adapted to students' needs. Otherwise they will not use it. Two different methodologies were used one for the interface development and another for software development. The first one was a User-Centered Design (UCD) approach and the second one, an Agile Software Development.

The UCD is an interface development process where end-users are in the center of the development and consequently influence the application design. The main purpose of UCD is to build interfaces adapted to end-users needs, i.e., an interface that end-users can use it with efficiency, efficacy and satisfaction [9]. This methodology is cyclical and consists in four phases: conception, prototype development, tests and results analysis. The phases are repeated until end-users are satisfied with the interface. The prototype englobes two types a low-fidelity and a high-fidelity prototype. A low-fidelity prototype permits allows verify if all the functionalities are in the interface, generally are drawn on paper. After the prototypes are tested and accepted by the end-user it is stated a new cycle where high-fidelity prototypes are created with the aim to test the visual appearance and usability issues.

The software development process was based in an Agile Software Development methodology which is incremental and iterative (Fig. 1). This methodology approach enables the constant change of requirements along the project as well as a constant feedback from customers. In the context of Agile Software Development methodologies, it was elected Extreme Programming (XP) [10]. Since, XP is an Agile Software Development methodology which intends to improve software quality and responsiveness to customers' requirements changes [11]. XP shares the values espoused by the Agile Manifesto for Software Development but goes further to specify a simple set of practices [12].

Fig. 1. Extreme programming cycle (from [12])

Extreme Programming is based on values of simplicity, communication, feedback, and courage [12, 13].

3.2 Requirements Identification

The process stated with interviews to students, so their needs could be attended. Moreover, teachers were also listened. The institution management systems was also analysed, their functionalities and their failures. From this point a set of requirements were defined.

3.2.1 Desktop Application Requirements

From meetings with teachers a requirement list was gathered aiming to improve the communication between teachers and students. Teachers appreciated the idea and considered the system useful and highlighted the importance of it to improve their job regarding paperwork. Since, there were many requirements that teachers would like to implement in the application, researchers identified the most important and assigned priorities to each one.

In the desktop application, teachers should be able to upload the students' grades. Later the system automatically notifies each student individually about grades. Teachers also will be able to visualise and control the students' absence to classes. Moreover, it will be possibly to make available in the mobile application the assessment schedules. Besides, teachers can manage their schedule tutorial orientation, which will be visible in the students' mobile application. Furthermore, the course head master will have dashboards for data analysis concerning courses, subjects and students' performance.

There are different profiles accessing the desktop application, namely, the administrator, the course head master and the teacher. The administrator is responsible for users' management. The course head master is responsible for the management of the course subjects and respective teaching staff. Teachers are responsible for their subject management. In Table 1 are listed the desktop application requirements for each profile and the respective priority.

Table 1. Desktop application requirements

Profile	Requirement	Priority
Administrator	Manage users	High
Head master	Manage teachers	High
	Manage subjects	High
	Analyse students performance	High
	Analyse students attendance to classes	High
	Turn available to students' mobile application the course assessment schedule	High
	Turn available to students' mobile application the course schedule	High
	Obtain Reports about students' performance and attendance to classes	Medium
	Schedule individual appointments with students	Low
Teacher	Import students' marks from Excel Worksheets	High
	Notify individually students about their marks	High
	Automatically confirm which students are attending to classes	High
	Analyse the students' attendance and notify the ones which are in danger to fail because the number of absences to classes	High
	Manage their tutorial schedule	Medium
	Receive emails about students' messages	Medium

3.2.2 Mobile Application Requirements

The students appreciated the idea of having academic data in their mobile devices, specially the fact they can easily verify the assessment schedule and the respective classroom. In addition, they consider important to observe the number of absences to each subject. Currently, to know this information they must contact each teacher who count presences manually. The requirements list to implement in the students' mobile application are present in Table 2.

Table 2. Mobile application requirements

Profile	Requirement	Priority
Student	Visualise each subject grade	High
	Visualise the course final grade	
	Visualise each class absence number	
	Visualise the course schedule	
	Visualise the assessment schedule	
	Visualize teachers' tutorial orientation schedule	
	Schedule tutorial classes with a specific teacher	
	Notification of subjects' alerts	
	Notify the system about their presence in the classroom	

3.3   System Architecture of the SAM System

The SAM system architecture consists of two applications, a desktop application and a mobile application interconnected through the web (Fig. 2). The system logical architecture follows the Client/Server model with multiple layers, namely: client, middleware and server. The client is represented by the two aforementioned applications. The middleware layer is responsible for the implementation of services which are available for both applications. Finally, the server layer represents both, the web server and the database server.

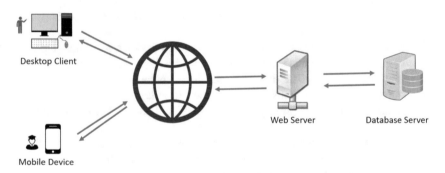

Desktop Client

Web Server   Database Server

Mobile Device

Fig. 2. SAM system architecture

4   Prototype Development

The next step of development consists on the creation of several Mockups, namely low-fidelity prototype, to test and analyse the user interface. The Mockups were developed for both the Desktop and Mobile applications. The Mockups were designed with the purpose to find the most suitable solution. And they were evaluated by end-users.

Later, a high-fidelity prototype was built with the purpose to assess the interface usability.

In Fig. 3 the reader can observe screens of both high-fidelity prototypes, the Desktop and Mobile applications. On the left a screen for the head master to manage subjects and on the right the main screen of the student mobile application (Fig. 3).

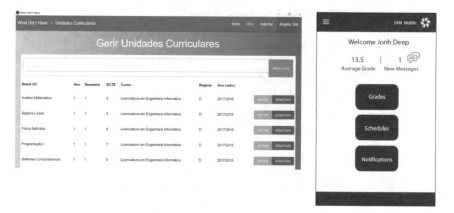

Fig. 3. Prototype desktop application (left) and the mobile application (right)

4.1 Low Fidelity Prototype Tests

The desktop low-fidelity prototypes were tested by end-users, who considered them confused and not balance with many empty spaces. Moreover, in the head master area end-users had difficulties in associating subjects to teachers. Consequently, these prototypes were redesign.

In the teacher area the most problematic layout was associated to the grades excel import files process. Since this is a main feature for teachers, the interface must be both intuitive and user friendly. The difficulty was the adaptation to a wide range of excel file structures. As so, it was considered important to redesign it.

The mobile application prototype was tested, and the end-users did not made considerations, so it was considered acceptable.

After this evaluation phase new prototypes were developed, considering the end-users feedback and, later, they were tested again. The results showed that the prototypes were better, so they went to the next phase, developing the high-fidelity prototype with the aim to test the application in terms of visualization and aesthetics.

4.2 Usability Tests

Usability is a quality attribute that assesses how easy user interfaces are used [14]. The usability is measured by 5 quality components:

- Learnability: How easy it is for users to accomplish the basic tasks in the first time they use the interface.
- Efficiency: How quickly users can perform tasks.
- Errors: How many errors do users make.

– Satisfaction: How pleasant is it to use the design?
– Efficacy: How users produce a desired or intended result.

According to Nielsen [15], to identify the most important usability problems, typically 5 users are enough. Considering this, the usability tests were made by individual users, namely, five teachers and five students, that had to solve several tasks to test all the applications functionalities.

Teachers tests
As stated the desktop application has three profiles, administrator, head master and teacher. Yet, the tests focus mainly in the head master and teacher profiles, since those were considered the most important.

With the head master profile, in general all the tasks were performed easily, with exception the task regarding the association of a subject to a teacher. Although, in the previous tests these issues were overcome. These new end-users considered this interface not intuitive and were not able to use it. Since this is a main task in the desktop application, it was necessary to rethink the interface.

The interface for teachers was considered confuse. During the tests end-users made several mistakes and took some time to accomplish tasks. Based in these results it was considered important redesign the interface.

After this tests phase, new prototypes were created (Fig. 4.) based in users' opinions and new usability tests performed. The problems observed in the previous tests were overcome.

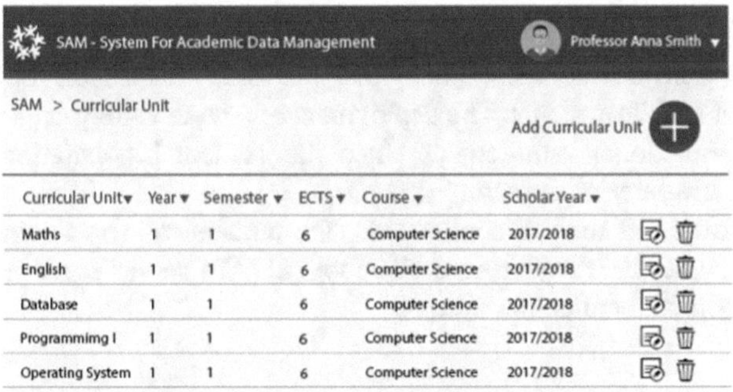

Fig. 4. Final prototype desktop application (Head master)

Students tests
Regarding the mobile application for students it was considered by the end-users, who tested it, easy to use, intuitive and very useful. Thus, the mobile application interface was considered well done and adapted to the user's needs.

To sum up, researchers were concerned in obtaining a user-friendly interface for both teachers and students to guarantee the interest in the application. Therefore, the researchers concluded the prototype phase and started the system development.

5 Development Process

The desktop application was developed using electron [16], which is a cross-platform framework for creating applications with web technologies such as JavaScript, HTML, and CSS. This means that the desktop application can run in multiple platforms such as Windows, Linux and Mac OS. Moreover, a REST (Representational State Transfer) [11] API (Application Programming Interface) was developed enabling the communication of data to mobile devices (Fig. 5).

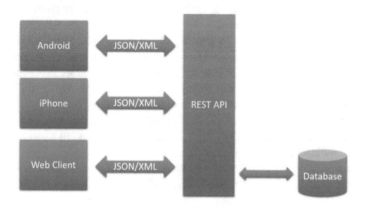

Fig. 5. Interoperability between layers

The mobile application is implemented using Ionic, an open source, and hybrid solution for mobile development [17]. This platform enables the deployment of a diverse range of mobile systems. The deployment of a wide variety of mobile devices is a system non-functional requirement since the students' population have mobile devices with a diversity of operating systems.

In the desktop and mobile applications, the authentication is done by using the school Lightweight Directory Access Protocol (LDAP) server avoiding the development of a specific authentication system.

6 Conclusion and Future Work

Nowadays, students consider mobile devices as the main communication mean. Academic Institutions should accomplish the technological enhancements. In this study, it is described a system capable of improving the communication between teachers and students for academic data and automate the presences management in classes.

For teachers, this software will reduce the time consuming in bureaucratic issues and allow them to focus on teaching and investigation. For students it is a quickly way to have access to the academic information.

The researchers consider this application user-friendly since it was planned and designed with a user-centered approach which is the best way to satisfy the end-users needs.

The system is under construction but until now the feedback obtained from end-users is positive, they like the application and highlight its usefulness, specially students.

As future work the researchers intend to insert more functionalities in the application such as the possibility to identify the students that are at risk of dropping out. Since this is a big concern in Portuguese Higher Education Institutions and having a way to identify the students that are at risk it is an important functionality to prevent it.

References

1. Considine, D., Horton, J., Moorman, G.: Teaching and reaching the millennial generation through media literacy. J. Adolesc. Adult Literacy **52**(6), 471–481 (2009)
2. Esteves, M., Pereira, A.: YSYD-you stay you Demand: user-centered design approach for mobile hospitality application. In: International Conference on Interactive Mobile Communication Technologies and Learning (IMCL). pp. 318–322. IEEE (Nov 2015)
3. Pereira, A., Esteves, M., Weber, A.M., Francisco, M.: Controlling diabetes with a mobile application: diabetes friend. In: Interactive Mobile Communication, Technologies and Learning. pp. 681–690. Springer, Cham (Nov 2017)
4. Esteves, M., Pereira, A., Veiga, N., Vasco, R., Veiga, A.: The use of new learning technologies in higher education classroom: a case study. In: Teaching and Learning in a Digital World, ed. Michael E. AuerDavid GuralnickIstvan Simonics, pp. 499–506. Springer International Publishing, Cham (2018). ISBN: 978-3-319-73209-1
5. Shah, M.: Impact of management information systems (MIS) on school administration: what the literature says. Procedia-Social Behav. Sci. **116**, 2799–2804 (2014)
6. Ryan, D.: Understanding Digital Marketing: Marketing Strategies for Engaging the Digital Generation. Kogan Page Publishers (2016)
7. Sago, B.: The Influence of Social Media Message Sources on Millennial Generation Consumers. Int. J. Integr. Mark. Commun. **2**(2) (2010)
8. Al-Emran, M., Elsherif, H.M., Shaalan, K.: Investigating attitudes towards the use of mobile learning in higher education. Comput. Hum. Behav. **56**, 93–102 (2016)
9. Abras, C., Maloney-Krichmar, D., Preece, J.: User-centered design. Bainbridge, W. Encyclopedia of Human-Computer Interaction. Thousand Oaks: Sage Publications **37**(4), 445–456 (2004)
10. Beck, K., Gamma, E.: Extreme programming explained: embrace change. Addison-wesley professional (2000)
11. Erickson, J., Lyytinen, K., Siau, K.: Agile modeling, agile software development, and extreme programming: the state of research. J. database Manage. **16**(4), 88 (2005)
12. Lindstrom, L., Jeffries, R.: Extreme programming and agile software development methodologies. Inf. syst. manage. **21**(3), 41–52 (2004)
13. Jeffries, R.: Extreme Programming and Agile Software Development Methodologies (2003)
14. Paz, F., Pow-Sang, J.A.: A systematic mapping review of usability evaluation methods for software development process. Int. J. Softw. Eng. Its Appl. **10**(1), 165–178 (2016)
15. Nielsen, J.: Heuristic evaluation. In: Nielsen, J., Mack, R.L. (Eds.) Usability Inspection Methods. John Wiley & Sons, New York, NY (1994)

16. Electron - Build cross platform desktop apps with JavaScript, HTML, and CSS (2018). Search in 5 of January of 2018. Available at https://electronjs.org/
17. Ionic - Build amazing apps in one codebase, for any platform, with the web. Search in 5 of January of 2018. Available at https://ionicframework.com/

Comparison of Reaction Times of a Visual and a Haptic Cue for Teaching Eco-Driving

An Experiment to Explore the Applicability of a Smartwatch

Matthias Gottlieb[1]($\boxtimes$), Markus Böhm[1], Matthias Utesch[1,2], and Helmut Krcmar[1]

[1] Information Systems, Technical University of Munich (TUM), Munich, Germany
{gottlieb,markus.boehm,utesch,krcmar}@in.tum.de
[2] Staatliche Fachober- und Berufsoberschule Technik München, Munich, Germany

Abstract. Climate change increases the interest in Green Information Systems (IS) research. Green IS technologies enable the reduction of energy consumption by teaching eco-driving to make the driver aware of inefficient fuel consumption. To teach eco-driving (energy-efficient human behavior), car manufacturers applied in-car information systems. Due to the increasing visual in-car information, alternative solutions are necessary. However, the haptic interaction channel has a short reaction time for a given cue and seems to be an applicable solution to replace visual in-car information. The examined studies investigating haptic cues in the context of energy consumption and automotive applications have used the gas pedal—with lower user acceptance—to interact with the driver. For this reason, we investigate a smartwatch as the haptic cue in a laboratory experiment to teach eco-driving.

Keywords: Haptic · Visual · Cue · Reaction Time · Experiment

1 Introduction

Climate change increases the interest in Green Information Systems (IS) research [1]. Green IS technologies enable the reduction of energy consumption by teaching eco-driving [2–4]. Teaching eco-driving can decrease fuel consumption by up to 20% depending on driving behavior [4, 5]. Teaching eco-driving with the intent of changing driving behavior to a more energy-efficient way of driving, we identified six eco-driving rules [6, 7]:

(i) Shift up as soon as possible
(ii) Use the highest gear possible and driving at low engine speed
(iii) Downshift late

M. E. Auer and T. Tsiatsos (Eds.): ICL 2018, AISC 917, pp. 185–196, 2019.
https://doi.org/10.1007/978-3-030-11935-5_18

(iv)   Maintain a steady speed by anticipating traffic flow
(v)   Accelerate swiftly
(vi)   Decelerate smoothly while leaving the car in gear.

Green IS **eco-feedback** mechanisms such as dashboards can teach eco-driving rules while driving.

Due to increasing visual in-car information [8, 9], the driver might be distracted from his driving task. However, an appropriate design of information is crucial for keeping the driver's attention on the primary task of driving and not distracting the driver with secondary or tertiary tasks such as eco-driving [8, 10]. The **haptic inter-action channel** has a short reaction time for a given cue and seems to be an applicable solution [11–13].

All examined studies [14] investigating haptic cue in the context of energy consumption have used the gas pedal to interact with the driver [11, 12, 15–19]. In this paper, we applied the didactic framework of Go4C [20] —which has been well established and proven for more than 10 years—in the first stage called *Preparation* to explore the applicability of a smartwatch for teaching eco-driving. Therefore, the purpose of this study is to discover in a laboratory experiment whether a **smartwatch** is applicable for further studies in teaching eco-driving to the driver while driving [21]. As an **outcome**, we measure the **reaction time** and consider **latency issues**, due to the distributed system of the smartwatch. Figure 1 illustrates the concept-map of our approach for teaching eco-driving.

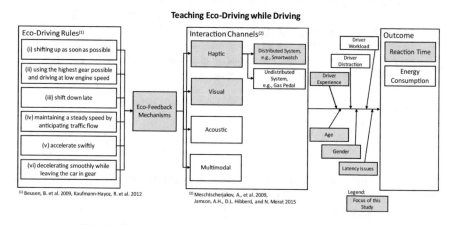

Fig. 1. Concept Map of Teaching Eco-Driving while Driving

2   Related Work

Studies have dealt with a visual cue [8, 9, 11, 12, 19, 22], but the use of a haptic cue is focused on the gas pedal [11, 12, 17, 19]. No significant difference between the gas pedal—with low user acceptance—and a visual cue is found [11, 19]. Figure 1 illustrates the focus of this study on teaching eco-driving while driving.

The haptic interaction reduces driver workload [23] and is the most effective cue variant of the interaction channels (s. Fig. 1) [15]. A suitable device to quantify the quality of the haptic cue is a smartwatch [24]. Studies examined a smartwatch in different contexts such as driver drowsiness [25–27]. None of them measured the latency within distributed systems. The smartwatch is a distributed system (not directly connected to the car such as gas pedals are). We compare the reaction times of a haptic and visual cue. Experimental research facilitating multimodal cues depict that haptic cues have a lower reaction time compared to visual cues [28]. However, they examined the cue on the finger [28]. Stimulating both senses resulted in even better reaction times [29]. Both experiments examine non-driving tasks. Rydström et al. [22] implemented a multisensory system with four treatments to examine the driver's ability to conduct secondary tasks while driving. The results indicated that the participants with haptic-only treatment expend the least amount of effort while spending the most time on the primary driving task [22]. Richter et al. [30] and Pitts et al. [31] featured an enhanced center console touchscreen with different vibrotactile patterns. In their experiments, a driver was supposed to conduct various tasks—mostly choosing the menu items on the screen—while driving. Aside from having no impact on the driving, the use of haptic cue in both experiments has enabled a significant increase in performance for solving the secondary tasks.

None of the examined studies has investigated a smartwatch as haptic cue while driving. Subsequently, a haptic cue is a promising interaction channel due to its short reaction time for a given cue [11–13]. In addition, the smartwatch is a distributed system and we consider latency issues. Therefore, we extend prior research [11, 12, 17, 19] and examine the reaction time at the example of a smartwatch as cue sender to explore whether we can apply a smartwatch for teaching eco-driving while driving.

3 Hypothesis

This research focuses on the difference in reaction times between a visual and a haptic cue. In an extension, we investigate a driving context that is more complex [32] than non-driving tasks. The additional driving context might have similar findings because of the visual and haptic cue comparisons. The receptor on the wrist as compared to the receptor on the finger has a higher bandwidth [33]. This might cause a lower reaction time. A haptic cue is more efficient than a visual cue [15]. While the results confirm the theory of Forster et al. [28], although Landau et al. [15] have not tested the signal with a smartwatch. Due to the purpose of this research, we want to examine whether the haptic cue can replace a visual cue. The replacement performs under the prerequisites that latency issues are considered. Consequently, we assume the reaction time of a haptic cue is less or equal compared to the visual cue. We assume the following hypotheses H_0 and H_1.

H_0: The reaction time of a haptic cue on the wrist is less than or equal to the reaction time of a visual cue on the screen.

H_1: The reaction time of a haptic cue on the wrist is greater when compared to the reaction time of a visual cue on the screen.

4 Methodology

Figure 2 illustrates the focus of the study and the application of the framework from Utesch et al. [34], we first describe the organization (experiment design and pre-experiment) and environment (driving simulator) of the experiment in more detail. Second, we present the introduction of execution from the experiment (instructions and usage-explanation of the simulator). Third, we explain the implementation (data collection and data analysis).

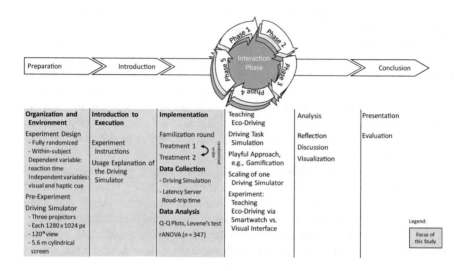

Fig. 2. Adoption and application of the framework of Utesch et al. [34]

In terms of **organization,** before, we first conducted a fully randomized within-subject design experiment. We performed a pre-experiment to eliminate possible biases [35]. To measure the dependent variable reaction time in milliseconds (ms), we applied the ReactionTask from the driving simulator software OpenDS. The independent variable consists of two treatments to effect braking in different ordered rounds: (1) a red visual cue on the screen and (2) a haptic cue, the vibration of the smartwatch. After the experiment, the participants had to complete a questionnaire with demographical data to identify differences among them such as gender or age. As an incentive for participation, we raffled off three Amazon vouchers for 20 EUR each. The **environment** is a high-fidelity driving simulator (s. Fig. 3a) that uses the OpenDS software. The simulator cockpit consists of a Mercedes dashboard with tacho- and speedometer replaced by a Nexus 9 tablet that illustrates these. The projection system consists of three projectors with a resolution of 1280×1024 pixels each. The images are blurred at the edges to provide a total horizontal view of $120°$ with a 5.6 m radius on a cylindrical screen at $45°$ to each other in front of the driver. To provide a fluent illustration of the tachometer, the data are updated if necessary every 20 ms.

Introduction to execution: We first present the experiment instructions to the participants on the screen. As the first instruction appeared, the experimenter began to read the instructions adapted from the original ReactionTest to the haptic cue. Afterward, the participant drove a familiarization round to get accustomed to the simulator.

Implementation: To randomize the experiment, each participant exposes both treatments in different orders. The participant had to react to the cue by pressing the brake pedal when the red visual cue (treatment 1, s. Fig. 3a) or the vibration (treatment 2, s. Fig. 3b) occured. In total, ten times per round. A vibrating LG Urban smartwatch provided the haptic cue for braking. For the haptic cue, we utilized a vibration pattern. Android programming only allows two states of vibration: on and off. The vibration was set to on for a duration of 100 ms, followed by a break of another period of 100 ms. The vibration pattern repeated five times. The vibration has no lead-time to achieve comparability between the haptic and visual cue. Each participant drove two rounds on a five-lane highway with a speedometer and a landscape background. In both rounds, we used an additional green visual cue in between the treatments in random order to prevent the drivers from anticipating the cue (s. Fig. 3c). The green visual cue appeared on the gantry and informs the driver to change the lane. It also occurred ten times per round.

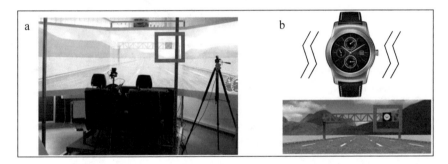

Fig. 3. a High-fidelity driving simulator with visual cue (treatment 1), **b** Smartwatch (treatment 2), **c** Instruction screen with visual cue to prevent guessing

To **collect** the *reaction time* in ms, we use the internal routines from OpenDS. A timer was set automatically as soon as the corresponding cue of a highway bridge (gantry) triggers. For the treatments (s. Fig. 2a, b), the timer stopped when the driver pressed the brake pedal. When the driver depressed the brake for less than two seconds, the ReactionTask marked the response as invalid with a value of 10 s. OpenDS calculated and saved the difference between the time stamps: start and end per treatment. We use the term "reaction time" in a simplified manner that does not *precisely describe* the actual psycho-motoric measurement we are conducting. According to Klebelsberg [36], we are measuring the driver's reaction time *plus* the brake's operating time. The operating time of the simulation is the input lag between the brake pedal and OpenDS. To determine the latency, we apply a *LatencyServer*. The Server received the acknowledgment messages from the smartphone application in the form of round-trip time (RTT) values described by the Cristian's algorithm [37]. For computing the adjusted reaction time values for latency, we utilize the following equation:

$$t_{hap,adj} = t_{hap} - RTT / 2 \qquad (1)$$

The RTT/2 approach requires *symmetric* RTT across the network. To calculate the latency, we measure every time stamp from each device for record purposes; the time stamps t_1 and t_8 are the essential ones. The server record the time stamp (eight in total) when the message arrives at the latency server (t_1) and the time stamp when it arrives back at the latency server after transmitting the message to the smartphone and the smartwatch (t_8).

We use a separated Wi-Fi network with IEEE 802.11b/g standard to ensure that we do not cause biases, such as quality-of-service rules. The smartwatch communicates via a low-energy Bluetooth. Figure 4 illustrates the separate communication via smartphone and smartwatch via Bluetooth by using the Google cloud messaging service.

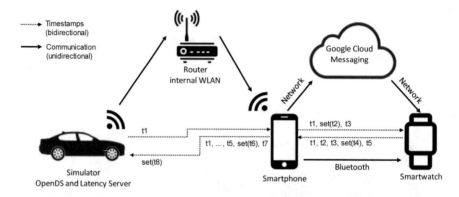

Fig. 4. Developer's view on the android wear networking and latency message exchange. *Source* Own Illustration; Google and Bluetooth Logos are in Public Domain

Data analysis: since the Shapiro-Wilk test is biased for large sample sizes, we compared Q-Q plots to test normal distribution [38] and determine heteroscedasticity with Levene's test. Reaction time values follow the convolution of exponential and normal distribution called ExGaussian distribution [39]. Due to these normality issues, we follow the approach from Ratcliff [40] and Whelan [39] for reaction time analysis. We transform the data with common logarithm to get rid of the biased data. For statistical analysis, we use a repeated measure, ANOVA (rANOVA), because of stochastically dependent measures for the visual and haptic reaction times from each subject [41]. We inspect each reaction time value (10 visual and 10 haptic) per participant—in total a data set of 350 values. The reaction test marks outliers with a value of 10,000 ms. Ratcliff [40] recommends dropping the outliers to conduct the rANOVA.

5 Results

The population of the experiment is 35 (13 female and 22 male) with ages ranging from 19 to 42 ($M = 25.97$, $SD = 4.65$). By the end of the experiment, we received 350 visual and haptic reaction test values. We removed outliers ($\geq 10,000$ ms) from the dataset by using an inter-quartile range analysis with a factor of k = 3 because of the skewness of the data [46], resulting in 347 pairs. The driving experience of test persons ranges from 0 to 45,000 km/year (M = 10,220 km/year, SD = 11,573 km/year). Participants have on average had a driver's license for 8.26 years (SD = 4.74). The results revealed no significant influence on the moderators that were used. We analyze the co-variances with repeated measures analysis of co-variances of age, mileage in km/year, years in possession of a valid driver's license and gender. The effect size is small for $\eta p^2 > 0.02$, medium for $\eta p^2 > 0.13$, and large for $\eta p^2 > 0.26$ [44]. The age and years in possession of a valid driver's license had no or a medium effect on the reaction time whereas the other moderators had any or a small effect. The mean of haptic adjusted 1290.99 ms compared to the latency 281.89 ms is 21.8%. The mean of the visual cue is similar to haptic but is adjusted (1276.97 ms). Comparing the standard deviation of the latency (87.43 ms) to visual reaction times (292.28 ms), we find that the haptic cue (283.57 ms) is lower. Figure 5 illustrates the comparison between mean and median per treatment grouped by gender.

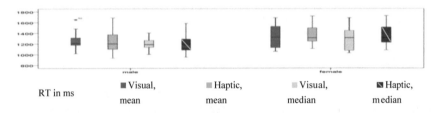

Fig. 5. Comparison of mean and median per treatment grouped by gender

Because reaction times have a positive skewness [39, 42], we examine the skewness on the mean and median values, cumulated for each participant. Male participants had a greater positive skewness than female. To get rid of the normality issues, we transformed the data using the common logarithm. We compared the linearity of points within Q-Q plots for assuring normality, with transformed data delivering better results.

We used a level of significance α = .05. We found no significant difference between the visual and the haptic cue, reciprocal: $p > .5$ and common logarithm: $p > .2$. The results of the applied t-test for paired samples on the mean ($t(34)$ = −.459, M = −12.85, SD = 165.67, SE = 28.00, p = .6) and the median ($t(34)$ = 1.705, M = −34.6, SD = 120.07, SE = 20.30, $p > .09$) indicate no significant difference. We use the Wilcoxon-test to clarify whether there was an influence exerted by the transformation but we found no significant differences between the treatments, $p > .05$ (arithmetic mean: z = −.860, p = .397, median: z = −1.482, p = .141). These confirm the findings from André [43]. In addition, the execution of a rANOVA resulted in no statistically significant difference between the transformed reaction time values of the treatments adjusted for latency, $F(1,346)$ = 1.010, p = .316, ηp^2 = .003. The Bonferroni-corrected pairwise comparison reflect no significant higher reaction time for the visual feedback (M = 1276.97, SD = 292.28) compared to the haptic feedback (M = 1290.99, SD = 283.57). Hence, we fail to reject H₀ for visual and adjusted haptic treatment. By contrast, a statistical difference is between the visual and unadjusted haptic reaction times indicated by rANOVA, $F(1,346) = 216.597$, $p < .0005$, $\eta p^2 = .385$. Bakeman [44] depicts a large effect size with $\eta p^2 > .26$. The calculated effect size shows a strong effect, $f = 3.129$ [45]. The Bonferroni-corrected pairwise comparison illustrates that the unadjusted haptic feedback ($M = 1572.88$, $SD = 313.16$) is significant ($p = .000$) higher as the visual feedback ($M = 1276.97$, $SD = 292.28$).

Figure 6 depicted the estimated marginal means of the reaction time data. There is enough evidence to reject the H_0 for visual and the unadjusted haptic treatments.

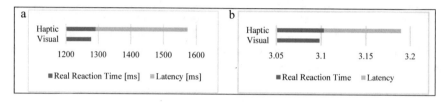

Fig. 6. Estimated Marginal Means: **a** Original and **b** $\log_{10}$-transformed Data

6   Discussion

Our reaction time values adjusted for latency were in line with the previous research on unexpected but common braking times presented by Green [47]. The results indicate no significant deviation from the population between visual and haptic cue with a mean of 1250 ms. Consequently, we depict that a haptic cue as a smartwatch can replace a visual cue to teach eco-driving. However, the reaction time results adjusted for latency are inconsistent with the groundwork on multimodal facilitated reaction times [28, 29, 48]. Our optimistic prediction based on these results implied significantly shorter reaction times triggered by a haptic cue in comparison to a visual cue. Despite an extensive inspection by transformed and non-transformed data, our statistical analysis results in no significant difference between the treatments. Therefore, we can apply the smartwatch to teach eco-driving while driving.

Forster et al. [28] and Ng and Chan [48] did not elaborate on data skewness, normality violations, and the ExGaussian distribution of the reaction time values. We extended the findings by elaborating our data on the normality issues and transforming the data to retain the conclusion validity of hypothesis testing. Ng and Chan [48] examine two different sample sizes, the visual with 69 participants (32 males) and the haptic with 25 participants (14 females). The experimental design is a between-subject design, but Forster et al. [28], Bauer, Oostenveld [29] and we applied a within-subject design to get rid of possible bias by a possible individual variability. In contrast to our experiment, the participants who underwent Ng's and Chan's [48] experiment had different tasks for the visual treatment. The participants had to decide between different buttons, whereas the haptic treatment only consists of one button [48].

We extend the findings from Ng and Chan [48], Forster et al. [28], and Bauer et al. [29] by examining a driving context explicitly. This difference contributes to new kinds of designing in-car interfaces to teach eco-driving.

The difference between our results and previous work on multimodal reaction times can probably be attributed to the complexity of the tasks that participants had to perform in our reaction test. The experiment conducted by Forster et al. [28] featured simple tasks and was a non-driving situation, which consisted of responding to dichotomous stimuli and pressing a button positioned directly under the participant's finger. Nevertheless, a driver needs his fingers to grab the steering wheel and steer the car.

The experiment by Ng and Chan [48] featured a low degree of uncertainty with a simple motoric task of pressing buttons on a numeric keypad with fingers that were already resting on it. Hence, this setup required only minor motoric and cognitive effort. By contrast, our task consisted of a complex simulation environment, which required the recognition of multiple treatment conditions (green or red sign, the position of the sign on the gantry), as well as the comparatively long-lasting movement of the right foot from gas to brake pedal. Driver motoric braking time may have outweighed the actual cognitive differences between both treatments. The mean deviation presented by Forster et al. [28] was approximately 20 ms between the visual, haptic, and the visual-haptic treatments. Such a difference is likely to be statistically insignificant for the sample size of 35 subjects—reflected in the minimal effect size. The underlying assumption is that the stimuli do not affect the motoric capabilities.

The latency issue features a statistically significant difference between the two tested treatments. While exposed to the haptic treatment, participants performed worse—exactly by the latency value. That implies the unsuitability of the current haptic cue implementation for reaction time-critical tasks.

Although we can confirm H_1 only for replacing the visual cue with a haptic cue, the fact that haptic cue is not inferior to a visual cue, which is the traditional way of driver sensation, for solving driving tasks needs further research. Hence, the haptic cue may be used for facilitating secondary driving tasks such as eco-driving in place of visual indicators for facilitating the primary and undertaking the secondary (and also tertiary) tasks [49–51]. Previous research on haptics for driving tasks has shown positive results, in particular, drivers were able to perform the secondary tasks more efficiently and were partially able to increase visual contact with the road [30, 31]. Additionally, the time-period for the haptic cue is essential because too-differentiated signals, such as different types of vibration, affect the time until a person can react, and as we learned, 200 ms is significant.

7   Conclusion, Limitations and Future Work

Our research adds a smartwatch successfully as an approach for further research in teaching eco-driving. In addition, haptic cues can use the watch as an advanced driver assistant system to create an effective driver support interface [21] for situations other than eco-driving, i.e., for the smartwatch to act like an alarm clock to wake up the driver in case she or he falls asleep. We can confirm the findings from Azzi et al. [19]. Hence, research can use a smartwatch for a haptic cue instead of a gas pedal and this extends the findings from Meschtscherjakov et al. [11] by implementing a smartwatch.

Also, it extends Forster et al. [28] findings by using a haptic cue not only on the wrist but also for a driving task. Nevertheless, we recommend considering latency for distributed systems, especially while driving.

We performed the reaction test without counterbalancing the independent variable levels among the participants. This results in a reduction of the robustness of the test against the habituation and sensitization effects. The correctness of the adjusted reaction times might affect the asymmetrical latency values. In other words, the latency between "there-and-back" phases may differ.

In future studies, we will analyze human behavior on energy consumption by teaching eco-driving with a smartwatch while driving by adopting the playful approach of teaching put forth by Utesch et al. [34]. Therefore, a specific challenge in teaching is how to scale one driving simulator for a certain number of participants. Additionally, the practitioner can use haptic cues on a smartwatch for secondary or tertiary driving tasks to reduce the visual information overload. We are going to research teaching eco-driving. Moreover, research can study longtime effects by teaching eco-driving. We suggest conducting experiments on driver distraction caused by a smartwatch and to make sure the driver is not interpreting the haptic signal as a phone call.

Acknowledgements. We thank Roman Trapickin for his contribution in the course of his student thesis.

References

1. Watson, R.T., Boudreau, M.-C., Chen, A.J.: Information systems and environmentally sustainable development: energy informatics and new directions for the IS community. MIS Q. **34**(1), 23–38 (2010)
2. Jamson, S.L., Hibberd, D.L., Jamson, A.H.: Drivers' ability to learn eco-driving skills; effects on fuel efficient and safe driving behaviour. Transp. Res. Part C Emerg. Technol. (2015)
3. Inbar, O., Tractinsky, N., Tsimhoni, O., Seder, T.: Driving the scoreboard: motivating eco-driving through in-car gaming. In: CHI Gamification Workshop. Vancouver, BC, Canada (2011)
4. Barkenbus, J.N.: Eco-driving: an overlooked climate change initiative. Energy Policy **38**(2), 762–769 (2010)
5. Gonder, J., Earleywine, M., Sparks, W.: Analyzing vehicle fuel saving opportunities through intelligent driver feedback. SAE Technical Paper (2012)
6. Beusen, B., Broekx, S., Denys, T., Beckx, C., Degraeuwe, B., Gijsbers, M., Scheepers, K., Govaerts, L., Torfs, R., Panis, L.I.: Using on-board logging devices to study the longer-term impact of an eco-driving course. Transp. Res. Part D Transp. Environ. **14**(7), 514–520 (2009)
7. Kaufmann-Hayoz, R., Lauper, L., Fischer, M., Moser, S., Schlachter, I., Meloni, T.: What makes car users adopt an environmentally friendly driving style? In: Proceedings of INTER-NOISE. New York, NY, USA (2012)
8. Ablaßmeier, M., Poitschke, T., Wallhoff, F., Bengler, K., Rigoll, G.: Eye gaze studies comparing head-up and head-down displays in vehicles. In: IEEE International Conference on Multimedia and Expo. IEEE, Beijing, China (2007)
9. Froehlich, J., Findlater, L., Landay, J.: The design of eco-feedback technology. In: Proceedings of the SIGCHI 2010 Conference on Human Factors in Computing Systems. Atlanta, GA, USA: ACM (2010)

10. Kern, D., Schmidt, A.: Design space for driver-based automotive user interfaces, In: 1st International Conference on AutomotiveUI '09. pp. 3–10, Essen, Germany: ACM (2009)
11. Meschtscherjakov, A. et al.: Acceptance of future persuasive in-car interfaces towards a more economic driving behaviour. In: 1st International Conference on AutomotiveUI '09. Essen, Germany: ACM (2009)
12. Jamson, A.H., Hibberd, D.L., Merat, N.: Interface design considerations for an in-vehicle eco-driving assistance system. Transp. Res. Part C Emer. Technol. **58**, 642–656 (2015)
13. Staubach, M., Kassner, A., Fricke, N., Schießl, C.: Driver reactions on ecological driver feedback via different HMI modalities. In: Proceedings of the 19th World Congress on ITS. Vienna, Austria (2012)
14. Gottlieb, M., Böhm, M., Krcmar, H.: Analyzing measures for the construct "energy-conscious driving": a synthesized measurement model to operationalize eco-feedback. In: 22nd Pacific Asia Conference on Information Systems. Yokohama, Japan (2018)
15. Landau, M., Loehmann, S., Koerber, M.: Energy flow: a multimodal 'ready' indication for electric vehicles, In: Adjunct Proceedings of the 6th International Conference on Automotive User Interfaces and Interactive Vehicular Applications. pp. 1–6, Seattle, WA, USA, ACM (2014)
16. Coughlin, B.: Haptic Apparatus and Coaching Method for Improving Vehicle Fuel Economy. Ford Global Technologies Llc, USA (2009)
17. Birrell, S.A., Young, M.S., Weldon, A.M.: Vibrotactile pedals: provision of haptic feedback to support economical driving. Ergonomics **56**(2), 282–292 (2013)
18. Mulder, M., Abbink, D.A., van Paassen, M.M., Mulder, M.: Design of a haptic gas pedal for active car-following support. Intell. Transp. Syst. IEEE Trans. **12**(1), 268–279 (2011)
19. Azzi, S., Reymond, G., Mérienne, F., Kemeny, A.: Eco-driving performance assessment with in-car visual and haptic feedback assistance. J. Comput. Inf. Sci. Eng. **11**(4), 181–190 (2011)
20. Utesch, M.C.: A successful approach to study skills: Go4C′s projects strengthen teamwork. Int. J. Eng. Pedagogy (iJEP) **6**(1), 35–43 (2016)
21. Mulder, M., Mulder, M., van Paassen, M.M., Abbink, D.A.: Haptic gas pedal feedback. Ergonomics **51**(11), 1710–1720 (2008)
22. Rydström, A., Grane, C., Bengtsson, P.: Driver behaviour during haptic and visual secondary tasks. In: Proceedings of the 1st International Conference on Automotive User Interfaces and Interactive Vehicular Applications. ACM, Essen, Germany (2009)
23. Birrell, S.A., Young, M.S., Weldon, A.M.: Delivering smart driving feedback through a haptic pedal. In: Proceedings of the International Conference on Contemporary Ergonomics and Human Factors 2010. Taylor & Francis Ltd (2010)
24. Miller, M.: The Internet of Things: How Smart TVs, Smart Cars, Smart Homes, and Smart Cities are Changing the World. 1 edn, Indianapolis, IN, USA: Pearson Education, Indianapolis, Indiana (2015)
25. Ríos-Aguilar, S., Merino, J.L.M., Sánchez, A.M., Valdivieso, Á.S.: Variation of the heartbeat and activity as an indicator of drowsiness at the wheel using a smartwatch. Int. J. Artif. Intell. Int. Multimedia **3**(3), 96–100 (2015)
26. Li, G., Lee, B.-L., Chung, W.-Y.: Smartwatch-based wearable EEG system for driver drowsiness detection. IEEE Sens. J. **15**(12), 7169–7180 (2015)
27. Liu, L., Karatas, C., Li, H., Tan, S., Gruteser, M., Yang, J., Chen, Y., Martin, R.P.: Toward detection of unsafe driving with wearables. In: Proceedings of the 2015 Workshop on Wearable Systems and Applications. pp. 27–32, ACM: Florenz, Italy (2015)
28. Forster, B., Cavina-Pratesi, C., Aglioti, S.M., Berlucchi, G.: Redundant target effect and intersensory facilitation from visual-tactile interactions in simple reaction time. Exp. Brain Res. **143**(4), 480–487 (2002)

29. Bauer, M., Oostenveld, R., Fries, P.: Tactile stimulation accelerates behavioral responses to visual stimuli through enhancement of occipital gamma-band activity. Vision. Res. **49**(9), 931–942 (2009)

30. Richter, H., Ecker, R., Deisler, C., Butz, A.: HapTouch and the 2 + 1 state model: potentials of haptic feedback on touch based in-vehicle information systems. In: Proceedings of the 2nd International Conference on Automotive User Interfaces and Interactive Vehicular Applications. ACM, Pittsburgh, PA, USA (2010)

31. Pitts, M.J., Burnett, G., Skrypchuk, L., Wellings, T., Attridge, A., Williams, M.A.: Visual-haptic feedback interaction in automotive touchscreens. Displays **33**(1), 7–16 (2012)

32. Pfleging, B., Broy, N., Kun, A.L.: An introduction to automotive user interfaces. In: Proceedings of the 2016 CHI Conference Extended Abstracts on Human Factors in Computing Systems. ACM, San Jose, CA, USA (2016)

33. Balakrishnan, R., MacKenzie, I.S.: Performance differences in the fingers, wrist, and forearm in computer input control. In Proceedings of the ACM SIGCHI Conference on Human Factors in Computing Systems, ACM (1997)

34. Utesch, M.C., Seifert, V., Prifti, L., Heininger, R., Krcmar, H.: The playful approach to teaching how to program: evidence by a case study. In: 20th International Conference on Interactive Collaborative Learning. Budapest, Hungary, Springer (2017)

35. van Teijlingen, E.R., Hundley, V.: The Importance of Pilot Studies. vol. 35, Social Research UPDATE, University of Surrey (2001)

36. Klebelsberg, D.: Verkehrspsychologie, vol. 308, Springer, Berlin, Heidelberg (1982)

37. Cristian, F.: Probabilistic clock synchronization. Distrib. Comput. **3**(3), 146–158 (1989)

38. Field, A.: Discovering Statistics Using IBM SPSS Statistics. Sage Publications, London, Thousand Oaksm New Delhi, Singapore (2013)

39. Whelan, R.: Effective analysis of reaction time data. Psychol. Rec. **58**(3), 475 (2008)

40. Ratcliff, R.: Methods for dealing with reaction time outliers. Psychol. Bull. **114**(3), 510 (1993)

41. Dytham, C.: Choosing and Using Statistics: A Biologist's Guide. 3rd edn, Wiley-Blackwell, Hoboken, NJ (2011)

42. Baayen, R.H., Milin, P.: Analyzing reaction times. Int. J. Psychol. Res. **3**(2), 12–28 (2015)

43. André, M.: The ARTEMIS European driving cycles for measuring car pollutant emissions. Sci. Total Environ. **334–335**, 73–84 (2004)

44. Bakeman, R.: Recommended effect size statistics for repeated measures designs. Behav. Res. Methods **37**(3), 379–384 (2005)

45. Cohen, J.: Statistical Power Analysis for the Behavioral Sciences. 2nd edn, Lawrence Erlbaum Associates, Hillsdale, NJ, USA (1988)

46. Hoaglin, D.C., Iglewicz, B.: Fine-tuning some resistant rules for outlier labeling. J. Am. Stat. Assoc. **82**(400), 1147–1149 (1987)

47. Green, M.: How long does it take to stop? Methodological analysis of driver perception-brake times. Transp. Hum. Factors **2**(3), 195–216 (2000)

48. Ng, A.W.Y., Chan, A.H.S.: Finger response times to visual, auditory and tactile modality stimuli. In: Proceedings of the International MultiConference of Engineers and Computer Scientists, Hong Kong, China (2012)

49. Griffiths, P.G., Gillespie, R.B.: Sharing control between humans and automation using haptic interface: primary and secondary task performance benefits. Hum. Factors J. Hum. Factors Ergon. Soc. **47**(3), 574–590 (2005)

50. Lefemine, G., Pedrini, G., Secchi, C., Tesauri, F., Marzani, S.: Virtual fixtures for secondary tasks. In: Human-Computer Interaction Symposium. Springer (2008)

51. MacLean, K.E.: Haptic interaction design for everyday interfaces. Rev. Hum. Factors Ergon. **4**(1), 149–194 (2008)

IQ and EQ Enhancement for People with Mental Illness

Samir Abou El-Seoud[1(✉)] and Samaa A. Ahmed[2]

[1] Faculty of Informatics and Computer Science (Computer Science), The British University in Egypt (BUE), Cairo Governorate, Egypt
samir.elseoud@bue.edu.eg
[2] Faculty of Informatics and Computer Science (Software Engineering), The British University in Egypt (BUE), Cairo Governorate, Egypt
samaa131048@bue.edu.eg

Abstract. This paper focuses on IQ and EQ enhancement for people with mental challenges such as Autism, Asparagus syndrome and Down syndrome. People with autism and asparagus syndrome develop a relatively high IQ but below average EQ. On the other hand, people with Down syndrome and intellectual disability have low IQ and relatively high EQ. The system uses speech recognition and detection to listen and understand the user. The application uses text to speech (TTS) in addition to an artificial intelligent chat bot to maintain conversations with the user. Furthermore the system uses image processing to detect users emotions and facial expression. The application uses the methodologies used in several medical institutes and psychiatrist. In addition to IQ and EQ tests that are performed to detect the enhancement of the users IQ and EQ. There are two types of Intellectual disability previously known as mental retardation, namely: mildly mentally retarded (MIMR) and severely mentally retarded (MOMR). Intellectual disability according to IQ is rated by scores. An IQ of 130 and above means Very Intelligent, an IQ of 120–129 indicates intelligent, and an IQ 110–119 indicates High Average, an IQ of 90–109 indicates Average, and an IQ of 80–89 indicates Low Average. An IQ of 70–79 means MIMR and an IQ 69 and lower means MOMR [1]. In Emotional intelligence, the average human rate is from 90 to 110. A person with an IQ below 90 is considered to have low EQ. The main goal of this paper is to help users with below average IQ and EQ to seek normal rates. The developed program uses machine learning, speech recognition, image processing, and best IQ methodologies available to perform an EQ test repeatedly until the user reaches the normal EQ rates. That will allow users with autism to participate in their society and have a normal life. Many parents cannot afford to take their children to a specialist or send them to special schools. Consequently, parents should no longer suffer from self-taking care for their disable kids. In developing countries, children in Orphanages would not be diagnosed for their disability nor would have support from the staff. There are many programs and therapy institutions exist that is specialized in helping those disable people. However, mostly there are no computerized programs nor enough personal to support such children, adults or even parents. The combination of methodologies and the repetitive EQ and IQ tests ensures the effectiveness and efficiency of the product.

Keywords: Autism · IQ improvement · EQ improvement · AI · Intellectual disabilities

© Springer Nature Switzerland AG 2019
M. E. Auer and T. Tsiatsos (Eds.): ICL 2018, AISC 917, pp. 197–209, 2019.
https://doi.org/10.1007/978-3-030-11935-5_19

1   Introduction

The National Alliance on Mental Illness (NAMI) surveys in the United States of America showed that one in five adults suffer from mental illness. It indicates that forty-three point eight million Americans suffer from a mental illness. Moreover, one in five youth with age between 13 and 18 suffers from a severe mental disorder. In Scotland, it was estimated that seventeen percent of Scottish women and fourteen percent of Scottish men tend to suffer a mental illness [2]. Mental illness is defined as a health condition that causes the change in person's behavior, emotion, and thinking. Mental illness affects the social functionalities, family, and work activities of its patient [3]. There are several diseases that affect the human's brain such as Autism, Asperger syndrome, Down syndrome, and Intellectual disability etc.

Worldwide many people suffer from mental illness such as Autism, Down syndrome, Asparagus syndrome, Intellectual disability. An intelligence quotient (IQ) is a total score expressed in a number and is derived from several standardized tests designed to measure human intelligence. The higher the score, the greater that person's reasoning ability. IQ scores are used for educational placement, assessment of intellectual disability, and evaluating job applicants. With the help of a professional, therapy can help to get out of an unhealthy cognitive, emotional, and behavioral pattern. Different tests have been deigned to evaluate and measure individual's human intelligence and can be used to assist in the diagnosis of mental health disorders.

Emotional Intelligence (EI), which is measured using Emotional Quotient (EQ), is known as the capability or ability of individual to recognize, evaluate and regulate their own emotions as well as those of others. It is also individual's ability to differentiate between different feelings, label them appropriately, and use emotional information to guide his/her thinking and behavior. Moreover, it is the individual's ability to adjust emotions and adapt them to environments. Mental illness is categorized into three types. Low intelligence (IQ), low emotion intelligence (EQ), and physiological illness such as paranoia, kleptomania, and Obsessive Compulsive Disorder OCD.

The aim of the proposed project is to enhance and improve the mental ability of people with mental illness.

1.1   Overview and Background

There are several kinds of mental illness: some affects the physical abilities of patient, some affects the thinking capabilities, some affects the emotional capabilities, some affect both thinking and emotional, and some tends to affect all. It is intended to focus in this paper on both the emotional and thinking abilities of the user. Patients with Down syndrome and intellectual disability tends to suffer a below IQ. This is due to a disease that affects patient's brain. Patients with Down syndrome or intellectual disability previously known as (Mental Retardation) show a graduate development. Which means that a 10 years old patient's brain is in fact way younger than any normal 10 years old kid. This actually depend on the degree of infection. As we mentioned above, there are two types of Intellectual disability previously known as mental retardation, namely: mildly mentally retarded (MIMR) and severely mentally retarded (MOMR). Intellectual disability according to IQ is rated by scores. An IQ of 130 and above means Very Intelligent, an IQ of 120–129 indicates intelligent, and an IQ 110–119 indicates High Average, an IQ of 90–109 indicates Average, and an IQ of 80–89 indicates Low Average. An IQ of 70–79 means MIMR and an IQ of 69 and lower means MOMR [4]. Although people with syndrome and intellectual disability tends to suffer a below IQ they tend to develop a high EQ. They are always willing and loving to communicate.

People with autism and asparagus syndrome develop high IQ but below average EQ. In Emotional intelligence, the average human rate is from 90 to 110 [2]. A person with an EQ below 90 is considered to have low Emotional Intelligence. Patients with autism spectrum disorder and asparagus syndrome show aggressive responses when their daily routine is changed. They tend to have an order tune and a rude attitude. They tend to avoid eye contact. In addition they suffer aggressive reaction when physically contacted [5].

It has been noticed that patients who suffer mental disorders have in general lower emotional intelligence. On the other hand, depression is considered as a mode of disorder. Patients who have problems to conceive a pleasure and positive feeling, show low levels of positive affect (mood) and often report states of sadness or fear.

Most of studies carried out indicate that brain capabilities are tidily related to wellbeing and psychological health. It is also related to get involved in self-destructive acts and in deviant behavior.

1.2 Problem Statement

About 1% of world population suffer from autism spectrum disorder [6]. In two thousand and ten the percentage of US children who have autism increased by one hundred and nineteen point four (119.4) percent [6]. One in every seven hundred born babies tend to have Down syndrome [7].

Despite the fact that large amount of people around the world suffer from a development disability, there is no automated service provided to support and improve their cases. Many Institutes are available worldwide, yet in Egypt and other developing countries, the efficiency of the programs provided by most of the institutes are very inefficient and very expensive.

2 Methodologies

2.1 EQ Methodologies

As mentioned above, Emotional Intelligence (EI) is defined as the capability to classify, control, and evaluate ones emotion. There are two types of emotional intelligence. The first is ability EI (cognitive emotional ability). The second is trait EI (trait emotional self-efficiency). The difference between them is, that the first can be measured and tested and is more related to the cognitive ability. The second is more related to how one measure himself, and how capable he/she is in expressing their emotions, handling themselves, and others emotions [8]. Professor Emeritus of Psychology at UCLA,Â Albert Merhabian, states that [9]:

- 55% of our correspondence is absorbed from non-verbal communication, such as body language.
- 38% is para-etymological (doing with the way that you say something: tone, delays, pace, and so on).
- An insignificant 7% relates to the genuine words talked.

There are many therapists, institutes and private schools exist and help people who suffer from mental illness. In the developing countries children and adults with mental illness tends to be treated inappropriately. Furthermore, not all parents are capable of sending their children to a private school, university and institutes. That is why E-therapy and/or software is needed to help people who cannot afford a therapist. Some games where provided for people with autism such as the online game Ron gets dressed, which shows the user/patient that it is fine to change his/her daily routine [5]. As mentioned above there are computer games that help people with autism. Yet these games tend to help patients cope not to increase their intelligence. Most of the Institutes in England, United States, and Australia tend to use this five way methodology to increase someone's emotional intelligence. This five-way methodology are [10]:

1. Focus on his/ her emotions
2. Improve his/her body language skills and recognize that of others
3. Exercise empathy
4. Practice self-direction
5. Increase his/her social skills.

These methodologies could be illustrated in Fig. 1.

Fig. 1. Five-way methodology

Some institutes use other methodologies, which are called the nine-way methodology. It includes [11]:

1. Acknowledge the feelings and accept they are difficult to manage
2. Calm the patient mind down and speak to them. The moment they pay attention is the moment they tend to learn
3. Ask the patient questions that helps them get in the mode or to use the correct strategies
4. Analyze what is true and what is false from the patients point of view and correct it to what is known to be right or wrong
5. Always keep patient in a situation where he/ she need to search for a solution
6. Show patient that his therapist is not judgmental and is respectful
7. Be diplomatic and do not force consequences unless it is necessary
8. Reflect on situation carefully
9. Learn about the verbal and body language clues.

2.2   IQ Methodologies

There are different perspectives to increase persons IQ. It is believed that being on a good diet and daily exercise will increase one's Intelligence. It is known that a healthy brain is in a healthy body. Despite that, institutes do include training exercise; it is known that challenging the brain every day by solving a puzzle or a riddle playing intelligence game like Sudoku and Chess increases humans intelligence. Most of the therapist uses the following steps to boost intelligence [12].

1. Physical Exercise
2. Training mental capabilities
3. Follow a specific diet
4. Resolve IQ test questions.

Some institutes use the following methodologies as well [13].

1. Break a big problem into smaller problems so he/she can solve it
2. Using coping strategies
3. Seek support
4. Using motivation
5. Providing feedback.

As mentioned above it is believed that resolving IQ test helps boosting intelligence. However, it is important to clarify different types of IQ test. Here below are different types of such tests.

1. Verbal reasoning: which shows the capability of person in understanding
2. Numerical reasoning: which shows the person's ability to calculate
3. Logical reasoning: shows the person's ability to give a reasonable conclusion, show evidence to support his reasons, and to find the most simplest explanation to a situation
4. Spatial reasoning: shows the ability to solve spatial and complicated shapes, and differentiate between them.

3 Proposed Solution

The proposed Idea is a program that preforms an EQ test, detecting points of weakness and strength. The above five methodologies in Fig. 1 will be used to train the user. The developed program provides training on how to act in specific situations. Moreover, the developed program trains the user to learn how to behave, and how to act according to the way the program is repetitively proposing. Thereafter, the program will perform an EQ test repeatedly until the user reaches the normal EQ rates. The goal of this work is to test the Emotional intelligence using EQ tests. The program detect emotions using face detection and check the attitude of user. The program trains the users on how to react in specific conditions, in order to gather measurable information and build up an information base for disable people. As mentioned above the developed program should:

1. Scan the user's face
2. Analyze user's emotion
3. Recognize user's voice
4. Detect users order tune
5. Be Capable of answering the user
6. Detect users Eye to make sure user keeps Eye contact
7. Train the user to enhance his IQ and EQ.

The program code for the interactive dialogue that take place between the computer and user has been written in C#.

The developed program performs the following tasks:

1. Preform an EQ Test.
2. Preform Emotion detection using Image recognition. This has been achieved using the so-called library OpenCV in Matlab, C#, python, java.
3. Preform sound recognition. This has been done using C# by Microsoft.net library. One could also use java by sphinx. However, it is preferable to use C# or python to increase the Efficiency.
4. Preform text to speech conversion. This could be performed using C# or python. Although efficiency in java is higher if Mobrola (a java program application) is used.
5. Preform the algorithm which uses the above five features appearing in Fig. 1. This will combine the methodologies appearing in Fig. 1 and the Methodologies addressed above for further illustration as an example Fig. 2 provides an explanation.
6. Preform an EQ test again to detect the improvement rates and continue on improving until user reaches normal EI rates.

Here bellow is a brief explanation for the steps illustrated in Fig. 2:

1. User will open the program
2. The user will start dealing with the program
3. The user shows an unfriendly attitude
4. Program detects what was wrong

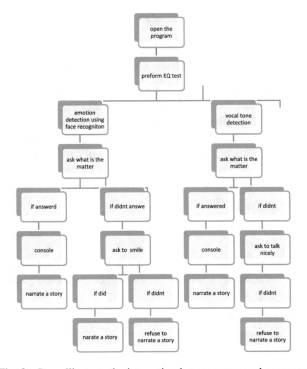

Fig. 2. Steps illustrate the interaction between user and computer

Fig. 3. Home Page UI

5. Program asks the user to change attitude
6. User changes his attitude
7. The program narrates a story regarding the user's misbehavior
8. Program will continue dealing with the user first after he changes his attitude.

In step 7 the program reacts like a human. It resamples a situation that the user will face in real life. Therefore, user will learn how to apologize.

The entity "console" in Fig. 2 means that the program will show empathy for the user. Such kind of empathy could be done using different answers, and machine learning algorithm. It will check its knowledge base to find a suitable answer depending on the situation.

3.1   System Constraints

There is only one admin for the system user cannot register as an admin (there is no admin registration).

System does not allow users to have same user name. Each user has to have a unique user name for security. User must enter his/her username, password, age when registering.

Password must be secured
System must refuse any insults
System must respond to any insult
Questions/stories/riddles are added ones (same question cannot be added twice).

3.2   System Details

The EQ, IQ tests, Stories and riddles are added by the admin to the systems Database. The admin has access to the application using a specific user name and password. When the admin log's in the system a page appears that shows the interface, where EQ, IQ tests, stories, and riddles are added refer to Fig. 4.

The Database is an object oriented database developed by adding DB4O libraries to the C# source code file. The database contains all users' questions, answers, stories, riddles and user grades. Grades are calculated depending on the users answer. The questions asked by the program has a Boolean answer either yes or no. the questions are added as a situation that the application had in a specific time for example the program will ask: "I had a colleague at work who was sick yesterday. I went to her mentioning that I am sorry and I asked her if she needed help. Do you think what I did was right". The user then needs to say yes as "Yes" is the right answer, if the user said yes then the grade for the user is 10 out of 10. Each question the application asks has a 10 marks score this scores are added. Each time the user accesses the application he/she will be asked different questions, told different stories and riddles. In case the user said no and the answer is no the application will tell the user, "I believe my friend that was the right answer when someone is sick we need to offer help". Using such method-ologies the user will not feel that he/she is in a test. The user will have the ability to register and login the user's information will be saved as well in the database.

Fig. 4. Admin UI

Sound recognition is maintained by adding oxford dictionary and Bing Speech API to enhance the accuracy of STT. Microsoft speech API is used for TTS. AIML Bot is used for conversation but the AI libraries has been enhanced.

System Security is maintained by hashing the user's password using C# SHA1Managed function.

3.3   Survey Analysis

A Google Survey was performed and the results are calculated as following:
In Q1, 2, 3, 6, 7, 8:

Pros: is determined with 7 or more
Cons: is determined with 4 or less
Neutral: is determined with 5 and 6.

In Q4, 5:

Pros: every day, because they are helpful
Cons: never use them, rarely use them
Neutral: every now and then, because I'm bored.

Refer to the references (Ref.* [14]*) to find the link for the survey (Table 1).

Table 1. Survey analysis

Question	Pros (%)	Cons (%)	Neutral (%)
Q1—Did you ever feel that you need to increase your intelligence?	68	12	20
Q2—Did you ever feel that you are not a sociable person?	24	52	24
Q3—Did you ever wish to become a more sociable person?	64	20	16
Q4—How often do you use your smart phone AI agent (siri/Cortana.. etc)?	40	48	12
Q5—What makes you use your phone Agent?	36	36	28
Q6—Would you like to have a program that increases your IQ and EQ?	80	4	16
Q7—Would you be excited to use such a program?	96	4	0
Q8—Do you think that such a program may effect user behavior?	84	4	12

4   Design and Implementation

Here below are some snapshots previewing the accomplished work. The snapshots reflect: home page, login and registration pages. Also the admin page and the screens where admin could add EQ and IE questions. The snapshots show also the screen for the admin to add stories and view users (Fig. 3).

The screens below show sample snapshots of the preformed conversation between user and program (Figs. 5, 6 and 7)

Fig. 5. Impolite interactive communication

Fig. 6. Polite interactive communication

Fig. 7. Polite interactive communication

5   Results

The Result of the Execution of the program indicate:

- High rates of accuracy in speech recognition. Almost 99% of speech is detected correctly
- High rates of Text to speech Conversion
- High rates of image processing results

- High effectiveness when used on users
- Slight Delay in speech recognition but for the sake of accuracy delay is tolerable.

6 Conclusion

The paper objective is a one-step forward to make a word free of discrimination, facilitate equal chances, and enhance the social and mental abilities of people with autism, Down syndrome, asparagus syndrome. In addition, the project helps parents and patients with low income to access a free developed program and have their therapy. The main scope of the paper is to use machine learning, speech recognition, image processing, and best methodologies available to support people with below average IQ and EQ. The developed program intend to increase EQ of autism users and the IQ of Down syndrome users. The program performs an EQ test and IQ test repeatedly until the user reaches the normal Intelligence rates in terms of IQ and EQ. The developed program acts like human during the interaction with the user. It will not tolerate the unusual or inappropriate attitude or unacceptable reaction of the user. Moreover, it will try to improve user's reaction by repetitively asked questions until the user's answer is satisfactory. Furthermore, it detects user emotions using face detection and check the user attitude. The program trains the users on how to react in specific conditions. In addition, the program instructs user on how to make friends in real life. The combination of methodologies and the repetitive EQ and IQ tests ensures the effectiveness and efficiency of the product.

7 Further Work

- Increase the Efficiency of Emotion Detection
- Increase the Performance
- Enhance the Graphical User Interface
- Enhance the Avatars Voice.

References

1. NAMI. (n.d.): Retrieved 27 Feb 2018, from: https://www.nami.org/Learn-More/Mental-Health-By-the-Numbers
2. Mental Retardation: the Background and the Issues: Retrieved 27 Nov 2017, from: http://www.achildwithneeds.com/disabilities/intellectual-disability/mental-retardation-the-background-and-the-issues/ (2012)
3. Parekh, R.: What Is Mental Illness? Retrieved 24 Mar 2018, from: https://www.psychiatry.org/patients-families/what-is-mental-illness (2018)
4. Dr. Poirier, K.: Teaching Emotional Awareness in Autism. Retrieved 27 Jan 2018, from: https://drkarinapoirier.com/teaching-emotional-awarness-in-autism (2015)
5. Transactions. (n.d.): Retrieved 5 Jan 2018, from: http://www.autismgames.com.au/game_trans.html

6. Facts and Statistics. (n.d.): Retrieved from: http://www.autism-society.org/what-is/facts-and-statistics/
7. Birth Defects: Retrieved from: https://www.cdc.gov/ncbddd/birthdefects/downsyndrome/data.html (2017)
8. Swenson, S.: Asperger's Syndrome and Emotional Intelligence. Retrieved 27 Feb 2018, from: https://www.goodtherapy.org/blog/aspergers-syndrome-emotional-intelligence-1002124 (2012)
9. Speaks, A.: Researchers Focus on Non-Verbal Autism at High Risk High Impact Meeting. Retrieved 28 Feb 2018, from: https://www.autismspeaks.org/science/science-news/researchers-focus-non-verbal-autism-high-risk-high-impact-meeting (2012)
10. Autism and Child Developmental Disorders in Africa. (n.d.): Retrieved 27 Feb 2018, from: www.myaspergers.net/what-is-aspergers/5-steps-to-emotional-intelligence/
11. Eskridge, R. (n.d.). Using the High IQ of ASD to Foster Emotional Intelligence. Retrieved 27 Feb 2018, from: https://aspergers101.com/using-high-iq-asd-foster-emotional-intelligence/
12. Meditation, O. (n.d.).: How to Increase Intelligence? 4 Powerful Methods. Retrieved 25 Jan 2018, from: http://operationmeditation.com/discover/how-to-increase-intelligence-4-powerful-methods/
13. Hartley, S. L., MacLean, W.E.: Retrieved 28 Mar 2018, from: https://www.ncbi.nlm.nih.gov/pmc/articles/PMC2838717/ (2008)
14. https://docs.google.com/forms/d/1rAI8OWvDzK_RafVJqdseyDtt9xX-yDWJt0Gt0sjvbDA/edit#responses
15. Burton, N.: A Brief History of Psychiatry. Retrieved 27 Nov 2017, from: https://www.psychologytoday.com/blog/hide-and-seek/201206/brief-history-psychiatry (2012)
16. IQ TEST TYPES YOU MUST KNOW. (n.d.): Retrieved 28 Mar 2018, from: https://www.iq-test.net/4-iq-test-types-you-must-know-16.html

Poster: From Unification to Self-identification of National Higher School in the World Educational Space: Comparative-Legal Research

Aleksandrov Andrei Yuryevich[1], Barabanova Svetlana Vasilievna[2],
Vereshchak Svetlana Borisovna[1], Ivanova Olga Andreevna[1(✉)],
and Aleksandrova Zhanna Anatolyevna[3]

[1] I.N. Ulyanov Chuvash State University, 15, Moskovsky Prospect,
Cheboksary 428000, Russia
{alexandrov_au, veres_k, public_law}@mail.ru
[2] Kazan National Research Technological University, 10, Popov Street,
420029 Kazan, Tatarstan, Russia
sveba@inbox.ru
[3] Chuvash State Agriculture Academy, 29, K. Marks Street,
Cheboksary 428000, Russia
zhanna978@mail.ru

Abstract. The relevance of the research topic is the urgent need for studying the higher education systems of countries having successfully retained authentic higher school, which is in high demand both in the educational services market and in the world labor market, considering their positive experience as a basis for further reform of the Russian higher education. The experience of countries that consistently fulfilled the Bologna process requirements, having lost their higher education self-identification is of interest for research, too. Working out a new concept of public policy towards Russian higher educational institutions and its place in the international educational space is the main goal of conducting a comparative legal study of higher education systems in the international educational space. The scientific novelty of the study is that the authors substantiate the desire for self-identification of higher education systems of states being a sign national and cultural identity. Loss of authenticity of higher education not only hinders the promotion of educational product on the world market, but it can also negatively affect its quality on the national markets. The authors put forward the evidence that preservation of national self-identification in education provides a variety of approaches to solving fundamental and applied scientific problems, provides the right of choice, gives way to extraordinary ideas and views.

Keywords: World educational space · State educational policy
Educational reform · Higher education institutions

M. E. Auer and T. Tsiatsos (Eds.): ICL 2018, AISC 917, pp. 210–216, 2019.
https://doi.org/10.1007/978-3-030-11935-5_20

1 Introduction

Consistent educational reform at all levels of education, which has been carried out in Russia for almost two decades, is aimed at integrating Russian education into the international educational space. The current situation is such that the place of Russian higher school on the world market of educational services continues to remain substantially below its real potential.

The current educational legislation establishes a sufficiently wide framework for conducting educational activities in the sphere of higher education with the preservation of academic freedom and authenticity of the national higher school. At the same time the diverse nature of the tasks set by the state, boosting formal criteria indicators for assessing the quality of educational services often have a negative impact on the development of higher education. Nearly all member-states of the Bologna agreement have faced these consequences. The founders of a single educational space on the territory of Europe quite easily abandon the established criteria for the unification of national higher education systems in favor of preserving their identity, while they insist on their implementation by other participants in the Bologna process. The relevance of the research lies in the urgent need to study the legal and regulatory framework for higher education systems of states that have successfully retained authentic higher school, which is in high demand both on the educational services market and on the world labor market, considering their positive experience as a direction for further reform of the Russian higher education. At the same time the experience of States that have consistently fulfilled the requirements of the Bologna Process and consequently lost authenticity of higher education, not having received recognition on the educational market, technology, skilled labor, is also of interest for research.

Development of a new concept of Russian state educational policy and higher education institutions' place in the international educational space is the main goal of conducting a comparative legal study of higher education systems of subjects of the international educational space.

2 Approaches and Methods

General scientific methods (dialectical materialistic method and methods of formal logic) and specific cognitive methods (comparative legal, formal legal, statistical, structural and functional methods) make up methodological basis for research. Interdisciplinary and systematic approach, historical principle and the research methods mentioned above allowed correlating Russian and foreign experience in implementing educational policy within the framework of the Bologna process and revealing effective practices for developing new approaches to the concept of Russia's educational policy aimed at developing educational system, boosting its efficiency and preserving best educational traditions.

3 Discussions

The authors reveal that foreign and domestic researchers show keen interest in organizational and legal support for the process of globalization of educational services. Thus, the problems of higher education in the context of globalization are considered in the works of foreign and domestic scientists—F. Altbach (Boston, 2009) [1], S. Marginson (2014) [7], I. V. Lazutina (Moscow, 2014) [6], K. Yu. Burtseva (Togliatti, 2014) [3], A. B. Ordabayeva (Kazan, 2013) [10], V. S. Senashenko (Moscow, 2013) [12] and others. The works of scientists in various branches of science devoted to the problems of increasing competitiveness of Russian universities on the world market of educational services are also noteworthy: G. F. Tkach, V. M. Filippov (Moscow, 2014) [13], V. V. Nasonkin (St. Petersburg, 2015) [9], by introducing ratings in the system of external and internal assessments of the quality of university—collections of research papers edited by A. L. Aref'iev (Moscow, 2010) [2]. The following research is devoted to the social consequences of educational migration to Russia: S. V. Dementieva (Tomsk, 2011) [4]. The peculiarities of educational services export as forms of international economic relations, as well as the regional tendencies of development of educational services export are examined in the work of A. V. Kosevich (Moscow, 2006) [5], T. M. Rogova (Rostov-on-Don, 2013) [11] and others.

Educational reform pace makes new demands and suggests criteria for assessing the effectiveness of higher education institutions in the context of determining its place on the national and world market of educational services. Therefore, it is often complicated to implement, and even more so, to evaluate practical recommendations of the reviewed scientific concepts based on long-term results. This trend is typical for both foreign countries and Russia. The negative consequences of total globalization leading to the destruction of authenticity of national higher education were revealed both by Russian and foreign researchers, who substantiate a drastic change in the legal support for educational reform.

The peculiarity of the resource base at the present stage, within the last three years (2014–2017), is the absence of a comprehensive study devoted to solving the problems of creating an optimal model of the university, capable of rapid change and practically applicable both within a large metropolitan university complex and regional classical university. Thus, the need for a comprehensive comparative legal research of the dynamics of educational legal relations within the framework of higher education remains relevant with every new stage of the international and domestic educational reform.

4 Results

As a result of legal research into the development of higher education system in Russia carried out by the authors for several years, the following ideas have been formulated and triggered the need to update state educational policy.

4.1 Competition of Different Educational Systems

Competition of different educational systems has become a key element of global competition, requiring constant updating of technologies, accelerated innovation, rapid adaptation to the demands and requirements of a dynamically changing world.

4.2 Combination of Domestic and Foreign Educational Practices

In Russia, an authentic system for training professional specialists in higher education has been created and is successfully functioning, the final product of which, especially in high-tech industries and heavy industry, is in demand and competitive on the domestic and world labor market. The primary objective in the field of higher education is not to replace authentic higher education system with foreign ones, but to adopt best practices of teaching techniques, extend sources of obtaining theoretical knowledge and practical skills, and to make students acquire a truly valuable ability to constantly update their knowledge throughout their career.

4.3 Competitive Advantages of Russian Educational Institutions in the International Market of Educational Services

It is necessary to note the significant competitive advantage of Russian educational institutions on the international market of educational services: that is the traditionally inherent fundamental nature of Russian education, high intellectual potential of students; comprehensive education, focused specializations alongside with comprehensive professional training practical focus of educational process in certain educational areas, developed methodology of educational work aimed at bringing up patriotism and citizenship in university graduates being the vanguard of society, etc.

4.4 Convergence of National Systems of the States Participating in the World Educational Market

The Bologna Process takes the lead in unification and harmonization of the world educational space in general and the European educational space in particular. Analysis of the reform of educational systems in foreign countries allows us to conclude that harmonization of educational legislation does not deprive educational systems of the Bologna member-states of their authenticity and national identity. This reform must be given positive evaluation and be used to modernize Russian educational space.

4.5 The Combination of the Harmonization of the Educational Environment and the Preservation of National University Traditions

In the process of implementing the provisions of the Bologna Process to unify educational systems of member-states, a tendency has emerged either to persistently preserve national university traditions or to return to them after undergoing the Bologna reforms. Thus, while acting as initiators of the Bologna process for the rapprochement and subsequent unification of educational systems to provide the freedom of labor movement within the European common market, core states of Western Europe have significant differences in legal regulation of higher education national systems.

E.g., British educational legislation is a set of codified and consolidated acts on education, which is a clear sign of strengthening the role of state in regulating educational relations. Changing the system of funding universities operating in the form of public corporations, the British Cabinet gradually limits their autonomy, which used to be the academic tradition of the English higher school. The system of higher education in the UK is represented by universities and polytechnic colleges. Universities can be collegiate and unitary. Collegiate universities are Oxford and Cambridge universities, which include 39 and 29 colleges respectively with a high degree of autonomy. British academic education is focused on fundamental scientific research. The adherence of universities to traditional academic work that runs counter to the ideas of the Bologna Process was manifested in the fact that internship is either not included in the curriculum at all or occupies insignificant place, although many universities require students to acquire practical experience on their own.

The most conservative of European core states is German education system. West Germany is a state with a decentralized government system in the sphere of education, in which the subjects of governance interact closely at different levels: the federation, the lands and communities (local authorities) performing various legislative, executive, coordinate and fiscal functions. The legal basis for higher education is the Federal Framework Law on Higher Education (Hochschul + rahmengesetz, HRG), adopted in 1998 following the Framework Law on Higher Education of 1976. It is characteristic of German legislation to vest federal legislation with working out educational standards and qualification requirements to staff involved in education process. In Germany, the principle of "academic freedom" has been proclaimed, according to which freedom is granted not only to higher education institutions, but to every student. In Germany, there is no rigid system of compulsory education. A German university student does not attend classes with his fellow students, but in accordance with the chosen specialty and its curriculum, he makes up his curriculum and organizes his time to meet the general requirements for this specialty. The term of study in public universities is from 4 to 10 years, in private universities—from 3 to 5 years.

A number of European states, such as France, Italy, or the Netherlands, abstain from centralizing the legal regulation of educational relations in their higher education systems, transferring it to local legal regulation, which is the basis for variability of approaches to university education and the expansion of autonomy in the scientific, pedagogical, administrative and financial spheres, increasing mobility on the labor market and educational services. It is obvious that in the core countries of Western Europe there is a difference in approaches to the legal regulation of educational relations, depending on the form of the state (territorial) structure, which is also characteristic for the regulation of other private and public-private spheres of relations. The specific nature of the legal system affects the system of education and management of higher education institutions. As a rule, in the Romano-Germanic legal systems there is a minimum degree of autonomy of universities.

4.6 Modern Directions of Development of Higher Education in Russia

Higher education system of the Russian Federation makes up for the shortcomings of the Bologna reform in the following areas: (1) introduction of three levels of higher education (bachelor's—master's degree—highly qualified staff training "Researcher. Research Instructor") while preserving traditional for the Russian science academic degrees—Ph.D. and Doctor of Science, as well as academic ranks—associate professor and professor; (2) preservation and restitution of specialty in the field of personnel training in strategically important sectors: defense, medicine, pharmacology, law enforcement, national security; (3) provision of mandatory disciplines in the Federal State Educational Standards; (4) development and implementation of exemplary educational programs ensuring uniformity in the implementation of standards and equal opportunities for consumers of higher educational services; (5) unification of educational and professional standards in order to meet the labor market demand and ensure employers' expectations for graduates of higher education institutions; (6) reinstitution of state (municipal) order for academic training in socially significant sectors.

5 Conclusion and Recommendations

In the course of the study the authors substantiated the desire for self-identification of higher education systems of states being a sign national and cultural identity. Loss of authenticity of higher education not only hinders the promotion of educational product on the world market, but it can also negatively affect its quality on the national markets.

Preservation of national self-identification in education provides a variety of approaches to solving fundamental and applied scientific problems, provides the right of choice, gives way to extraordinary ideas and views. We believe that it is necessary to update the concept of state educational policy aimed at preserving and promoting authenticity of the Russian educational product on the world market of educational services, science and technology.

References

1. Altbach, P.G.: Trends in Global Higher Education: Tracking an Academic Revolution, p. 149. Harvard Business School Press, Boston (2009)
2. Arefiev, A.L.: Current Condition and Prospects of Russian Education Export. Moscow (2010)
3. Burtseva, K.Yu.: Integration and Cooperation of Universities in Global Economy. The Vector of Science of TSU, vol. 2, no. 28, pp. 63–67 (2014)
4. Dementieva, S.V.: Social and Legal Aspects of Educational Migration in the Context of the Russian Educational Legislation Reform. Izvestia of Tomsk Polytechnic University, vol. 319, no. 6, pp. 166–172 (2011)
5. Kosevich, A.V.: Higher educational services export: international experience and Russian praxis. Abstract of diss. Ph.D. in Economics, Moscow (2006)

6. Lazutina, I.V.: Priorities and tools for Russia's international cooperation in science and education. Bulletin of International Organizations: Education, Science, New Economy, 1 (2014)
7. Marginson, S.: Russian science and higher education in the context of globalization (Russian science and higher education in a more global era). Education Issues **4**, 8–35 (2014)
8. Mitina, N.A.: International cooperation as a condition for improving the quality of higher education in Kazakhstan. Young Scientist **2**, 797–799 (2014)
9. Nasonkin, V.V.: National and regional dimension of state educational policy in the context of globalization (the EU's experience): dissertation. Dr. of Political Sciences, St. Petersburg, 387 p (2015)
10. Ordabayeva, A.B.: International legal regulation of cooperation between states in the field of education. Bulletin of KazNPU. URL: http://articlekz.com/article/11095 (2013)
11. Rogova, T.M.: Export of educational services of Russian universities: barriers and perspectives: dissertation. Ph.D. in Economics. Rostov-on-Don, 200 p (2013)
12. Senashenko, V.S.: On the participation of the Russian education system in international integration. Alma mater (Bulletin of Higher School) **3**, 7–13 (2013)
13. Tkatch, G.F., Filippov, V.M.: Administrative and Legal Practical Mechanisms for Ensuring Academic Mobility and Expanding the Export of Educational Services: Monograph, p. 288. PFUR, Moscow (2014)

Be Innovative or Fail

Rupert Erhart[✉], Günther Laner, and Helmut Stecher

Higher Technical College Innsbruck Anichstraße, Anichstraße 26-28,
6020 Innsbruck, Austria
{erhart,laner,stecher}@htlinn.ac.at

Abstract. A technician is mainly a technician but should also know about methods of marketing, innovation and how getting started with his idea in a short and practical way. A higher technical college with all its possibilities must be a place to learn. For technicians, the most important part is to try and make mistakes in the process. A mistake for a technician means that there is work for him, in other words an error gives a meaning to his oeuvre. The process of learning, searching for solutions and implementing these solutions is part of the typical work description of a technician. Standardization is another important part of a technician's work. Value Management for example as a method of innovation is standardized in EN 12973 [1]. Standardization with comprehensible methods is needed if you want to be certified for ISO 9000 (EN 27000) [2] and is also necessary in the case of product liability. The literature on the market doesn't give much material to spend time with, neither does it accord to the curricula of the higher technical college format. By creating two books we wanted to facilitate the teaching of our students in the fields of the newest market research methods, correct processes of strategy creation and modern methods of innovation such as the process of looking for problems, solving problems, analysing problems or the ideas portfolio, the Canvas Business plan—every step in the book has been tested and is now standardized material for education. We can teach the necessity of marketing and innovation in a practical manner. The outcome of the process can be implemented directly either with a start-up company of your own or as a further development in an existing company. The second advantage of the books is that they are not only books, but also advertising material for changing behavioural patterns.

Keywords: Innovation · Innovation methods · Ideas portfolio
Junior company · Marketing for technicians · Consumer behaviour
Standardized teaching methods

1 Introduction

Nearly everything we use in our daily life is invented and produced by human beings. For example, we use spoons for eating, we wear clothes and many other things which are designed and produced. Each of these goods originates from a technician's brain or is made by a machine. The question to ask is: is creating and assembling already everything? Is this the key to success? If we look into the past we can see that there are inventions that changed the world but the inventors passed away inglorious and did not

© Springer Nature Switzerland AG 2019
M. E. Auer and T. Tsiatsos (Eds.): ICL 2018, AISC 917, pp. 217–226, 2019.
https://doi.org/10.1007/978-3-030-11935-5_21

set up wealth with their discoveries,[1] others invented by coincidence (for instance penicillin[2]) and other inventions did not even come to the market. Also, there are inventions that came to the market and big players missed the chance and lost their position on the market (for instance Nokia) [7] as others jumped on the train.

It looks like the line between failure and success following innovation can be very thin and unpredictable. Still there must be people who can recognize a problem, find a solution for the problem and sell the solution.

2   The Duty Stapler

The questions about success posted above is comes natura with work of a technician. We also wanted to use this approved way in finding a solution for our problem with standardized methods for innovation in a successful way for the 21st century.

The second requirement in our duty stapler was to find methods which can be used for every field a technician can work on.

The methods must purport our curriculum.

The methods must relate to the education. A Technician is, first of all, a technician and should be an expert in his field and only secondly an expert in marketing, calculation Canvas, Business Plan, strategy, …

In the past, an engineer was a person in a white or blue coat that solves a problem, but no one thought that he could also be able to think about design, the costs or how to sell a solution. Times have changed and so did the teaching methods and the curriculum, but the thinking process of the people stayed the same. The sad truth is that—to most people—an engineer is a person who solves a problem and knows nothing about modern marketing, innovation and innovation methods. The solution proposed here also includes changing the current consumer pattern by eliminating these old-fashioned stigmas.

3   Initial Situation

We are a modern state in the European Union and still, not everything is perfect. We have nearly no goods that can be used infinitely. In consciousness of this fact it is essential to take care of the resources that have been given to us. This requires creative people that have been educated in innovation methods, finding new ideas, giving old things a new life and a new purpose. According to a statement of Antoine de Saint-Exupéry "If you want to build a ship, don't drum up people to collect wood and don't assign them tasks and work, but rather teach them to long for the endless immensity of the sea." [8] we have to follow and also teach this philosophy to the next generations.

---

[1] Michael Kelly submitted his idea of a thorn wire for a patent. It was patented with US. No. 84.062 improvement in metallic fences. The company which produced the thorn wire was called Thorn Wire Hedge Company [3, 4]. The production was to expensive. Glidden and Haish patented in 1874 a cheaper variant of Kellys Idea and are considered since then as the inventors of the barbed wire [5]

[2] Alexander Flemming invented 1928 penicillin [6].

As the life changed tremendously over time due to innovations as for example the big changes with motorcycles and, mobile phones, now digitalization urges us to educate people in creativity—also described by the circles of Kondratjew [9]. They don't have to make great inventions themselves (it is perfect if they do, but not necessary), it is useful for all of us if they think of the little things in our daily life. They must recognize problems and implement standardized methods as a solution. They must have methods with which they are able to quickly check if there is a chance in a market and a possible start-up which could lead the project to success. They don't need to be experts in cost calculation, marketing, strategy or have undergone a specialized university degree because we don't have time to teach all these things to everybody. They simply should be experts in their fields.

4   Solution

For this purpose, we created two textbooks [10, 11] according to the Austrian curriculum for higher technical colleges and tested the described methods in our possible space within Junior Companies and other projects.

4.1   Testing the Solution

4.1.1   Case Study: Digitalization in the Classroom

For the digitalization in the classroom we took 4 classes and tried to find answers how teaching will look like in 2025.
Facts:

We have standardized curriculums in mathematics, German, English and many other subjects.
There are many lessons every day with the same contents.
Teaching is manpower and so it is very expensive.

Going into digitalization will reduce the cost of manpower, the cost for the buildings, the cost for the parents. We did not go further at this time because we had to look how the system partners will react and if the system has a chance of being executed. If our analysis would state that there is a chance of being achieved, we can go on with the next step of the cost analysis. But at this time, it might be wasted money.

We took the Brainwriting method [12] for finding ideas how digitalization could change the world of teaching and how real the chance for digitalized teaching is including the influence of the system partners. The system partners are the student representatives, the parent representatives, the labor union of the teachers, the Ministry of Education. With the method suggestibility/influencing matrix invented by Malik [13] we analyzed the system partners we have to include, and which systems partners we can neglect.

As shown in Fig. 1 the active partner is the Ministry of Education. The costs for the manpower will go down and the government can spend saved money for other projects. But, how real is the realization of the project? The labor union will not agree because many teachers will lose their jobs. The parents will not agree because they don't know what their children are doing? Are they really learning, or watching TV, or playing or sleeping the whole day, or drinking alcohol, or … The parents have nearly no opportunity to control their children's behavior. The student representatives will not agree because there will be no more friends from and at school—no social interaction. Nobody will dare to change a running system. Every change in an organisation leads to distrust and blocking.[3] The only way for being successful is a long period of discussion and the integration of all the system partners. In order to successfully implement a new system in an already working and running system, in other words replacing a running structure, you need representatives that oversee the whole process. In Austria neither the legislative period nor the period of the parent's and/or the student's representatives bodies is long enough to go through with a change that needs time and patience to implement.

The ministry will not agree robot teachers because of the lack of social interaction, even considering the huge progresses that computer scientists have made in programming life-like robots. Other issues are the solution for conflicts which are a very important part in living together in peace and by todays understanding cannot be replaced by a written code. Teaching is more than presenting methods. Teaching is a philosophy. If the ministry had wanted to go this way, it could already have done so. It would be easy to learn from books and to come for exams to a special place and to learn at home and to eliminate the teachers. When students come to school for their first day they are so happy that they have the possibility to learn. If you look into their eyes it is like Christmas for them. The task for modern teaching, the philosophy of teaching with the words of marketing the formula of AIDA[4] [16] is to keep their interest, to take them by hand, to see through their eyes. This philosophy can't be taught by a book. A book can be a good tool but cannot replace a teacher.

The suggestibility/influencing method shows that there will be nearly no changes in the school system in the next years. By using this method to find out if there is a chance for a new product in a new market you will need one or in maximum 2 h and you will have the result. Should you spend time or money into this idea?

A company which wants to invest in technologies for digitalized classrooms and has made detailed studies about the cost advantages will think it will be a big market. They have only looked at one side and forgot the other. So, they took the wrong approach leading to the wrong decision. In fact, they had to close down this sector of their market after realizing that there is no potential anymore.

4.1.2   Case Study: Power of Customers

The power of customers can be seen in a project called 'HTL is(s)t gesund' (double meaning in German 'HTL is healthy' and 'HTL eats healthy'—We played with the words like modern advertising companies do all the time). All the foods considered

---

[3] Erhart (1993) showed in his work CIMMO (a model for integrating Computer Integrated Manufacturing in organizations successfully) the power of the system partners and how they can block changes [14]

[4] Advertising is made in 4 steps. A customer should run through these 4 steps in the hope that the product will be sold at the end of the process. These process is called A (Attention) I (Interest) D (Desire) A (Action) and was invented 1898 by Elmo Lewis [15]

Influence From	On	A	B	C	D	E	Sum(influencing matrix)
A		X	3	0	1	2	6
B		2	X	2	1	2	7
C		2	0	X	1	2	6
D		3	0	2	X	3	8
E		1	2	2	1	X	6
Sum (suggestibility)		8	5	6	4	9	

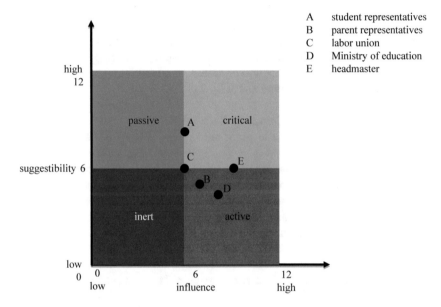

A   student representatives
B   parent representatives
C   labor union
D   Ministry of education
E   headmaster

Fig. 1. Example suggestibility/influencing matrix (For each other system partner you can give points from 0 to 3.0 means no or little effect, 3 means strong effect. The disadvantage of this method is that subjective ratings are given. The advantage is that the result is done within a few hours)

unhealthy were removed from our college's canteen since there simply was no demand for them from the student' sides. It was their decision.

The idea was created by using the method of brainstorming.[5]

4.1.3   Case Study: Diving as School Sports

Everything which is done at school must be backed by law. Diving is a dangerous sport. We wanted to test the "Lead User method" in getting rules for diving as school sports. We did not have the know-how at the college what can happen and what do we have to look for. We invited the Austrian Water Rescue Unit and a local Diving Centre to create with us the rules in compliance with the Austrian law. We set the rules and followed these during a week of diving in Croatia with our students.

---

[5] The method of brainstorming was invented by Alex Faickney Osborn and renewed by Clark [12]

4.1.4   Case Study: Cash App

The Austrian Federal Ministry of Finance made a law fighting fraud as most other countries have as well. Every company must have a machine which registers every single cash payment. If a company has a fixed location, you can elect from a variety of many solutions. But if you have no fixed location, for instance a taxi driver, you still have no solution. We tried to find a solution for this group of companies by an app on mobile phones. We used the method P (Problems) I (Ideas) S (Searching) A (Analysing). Looking for the ideas we made a duty stapler. The data must be stored safe and once the data has been put in into the mobile phone, there must be no way that it can be manipulated. Still there must be a possibility to change if it the user made a mistake during insertion of data. If the data is changed there must be a possibility that the Finance Ministry can follow these changes. The solution was an app which can be used on every mobile device. The data will be stored on a Host through WIFI. A functional prototype of this idea was constructed (Fig. 2).

4.1.5   Case Study: Junior Company Zero Waste

The Junior Company Zero Waste produced organic waste baskets from used paint buckets. Normally the paint buckets are getting disposed. For us, the used paint buckets were our raw material. For one organic waste basket we needed 2 used paint buckets and one corresponding lid. The basket below got a valve and the upper basket got holes drilled into. The two baskets were put together (shown in Fig. 3).

Then the containers got filled with organic waste and ferments. At the end of the process we got earth and fertilizer. We did not know if the idea had a chance for being successful on a market. We tried to get the answer with the method of the ideas portfolio.

As seen in Fig. 4: result can be clearly seen in the green section, thus investing is the right choice.

All baskets we produced could be sold in a brief period of time.

5   Result

There must be a market for the product, it can be placed in. At higher technical colleges in Austria we try to teach our students the process of searching and solving problems with different methods and to check if there is a chance of being successful in a market. We teach our students to critically question to the customer's position, to look through the customer's eyes, to understand the needs of the customer and to try to find a solution that solves the problem of the customer and to sell him the solution. We show them how to find the unique selling point and to build a border to the competition so that the main advantage of the product, the differentiation, is hard to copy. We teach them how to communicate the advantage in an effective way with marketing instruments. Modern teaching of technicians needs special methods. They cannot only be taught the theory and methods. They must also have the chance to try out the methods in projects in a simulated space. For this purpose, we have several units where a small business plan can be created within a few hours, ideas can be found within a small group, a strategy can be checked without long planning and research. Technicians must

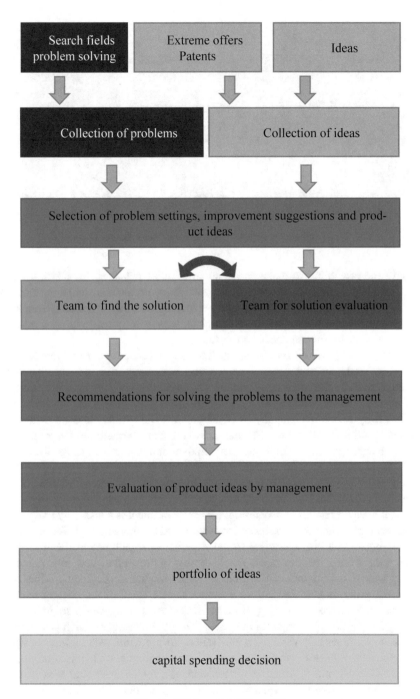

Fig. 2. PISA model [10, p. 55]

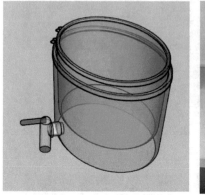

Fig. 3. Model and prototype of the organic waste basket [10, p. 24]

also have the possibility to make variants and to make mistakes. Mistakes, to technicians, mean that there is something missing, which means there is something to do. We need curricula which allow to teach and try standardized methods. Then we are able to undergo a quick check whether there is a chance for being successful in a market or not. That's what technicians need in the future.

Erhart [10, 11] created two books to teach our students about the newest market research methods, correct process of creating a strategy and modern methods of innovation such as the process of looking for problems, solving problems, analysing problems or the ideas portfolio, Canvas and Business plan—every step in the book has been tested by a Junior Company or other projects from different fields and is standardized material for education. Standardization is an important part for a technician. Value Management for example as a method of innovation is standardized in EN 12973. Standardization with comprehensible methods is needed if you want to be certified for ISO 9000 (EN 27000) and is also necessary in the case of product liability.

Using the facts that have been mentioned earlier, one can tell that there are only few things that have no relation to technology. Because of that it is a necessity that we teach technicians more about the process of innovation and occupy this topic for us. Now we have two books that are, according to the Austrian Curriculum, specifically for technicians with whom we can teach the necessity of marketing and innovation in a practical way. The outcome of the process can be implemented directly either with a start-up company of your own or as a further development of an existing company. These two books are not only textbooks for students, they can also be useful in changing the picture of technicians in society. The attitude theory, which is a part of the theories of consumer behaviour, says that rational instruments with arguments will fail in this case. The advantage in this case of the "marketing book" is that if anyone sees the book with the title "marketing for technicians", looks inside and sees that it's a textbook approved by the Austrian Federal Ministry of Education, will think about his behaviour and will change it. The second book "Innovation for technicians" teaches all the methods of innovation management in compliance with the Austrian curriculum in a standardized way. We are prepared for the future.

Example: Case Study: Bokashi Basket

Potential for success:

Novelty level	4
Competitive situation	4
Strategic position of success	4
Market volume	1
Market growth	5
Contribution margin	2
Calculation: 4×4×4×4×1×5×2 =	640

Risk:

Investment amount: 0×0. 6 =	0
Know-how marketing: 3×1. 2 =	3. 6
Know-how technology: 2×1. 4 =	2. 8
Patent: 3×0. 8 =	2. 4
Calculation: 0 + 3. 6 + 2. 8 + 2. 4 =	8. 8

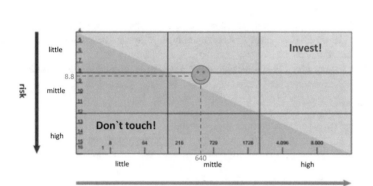

→ invest

Fig. 4. Example Ideas Portfolio [10, p. 60] (The ideas portfolio uses chances and risks of ideas and makes a comparison. The potential for success is divided in categories (Novelty level, Competitive situation, Strategic position of success, Market volume, Market growth, Contribution margin). For each category points are issued. For the "risk" categories (Investment amount, Know-how marketing, Know-how technology, Patent) also points are issued. The result is the position in the matrix. In the green field the chances are higher than the risks. The advice is to invest. When the idea is situated in the red field the advice is to stop the commitment [10, pp. 57–60])

References

1. EN 12973. (19-05-2018). Value Management
2. ISO 9000. (19-05-2018). Quality Management
3. Petrosky, H.: An Engineer's Alphabet: Gleanings from the Softer Side of a Profession, p. 217. Cambridge University Press, Cambridge (2011)
4. Reuss, M.L: The Illusory Boundary: Environment and Technology in History. University of Virgina Press, Charlottesville and London (2010)
5. https://historyrat.wordpress.com. (19-05-2018). https://historyrat.wordpress.com/2011/09/04/barbed-wire-barons-glidden-ellwood-and-haish/
6. https://de.wikipedia.org. (19-05-2018). https://de.wikipedia.org/: https://de.wikipedia.org/wiki/Antibiotikum#cite_note-8
7. https://www.ft.com. (19-05-2018). https://www.ft.com/content/e3cc3338-9449-11e1-bb47-00144feab49a
8. http://www.elise.com. (19-05-2018). http://www.elise.com: http://www.elise.com/quotes/antoine_de_saint-exupery_-_if_you_want_to_build_a_shipabgerufen
9. https://www.kondratieff.net. (19-05-2018). Von https://www.kondratieff.net: https://www.kondratieff.net/kondratieffzyklenabgerufen
10. Erhart, R.: Entrepreneurship and Innovation for Technicians. TÜV, Brunn (2018)
11. Erhart, R.: Marketing for Technicians. TÜV, Brunn (2018)
12. Clark, C.: Brainstorming. Moderne Industrie, Landsberg (1967)
13. Malik, F.: Führen. Lesiten, Leben: Wirksames Management für eine neue Zeit. Campus Verlag, Frankfurt/Main. p. 37 (2006)
14. Erhart, R.: CIMMO Computer Intelligent Manufacturing. Innsbruck (1993)
15. Schröder, S.: (19-05-2018). https://vorlesungen.info. https://vorlesungen.info: https://vorlesungen.info/node/1120
16. https://en.wikipedia.org/wiki. (19-05-2018). https://en.wikipedia.org/wiki/AIDA_(marketing)

Using the Jupyter Notebook as a Tool to Support the Teaching and Learning Processes in Engineering Courses

Alberto Cardoso[✉], Joaquim Leitão, and César Teixeira

CISUC, Department of Informatics Engineering,
University of Coimbra, Coimbra, Portugal
{alberto,jpleitao,cteixei}@dei.uc.pt

Abstract. Teaching and learning processes can benefit from the use of online resources, enabling the improvement of teachers and students productivity and giving them flexibility and support for collaborative work. Particularly in engineering courses, open source tools, such as Jupyter Notebook, provide a programming environment for developing and sharing educational materials, combining different types of resources such as text, images and code in several programming languages in a single document, accessible through a web browser. This environment is also suitable to provide access to online experiments and explaining how to use them. This article presents some examples of online resources supported by Jupyter Notebook, in subjects of an Informatics Engineering course, seeking to contribute to the development of innovative teaching methodologies.

Keywords: Online resources · Jupyter notebook · Teaching · Learning Engineering courses

1 Introduction

Nowadays, the Internet and the Information and Communication Technologies offer an appropriate environment to improve the teaching and learning processes, supported by innovative teaching methodologies and online resources. Collaborative work is of great importance, especially in engineering courses, where the teamwork and the interaction between teachers and students are very frequent.

In this context, the necessary tools are available to develop and share resources online, providing teaching and learning conditions that respond to current societal challenges [1]. Thus, teachers should feel challenged to tailor their classes to address the new student behavior, leveraging existing tools and technologies to prepare educational materials that can contribute to developing collaborative resources and enhancing interaction among different actors.

Taking into account that several subjects are taught using a particular programming language, there are different tools that can be very useful for the development of online resource for theoretical and practical classes.

© Springer Nature Switzerland AG 2019
M. E. Auer and T. Tsiatsos (Eds.): ICL 2018, AISC 917, pp. 227–236, 2020.
https://doi.org/10.1007/978-3-030-11935-5_22

Jupyter[1] is an open-source project that can be used to support interactive data science and scientific computing across several programming languages as, for example, Python and MATLAB. This project grew out of the IPython project [2], which initially provided an interface only for the Python language and continues to make available the canonical Python kernel for Jupyter.

Therefore, Jupyter notebooks give the necessary and adequate support to implement the concept of "Literate Computing" and "Reproducible Research", providing tools to develop and make available narratives anchored in a live computation, which offer the possibility of communicating knowledge and research based on data and results in a readable and replicated way. Notebooks are accessed through a web browser and are designed to support the workflow of scientific computing, from an initial interactive exploration phase to publishing a comprehensive record of computation [3].

A Jupyter notebook is organized into cells, which can include text, video, images and code or math operations, which weave together to produce an interactive document. This approach corresponds to an evolution of the interactive shell or REPL (Read-Evaluate-Print Loop) that has long been the basis of interactive programming [4, 5]. Thus, teachers can provide students with self-contained Jupyter notebooks, representing educational resources that can improve teaching and learning, providing a suitable environment for improving student engagement and motivation.

Knowing that laboratory classes using experimental environments play a crucial role in engineering courses, teachers can take advantage of using this collaborative tool to share educational materials and to provide access to remote and virtual experiments. The use of online experimentation represents a great opportunity to support teaching and learning activities complementing the laboratory activities and motivating students to perform practical works, acquiring knowledge, understanding the concepts and achieving experimental skills in a flexible way [6].

This article presents some examples of online educational resources supported by Jupyter Notebook in subjects of an Informatics Engineering BSc course, seeking to contribute to the development of innovative teaching methodologies.

2   The Jupyter Notebook

Considering the Jupyter project, notebooks are open with a web browser, making practical the use of the same interface running locally like a desktop application, or running on a remote server. Thus, notebooks provide a programming environment that can be shared and offers many advantages for students and instructors as these are free and open source software, as well as it can be used to promote teaching/learning based on innovative approaches and reproducible research [7]. For example, a teacher can provide the notebooks on a web server and easily give students access. Each notebook file is documented in the JSON format with the extension '.ipynb', making easier the process to write, manipulate and share these files.

---

[1] http://jupyter.org/ (last accessed: July 20, 2018).

Therefore, notebooks can be used to record a computational piece of code in order to explain it or a given subject in detail to others, and a variety of tools help users to conveniently share notebooks [3].

In association with Jupyter, there are several services available, such as "*nbconvert*", "*nbviewer*", "*binder*" or "*nbgrader*", which complement the project tools and make it more powerful. *Nbconvert* converts notebook files into different file formats, including HTML, LaTeX and PDF, making them accessible without needing any Jupyter software installed. *Nbviewer* (https://nbviewer.jupyter.org/) is a hosted web service that provides an HTML view of notebook files published anywhere on the web. These HTML views have a major advantage over publishing converted HTML directly because they link back to the notebook file, given the possibility to interested readers to download, run and modify it themselves.

Binder (http://mybinder.org/) allows sharing of live notebooks including a computational environment in which users can execute the code. With this, authors can publish notebooks, for example on GitHub, along with an environment specification in one of a few common formats in an interactive and immediately verifiable form [3].

nbgrader, is a tool designed to facilitate the grade process of notebooks, by providing an interface that blends the autograding of notebook-based assignments with manual human grading. Additionally, it streamlines and simplifies the process of assignment creation, distribution, collection, grading, and feedback [8].

Considering this set of tools, among others, the scientific code, its computational environment, data and execution conditions can be maintained in a *git* repository, enabling its maintenance and reusability. Consequently, numerous papers have been published supported by notebooks to reproduce the analysis or the creation of key results. Moreover, there are already available various examples of books as a collection of Jupyter notebooks, such as "Python for Signal Processing" [9] or "QuTiP Lectures as IPython Notebooks" [10].

This work aims to show that Jupyter notebooks can be used to provide educational materials using two different programming languages, Python and MATLAB, and to support the interaction with remote and virtual experiments, taking advantage of their services and functionalities.

3 Data Analysis and Transformation Subject

Data Science has become a great research topic and a very important field in every engineering subject. It is an interdisciplinary field that uses scientific methods, processes, algorithms and systems to extract knowledge and insights from data in various forms, both structured and unstructured, similar to data mining [11, 12].

Data Science employs techniques and theories, drawn from many fields within the context of mathematics, statistics, information science, and computer science, to analyze and understand phenomena with data, trying to unify diverse areas, such as statistics, data analysis, machine learning and their related methods.

The resources presented in this article were considered in the subject "Data Analysis and Transformation" (DAT) of the 2nd year (2nd semester) of the Informatics Engineering B.Sc. course at the University of Coimbra, Portugal.

This curricular unit corresponds to a broad and integrative view of the tools of data analysis, modeling and transformation for the computational treatment of phenomena and dynamic processes of the material and immaterial world. This subject is presented in specific contexts of Informatics Engineering and its interconnection with other domains.

The DAT subject aims to develop the following competencies:

- Understanding and interpreting dynamic and real-world phenomena and processes;
- Ability to analyze data supported by computational tools;
- Deepening the mathematical reasoning for extracting information;
- Ability to solve specific problems in the analysis and transformation of uni and multidimensional data in the field of time, space and frequency.

DAT also contributes to the development of skills in group work, critical thinking, argumentative capacity, research and self-learning.

The program of the DAT subject includes the following topics:

- Chapter 1: Introduction
  - Reasons for Data Analysis and Transformation
  - Analysis of dynamic phenomena
  - Environment-computer interaction
  - Signals and their properties
- Chapter 2: Time Series Analysis
  - Introduction
  - Stationary and non-stationary linear processes
  - Models identification
  - Estimation and forecasting
- Chapter 3: Linear Systems
  - Discrete-Time Systems
  - Z Transform
  - Properties of systems (causality, convolution, stability, ...)
  - Systems Analysis (feedback, transfer function, impulse response)
- Chapter 4: Fourier Transforms
  - Fourier Transforms (FS, FT, DTFT, DFT)
  - Frequency response and sampling theorem
  - Resolution and Noise
  - Digital filters
- Chapter 5: Time-Frequency Analysis
  - Short-Time Fourier Transform (STFT)
  - The dilemma of the uncertainty principle
  - Wavelet Transform (multi-resolution analysis and scaling function)
  - Analysis of non-stationary time series
- Chapter 6: Other Transforms
  - Karhunen-Loéve Transform
  - Other transforms of orthogonal basis
  - Transformations for data compression
  - Examples in various application domains.

The teaching activities comprise theoretical and practical classes, in which students have to carry out various practical works. Some of them involve solving theoretical-practical problems and carrying out simulation exercises and some others include experimental tasks, namely through the interaction with lab systems.

In one of these practical works, students should interact with a remote lab, available online at the Laboratory of Industrial Informatics and Systems of the Department of Informatics Engineering of the University of Coimbra, to acquire data from a real setup. Thus, they can use real data to analyze a given phenomenon and extract knowledge.

4 Examples of Jupyter Notebooks

All practical works of the DAT subject include solving exercises using the MATLAB or Python programming languages. A simple exercise is as follows:

– Consider the following continuous-time function $f(t) = \sin(2\pi t) + \sin(\pi t)$.

   (a) Create a time vector t from -10 to 10 s that gives values for $f(t)$, considering a sampling period of 0.01 s. Indicate the size of the time vector.
   (b) Plot the evolution of $f(t)$ as a function of time.

Figure 1 presents a cell of a Jupyter notebook programmed in MATLAB to illustrate the resolution of the exercise using the symbolic calculus. Figure 2 illustrates the resolution of the same exercise using Python as the programming language.

Other exercises consider the analysis of time series based on data acquired remotely. As an example, a remote lab to measure precipitation by a udometer can be considered to generate time series. In order to remotely monitor the precipitation, a web server is running in a Raspberry Pi computer, which sends data to a remote computer that acts as station manager, using an Internet connection. This computer is used to manage the system and to store and deliver data through web services. For this, an API with various endpoints was developed offering different possibilities of interaction using RESTful web services, configured as URL commands. These web services include commands to fetch data in real time, in a given period of time or between two specific dates [6].

The following commands exemplify the way the manager station can be inquired to get data about the amount of precipitation, acquired by the udometer connected with a given web server (identified by <node>):

• Fetch data registered in a given hour:
   – .../api/data/<node>/<year>/<month>/<day>/<hour>/
• Fetch data registered between two specific dates, specifying the sampling period (<search_type> as hour or minute):
   – .../api/data/search/<search_type>/<node>/<date1>/<date2>/

To use the available web services, it is necessary to send requests (commands) to the web service endpoint URL and receive responses from it. The services receive and

```
In [9]:  syms t  % t=sym('t')
         ft=sin(4*pi*t)+sin(pi*t)        % ft=sym('sin(4*pi*t)+sin(pi*t)')
         t=-10:0.01:10;
         f1=double(subs(ft));
         plot(t,f1,'-or')
         xlabel('tempo [s]')
         ylabel('f(t)')
         title('Grafico de f(t)')

         ft =
         sin(pi*t) + sin(4*pi*t)
```

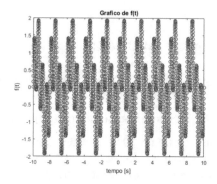

Fig. 1. A cell from a Jupyter notebook to solve an exercise using MATLAB

send data in JavaScript Object Notation (JSON). The requests are sent in a web browser and the responses are displayed within a webpage.

Considering the environment provided by the described remote lab, the use of a Jupyter notebook to interact with the remote udometer arises as a natural and very interesting option, where it is possible to show as the web services can be used to obtain data about the precipitation in the location where the udometer is positioned.

Moreover, the Jupyter notebooks offer the ideal setting to describe the concepts related with the observed time series, using contents as text or images, and program different approaches and methodologies to analyze, process and visualize data fetched from the remote system, using one of the several programming languages available.

To give an example of how to get and visualize data (precipitation in mm) fetched from the remote udometer presented in the previous section, Fig. 3 presents two cells of a Jupyter notebook (available at GitHub[2]) programmed in Python 2.7 (it can also be programmed in other versions, such as Python 3.x).

These cells are simple but exemplary instances of code in Python that use the "get" function from the "requests" module to get the JSON object returned by the request defined by the web service endpoint URL and the "matplotlib" module for data visualization purposes.

---

[2] https://github.com/albjlcardoso/python_examples/blob/master/Udometer_online_v2.ipynb   (last accessed: July 20, 2018).

```
In [8]:  import sympy as sp
         import matplotlib.pyplot as plt

         t=sp.symbols('t')
         ft=sp.sin(2*sp.pi*t)+sp.sin(sp.pi*t)

         tt=np.arange(-10,10.01,0.01)
         f=sp.lambdify(t,ft,"numpy")
         f1=f(tt)

         plt.plot(tt,f1,'-or')
         plt.xlabel('tempo [s]')
         plt.ylabel('f(t)')
         plt.title(u'Gráfico de f(t)')
         plt.show()
```

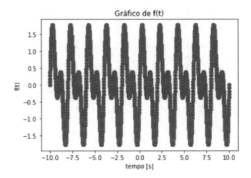

Fig. 2. A cell from a Jupyter notebook to solve an exercise using Python

```
In [1]:  from requests import get
         import matplotlib.pyplot as plt
         r=get("http://hydra.dei.uc.pt/station/manager/api/data/DEC1/2018/01/")
         d=r.json()
         data=d.get('data')
         e=data.get('events')
         days=range(1,len(e)+1)
         plt.figure(figsize=(20,5))
         plt.plot(days,e,'-o', label="January, 2018")
         plt.xlabel('Days')
         plt.ylabel(u'Precipitation [mm]')
         plt.title(u'Diary precipitation in Pólo II of UC, Coimbra, Portugal')
         plt.legend()
         plt.show()
```

```
In [2]:  from requests import get
         import matplotlib.pyplot as plt
         r=get("http://hydra.dei.uc.pt/station/manager/api/data/search/hour/DEC1/2018-03-01 00:00/2018-03-0
         1 23:00/")
         d=r.json()
         data=d.get('data')
         e=[]
         for elem in data:
             e.append(elem.get('value'))
         hours=range(0,len(e))
         plt.figure(figsize=(20,5))
         plt.plot(hours,e,'-o', label="March 1, 2018")
         plt.xlabel('Hours')
         plt.ylabel(u'Precipitation [mm]')
         plt.title(u'Hourly precipitation in Pólo II of UC, Coimbra, Portugal')
         plt.legend()
         plt.show()
```

Fig. 3. Example of two cells of a Jupyter notebook to acquire data from the remote lab

Figure 4 shows the plot representing the diary precipitation, in mm, occurred during January, 2018, at the "Pólo II" campus of University of Coimbra. Figure 5 shows the hourly precipitation, in mm, occurred on March 1, 2018, at the same place.

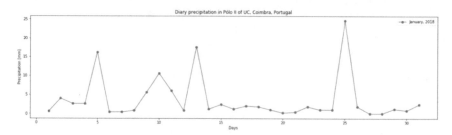

Fig. 4. Result of running the first cell of the Jupyter notebook in Fig. 3

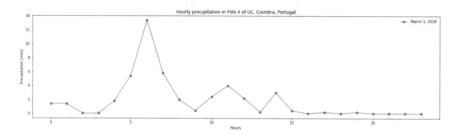

Fig. 5. Result of running the second cell of the Jupyter notebook in Fig. 3

Therefore, the notebooks represent a very useful tool that can be used in different teaching and learning contexts. Given the large flexibility to structure the content of the notebooks, as well as their characteristics that favor collaboration and sharing, they can support different teaching/learning approaches and strategies, taking advantage of the interaction with the remote/virtual laboratories, which can motivate students and improve their performance.

Based on these illustrative examples, several other notebooks can be developed with contents adjusted to the specific learning context, including text, images and code using data fetched from the remote system and to present and exemplify the use of different objects in Python and to develop various algorithms for data analysis and processing.

The internet-based resources considered for the DAT subject include also tutorial information, guidelines to carry out the experiments using the experimental setups and a set of quizzes for self-evaluation and assessment of students. They are implemented using the Moodle platform and each quiz contains different types of questions (multiple choice, matching, true/false, numerical and short answers, etc.), obtained in a random manner from a database with questions grouped by topics. For each practical work, students can perform a quiz with different questions about the topics of each work for self-assessment. Furthermore, they have to accomplish online other quizzes for evaluation purposes.

5 Conclusion

This article presents the use of Jupyter notebooks as tools that provide a programming environment to develop and share scientific contents and that can promote the access to remote and virtual labs. The use of this type of resources, where text, images and code can be combined in a harmonious and comprehensible manner, can contribute to explore innovative approaches and improve teaching and learning activities in different high education courses, especially in engineering subjects.

In particular, this article presents some examples of Jupyter notebooks using the programming languages Python and MATLAB to solve exercises and to interact with a remote laboratory, which can be useful in different learning contexts, namely in engineering subjects.

The use of notebooks can represent an important and very interesting tool in theoretical and practical classes, contributing for the enhancement of the students' learning process and their experimental skills.

This approach is being used in subjects of some engineering courses of the Faculty of Sciences and Technology of the University of Coimbra and the preliminary results, which show increased student participation and improved performance, are very encouraging to continue to use and to develop this type of educational resources for teaching activities in different areas of engineering courses and to share research results.

Acknowledgements. This work has been partially supported by the Portuguese Foundation for Science and Technology (FCT) under the project UID/EEA/00066/2013 and the Ph.D. grant SFRH/BD/122103/2016.

References

1. Garrison, D.R., Vaughan, N.D.: Blended Learning in Higher Education: Framework, Principles, and Guidelines. Wiley (2008)
2. Pérez, F., Granger, B.E.: IPython: a system for interactive scientific computing. Comput. Sci. Eng. **9**(3), 21–29 (2007)
3. Kluyver, T., Ragan-Kelley, B., Pérez, F., Granger, B., Bussonnier, M., Frederic, J., Kelley, K., Hamrick, J., Grout, J., Corlay, S., Ivanov, P., Avila, D., Abdalla, S., Willing, C.: Jupyter development team: Jupyter notebooks—a publishing format for reproducible computational workflows. In: ebook "Positioning and Power in Academic Publishing: Players, Agents and Agendas", pp. 87–90 (2016)
4. Iverson, K.E.: A Programming Language, New York, NY. Wiley, USA (1962)
5. Spence, R.: APL Demonstration, Imperial College London (1975). Available from: https://www.youtube.com/watch?v=_DTpQ4Kk2wA. Last Accessed 20 July 2018
6. Cardoso, A., Leitão, J., Gil, P., Marques, S.M., Simões, N.E.: Using IPython to demonstrate the usage of remote labs in engineering courses—a case study using a remote rain gauge. In: Proceedings of the 15th International Conference on Remote Engineering and Virtual Instrumentation (REV2018), pp. 683–689 (2018)

7. Raju, A.B.: IPython notebook for teaching and learning. In: Natarajan R. (eds) Proceedings of the International Conference on Transformations in Engineering Education. Springer, New Delhi (2015)

8. Hamrick, J.B.: Creating and grading IPython/Jupyter notebook assignments with NbGrader. In: Proceedings of the 47th ACM Technical Symposium on Computing Science Education—SIGCSE'16, pp. 242–242 (2016)

9. Unpingco, J.: Python for Signal Processing. Springer (2014). Available from: https://github.com/unpingco/Python-for-Signal-Processing. Last Accessed 20 July 2018

10. Johansson, R.: QuTiP Lectures as IPython Notebooks. Available from: https://github.com/jrjohansson/qutip-lectures. Last Accessed 20 July 2018

11. Dhar, V.: Data science and prediction. Commun. ACM **56**(12), 64 (2013)

12. Leek, J.: The key word in "data science" is not data, it is science. Simply Stat. (2013)

Gruendungsgarage

A Five-Year-Experience at Graz University of Technology

Martin Glinik[✉]

Institute of General Management and Organization, Graz University of
Technology, 8010 Graz, Austria
`martin.glinik@tugraz.at`

Abstract. This semester the Gruendungsgarage starts its 10th volume, an
occasion to analyze the development of this special education format in
entrepreneurship. The Gruendungsgarage is a cooperation between the
University of Graz (KFU) and the Graz University of Technology (TUG).
Originally it was designed as an interdisciplinary and interuniversity teaching
format offered as an elective course where students were able to work on their
own real business ideas while getting assistance from university experts as well
as from successful practitioners with a wide range of experience. Because of its
popularity and constant improvement, Gruendungsgarage became more and
more professional and over time changed its format into an academic start-up
accelerator. Setting up university start-up accelerator programs has been a
worldwide proven model that enables students to intensively develop their
business idea into a marketable product or service in just a few months. Since
the winter semester 2017/18 Gruendungsgarage has admitted not only students
but also scientific staff. University and postdoctoral researchers of the engineering faculties benefit from a wide range of Gruendungsgarage services, as
technical scientists with high-tech innovation get supported by experts and
entrepreneurs, creating a business model suitable to their scientific discoveries.
Because of this comprehensive service offer, one of the two located universities
of applied science recently showed interest in becoming a partner of the
Gruendungsgarage.

Keywords: Entrepreneurship lecture · Best practice
Academic start-up accelerator

1 Introduction

Five years ago, the Gruendungsgarage started as an interdisciplinary teaching format
offered as an elective course in order to provide support for preparing innovative and
knowledge-based start-up projects developed by students of the University of Graz
(KFU). A volume extends over one university semester and is divided into several
teaching phases. After two successful volumes, the University of Technology
(TUG) became a partner of the Gruendungsgarage creating a platform for students with
different fields of interest and encouraging exchange among students. Because of this

M. E. Auer and T. Tsiatsos (Eds.): ICL 2018, AISC 917, pp. 237–244, 2019.
https://doi.org/10.1007/978-3-030-11935-5_23

partnership, engineering students became aware of the Gruendungsgarage which increased the number of technology-oriented start-up projects hosted by the Gruendungsgarage. This hands-on teaching format reflects the entrepreneurship reality in a controlled environment. Practical tasks are mainly used to teach entrepreneurship, rather than a reliance on traditional classroom teaching. Program leaders originate from both academia and industry, with the aim of blending theoretical and practical knowledge [1]. For this reason, the teaching units are always outside a typical lecture room to enable a creative work climate. As a result, the participants can experiment with their business ideas get them ready for market and prepared for founding a start-up company. Another characteristic is the group-based learning method, which allows the teams to get an idea about how it is to work with a new venture team [2]. This experience is particularly important for each individual to define different roles inside the team, understand the mindset of the others, and learn on how to handle conflicts when they occur. As a part of the strategic project "entrepreneurial university" at the TUG, the Gruendungsgarage expanded its target audience—this means that services are not for students exclusively. Since the winter semester 2017/18, scientific staff is also eligible to apply proposing their business ideas as well as benefiting from the same kind of support and network use. To draw the attention of the scientific staff, the Gruendungsgarage was promoted in the staff magazine "TU Graz *people*" and integrated into the postgraduate training catalogue of the TUG to make it more visible for a new target audience. As an incentive, the scientific staff gets an official confirmation when they attend any program that is credited to them as a further education. Recently the Gruendungsgarage also addressed its services to scientists who are admitted to the national funding program and called "Spin-off Fellowship". The fellows get financial support to exploit and commercialize their intellectual property for up to 18 months. Besides the further development of their research results, the scientists are required to develop a business plan that describes their strategy to start a business after completing the fellowship program. The program leaders of the Gruendungsgarage prepared a special training catalogue for the fellows in order to teach them entrepreneurial skills.

The ongoing success of this format is based on special characteristics that make the Gruendungsgarage more than just a lecture. One of the features represents the opportunity for students and scientific staff to work on their own business idea which increases their motivation and willingness to participate. After their successful participation, all members get the Gruendungsgarage certificate, and students also benefit from it by receiving 2 ECTS credits. The latter is more a symbolic reward in consideration of the effort they provide during the semester.

Beyond that, the teams are supported by university lectors with an expertise in Entrepreneurship as well as successful mentors with an experience in business consultancy. Mentors play a significant role as they guide the teams during the whole semester and provide feedback at an early stage where they haven't already started a business. In that way, the teams may easily rearrange or modify their business ideas without losing their precious time. Basically, each team is supported by two mentors, who can identify professionally with their business idea, from complementary fields. With a total of 26 active mentors, the consulting services range from business modelling, design-thinking, inbound logistics and online marketing, intellectual property right, and software development to legal and tax advice. In addition to the assigned mentors, the teams have also the opportunity to seek the advice from all the mentors by arranging a private appointment or during so-called "mentoring days." These "mentoring days" are an organizational improvement: in the last few volumes, the organizers of the Gruendungsgarage realized that it is particularly difficult for teams and mentors to arrange a suitable date for everyone. Since the last two volumes, a doodle survey was performed: mentors are asked to arrange a date when the actual majority of participants are available for several hours at a time. As a result, the teams are provided with a great opportunity to book a time slot with each mentor, allowing participants to receive a comprehensive consultation from different experts in only one session. This organizational change saves a lot of time to mentors and teams as well as leading to a more intensive exchange between these two parties.

A further improvement is the stronger involvement of the alumni of the Gruendungsgarage in the teaching process. Every semester, one teaching unit is reserved for a so-called "alumni-talk," where the current teams get the chance to speak to a couple of alumni who successfully founded a start-up company after their participation at the Gruendungsgarage. The alumni introduce themselves briefly, present an outline of their career, and share their experiences with the teams of the current volume. Usually, the alumni not only discuss the aspects of their successes—they also share their failures and what they have learnt from them. These talks provide the current members of the Gruendungsgarage with an authentic insight into young academic entrepreneurs' lives, actually showing them what it truly takes to start a business.

Besides the mentors and the alumni, the network of the Gruendungsgarage also includes investors and business angels. It offers to the promising teams start-up funding to facilitate the realization of their own business ideas. Although the teams get an overview about the regional and national funding agencies in Austria, private investors are essential at the certain point, as it makes them important partners of the Gruendungsgarage.

2 Structure of the Gruendungsgarage

At the beginning of each university semester, interested students and scientific staff (TUG, KFU, and other universities) have to summarize their business idea and briefly describe it in a very short business proposal. After the submission deadline, the ideas are rated by an expert team. The experts check the business idea, the status, the roadmap, the innovation character and the potential and accessible scalability of the idea. Furthermore, the project is evaluated regarding its status for not having projects in line that are already ready for market access. Idea projects are situated in the (pre-) seed phase, where the input from university experts and the mentors could be fertilized. Because of the resource-intense support of the business ideas/teams, a maximum of 10 teams can participate in the course. During each semester, the teams receive training in class-room-lessons, workshops, exercises, and discussions within the group, and also with alumni teams who have already founded a business. The didactic concept is set up in the following way (Fig. 1): After the submission, the ideas of the students and the scientific staff are reviewed, and the selection process starts. A team of experts evaluates the ideas, chooses the 10 best rated ones and links the ideas to the mentors whose field of expertise best matches the idea. The selected teams are informed and invited to a kick-off meeting. At this meeting, the teams, the mentors and the university staff meet in person for the first time. During the kick-off meeting, the teams pitch their idea for all participants to have an overview. The meeting also helps people interested in start-ups to get to know each other and start networking. As seen in Fig. 1, after the kick-off meeting a start-up phase begins. Over the next few weeks, the students are taught to use a couple of helpful entrepreneurial tools as well as receive primary start-up knowledge. Then the workshop phase starts. Teams learn how to create business plans, learn the methods for human centered design, and gain an insight into the economy and the tax situation as well as create their business model [3].

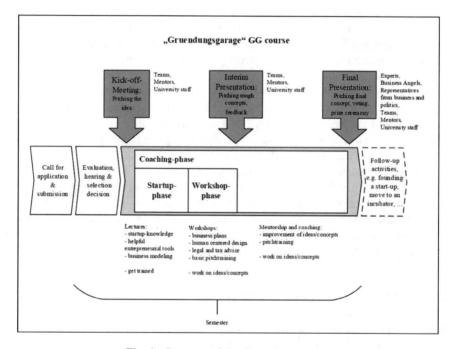

Fig. 1. Structure of the Gruendungsgarage

Over the numerous volumes, this phase has changed in terms of period and frequency in which the workshops are offered. The program leaders of the Gruendungsgarage realized the need for workshop-days instead of several workshop-units, which usually used to take two hours. The disadvantage of the previously offered workshop-units is that, just when the teams reach a productive working phase, they have to stop because of the short lesson. It turned out that the teams are much more productive if they complete three full workshop-days where they get time to focus on certain tasks and acquire the skills to start a business. In the middle of the semester, the teams have to pitch their idea and show their results to all the teams/mentors again at the interim presentation. At this intermediary presentation, they again receive a feedback from all the mentors and university staff. Then the teams eventually continue to work on their idea, gradually improving it: it is all based on the feedback and certainly on the mentorship as well as on coaching advice.

Beside the workshop-phase, the procedure of the coaching-phase has also changed over time. In former volumes, the participants were advised to make individual appointments with mentors from businesses and university to get feedback on their elaborated results. However, the experience has shown that it is difficult to find common dates where both, the team members and the mentors, would have time. Because of this reason, the so-called "mentoring-days" were introduced to offer the teams one to two days where the majority of mentors are available for several hours. Especially for busy mentors, this organizational change represents an enormous relief. The course concludes with a final presentation at the end of the semester to provide the teams a platform to share their results. A lot of experts, entrepreneurship affine persons, business angels, and representatives from business and politics do attend the presentation. At this final presentation session, the audience may vote for the best team, which is later awarded with prizes from sponsors, e.g. tickets for entrepreneur festivals, vouchers for co-working-spaces and legal advice [3].

3 Facts and Figures

As seen in Fig. 2, a total of 186 students and scientific staff took part at the Gruendungsgarage during the last five years, which led to a support of 91 projects/teams during this time. 27 out of these 91 teams actually started a business after attending the academic start-up accelerator, and 16 teams are currently in the start-up process or shortly before founding a company.

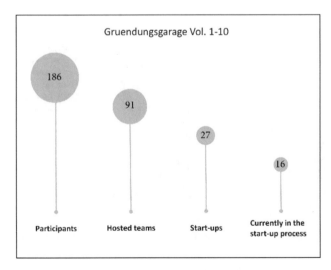

Fig. 2. Key figures of the Gruendungsgarage Vol. 1–10

The percentage of women among the 27 start-ups of the Gruendungsgarage is 25.42% which is well above the Austrian and the European average. According to a study by the European Startup Monitor (ESM), the percentage of female start-up founders in Austria is only 7.1% and in Europe it is 14.8% on average [4]. Thus, the Gruendungsgarage represents a true diversity format through its percentage of female start-up founders.

In the last three volumes (volume 8–10), the data collection was systematized to allow a more detailed analysis of key figures such as:

- university affiliation of the teams,
- team composition and,
- educational background of the participants.

A notable aspect of the data analysis was the number of applied and selected teams of the Graz University of Technology. While initially, most of the applicants were students or members of the KFU, the number of applied and selected teams from the TUG has increased during the last three volumes. More and more teams with a technical background are interested in participating in the accelerator program to benefit from an authentic entrepreneurship experience, as it exposes academic participants to real world start-up problems and opportunities. Figure 3 compares the number of teams that applied for the Gruendungsgarage and those that have been selected, labeled by their university affiliation.

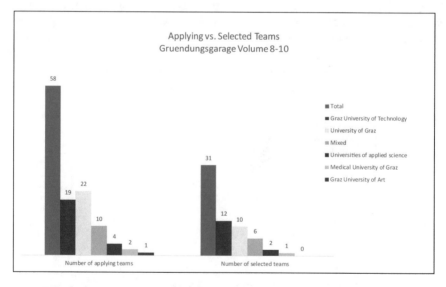

Fig. 3. Applying versus selected teams at the Gruendungsgarage Vol. 8–10

Although the number of participants from universities with a technical focus is increasing, there is still a lack of heterogeneous teams. In the last three volumes there were only six mixed teams out of 31 with a different field of study/discipline.

Furthermore, there is a trend towards digitalization in terms of the products, processes, and business models of the selected teams and those currently in the start-up process. For the development of digital business models, programming skills are an advantage that qualifies students and scientists of the TUG and universities of applied science to participate at the Gruendungsgarage [5].

To extend the investigation, the e-mail correspondence between the program leaders and the teams of the Gruendungsgarage were analyzed to get an idea of the content exchanged via this medium. It turned out that there is little exchange regarding subject-specific issues. Instead, mainly organizational topics are discussed via e-mail in communication. Frequently, teams contact program leaders of the TUG with technical questions, with the intent to get access to specific institutes at Graz University of Technology that can offer them technical support for their products. However, strategic questions concerning the business models are typically discussed in person during one of the numerous workshops, meetings, or events of the Gruendungsgarage.

4 Discussion and Outlook

After five exciting years and more than 25 successfully realized projects, the Gruendungsgarage has demonstrated its valuable contribution for students and scientific staff interested in starting their own business. Currently, the Gruendungsgarage is represented only in a few faculties, which results in a high number of applicants from these scientific fields. This includes business administration, mechanical engineering and software development. In the future, the Gruendungsgarage should be integrated in other curricula to reach a wider audience. Meanwhile only one curricula at the TUG contains the Gruendungsgarage as an elective course, which is why the majority of the students counts the accelerator program as an optional subject with 2 ECTS. Through a rollout to other faculties and curricula, more students and scientific staff will become aware of entrepreneurship as a career option. A further possibility for optimizing the teaching format refers to the diversity within the teams. Diverse people are sorely underrepresented in the business world—particularly in start-ups—and the Gruendungsgarage is no exception. Until now, many teams consisted of students or scientists from the same university or the same discipline, which limited their mindset and reduced their creativity [6]. To remedy such unsatisfactory tendencies, the program leaders thought about a change in the application process. For instance, the team-building process could take place at the beginning and not before the Gruendungsgarage starts. In a kind of matchmaking, potential founders with different fields of expertise would get to know each other and exchange ideas about possible partnerships before they start the regular program. A second modification of the teaching format aims to adopt an approach where universities exploit their intellectual property generated through theoretical and applied research. According to this approach, the Gruendungsgarage could offer students and scientific staff access to the patent archive where they analyze all the patents, pick an idea, and consider how they can commercialize it. This setup could lead to a format where participants of the Gruendungsgarage are no longer forced to apply with a business idea but still prove their entrepreneurial skills.

Five years of experience have shown that the teaching format as the Gruendungsgarage is essential for the academic start-up scene in Graz in order to get a comprehensive support at a very early start-up phase. The numerous applications verify the interest of students and scientific staff in the topic entrepreneurship which can be used as an evidence that there is a remarkable demand for more practical entrepreneurship education.

References

1. Nieuwenhuizen, C., Groenewald, D., Davids, J., Rensburg, L.J., Schachtebeck, C.: Best practice in entrepreneurship education. J. Prob. Perspect. Manage. **14**(3), 528–536 (2016)
2. Harms, R.: Self-regulated learning, team learning and project performance in entrepreneurship education: learning in a lean startup environment. J. Technol. Forecasting Soc. Change **100**, 21–28 (2015)
3. Vorbach, S.: Lecturing entrepreneurship at Graz University of Technology—the case of "Gruendungsgarage". In: Auer, M.E., Guralnick, D., Simonics, I. (eds.) Interactive Collaborative Learning. Proceedings of the 20th ICL Conference (in press)
4. Kollmann, T., Stöckmann, C., Hensellek, S., Kensbock, J.: European Startup Monitor, p. 39 (2016)
5. Poandll, E., Glinik, M., Taferner, R.: Digitalisierung in Academic Startups—Trends bei Gründungsprojekten der Gründungsgarage Graz. G-Forum 2018, 22. In: Interdisziplinären Jahreskonferenz zu Entrepreneurship, Innovation und Mittelstand (in press)
6. Novellus, R.: Why Diversity Is Needed in Startups. https://www.forbes.com/sites/yec/2017/06/15/why-diversity-is-needed-in-startups/#302a54fe76ea. 1st June 2018

Digitization and Visualization of Movements of Slovak Folk Dances

Matus Hajdin[1], Iris Kico[1(✉)], Milan Dolezal[1], Jiri Chmelik[1],
Anastasios Doulamis[2], and Fotis Liarokapis[1]

[1] Faculty of Informatics, Masaryk University, Botanická 68A, 60200 Brno,
Czech Republic
{445580,479048,396306,jchmelik}@mail.muni.cz,
liarokap@fi.muni.cz
[2] National Technical University of Athens, Athens, Greece
adoulam@cs.ntua.gr

Abstract. Folk dances as a part of intangible cultural heritage are important for
the cultural identity of human society. In this paper, we present a novel application for interactive, 3D visualization of folk dances based on motion capture
data set. A pilot user study was conducted, comparing the educative potential of
the application with a classic video recording of dance performance. After the
training phase, the dance performances of ten participants were ranked by
the professional dancer. The results of the study presented in this paper indicate
the potential of our approach for learning purposes.

Keywords: Motion Capture · Dance Teaching · Unity · Cultural Heritage ·
Folk Dances

1 Introduction

Folk dances are an important part of cultural heritage and they are deeply integrated
into society. As a result, preservation and dissemination of this heritage should be a
high priority for every nation and ethnicity. Folk dances are learned informally, and
they are passed on from one generation to the next [1]. Nowadays, we have available
technologies that can be used to achieve these goals. To better preserve this kind of
intangible cultural heritage (ICH), dances could be "digitized". Digitization could
include a semantic description of dance, audio, and video recordings or a computer
animation—the technique which forms the basis of our approach. A widely used
approach for animation is a "motion capture" technique—the process of digital
recording motion of a real-world object, most usually a human performer. Several
technologies of data acquisition can be utilized for motion capture, including magnetic,
optical, ultrasound, etc., see [2] for a comprehensive review of this field.

The main advantage of this approach is the digitization of the performer
(i.e. dancer) with high precision. The recorded data can be processed afterwards to
produce high quality, realistic animations in a relatively short time. This enables the use
of motion capture for the recording of an art performance and specifically folk dances.
Digitization itself is not sufficient to forward folk dances to the next generations.

© Springer Nature Switzerland AG 2019
M. E. Auer and T. Tsiatsos (Eds.): ICL 2018, AISC 917, pp. 245–256, 2019.
https://doi.org/10.1007/978-3-030-11935-5_24

Three-dimensional (3D) applications and dance learning platforms can help with this step. The current state of computer graphics, motion capture has strong advantages in the digitization of a dance, compared to other methods such as a video recording [3]. There have been numerous European projects committed to preservation of ICH and folk dances using motion capture technique [4–7].

This paper focuses on the development of an interactive application for visualization of the digitized dances. The goal is to represent the recorded dance movements in a way that every part of the performance can be closely examined. After using the application, the adequately skilled dancer should be able to reproduce the visualized dance. This is achieved by using the recorded data to animate a virtual 3D character and by providing interactive control for the user. The second part of this paper provides the design and description of the workflow for the digitization process. The application currently contains four folk dances that were acquired using this workflow. The main novelty of this work is that the application has an interactive approach, allowing the user to personalize the visualization. To assess the usefulness of the application a pilot study was carried out with ten participants. The objective was to experimentally compare our application with a classical approach (a video recording of a dance). User's experience and the usefulness of various features of the application were evaluated and initial results are promising.

The rest of the paper is organized as follows. Section 2 represents work related to digitization of dances, and Sect. 3 illustrates the system architecture. In Sect. 4 the experiment is described, and results are represented in Sect. 5. Conclusions and future work are presented in Sect. 6.

2   Related Work

The European Union (EU) project Open Dance proposed the learning framework for folk dances using the web 3D platform [4]. Optical motion tracking was applied using the VICON system. The user could watch a dancing model, choose the point of view and zoom level and control the speed of the 3D animation. Evaluation included questions quality and presentation of content, pedagocial quality and interactivity and usability aspects. Interactions in dances are something common. WhoLoDance [5] is another EU project focusing on developing and applying technologies to dance learning to achieve results with impact on researchers and professionals, but also dance students and interested public. Among objectives is to preserve cultural heritage by creating a proof-of-concept motion capture repository of dance motions. One session was recorded using unique platform that integrates a Qualisys motion capture system with RGB and RGBD video cameras, accelometers, biometric sensors, microphones, smartphones. Other sessions were recorded using passive optical motion capture system, VICON.

The WebDANCE project [6] was a pilot project to experiment with the development of web-based learning environment of traditional dances. The final tool included teaching units and three-dimensional (3D) animation for Greek and English dance. The users could choose different lectures, e.g. costumes, music, the role of the participant, etc., depending on the knowledge they want to get. Evaluation had the important role.

For the evaluation process interviews, questionnaires, observations, face-to-face meetings, electronic discussion, review meetings and focus group were used. The evaluation group included experts, teachers, students and members of traditional dance organizations. This project showed potential for teaching folk dances to young people. Furthermore, the i-Treasures Intangible Cultural Heritage EU project [7] is committed to the preservation of ICH. They recognized the need for multimodal ICH datasets and digital platforms for various multimedia digital content. This includes folk dances as well. The main objective of the project is to develop an open and extendable platform to provide access to ICH resources. Several folk dances have been recorded using multiple-Kinect approach and the Qualisys optical motion capture system. Educational game-like applications have been implemented for visualization of these recordings.

In [8], a framework was presented based on the principles of Laban Movement Analysis. A virtual reality simulator for teaching folk dances was implemented and tested with Cypriot folk dances. Dance segments performed by 3D avatar were presented to users. Data were captured using an optical motion capture system with active markers, PhaseSpace's Impulse X2. In [9] a game was presented where the human dancer can interact with computer dancer. For capturing the real-time movement of the human dancer, optical 3D motion capture system was used. For continuous recognition of real-time dance moves Progressive Block Matching Algorithm was used. The use case was a go-go dance. In [10] a method was proposed where passive optical motion capture system was used for motion capture. 3D models were made using Blender and MakeHuman. Five Indian dances were documented using 3D dance animation videos.

Other researchers used Microsoft Kinect, since it can achieve real-time 3D skeleton tracking and at the same time is cheap and easy to use [11]. To capture and record the performer's moves, the Kinect II depth sensor in combination with ITGD module [7] was used in [11]. The recorded data were used for identifying Greek folk dances by matching recorded sequences to a database of characteristic dance instances. In [12], the researchers proposed a Kinect-based human skeleton tracking for the acquisition of performer's motion. The performer gets the visual feedback in the 3D virtual environment. There is a real online interaction between teacher and performer. Moreover, a framework for real-time capture, assessment, and visualization of ballet dance movements was proposed [13]. For the acquisition of motion data was used Microsoft Kinect camera system. 3D visualizations and feedback for performance evaluation are provided through the CAVE virtual environment.

In [14], an expert was recorded using high-precision motion capture system (Qualisys) and Kinect V2 sensor. The expert performed different basic steps of the traditional Belgian dance, Walloon. Game-like application was developed for learning purposes. Similar approach using Kinect and game-like application can be found in [15]. Multiple-Kinect approach was used in [16, 17]. For example, in [16] a 3D game environment was presented with the natural human-computer interface, which is based on the fusion of multiple depth sensor data. Skeletal data from different sensors are fused into one robust skeletal representation used for user's avatar in the 3D environment and for evaluation.

Our approach presents a similar approach to [4], in the way that users can choose between several options including: playback control, viewport, dance picker. However, our approach extends this functionality by incorporating trail control, character picker and sound. Users can interact with the application and select the best options aiming in making the learning process easier.

3   System Architecture

Movements during any dance act are susceptible to the performer's mentality and body type. These variations should be appropriately handled by the system. Method without significant effect on dancer's performance has to be prioritized. A high degree of precision must be maintained as dances may contain large numbers of intricate movements. Since high precision was required the optical approach was selected. Optical systems with active markers can minimize marker swapping and identity, but adding wires in the system can limit the freedom of the movement. Even though optical systems with passive markers are very expensive and they may not be able to uniquely identify each marker, these systems have no need for obtrusive cables and can provide very accurate data. For recording the motion data, in our research, an optical system with passive markers was used.

After recording the data, visualization is the next step. The main aim of the application, called MotionDance, is to represent the recorded dances in a three-dimensional (3D) environment and use it for learning purposes. The application was developed in Unity 2017.3,[1] using C# language for all scripting purposes. Firstly, the motion data are visualized by animating a virtual 3D character. Secondly, the application allows interactive approach for the users. This consists of: pausing and resuming the playback of the dance, adjusting the playback speed, controlling the angle and distance of the camera, displaying the comments linked to the specific time in the dance and toggling the trails displaying the trajectory of each limb. Users can interact with the application through the interface shown in Fig. 1. In the first box, it is possible to choose one of three avatars. In the current version of the application generic avatars are presented. In the second box, the trail for each limb can be toggled with the associated checkbox. The third box enables the choice between four Slovak dances.

For trail control, users can select separately Left Arm, Right Arm, Left Leg, Right Leg or any combination of these. Limb trails are present for three seconds and then they disappear. Figure 2 shows trails of one arm and trails for one arm and one leg.

The whole process of digitization of folk dances will be described as follows. Digitization was done at Human-Computer Interaction Laboratory (HCI Laboratory) using a motion capture system with passive markers and 16 cameras Prime13W with proprietary software Motive used for the recording. Figure 3 shows the workflow of the digitization and visualization process.

3.1   Motion Capture

Most motion capture systems, especially optical ones with multiple cameras require some sort of calibration to ensure accurate recording of motion. The purpose of calibration is to

---

[1] www.unity3d.com

Fig. 1. The interface of the application

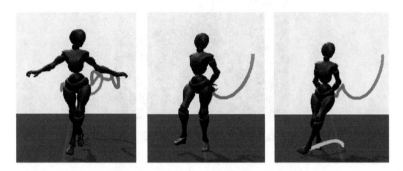

Fig. 2. Dance movements with trails

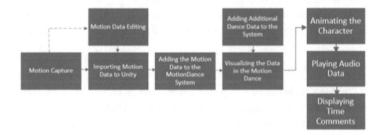

Fig. 3. Digitization and visualization workflow

exactly determine the positions of the cameras in relation to each other in space, as well as the orientation and position of the ground. Before the calibration, it is required to remove any objects that can be incorrectly identified as motion capture markers by the system. The next step consists of properly identifying skeleton of the dancer wearing marker suit. Markers are placed on specifically determined positions. After aligning the markers and proper identification by the software, the recording can start.

For the recording, it is important to start and stop with the dancer in a T-pose. T-pose is also known as reference pose in the unanimated state of a model in 3D graphics. This is important for the later use of the recorded data. It is a good practice to record multiple takes of the same performance even in case of an apparently successful case because some errors can be discovered later. Figure 4 shows the dancer wearing marker suit during the motion capture process. On the projector screen dancer can track their movements.

Fig. 4. Motion capture of dance moves

Post processing is done using Motive. Positions in space and recordings of all markers are displayed. Any gaps in the recordings can be filled using interpolation. This solution produces solid results, especially when the gap size is relatively small. An additional function for post-processing of the motion data is the smoothing of all position data. In our case, the cutoff frequency for a low-pass filter was 6 Hz.

3.2 Motion Data Editing

Some data can need additional processing and editing beyond the Motive software. Suitable and often used as an industry standard is MotionBuilder.[2] It is possible to edit the animation, fix errors or edit the performance for more truthful representation of the dance. First, data clean-up and re-targeting have to be done, e.g. finding errors in dancer's performance, noisy motion, flipped joints, unnatural limb positions, strange movements. The next step is to identify technical issues and fix them. Some technical issues include character's alignment, motion paths and trajectories, foot floor contacts, mesh intersection, object interactions. The final step is to enhance the performance. This can be done by exaggerating poses, refining motion, adjusting weight and balance, adding missing details.

3.3 Importing and Adding Motion Data to Unity

It is necessary to set the import settings for the recorded movements to work properly inside the MotionDance. In Muscle and Setting Menu, the limits on Per-muscle settings that control the range of possible rotation of individual limbs have to be set up properly. As the default ranges are too small for many dances a full range of motion

---

[2] www.optitrack.com/products/motive/

should be enabled. This solves issues that cause limbs switching rotation during an animation. Adding the imported motion data to the MotionDance is done through the Unity editor. Sound clip was also added. Synchronization of sound and dance moves using the application is still an open question.

3.4 Visualizing the Data in MotionDance

The application allows additional virtual characters. Any fully-rigged humanoid 3D characters should be suitable. Unity supports 3D file formats, e.g. fbx and obj, and 3D application file format, e.g. c4d and max3D. After the character is imported to the MotionDance, it can be added to the visualization system. It is necessary to create a Prefab from the character model. This Prefab is created by adding the Animator component which is responsible for applying any animation to the object. The Animator component uses Controller, which is a state machine and determines the animation applied to the object. Time comments in the application are added to help in the learning process. The comments contain important tips for the user and they are displayed at the specific time moments.

4 Pilot Study

A pilot study was conducted to test the learning effectiveness learning steps from dances as well as test the overall usability of the application. The hypothesis was that the application would be more useful for dance learning then the video recording. We also expected that the overall usability of the application would be at least at a satisfactory level. Ten subjects participated that are members of a Slovak folk ensemble Poľana in Brno, Czech Republic. All of them are aged between 18 and 30 years old and had years of experience with folk dancing.

The first part of the experiment aimed to compare the application and the video recording. First, a professional dancer who is also a member of Poľana was recorded in both ways during the performance of different folk-dances (video can be found on https://youtu.be/lT2WdOLVUr4). Movements of the dances recorded for this purpose are from different regions of Slovakia: Horehronie, Abov, Podpoľanie and Horné Považie. Recorded data were added to the application (the application can be found on https://youtu.be/ccctHl4oBKk). These two digital representations of the same dance were used for the experiment.

At the beginning of the experiment, a brief tutorial was provided to the participants. Participants were randomly split into two even groups. Members of one group were supposed to use the application for dance learning while the members of the other group were using the video recording. The members, one from each group were individually led into separate spaces where the corresponding approach for dance learning was prepared. For the application, a quick explanation of its features and controls was given. The participants had five minutes to learn the dance.

Participants performed the dance one by one in front of a 'judge'. The 'judge' was the professional dancer recorded at the beginning and assessed how well participants performed the dance on a scale of 1–10 (where 10 was the best possible score).

Participants who used the application were asked to fill out a questionnaire after their performance in front of the judge. The participants had to rate the experience of using the application and the application functionalities. The work-flow of the pilot study for both groups is shown in Fig. 5.

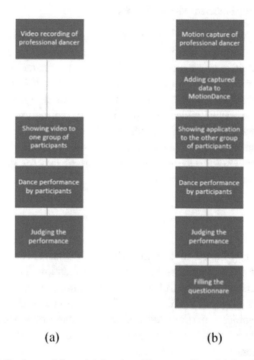

(a)   (b)

Fig. 5. Pilot Study workflow (**a**) for the video recording (**b**) for the application

5   Results

Initial results of the experiment are shown in Table 1. In particular, the results show the slightly better effectiveness of the application for learning a dance in comparison to video recording. The average score for the participants using application was 6.2 while the score for participants using video was 5.8, but since the difference is not significant generalization cannot be made.

Table 1. Dance learning results

Method	Dancers score					Average
Application	5	4	8	7	7	6.2
Video	5	6	4	7	7	5.8

The results from the questionnaire, shown in Tables 2 and 3 were also positive but not significant again. All different parts of the program, the possibility of learning a dance using the application and the overall experience were all ranked positively. The

Table 2. Numbered rating of the application from the questionnaire

Question	Users answer					Average	Median
How do you rate the ability to learn the dance using the software?	8	7	7	7	10	7.8	7
How do you rate the overall experience with the software?	8	9	7	9	9	8.4	9
How do you rate the user interface?	9	9	5	6	9	7.6	9
How do you rate the ability of interaction with the displayed dance?	10	9	6	10	10	9	10
How do you rate the quality of the dance visualization?	9	9	5	10	10	8.6	9

Table 3. Picking the suitable functionalities

	Users choices					Total
Which functionality did you use while learning the dance?						
Pausing	✓	✗	✓	✓	✗	3
Seeking	✓	✓	✓	✗	✓	4
Change of playback speed	✓	✗	✗	✓	✗	2
Camera control	✓	✗	✗	✗	✓	2
Limb trails	✗	✗	✗	✗	✗	0
Comments on the timeline	✓	✗	✗	✗	✓	2
Which functionality do you think is useful for learning a dance?						
Pausing	✓	✓	✓	✗	✓	4
Seeking	✓	✓	✓	✓	✓	5
Change of playback speed	✓	✗	✓	✓	✓	4
Camera control	✗	✗	✗	✗	✓	1
Limb trails	✗	✗	✗	✓	✓	2
Comments on the timeline	✓	✓	✗	✗	✓	3
Which functionality do you think is useful for viewing a dance without learning?						
Pausing	✗	✗	✓	✗	✓	2
Seeking	✗	✗	✓	✗	✓	2
Change of playback speed	✗	✗	✗	✓	✓	2
Camera control	✓	✓	✓	✓	✓	5
Limb trails	✗	✓	✗	✗	✗	1
Comments on the timeline	✓	✗	✓	✓	✗	3

lowest average rating was 7.6 of 10 for the user interface and the highest average rating was 9 for the ability of interaction with the visualized dance.

The additional part of the questionnaire showed that all functionalities of the application were used by some participants during their learning process except for trails display. However, 2 of 5 participants still thought that it could be useful for dance learning even though they did not use it. The most commonly used functionality was the seek functionality, 4 of 5 participants were using it. It was also the most useful functionality for dance learning where 5 of 5 participants chose it. Only 2 participants used the camera control during the dance learning and only 1 of them found it useful for dance learning. All 5 participants found this functionality as useful for viewing a dance without learning. Limb trails display functionality got an overall low score.

6   Conclusions

The goal of this research was to create an interactive application for visualization of folk dances recorded using motion capture. The resulting software provides an interactive visualization of the dance that enables the user to view any part of the dance. Dance movements from different dances from several regions of Slovakia have been digitized and added to the visualizing software using this process. Initial results show the potentials of the application for dance learning, although, there is no statistical significance. The small difference between the effectiveness for learning purposes using the application in comparison to a video recording can be partially attributed to the short 5-minute interval that participants had for learning the dance. The short time limit may also be the reason why some of the functionalities were not used or not found useful by some users, as they did not have time to properly try and evaluate them. Participants were rated by the professional dancer and this is a very subjective approach. The problem with the performance rating is challenging and will be further explored in the future. The comparison of dance movements is not a straightforward process. Different people can perform the same dance correctly but due to the differences in shapes of their bodies and their personal touch, movements can look different. A large-scale evaluation focusing on different parts and uses of the application will be performed in the future.

Acknowledgements. This work is supported under the H2020 European Union funded project TERPSICHORE project: Transforming Intangible Folkloric Performing Arts into Tangible Choreographic Digital Objects, under the grant agreement 691218.

References

1. Aristidou, A., Stavrakis, E., Chrysanthou, Y.: Motion analysis for folk dance evaluation. In: Proceedings of the Eurographics Workshop on Graphics and Cultural Heritage, pp. 55–64. Eurographics Association Aire-la-Ville, Switzerland (2014)
2. Nogueira, P.: Motion capture fundamentals—a critical and comparative analysis on real world applications. In: Oliveira, E., David, G., Sousa, A.A. (eds.) Proceedings of the 7th

Doctoral Symposium in Informatics Engineering, pp. 303–314. Porto, Portugal, 26–27 Jan 2012. Faculdade de Engenharia da Universidade do Porto

3. Smigel, L.: Documenting Dance: A Practical Guide. Dance Heritage Coalition, Incorporated (2006)

4. Magenant-Thalmann, N., Protopsaltou, D., Kavakli, E.: Learning how to dance using a web 3D platform. In: Leung, H., Li, F., Lau, R., Li, Q. (eds.) Advances in Web Based Learning—ICWL 2007. LNCS, vol. 4823, pp. 1–12. Springer, Berlin, Heidelberg (2008)

5. Camurri, A., El Raheb, K., Even-Zohar, O., Ioannidis, Y., Markatzi, A., Matos, J.M., Morely-Fletcher, E., Palacio, P., Romero, M., Sarti, A., Di Pietro, A., Viro, V., Whatley, S.: WhoLoDancE: towards a methodology for selecting motion capture data across different dance learning practice. In: MOCO'16 Proceedings of the 3rd International Symposium on Movement and Computing, Article 43, pp. 43:1–43:2. ACM, New York, NY, USA (2016)

6. Karkou, V., Bakogianni, S., Kavakli, E.: Traditional dance, pedagogy and technology: an overview of the WebDANCE project. Res. Dance Educ. 9(2), 163–186 (2008)

7. Grammalidis, N., Dimitropoulos, K., Tsalakanidou, F., Kitsikidis, A., Roussel, P., Denby, B., Chawah, P., Buchman L., Dupont S., Laraba, S., et al.: The i-treasures intangible cultural heritage dataset. In: MOCO'16 Proceedings of the 3rd International Symposium on Movement and Computing, Article 23, pp. 23:1–23:8. ACM, New York, NY, USA (2016)

8. Aristidou, A., Stavrakis, E., Charalambous, P., Chrysanthou, Y., Loizidou Himona, S.: Folk dance evaluation using Laban movement analysis. ACM J. Comput. Cult. Herit. 8, 20:1–20:19 (2015) Article 20

9. Tang, J.K.T., Chan, J.C.P., Leung, H.: Interactive dancing game with real-time recognition of continuous dance moves from 3D human motion capture. In: Proceedings of the 5th International Conference on Ubiquitous Information Management and Communication. ACM, New York, NY, USA (2011)

10. Hegarini, E., Dharmayanti, Syakur, A.: Indonesian traditional dance motion capture documentation. In: 2016 2nd International Conference on Science and Technology—Computer (ICST), pp. 108–111. IEEE, New York (2016)

11. Protopapadakis, E., Grammatikopoulou, A., Doulamis, A., Grammalidis N.: Folk dance pattern recognition over depth images acquired via Kinect sensor. The International Archives of the Photogrammetry, Remote Sensing and Spatial Information Sciences. vol. XLII-2/W3, pp. 587–593 (2017)

12. Alexaidis, D.S., Kelly, P., Daras, P., O'Connor, N.E., Boubekeur, T., Moussa, M.B.: Evaluating a Dancer's Performance Using Kinec-based Skeleton Tracking. In: Proceedings of the 19th ACM International Conference on Multimedia, pp. 659–662. ACM, New York, NY, USA (2011)

13. Kyan, M., Sun, G., Li, H., Zhong, L., Muneesawang, P., Dong, N., Elder, B., Guan, L.: An approach to ballet dance training through MS Kinect and visualization in a CAVE virtual reality environment. ACM Transactions on Intelligent Systems and Technology (TIST)—Special Section on Visual Understanding with RGB-D Sensors, vol. 6, Article 23, pp. 23:1–23:37 (2015)

14. Laraba, S., Tilmanne, J.: Dance performance evaluation using hidden Markov models. Comput. Animat. Virtual World. 27, 321–329 (2016)

15. Stavrakis, E., Aristidou, A., Savva, M., Loizidou Himona, S., Chrysanthou, Y.: Digitization of cypriot folk dances. In: Ioannides, M., Fritsch, D., Leissner, J., Davies, R., Remondino, F., Caffo, R. (eds.) EuroMed 2012: Progress in Cultural Heritage Preservation. LNCS, vol. 7616, pp. 404–413. Springer, Berlin, Heidelberg (2012)

16. Kitsikidis, A., Dimitropoulos, K., Ugurca, D., Baycay, C., Yilmaz, E., Tsalakanidou, F., Douka, S., Grammalidis, N.: A game-like application for dance learning using a natural human computer interface. In: Antona, M.; Stephanidis, C. (eds.) UAHCI 2015. LNCS, vol. 9177, pp. 472–482. Springer International Publishing, Switzerland, 2015
17. Kitsikidis, A., Dimitropoulos, K., Yilmaz, E., Douka, S., Grammalidis, N.: Multi-sensor technology and fuzzy logic for dancer's motion analysis and performance evaluation within a 3D virtual environment. In: Stephanidis, C., Antona, M. (eds.) UAHCI 2014. LNCS, vol. 8513, pp. 379–390. Springer International Publishing Switzerland (2014)

Design, Development and Evaluation of a MOOC Platform to Support Dual Career of Athletes (GOAL Project)

Panagiotis Stylianidis[(⊠)], Nikolaos Politopoulos,
Thrasyvoulos Tsiatsos, and Stella Douka

Aristotle University of Thessaloniki, Thessaloniki, Greece
{pastylia,npolitop,tsiatsos}@csd.auth.gr,
sdouka@phed.auth.gr

Abstract. This paper presents the expert evaluation results of the alpha version of an e-learning platform for supporting dual career of athletes. The platform has been designed and implemented in the context of GOAL (Gamified and Online Activities for Learning to support dual career of athletes) Erasmus+ project. More specifically, the paper presents briefly the structure and content of the platform along with the technological solution, which is based on the well-known Moodle platform. According to the evaluation results, GOAL platform is easy to use, consistent, with a simple user interface and the overall attitude of the evaluators is positive.

Keywords: MOOCS · Dual career · Sports

1 Introduction

Nowadays, elite sport has reached a high level of professionalism (Brackenridge 2004). Athletes' training hours have increased resulting in more than 40 h of work when considering training hours, competition travel time, and study requirements (Amara et al. 2004). The elite sport career entails five to ten years dedicated to sport (Alfermann and Stambulova 2007). Along these lines, balancing studies along with the sport career allows the athlete to better prepare for future employment (Aquilina 2013). A significant body of research has been accumulated on athletic career, examining career development, transitions, and especially athletic retirement (Stambulova et al. 2009). Contemporary research emphasizes the need for "whole career" and "whole person" approach, highlighting that athletes go through several transitions in sport, education, and psycho-social development simultaneously (Wylleman and Lavallee 2004). Dual career research is a response to the call for this holistic approach and it has become a growing area of study (Burnett 2010).

The combination of education and training often becomes a challenge for athletes. Transitions are taking place often at this stage when athletes are changing homes, sports clubs and have to make new training and sports arrangements. Sports and Physical Education faculties in Europe are focused only on sports training without offering any flexible courses, predominantly through distance learning. This kind of

© Springer Nature Switzerland AG 2019
M. E. Auer and T. Tsiatsos (Eds.): ICL 2018, AISC 917, pp. 257–266, 2019.
https://doi.org/10.1007/978-3-030-11935-5_25

learning may provide to athletes the flexibility in terms of the timing and location of their sporting and academic activities.

On the one hand, distance learning programmes in Europe for supporting the dual career of athletes have not been convincing, in terms of quality, level, accessibility and interactive character.

On the other hand, EU supports the development of more flexible, outcome-driven learning systems to allow the validation of competencies acquired outside of formal education. This effort relies heavily on the widespread use of digital technology in education, to unlock and exploit freely available knowledge (Rethinking Education: Investing in skills for better socio-economic outcomes 2012). All of these policy goals are relevant to sport and for Dual Career. For example, student mobility facilitates competition and training in other member states, and flexible learning systems are supportive of combining sport and education.

Therefore, there is a clear need to support dual career of athletes through flexible learning activities based on digital technologies.

This paper is supporting this need by presenting expert evaluation results of an e-learning platform implemented to deliver services and online courses for dual career of athletes. The platform has been designed and implemented in the context of GOAL (Gamified and Online Activities for Learning to support dual career of athletes, http://goal.csd.auth.gr/elearning/) Erasmus+ project.

This paper is structured as follows. The next section presents the background, previous work done and objectives in the context of the GOAL Erasmus+ project. Afterwards the development of the platform is presented. The evaluation methodology, participants procedure and results are presented. The last section presents the concluding remarks and our vision for the next steps.

2 Background and Objectives

GOAL aims to support active and non-active athletes in the development of their professional endeavours, after the end of their athletic career. By active and non-active, we mean athletes that are already doing a sport and wish to be prepared on achieving a smooth transition between the end of their professional careers and further educational/business endeavours and those athletes who discontinued their professional athletic career and experience challenges with their integration in education, training and the open labour market.

In particular GOAL is identifying and testing gamified learning and training activities to form best practices for supporting dual careers of athletes using digital technology such as games/gamification in sports. A set of interactive ICT-based tools will be offered to active and non-active athletes for acquiring skills and competencies necessary to consciously discover, plan and determine their future career goals once they complete their competitive sports career. Such skills are critical in developing athlete's continuous professional career development including efforts of coping with transition and change both as individual personalities being part of a wider community (e.g. active citizenship) as well as professionals that will be following a career after sports competition, and thereby preparing them for a new job.

As it was presented before (Tsiatsos et al. 2018) a state of the art was conducted about the e-learning platforms and the result was that the most appropriate is Moodle (https://moodle.org/).

This platform will contain the six courses of the curriculum:

- Cycle 1
  - Entrepreneurship
  - Personal Skills Development—Teamwork
  - Personal Skills Development—Decision Making Skills.
- Cycle 2
  - Sports Management
  - Sports Marketing
  - Coaching in Sports.

It will also contain the online services of the project which include:

- E-Mentoring
- Online Psychological Support.

Finally the will support gamified activities and two serious games aimed to improved users' personal soft skills.

In this study will be presented the MOOC development and the first technical evaluation. This will be the Alpha-testing of the platform. In order to ensure the highest quality of the evaluation, it was decides to be conducted an expert heuristic evaluation.

3 Development of the Platform

3.1 MOOC Platform

As it was described before GOAL team developed a MOOC platform using the LMS Moodle. Moodle provides a wide range of plugins and solutions to enable gamification and create a friendlier and easy customizable, from the user perspective, environment. GOAL team used plugins such us "Level up" to create a gamified environment or "Accessibility" for users with disabilities (Fig. 1).

3.2 Services

Psychological assistance due to sudden terminations of the athletic career is a key aspect for supporting athlete's career transitions serving as a crucial supporting service. Education and preventive interventions can help athletes of becoming better aware of forthcoming transitions and develop resources in coping them. Psychological assistance as part of supporting services will be provided by the GOAL project, aimed at helping athletes to overcome transitions in their careers inside and outside of sports. GOAL aims to introduce its comprehensive coaching/mentoring psychological intervention services. Exploiting these services, athletes may balance life-style to reduce stress and enhance wellbeing as part of teaching life skills for a holistic life-career.

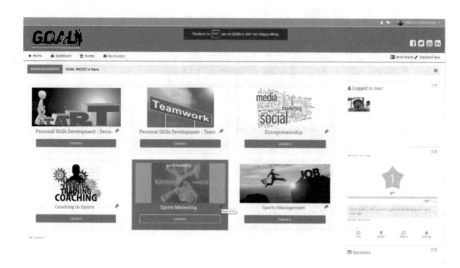

Fig. 1. GOAL MOOC prototype

Those services will be provided both synchronous and asynchronous methods through the platform and the consortium experts will organize the meetings.

4 Evaluation

In this part a review of the evaluation methodology is presented as well as the evaluator profiles, the heuristic rules used and the evaluation process.

After that for every group of interactions that a user can engage in:

- The results of the heuristic evaluation are presented as well as the relevant comments from the evaluators
- The positive points of usability of the system are mentioned
- A brief description of the problems and suggested solutions are described.

4.1 Methodology

For the evaluation of applications in which user interactions cannot be accurately predicted and the user is not a novice, general practices in the lab for the measurement of usability and design control are required. This requirement endeavors to be satisfied by the heuristic evaluation method (HE). This particular method is not analytical, but has a subjective character and is based on empirical rules and findings that are well known and are related to good interface design. Another term used to describe a heuristic evaluation is Usability Inspection (UI), since in practice it is an inspection process by experts on interface characteristics based on heuristic rules.

4.2   Evaluators

The heuristic evaluation, (as mentioned in the bibliography), was conducted by external evaluators and not by the system designers as to ensure impartial judgement and a second opinion on the design. Furthermore as mention in the bibliography, an optimal number of evaluators has been adopted (5–6), whose opinions accumulate cumulatively. In Table 1 the demographics of the evaluators (sex, age and educational background) are being presented.

Table 1. Demographics of evaluators

	Size	Percentage (%)
Gender		
Female	1	14.29
Male	6	85.71
Age		
25–34	3	42.8
35–44	2	28.6
>45	2	28.6
Educational level		
M.Sc. graduate	5	71.4
Ph.D. graduate	2	28.6

The rules that can be used for a heuristic evaluation are not strictly specified, different experts might consider some rules more important than others. Furthermore some problems require that greater emphasis is attributed to some of the rules. A widespread set of heuristic evaluation rules is presented in Table 2.

4.3   Heuristic Rules

The heuristic evaluation method focuses on two key points:

- The general design of system screens
- The flow of dialogues, messages and actions required by the user to perform a specific task.

The rules that can be used for a heuristic evaluation are not strictly specified, different experts might consider some rules more important than others. Furthermore some problems require that greater emphasis is attributed to some of the rules. A widespread set of heuristic evaluation rules is presented in Table 2.

4.4   Evaluation Procedure

The user interface characteristic of a GOAL MOOC platform user that was evaluated with heuristic evaluation was:

Table 2. Heuristic evaluation rules

A/I	Heuristic rule	Description
1.	Aesthetic and minimalist design	Dialogues should not contain information which is irrelevant or rarely needed. Every extra unit of information in a dialogue competes with the relevant units of information and diminishes their relative visibility
2.	Match between system and the real world	The system should speak the users' language, with words, phrases and concepts familiar to the user, rather than system-oriented terms. Follow real-world conventions, making information appear in a natural and logical order
3.	Recognition rather than recall	Minimize the user's memory load by making objects, actions, and options visible. The user should not have to remember information from one part of the dialogue to another. Instructions for use of the system should be visible or easily retrievable whenever appropriate
4.	Consistency and standards	Users should not have to wonder whether different words, situations, or actions mean the same thing
5.	Visibility of system status	The system should always keep users informed about what is going on, through appropriate feedback within reasonable time
6.	User control and freedom	Users often choose system functions by mistake and will need a clearly marked "emergency exit" to leave the unwanted state without having to go through an extended dialogue. Support undo and redo
7.	Flexibility and efficiency of use	Accelerators—unseen by the novice user—may often speed up the interaction for the expert user such that the system can cater to both inexperienced and experienced users. Allow users to tailor frequent actions
8.	Help users recognize, diagnose, and recover from errors	Error messages should be expressed in plain language (no codes), precisely indicate the problem, and constructively suggest a solution
9.	Error prevention	Even better than good error messages is a careful design which prevents a problem from occurring in the first place. Either eliminate error-prone conditions or check for them and present users with a confirmation option before they commit to the action
10.	Help and documentation	Even though it is better if the system can be used without documentation, it may be necessary to provide help and documentation. Any such information should be easy to search, focused on the user's task, list concrete steps to be carried out, and not be too large

- Navigation and accessing course content
- In the next section are presented:
- The heuristic evaluation results
- Good usability practices on the platform
- Problems and suggested solutions.

Heuristic Evaluation Results

Tables 3 and 4 present the heuristic evaluation results and the evaluator comments for each rule.

Table 3. Heuristic evaluation results

No.	Heuristic rule	Ev. 1		Ev. 2		Ev. 3		Ev. 4		Ev. 5		Ev. 6		Ev. 7	
		Sev.	Fq.	Sev.	Fq.	Sev.	Fq.	Sev.	Fq.	Sev.	Fq.	Sev.	Fq.	Sev.	Fq.
								(0-4)							
1	Aesthetic and minimalist design	1	2	0	1	0	1	0	0	1	0	1	1	0	2
2	Match between system and the real world	0	0	0	0	0	0	0	0	1	0	1	1	0	0
3	Recognition rather than recall	0	0	0	1	0	0	0	0	2	50	0	0	1	1
4	Consistency and standards	0	0	0	0	0	0	0	0	1	0	2	1	0	0
5	Visibility of system status	0	0	1	1	1	1	1	2	2	2	0	0	1	2
6	User control and freedom	2	1	0	0	1	2	0	0	1	0	0	0	0	0
7	Flexibility and efficiency of use	2	2	0	0	2	2	1	0	2	0	0	0	1	0
8	Help users recognize, diagnose, and recover from errors	0	0	0	1	0	0	0	0	0	0	1	1	0	0
9	Error prevention	0	0	2	1	0	0	0	0	1	0	0	0	0	3
10	Help and documentation	2	2	0	0	0	0	0	0	1	0	2	1	1	0

Table 4. Evaluator comments by heuristic rules

A/I	Heuristic rule	Comments
1.	Aesthetic and minimalist design	**Evaluator 6**: Starting screen Pictures could have a more uniform theme/style
2.	Match between system and the real world	**Evaluator 6**: Intents could be improved to better represent the various headings and subheadings displayed in each course page
3.	Recognition rather than recall	–
4.	Consistency and standards	**Evaluator 6**: There are some issues with the different font sizes and styles used. Additionally, some bibliography links appear to follow different formats
5.	Visibility of system status	–
6.	User control and freedom	**Evaluator 1**: Some screens are missing the cancel button to go back **Evaluator 3**: Exit Button on the slideshows does nothing
7.	Flexibility and efficiency of use	–
8.	Help users recognize, diagnose, and recover from errors	**Evaluator 7**: The text and structure of the 404 page could be further improved (e.g., try visiting a non-existent URL like "http://goal.csd.auth.gr/elearning/course/lost" there are some font and alignment issues)
9.	Error prevention	–
10.	Help and documentation	**Evaluator 1**: I could not find a manual on the site **Evaluator 4**: Support section is missing **Evaluator 6**: Is there a help section? (if so I could not find it easily). The search UI could be improved

Good Usability Practices on the Platform

The main positive points of the platform are the following, according to the evaluators' comments:

- Good information categorisation without elements that distract a user from his task
- Understandable use of language without difficult technical terms that might confuse the user
- Every course is organised the same way. There is consistency
- The breadcrumbs appearing at the top of the UI are helpful for the navigation.

Problems and Suggested Solutions

Most of the comments indicate a lack of a tutorial guide and help section on the platform and some minor user interface problems such as the omission of cancel buttons on some screens.

5 Conclusions—Future Work

This paper presents the heuristic evaluation of the alpha version of an e-learning platform for offering MOOCs supporting the dual career of athletes. This platform has been deployed in the context of GOAL Erasmus+ Sport project. The development of the alpha version of the platform has taken into account the needs analysis and the design conducted in previous steps of the GOAL project and presented at IMCL Conference 2017 (Tsiatsos et al. 2018).

The heuristic evaluation results show that GOAL MOOC platform is easy to use, consistent, with a simple user interface and the overall attitude of the evaluators is positive.

It is worth mentioning that in the expert evaluation most of the rules had few problems. Most of the comments were on rule 10 "Help and documentation" and were about the lack of a help section and user guide on the platform. This can be remedied by creating a help section and user guide that will be easily accessible to the end users. There were also some minor comments on rule 8 "Help users recognize, diagnose, and recover from errors" about the structure of the error screen on the platform in case a user tries to access a page that does not exist. A solution to that problem is to change the error message to be more informative and present solutions to the user on how to proceed.

The next step is to resolve any issues that rose from the heuristic evaluation and conduct a technical evaluation of the platform with a larger user base.

Acknowledgements. This project has been funded with support from the European Commission. This publication reflects the views only of the author, and the Commission cannot be held responsible for any use which may be made of the information contained therein.

The authors of this research would like to thank GOAL team who generously shared their time, experience, and materials for the purposes of this project.

References

Alfermann, D., Stambulova, N.: Career transitions and career termination. In: Tenenbaum, G., Eklund, R.C. (eds.) Handbook of Sport Psychology, pp. 712–736. Wiley, New York (2007)

Amara, M., Aquilina, D., Henry, I., PMP: Education of young sportspersons (Lot 1). European Commission, Brussels (2004)

Aquilina, D.: A study of the relationship between elite athletes' educational development and sporting performance. Int. J. Hist. Sport **30**(4), 374–392 (2013). http://dx.doi.org/10.1080/09523367.2013.765723

Brackenridge, C.: Women and children first? Child abuse and child protection in sport. Sport Soc. **7**(3), 322–337 (2004)

Burnett, C.: Student versus athlete: professional socialisation influx. Afr. J. Phys. Health Educ. Recreat. Dance, 193–203 (2010)

Rethinking Education: Investing in Skills for Better Socio-Economic Outcomes. European Commission, Strasbourg (2012)

Stambulova, N., Alfermann, D., Statler, T., Côté, J.: ISSP position stand: career development and transitions of athletes. Int. J. Sport Exerc. Psychol. **7**, 395–412 (2009)

Tsiatsos, T., Douka, S., Politopoulos, N., Stylianidis, P., Ziagkas, E., Zilidou, V.: Gamified and online activities for learning to support dual career of athletes (GOAL). In: Auer, M., Tsiatsos, T. (eds.) Interactive Mobile Communication Technologies and Learning. IMCL 2017. Advances in Intelligent Systems and Computing, vol. 725. Springer, Cham (2017)

Wylleman, P., Lavallee, D.: A developmental perspective on transitions faced by ath-letes. In: Weiss, Maureen R. (ed.) Developmental Sport and Exercise Psychology: A Lifespan Perspective, pp. 507–527. Fitness Information Technology, Morgantown, WV, US (2004)

Skeleton Extraction of Dance Sequences from 3D Points Using Convolutional Neural Networks Based on a New Developed C3D Visualization Interface

Ioannis Kavouras, Eftychios Protopapadakis, Anastasios Doulamis[✉],
and Nikolaos Doulamis

National Technical University of Athens, 15773 Zografou, Athens, Greece
johncrabs1995@gmail.com,eftprot@mail.ntua.gr,
{adoulam, ndoulam}@cs.ntua.gr

Abstract. A combined approach, involving 3D spatial datasets, noise removal prepossessing and deep learning regression approaches for the estimation of rough skeleton data, is presented in this paper. The application scenario involved data sequences from Greek traditional dances. In particular, a visualization application interface was developed allowing the user to load the C3D sequences, edit the data and remove possible noise. The interface was developed using the OpenGL language and is able to parse aby C3D format file. The interface is supported by several functionalities such as a pre-processing of the 3D point data and noise removal of 3D points that fall apart from the human skeleton. The main research innovation of this paper is the use of a deep machine learning framework through which human skeleton can be extracted. The points are selected on the use of a Convolutional Neural Network (CNN) model. Experimental results on real-life dances being captured by the Vicon motion capturing system are presented to show the great performance of the proposed scheme.

Keywords: Deep learning · C3D files · Skeleton estimation

1 Introduction

Intangible cultural heritage (ICH) is a major element of peoples' identities and its preservation should be pursued along with the safeguarding of tangible cultural heritage. In this context, traditional folk dances are directly connected to local culture and identity [1]. For this reason, recently, European Union has been funded research projects for preserving, documenting and analyzing intangible cultural heritage aspects and folkoric performing arts [2,3]. The current technological achievements, in the area of software and hardware engineering, have emerged the use of efficient 3D motion capturing interfaces for digitizing human

© Springer Nature Switzerland AG 2019
M. E. Auer and T. Tsiatsos (Eds.): ICL 2018, AISC 917, pp. 267–279, 2019.
https://doi.org/10.1007/978-3-030-11935-5_26

kinesiology. Examples include the low cost Kinect sensor [4] and the most professional Vicon motion capturing interface [5]. However, these motion capturing interfaces mainly focus on the mechanisms for acquiring raw human data in the form of 3D point clouds, instead of an intelligent 3D oriented processing and visualization methodology.

Towards this direction, methods for estimating human 3D skeleton from the acquired 3D point clouds have been performed. For this reason, the single depth acquired image, taken for example from the Kinect sensor, are processed in order to extract 3D human skeleton data in a real time constraint framework [6]. In other words, the unstructured 3D point clouds are processed, using machine learning paradigms, in order to derive a compact representation regarding human kinesiology. Using this information provided by the Kinect sensors, research methods have been proposed in the literature for analyzing a choreographic pattern. In this context, the work of [7] exploits multiple Kinect sensors for dance analysis. Additionally, the work of [8] classifies pop dances based on skeleton information provided by a Kinect sensor. Other methods exploit a set of cameras for digitizing human activities such as the work of [9] regarding Japanesse dances or the work of [10] for cypriot performance art.

In the area of choreographic data analysis and processing dance summarization methods are recently investigated by the research community. Charasteristic examples are the works of [11,12] where a k-means algorithm is proposed for the estimation of the most characteristic choreographic patterns (as in [11]) or an hierarhical spatial-temporal algorithm exploiting principles of Sparse Modeling Representative Selection (SMRS) [13] (as in [12]). Additionally, the work of [12] deals also with kinesiology data structures in order to get the most salient (key) human movements.

Recently, deep learning architectures have been proposed for human action recognition exploiting skeleton information [14]. More specifically, Convolutional Neural Networks (CNNs) Models have been proposed in [15] to classify different type of dances, while in [16] a deep learning pipeline is introduced for dance style classification. Other approaches exploit Grassmannian point structures instead of deep learning towards the classification of time varying data [17].

In this paper, a method for better organization, structuring and visualization of 3D choreographic data is proposed. In particular, as far as choreographic data structuring is concerned, we have investigated the use of deep machine learning in extracting dancer's 3D skeletons from raw point clouds. For this purpose a Convolutional Neural Network model is introduced [18], with the aim of transforming 3D information, fed as input to the neural network, to discrete 3D human joints, produced as output of the network. In this way, we derive a more semantic description and organization of the raw 3D points, captured from depth sensor interfaces, the Kinect in our case. In addition, regarding the visualization of the choreographic movements, we describe an efficient interface developed in openGL framework [19] for editing and manipulating choreographies. The developed editing interface complies with the Coordinated 3 Dimensional (C3D) file format [20], which is a data structure representation for representing 3D moving objects [20]. The C3D data format has been deployed in various application scenarios, such as visualization of gaits [21] or for bio-mechanical data [22]. The developed application is a first step for creating a full featured package for visualizing and editing C3D files with emphasis on dance analysis.

This paper is organized as follows: Sect. 2 presents the new developed OpenGL based visualization interface based on C3D data. The interface is enriched with some editing and noise removal functionalities. Section 3 presents the skeleton extraction from the 3D points using deep learning Convolutional Neural Networks. Experimental results are presented in Sect. 4 while Sect. 5 concludes the paper.

2   Functionalities of the Editing Interface for Choreographic Representation

In this section, we describe the key functionalities of the proposed openGL interaface for choregraphic representation, analysis and editing. The proposed methodology imports data in C3D format and includes methods for (i) noise removal, (ii) unsupervised clustering and supervised classification, (iii) extracting human 3D skeletons (using a CNN model) and finally (iv) exporting capabilities to CSV files. Skeleton extraction includes the estimation of human body parts/joints, including head, left & right palm, upper & lower torso, knees, ankles and shoulders.

2.1   The C3D Format

The C3D data file format is originally developed by AMASS photogrammetric software system during 1986–1987. The C3D format provides a convenient and efficient means for storing 3D coordinate and analog data, together with all associated parameters [20]. C3D files are binary data types, consisting of three section blocks, i.e. header, parameters, and data. A short description for each of the blocks is provided bellow.

The Header section covers the first 512-bytes of the file. In this section, the parameter values are saved which are needed for reading the data section followed. It is not recommended to use these values from the header section, because the rational of a C3D file format is the connection between the data and their meanings (parameters). However, for simple projects, where only data are needed, it is easier to program a code, which reads the data without reading the whole parameter section.

In the parameter section, metadata information is saved. Without the parameter section, the C3D file would be just another simple numeric file format, like CSV. The parameter section is divided into the group part and the parameter part. The latter is laid inside each group. In this way, we can have more parameters with the same name, each pointing to a different description. For example, the SCALE parameter in the POINT and the ANALOG group with a different meaning in each group.

Data section includes all numeric values of the data. The numeric values are saved in frames and each frame is described by spatial and analog information.

2.2   The Developed User Interface

The application's interface developed consists of a menu toolbar, the visualization widget and the cluster list widget as shown in Fig. 1. Especially the menu toolbar subdivided into the following categories:

1. **File:** It includes commands for import and export data files. The application can only import C3D file and can export TXT and CSV files.
2. **Edit:** It includes commands for simple editing the visualization scene, such us metric scaling, and clustering editing.
3. **View:** User can use the commands of this category to change the viewpoint of the scene.
4. **Tools:** It includes all the processing commands for noise removal. A k-means clustering is exploited towards this.
5. **About:** It includes information about the application.

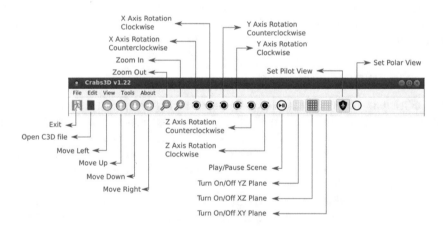

Fig. 1. The Menu developed in our case analyzing, presenting and visualizing 3D points data of a dancer

2.3   Data Editing, Preprocessing and Noise Removal

The 3D point data are often noisy. This is due to the errors of the motion capturing procedure or due to inferent low resolution of the depth sensors (see Fig. 2). In general, three types of noise is observed: (i) the cloud points are far away from the actual human body, (ii) 3D points of the human body that

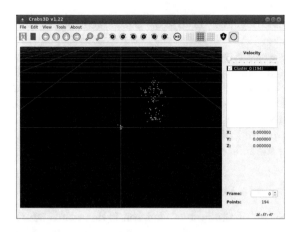

Fig. 2. Perspective view of a C3D file. In this figure, we depict the noise parts

suddenly differentiate from the body cloud and (iii) body points that suddenly gather around point of origin.

As far as the cloud points that are located far away from the actual body are concerned, the developed functionality of our interface calculates the Euclidean distance between the first frame and the other ones. If this distance is lower than a given threshold value, set by the user, then this point is recognized as immobilized. The second parameter of this algorithm is the frame rate or the tolerance. If the number of frames, where a point has been declared as immobilized, is higher than the frame rate, then this point is set invisible.

The 3D points being located far away from the actual human body are points, which behave either as object points at same frames or as noise points on other frames. To filter this kind of noise, our interface exploits the k-nearest neighbor algorithm. The user set a maximum radius and a tolerance value so that the k neighbors are specified. Then, for each point, the distance to the other points is calculated. If this point has a neighbor value at least equal to the tolerance value for all frames, then this point is declared as an object point. In any other cases, this point is considered as noise and is set to be invisible in the display mode of our interface. The main drawback of this algorithm occurs when the noise points are too close. Then the application declares them as object points since they cannot be distinguished from the other human points.

To correct the body points that suddenly gather around point of origin, we need to create a fix cloud command. This command uses the tendency of the point and predicts a probability location for the frame, considering this point as noise. This command rewrites the coordinates of a point and the result is irreversible.

3   Deep Learning for Human Skeleton Extraction

The goal of this section is to develop a new deep learning based framework for extracting human skeletons from the 3D data. The algorithm receives as inputs the 3D points of the C3D format as well as a pre-processed data filtering on the 3D points so as to remove the noise. The main steps of the algorithm are: (i) the estimation of a rough human skeleton through the application of a clustering algorithm and (ii) the application of a deep machine learning scheme for implementing the final refinement of the data.

3.1   Rough Skeleton Creation

The rough estimation of the human skeleton is done through the use of a clustering scheme. In this paper, the simple k-means algorithm is implemented. The goal of the k-means is to categorize the data into k clusters with respect to their position and orientation in the 3D space. The main disadvantage of the k-means algorithm is that fact that it is sensitive to the selection of the initial values of the cluster centers. Upon a different initial cluster center selection, different clusters are estimated and the categorization of the data may be different. Anther drawback point of the k-means clustering scheme is that the number of clusters should be a priori known. To address these limitations in this paper, and as functionality of our developed interface, a modification of the k-means is adopted called k-means++ [23]. The k-means++ can work more robustly with respect to the number of clusters than the conventional k-means algorithm.

3.2   Deep Learning Models-The Convolutional Neural Network Scheme for Skeleton Extraction

Having derived a rough human skeleton through the clustering approach proposed in the aforementioned section, we proceed with the final refinement of the model. To do this, we apply a deep machine learning scheme via the use of a

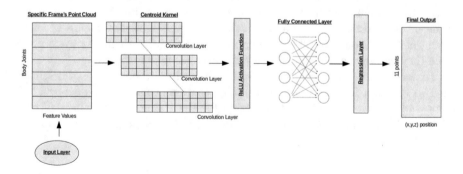

Fig. 3. The topology of the adopted Convolutional Neural Network for human skeleton extraction

Convolutional Neural Network (CNN) model. A CNN typically comprises three main types of layers, namely, (i) Convolutional Layers, (ii) ReLU Layers, and (iii) Fully Connected Layers. Each layer type plays a different role in the analysis f the data. Figure 3 shows a CNN topological architecture. This topology is useful for object detection from imaginary data. In particular, every layer of the CNN transforms, via convolutions, the inputs into more semantic meanings (descriptions) and eventually leads to the final fully connected layers. This layer performs the actual classification of the skeleton data.

In contrast to conventional neural networks structures, CNNs work well with images or even 3D volumes. This allows us to encode certain properties of the architecture and therefore a CNN can be used as a feature extractor module so that the most suitable features of the input data can be exploited.

In this paper, a modification of the classical CNN structure is proposed using the 3D scheme of [24]. The goal is to exploit a 3D CNN structure which will be useful for detecting objects. This structure is very important in our case since 3D points are received as inputs.

The Convolutional Layer. In the convolutional layers, a CNN utilizes various kernels to convolve the whole image as well as the intermediate feature maps, generating various feature maps. Because of the advantages of the convolution operation, several works (e.g., [25, 26]) have proposed it as a substitute for fully connected layers with a view to attaining faster learning times.

ReLU Layers. The units of this layer are called Rectified Linear Units (ReLU). These units apply non-saturating activation function of the form,

$$f(x) = \max(0, x) \tag{1}$$

The goal of this layer is to increase the nonlinear properties of the overall network without affecting the receptive fields of the convolution layer [27].

The Fully Connected Layer. Following several convolutional and pooling layers, the high-level reasoning in the neural network is performed through the fully connected last layer. The neurons in the fully connected layer have full connections to all nodes of the previous layer, as name of the layer implies. The neurons' output, in this layer, can hence be computed as a matrix multiplication followed by a bias offset. Fully connected layers eventually convert the 2D/3D feature maps into a 1D feature vector. The derived vector could either be fed forward into a certain number of categories for classification [28] or could be considered as a feature vector for further processing [29].

4   Experimental Setup

The application has been developed using C++ programming language. For developing the interface the QtCreator and the OpenGL are adopted.[1]

4.1   Dataset Description

For the purpose of this paper, two C3D files are used. These files have been recorded using the Vicon motion Camera System and describe the Greek traditional dance Syrtos. In each dataset a dance is executed by different dancers, one male and one female. The data of the female dancer are used for training inputs, while the data of the male dancer are used for testing. Most specifically, the training dataset consists of 1000 frames, each of which compose of 50 3D points. On the other hand, the test data set consists again of 1000 frames, in each of which we have extracted eleven rough 3D skeleton points using the clustering algorithm described in Sect. 3.1. We have also created another test set consisting of 500 image samples of a different dancer.

4.2   Performance Metrics

For the estimation of the accuracy of convolution neural network, the following four objective criteria have been taken into account.

Minimum Deviation (MIN). The minimum deviation is the lowest absolute value of the difference between the estimated values and the real measurements.

Maximum Deviation (MAX). The maximum deviation is the higher absolute value of the difference between the estimated values and the real measurements.

Mean Absolute Estimation (MAE). The mean absolute error is defined as:

$$MAE = \frac{1}{n} \sum_{i=1}^{n} |x_i - m(X)| \tag{2}$$

where x_i is the estimated value and $m(X)$ the respective central value.

Root Mean Square Estimation (RMSE). The root mean square error is defined as:

$$RMSE = \sqrt{\frac{\sum_{i=1}^{n} (x_i - m(X))^2}{n}} \tag{3}$$

where again x_i is the estimated value and $m(X)$ the respective mean operator.

---

[1] The developed user interface can be found in https://github.com/JohnCrabs/Crabs3Dv122.

Table 1. Convolutional neural network errors

Training Test for 1000 Epochs

	X Axis				Y Axis				Z Axis				Distance	
	Min [m]	Max [m]	MAE [m]	RMSE [m]	Min [m]	Max [m]	MAE [m]	RMSE [m]	Min [m]	Max [m]	MAE [m]	RMSE [m]	MAE [m]	RMSE [m]
Right Palm	0.0903	0.4682	0.3189	0.3325	0.2612	0.4391	0.3477	0.3500	0.0003	0.2938	0.1346	0.1576	0.4906	0.5078
Left Palm	0.1120	0.5133	0.3088	0.3216	0.1170	0.4458	0.2655	0.2742	0.0004	0.2866	0.1372	0.1557	0.4297	0.4504
Head	0.1944	0.5034	0.3793	0.3876	0.0002	0.1252	0.0404	0.0486	0.0016	0.2589	0.1110	0.1264	0.3973	0.4106
Upper Torso	0.1873	0.4737	0.3643	0.3706	0.0001	0.2002	0.0525	0.0601	0.0004	0.3687	0.1807	0.2064	0.4100	0.4284
Lower Torso	0.3648	0.6569	0.5204	0.5261	0.0364	0.2058	0.1256	0.1304	0.0002	0.2137	0.1127	0.1247	0.5471	0.5562
Right Shoulder	0.0008	0.5644	0.2305	0.2668	0.0001	0.2544	0.0689	0.0840	0.0003	1.0740	0.2645	0.3513	0.3575	0.4491
Left Shoulder	0.1680	0.8349	0.4355	0.4666	0.0001	0.2746	0.0653	0.1032	0.0002	0.4657	0.2459	0.2746	0.5044	0.5512
Right Knee	0.3059	0.8789	0.6361	0.6554	0.0557	0.2345	0.1435	0.1486	0.0016	0.4408	0.1913	0.2189	0.6796	0.7068
Left Knee	0.3693	0.7366	0.5597	0.5675	0.2087	0.5211	0.3584	0.3657	0.0001	0.1639	0.0768	0.0894	0.6690	0.6810
Right Ankle	0.0001	0.1639	0.0768	0.0894	0.1292	0.6807	0.4112	0.4343	0.0001	0.0900	0.0283	0.0361	0.4193	0.4449
Left Ankle	0.2071	0.6947	0.4906	0.5016	0.0682	0.2191	0.1330	0.1371	0.1329	0.2071	0.2346	0.2439	0.5598	0.5744

Training Test for 2000 Epochs

	X Axis				Y Axis				Z Axis				Distance	
	Min [m]	Max [m]	MAE [m]	RMSE [m]	Min [m]	Max [m]	MAE [m]	RMSE [m]	Min [m]	Max [m]	MAE [m]	RMSE [m]	MAE [m]	RMSE [m]
Right Palm	0.0002	0.4353	0.1787	0.2280	0.2268	0.5457	0.3758	0.3834	0.0019	0.4412	0.2509	0.2694	0.4859	0.5211
Left Palm	0.0001	0.3698	0.1916	0.2167	0.2224	0.5088	0.3353	0.3420	0.0002	0.2595	0.1159	0.1354	0.4032	0.4269
Head	0.0005	0.5309	0.3000	0.3253	0.1518	0.4070	0.2521	0.2596	0.0001	0.1751	0.0551	0.0732	0.3957	0.4226
Upper Torso	0.0002	0.4093	0.1831	0.2205	0.0065	0.3475	0.2300	0.2359	0.0006	0.6280	0.2621	0.3166	0.3939	0.4522
Lower Torso	0.0003	0.3279	0.1812	0.1955	0.1466	0.3421	0.2185	0.2218	0.0003	0.4234	0.1498	0.1879	0.3210	0.3503
Right Shoulder	0.0015	2.6338	0.8822	1.1857	0.0032	0.6290	0.4075	0.4212	0.0012	2.6413	0.8618	1.1712	1.2989	1.7190
Left Shoulder	0.0002	0.4799	0.1630	0.2280	0.0224	0.3284	0.1561	0.1745	0.0007	0.6828	0.3144	0.3411	0.3870	0.4459
Right Knee	0.0004	0.5411	0.2341	0.2721	0.1595	0.3374	0.2190	0.2218	0.0001	0.4506	0.2057	0.2436	0.3809	0.4273
Left Knee	0.0001	0.3307	0.1266	0.1612	0.3838	0.7579	0.5526	0.5600	0.0001	0.2641	0.0861	0.1086	0.5734	0.5928
Right Ankle	0.0001	0.2641	0.0861	0.1086	0.0339	0.6287	0.3047	0.3432	0.0001	0.1351	0.0354	0.0506	0.3186	0.3635
Left Ankle	0.0912	0.6240	0.3557	0.3827	0.0573	0.2639	0.1565	0.1643	0.1409	0.5429	0.3781	0.3918	0.5422	0.5718

Training Test for 3000 Epochs

	X Axis				Y Axis				Z Axis				Distance	
	Min [m]	Max [m]	MAE [m]	RMSE [m]	Min [m]	Max [m]	MAE [m]	RMSE [m]	Min [m]	Max [m]	MAE [m]	RMSE [m]	MAE [m]	RMSE [m]
Right Palm	0.0515	0.6427	0.2697	0.3175	0.4703	1.0074	0.7712	0.7843	0.0025	0.4389	0.2624	0.2816	0.8581	0.8918
Left Palm	0.0170	0.3021	0.1112	0.1283	0.3348	0.8011	0.6197	0.6324	0.0623	0.3231	0.2046	0.2167	0.6620	0.6807
Head	0.2290	0.7120	0.4461	0.4724	0.3330	0.7028	0.5570	0.5662	0.2460	0.5291	0.3363	0.3450	0.7889	0.8141
Upper Torso	0.0466	0.5526	0.2991	0.3150	0.0156	0.5784	0.4380	0.4494	0.0128	0.5129	0.2759	0.3008	0.5979	0.6258
Lower Torso	0.2496	0.6172	0.4612	0.4673	0.1730	0.4548	0.3216	0.3288	0.0001	0.1664	0.0350	0.0469	0.5633	0.5733
Right Shoulder	0.0018	1.5806	0.6349	0.6859	0.0003	0.9963	0.6150	0.6936	0.0043	1.5896	0.5176	0.7348	1.0243	1.2213
Left Shoulder	0.0003	0.4796	0.1775	0.2427	0.1434	0.5412	0.3963	0.4149	0.0038	0.3205	0.3205	0.3429	0.5397	0.5904
Right Knee	0.3267	0.9705	0.6378	0.6474	0.0740	0.2828	0.1690	0.1745	0.0004	0.3357	0.1085	0.1432	0.6687	0.6856
Left Knee	0.0001	0.3472	0.1487	0.1744	0.4003	0.8814	0.6918	0.7022	0.0561	0.3591	0.2192	0.2346	0.7408	0.7606
Right Ankle	0.0561	0.3591	0.2192	0.2346	0.3417	0.7271	0.5357	0.5465	0.0417	0.2033	0.1320	0.1380	0.5937	0.6105
Left Ankle	0.0001	0.2673	0.1080	0.1248	0.1728	0.3148	0.2335	0.2362	0.0758	0.6393	0.4517	0.4640	0.5198	0.5354

Training Test for 4000 Epochs

	X Axis				Y Axis				Z Axis				Distance	
	Min [m]	Max [m]	MAE [m]	RMSE [m]	Min [m]	Max [m]	MAE [m]	RMSE [m]	Min [m]	Max [m]	MAE [m]	RMSE [m]	MAE [m]	RMSE [m]
Right Palm	0.0010	0.4986	0.2828	0.3073	0.2629	0.6454	0.4464	0.4551	0.0002	0.5030	0.2540	0.2937	0.5863	0.6227
Left Palm	0.0003	0.5658	0.2140	0.2137	0.4038	0.6819	0.5455	0.5494	0.0003	0.2927	0.1136	0.1352	0.5969	0.6048
Head	0.0002	0.5493	0.3645	0.3909	0.2055	0.4214	0.2589	0.3026	0.0023	0.4572	0.1871	0.2220	0.5072	0.5419
Upper Torso	0.0005	1.1522	0.5110	0.6128	0.1978	0.5850	0.3579	0.3691	0.0007	1.1237	0.4745	0.5305	0.7838	0.8906
Lower Torso	0.0019	0.4606	0.2819	0.3071	0.1560	0.3496	0.2440	0.2481	0.0010	0.5580	0.2403	0.2915	0.4436	0.4908
Right Shoulder	0.0006	5.8187	1.7217	2.6620	0.1786	2.2193	0.8004	1.0930	0.0003	5.5726	1.6070	2.5340	2.4874	3.8343
Left Shoulder	0.0001	0.4768	0.3057	0.3286	0.1968	0.4898	0.3460	0.3528	0.0015	0.6745	0.3614	0.4114	0.5863	0.6338
Right Knee	0.0024	0.6906	0.3877	0.4458	0.2336	0.4173	0.3282	0.3310	0.0005	0.5148	0.2212	0.2641	0.5540	0.6149
Left Knee	0.0029	0.4495	0.2284	0.2589	0.0984	0.6692	0.3715	0.4082	0.0031	0.8519	0.3435	0.4144	0.5551	0.6367
Right Ankle	0.0031	0.8519	0.3435	0.4144	0.0004	0.5885	0.3441	0.3943	0.0001	0.1210	0.0492	0.0581	0.4887	0.5750
Left Ankle	0.0056	0.3608	0.2369	0.2467	0.0001	0.1577	0.0623	0.0765	0.0001	0.3914	0.1841	0.2147	0.3064	0.3359

Training Test for 5000 Epochs

	X Axis				Y Axis				Z Axis				Distance	
	Min [m]	Max [m]	MAE [m]	RMSE [m]	Min [m]	Max [m]	MAE [m]	RMSE [m]	Min [m]	Max [m]	MAE [m]	RMSE [m]	MAE [m]	RMSE [m]
Right Palm	0.0532	0.5253	0.2693	0.2995	0.1912	0.6608	0.3824	0.3959	0.0007	0.4666	0.2908	0.3141	0.5507	0.5874
Left Palm	0.0003	0.3306	0.1316	0.1622	0.0003	0.4748	0.2175	0.2540	0.1097	0.4872	0.3174	0.3357	0.4067	0.4511
Head	0.0006	0.5617	0.3068	0.3451	0.0037	0.4176	0.2500	0.2688	0.0001	0.1962	0.0866	0.1026	0.4051	0.4493
Upper Torso	0.0640	0.5373	0.3141	0.3520	0.0001	0.2765	0.1252	0.1537	0.0001	0.3749	0.1293	0.1587	0.3620	0.4156
Lower Torso	0.0007	0.6104	0.3843	0.4031	0.0001	0.2654	0.1126	0.1397	0.0006	0.5303	0.3511	0.3630	0.5326	0.5602
Right Shoulder	0.0069	1.6632	0.6372	0.7003	0.0001	0.5389	0.1091	0.1628	0.0001	1.4517	0.5098	0.5770	0.8233	0.9219
Left Shoulder	0.0235	0.4891	0.2197	0.2633	0.0001	0.3267	0.1540	0.1812	0.0001	0.4595	0.2356	0.2715	0.3571	0.4194
Right Knee	0.0036	0.8403	0.4710	0.5172	0.0001	0.2285	0.1105	0.1253	0.0027	0.4644	0.3158	0.3275	0.5777	0.6249
Left Knee	0.0004	0.4307	0.1821	0.2147	0.0002	0.2383	0.0715	0.0917	0.0658	0.6808	0.4157	0.4411	0.4594	0.4991
Right Ankle	0.0658	0.6808	0.4157	0.4411	0.0585	0.8037	0.5383	0.5940	0.0027	0.2006	0.1240	0.1304	0.6913	0.7513
Left Ankle	0.0001	0.1659	0.0795	0.0890	0.0006	0.2574	0.1064	0.1206	0.1384	0.7087	0.4468	0.4754	0.4661	0.4985

Training Test for 2000 Epochs with 500 frames

	X Axis				Y Axis				Z Axis				Distance	
	Min [m]	Max [m]	MAE [m]	RMSE [m]	Min [m]	Max [m]	MAE [m]	RMSE [m]	Min [m]	Max [m]	MAE [m]	RMSE [m]	MAE [m]	RMSE [m]
Right Palm	0.0005	0.6301	0.3165	0.3665	0.0827	0.1969	0.1326	0.1358	0.0155	0.2825	0.1784	0.1977	0.3868	0.4380
Left Palm	0.0008	0.7446	0.3310	0.4119	0.1791	0.3081	0.2412	0.2433	0.0785	0.3187	0.1874	0.2004	0.4504	0.5187
Head	0.0003	0.5994	0.3360	0.3821	0.0497	0.1373	0.0855	0.0873	0.0892	0.2604	0.1686	0.1750	0.3855	0.4292
Upper Torso	0.0009	0.5671	0.3374	0.3816	0.0039	0.1041	0.0473	0.0516	0.3032	0.5148	0.3892	0.3946	0.5173	0.5514
Lower Torso	0.0005	0.6655	0.3384	0.3942	0.0664	0.1474	0.0958	0.0972	0.2230	0.4908	0.3287	0.3393	0.4814	0.5291
Right Shoulder	0.0006	0.6393	0.3420	0.3930	0.0223	0.1106	0.0581	0.0608	0.1213	0.3555	0.2383	0.2495	0.4209	0.4695
Left Shoulder	0.0011	0.6299	0.3369	0.3862	0.0254	0.1277	0.0695	0.0725	0.3141	0.5410	0.4099	0.4153	0.5351	0.5717
Right Knee	0.0003	0.6497	0.3588	0.4078	0.1580	0.2771	0.1985	0.1998	0.0290	0.3530	0.2049	0.2288	0.4584	0.5085
Left Knee	0.0005	0.7429	0.3560	0.4303	0.1265	0.2615	0.1757	0.1788	0.4147	0.7033	0.5248	0.5310	0.6580	0.7065
Right Ankle	0.4147	0.7033	0.5248	0.5310	0.0022	0.5565	0.3746	0.4142	0.0079	0.0951	0.0381	0.0438	0.6459	0.6749
Left Ankle	0.0032	0.6513	0.3033	0.3544	0.0198	0.1398	0.0649	0.0737	0.1112	0.4546	0.2783	0.2885	0.4167	0.4629

4.3   Experimental Results

In order to evaluate the performance of the CNN model in estimating human skeleton from 3D points, the aforementioned objective criteria are used. Table 1 shows the results regarding the eleven targeted skeleton points for different values of training epochs of the CNN model. The results are depicted for the two considered data sets; the one of the 1000 image frames and the other of the 500 image samples. Both test sets comprise different dancers against the one of the training set. In this table, the error over the three axes (i.e., xyz) are depicted.

Figure 4 shows the RMSE metric for the eleven skeleton human points over all epochs that the CNN model has been trained on. The figure considers the first test set of the 1000 image samples. As is observed, the majority of the human skeleton joints are well identified by the CNN model.

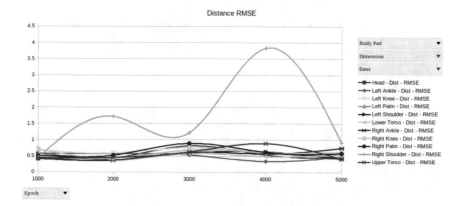

Fig. 4. Test results for all epochs

Figure 5 depicts performance evaluation between the two test sets; the one of 500 frames and the one of 1000 frames, in case that 2000 epochs are used. Epochs. As is observed, the test set of 500 image samples are slightly better than that the one of the 1000 image samples.

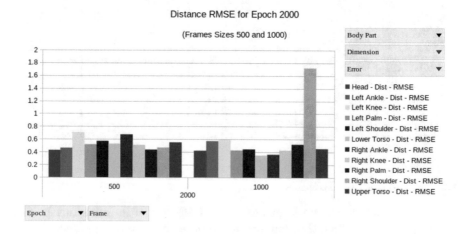

Fig. 5. Comparison between datasets 500 and 1000 frames for epoch 2000

5   Conclusions and Future Works

An advanced user interface for visualizing and editing C3D dataset was proposed in this paper. The application considers the case of editing traditional Greek dance sequences, obtained using VICON sensors. The interface allows the user to monitor the dance choreography, step-by - step, filter the noise using clustering approaches and create a rough estimation of the dancer's body joints. Different types of functionalities are supported such as noise removal of 3D points that do not correspond to human skeletons and clustering. The developed interface was built on C++ programming on the exploitation of the OpenGL language. The interface is extensible in the sense that it can parse any C3D type following the precise instructions and recommendations of the standard.

Another key innovation of this paper is the use of a novel deep machine learning framework for the extraction of human skeleton. Deep machine learning is implemented through Convolutional Neural Networks. Initially a rough human skeleton is extracted using clustering approaches.

Experimental results and validation on real-life dance 3D point video sequences are conducted. The results show the excellent performance of the proposed method and the ability of the developed interface to support any type of dance sequence. Semi-supervised and adaptive methods can be also exploited, as the ones in [30,31], for improving the skeleton modelling performance.

Acknowledgments. This work was supported by the EU H2020 TERPSICHORE project "Transforming Intangible Folkloric Performing Arts into Tangible Choreographic Digital Objects" under the grant agreement 691218.

References

1. Barbara, S.-Y., Shay, A.: The Oxford Handbook of Dance and Ethnicity. Oxford University Press, Oxford (2016)
2. Dimitropoulos, K., Manitsaris, S., Tsalakanidou, F., Denby, B., Buchman, L., Dupont, S., Nikolopoulos, S., Kompatsiaris, Y., Charisis, V., Hadjileontiadis, L., Pozzi, F., Cotescu, M., Ciftci, S., Katos, A., Manitsaris, A., Grammalidis, N.: A multimodal approach for the safeguarding and transmission of intangible cultural heritage: the case of i-treasures. IEEE Intell. Syst. 1–1. https://doi.org/10.1109/MIS.2018.111144858 (2018)
3. Doulamis, A.D., Voulodimos, A., Doulamis, N.D., Soile, S., Lampropoulos, A.: Transforming intangible folkloric performing arts into tangible choreographic digital objects: the terpsichore approach. In: International Conference on Computer Vision, Theory and Applications (VISIGRAPP), Porto, Portugal, pp. 451–460 (2017)
4. Zhang, Z.: Microsoft kinect sensor and its effect. IEEE Multimed. **19**, 4–10 (2012)
5. Windolf, M., Gtzen, N., Morlock, M.: Systematic accuracy and precision analysis of video motion capturing systems-exemplified on the Vicon-460 system. J. Biomech. **41**, 2776–2780 (2008)
6. Shotton, J., Fitzgibbon, A., Cook, M., Sharp, T., Finocchio, M., Moore, R., Kipman, A., Blake, A.: Real-time human pose recognition in parts from single depth images, pp. 1297–1304 (2011)

7. Kitsikidis, A., Dimitropoulos, K., Douka, S., Grammalidis, N.: Dance analysis using multiple kinect sensors. In: VISAPP 2014—Proceedings of the 9th International Conference on Computer Vision Theory and Applications, vol. 2, pp. 789–795 (2014)

8. Kim, D., Kim, D.-H., Kwak, K.-C.: Classification of k-pop dance movements based on skeleton information obtained by a kinect sensor. Sensors **17**, 1261 (2017)

9. Hisatomi, K., Katayama, M., Tomiyama, K., Iwadate, Y.: 3D archive system for traditional performing arts: application of 3D reconstruction method using graph-cuts. Int. J. Comput. Vis. **94**, 78–88 (2011)

10. Stavrakis, E., Aristidou, A., Savva, M., Himona, S., Chrysanthou, Y.: Digitization of cypriot folk dances. Lecture Notes in Computer Science (including subseries Lecture Notes in Artificial Intelligence and Lecture Notes in Bioinformatics), vol. 7616, pp. 404–413 (2012)

11. Rallis, I., Georgoulas, I., Doulamis, N., Voulodimos, A., Terzopoulos, P.: Extraction of key postures from 3D human motion data for choreography summarization. In: Proceedings of the IEEE 9th International Conference on Virtual Worlds and Games for Serious Applications, (VS-Games), pp. 94–101 (2017)

12. Rallis, I., Doulamis, N., Doulamis, A., Voulodimos, A., Vescoukis, V.: Spatio-temporal summarization of dance choreographies. Comput. Graph. **73**, 88–101 (2018)

13. Elhamifar, E., Sapiro, G., Vidal, R.: See all by looking at a few: sparse modeling for finding representative objects. In: Proceedings of IEEE Conference on Computer Vision and Pattern Recognition, pp. 1600–1607 (2012)

14. Wang, H.: A survey on deep neural networks for human action recognition based on skeleton information. In: Advances in Intelligent Systems and Computing, vol. 541, pp. 329–336 (2017)

15. Protopapadakis, E., Voulodimos, A., Doulamis, A., Camarinopoulos, S., Doulamis, N., Miaoulis, G.: Dance pose identification from motion capture data: a comparison of classifiers. Technologies **6**(1), 31 (2018)

16. Dewan, S., Agarwal, S., Singh, N.: A deep learning pipeline for Indian dance style classification, vol. 10696 (2018)

17. Dimitropoulos, K., Barmpoutis, P., Kitsikidis, A., Grammalidis, N.: Classification of multidimensional time-evolving data using histograms of Grassmannian points. IEEE Trans. Circuits Syst. Video Technol. **28**, 892–905 (2018)

18. Voulodimos, A., Doulamis, N., Doulamis, A., Protopapadakis, E.: Deep learning for computer vision: a brief review. Comput. Intell. Neurosci. **2018**, 13 pages (2018)

19. McReynolds, T., Blythe, D.: Advanced Graphics Programming Using OpenGL. Morgan Kaufmann Publishers Inc., San Francisco (2005)

20. Motion Lab systems Inc.: The C3D File Format User Guide. United States of America, 1997–2008

21. Alfalah, S., Chan, W., Khan, S., Falah, J., Alfalah, T., Harrison, D., Charissis, V.: Gait analysis data visualisation in virtual environment (GADV/VE). In: Proceedings of 2014 Science and Information Conference, SAI 2014, pp. 742–751 (2014)

22. Barre, A., Armand, S.: Biomechanical toolkit: open-source framework to visualize and process biomechanical data. Comput. Methods Programs Biomed. **114**(1), 80–87 (2014)

23. Nguyen, T.-H., Huynh, V.-N.: A k-means-like algorithm for clustering categorical data using an information theoretic-based dissimilarity measure. In: Lecture Notes in Computer Science (including subseries Lecture Notes in Artificial Intelligence and Lecture Notes in Bioinformatics), vol. 9616, pp. 115–130 (2016)

24. Makantasis, K., Doulamis, A., Doulamis, N., Psychas, K.: Deep learning based human behavior recognition in industrial workflows. In: Proceedings— International Conference on Image Processing, ICIP, August 2016, pp. 1609–1613 (2016)
25. Laptev, I., Oquab, M., Bottou, L., Sivic, J.: Is object localization for free? - weakly-supervised learning with convolutional neural networks. In: Proceedings of the IEEE Conference on Computer Vision and Pattern Recognition, CVPR 2015, pp. 685–694, June 2011
26. Jia, Y., Szegedy, C., Liu, W.: Going deeper with convolutions. In: Proceedings of the IEEE Conference on Computer Vision and Pattern Recognition (CVPR 2015), Boston, Mass, USA. p. 19, June 2011
27. Wikipedia. Convolutional neural network. https://en.wikipedia.org/wiki/ Convolutional_neural_network
28. Sutskever, I., Krizhevsky, A., Hinton, G.E.: Imagenet classification with deep convolutional neural networks. In: Proceedings of the 26th Annual Conference on Neural Information Processing Systems (NIPS 2012), Lake Tahoe, Nev, USA, pp. 1097–1105, December 2012
29. Darrell, T., Girshick, R., Donahue, J., Malik, J.: Rich feature hierarchies for accurate object detection and semantic segmentation. In: Proceedings of the 27th IEEE Conference on Computer Vision and Pattern Recognition, pp. 580–587, June 2015
30. Doulamis, N., Doulamis, A.: Semi-supervised deep learning for object tracking and classification, pp. 848–852 (2014)
31. Doulamis, N., Doulamis, A.: Fast and adaptive deep fusion learning for detecting visual objects. In: Lecture Notes in Computer Science (including subseries Lecture Notes in Artificial Intelligence and Lecture Notes in Bioinformatics), vol. 7585, no. PART 3, pp. 345–354 (2012)

The Effects of 8-Week Plyometric Training on Tennis Agility Performance, Improving Evaluation Throw the Makey Makey

Efthymios Ziagkas[1](✉), Vassiliki I. Zilidou[1], Andreas Loukovitis[1],
Nikolaos Politopoulos[2], Styliani Douka[1], and Thrasyvoulos Tsiatsos[2]

[1] School of Physical Education and Sport Science,
Aristotle University of Thessaloniki, Thessaloniki, Greece
{eziagkas, sdouka}@phed.auth.gr,
vickizilidou@gmail.com, Louko-vitis@hotmail.com
[2] School of Informatics, Aristotle University of Thessaloniki,
Thessaloniki, Greece
{npolitop, tsiatsos}@csd.auth.gr

Abstract. Tennis is an intermittent sport characterized by repeated high-intensity efforts during a variable period of time. Aiming to be competitive and successful, tennis players need a mixture of speed, agility, and power combined with high aerobic fitness. Plyometrics are training techniques used by athletes in all types of sports to increase strength and explosiveness. The objectivity of the performance measure varies through measures such as: time, checklists, or established criteria. Makey Makey is an invention kit for everyone, is an electronic invention tool and toy that allows users to connect everyday objects to computer programs. The aim of this study was twofold. Firstly the purpose of this study was to examine the effects of a 8-week plyometric training program on tennis specific agility. Secondly, this study was directed towards examining objectivity evaluation among three different evaluators/coaches and highlight the need to use new technologies in order to evaluate performance on agility objectively. 24 male amateur tennis players participated in this study. Subjects separated randomly in two groups, the first group underwent an intervention of a 8 week plyometric program. In order to examine agility we used two tennis specific agility test. 3 independent coaches were used in order to measure participants score on each test. When we use the mean scores of coaches we show that the experimental group showed significant improvement on doth test. We also examined the differences and the correlations among coach's test in order to estimate the objectivity. Our findings highlight the need of mew technologies in order to improve objectivity in laboratory and field measurements. Here we suggest the use, and the development of suitable software for the Makey Makey device.

Keywords: Plyometrics · Tennis · Agility · Informatics · Objectivity

© Springer Nature Switzerland AG 2019
M. E. Auer and T. Tsiatsos (Eds.): ICL 2018, AISC 917, pp. 280–286, 2019.
https://doi.org/10.1007/978-3-030-11935-5_27

1 Introduction

Over 75 million people worldwide play tennis [1] and 200 nations are associated with the International Tennis Federation [2]. The health benefits of this sport are well recognized; tennis players have a higher level of aerobic fitness, a decreased risk of cardiovascular disease, a lower body fat percentage, and improved bone health compared with less active individuals [3]. Also tennis is an intermittent sport characterized by repeated high-intensity efforts (i.e., accelerations, decelerations, and changes of direction and strokes) during a variable period of time. Aiming to be competitive and successful, tennis players need a mixture of speed, agility, and power combined with high aerobic fitness [4]. Players must be able to react as fast as possible to actions performed by the opponent, where reaction time, initial acceleration, and agility play an important role [5]. Initial acceleration can be referred to as the first 10 m of a sprint [6], while agility can be recognized as the ability to change direction by starting and stopping quickly during points [7]. The mean sprint distance performed in tennis court is 4–7 m, with an average of 4 changes of direction per point [4]. Based on those facts, tennis players need to possess exceptional dynamism in multidirectional movements during matches.

Plyometrics are training techniques used by athletes in all types of sports to increase strength and explosiveness [8]. Plyometrics consists of a rapid stretching of a muscle (eccentric action) immediately followed by a concentric or shortening action of the same muscle and connective tissue [9]. Plyometric drills usually involve stopping, starting, and changing directions in an explosive manner. These movements are components that can assist in developing agility [10–14]. Agility is the ability to maintain or control body position while quickly changing direction during a series of movements [15].

Objective and subjective performance measures are used to classify the various different types of performance measures. Objective performance measures are independent of the observer. That means the measurement is done using something other than the person observing. This independent measure can include: a stop-watch, measuring tape or record of goals. The objectivity of the performance measure is increased through measures such as: time, checklists, or established criteria. In contrast subjective performance measures are dependent on the observer and based on opinions, feelings, and general impressions. Subjective measures rely more on the observer than independent measures [16].

Makey Makey is an invention kit for everyone, is an electronic invention tool and toy that allows users to connect everyday objects to computer programs [17]. Using a circuit board, alligator clips, and a USB cable, the toy uses closed loop electrical signals to send the computer either a keyboard stroke or mouse click signal. This function allows the Makey Makey to work with any computer program or webpage since all computer programs and webpages take keyboard and mouse click inputs.

Therefore, the aim of this study was twofold. Firstly the purpose of this study was to examine the effects of a 8-week plyometric training program on tennis specific agility. Secondly, this study was directed towards examining objectivity evaluation among three different evaluators/coaches and highlight the need to use new technologies in order to evaluate performance on agility objectively.

2 Methods

2.1 Participants

24 male amateur level tennis players participated in this study. The participant's mean age was 20.9 (SD = ±0.66). The training years in tennis were between 1 and 3 years. Participants separated randomly in two sub groups. The experimental group including 12 players and the control group consisted of the same number of players.

2.2 Training Program

The plyometric training program consisted of a combination of upper body and lower body exercises. Three coaches organized weekly meetings to assign similar tennis training loads to both the experimental and the control group (i.e., number of exercises, technical/tactical aims). A program of 8 week exercises, performed at maximal intensity, with 2–4 sets and 10–15 repetitions each was applied. The rest period ranged between 15s and 90s depending on the exercise and number of sets performed during the trials. Proper technique was ensured via verbal cues and demonstration by the strength and conditioning coaches. plyometric sessions were conducted within the tennis training sessions (i.e., as a substitute for some tennis training within the usual 90-min practice), lasted from 30 to 60 min, and were followed by a 5-min cool-down protocol (e.g., general mobilization).

2.3 Testing Protocol

Hexagon Test: The hexagon test measures the time it takes to jump over six sides of a 24-in. taped hexagon three times around. The score is in seconds. It measures foot quickness in changing direction backward, forward, and sideways while facing in one direction. (Facing in the same direction during the test simulates facing the net during play.) The hexagon also tests the ability to stabilize the body quickly between those changes of direction because the body needs to be stabilized before the next jump can be performed. If the body is not stabilized, players will lose their balance. This causes them to either touch one of the lines and receive a time penalty of one half of a second or causes to jump out of order and receive a time penalty of one second.

Spider Test: The spider test measures the time it takes you to pick up five tennis balls and return them individually to a specified zone. The score is in seconds. The spider test includes agility and speed. In this test it is allowed to face in any direction and move in whatever direction possible. The spider test includes stopping; starting, and changing direction.

2.4   Procedure

All subjects participated in the semester's tennis lesson at the Department of Physical Education and Sport Science at Aristotle university of Thessaloniki. As an intervention the experimental group followed the 8 weeks plyometric program biweekly as described by Fernandez et al. in 2016 [18]. while the control group at the same time watched tennis matches on the television. Two agility tests were carried out before (pretest) and after (posttest) the training period, including (a) the hexagon test, and (b) the spider test. Three experienced, independent tennis coaches measured the time on those test of each participant using a digital chronographer.

2.5   Statistical Analysis

For the data analyses we used descriptive statics to present mean scores in the examining protocol. In order to test the first hypothesis one way Anova was used. As depended variable for this analysis we used the estimated mean time of the three observers/coaches. Especially, one way Anova was used in order to show that there were no significant differences between two groups at baseline. Also this analysis was used in order to test performance on agility after intervention between two groups.

To test the second hypothesis we performed paired samples t-test and Pearsons r correlation in order to evaluate objectivity among the three observers/coaches. All analysis performed using SPSS 24.0 IBM and the p value was set at 0.05.

3   Results

When we used the mean score of three observers/coaches we show that, at baseline, no significant differences were found between groups on both agility tests ($F_{1,22} = 0.574$, $p = 0.05$). The mean score of three observers for the experimental group was 12.6 ± 1.67 s on the hexagon test and 16.4 ± 2.28 s on the spider test, while in the control group the mean score was 12.3 ± 1.85 s on the hexagon test and 16.6 ± 2.18 s on the spider test.

After the intervention control group shown no significant differences although has achieved better scores on both tests (11.5 ± 1.83 s on hexagon test $F_{1,23} = 0.728$ $p > 0.05$ and 15.2 ± 2.22 s on spider test $F_{1,23} = 1.264$ $p > 0.05$). As regards the experimental group, significant differences have shown better improvements in both hexagon (9.8 ± 1.69) and spider test ($11.7 \pm s2.32$) (Table 1).

Analyzing the mentioned score of each observer/coach we found significant differences between the observers in all variables. In Figs. 1 and 2 the mean score of each observer is presented pre and post testing on hexagon test (Figs. 1 and 2).

Also the Pearson r correlation showed moderate or low correlations between observers' scores as presented in Table 2.

Table 1. Pre and post intervention mean scores on agility tests of the two groups

	Hexagon test		Spider test	
	Experimental group mean (s), SD	Control group mean (s), SD	Experimental group mean (s), SD	Control group mean (s), SD
Pre test	12.6 ± 1.67	12.3 ± 1.85	16.4 ± 2.28	16.6 ± 2.18
Post test	9.8* ± 1.69	11.5 ± 1.83	11.7* ± 2.32	15.2 ± 2.22

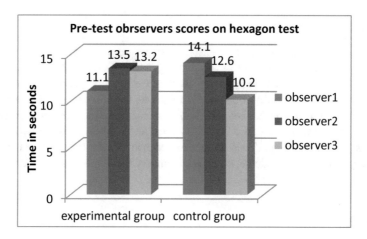

Fig. 1. .

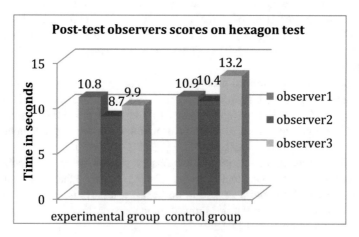

Fig. 2. .

Table 2. Pearson r correlation between observers/coaches scoring

	Hexagon test						
	Pre test			Post test			
	Observer 1	Observer 2	Observer 3		Observer 1	Observer 2	Observer 3
Observer 1	1			Observer 1	1		
Observer 2	0.53	1		Observer 2	0.63	1	
Observer 3	0.68	0.47	1	Observer 3	0.65	0.52	1
	Spider test						
	Observer 1	Observer 2	Observer 3		Observer 1	Observer 2	Observer 3
Observer 1	1			Observer 1	1		
Observer 2	0.58	1		Observer 2	0.49	1	
Observer 3	0.45	0.32	1	Observer 3	0.60	0.61	1

4 Discussion

The aim of this study was twofold. Firstly the purpose of this study was to examine the effects of a 8-week plyometric training program on tennis specific agility. Secondly, this study was directed towards examining objectivity evaluation among three different evaluators/coaches and highlight the need to use new technologies in order to evaluate performance on agility objectively.

As regards the first hypotheris, findings showed that the 8 week plyometric training program occurred significant differences on tennis specific agility in the experimental group. Those findings are on line with recent bibliography supporting that plyometrics seems to affect agility positively [18]. It has been suggested that increases in power and efficiency due to plyometrics may increase agility training objectives [19] and plyometric activities have been used in sports such as football, soccer or other sporting events that agility may be useful for their athletes [12, 20, 21].

Concerning the second hypothesis, our data showed that observers/coaches score measurements seems to significantly differ among them and we also found moderate to low correlation among their scores. Those findings support that objectivity of the performance measure may vary through measures such as: time, checklists, or established criteria. Those performance measures are dependent on the observer and based on opinions, feelings, and general impressions. Subjective measures rely more on the observer than independent measures [16].

Consequently, its necessary for new technologies and informatics to develop hardware and software in order to improve objectivity in laboratory and field measures in sports.

5 Conclusion

The findings of this study shows that the 8 week plyometric training program induced significant differences on tennis specific agility, although the low degree of objectivity among observers/coaches. Based on these findings, we recommend the use of Makey Makey as hardware as it is a low cost equipment with a wide range of uses in order to improve the necessary software aiming to more reliable and objective measurements in the laboratory in field measures in sport.

References

1. Pluim, B.M., Miller, S., Dines, D., Renström, P.A., Windler, G., et al.: Sport science and medicine in tennis. Br. J. Sports Med. **41**, 703–704 (2007)
2. Abrams, G.D., Renstrom, P.A., Safran, M.R.: Epidemiology of musculoskeletal injury in the tennis player. Br. J. Sports Med. **46**, 492–498 (2012)
3. Pluim, B.M., Staal, J.B., Marks, B.L., Miller, S., Miley, D.: Health benefits of tennis. Br. J. Sports Med. **41**, 760–768 (2007)
4. Fernandez-Fernandez, J., Sanz-Rivas, D., Mendez-Villanueva, A.: A review of the activity profile and physiological demands of tennis match play. Strength Condition. J. **31**(4), 15–26 (2009). https://doi.org/10.1519/SSC.0b013e3181ada1cb
5. Reid, M., Sibte, N., Clark, S., Whiteside, D. (eds.): Tennis players. In: Physiological Tests for Elite Athletes, 2nd edn. Australian Institute of Sport. Human Kinetics, Champaign, IL (2013)
6. Nagahara, R., Matsubayashi, T., Matsuo, A., Zushi, K.: Kinematics of transition during human accelerated sprinting. Biol Open. **3**(8), 689–699 (2014). PubMed https://doi.org/10.1242/bio.20148284
7. Sheppard, J.M., Young, W.: Agility literature review: classifications, training and testing. J. Sports Sci. **24**(9), 919–932 (2006). PubMed https://doi.org/10.1080/02640410500457109
8. Chu, D.A.: Jumping into Plyometrics. Human Kinetics, Champaign, IL (1998)
9. Baechle, T.R., Earle, R.W.: Essentials of strength training and conditioning, 2nd edn. National Strength and Conditioning Association, Champaign, IL (2000)
10. Craig, B.W.: What is the scientific basis of speed and agility? Strength Condition. **26**(3), 13–14 (2004)
11. Miller, J.M., Hilbert, S.C., Brown, L.E.: Speed, quickness, and agility training for senior tennis players. Strength Condition. **23**(5), 62–66 (2001)
12. Parsons, L.S., Jones, M.T.: Development of speed, agility and quickness for tennis athletes. Strength Condition. **20**(3), 14–19 (1998)
13. Yap, C.W., Brown, L.E.: Development of speed, agility, and quickness for the female soccer athlete. Strength Condition. **22**, 9–12 (2000)
14. Young, W.B., McDowell, M.H., Scarlett, B.J.: Specificity of spring and agility training methods. J. Strength Condition. Res. **15**, 315–319 (2001)
15. Twist, P.W., Benicky, D.: Conditioning lateral movements for multi-sport athletes: practical strength and quickness drills. Strength Condition. **18**(5), 10–19 (1996)
16. Rasidagić, F.: Objectivity in the evaluation of motor skill performance in sport and physical education. Homo Sporticus **16**, 10–16 (2014)
17. MaKey MaKey: How would YOU interact with your computer? on YouTube
18. Fernandez-Fernandez, J., Saez de Villarreal, E., Sanz-Rivas, D., Moya, M.: The effects of 8-week plyometric training on physical performance in young tennis players. Pediatr. Exerc. Sci. **28**, 77–86 (2016)
19. Stone, M.H., O'Bryant, H.S.: Weight Training: A Scientific Approach. Burgess, Minneapolis (1984)
20. Renfro, G.: Summer plyometric training for football and its effect on speed and agility. Strength Condition. **21**(3), 42–44 (1999)
21. Robinson, B.M., Owens, B.: Five-week program to increase agility, speed, and power in the preparation phase of a yearly training plan. Strength Condition. **26**(5), 30–35 (2004)

Integrating Technology into Traditional Dance for the Elderly

Vasiliki I. Zilidou[1,2]([⊠]), Efthymios Ziagkas[1],
Evdokimos I. Konstantinidis[2], Panagiotis D. Bamidis[2],
Styliani Douka[1], and Thrasyvoulos Tsiatsos[3]

[1] School of Physical Education and Sport Science,
Aristotle University of Thessaloniki, Thessaloniki, Greece
vickyzilidou@gmail.com,
{e.ziagkas,sdouka}@phed.auth.gr
[2] Laboratory of Medical Physics, Medical School,
Aristotle University of Thessaloniki, Thessaloniki, Greece
{vickyzilidou,evdokimosk}@gmail.com,
bamidis@med.auth.gr
[3] School of Informatics, Aristotle University of Thessaloniki,
Thessaloniki, Greece
tsiatsos@csd.auth.gr

Abstract. The increased life expectancy is associated with an increase in multiple chronic conditions, such as functional disability, need for help, reduced mobility, depression, isolation and loneliness. It is considered necessary to give particular attention in order to observe changes to services and facilities aimed at the elderly, to design and to implement appropriate health programs initially on prevention and continuously on treatment to achieve a satisfactory level of quality of life and a significant improvement and facilitation in their daily life. The traditional dance which is a health program that is very close to the interests of the elderly, could promote their health and contribute to the maximum delay of aging. The aim of this work was to investigate whether the elderly can participate in traditional dance programs using new technologies, effectively promoting active aging and pursuing an autonomous and independent living. Fifty-one (51) elderly women participated in this study and they were divided into two groups, the first group was the traditional classical dance intervention [26 elderly, age 66.23 (SD = 4.46), years of education 7.38 (SD = 2.15)] and the second was the dance intervention group using the technology [25 elderly, age 67.12 (SD = 4.43), years of education 11.08 (SD = 4.24)]. The interventions lasted 24 weeks with a frequency of 2 times per week in sessions of 75-min at a Day Care Centers. Both groups improved their aerobic capacity, but the Dance group also improved their sociability, preventing falls, while reducing their depression. The technology exists in the lives of most people but for elderly the last years become an attempt to approach this situation through health or education programs to familiarize them with technology and leaving social exclusion.

Keywords: Traditional dance · Technology · Quality of life · Elderly · Physical activity

© Springer Nature Switzerland AG 2019
M. E. Auer and T. Tsiatsos (Eds.): ICL 2018, AISC 917, pp. 287–296, 2019.
https://doi.org/10.1007/978-3-030-11935-5_28

1 Introduction

In recent years, in our country and worldwide the phenomenon of aging is strongly observed, either because of restriction of birth or the prolongation of life expectancy. Elderly people are facing several changes such as physical (motor, cardiovascular, respiratory, secretory, autonomous, reproductive system, reduced acidity sensors), changes in cognitive functions (information processing, memory, intelligence), changes in their family, or the feeling of marginalization from the family side. The increased life expectancy is also associated with an increase in multiple chronic conditions, such as functional disability, need for help, reduced mobility, depression, isolation and loneliness [1].

These changes, when they occur, can have positive or negative effects on their lives and are a consequence of all the attributes they have acquired in their earlier lives, such as health, education, employment, financial situation [2]. According to the World Health Organization (WHO), it is estimated that the world's population will be more than two billion people over the age of 60 by 2050, where 80% will live in developing countries [3]. For all these elements in recent years, there is growing concern that the quality of life deserves attention in order to bring prosperity to these individuals, so many countries are moving in new directions that will promote active and healthy aging [4].

It is considered necessary to give particular attention in order to observe changes to services and facilities aimed at the elderly, to design and to implement appropriate health programs initially on prevention and continuously on treatment to achieve a satisfactory level of quality of life and a significant improvement and facilitation in their daily life [5]. Increasing physical activity is a supportable strategy for improving both health and quality of life in elderly [6]. The traditional dance which is a health program that is very close to the interests of the elderly, could promote their health and contribute to the maximum delay of aging.

The Greek traditional dance helps to improve physical and mental wellness, leads to spiritual benefits as an exercise program and inseparably bound with the quality of life. Elderly people who participate in the Greek traditional dance program have a better view of their body and function and are more satisfied with their lives [7]. In a relevant research, about the effectiveness of a 24-week program of Greek traditional dance and its effects on the physical and mental health variables for healthy elderly and elderly with Mild Cognitive Impairment (MCI), observed statistically significant improvement in their physical and mental health, such as their social interactions and environment [8]. In another study, which was designed to assess the effect of a 10-week dance program on the balance of the elderly, the results supported the use of dance as an effective physical exercise tool to improve the control of static and dynamic balance in the elderly [9]. In Meekums study [9], investigating the effectiveness of dance therapy, showed that the intervention group had statistically reduced body image anxiety and increased self-esteem as compared to the control group [10]. It also showed that Greek traditional dance can affect other parameters of physical and mental health of people with health problems, as it was found that in women with breast cancer, it has contributed to the improvement of physical functioning, life satisfaction, while the depression symptoms were reduced [11].

Despite limited research into the elderly using new technology, it is considered to be a source of social support and improvement of elderly lives and, more generally, a means of integrating into society as citizens [12]. Research groups are studying and adapting ways of utilizing Information and Communications Technology (ICT) to provide elderly with new services, specifically social networking services, to not feel socially excluded, as well as services that will interact and improve active aging. The need for updating the training and education programs in the data that create new trends in society, have made it necessary to use ICT and especially those of the Internet. The use of ICTs can provide the quality of elderly's life, making them more independent, less isolated as it gives them more opportunities. It is considered that there is a link between quality of life (QOL) and ICT in terms of skills, health, relationships, self-confidence and connection with family and the general environment. Until today, there are a lot of studies showing that ICT improves QOL [13]. In previous research, we have shown that serious games focusing on engaging elderly people in physical activity are designed to encourage them to participate in order to promote a more active lifestyle and improved quality of life. Moreover, it has been shown that serious games promote and allow the socializing of the elderly [14].

The aim of this work was to determine whether the elderly can participate in traditional dance programs using new technologies, effectively promoting active aging and pursuing an autonomous and independent living.

2   Methods

2.1   Intervention

Dances were selected from all over Greece and divided into three categories depending on the position of hand movement, the intensity (slow speed) and complexity and number of steps. Dances were further classified into three categories: mild, moderate and high intensity. Most of them were moderate, progressive and with increasing intensity, indicative of the age and the physical abilities of the participants. Two interventions were designed, one of them was the traditional classical dance intervention (Dance), i.e. with the presence of a physical person, the instructor, and the second was the (webDance) intervention with the use of technology. The dances that were presented were the same for both groups.

In particular, in the Dance group, the participants tried to train with the physical presence of the instructor, who showed the steps of each dance and then they try to follow him. Initially, the steps were done without music and then with music (see Fig. 1). A computer and a projector were used to train the webDance group. The dances were presented by a model dancer through video in the wall, initially presenting and measuring the steps of each dance and then with music. At the end, a group of dancers presented the dance as a performance. The participants of this group tried to perform individually the steps as they watched them from the model dancer and then they followed as a group the video performance group (see Fig. 2).

Participants with heart failure, hypertension and respiratory failure were excluded from the program, as well as these participants who did not complete at least 80% of the

Fig. 1. Dance group in day care center in Pella

Fig. 2. Dance groups in day care centers in Pylaia and Thessaloniki

total attendance hours. At the beginning and after the intervention, their fitness and functional capacity was evaluated by a fitness instructor, and their quality of life was assessed through appropriate questionnaires. Written consent was requested from them for their participation in this study, after being given the necessary explanations for its purpose. Ethical and Scientific Committee of GAARD approved the protocol of this study.

2.2   Physical Evaluation

The assessment of their physical status was held with the following tests: Fullerton Senior Fitness Test is contained six tests that measure the overall physical status for basic activities of daily living. These are: Chair stand, Arm Curl, Two-Min Step, Back Scratch, Chair Sit and Reach and 8 Foot Up and Go [15], the Berg Balance Scale [16] for the balance and risk of falls, the Tinetti Test [17] for walking and risk of falls and the Stork Balance Stand Test [18] for assessing the balance when standing on one leg.

2.3   Quality of Life Evaluation

The questionnaire was chosen to evaluate the quality of life is WHOQOL [19] which developed by the World Health Organization. It aims to promote an intercultural Quality of Life assessment system and the use of this questionnaire in the wider health sector. It includes 26 questions and is divided into four thematic sections: physical health; mental health; social relations; environment. The two questions, offer an overall assessment of Quality of Life and health status [20, 21]. The results with the highest outcomes are indicative of a better quality of life. Other tests used are the: Risk of falls (it concerns the Fall Prevention Checklist) [22], Friendship Scale (it concerns the Socialization) [23], Beck Anxiety Inventory-BAI (is a self-report measure of anxiety) [24], Patient Health Questionnaire-PHQ-9 (is a depression scale) [25], Medical Outcomes Study—Short Form, SF-12 (it concerns the physical and mental health) [26].

2.4 Usability Evaluation

The System Usability Scale—SUS was selected to measure the usability of the system. It is a Likert scale which includes 10 questions and measures the effectiveness (the ability of users to complete tasks using the system and the satisfaction by evaluating subjectively the reactions to using the system [27].

2.5 Participants

Fifty-one (51) elderly women participated in this study and they had a good functional and emotional state. They were divided into two groups; the first group was the traditional classical dance intervention (Dance) and the second was the dance intervention group using the technology (webDance). Twenty-six (26) of them were in the (Dance) group with an average of age 66.23 (SD = 4,46) and average years of education 7.38 (SD = 2.15) and the twenty-five (25) elderly were in (webDance) group with an average of age 67.12 (SD = 4.43) and average years of education 11.08 (SD = 4.24). (see Table 1). Nobody of them participated in another Greek Traditional Dance program. The interventions took place at the Day Care Centers of Municipalities of Pella, Thessaloniki and Pylaia. The interventions lasted 24 weeks with a frequency of 2 times per week in sessions of 75-min.

Table 1. Discriptives statistics for both groups

Group	#Participants	Age	Education	Use of technology
Dance	26	66.23 ± 4.46	7.38 ± 2.15	–
Web dance	25	67.12 ± 4.43	11.08 ± 4.24	√

3 Results

The statistical analysis was conducted via a 2 × 2 Mixed Model ANOVA. The groups of intervention (Dance and webDance) served as between-subjects factor whereas the time (pre-post) as within-subject factor.

Table 2. 2 × 2 ANOVA results

Variables	F(1,49)	Time/intervention	Between intervention
Two-minute step test	4.604	0.037	0.176
PHQ-9 test	5.926	0.019	0.005
Friendship scale test	4.146	0.047	0.005
Risk of falls test	6.399	0.015	0.008

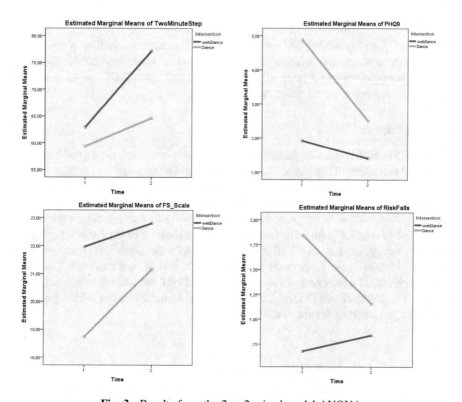

Fig. 3. Results from the 2 × 2 mixed model ANOVA

Results revealed an interaction between time and intervention for the Two Minute Step test, $F(1,49) = 4.604$; $p = 0.037$, the PHQ-9 test, $F(1.49) = 5.926$; $p = 0.019$, the Friendship Scale test, $F(1,49) = 4.146$; $p = 0.047$, as well as the Risk of Falls test, $F(1,49) = 6.399$; $p = 0.015$ (see Table 2, see Fig. 3).

In Fig. 4, we observe the evaluation of the questionnaire on the usability and effectiveness of the system with an average value of 74.4.

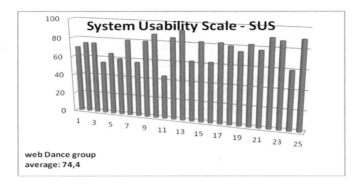

Fig. 4. System usability scale—SUS for webDance group

4   Discussion

Greek traditional dance is associated with the tradition of our country and at the same time it is a suitable type of exercise for all ages, aiming at the physical, mental and social well-being of the participants. In the present study, we desired to investigate whether the unrelated elderly people to the technology can use it for learning traditional dances and to be able to perform them, but also to consider whether the use of technology could offer them some benefits and especially quality in their lives.

Following the completion of the evaluations for the two interventions, the results showed that in the four of the tests were performed statistically significant results. Dance is considered to be a type of aerobic activity, so in the Two-Minute Step test which evaluates the aerobic capacity of individuals, expecting statistically significant results for both groups. It is important that the improvement in the webDance group was found to be more effective for this specific result.

The statistical analysis showed that the Dance group proved to be the best way for improving depression, sociability and the prevention of falls in relation to the web-Dance group. It seems that the use of technology in people who have not been used so far in their lives, requires more time to familiarize themselves with it. The sociability for the Dance group has shown better results, probably because in this group the time for social contacts is greater than that of the other group where it takes time for someone to understand the learning process from a model instructor without having the opportunity to discuss together for difficulties in implementation. Their attention is tense to be able to respond to this challenge.

The reduced risk of falls that proved at the Dance group, probably because this group feels more confident in all of them that they have to perform in relation to the webDance group, which has the difficulty of thinking, understanding and performing the dance properly.

Many elderly people are not familiar with new technology and this is not only due to individual factors, but also in complexity of the technology itself, the non-supply remarkable programs and the lack of funds and funding from the relevant one's government agencies. Therefore, they need appropriate support and guidance, in order to satisfy the elderly in the demands of daily life.

The System Usability Scale—SUS was used in a wide range of user interfaces. Initially, it was used to determine a unique usability and satisfaction rating for a specific product or service. The last 10 years the SUS can be used to complete a positive test program and usability assessment. The average of System Usability Scale score is 68. If the total score is below 68, then probably there are serious problems with the usability of system that need to be addressed. If the score is above 68, then we consider that usability is better, because the problems could be addressed more easily. The score of this study was 74,40 so we believe that there aren't any serious usability problems and there are margins for improvement to make the system more acceptable and effective.

5   Conclusion

Different groups of people use technology in their daily lives utilizing various tools for rehabilitation purposes, cognitive or physical training in order not to be isolated. Dance is considered to be a simple physical activity using a minimum equipment with low cost. Moreover, Greek traditional dance offers cultural heritage to a different percentage of population being a pleasant and interesting condition that allows participants to commit themselves in the long term. Consequently, dance may enhance patient compliance with treatment, which is usually a problem for many rehabilitation training programs.

Acknowledgements. This work was supported in part by the UNCAP Horizon 2020 project (grant number 643555), as well as, the business exploitation scheme of the ICT-PSP funded project LLM, namely, LLM Care which is a self-funded initiative at the Aristotle University of Thessaloniki (www.llmcare.gr).

References

1. Ory, M.D., Cox, D.M.: Forging ahead: linking health and behavior to improve quality of life in older people. In: Romney, D.M., Brown, R.I., Fry, P.S. (eds.) Improving the Quality of Life, pp. 89–120. Springer Publishing Company, Dordrecht, The Netherlands (1994)
2. Kostaridou-Efkleidi, A.: Topics of psychology and gerontology. Greek Letters (1999)
3. World Health Organization: Active Ageing. A Policy Framework. Geneva 2002, http://whqlibdoc.who.int/hq/2002/WHO_NMH_NPH_02.8.pdf
4. Drewnowski, A., Evans, W.J.: Nutrition, physical activity, and quality of life in older adults: summary. J. Gerontol. A Biol. Sci. Med. Sci. **56**(2), 89–94 (2001)
5. Dardavesis, T.I., Chousiadas, L.B, Kostaridou-Efkleidi, A., Nouskas, I., Chatzixristou, D., Kosta-Tsolaki, M., Benos, A.: Topics psychology and gerontology. Publications Field (2011)
6. Dechamps, A., Diolez, P., Thiaudière, E., Tulon, A., Onifade, C., Vuong, T., et al.: Effects of exercise programs to prevent decline in health-related quality of life in highly deconditioned institutionalized elderly persons: a randomized controlled trial. Arch. Intern. Med. **170**(2), 162–169 (2010). https://doi.org/10.1001/archinternmed.2009.489. [Medline: 20101011]

7. Konstantinidou, M., Charachousou, Y., Kabitsis, C.: Dance movement therapy effects on life satisfaction of elderly people. In: Proceedings of the 6th World Leisure Congress, Bilbao, 207 pp. I (2000)
8. Zilidou, V., Douka, S., Tsolaki, M.: The results of an intervention program of traditional dances, as a recreational activity, in the improvement of the quality of life of elderly people in day care centers of the municipality of Thessaloniki. Hellenic J. Sports Recreat. Manage. **12**(1), 13–25 (2015)
9. Sofianidis, G., Hatzitaki, V., Douka, S., Grouios, G.: Effect of a 10-week traditional program on static and dynamic balance control in elderly adults. J. Aging Phys. Act. **17**, 167–180 (2009)
10. Meekums, B.: Responding to the embodiment of distress in individuals defined as obese: Implications for research. Couns. Psychother. Res. Link. Res. Pract. **5**(3), 246–255 (2005)
11. Kaltsatou, A., Mameletzi, D., Douka, S.: Physical and psychological benefits of a 24-week traditional dance program in breast cancer survivors. J. Bodywork Mov. Ther. **20**, 1–6 (2010)
12. Czaja, S.J., Charness, N., Fisk, A.D., Hertzog, C., Nair, S.N., Rogers, W.A., Sharit, J.: Factors predicting the use of technology: findings from the center for research and education on aging and technology enhancement (CREATE). Psychol. Aging **21**(2), 333 (2006)
13. Boulton-Lewis, G.M., Buys, L., Lovie-Kitchin, J., Barnett, K., David, L.N.: Ageing, learning, and computer technology in Australia. Educ. Gerontol. **33**(3), 253–270 (2007). https://doi.org/10.1080/0360127060116249
14. Zilidou, V., Konstantinidis, E., Romanopoulou, E., Karagianni, M., Kartsidis, P., Bamidis, P.: Investigating the effectiveness of physical training through exergames: focus on balance and aerobic protocols. In: 1st International Conference on Technology and Innovation in Sports, Health and Wellbeing (TISHW), Vila Real, Portugal (2016)
15. Jones, C.J., Rikli, R.E.: Measuring functional fitness of older adults. J. Act. Aging, 24–30 (2002)
16. Berg, K., Wood-Dauphine, S., Williams, J.I., Gayton, D.: Measuring balance in the elderly: preliminary development of an instrument (2009) http://dx.doi.org/10.3138/ptc.41.6.304
17. Tinetti, M.E.: Performance-oriented assessment of mobility problems in elderly patients. J. Am. Geriatr. Soc. **34**(2), 119–126 (1986)
18. Johnson, B.L., Nelson, J.K.: Practical Measurements for evaluation in physical education (1969)
19. T. W. Group: The world health organization quality of life assessment (WHOQOL): development and general psychometric properties. Soc. Sci. Med. **46**(12), 1569–1585 (1998)
20. WHOQOL Group: The world health organization's WHOQOL-BREF quality of life assessment: psychometric properties and results of the international field trial. A report from the WHOQOL group. Qual. Life Res. **13**, 299–310 (2004)
21. Ginnieri-Kokkosi, M., Triantafyllou, E., Antonopoulou, V., Tomaras V., Christodoulou: Life Quality Manual with Axis Questionnaire WHOQOL-100. Publications Beta, Athens (2003)
22. Personal Risk Factors: Fall Prevention Checklist. Adopted from the Minnesota Safety Council Fall Checklist Personal Risk Factors and Hennepin County Community Health Department with Permission. https://fallpreventiontaskforce.org
23. Hawthorne, G., Griffith, P.: The friendship scale: development and properties centre for health program evaluation. The University of Melbourne (2000)
24. Beck, A.T., Epstein, N., Brown, G., Steer, R.A.: An inventory for measuring clinical anxiety: psychometric properties. J. Consult. Clin. Psychol. **56**, 893–897 (1988)

25. Kroenke, K., Spitzer, R.L., Williams, J.B.: The PHQ-9: validity of a brief depression severity measure. J Gen Intern Med. **16**(9), 606–613 (2001)
26. Kontodimopoulos, N., Pappa, E., Nikas, D., et al.: Validity of SF-12 scores in a Greek general population. Health Qual. Life Outcomes **2007**(5), 55 (2007)
27. Kirakowski, J., Corbett, M.: Measuring user satisfaction. In: Jones, D.M., Winder, R. (eds.) People and Computers IV. Cambridge University Press, Cambridge (1988)

Project Based Learning

Implementing Digital Methods into Project-Based Engineering Courses

Peter Vogt[1(✉)], Uwe Lesch[2], and Nina Friese[3]

[1] Ruhr West University of Applied Sciences, Department of Civil Engineering, Mülheim, Germany
peter.vogt@hs-ruhrwest.de
[2] Ruhr West University of Applied Sciences, Department of Mechanical Engineering, Mülheim, Germany
uwe.lesch@hs-ruhrwest.de
[3] Ruhr West University of Applied Sciences, Higher Education, Mülheim, Germany
nina.friese@hs-ruhrwest.de

Abstract. Digital transformation increasingly affects all fields of life. Today's students need a vast variety of collaboration skills and have to be able to work in interdisciplinary teams and in remote workspaces by applying future-oriented digital instruments at the same time. It is the role of academic education to impart technical and social competences, stimulating the courage and the curiosity for a lifelong learning. Against this background, academic courses require contemporary content and didactical methods. In fact, academic teaching in engineering disciplines today predominantly consists of conventional teaching methods such as lectures, exercises or laboratory tutorials. This paper outlines two project-based concepts, including digital instruments and collaborative forms of teaching and learning. Students, who participate in pilot projects since 2017 study either mechanical or civil engineering at Ruhr West University of Applied Sciences.

Keywords: Teaching in engineering · Teamwork · Academic-industry partnership · Digital strategy · Intrinsic motivation

1 Purpose of the Teaching Concept

1.1 Motivation

Future generations of engineers will apply a wide range of digital working methods. The digital transformation will influence how people work and live together and how they think and solve problems. Graduates in engineering disciplines will be responsible for developing and influencing technical instruments and social aspects within their working environment. Therefore, it is crucial that students acquire sufficient competences to act and react in a digitized world [1].

In order to achieve this objective, the focus is on creating new course concepts. Furthermore, the students learn how to work and communicate in interdisciplinary teams.

M. E. Auer and T. Tsiatsos (Eds.): ICL 2018, AISC 917, pp. 299–310, 2019.
https://doi.org/10.1007/978-3-030-11935-5_29

The implementation of digital tools pursues the goal to increase the attractiveness of the future field of work and to lead to a rise in the intrinsic motivation of the participants. Moreover, an evaluation shows the students´ abilities to link classic with newly developed methods. The knowledge about basic principles forms the foundation to assess critically new methods. Finally, the optimization potential associated with the question of meaningfulness of the applied technologies will be taken into account.

1.2   Project-Based Approach

Academic-industry partnerships represent the basis to carry out two pilot projects in a collaborative learning environment. Of course, basic theoretical knowledge needs to be part of the curriculum, but it is reasonable to apply new trends and developments into each study program once students enter an advanced stage of their study.

1.3   Implementation of Innovative Technology

Innovative technologies increasingly influence daily work routines. Especially digitization changes working processes sustainably by promoting the human-technology interaction or by linking machine control systems. Professionals predict that ignoring important trends might result in the fact that small and medium sized enterprises vanish and lose touch with latest technologies and competitors. Universities must meet this demand, especially by identifying these trends as early as possible associated with offering specific study programs. Furthermore, today's students expect the application of innovative technologies in courses, because they grew up in a partially digitalized environment.

Another aspect concerns legal issues, which relate closely to the application of new technologies. Operating processes need to be transparent, precise and approved by all project partners. In that context, students need to be aware of potential obstacles and formal constraints.

The two course concepts, described in this paper deal with exemplary digital process chains in engineering disciplines. Of course, these examples only demonstrate digital facets, but offer significant potential for further trends. To sum it up, developments cycles become shorter, which requires a continuous monitoring of market trends. This fact sustainably underlines the importance of lifelong learning.

1.4   Working Environment

The tasks, an engineer is faced with almost every day predominantly consists of project-based work. Most of these projects are complex, have a big impact on the company, and demand self-employed action in interdisciplinary teams. Classical education methods in engineer courses such as lectures and exercises do not prepare adequately for this kind of challenge. "Real life projects" fulfil these requirements. Therefore, two courses concepts are based on real industry projects in cooperation with local companies.

2 Target Group Analysis of Potential Course Participants

2.1 History and Technical Orientation of Ruhr West University of Applied Sciences

Ruhr West University of Applied Sciences (HRW) was officially founded as a science and engineering institution on May 1st, 2009. The state-run University is situated in the federal state of North Rhine-Westphalia in general and the Ruhr region in particular. The Ruhr region used to be a manufacturing region. Since the beginning of the 20th century, plenty of people were employed in the heavy metal industry and in coal mines. During that period, workers from foreign countries as well as from rural arias settled down here. Traditionally, the Ruhr region has both, well-qualified employees and residents with a low educational background. Currently, a great shift to a knowledge and educational region is recognizable.

The charter of the HRW founders announces to support the structural transformation by transferring knowledge and innovation from the University to the society and vice versa. Because of the highly technologized facilities and innovative principles of teaching, Ruhr West University pursues the goal to get the most modern University of Applied Sciences in the Ruhr region.

2.2 Students

Due to the history of the Ruhr region, the amount of students, who attend University as a first family representative is high. Additionally, the number of students, who have a migration background, is the highest in Germany. As shown in Fig. 1, at Ruhr West University 72.4% of the students' parents do not have an academic degree [2]. Compared with other Universities of Applied Sciences in the Ruhr region, this percentage is about 15.1% higher [3].

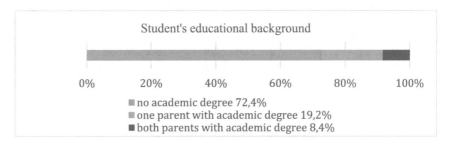

Fig. 1. Background of students at HRW [2]

In North Rhine-Westphalia pupils achieve the general higher education entrance either from a *Gymnasium* (academic high school), from a *Gesamtschule* (comprehensive school), or from a *Berufskolleg* (vocational school). Only the high school degree directly allows students to enroll at a University. All other degrees enable the registration at Universities of Applied Sciences. Courses at Universities are generally more theoretical than at Universities of Applied Sciences.

Figure 2 shows the heterogenic ways for students to access Ruhr West University of Applied Sciences. Figures 1 and 2 illustrate that students begin studying with different educational backgrounds and with a wide range of competences.

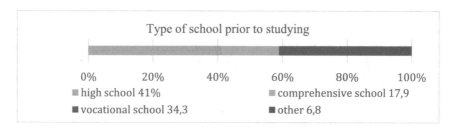

Fig. 2. Visited types of school [2]

During winter semester 2017/18 at total number of 5870 students was registered at HRW.

2.3   Study Programs

The Bachelor programs for Mechanical Engineering (start in 2009) and Industrial Engineering (2013) each have a standard period of study of seven semesters including an internship semester. Both programs have an industry based advisory board that emphasizes the need for basic theoretical knowledge, most recent technologies and trends as well as practical projects.

The consecutive Master's program Civil Engineering at Ruhr West University of Applied Sciences with a standard period of study of three semesters is open to all applicants, who have received a Bachelor´s degree in the field of construction or architecture. The accredited program started in summer 2017. Prior to enrolling, students decide if they specialize in structural engineering or construction management.

During the development phase of the study programs, industry representatives contributed their points of view concerning course contents and soft skills reflecting the requirements of the labor market. Respondents often request well-founded theoretical knowledge taking into account the fast change of the working environment under the application and dissemination of digital methods. In summary, it is the overall aim to intensify project-based teamwork under contemporary and complex boundary conditions.

3   Didactic Concept

3.1   Lectures, Simulation and Project-Based Learning

Project-based learning is a dynamic approach, which illustrates how to connect academic work with real-life issues. Students explore real-world problems and challenges while simultaneously developing cross-curriculum skills. Project-based learning has a positive effect on the student's attention and their improvement of competences [4, 5]. Especially economically disadvantaged students and students with a non-academic background benefit from project-based learning [6, 7]. By confronting students with authentic learning situations like real-world problems, collaboration, inquiry, writing, analysis and effective communication, they start to think and reflect autonomously and critically. Barron and Darling-Hammond [8] note, by engaging in authentic projects that draw subject knowledge to solve real-world problems, students learn at deeper levels and perform better on complex tasks.

Project-based learning is not a new approach. However, the merger of real and virtual world in the context of the course concept is innovative. In Germany, the professional associations in engineering require a fast and professional education for the digital transformation in all types of education [9].

Biggs and Tang point out, that in problem-based learning environments students use higher-level strategies for understanding and applying self-directed learning. Compared with students, who attend traditional lectures or seminars, they get progressively deeper into the approach to learn [10].

The course concepts comprise three steps, whereas one of them is project-based learning. In the first step, students get the basic theoretical knowledge in traditional lectures. Lectures are restricted to those theories they need to start into the project-phase. This means that they have to study more than the impact of the lectures and they have to learn how to find relevant information for themselves. In addition, students learn to ask, "Which information is relevant?" In the second step, they try out the new knowledge by applying it. For example, course members digitally capture a part of the University campus in order to do the same in the real-life project with industry partners later. In the second phase, students work in teams, share their knowledge and learn how to apply digital tools in a real-life simulation context. In this phase, they also develop a concept on how to communicate with the industry partner. Finally, they are well prepared to start the third phase: the real project!

3.2 Industry Project

The first project addresses students of Mechanical and Industrial Engineering and deals with the planning of a new factory for a company, whose facilities became too small. The second project supplies a set of data for a construction company, which in 2017 and 2018 erects a building close to the University campus.

Students work in teams with 4–5 members, whereas the team compilation is random and independent from the students' wishes. As soon as the teams have formed, the team roles are being distributed. It is very important to assign responsibilities, e.g. the clarification of technical details, the operation of hard- and software requirements, scheduling, documentation of important contact data etc. If needed, the module manager advises the students concerning team building and team preparation.

3.3 Presentation of Results in Front of Professionals

It is part of the project-based concept to prepare a project report, which stresses all aspects of the problem definition. The report includes all processing steps and critically evaluates the approach. Students are aware of the fact that representatives of the practice partners also get an insight into the final report.

At the end of the project term all groups present their results in front of professionals and provide as many information as necessary.

4   Approach

4.1   Mechanical Engineering Course "Factory Planning"

The course "Factory Planning" is an elective course in the 6th semester of the Bachelor curriculum, open for mechanical engineers, industrial engineers and business administration students, who specialize on international trade management and logistics. The course combines project-based learning with a collaborative learning approach in a partial virtual environment.

Subsequent to basic lectures about the methods and procedures of factory planning and a practical introduction into the principles of laser scanning, the students visit the existing factory of a local company in order to familiarize with products, processes, and the equipment (Fig. 3). At the same time, each team learns more about the future demands of the company.

Fig. 3. Visit of the company "EME" on April 25th, 2017

The information collected during this visit serves as a basis for the development of a new factory layout. Using approximately 20 point-clouds, taken from laser scans at different positions, the students create a digital 3D image by combining these point-clouds to a 3D-model of the whole factory (Fig. 4). The application of the laser scanning technology lies within the responsibility of each team. A manual, providing basic instructions is available for each team.

The 3D-model of the existing factory allows the inspection of details of machines, equipment and the building, as exemplarily illustrated in Fig. 5.

Fig. 4. 3D-model of the existing factory

Fig. 5. Point-cloud of a machine with an enlarged detail

In the following, the teams develop a 2D-concept for a new factory based on the collected data and the methods taught during the initial lectures. As a final step, the students extract the machines from the 3D scan and use them to create the 3D-layout of the new factory (Fig. 6). If new machines are needed, 3D-models of each machine might be uploaded from the data bases of the suppliers.

A documentation of the project, including the applied planning methods and the propose layout combined with a presentation of the new factory in front of the management board of the company finalize the project.

4.2 Civil Engineering Course "Digital Design and Construction"

The course "Digital Design and Construction", which is part of the Master's degree Program civil engineering, is a mandatory course for students of both specializations.

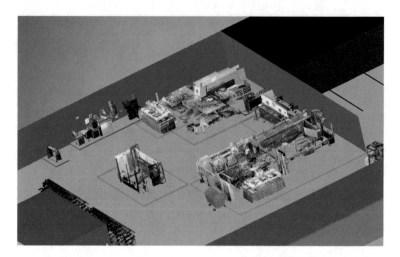

Fig. 6. 3D proposal for the layout of the new factory

It aims to build up a digital planning chain during the development phase for a new building. In this respect, the students have to apply skills they have learned during the Bachelor's degree program.

The development of the chain is based on the BIM-technology, whereas BIM stands for Building Information Modeling. It is the idea of BIM to create an entire virtual model prior to starting the building process itself on site. By this means, the extent of planning details increases. At the same time, due to this high level of detail BIM-projects will be on time and meet the proposed budget. In the end, the digital planning method and its interconnectivity leads to a new standard, exceeding the quality of conventionally planned projects by far. Visualization technics also help to investigate different execution phases and the overall feasibility.

During the conceptual planning stage of the course for the winter semester 2017/18, the module coordinator concluded a cooperation with the project managers of a construction project next to the University campus. The project "StadtQuartier Schlossstraße" (SQS) is a combined residential and commercial building in the city center. The structural works, mainly consisting of reinforced concrete elements, began in early 2017. The complex itself is characterized by a clear structure of concrete elements like base plates, walls, columns, and ceilings.

On Friday, October 20th, 2017, the 3D laser scanner was used to conduct measurements in the basement of the building (Fig. 7). In advance, the students formed teams of four to five persons, so that all groups had approximately two hours to scan a part of the basement.

The students applied the technics, they familiarized with in the previous lectures and seminars, respectively. Every team produced six to eight scans from different locations, taking into account that adjacent scans must have overlapping sections. In addition, it was important to document the boundary conditions for each scan thoroughly.

Fig. 7. Application of the 3D laser scanner in the basement of the SQS project

After the scan process has been finished, the obtained data referred to as point-clouds, have been processed in the BIM-lab at the University. Here, the students applied a number of licensed software programs. The first step consisted of transferring the point-cloud into a 3D-model by digitally combining all recorded data. While capturing the points by using laser technology, the scanner simultaneously takes a digital all-round photo in order to ensure an adjustment to reality. Figure 8 visualizes a point-cloud consisting of data taken from approximately 15 scanner positions. Wall and floor surfaces, columns and openings are clearly identifiable. However, the conversion of the point-cloud into a technical model requires a previous cleanup of the cloud data. Therefore, the students had access to the technical construction drawings of the basement area.

Fig. 8. Visualization of the 3D point-cloud of the basement

In the next step, the prepared point-cloud was exported to a CAD program (CAD = Computer Aided Design). As Fig. 9 shows, all components have been converted into three-dimensional structural elements. Likewise, the geometry data of the virtual structure were enriched with the corresponding information, for instance the concrete strength or the amount of reinforcing steel.

The last step within the BIM process involved the implementation of the proposed construction period as well as the cost planning. By doing so, the former 3D-model is transferred into a five-dimensional BIM model.

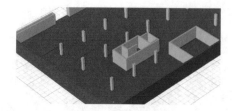

Fig. 9. View of the three-dimensional model with attributed components

5   Outcomes

5.1   Feedback and Evaluation by Students

In order to measure the performance, each project team documents its approach combined with the amount of work hours in individual reports. Additionally, the students fill in evaluation questionnaires during the last third of the semester.

For both projects in 2017, the students got a detailed problem definition and an outline of the anticipated steps at the beginning of the semester. Additionally, the expectations concerning the documentation of the results have been communicated transparently. Almost every week, the students presented and discussed intermediate results with other participants. Communication and collaboration are integral factors that contribute to the success of project-based teamwork.

For the course "Digital Design and Construction", the students provided feedback as summarized in Table 1. Table 2 includes the feedback concerning the course "Factory Planning".

Table 1. Feedback for the course "Digital Design and Construction"

Positive	Potential for improvement
• Well-equipped computer lab • Use of new and innovative equipment on a real construction site • Introduction to many different software programs • The interactive work during the seminars in the computer lab • The high degree of freedom to work on the project	• Five students per group are too much • Too many students with too many opinions • Elimination of malfunctions in the programs • Laboratory accessibility also on weekends

Table 2. Feedback for the course "Factory planning"

Positive	Potential for improvement
• Use of 3D laser scanning • Real life project	• Should be run in a full semester

It is the standard at Ruhr West University of Applied Sciences, that during the last two weeks of the semester, the students get an insight into the results of the evaluation followed by a discussion. The valuable comments and the feedback will be used for the enhancement of the course concepts.

5.2   Conclusion

During the project realization, a high motivation of the students was recognizable. Additionally, students handle digital tools much easier than expected beforehand. A significant intrinsic motivation was apparent during self-study periods, for example when familiarizing with software applications. The students are very experienced in browsing social media for detailed troubleshooting or online tutorials. In each lecture or discussion, the students applied scientific vocabulary, showing a broad comprehension of the overall topic.

The "real life aspect" of the project and the responsibility to present results in front of the company management is a strong motivation for the students. The teamwork enables them to solve tasks, which were largely unknown to them in the beginning. Team coaching helps to minimize uncertainties in the beginning of the project. Moreover, the students learn that only well-functioning teams achieve good results.

Students with an engineering background gained an easy access to 3D-modelling and augmented reality even though they experienced that it is additional work to create and use it. On the other hand, some students with a business administration background were discouraged from the technical aspects and the additional work, so that they left the course.

To sum it up, the objective of solving a real life problem independently has been fully achieved by both projects.

Generally, the implementation of digital tools in academic teaching should be standard in engineering programs at an advanced study level. Digital competence and intrinsic motivation help students to work on projects in small groups (e.g. 4 students per group). For that reason, an appeal summarizes the authors' experiences: Try out, even though some uncertainties remain!

Experience has taught that clear expectations concerning the project process and the results have to be provided in advance. At least one contact person should be available during the project execution. Be aware, that students request clear answers to urgent problems. Pragmatic solutions because of former discussions help students to continue work without losing their motivation.

5.3   Outlook

The further development of the course concepts will consist of two steps. The first step pursues the goal to transfer the 3D-model from the computer screen into virtual and augmented reality. This technique allows the students as well as the company management to walk around within the boundaries of the new planned environment. This step will even more increase the motivation of the students and enable them to transfer the knowledge into other fields of business like industry 4.0. The second step covers the development of a subsequent course, open to mechanical and civil engineering students. This course will offer the possibility for civil engineers to continue the work with the 3D-model created by industrial engineers. It covers the entire digital process-chain from the idea of a new factory/building to its virtual 3D layout. The result will contain a fully equipped BIM-model for a factory.

References

1. Hochschul-Bildungs-Report.: Hochschulbildung für die Arbeitswelt 4.0. Stifterverband für die deutsche Wissenschaft. Essen (2016)
2. Erstsemesterbefragung, H.R.W.: Ergebnisse der Befragung von Studierenden im ersten Hochschulsemester im Zeitraum 2016/17. Mülheim (unpublished), Hochschule Ruhr West (2017)
3. RuhrFutur.: Studieren im Ruhrgebiet heute. Erste Ergebnisse der gemeinsamen Studienein-gangs- und Studienverlaufsbefragung an den RuhrFutur Hochschulen. Essen (2018)
4. Kolb, D.A.: Experiential Learning: Experience As The Source Of Learning And Development. Prentice Hall, Englewood Cliffs, NJ (1984)
5. Kolb, A.Y., Kolb, D.A.: Learning styles and learning spaces: Enhancing experiential learning in higher education. Acad. Manage. Learn. Educ. 4(2), 193–212 (2005)
6. Creghan, C., Adair-Creghan, K.: The Positive impact of project-based learning on attendance of an economically disadvantaged student population: a multiyear study. Interdisc. J. Prob. Based Learn. 9(2) (2015). Available at: https://doi.org/10.7771/1541-5015.1496
7. Altieri, M., Schirmer, E.: Learning the concept of eigenvalues and eigenvectors in a problem-based learning environment: A qualitative and quantitative comparative analysis of achieved learning depth using APOS theory among students from different educational backgrounds. ZDM Mathematics Education, special issue 6/2019 (submitted 2018)
8. Barron, B., Darling-Hammond, L.: Teaching for meaningful learning. In: Hammond, D.H., Barron, B., Pearson, P., Schoenfeld, A., Stage, E., Zimmerman, T., Cervetti, G., Tilson, J. (eds.) Powerful Learning: What We Know About Teaching for Understanding. Jossey-Bass, San Francisco, CA (2008)
9. VDI Verein Deutscher Ingenieure e.V.: Ingenieurausbildung für die digitale Transformation. Diskussionspapier zum VDI-Qualitätsdialog, 1 Mar 2018
10. Biggs, J., Tang, C.: Teaching for quality learning at university. The Society for Research into Higher Education. 3rd edn. NY. Mc Graw Hill (2007)

Technique of Design Training When Performing Laboratory Works

Khatsrinova Olga$^{(\boxtimes)}$, Mansur Galikhanov, and Khatsrinova Julia

Kazan National Research Technological University, Kazan, Russia
{khatsrinovao, khatsrinoval2}@mail.ru,
mgalikhanov@yandex.ru

Abstract. The article is devoted to use of a design method of training at laboratory researches in discipline "Division of multicomponent systems" where the complex problem of design of the optimum scheme (device) of division multicomponent mixes is solved step by step on the basis of the solution of subtasks.

Keywords: Laboratory research · Design training · Professional competences · Problems of design

1 Introduction

Development of economy and industrial production depends on quality of the engineering education demanded by employers. Modernization of engineering education is based on transition from traditional disciplinary model of training to the competence-based format which is substantially focused more on logic of future professional activity. The competence is shown, first of all, by employers and society in the form of some expectations connected with professional activity of the graduate. Level of compliance of individual indicators of future expert to expectations of the employer and to the social order of society is necessary as the main criterion of competence—readiness for performance of professional tasks. In this regard training, development of information and professional skills (it was always), but, the main thing, readiness formation independently and productively to work in real life and production situations, to be able to diagnose them and to make expedient decisions becomes a dominant in the Russian higher education not so much today. Necessary is a transfer of emphases on the developing education function, on acquisition of knowledge during all life and harmonious personal and professional development of the student.

1.1 The Purpose

Quality of training of engineers—one of the most important indicators of work of engineering higher education institutions. Engineering education has to be directed to creation of design teams which even during training or right after release will be capable to analyze the market, to develop new products and to create own technological enterprises or to join entirely larger corporations. The graduate needs to be guided well

© Springer Nature Switzerland AG 2019
M. E. Auer and T. Tsiatsos (Eds.): ICL 2018, AISC 917, pp. 311–320, 2019.
https://doi.org/10.1007/978-3-030-11935-5_30

in current trends, to represent what happens in branch in Russia and abroad where there are productions whether it is possible to give some processes on outsourcing with whom it is favorable to cooperate also what real practice of use of the mastered technologies in Russia and in the world [1].

The quality of training engineers is one of the most important indicators of the work of engineering universities. Practice shows that the professional competencies of future engineers are most effectively formed within the framework of a laboratory workshop. The forecast of development of laboratory works as a form of training will be associated with the integrated application of full-scale experiments and the use of active teaching methods.

Main objectives of performance of laboratory works are: fixing and increasing knowledge of theoretical material and also development by students of abilities of the choice of optimum methods of division of multicomponent substances and skills of performance of calculations of devices for division of multicomponent mixes taking into account economic efficiency. In a traditional laboratory practical work students, acquiring subject concepts, carrying out calculations are not capable to correlate this knowledge to future professional activity, there is a lack of mutually assistance of students, there is no exchange of any results. More expedient to offer the student a set of works which correspond to the through design task broken into subtasks which students need to solve when performing laboratory works. After acquaintance of students with the practical contents reflecting real engineering practice independent work with various sources of information, their processing and systematization for the purpose of drawing up research problems is offered to them: to carry out literary search and to choose a research method; to study theoretical questions and to offer tools for calculation of quantitative characteristics of process of division of substances and its thermodynamic parameters; to process the received results and to interpret them in the form of tables and schedules; to draw conclusions on mutual influence of ions in exchange processes; to prepare the report and the presentation of the project.

Search and research stage. For support of process of training information technologies are used. Students make hypotheses of a solution, formulate tasks, choose an optimal variant of carrying out calculations. In the course of work the magazine of the project is kept. Members of the team determine by discussion a position which they will take in the project,—the most active become project managers, and someone from students is inclined to work at a position of the member of the team. The teacher accompanies classroom and independent implementation of the project through internal and virtual consultations and gradually forms the culture of project work.

Technological stage. Students carry out calculations, analyze results, process data, form the general report of group. At this stage function of the teacher—to direct, coordinate work, to create the conditions optimum for manifestation of creative potential of students, to help to overcome difficulties which can arise.

Final stage. It is a stage of protection of the project and assessment of its results. It includes the expert analysis of the received results, the analysis of performance of a goal, preparation of the report and the presentation. Students and the teacher take part in assessment. Thus, in process of implementation of the project the necessary level of professional knowledge, abilities to organize a research, to work in team, to represent results is reached. All this fills up an arsenal of abilities of students, forming their personal and activity potential.

The stage-by-stage solution of subtasks as a result of performance of all modules gives integrated knowledge of all process and influence of various factors on regime and design parameters of process of division. Use of a design method allows to achieve in the most optimum way the objectives of discipline and to create necessary competences at students and also contributes to the development of research skills in students. Despite the high importance of a problem, its complex decision does not exist still. The organization of methodological orientation of educational process for discipline in that part which concerns support of the experiment significantly expanding scope of training and educational researches of students [2, 3] is required. More concrete methodical receptions and the corresponding model of application of laboratory works are necessary for creation on their base of the design environment. Such environment represents also a basis of the organization of the independent educational research of students reflecting the implementation nature of future professional activity [4]. The objective can be carried out by means of carefully selected receptions and standards by types of laboratory works in the conditions of the developing computer technologies.

The complexity of realization of this approach is that it is required to develop innovative educational tasks and to organize collective work of students already on younger courses. At the same time it is necessary to keep within the hours planned by the program allotted in the curriculum on studying of discipline. In these conditions for introduction in educational process of project-oriented educational tasks big methodical and technological preparation of the corresponding maintenance of such integrative activity of students, including optimum selection of tasks, creation of electronic base and a reference information, the organization of study of students and control of the developed projects is necessary. The problem of introduction of design training in educational process has two main aspects connected readily to its application, both teachers, and students.

From the teacher prerequisites of successful realization of the considered method are: psychological readiness to change own behavioural model of professional activity; susceptibility to the creative ideas as the project surely assumes creation something new; knowledge of techniques of the organization of design activity of students; experience of development and implementation of projects in the sphere of future professional activity of trainees. From the student existence of skills of independent search and selection of information, interest in the final product, knowledge of a conceptual terms framework and essence of the processes which are object of transformation are required.

2 Approach

Main for the higher engineering school is a task—to provide compliance of content of education to new prospects and priorities of scientifically technical, economic and social development. At the same time the important role is carried out by predictive information on future condition of nature of work and conditions in which university graduates will work. From this requirements to the level of professional competences, to abilities, to professional significant personal qualities of future expert follow. Rational use of laboratory works is connected with modeling of real production activity of future experts. V.G. Tchaikovsky came to a conclusion that holding complex laboratory researches promotes increase in activity of students on occupations since they allow to use results of experiments in academic year projects [5]. N.N. Semashko, V. P. Kobelev, analyzing experience of many higher education institutions, came to a conclusion that quite good results in increase in efficiency of laboratory practical works are yielded by use of a frontal method of its carrying out and the unified blocks of the equipment [6]. If to speak about a method of projects as about pedagogical technology, then this technology represents set of research, search, problem methods, creative on the most essence [7]. According to the German researcher R. Dreer, updating of design training is directly connected with transition of system of professional education to the two-level model "bachelor degree-a magistracy". The scientist believes that introduction of a method of projects should be begun not only with training not of students, but also teachers, each of which has to prove the readiness for a similar type of training, having presented as the project own subject matter [8]. E.M. Turlo considers too that introduction to practice of higher education institutions of design training is interfered by lack of special training of teachers, their orientation not on essence, and on formal signs of the corresponding method [9]. The main goals of the laboratory work are: consolidation and deepening of the knowledge of theoretical material, as well as the development by students of the skills of choosing the best methods and devices for separation of substances and skills in performing design and verification calculations of devices for separation of multicomponent mixtures taking into account economic efficiency. The set of settlement works corresponds to a cross-cutting project task, divided into subtasks, which must be solved by students in the performance of laboratory work. The step-by-step solution of the subtasks as a result of the execution of all modules gives an integral knowledge of the entire process and the influence of various factors on the regime and design parameters of the separation process. The use of the project method allows to achieve the goals of the discipline in the most optimal way and to form the necessary competences for students, and also promotes the development of research skills among students. This method allows, as a result, to solve a complex design problem that does not have an obvious solution, and in the event of a change in conditions, a completely different schematic or regime solution may turn out.

3   Actual Results

On discipline "Division of multicomponent systems" us is developed the content of laboratory occupations with the large volume of independent work. The discipline is studied by students on the 3rd course that assumes presence at them of the big list of competences (Tables 1 and 2).

The purpose of discipline is: development of specifics of the mathematical description of a mass transfer in multicomponent mixes, generalization of methods of calculation of devices for division of binary mixes in relation to multicomponent systems, studying of special types of division of binary mixes due to introduction of the

Table 1. Structure and content of laboratory works

No.	Hours	Name of laboratory work	Summary	The formed competences
1	2	Finding of matrixes of coefficients of multicomponent diffusion in steam and liquid phases.	To calculate Einstein coefficients of diffusion and practical matrixes of coefficients of multicomponent diffusion for three-component mix in steam and liquid phases	PK-2, PK-4, PK-5, PK-9, OK-1, OK-10,
2	8	Modeling of absorbers for division of multicomponent mix	To make design calculation of a nozzle and dish-shaped absorber for extraction of two components of three-component gas mix, to compare the calculated devices	OK-4, OK-5, OK-10, PK-1, PK-2, PK-9,
3	4	Design and testing calculations of rectifying columns for continuous division of multicomponent mix, the choice of an optimal variant of the scheme of installation	Approximate design calculation of four rectifying columns of accurate rectification of three-component mix for two schemes of division; the testing specified potarelochny calculation of all columns, the choice of the optimum scheme of division; design kinetic calculation of columns with various nozzle. The comparative analysis of the received results	OK-1, OK-10, OK-12, PK-2, PK-4, PK-5, PK-9,

dividing agents, training in methods of application of the gained knowledge for the solution of practical tasks. The main form of laboratory researches—"Work on the project". This kind of activity is complex creative work (a task with many unknowns, individual work with students, work on a resulting effect, etc.). It should be noted that the presented set of settlement works corresponds to the through design task broken into subtasks which students need to solve when performing laboratory works. The stage-by-stage solution of subtasks as a result of performance of all modules gives integrated knowledge of all process and influence of various factors on regime and design parameters of process of division. The work result—allows to solve as a result a difficult design task which has no obvious decision and in case of change of conditions other schematic or regime decision can turn out absolutely.

The team—each group working on an individual task consists of 4–6 students (usually—the head of group, engineers—researchers, "programmer"), the head of group is responsible for the general organization of work, cast, the analysis of the scientific and patent information, the analysis of correctness of calculations, registration of results of work in the form of the report, the scientific article and theses, protection of the project, engineers—researchers are obliged to project and collect simple

Table 2. The competences formed on project stages

Stage	Competences	Works		
		1	2	3
Informational	OK1—possession of the culture of thinking, ability to generalization, the analysis, perception of information, statement of the purpose and the choice of ways of its achievement	+	+	
	OK10—possession of the main methods, ways and means of receiving, storage, processing of information	+	+	
	OK12—possession of ability to work with information in global computer networks	+		
	PK-1—ability to the analysis and synthesis	+	+	+
Personal	OK-5—ability to self-organization and self- education	+	+	+
	OK-3—to find organizational and administrative solutions in unusual situations and readiness to bear responsibility for them;	+	+	+
	OK-4—ability to work in team, tolerantly perceiving social, ethnic, confessional and cultural distinctions	+	+	+
Professional and activity	PK-2—ability to choose research methods, to plan and make necessary experiments, to interpret results and to draw conclusions	+		+
	PK-4—readiness to use the basic concepts, laws and models of chemical reactions	+	+	+
	PK-5—ability to choose and apply the corresponding methods of modeling of physical, chemical and engineering procedures	+	+	+
	PK-9—ability to prepare basic data for the choice and justification of scientific and technical and organizational decisions on the basis of economic calculations	+		+

installations, to work on the existing devices and devices, to carry out calculations, "programmer"—is responsible for search of the scientific and patent information in the Internet, carrying out calculations, for statistical data processing, development of the presentation, the exhibition stand (Russian-English option).

This settlement work consists of two stages: calculation of Einstein coefficients of diffusion and a matrix of coefficients of multicomponent diffusion in steam and liquid phases; check of the received results with computer calculation. Calculations of processes and devices of a mass transfer are based on the main equation of a mass transfer, balance of phases and kinetics of a mass transfer. The kinetics of transfer is described by means of a matrix of coefficients of diffusion and models of a mass transfer. Therefore the following element with which students get acquainted during calculations is definition of balance of multicomponent mix. For the description of balance of phases there is a set of models: NRTL, Wilson, and others. For vapor-liquid balance are generally used the first two NRTL models and Wilson. Therefore during the calculations students independently choose one of these models. Carry out calculation of a nozzle and dish-shaped absorber for division of three-component mix. Then by means of the software package of IVC-SEP developed in Technical University of Denmark and provided for application in the educational purposes balances and a dish-shaped absorber as in this package only potarelochny calculation of a column is put verify the calculations. As a result data sets on two types of devices turn out and students have to prepare the comparative analysis of devices for protection of settlement work directly for their case. For implementation of process of division in mass-exchanged devices select a certain type of the contact device depending on specific conditions. The choice of internal devices of mass-exchanged devices is a difficult task which solution demands accounting of many factors. Performance data of various devices can change depending on properties of the divided system, operating conditions and the geometrical characteristic of the contact device. Therefore comparison of contact devices under various conditions without tests can lead to wrong conclusions. Carrying out researches taking into account various factors demands big capital expenditure and time. It is theoretically possible to allocate merits and demerits of certain contact devices and to carry out the choice by the main criteria. However the multicomponent of a task and mutually exclusive criteria of contact devices do not allow to carry out the unambiguous choice, carrying out technical and economic calculation is required. For this purpose future engineers have to own methods of calculation of columns with various types of contact devices and be able to carry out comparison and the choice by the set criteria. This aspect of future profession is also mastered by students during performance of laboratory work. It allows to increase the accuracy of computing manipulations of students and to remove routine work of check from the teacher. In addition, use of the computer program induces motivation of students to research activity. During interactive interaction with the program students manage to estimate much more aspects of influence of various factors (properties of substances, regime parameters) on results of the project. The list of the competences developed within the educational and research laboratory project as the strategy of scientific practical activities of students is provided in the table.

The analysis showed that a number of competences (PK-1, OK-5, OK-3, OK-4, PK-9) are through. PK-1 (ability to the analysis and synthesis) is one of the most important competences of research. In the context of the considered project the PK-1 components can be described: motivational—understanding of value of design preparation in the context of future profession of the process engineer; cognitive—development of basic concepts, operations, methods, sections of processes and devices of chemical technology and a set of practical tasks for their application; activity—ability to set the purposes, to plan activity, to apply knowledge to achievement of the planned result; personal—a subject position, ability to work in team, responsibility, creative and thinking; reflexive and estimated—ability to an adequate self-assessment of participation in an experiment and interpretation of results. At the final stage of the project important are criteria for evaluation of design activity of students in design and research laboratory work. Each criterion is estimated from 1 to 3 points: 1—does not correspond, 2—corresponds insufficiently, 3—completely corresponds. Assessment is carried out with attraction as experts of teachers and students, estimated sheets are made in the form of a matrix with the detailed description of level of compliance of

Table 3. Distribution of criteria for evaluation of competences

Competence	Criterion		
	Stage 1	Stage 2	Stage 3
(IK) OK-1, OK-10, OK-12, PK-1	Compliance of structure of work to stages (methodology) of scientific research. Correctness of determination of relevance, formulation of a problem, purposes, tasks. Quantity and quality of the used sources. Detailed scheduling of the project	Quality of results of a research	Structure and quality of contents of the written and oral report, the presentation, article on a conference
(LK) OK-5, OK-3, OK-4	Ability to goal-setting, planning, creative relation to the project. Creation of team of the project (speed of passing of stages of development of group/team), statement of the command purposes, coordination of the individual and command purposes, distribution of tasks, creation of open climate of communication	(data and information). Correction of the purposes, tasks, plan of implementation of the project	Compliance of the written report to the required style of statement. Compliance of the presentation of results of design activity to requirements to the presentation
(P-DK) PK-2, PK-4, PK-5, PK-9,	Compliance of the chosen methods of a research and cross-disciplinary tools of the analysis to requirements of the project	Work in team, cast in team. Participation in discussion of results at seminars	Cross-disciplinary and professional context of the report and protection

criterion to the appropriated point. You may mention here granted financial support or acknowledge the help you got from others during your research work (Table 3).

For the qualitative simulation of mass-exchange processes, it is necessary to know the diffusion coefficients. From the accuracy of reproduction of transport properties of substances, the final results of modeling largely depend. Despite the study of the phenomena of diffusion in gases and liquids, the formalization of the description of the matrices of the coefficients of multicomponent diffusion is very complicated for students to perceive the nature of cumbersome formulas and calculations. Therefore, the design task was divided into two stages: calculation of the Einstein diffusion coefficients and the matrix of multicomponent diffusion coefficients in the vapor and liquid phases; checking the results with computer calculations. This makes it possible to improve the accuracy of students' computing manipulations and to remove the routine work of verification from the teacher. In addition, the use of interactive techniques (computer program) motivates students to engage in research activities. During interactive interaction with the program, students are able to evaluate much more aspects of the influence of various factors (properties of substances, regime parameters) on the results of calculations. Therefore, the next element that the students are familiar with during calculations is the determination of the equilibrium of a multicomponent mixture.

4 Conclusion

The laboratory practical work of the RMS discipline is realized within the framework of the design method. Students solve individual tasks during laboratory work, which increases personal responsibility and motivation. The use of information technology tools in the project method promotes the formation of certain knowledge, skills and abilities for students to carry out information activities with computer equipment; development of visually-figurative, intuitive, creative, creative types of thinking; the motivation of the use of information technology in educational activities; development of aesthetic perception of any objects; the formation of skills to make the best decision or to find solutions in a difficult situation; development of skills to carry out experimental activities; development of spatial imagination and spatial representations of students. As a result of a step-by-step solution of the cross-cutting design problem for the separation of a multicomponent mixture, students determine the optimal design and design parameters of the column for the process. In the course of the solution, the students develop an integral picture of the interrelationship of various aspects of the process of separation of multicomponent mixtures. In groups within two years studied by this technique of 136 students. The conducted researches allowed to determine parameters of increase in effectiveness of activity: growth of cognitive interest of students to discipline and in general to future professional activity—for 52%; positive dynamics of growth of level of proficiency on discipline—for 27%; growth of motivation of educational activity—first place at students was won by group of communicative motives (67%); In general at respondents higher importance of educational informative and professional motives which can be referred to internal motivation is observed. Also social motives gain bigger value, understanding of value of creative opening (56%); increase in level of satisfaction of students with quality of education (75%); annual participation of students in scientific and practical conferences and competitions with good results (56%). Throughout all work students use the developed educational reference materials necessary for the analysis of an objective and its implementation, they are available as in electronic, and in printing. When performing of the represented task students plunge into the atmosphere close to real design activity. Similar practice increases the level of systematization of knowledge, promotes increase in professional orientation of discipline, forms the steady design competences [8] demanded in future engineering creativity when modeling processes at students. Public protection of the project in the presence of students of other groups (the project manager's presentation in English, submission of educational scientific articles, representation of the exhibition stand, demonstration of the commercial of the settlement equipment) also motivates students. Design training allows to connect theoretical knowledge to practical experience of their application, brings closer process of training of specialists to real professional activity that increases quality of the acquired knowledge and skills and competence of graduates. Thanks to implementation of projects students show big interest in results of the education, than at traditional training, understand integrity of design process and a role of each of its stages more distinctly, gain skills of independent search, selection and information processing, necessary for achievement of a goal. Positive sides of use of a design method are also that the principle of mutual training is implemented; restrictions of school hours which are allowed for studying of discipline are overcome. The most perspective innovative projects can be implemented at the industrial enterprises.

References

1. Khatsrinova O., Ivanov W.: Career-building training as a component of talent management. In: Proceedings of 2015 International Conference on Interactive Collaborative Learning (ICL). IEEE 978-1-4799-8706-1/15/$31.00 ©2015 Florence, Italy, 20–24 Sep 2015
2. Brykova O.V., Gromova T.V.: Design Activity in Educational Process. Clean Ponds, Moscow, 2006.3s
3. Trishchenko D.A.: Experience of design training: attempt of the objective analysis of achievements and problems Science and education. 2018. 22(4), 132–152. https://doi.org/10.17853/1994-5639-2018-4-132-152. Experience of project-based learning: an attempt at objective analysis of results and problems. D. A. Trishchenko Belgorod University of Cooperation, Economics and Law, Belgorod, Russia. E-mail: gastronom-tv@yandex.ru © Д. А. Трищенко Образование и наука. Том 20, № 4. 2018. Educ. Sci. J. 20(4) (2018)
4. Zeer E.F., Lebedeva, E.V., Zinnatova M.V.: The methodological bases of realization of process and design approaches in professional education. Sci Educ. 7(136), 40–56 (2016)
5. Tchaikovsky, V.G., Anokhina, L.V.: Design technologies in realization of praktikooriyentirovanny approach to training in higher education institution. Curr. Trends Develop. Sci. Technol. 6, 61–63 (2015)
6. Semashko, N.N., Kobelev, V.P.: Some aspects of the organization of design training in higher education institution. Inf. Commun. Technol. Pedagogical Educ. 3(48), 97–100 (2016)
7. Zaynulina F.K.: A design method of training in formation of motivation of educational process of students. Messenger Kazan State Univ. Cult. Arts. 4, 164–167 (2016)
8. Dreer R.: Use of the principles of design education in program—a bachelor degree move. The Higher Education in Russia, vol. 2, pp. 46–49 (2013)
9. Turlo E.M.: Design training at the higher school. Problems and Per-spektiva of Development of Education in Russia, vol. 19, pp. 79–84 (2013)
10. Krotova E.A., Maksheeva A.I.: Design training as development tool of creative activity. Mod. High Technol. 1, 120–123 (2016)

The Transformative Role of Innovation in the Higher Education

A Case Study

Ferenc Kiss[1] and Vilmos Vass[1,2(✉)]

[1] Budapest Metropolitan University, Budapest, Hungary
{fkiss,vvass}@metropolitan.hu
[2] J. Selye University, Komarno, Slovakia

Abstract. The main objective of this paper is to explore the relationship between transformation and innovation in the context of higher education. It is argued in academic theory that the process of changing role of the universities has two dimensions: interdisciplinarity and competency-based development is an integral part of transformation and innovation. In order to explore this relationship, the basic statement of the paper, that innovation has enormous transformative role in the higher education at different levels (curriculum, teaching methodologies, assessment). The contextual background of the paper is a challenge full "skill gap", which is increasingly strengthening at the Age of 4th Industrial Revolution. This paper adopts a case study approach to analyze the horizontal dimensions of competency-based curriculum planning. The conclusions are drawn from the main analytical results of the theoretical part and the case study approach.

Keywords: Higher education · Transformation · Interdisciplinarity · Curriculum map · Competency-based curriculum · Hungary

1 Introduction

This paper has focused on the experience and best practices of Business Studies (Crisis and Change Management, International Business Culture, Management and Organization, Project Management, Information Management) at Budapest Metropolitan University, especially focusing on the innovation-focused competency-based curriculum planning, learning and teaching methodologies, diagnostic and formative assessment. The basic statement of the paper, that innovation has enormous transformative role in the higher education at different levels. At the curriculum development level, with increasing interdisciplinary cross-impacts among the courses, we use the structural planning process, which is based on the

© Springer Nature Switzerland AG 2019
M. E. Auer and T. Tsiatsos (Eds.): ICL 2018, AISC 917, pp. 321–330, 2019.
https://doi.org/10.1007/978-3-030-11935-5_31

1. needs analysis from the market and our business partners, e.g. TATA Consultancy Services, Exxon Mobile, IBM, Prezi, etc.,
2. competency standards focusing on the 4C's [10], namely critical thinking and problem solving, creativity and innovation; communication, collaboration, and the 21st century skills.
3. the results on mapping and diagnosing the students' prior knowledge on business. During the curriculum planning process, we use the webbing techniques: sharing the aims and expectations, formulating the key questions, planning the students' activities and teaching methods, planning the key topics and deciding the requirements on the based on the revised Bloom's taxonomy [1].

There are some horizontal dimensions at the curriculum planning process: new meaning of motivation, the systematic change, creativity and innovations. These horizontal dimensions have overlapped the curriculum web in order to strengthen the interdisciplinary connections on the Business Studies competency-based curriculum stressing the transformative role of innovation. On the base of the structural curriculum planning process, at the learning and teaching methodological level, basically we use the project method [11] at some courses. Especially, focusing on the social constructive learning philosophy, in the beginning of the courses, mapping the students' prior knowledge using brainstorming, mind mapping and question cards in order to sharing the aims and expectations, building the collaborative experience, beliefs and conceptions. The next step (expected outcome 1) is making personalized competency portfolio comparing the planned competency standards. This portfolio contains personal and professional competencies (knowledge, skills and attitudes), learning strategies and the phenomena on self-directed and deep learning in order to follow the progression on the courses. From the above-mentioned horizontal points of view, the new meaning of motivation [13] the systematic change [4], creativity and innovation [3], we use these horizontal topics as the tools for developing the students' competencies and introducing the projects, which are based on these dimensions (expected outcome 2). The students form the groups, choose the topics, practice the research-based information processing and project management, later the presentation skills. Parallel the planning and development of the projects, they need to use the personalized portfolios and the project documentation framework formulating the aims, research questions and problems, describing the topic, planning the collaborative work, deciding the responsibilities and the tasks, managing the time and planning the expected outcomes of the project (expected outcome 3).

2 Contextual Background

Generally, higher education is under the pressure to change all over the world. Under the umbrella of globalization and internationalization, the growing competition of the international higher education arena and the prioritizing the world class universities require to renew the concept of quality and accountability from theoretical and strategic perspectives [14]. The one side of the coin, this is the huge challenge to the expansion of higher education, but it has just been one side of the coin. The other side of the coin is a growing need for closing the gap between the academic and the work sector. As Egron-Polak and Marmolejo stated: "In globalization, the creation (or perception) of a single worldwide market (for students or faculty members in HE, for example) also demonstrates the extent to which the central drivers of globalization are economic, no matter what sector is being transformed" [2]. This transformation is a paradigm shift from political to economic dimension, which has enormous impact to change from policy to action to the higher education. Firstly, it has been resulted the competency-based higher education, where the required coherency between changes and innovation focus on new meaning of learning and knowledge. As a result of new meaning of learning and lifelong learning strategy, higher education are growing demand for developing self-directed, active, meaningful and constructive learning. The other type of transformation has interdisciplinary characteristics, which is based on active inter-activity and interdisciplinary knowledge—said Holley [6]. In fact, there are several scientific-based definitions of interdisciplinarity and "the absence of widespread consensus on terminology" or "lack of common definitions", from the relevancy of the paper I emphasize three dominant concepts:

Interdisciplinary: A knowledge view and curriculum approach that consciously applies methodology and language from more than one discipline to examine a central theme, issue, problem, topic, or experience.—suggested Jacobs [7]
INTERDISCIPLINARITY is usually defined in one of four ways:

1. *by example, to designate what form it assumes*;
2. *by motivation, to explain why it takes place*;
3. *by principles of interaction, to demonstrate the process of how disciplines interact;and*
4. *by terminological hierarchy, to distinguish levels of integration by using specific labels.—* summarizes Klein [12]

Interdisciplinarity is, I suspect, a path to be taken when we are confronted with phenomena that cannot be understood from one or another discipline alone and only yield their secrets and fascinations when approached with new tools and from new perspectives that derive their methods from more than one discipline.—concludes Renyi [15]

Basically, the similarities of the definitions are the interactions of the different disciplinary areas and interdisciplinarity is a challenge of curriculum planning and teaching methodology as well. The differences of the above-mentioned terminology are on the one hand the basic pillar of interdisciplinarity is "a knowledge view" related to revised concept of knowledge (see above). On the other hand, interdisciplinarity can promote the process of understanding among the different disciplines. From the perspective of higher education, growing demand of interdisciplinary at the curriculum planning level relates to more effective collaboration among the teachers organizing interdisciplinary courses and projects. This is the structural curriculum planning process, where horizontally the teachers discuss about the overlapped themes and the redundancies of the disciplinary courses.

3 Curriculum Mapping Definitions and Use

Curriculum mapping is a conscious structuring fosters interdisciplinarity in higher education. Regarding the concept of mapping, Jacobs cited Fenwick English, "a prominent curriculum leader and powerful theoretician's' definition: "Mapping is a technique for recording time on task data and then analyzing this data to determine the 'fit' to the officially adopted curriculum and the assessment/testing program" [8]. Jacobs described seven phases of curriculum planning process:

1. Collecting the Data: "each teacher describes three major elements that comprise the curriculum on the curriculum map" (processes, skills, content, products)
2. The First Read-Through: "each faculty member should become familiar with his or her colleagues' curriculum" (teacher-as-editor, new content, skills and assessment, looking for repetitions, potential areas for integration, critical evaluation, interdisciplinary team)
3. Mixed Group Review Session: "each teacher shares his or her findings from the individual review of the maps" (delaying judgment, sheet listing, six-eight staff members)
4. Large Group Review: "the facilitators of each small group review session have reported on the findings of the small group sessions" (overall findings, critical decisions, editing, revising and developing mode)
5. Determine Those Points That Can Be Revised Immediately: "the faculty starts to sift through the data and determine areas that can be handled by faculty members, teams and administrations" (exchange the ideas, decision about the curriculum)
6. Determine Those Points That Will Require Long-Term Research and Development: "the implications likely will include structural decisions" (professional discussions, internal needs and external best practices, large-scale steps)
7. The Review Cycle Continues: "curriculum review should be active and ongoing" (ongoing, systematic planning, increasing communication).

In fact, Jacobs' curriculum mapping model used in public education, but there are several opportunities to implement this structuring into the higher education. Basically, this is a visual representation of curriculum based on fostering interdisciplinarity and collaborative curriculum planning [8, 9]. From the higher education dimension, Uchiyama and Radin did a qualitative study on the implementation of curriculum mapping in higher education. They stated: 'The curriculum maps are aggregated first horizontally by course and then vertically across all courses in sequence" [16]. They described the same "cyclical circle" of curriculum mapping than Jacobs's model stressing the horizontal and vertical process. Summarizing the advantage of curriculum mapping, they raised the point as: "The result is a curriculum that is fluid and adaptable as the needs of students, policies and new research findings change over time" [16]. Obviously, this is a positive perspective of the structural curriculum planning. Flexibility and adaptability is the strength of curriculum planning at the department or faculty level (micro level curriculum planning and implementation). But at the university level, "collaboration and collegiality" and "increasing interactivity" basically can change the traditional organizational culture. As Uchiyam and Radin concluded their research: "Curriculum mapping is an ongoing, dynamic process. Our faculty recognizes that, by accepting this as an ongoing process, we will continue to grow as a collaborative community, to connect with each other to decrease isolation, to consider curricular changes carefully, and to promote collegiality" [16].

4 Overview of the Transformative Role of Innovation in the Higher Education

As most of the phenomena of curriculum mapping mentioned above are strongly linked to revised concept of knowledge and teaching, the organizational change (see cooperation and collegiality). Basically, this a complex process at micro and macro level, where continuous innovation is a vehicle of change. Halász stated: "An innovative university that is constantly on the lookout for new or original solutions to improve its activities in research, teaching and community services is a university where the notion of change has a positive meaning and where the culture of the organization encourages innovation" [5]. There is no doubt; in this "systematic triangle" (innovation-change-culture) curriculum mapping plays an essential role. But as we know: "One swallow doesn't make a summer." It is therefore that the other "important triangle" (curriculum planning-teaching methodology-assessment), which I think worth pointing out. It is apparent that these two triangles have strong consistency, the elements are mutually reinforcing. Curriculum mapping is in close correlation with teaching methodology and assessment fostering project-based courses and diagnostic-formative assessment in higher education. One of the worldwide and most popular definitions of innovation has the following economic and social connections:

"The process of translating an idea or invention into a good or service that creates value or for which customers will pay. To be called an innovation, an idea must be replicable at an economical cost and must satisfy a specific need. Innovation involves deliberate application of information, imagination and initiative in deriving greater or different values from resources, and includes all processes by which new ideas are generated and converted into useful products. In business, innovation often results when ideas are applied by the company in order to further satisfy the needs and expectations of the customers. In a social context, innovation helps create new methods for alliance creation, joint venturing, flexible work hours, and creation of buyers' purchasing power. Innovations are divided into two broad categories:

1. Evolutionary innovations (continuous or dynamic evolutionary innovation) that are brought about by many incremental advances in technology or processes and
2. Revolutionary innovations (also called discontinuous innovations) which are often disruptive and new" [18].

How can we transfer this definition into the higher education? Admittedly, the above-mentioned definition contains some remarkable approaches to the higher education. Firstly, "application of information, imagination and initiative" are central components of the transformation in the higher education. New ideas are the starting point of innovation. Secondly, students' and teachers' needs and expectations as important as costumers' satisfaction. Obviously, sharing the needs and expectations before starting curriculum mapping is a fundamental task. This is the significant part of the diagnostic assessment as well. Thirdly, the two broad categories of innovations are forcefully characterizing higher education; especially revolutionary innovations determine the processes. The conscious coherency between the above-mentioned two triangles: "systematic triangle" (innovation-change-culture) and "important triangle" (curriculum planning-teaching methodology-assessment) that continuous or dynamic evolutionary innovation can produce.

5   A Case Study of Budapest Metropolitan University

Budapest Metropolitan University (METU) is a unique institution as being the most national private one with creative industry studies with 4 basic faculties:

- Communication: communication and media studies, development of competencies, protocol and event organization, speaker studies, mediation, etc.
- Business: HR, Management and Economics, Commerce and Marketing, Financial studies, Development of Entrepreneurship, etc.
- Tourism: Tourism—Hospitality, Tourism—Management, Wine Tourism, Health Tourism, Gastronomy, Organization and MICE Tourism, International Event and Hospitality Management, etc.
- Art: Animation, Design Culture, Photography, Craftsmanship, Graphic Design, Filmmaking, etc.

Up to now we have 6000 students from 90 different countries, 80 granted projects have been carried out, and currently, several multicultural BA and MA courses and projects are being developed, and new proposals are to be submitted with our best partners in each relevant field. In addition, we are about to keep our knowledge up-to-date and develop or acquire new skills, learn new methods and technics. Our main priorities:

- Tourism and wellbeing
- Promoting STEM involving the creative industry (including Arts) as STEM → STEAM
- Developing and strengthening of start-ups, launching spin-offs
- Heritage protection: preserving and promoting national and common European tangible and intangible natural and cultural heritage
- Developing non-formal learning and education:
  - developing training materials, modules and curriculums;
  - developing competencies, carrier planning;
  - launching joint platforms, online fora;
  - gamification, edutainment and applying their means to increase the motivation of the Youth and (university) students
- Future research and the Grand Societal Challenges.

The most important purpose of Budapest Metropolitan University is to provide its' students with valuable competencies, make them successful in their jobs and professions with their capabilities and abilities. Inspire them to become creative, independent thinkers, exceptional experts, who are committed to life-long learning, and sense responsibility for the community. [19] Yin underlined the importance of case study as a social science research method. He noted: *"In general, case studies are the preferred strategy when "how" or "why" questions are being posed, when the investigator has little control over events, and when the focus is on a contemporary phenomenon within some real-life context"* [17]. I choose case study as a relevant research methods of my paper, because via the experience-based examples and best practices it can foster relevancy and validity of the transformative role of innovation in the higher education.

Our study question is: How and why do we need to foster the transformative role of innovation in the higher education? The proposition of the case study: Because of the traditional unidisciplinarity and isolation in higher education, it is difficult to foster the transformative role of innovation.

At the curriculum development level, we use curriculum mapping structuring strategy (see above) turning from the knowledge-based to the competency-centered approach. It means that in the case of Business Studies (Crisis and Change Management, International Business Culture, Management and Organization, Project Management, Information Management) there are some horizontal points which determine the curriculum mapping process, e.g. promoting intrinsic motivation, creativity and innovation. In order to strengthen these horizontal dimensions the interdisciplinary curriculum planning group made a competency map, especially focusing on general key competencies, for instance: critical thinking and problem solving, creativity and innovation; communication, collaboration, and the 21st century skills (entrepreneurship, digital competence etc.). On the base of this competence-map, at the first step of curriculum mapping (collecting data) the teachers need to find the consistency with the general competencies. They do not focus on knowledge or fragmented topic or content in the curriculum, but they plan to map students' prior knowledge (topic-relevant experience, beliefs, values etc.) and the methodology of sharing the needs and expectations, collecting key questions with question cards. Parallel with planning the introduction part of the course, they revise the assessment techniques as well in order to foster diagnostic and formative assessment. At the second and third phase of curriculum mapping (the first read-through, mixed group review session) we organized a workshop to the colleagues of Business Studies in order to discuss about the innovations (ideas, methods, techniques) and the potential areas for integration via defining the domain-specific competence web parallel to the general key competencies.

At the methodological level, we organize a pedagogical workshop on collaborative learning and project-method. Our hypothesis is these two methods can foster the transformative role of innovation in the higher education. On the base of this internal staff training, all members of the department attended the large group review in order to make comments and make critical decision of the first version of curriculum map. The expected outcome of the internal staff meeting and the large group review is to finalize the standard of the project plans: focused topics (e.g. e-business, start-up thinking, electronic vehicles, smart cities, silver economy, blockchain, application of artificial intelligence), formulating the research questions and defining the problem, project team's decision on the qualitative research methods (document analysis, participated observation, interview, Delphi-method), time and project management plan, success criteria. During the project work the formative assessment techniques and tools have fundamental role. For instance ongoing feedback of the project work based on weekly consultations including the presentation and discussion of the meeting minutes and the outcomes, portfolio is based on project documentations with some reflections and comments, final feedback circle in order to evaluate the progression and the shared aims and expectations.

Time management and creative problem-solving are the prioritized areas of the project plan, although, besides these, students and teachers face various multicultural challenges. Because of the students of the university are coming from 95 countries (2018 Spring) for full time or one semester mobility, the courses delivered in English language have to manage the implementation of team collaboration framework and assessment culture which can give the possibility to work effectively, overwhelm many difficulties originated for the different culture of the team member, as well as to compare the performances. This includes the common reporting and time management requirements, the understanding and acceptance of the assessment methods, the ways to access information, consultation and reflection resources (e.g. consultation with experts and teachers, market feedback, community). Teachers and trainers involved in these courses have regular discussion and development circles for sharing experiences and understanding the new multicultural phenomena among in the cohort.

Finally, at the end of the project there will be an Interdisciplinary Presentation Day, where the students present their research work with collaborative assessment. The main assessment criteria are: relevancy of the project, logical structure, innovative and creative parts of the project, qualitative research methods, quality of conclusion. Last, but not least, after the project presentations the review circle is starting in order to evaluate the strength and weaknesses of the project-based courses.

6 Conclusion

From the theoretical and research dimensions, no doubt, innovation as an important and integrated part of transformational higher education. Basically, innovationin an economic, social and educational context is based on creative ideas, effective application information and knowledge, and imagination. Creative ideas emphasize the importance of divergent thinking and flexible adaptivity. From the outcome-based dimension of curriculum planning, innovation has resulted after the successful transformation process, useful inventions and products. Firstly, this statement tightly fits to the project-method and project-based learning. Secondly, under the umbrella of professional accountability, the transformation role of innovation in the higher education has based on evidence-based, criteria-oriented, operationalized innovation approach. From the process-based dimension of curriculum planning, one of the most relevant and pragmatic answers to the challenging question of innovation in higher education is revising the content, teaching and learning methods, assessment techniques to focus on general and domain-specific competencies. From a practical point of view, organizing interdisciplinary curriculum planning team is an effective solution on the above-mentioned challenges. We propose that the process-based dimension of curriculum planning and interdisciplinary approach of curriculum development, can be integrated innovation in the transformation at two levels. The personal level of integration has enormous impact to the individuals developing competencies and increasing professionalism. The organizational level of integration has high added value via the transformative channels to change the traditional hierarchical structure of the universities moving forward to the professional learning communities. In addition, professional learning communities have some general characteristics: shared vision and mission, effective change management and conflict resolution and creativity, but on the base of the required coherency between innovation and transformation, interdisciplinary can play an integrated role among these fields.

Interdisciplinarity has enormous impact to foster collaboration and collegiality at organizational level. Basically, curriculum mapping is a vehicle for collaboration, an "ongoing, dynamic process" [16]. But there are some dilemmas about it. On the base of the research data, it is clear that it is important task to revise curricula and teaching methodologies, especially focusing on project-method and cooperative learning. The conscious coherency between the above-mentioned two triangles: "systematic triangle" (innovation-change-culture) and "important triangle" (curriculum planning-teaching methodology-assessment) that continuous or dynamic evolutionary innovation can produce. These two triangles can guarantee, on the one hand, the competency-based higher education. On the other hand, innovation is a part of the everyday life at the universities changing the higher education culture. As a result, the curriculum development level, curriculum mapping structuring strategy turning from the knowledge-based to the competency-centered approach. The teaching methodology level prioritizes the meaningful and active learning fostering the lifelong learning paradigm. The assessment level emphasizes diagnostic and formative function promoting self-directed learning and ongoing, continuous feedback and comments.

From the organizational level, there are some suggestions in order to move forward to the professional learning community:

- Sharing strategic vision and aims via workshops among the staff and non-staff members.
- Strengthening interdisciplinary project-based dissemination process on the base of shared vision and aims formulating some projects and teams.
- Fostering "intelligent accountability" finding the balance between summative and formative assessment, and internal and external evaluation.
- Promoting the culture of trust, creativity and innovation at the horizontal and vertical levels of the organization.

As a summary, innovation is important and play a dominant transformative role in the higher education in the context of growing demand of interdisciplinarity (see curriculum mapping). Parallel to this process, the case study of the Budapest Metropolitan University emphasized the importance of innovative teaching and learning methods, assessment techniques in the context of competency-based higher education.

References

1. Drake, M.S.: Planning Integrated Curriculum. The Call to Adventure. Association for Supervision and Curriculum Development, Alexandria, VA (1997)
2. Egron-Polak, E., Marmolejo, F.: Higher education internationalization: adjusting to new landscapes. In: de Wit, H., Gacel-Ávila, J., Jones, E., Jooste, N. (eds.) The Globalization of Internationalization. Emerging Voices and Perspectives, pp. 7–18. Routledge, London and New York (2017)
3. Florida, R.: The Rise of the Creative Class. Basic Books A member of the Perseus Books Group, New York (2011)
4. Fullan, M.: The Principal. Three Keys to Maximizing Impact. Jossey-Bass A Wiley Brand, San Francisco, CA (2014)
5. Halász, G.: Organizational Change and Development in Higher Education. EHMD Change Study (n.a.). http://halaszg.ofi.hu/download/EHEMD.pdf
6. Holley, A.K.: Interdisciplinary strategies as transformative change in higher education. Innov. High. Educ. **34**, 331–344 (2009)
7. Jacobs, H.H.: Interdisciplinary Curriculum: Design and Implementation. Association for Supervision and Curriculum Development, Alexandria, VA (1989)
8. Jacobs, H.H.: Mapping the Big Picture: Integrating Curriculum and Assessment K-12. Association for Supervision and Curriculum Development, Alexandria, VA (1997)
9. Jacobs, H.H.: Getting Results with Curriculum Mapping. Association for Supervision and Curriculum Development, Alexandria, VA (2004)
10. Jacobs, H.H. (ed.): Curriculum 21. Essential Education for a Changing World. Association for Supervision and Curriculum Development, Alexandria, VA (2010)
11. Kilpatrick, W.H.: The project method. Teach. Coll. Rec. **19**, 319–335 (1918). Klein, T.J.: Interdisciplinarity. History, Theory, and Practice, p. 55. Wayne State University Press, Detroit (1990)
12. Maringe, F.: The meanings of globalization and internationalization in HE: findings from a world survey. In. Maringe, F., Foskett, N. (eds.) Globalization and Internationalization in Higher Education, pp. 17–35. Continuum International Publishing Group, New York (2010)
13. Pink, H.D.: Drive: The Surprising Truth About What Motivates Us. Riverhead Books, New York (2011)
14. Rényi, J.: Hunting the Quark: interdisciplinary curricula in public schools. In: Wineburg, S., Grossman, P. (eds.) Interdisciplinary Curriculum. Challenges to Implementation, p. 14. Teachers College, Columbia University, New York and London (2000)
15. Uchiyama, P.K., Radin, L.J.: Curriculum mapping in higher education: a vehicle for collaboration. Innov. High. Educ. **33**(4), 271–280 (2009)
16. Yin, K.R.: Case study research design and methods. Second Edition. http://www.madeira-edu.pt/LinkClick.aspx?fileticket=Fgm4GJWVTRs%3D&tabid=3004
17. http://www.businessdictionary.com/definition/innovation.html
18. http://metubudapest.hu/about-us/81/why-metropolitan.html

Engineering Design of a Heat Exchanger Using Theoretical, Laboratory and Simulation Tools

Rabah Azouani[1(✉)], Rania Dadi[1,2], Christine Mielcarek[1], Rafik Absi[1],
and Abdellatif Elm'selmi[1]

[1] Process Department, Ecole de Biologie industrielle, 49 avenue des
Genottes CS 90009, 95895 Cergy Cedex, France
r.azouani@hubebi.com
[2] LSPM-CNRS, Laboratoire de Sciences des Procédés et des Matériaux,
Université Paris 13, Sorbonne Paris Cité, 99 Avenue Jean-Baptiste Clément,
93430 Villetaneuse, France

Abstract. Project based learning (PBL) is developed in this work, the main aim is the development of engineering and design skills of our graduate students for heat exchangers devices. We use a project pedagogy to enhance the understanding and demonstrate with different ways the effect of several parameters on the efficiency of heat transfer, showing the relation between theory, experiments and computer modeling. Pedagogical experience was carried out with 150 graduate students during heat and mass transfer course. This work shows the interest of working with several complementary educational tools. Results suggest that this engineering innovative pedagogy is an effective approach to improve motivation, implication, rigor, critical thinking and communication skills. Through PBL, connection with real problems of process industry is achieved.

Keywords: Project based learning · Heat exchangers · Heat transfer

1 Introduction

The transfer of heat to and from process fluids is an essential part of most industrial processes in different fields. The word "exchanger" is often used specifically to denote equipment in which heat is exchanged through an exchange surface between two process streams, where fluid is heated or cooled. Heat exchange is an important unit operation that contributes to the efficiency and safety of many processes [1]. Heat recovery of heat exchanger network is an important subject [2], numerous studies about heat exchanger efficiency and influence of fluid parameters on heat transfer have been reported [3–5]. The principal types of heat exchanger used in the industrial process are: double-pipe heat exchanger, shell and tube heat exchanger and plate exchanger. All these heat exchangers can be operated in both parallel and counter-flow configuration.

Our engineering school in industrial biology is implementing active learning pedagogy [6–8], to cultivate engineers' core competencies, practical skills and professional knowledge, this approach allows our future engineers to tackle innovation projects with a multitude of data from different sources and complexity. PBL in

© Springer Nature Switzerland AG 2019
M. E. Auer and T. Tsiatsos (Eds.): ICL 2018, AISC 917, pp. 331–341, 2019.
https://doi.org/10.1007/978-3-030-11935-5_32

engineering has been applied in several specialization [9–12]. It seems to be in accordance with students' profile evolution and industry demands on our engineers that are better prepared to support their organizations. In this work, we report the use of PBL in heat transfer engineering course with 150 third year graduate students. This project consists on engineering design of heat exchanger using theoretical calculations, experiments and computer modeling. We expect to develop skills of calculations, experimentation, simulation and results analysis. At the end of this course, students will be able to calculate the energy needs in any industrial systems and design the required exchanger in compliance with the standards of The Tubular Exchanger Manufacturers Association (TEMA) [13].

2 Methodology

2.1 Pedagogical Methodology

The methodology of this study is based on the project realization approaches of heat exchanger calculation, design and simulation. Heat exchange is performed between hot and cold water, the objectives of this project are:

1. To investigate the effects of the control parameters and the heat exchanger configuration on the rate of the heat transfer and the overall heat transfer coefficient
2. To check the validity of the theoretical calculation
3. To compare the results of theoretical calculations, with the experimentation on a pilot of heat exchanger and with a simulation on the same heat exchanger.

The students work is organized in different phases as shown in Fig. 1:

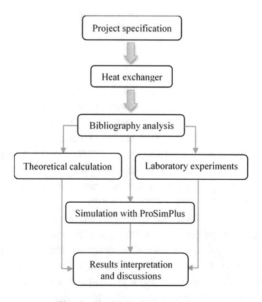

Fig. 1. Basic design procedure

2.2   Theoretical Calculations

This project part consists on calculation using several thermal formulas to estimate overall heat transfer coefficient (h_g), which is an indicator of the efficiency of heat transfer between two fluids. The general equation of heat transfer in exchanger is represented below:

$$\Phi = h_g.S.\Delta T_{lm} \tag{1}$$

With

Φ	heat transfer flux [W]
h_g	overall heat transfer coefficient [W/m^2 °C]
S	surface of exchange between fluids [m^2]
ΔT_{lm}	logarithmic mean temperature difference between hot and cold fluid [°C].

The calculation of h_g is required, student group use physicochemical and thermal proprieties of hot and cold water. Example of calculation of h_g for tubular heat exchanger is represented in the following equation:

$$\frac{1}{h_g} = \frac{r_2}{r_1}\frac{1}{h_{hot}} + \frac{r_2 ln \frac{r_2}{r_1}}{\lambda} + \frac{1}{h_{cold}} \tag{2}$$

With

r_1, r_2	internal and external radius of the tube [m]
h_{hot}, h_{cold}	coefficient of convection respectively of hot and cold fluid [W/m^2 °C]
λ	thermal conductivity of the tube [w/m °C].

To perform this calculation, students estimate convection coefficient of hot and cold water respectively h_{hot} and h_{cold}, taking into a count hydrodynamic propriety represented by Reynolds number (Re). They use experimental correlation with dimensionless numbers: Nusselt (Nu) and Prandtl (Pr), as shown in Eq. 3:

$$Nu = \frac{h_i D}{\lambda} = a\,Re^b Pr^c \tag{3}$$

With

D	diameter of the tube [m]
h_i	coefficient of convection of the fluid [W/m^2 °C]
λ	thermal conductivity fluid [W/m °C]
a, b, c	constants of correlation depending from experimental conditions.

The output data in this part of the project is h_g calculation.

2.3   Experiments: Heat Exchanger Unit

The experiment was conducted in a heat exchanger pilot PIGNAT-BME/3000 (Fig. 2), using as fluids cold and hot water. The influence of the flow rate of cold water on the temperatures of the fluids was performed. Students studied also the influence of the direction of flow (parallel flow and counter flow). The overall heat transfer coefficient was calculated with the collected experimental data using Eq. 1.

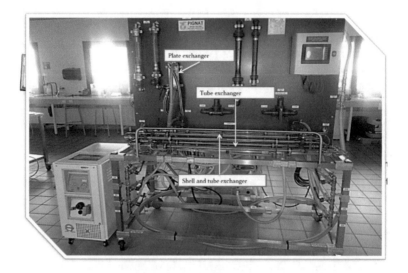

Fig. 2. Heat exchanger pilot

This experimentation doesn't allow the variation of all exchanger parameters: fluid proprieties, pressure, flow rate, geometry, specific heat capacity, thermal conductivity and fouling.

To improve the understanding and analyze the effect of each parameter on the efficiency of the heat exchanger, we introduce simulation with industrial software ProSimPlus.

2.4   Simulation with ProSimPlus

ProSimPlus is a process engineering software that performs rigorous mass and energy balance calculations for a wide range of industrial steady-state processes. It is used to design and optimize several unit operations. The steps to build the flowsheet are as follows:

(a) Select your components
(b) Select your thermodynamic model
(c) Describe your chemical reactions step (no reaction)
(d) Create your flow sheet with streams and exchanger
(e) Run the simulation

(f) Reports generated
(g) Analyze the results from the flowsheet.

We use this tool to simulate the three types of heat exchangers and to determine the influence of parameters on the efficiency of heat exchangers. For each parameter we measure 5 variations to obtain a profile. Figure 3 shows all the parameters studied.

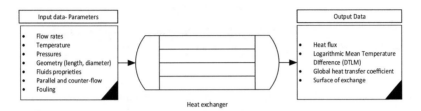

Fig. 3. Simulation methodology

Students create a flowsheet for all heat exchangers. Figure 4 presents a general overview of the shell and tube heat exchanger.

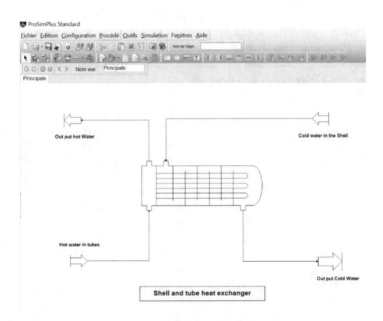

Fig. 4. Shell and tube flowsheet with ProSimPlus

3   Results and Discussion

3.1   Overall Heat Transfer Coefficient

The value of this parameter was obtained using the different approaches. Table 1 sum up the result:

Table 1. h_g values

Methods	Theorical calculation	Experimentation	ProsimPlus simulation
h_g (W/m^2 °C)	256	2354	2200

Students Results show a significative difference between theoretical calculation performed with empirical correlation compared to experimentation and ProsimPlus simulation. The explanation is the fact that empirical correlations are just valid in the same conditions in which they were obtained. This part expects to develop critical thinking of our engineers.

Influence of fluids parameters on the heat transfer
Temperatures of cold and hot water
Temperatures of fluids are an important parameter in heat transfer. Students used in their calculation logarithmic mean temperature difference method, between hot and cold fluid to estimate temperature gradient, it means that heat transferred is proportional to ΔT_{lm}. Equation 1 shows that the increase of this parameter implies increase of energy transferred from the hot to the cold fluid.

Figure 5a shows evolution of ΔT_{lm} with temperature of cold and hot water. Students obtain a linear relation between ΔT_{lm} and cold-water temperature, increase of this temperature decrease the value of ΔT_{lm}, because the temperature gradient is reduced with the hot fluid. The shape of the curve of hot water Fig. 5b is significantly different from cold water. They observe a proportional relationship with ΔT_{lm}; a sharp rise of ΔT_{lm} from 75 to 85 °C, then, a leveling off from 85 to 100 °C, this observation can be explained with the phase change occurred. They conclude that the temperature of hot fluid (primary fluid) is the most important one and cannot increased infinitely. They have to pay attention to phase change, instability, and risks.

Flow rate of cold and hot water
Same approach was used with flow rate of fluids. Figure 6 shows the evolution profile for each fluid. Students remark a slight increase of heat flux for secondary fluid (cold fluid) and sharp rise with primary fluid (hot fluid). This sensitivity to primary fluid is due to the fact that he has the highest heat capacity and transfers its energy to the secondary fluid. This physical phenomenon is limited by the ability of the secondary fluid to absorb the energy transmitted by the primary fluid. Heat capacity is a parameter of choice between heat transfer fluids.

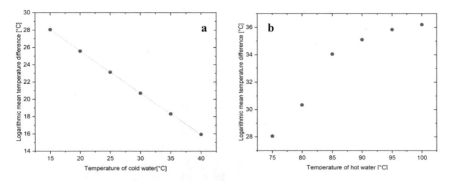

Fig. 5. Influence of fluid temperature variation on heat transfer

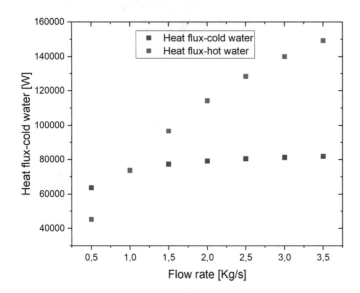

Fig. 6. Flow rate influence on heat flux exchanged

Nature of fluids

Thermophysical properties are available for many fluids. These data include: viscosity, thermal conductivity, specific heat capacity, enthalpy and density. These proprieties are specific for each fluid. If we change a fluid in heat exchanger, primary or secondary one, it generates properties change, and influence the heat transfer, because all the priorities are used in overall coefficient of heat transfer. Figure 7 shows the influence of the primary fluid change on heat transfer, water seems to be the best one. This fluid is very used as primary fluid for many raisons: large specific heat capacity (4.18 kJ/kg °C), toxicity, phase change, and cost.

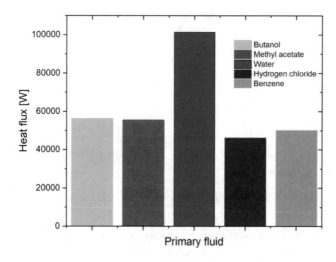

Fig. 7. Influence of the nature of fluids on heat tranfer

Influence of the surface of exchange on heat transfer
Number and length of tubes

Surface of exchange determine the cost and efficiency of heat exchanger; this parameter depend on geometry of heat exchanger: tubular or rectangular. In this work, students studied shell and tube heat exchanger, they have done a variation of tube dimension (diameter and length) and their number in exchanger as shown Fig. 8.

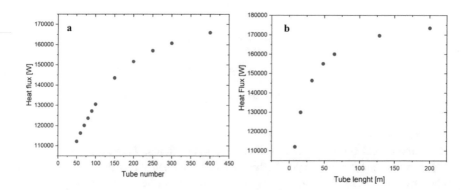

Fig. 8. Surface exchange influence

Heat transfer increase with the surface. In Fig. 8a, b, the evolution of heat flux has the same profile: sharp rise and leveling off, with more pronounced slope for tube length. Heat transfer reaches saturation, because beyond surface increase, it fails to consider fluid speed in tube, pressures losses, mechanical stresses and vibrations.

Flow pattern influence

There is a variety of flow configuration in heat exchanger, the most common are parallel flow and counter flow. In parallel flow hot and cold fluid have the same direction, where in the counter flow their direction is opposite. Figure 9 shows that counter flow configuration is more efficient, all his experimental points have a higher value of ΔT_{lm}. Explanation of this observation is due to temperature gradient evolution between primary et secondary fluids. In counter flow, temperature gradient still constant, but in parallel flow the gradient varied; higher in the beginning and lower in the end of transfer. Students conclude that counter flow configuration is more efficient in heat transfer.

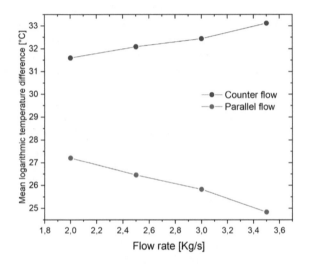

Fig. 9. Flow pattern influence

4 Pedagogical Evaluation

This global project offers opportunity to master design and calculation of heat exchangers related to industrial practice. It's a new pedagogical approach in engineering where students apply all exchanger aspects; theory, laboratory work and process simulation. All calculations and data analysis were done on Excel. With this method, students apprehend all the optimization levers of heat exchangers: fluid parameters, flow configuration and surface of exchange.

Furthermore, students group is composed of 5 students, they were active, motivated, and demonstrated good communications skills. Evaluation of students have shown, that this project allows them to develop critical thinking, application of their theoretical knowledge in practice, increase understanding of the module and acquire autonomy.

Nevertheless, it is necessary for the professor to accompany students in order to guide and to reassure them, emitting at the same time encouragements and alerts.

Hence, we found in project reports that the results interpretation of some students remains very superficial, we have to find a balance between time of the theoretical classes versus time devoted to the project. To overcome this limitation, we use the data of the project of some students as examples in the course.

5   Conclusion

In this work, we report a successfully PBL development in engineering. The methodology was implemented in heat and mass transfer module for the third-year graduate students. The experiment took part in a period of one semester simultaneously with course. Evaluation is based on project rapport and exam.

Comparing with traditional form of education, PBL capture students' interest and leads to serious implication and motivation to work on real process industry problems. Knowing the good practice and respecting the standards, which would allow them to work in an effective and structured approach on their real industrial projects as future engineers.

Students' evaluations show great interest for this project, for various reasons: variety of tools, experiments, concrete industrial problem, interaction between theory and practice, raising awareness of the economic issues/risks and project management.

We have seen a significant increase in student interest and marks, but at this stage we cannot perform a reliable statistical analysis, we need a return over several years.

References

1. Sinnott, R., Towler, G.: Chemical Engineering Design, 5th edn. Elsevier (2009)
2. Sadeghian Jahromi, F., Beheshti, M.: Extended energy saving method for modification of MTP process heat exchanger network. Energy **140**(Part 1), 1059–1073 (2007)
3. Noorollahi, Y., Saeidi, R., Mohammadi, M., Amiri, A., Hosseinzadeh, M.: The effects of ground heat exchanger parameters changes on geothermal heat pump performance. Appl. Therm. Eng. **129**, 1645–1658 (2018)
4. Hamidreza, S., Allen, M.J., Nourouddin, S., Benn, S., Faghri, A., Bergman, T.L.: Heat pipe heat exchangers and heat sinks: opportunities, challenges, applications, analysis, and state of the art. Int. J. Heat Mass Transf. **89**, 138–158 (2015)
5. El Gharbi, N., Blanchard, R., Absi, R., Benzaoui, A., El Ganaoui, M.: Near-wall models for improved heat transfer predictions in channel flow applications. J. Thermophys. Heat Transf. **29**(4), 732–736 (2015)
6. Absi, R., Lavarde, M., Jeannin, L.: Towards more efficiency in tutorials: active teaching with modular classroom furniture and movie-making project. In: 2018 IEEE Global Engineering Education Conference (EDUCON), 17–20 April 2018, pp. 774–778. Santa Cruz de Tenerife, Tenerife, Islas Canarias, Spain (2018)
7. Elm'selmi, A., Boeuf, G., Elmarjou, A., Azouani, R.: Active pedagogy project to increase bio-industrial process skills. In: Auer M., Guralnick D., Uhomoibhi J. (eds.) Interactive Collaborative Learning. ICL 2016. Advances in Intelligent Systems and Computing, vol. 544 (2017)

8. Absi, R., Nalpace, C., Dufour, F., Huet, D., Bennacer, R., Absi, T.: Teaching fluid mechanics for undergraduate students in applied industrial biology: from theory to atypical experiments. Int. J. Eng. Educ. (IJEE) **27**(3), 550–558 (2011)
9. Alaya, Z., Chemek, A., Khodjet El Khil, G., Ben Aissa, M., Marzouk, A.: An integrated project for freshmen students in a software engineering education. In: Auer, M., Guralnick, D., Uhomoibhi, J. (eds.) Interactive Collaborative Learning. ICL 2016. Advances in Intelligent Systems and Computing, vol. 544. Springer, Cham (2017)
10. Davis G.W.: Motivating students with bio-fuel student engineering competition projects. In: Auer, M., Guralnick, D., Uhomoibhi, J. (eds.) Interactive Collaborative Learning. ICL 2016. Advances in Intelligent Systems and Computing, vol. 544. Springer, Cham (2017)
11. Reinhardt A., et al.: Didactic robotic fish—an EPS@ISEP 2016 Project. In: Auer, M., Guralnick, D., Uhomoibhi, J. (eds.) Interactive Collaborative Learning. ICL 2016. Advances in Intelligent Systems and Computing, vol. 544. Springer, Cham (2017)
12. Seman, L., Hausmann, R., Bezerra, E.-A.: On the students' perceptions of the knowledge formation when submitted to a project-based Learning environment using web applications. Comput. Educ. **117**, 16–30 (2018)
13. Tubular Exchanger Manufacturer's Association: Standards-of-the-Tubular-Exchanger-Manufacturers-Association-TEMA-9th-Edition (2007)

HappyGuest—A System for Hospitality Complaint Management Based on an Academic-Industry Partnership

Angela Pereira[1,3(✉)], Micaela Esteves[2,3], Rosa Matias[2,3],
Alexandre Rodrigues[3], and André Pereira[3]

[1] CiTUR—Tourism Applied Research Centre—ESTM, Leiria, Portugal
angela.pereira@ipleiria.pt
[2] CIIC—Computer Science and Communication Research—ESTG,
Leiria, Portugal
{micaela.dinis,rosa.matias}@ipleiria.pt
[3] Polytechnic Institute of Leiria, Leiria, Portugal
{2150670,2150689}@my.ipleiria.pt

Abstract. Academic-Industry Partnership is important to increase knowledge exchange and boost the investigation. This partnership can be done in many ways, for instance, through: long-term projects investigation, academies or small projects to be developed by students. This paper presents an Academic-Industry Partnership case study between a Portuguese Higher Education Institution and a local Hospitality Industry. In this study the researchers analyse the benefits for students and the drawbacks results for both parts. The results show that for students there are many advantages since they can learn through real-world problems and deal with real customers. This way, students improve their communication skills and can observe the practical application of the theoretical concepts. In this case, the local Hospitality highlights the solution presented, even though considering that is crucial for this partnership to reduce the time consuming in production. In addition, the economy is constantly changing, and industries cannot wait a long time for products as they run the risk of being outdated.

Keywords: Academic-industry partnership · Experiential learning
Higher education · Hospitality industry · Hospitality complaint management

1 Introduction

Nowadays, it is important for Higher Education Institutions to establish Academic-Industry Partnerships to reduce the gap between the needs of the labour market and what is taught at universities [1, 2]. Therefore, striving for excellence helps universities career opportunities, research funding, awareness of industry trends and students' inspiration through the discussion of real problems [3]. On the other hand, the industry has access to extended networks of R&D obtaining innovative solutions [3]. This relationship is crucial to both partners in exchanging knowledge and making contacts.

© Springer Nature Switzerland AG 2019
M. E. Auer and T. Tsiatsos (Eds.): ICL 2018, AISC 917, pp. 342–352, 2019.
https://doi.org/10.1007/978-3-030-11935-5_33

Especially, in small regions, Higher Education Institutions can boot the local economy through partnerships with local companies. This way, the institutions graduate students with the profile needed locally. Usually, those companies have small problems which can be solved by higher education students. Moreover, local industries have access to R&D and personalized solutions to their problems.

Usually, the Portuguese Computer Science students learn through the development of academic projects. This is a limitation, due to these students are not prepared to face real projects constraints as well as interact with customers. As teachers, we have been observing students over the years and they are not keen on the project development. Moreover, the solutions presented by students are usually not very suitable for the problems. Through conversations with students the main reason pointed out is the absence of real projects as well as the need to interact with customers particularly during the requirements specification and test phases. Also, when students went to the internship, usually face a lot of integration problems, namely, they have difficulties to deal with the customers and therefore to understand what they want.

In this context, teachers from computer science department of Polytechnic Institute of Leiria established a partnership with local Hospitality Industry, with the aim to solve a specific problem regarding hospitality complaints management and to prepare students for labour market. With this study, researchers want to verify if the partnership can be of great value for both sides. Our goal is to change the academic project to real life ones, so students can address genuine business needs and real customers. The researchers want to study the students' motivation and commitment when facing real life problems.

The rest of this paper is structured as follow: first, related work and motivation; next the methodology and then, results and discussions are presented. Finally, conclusion is made.

2 Related Work and Motivation

It is widely recognized the importance of collaboration between Industry sector and Higher Education Institutions. Not only for knowledge exchange, but also for better prepare students to the labour market [2]. Since, the Industry has the practical know-how and the Higher Education Institutions have the academic theory and research orientation [4]. According to several authors [2, 4, 5] there are many advantages from the collaboration between Industry and Higher Education which are summarized in Table 1.

Meanwhile, the literature [6, 7] refers to three main types of partnership between Industry and Higher Education (Fig. 1), which are Industry—University Research; Industry—University Academy and Industry—University Experience Learning.

Industry—University Research refers to the development of research projects that are funded by enterprises. This is a way for universities to obtain external funding, providing opportunities for professors and graduating to work on ground-breaking research, vital inputs to keep teaching and learning on the cutting edge of a knowledge area [6]. For an effective collaboration between universities and industry, in this type of partnership, it is important to have individuals who understand both, academia and

Table 1. Academic-industry partnership advantages

Academic	Industry
Career opportunities	Access to extended networks
Research funding	Thinking outside the box
Awareness of industry trends	Training
Inspiration by application derived discussions	Ability to find new talent to hire
	Access to specialized
	World-leading resources
Making contacts and exchange of knowledge	

Fig. 1. Types of partnership between industry and higher education

business as refer Alan Begg [6]. An example well succeeded, it was the partnership in nanoscience between IBM and the Swiss Federal Institute of Technology (ETH Zurich) which spent $90 million in a nanotechnology centre [6]. Furthermore, the universities should orient their courses/programs to solve the scientific and technological problems that enterprises are interested [7].

In the Industry—University Academy partnership the most common is collaboration with big companies such as CISCO, Oracle, Microsoft and IBM [6], where the universities as well as other Higher Education institutions have academies. In these academies students can obtain a certificate in a specific knowledge area. Academicians believe that this type of partnership is more valuable for students since it transmits them extensive technology knowledge widely use in the labour market. Also, it is recognized that these academies attract students to Higher Education.

Industry—University Experience Learning consists of the development of small projects with the intention of helping students to learn by doing. With this partnership it is possible to create an environment where students learn through real-world problems.

Moreover, there are studies which indicate that recent graduates' students struggle with technical problems effectively when communicating with co-workers, customers and other soft-skills [8, 9]. In some cases, there is a gap between students' skills and employers' expectations that prevent students from getting a job [10]. Through the Experience Learning students can work on real life projects and learn to develop transversal abilities namely as soft-skills. In this methodology, students work in small groups and try to solve real problems that industries propose [11–13]. This is a strong education approach where students discuss ideas, exchange opinions and enquire solutions. In addition, students learn to communicate with the customers during the process and be better prepared for their future career.

In this context, the researchers decide to apply this type of collaboration between Industry and Higher Education to improve the students' skills.

3 Methodology

In this research was used a Case Study methodology which enables a researcher to closely examine the data within a specific context [14]. The case in study is the students' motivation to solve real projects and to verify the local industrial acceptance. Throughout the project phase, the teachers were also researchers. As teachers, they made the learning process easier and as researchers, observed and registered the results [15].

In this study it was used an Experiential Learning approach. This methodology means learning from experience or learning by doing. It is associated with the constructivist theory of learning and students play the main role in the process of their own learning. According to Wurdinger [16], students must think about problems to be solved rather than memorizing information.

In the 2017/2018 school year, within the scope of the final project of Computer Science course, it was used the Experiential Learning approach. The project outline was presented to students that choose, in group of two elements, the project they want to develop. The project subject has a period of four months in which students develop their project and are supervised by tutors. Students have regular meetings with tutors (at least one a week). During these meetings students explain what they have done and present their doubts. At the end of the forth month students must write a report and present their work to a jury of three teachers.

In this research it was used a control group formed by two students. In contrast, this group developed an academic project. The level difficulty of both projects is similar, and both have two applications, one for the web and another for mobile devices. The academic levels of all students in both groups are similar. Their score was in the range of twelve to fourteen.

3.1   Industry—Experience Learning University

The local Hospitality Industry wants to solve the guests' complaints in social media since it is a problem they face. Several guests instead of complaining during their stays prefer to express their opinion in social media. Most of the time, hotel managers are surprised with this complaints and feel helpless and frustrated, since they cannot solve the problem, referred by local manager. Moreover, opinions in social media leave a footprint which may affect the hotel reputation. In this context, the hotel manager wants an effective mechanism that allows the guests to easily explain and make complaints during their stay, avoiding spreading the undesired information in social media.

This problem was presented to students, so they had to meet with the local hotel manager. After it, researchers verified that students had some difficulty in acknowledge what the customer desires, because there is a gap between the students' technology language and the natural customer language. Also, students' inexperience in dealing with customers was notorious due to their nervousness and lack of doubts. To overcome these difficulties, teachers proposed students to investigate additional information by looking at comments on the social media channels. Also, they analysed the complaints presented in the hotel reception and as a result they understood the problem of the hotel manager.

Despite students having found out that complaints presented in social media channels were more than the complaints presented at the hotel reception, the hotel managers could not solve these problems as they not even knew who did them. After this, students were aware of the problem and tried to find a solution for it.

Until now, researchers had observed a change in students' behaviour. They are more motivated, committed and feel more the burden of responsibility.

Students' Solution

The system proposed by students has two applications that communicate over the internet, named HappyGuest. One is a web application for the hotel manager and a mobile application for guests.

Firstly, for the web application they defined two profiles: the receptionist and the hotel manager. Secondly, for the mobile application only one profile: the guest.

When guests arrive at the hotel they are informed about the existence of the HappyGuest application where they may submit complaints. It is also delivered a key which they can use as an authentication mechanism in the HappyGuest. When the guest has a complaint to report he can do it through the application and choose if he wants to remain anonymous or not. If the complaint is not anonymous the guests can follow the complaint state and observe all the steps of its resolution. The guests can also compliment the hotel by using the mobile application.

Through the web application the hotel manager can handle complaints made by the guest through the mobile application. The hotel manager takes notes of them and assigns priorities. Later, the hotel manager can elect an employee who is responsible for the complaint resolution. The web application has a dashboard for data analysis where hotel managers can visualize the most problematic hotel departments.

The students' solution was shown to the customer and for a better understanding it was developed a prototype. Firstly, the students created a low fidelity prototype with the aim to verify if the customer agrees with all the functionalities.

The customer enjoyed the solution and made some suggestions related with mobile and web applications interface to improve the usability of it. The next step was consisted in the development of a high-fidelity prototype (Figs. 2 and 3). Later, will be validate aspects such as colours, letter type, images, the navigation system and the way the two applications communicate with each other. These prototypes are in the final step of development and later it will be made usability tests with the end-users (guests and hotel staff). The researchers hope that at the end of the project a solid prototype is obtained to be implemented in the next year by another group of students.

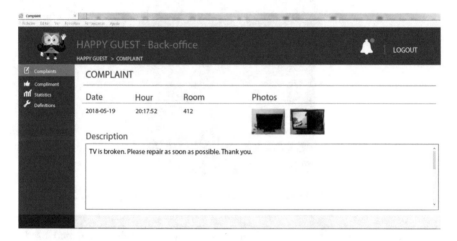

Fig. 2. Web application prototype high-fidelity prototype

During the prototype phase, students mentioned it was the first time they understood the importance of creating prototypes. In their course they had to make prototypes for the applications but did not considered that useful. This comment reinforces the researchers' opinion about the development of real project and partnership Industry-University.

3.2 The Group Control Problem

The academic project consisted in the development of a solution to facilitate the students' access to academic information and to improve the communication between students and teachers. Specifically, this project intends students to have access to their grades through a mobile application as well as information about schedules, assessments and absences.

The system is formed by a desktop application for teachers and a mobile for students. This project, besides being purely academic, has some interesting points, for

Fig. 3. Mobile application high-fidelity prototype

example if the solution is acceptable the system will be implemented and used by students and teachers. In this project, the tutors' made customers role.

The first steps consisted in students' understanding the problem and presenting a solution.

Students' Solution

The system proposed by students has two applications that communicate over the internet, named "What Did I Have?" One is a desktop application for teachers and the other one is a mobile application for students. The desktop application has two profiles the staff coordinator and teachers. The coordinator is responsible for assigning subjects to teachers and has a dashboard for data analysis. Teachers have the responsibility of importing the grades from excel sheets to the systems and notify students.

This group developed only the desktop application prototype (Fig. 4), with a few functionalities. Although, the supervisors' explanation about the prototype importance, it was not taken in consideration by students. The students' justification for this was: *"the application is simple and it is not necessary to create a prototype. We know what to do".*

The students did not understand the usefulness of prototyping, despite the supervisions' explanations.

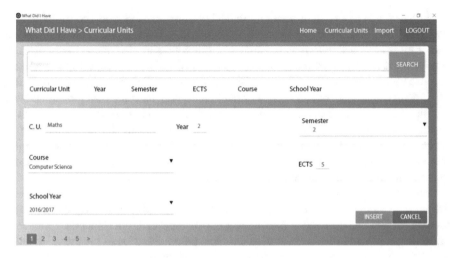

Fig. 4. Desktop application prototype

Regardless of the technical skills of this group, they did not have the ability to develop a coherent end-user interface. Moreover, the final solution did not have valid aspects such as colours, letter type, images, a clear search paradigm or a simple navigation system.

The academic project group did not understand clearly the system requirements and did not have engagement with teachers to find out a more suitable solution. Even though the first desktop application screen was just a window with a lonely button for authentication purposes and did not have a proper background colour.

4   Results and Discussion

This type of partnership between the local Industries and the Higher Education Institution has some risks for both parts. As researchers observed along this case study, the local Industry had high expectations about the solution for their problem. It is difficult to make them understand that projects are developed by students which are inexperience, so it is necessary more time to create a solution with quality. For the institution, it is a responsibility to accept projects, since these are entirely developed by students with the teachers' supervision, so it is hard to guarantee the quality and released date. This type of problems can compromise the institution image.

From the interview with the hotel manager, that participate in this project, it was notorious his satisfaction with the solution presented by students. Since, until now it is only a prototype, the hotel manager must wait for the complete project development. In hotel manager opinions, the time necessary for having a final product must be reduce if the university wants to have more projects and keep the partnership. He also mentions: *"we are proud in collaborating with students' education; we believe that it is our contribution to create a better society"*.

For students, this is an opportunity to deal with real case studies and customers, so they can learn inside a protected environment. Along the meetings between students and customers it was notorious the students' difficulties in understand the customers' requirements and the business processes. Students considered important these meetings because they could learn how to deal with customers and improve their communication skills, which they understood if did not have to be very technical. Moreover, students highlight the responsibility for having to develop a good solution and be proud of seeing their work recognized. These results are in line with the research made by several authors [17, 18], which refer the importance of public recognition to students' motivations. Along the project development it was observed improvements in students' transversal skills such as communication and autonomy.

The researchers compared this project with other projects which are mainly academic. The students' motivation was different. The students with real project were more motivated and committed in comparison with their colleagues that only wanted to finish the project and obtain the final grade. This group did not seek for a good solution and were not interested in overcoming their difficulties. Moreover, they were not receptive to supervisors' suggestions to improve the interface and usability of the system. In contrast, the other group accepted the customer suggestions and presented new prototypes to be tested, as can be observed in the following students' comments: "...this is what the customer wants", "... we need to ask the customer if this is all right".

All in all, researchers considered that it is important for students to have this kind of opportunities, so they can develop their transversal skills that will be important for their future living. Furthermore, students can see the practical application of the theoretical concepts. Although, it is necessary to inform the partnerships about the solutions, which may not be delivered in a short time and must, be aware, that they are created by students and not by professionals' expertise.

5   Conclusions

The Academic-Industry Partnership may be stablished in various formats. For instance, by involving investigators to solve complex industry problems, by creating academies in which students may get a certificate or by promoting projects with students.

In this study, it was analysed how this partnership could improve the students learning process. Moreover, the results show that, for students' motivation, it is crucial to work with real life projects due to their sufficient liability and public recognition of their work. It makes them proud and stimulus to work harder. Also, students acquire and reinforce their soft-skills and know-how in different areas.

The Academic-Industry Partnership has benefits and drawbacks. For industrial partners this is an opportunity to have R&D access and to obtain innovative solutions. However, one problem for the industry is the time they have to wait for a good solution.

For Higher Education Institutions it is a challenge since they have to deal with the industrial world and have to know the business processes to tutor students. It is also important to create a good relationship in order to protect the institutions' image once students are who are going to develop the projects and to create new ones.

Based on our experience we believe this type of partnership is important for both sides.

References

1. Blumenthal, D., Causino, N., Campbell, E., Louis, K.S.: Relationships between academic institutions and industry in the life sciences—an industry survey. New Engl. J. Med. **334**(6), 368–374 (1996)
2. Dooley, L., Kirk, D.: University-industry collaboration: grafting the entrepreneurial paradigm onto academic structures. Eur. J. Innov. Manag. **10**(3), 316–332 (2007)
3. Poyago-Theotoky, J., Beath, J., Siegel, D.S.: Universities and fundamental research: reflections on the growth of university–industry partnerships. Oxf. Rev. Econ. Policy **18**(1), 10–21 (2002)
4. Nakagawa, K., Takata, M., Kato, K., Matsuyuki, T., Matsuhashi, T.: A university–industry collaborative entrepreneurship education program as a trading zone: the case of Osaka University. Technol. Innov. Manag. Rev. **7**(6), 38–49 (2017)
5. Perkmann, M., Walsh, K.: University–industry relationships and open innovation: towards a research agenda. Int. J. Manag. Rev. **9**(4), 259–280 (2007)
6. Edmondson, G., Valigra, L., Kenward, M., Hudson, R.L., Belfield, H.: Making Industry-University Partnerships Work: Lessons from Successful Collaborations, pp. 1–52. Science Business Innovation Board AISBL (2012)
7. Ivascu, L., Cirjaliu, B., Draghici, A.: Business model for the university-industry collaboration in open innovation. Proc. Econ. Finance **39**, 674–678 (2016)
8. Riggio, R.E., Tan, S.J. (eds.): Leader Interpersonal and Influence Skills: The Soft Skills of Leadership. Routledge (2013)
9. Brunello, G., Schlotter, M.: Non-cognitive skills and personality traits: labour market relevance and their development in education & training systems (2011)
10. Radermacher, A., Walia, G., Knudson, D.: Investigating the skill gap between graduating students and industry expectations. In: Companion Proceedings of the 36th International Conference on Software Engineering, pp. 291–300. ACM, May 2014
11. Esteves, M., Fonseca, B., Morgado, L., Martins, P.: Improving teaching and learning of computer programming through the use of the second life virtual world. Br. J. Educ. Technol. **42**(4), 624–637 (2011)
12. Esteves, M., Antunes, R., Fonseca, B., Morgado, L., Martins, P.: Using Second Life in Programming's communities of practice. In: International Workshop of Groupware, pp. 99–106. Springer, Berlin, Heidelberg, Sept 2008
13. Pereira, Â., Morgado, L., Martins, P., Fonseca, B.: The use of three-dimensional collaborative virtual environments in entrepreneurship education for children. In: Proceedings of the IADIS International Conference WWW/Internet, pp. 319–322 (2007)
14. Yin, R.K.: Case Study Research: Design and Methods (Applied Social Research Methods). Sage, London and Singapore (2009)
15. Pereira, A., Martins, P., Morgado, L., Fonseca, B., Esteves, M.: A technological proposal using virtual worlds to support entrepreneurship education for primary school children. In: International Conference on Interactive Collaborative Learning, pp. 70–77. Springer, Cham, Sept 2017

16. Wurdinger, S.D.: Using Experiential Learning in the Classroom: Practical Ideas for All Educators. R&L Education (2005)
17. Ames, C.: Classrooms: goals, structures, and student motivation. J. Educ. Psychol. **84**(3), 261 (1992)
18. Linnenbrink, E.A., Pintrich, P.R.: Motivation as an enabler for academic success. Sch. Psychol. Rev. **31**(3), 313 (2002)

Short Collaboration Tasks in Higher Education: Is Putting Yourself in Another's Shoes Essential for Joint Knowledge Building?

Vitaliy Popov[1]([⊠]), Anouschka van Leeuwen[2], and Eliza Rybska[3]

[1] Jacobs Institute for Innovation in Education, School of Leadership and Education Sciences, University of San Diego, San Diego, CA, USA
vpopov@sandiego.edu
[2] Social and Behavioral Sciences, Utrecht University, Utrecht, The Netherlands
a.vanleeuwen@uu.nl
[3] Department of Nature Education and Conservation, Adam Mickiewicz University in Poznan, Poznań, Poland
elizary@amu.edu.pl

Abstract. In Higher Education, students are often grouped together for short collaborative assignments. The question is how to optimally group students in order to achieve the potential rewards of such collaboration. In this study, the relation between the ability to take each other's perspective, familiarity, experienced relatedness, and quality of collaboration was investigated. Thirty-three dyads of undergraduate students collaborated on a short genetics assignment. They first read information individually and then had to combine their knowledge to solve the task. By means of questionnaires, participants' prior knowledge, perspective taking ability, familiarity, and relatedness to collaborative partner were measured. Only dyads' average prior knowledge predicted the score on the group assignment. Implications for dyad grouping and further research are discussed.

Keywords: Collaborative learning · Perspective taking · Higher education

1 Introduction

1.1 Short Collaborative Assignments in Higher Education

In Higher Education, it is common practice to group students together for short collaborative assignments. There is a well-documented body of literature that demonstrates the positive relationship between collaborative learning and student achievement, persistence, and motivation (see Johnson and Johnson 2009; Slavin 1990; Kyndt et al. 2013 for reviews). However, just putting people together in groups does not necessarily guarantee the development of a shared understanding, better learning, nor motivation, especially when students are limited by the time period allocated to collaborate (Khosa and Volet 2013). Due to a range of individual differences (e.g., disciplinary/cultural backgrounds, socio-emotional intelligence, interests, skills) that members bring to a group, the so called web of intra-group dynamics becomes

© Springer Nature Switzerland AG 2019
M. E. Auer and T. Tsiatsos (Eds.): ICL 2018, AISC 917, pp. 353–359, 2019.
https://doi.org/10.1007/978-3-030-11935-5_34

more complex (Halverson and Tirmizi 2008). Group processes play an important role in determining how individual differences mesh together or compete for effective expression. If the arising differences and similarities are not properly managed or articulated, students might not be able to engage in high-level collaboration processes, and ultimately lose the potential learning effect of collaborating. Particularly important in this sense is the occurrence of transactive discourse —building on each other's reasoning to co-construct (new) knowledge (Roschelle and Teasley 1995).

Therefore, group formation is an important choice teachers have to make in Higher Education to establish a high-performing group—a group that capitalizes on its diversity rather than being constrained by it (Cruz and Isotani 2014). Among many factors influencing the effectiveness of collaboration covered by decades of research on group work, previous studies emphasized the importance of group members' ability to understand each other's viewpoints, which we hypothesize is related to the ability to connect emotionally with one another (Järvelä and Häkkinen 2002). The ability called perspective taking is defined as "the cognitive capacity to consider the world from another individual's viewpoint" (Galinsky et al. 2008, p. 378). From a cognitive perspective on collaborative learning (King 1997), gaining a new way to think about a particular topic/subject can lead to improved learning outcomes. For example, in a recent study, Kulkarni et al. (2015) found that high geographically diverse MOOC discussion groups demonstrated higher learning gains than low-diversity groups. Students were asked to reflect on lecture content by relating it to their local context and discussing it with their peers in an online forum. Differences in opinions between collaborative partners are considered valuable, whereas easily compromising, agreeing too quickly, or not making the effort to understand the point of view of others reduces the value of the discussion (Kulkarni et al. 2015).

Because learning is a social act, the willingness and ability to understand each other's viewpoints is hypothesized to be dependent on the ability to connect socio-emotionally with collaborative partners (Järvelä and Häkkinen 2002). The social-emotional connection can benefit collaboration especially in the first stages of group development (Kreijns et al. 2003). For example, how well students know each other prior to their collaboration (familiarity) may be an important factor that moderates the effectiveness of collaboration. Previous research found that familiar group members more efficiently regulate their task-related activities, but a clear relation between familiarity and group performance has not yet been demonstrated (Janssen et al. 2009). Another characterization of socio-emotional connection is the amount of social relatedness students experience (Deci and Ryan 2008). The basic psychological need to experience relatedness could be a prerequisite for being willing and able to take the other's perspective (Deci and Ryan 2008) and subsequently, for effective collaboration to occur.

1.2   The Present Study

To summarize, previous studies provide insights into the influence of perspective taking and socio-emotional factors that facilitate collaborative learning, but it often concerns studies in which students collaborate over longer periods of time. However, it remains unclear how these factors play out in short term collaborative assignments within one lecture or meeting, which are common in Higher Education. The question is how to group students in such a way that their background and their relation to each other is set-up to facilitate collaboration as effectively as possible. Therefore, in the present study, we explore the role of students' ability to take each other's perspective, as well as the role of familiarity and social relatedness in the effectiveness of collaboration. Teachers and educational designers can take into account these factors to enable more effective and enjoyable collaboration experiences for students. The study is performed in the context of a short collaborative assignment in the Biology domain in Higher Education. The following research question was posed: *what is the relation between university students' ability to take their collaborative partner's perspective, familiarity between group members, students' experienced relatedness, and group performance in the context of a short term collaborative assignment?*

2 Method

2.1 Design

We report on a correlational study in which the relation between familiarity, relatedness, perspective taking, and group assignment score was examined in the context of a collaborative assignment on genetics. The data collection was part of a quasi-experimental study, in which the effect of an intervention to stimulate relatedness was examined. As the intervention had no effect on any of the main variables in the present study (perspective taking and group assignment score), nor on the intended variables (relatedness), we treated the two groups as one dataset.

2.2 Participants

The sample of participants consisted of 66 undergraduate students from a large University in the Netherlands. Their mean age was 25.5 ($SD = 5.8$). Thirteen students were male. After filling in background questionnaires and the pretest (see Sect. 2.3), students were randomly divided into 33 dyads. Seven dyads consisted of one male and one female, three dyads consisted of two males, and the remaining 23 dyads consisted of two females.

2.3 Procedure

The procedure of the study is depicted in Table 1. Participants first filled in questionnaires concerning their age and sex, as well as about their ability to take other's perspective, followed by the pretest. Then, the participants were randomly divided into dyads. The experimental condition performed the 5-min activity aimed at stimulating relatedness (for which no effects were found). The members of each dyad were each handed 4 pages of information about genetics. The first part contained identical basic knowledge for both students, but the second part differed and offered each group member more in-depth knowledge on one of two types of genetic inheritance. After 15 min, the individual information was handed in and the dyads collaborated on an

Table 1. Procedure

1	2	3	4	5
Fill in questionnaires/pre-test (15 min)	Randomly divide into dyads. (5 min)	Read information individually. (15 min)	Collaborative assignment. (20 min)	Fill in questionnaires. (10 min)
Demographics, Perspective taking, Pretest	Experimental condition performs activity aimed at relatedness (5 min)		9 dyads are audiotaped	Relatedness, Familiarity

assignment that required them to integrate their knowledge. They had 20 min to collaboratively finish the assignment and then handed in their answers. The discussions of nine randomly selected dyads were audiotaped. Finally, the participants individually filled in questionnaires about experienced relatedness to their group member, and about how well they knew their group member prior to the collaboration (familiarity).

2.4   Materials

2.4.1   Questionnaire Perspective Taking

The ability to take other people's perspective was measured by means of the perspective taking scale by Davis (1980). The scale consisted of 7 items, which were judged on a 5-point scale ranging from 1 (do not agree at all) to 5 (completely agree). Example items are "I believe there are two sides to every issue, and I try to look at both" and "I try to understand my friends better by imagining what things look like from their perspective". Reliability of the scale in terms of Cronbach's α was 73. Following Galinsky et al. (2008), the score for perspective taking for each dyad was calculated by taking the average of the two group members.

2.4.2   Questionnaire Familiriaty

Familiarity between group members was measured by asking participants how well they knew their collaborative partner prior to the collaboration. Following Janssen, Erkens, Kirschner, and Gellof (2009), familiarity was measured on a 4-point scale, ranging from 0 (did not know him/her at all) to 3 (knew him/her very well). To prompt participants about how well they knew their collaborating partner, the item was preceded by four yes/no questions such as "I have collaborated with my group member before". The 4-point scale question was used for the familiarity score of a dyad, which was calculated by taking the average of the two group members.

2.4.3   Questionnaire Relatedness

Feelings of relatedness between group members were measured using the relatedness scale by Broeck et al. (2010). The scale consists of 6 items, which were judged on a

5-point scale ranging from 1 (do not agree at all) to 5 (completely agree). Example items are "I felt part of the group" and "I got along well with my group member". Reliability of the scale in terms of Cronbach's α was 60. The score for relatedness for each dyad was calculated by taking the average of the two group members.

2.4.4 Pretest
The pretest consisted of 6 multiple choice questions concerning genetics, each with 4 answer options. An example question is "How many X-chromosomes do males have?" Each correct answer was scored as 1 point, leading to a possible total of 6 points. In the analyses, the dyads' average pretest score was used.

2.4.5 Collaborative Assignment
Students collaborated in dyads on a genetics assignment. An inheritance tree was depicted, and the assignments were to determine which type of inheritance was the case (X-chromosomal versus mitochondrial inheritance), and to determine each person's genotype within the tree. To solve the assignment, information from both collaborating partners (which they read in the individual phase) was needed, thereby creating resource interdependence (Johnson and Johnson 2009). The group assignment was scored, with a maximum of 6 points.

2.5 Results

Table 2 shows the descriptive values for all included variables. Dyads scored relatively high on relatedness as well as perspective taking. Familiarity showed a low average but high standard deviation, which is probably a result of the random allocation of students within dyads. Stepwise regression analysis was performed with relatedness, perspective taking and familiarity (at dyad average level) as predictors and group assignment score as dependent variable (Table 3). A significant model was found ($F(1,31) = 5.97$, $p = 02$, $R^2 = 0.16$). Only the average pretest score within a dyad was a significant predictor of group assignment score ($B = 0 .59$).

Table 2. Descriptive values of included variables

N = 33	Pretest (0–6)	Relatedness (1–5)	Perspective taking (1–5)	Familiarity (0–3)	Group assignment score (0–6)
Dyad average	2.99	4.37	3.67	0.64	4.02
SD	0.99	0.30	0.34	0.84	1.44
Min	1.50	3.67	3.00	0.0	1.70
Max	5.00	4.83	4.29	3.0	6.00

Table 3. Correlation matrix

	Relatedness	Perspective taking	Pretest	Familiarity	Group assignment
Relatedness	1				
Perspective taking	0.019	1			
Pretest	0.26	0.16	1		
Familiarity	0.21	0.13	0.23	1	
Group assignment	0.28	−0.05	0.40	0.06	1

3 Conclusion

The findings of the present study showed that only average pretest scores were predictive of group performance, and that none of the socio-emotional factors we examined influenced group performance. Previous studies *did* show influence of socio-emotional factors in studies where students collaborated over longer periods of time. Our results could mean that in short term collaboration, perspective taking ability and socio-emotional connection do not play a large role for the collaboration outcome. It might be explained also by the nature of tasks that students were taking, that was directly connected to the biological content not to emotional or social aspects of genetics. One of the practical implications for both educators and instructional designers is to form groups based on students' prior knowledge rather than on group members' familiarity or students' perspective-taking abilities. Further research should examine these variables by looking at a variety of different time spans for group work, paying specific attention to how the period of time allocated for collaboration affects group process and outcomes. Furthermore, the explained variance of the regression model we found was rather low, indicating that indeed other factors than socio-emotional ones influenced the group assignment score.

There is a need to examine this relation also in other domains and collaborative task structures. The genetics assignment used for this study, which is in the science domain, had clear-cut wrong and correct answers. Other domains such as social sciences often have more ill-defined problems, in which the investigated variables might play a larger role. For example, the ability to take each other's perspective might be more important when it concerns societal controversial issues, such as nuclear energy or animal cloning.

In subsequent analyses, our aim is to examine the audiotapes from the subsample of 9 dyads whose discussions were recorded to see whether this explanation is observable in terms of a primarily cognitive (task-related) focus in the discussions. Additional qualitative analysis of the audiotapes could also shed light on how and to what extent the ability to take the other's perspective is demonstrated in collaborative discussions.

References

Broeck, A., Vansteenkiste, M., Witte, H., Soenens, B., Lens, W.: Capturing autonomy, competence, and relatedness at work: construction and initial validation of the work related basic need satisfaction scale. J Occup. Organ. Psychol. **83**(4), 981–1002 (2010)

Cruz, W.M., Isotani, S.: Group formation algorithms in collaborative learning contexts: a systematic mapping of the literature. In: Baloian, N., Burstein, F., Ogata, H., Santoro, F., Zurita, G. (eds.) Lecture Notes in Computer Science, vol. 8658. Springer, Cham (2014)

Davis, M.H.: A multidimensional approach to individual differences in empathy. JSAS Catalog Sel. Doc. Psychol. **10**(85) (1980) Retrieved from http://www.uv.es/ ~ friasnav/Davis_1980.pdf

Deci, E.L., Ryan, R.M.: Self-Determination Theory: An Approach to Human Motivation and Personality. University of Rochester, NY (2008)

Galinsky, A.D., Maddux, W.W., Gilin, D., White, J.B.: Why it pays to get inside the head of your opponent: the differential effects of perspective taking and empathy in negotiations. Psychol. Sci. **19**(4), 378–384 (2008)

Halverson, C.B., Tirmizi, S.A.: Effective Multicultural Teams: Theory and Practice. Springer, Netherlands (2008)

Janssen, J., Erkens, G., Kirschner, P.A., Kanselaar, G.: Influence of group member familiarity on online collaborative learning. Comput. Hum. Behav. **25**(1), 161–170 (2009)

Järvelä, S., Häkkinen, P.: Web-based cases in teaching and learning—the quality of discussions and a stage of perspective taking in asynchronous communication. Interact. Learn. Environ. **10**(1), 1–22 (2002)

Johnson, D.W., Johnson, R.T.: An educational psychology success story: social interdependence theory and cooperative learning. Educ. Res. **38**, 365–379 (2009)

Khosa, D.K., Volet, S.E.: Promoting effective collaborative case-based learning at university: a metacognitive intervention. Stud. High. Educ. **38**, 870–889 (2013)

King, A.: ASK to THINK – TEL WHY®©: a model of transactive peer tutoring for scaffolding higher level complex learning. Educ. Psychol. **32**(4), 221–235 (1997)

Kreijns, K., Kirschner, P.A., Jochems, W.: Identifying the pitfalls for social interaction in computer-supported collaborative learning environments: a review of the research. Comput. Human Behav. **19**(3), 335–353 (2003)

Kulkarni, C., Wei, K.P., Le, H., Chia, D., Papadopoulos, K., Cheng, J., Koller, D., Klemmer, S.R.: Peer and self assessment in massive online classes. In: Design Thinking Research, pp. 131–168. Springer International Publishing (2015)

Kyndt, E., Raes, E., Lismont, B., Timmers, F., Cascallar, E., Dochy, F.: A meta-analysis of the effects of face-to-face cooperative learning. Do recent studies falsify or verify earlier findings? Educ. Res. Rev. **10**, 133–149 (2013)

Roschelle, J., Teasley, S.D.: Construction of shared knowledge in collaborative problem solving. In: O'Malley, C. (ed.) Computer-Supported Collaborative Learning, pp. 69–97. Springer, New York (1995)

Slavin, R.E.: Cooperative Learning: Theory, Research, and Practice. Prentice-Hall (1990)

Educational Robotics—Engage Young Students in Project-Based Learning

Georg Jäggle[1], Markus Vincze[1], Astrid Weiss[3],
Gottfried Koppensteiner[2,4], Wilfried Lepuschitz[2], Zakall Stefan[2,4],
and Munir Merdan[2(✉)]

[1] ACIN Institute of Automation and Control, Vienna University of Technology,
Vienna, Austria
{jaeggle,vincze}@acin.tuwien.ac.at
[2] PRIA Practical Robotics Institute, Vienna, Austria
{koppensteiner,lepuschitz,zakall,merdan}@pria.at
[3] Institut of Visual Computing and Human-Centered Technology, Vienna
University of Technology, Vienna, Austria
astrid.weiss@tuwien.ac.at
[4] TGM, Vienna Institute of Technology, Vienna, Austria
{gkoppensteiner,szakall}@tgm.ac.at

Abstract. This paper reports on the impact of a cross-generational project, which links seniors' needs and high school students' expertise with the development of the student's self-efficacy, communication and collaboration skills as well as their interests. The project iBridge integrates in its framework several different out-school activities with the concrete development of a prototypical service assistant (a cuddly toy robot) for seniors. During the study, the self-efficacy is measured with questionnaires. Students' communication skill is demonstrated during interviews and in their interactions training elderly people in computer technology. Collaboration skills and responsible work efforts are manifested in their team efforts to develop a robotic prototype. The study also shows different outcomes for students participating in iBridge and students who are not involved in the project.

Keywords: Project-based learning · Vocational school · Educational robotic ·
Out-of-school-learning · Self-efficacy

1 Introduction

Engineering graduates require skills that extend beyond content knowledge. Several studies have indicated that they need to have communication and collaboration skills [1] to solve future technical challenges and therefore to be successful in their job. Project-based learning represents one teaching method in which high school students can develop these skills. This form of learning provides these students with a technical expert role to plan their own milestones, develop their own goals, and produce different products. In the project iBridge, we use the project-based learning approach to improve student skills and their interest in technology. The project has linked vocational schools and universities with older adults and students, enabling students to take the roles of co-researchers and trainers at the same time and learn how to approach social problems in their local community. On the one side, the students are learning about assistive robot technology development and applications by building a cuddly toy robot as a prototype of a service robot including the potential end users (seniors) in the development process. On the other side, the project goals are that students improve social and communication skills by introducing older adults to computer technology and work in teams to develop a sensitive cuddly toy to serve as a service robot (collaboration skills). The project approach seeks to get young people involved in the field of assistive technologies for senior citizens. This involvement requires addressing the needs and fears of the elderly while understanding their desire to use new technologies and participate in new developments. The project-based learning in this context also aims at motivating high school students to take on responsible work efforts linked to current technological problems; in fact, one impact of such learning is also improved self-efficacy and self-reliance [2].

We hypothesize that this kind of project-based learning linked with out-of-school-learning will increase students' self-efficacy during the educational robotic project (H1). We also hypothesize that this kind of the project seeks to engage adolescent students in technical content knowledge (H2), communication (H3), and collaboration (H4) while fostering their responsible work efforts and increasing their self-efficacy. Self-efficacy represents an important point because a high level of self-efficacy can impact students' willingness to perform a difficult task, to do exercises, and to expand their horizons [8]. The study within this paper will show different outcomes for students participating in iBridge and students who are not involved in the project. This leads to the Research Question (Q1): What did students involved in the project learn that other students did not? The study will measure what students learnt during the project using mind-mapping as a research method.

The paper is structured as follows: The following Sect. 2 briefly introduces the iBridge project-based learning environment. Section 3 presents the evaluation of the proposed hypotheses. Section 4 gives a short evaluation of the out-of-school learning activities. Finally, a conclusion is given in Sect. 5.

2   iBridge Project-Based Learning Environment

Project-based learning is a teaching method that involves students in problem-solving and gives them the opportunity to work autonomously over extended periods of time. In our case, it concludes with a realistic product and presentation of a robotic prototype [3]. In this context the technology plays a powerful role in enhancing student motivation to do projects, impacting student learning by attracting their interest [4]. Combined with out-of-school-learning, this teaching method offers more motivation and better learning and particularly influences self-efficacy. Especially, the novel situations and out-of-school workshops force an adaptation or adjustment of students, directing their behaviour towards the environment and away from structured learning activities [5, 6]. The assumption here is that learning, which occurs within the physical environment, is in fact, always a dialogue with the environment [5]. The project iBridge is a cross-generational project that engages the students in social and cross-cultural research topics through the application of robotics. The project aims to introduce students from vocational schools into the user-centered design process through different entry-points: workshops with experts, workshops with seniors as well as the project-based prototypical development of a cuddly toy robot for seniors.

(a) **Workshops with experts**

There are two main goals for the workshops at the Automation and Control Institute (ACIN) at Vienna University of Technology. The first one is to show students state-of-the-art robotic technology and current robotic projects. The second one is to instruct them in research methodologies. The students learn two qualitative methods: (1) the half-structured research interview and (2) keeping a research diary [10]. The design of the workshops is based on the didactic model of AVIVA, which is a model of classroom management and lesson planning in schools. This model links to the living world of the students and combines the instruction of knowledge with the application of knowledge [11].

The workshops at the Practical Robotics Institute Austria (PRIA) are intended to give students an effective introduction to robotics through hands-on activities. At the very beginning the students build their first simple robot in electronic lessons to get in touch with an introductory level of robotics. In a further stage, the workshop students learn about robot motors, microcontrollers sensors, the mechanical structure of a robot as well as how to build, program and test robots.

(b) **Workshops with Seniors**

The Information Technology Department of the Vienna Institute of Technology offers its students a social skills learning workshop with seniors. With the cooperation partner City of Vienna and its senior citizen clubs, high school students visit the seniors within a framework of six workshops to introduce them to using computers and to observe their needs and demands. On the one side seniors learn dealing with new media and on the other side students learn passing on their technology competence. It is also an excellent opportunity for students to approach a completely new "user-oriented view of things" that can help them in their future career. The workshop levels for seniors range from beginners who have never dealt with a computer in their lives to advanced users who want to learn how to deal with special programs or take care of official channels.

(c) **Prototypical Service Assistant (Cuddly Toy Robot)**

Within the framework of the project, the high school students develop a robot in the form of a sensitive cuddly toy to be used as a cute and friendly interaction partner for the seniors (generation 60+ years), who have a basic understanding of computer knowledge. Based on the requirements and wishes they have learned from the seniors, the students incorporate the following features (Fig. 1) in the cuddly toy:

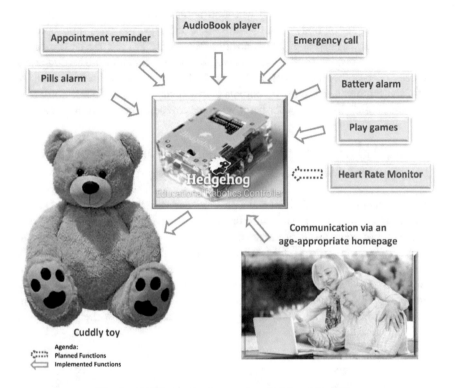

Fig. 1. Cuddly toy implemented and planned functions

A pills alarm function is implemented to remind the user to take the medication at a specified time. In order to help in the case of discomfort or emergency, an emergency function is implemented to send a distress signal to an emergency contact previously set via the website. When the emergency signal button is pressed, an LED trip indicator lights red and the emergency call is transmitted via the GSM mobile network. The cuddly toy can play audiobooks, which are stored in an internal memory and made available for anytime use. The books are permanently uploaded over USB port and seniors manage the audiobook library, with over currently 50 books, by using the webpage. The appointment reminder function is also integrated into the cuddly toy and serves as a reminder alarm for a personal calendar. It can read for example from a Google Calendar and speak out the reminders when the time comes. The seniors are besides notified by a vibration function that the cuddly toy battery level is low. In that case, the cuddly toy vibrates several times every few minutes until somebody starts the charging. The cuddly toy has tactile sensors integrated in each hand enabling to use a toy for playing interactive games. By pressing a sensor a user can start a game similar to the game "Simon says". The main functions of the cuddly toy are manageable via an age-appropriate webpage. The age-appropriate design is reflected through following characteristics: clarity, goal orientation, simple structure and readability. The data collected by the cuddly toy as well as the data on the website are securely stored and it is ensured that no one unauthorised can gain access to the system. The cuddly toy incorporates a series of body regions—the arms, the legs, the head, and body. The system integrates a Hedgehog [12] controller (Raspberry Pi A+), tactile sensors (left arm and right arm inside), light sensors as well as a wireless module and a power bank. In the head of the robot are several multicolor LEDs and a speaker in the mouth. As a basis for the prototype, the most favourable components and open source software are used. In the future, the function of heart rate measurement should be implemented.

3 Evaluation of the Students' Self-efficacy During Engagement with Project-Based Learning

To measure outcomes, we administered a pre-questionnaire to obtain students' scores at the beginning of the project. The questionnaire is a standard test from Jerusalem and Schwarzer [7] to get a result of students' self-efficacy. As the iBridge project started midterm, 32 students from TGM filled in the PRE-, POST-Questionnaires, and mindmaps (they are called Project-Group in the further text). The Control-Group encompassed 43 students, who were not involved in the iBridge project. Figure 2 shows the difference between Control- and Project-Group. The data represented similar group characteristics, which is important to compare the results of the groups.

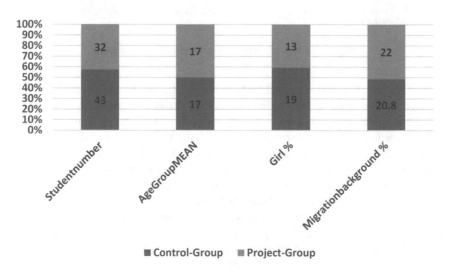

Fig. 2. Data Control- compare Project-Group

(a) **Increase of students' self-efficacy (H1)**

H1: The hypothesis 1 was that the project will increase the students' self-efficacy. The self-efficacy was measured with a standard test with 10 questions from Jerusalem and Schwarzer [7] at the beginning with PRE-Questionnaires and at the end with

POST-Questionnaires. Table 1 shows that the self-efficacy only slightly increased during the project. A Wilcoxon signed-rank test showed that pre- and post-questionnaire ratings on self-efficacy did not elicit a statistically significant change.

Table 1. Mean of students self-efficacy

Mean of students' self-efficacy				
	N	Mean	Std. Deviation	Std. Error Mean
POSTSelfEfficacy	32	29.7813	3.88325	.68647
PRESelfEfficacy	32	29.1563	3.64656	.64463

Engaging in technical content (Q1, H2)

Researcher Question Q1 was: what did students involved in the project learn that other students did not? This question is answered with the mindmap method. This method gives the results about the learning outcomes from students in the content knowledge of three contexts: Need of older people, Research methods and Robotics. These results are compared with the Control-Group.

The mindmaps were filled in by 85 students, who gave 772 answers in total. The Project-Group filled in a PRE-Mindmap at the beginning of the project and complemented their knowledge in a POST-Mindmap at the end of the project. The Control-Group got the Mindmap at the same time as the POST-Mindmap was given out. Table 2 shows that the Project-Group gave more answers than the Control-Group, although the Project-Group had fewer students.

Table 2. Number of answers in the different contexts

Context	Control-Group	Project-Group
Needs of older people	145	174
Research methods	67	82
Robotics	127	177
Total	**339**	**433**

When asked about ***needs of older people*** the Control-Group gave 145 answers and the Project-Group gave 174 answers in eight relevant categories: Activities/Entertainment; Basic Needs/Health/Medicine; Convenience; Interest Knowledge; Technicaly Skills, Security, Social Contact and Support/Help. Figure 3 shows that the students of the Project-Group named more Interest in Knowledge and Technically Skilled and not more Support and Help as needs of older people. This result shows that the students of the Project-Group learned more about the needs of the older people in the pensions club. They got a different view about the needs of older people. Older people do not need just support and help, they have interests and want to use technologies and perform activities.

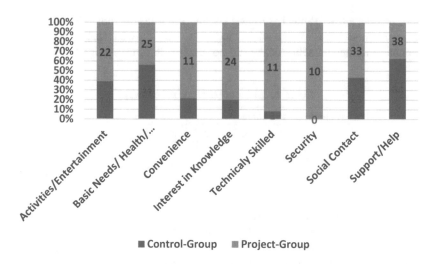

Fig. 3. Cluster needs of older people

When asked to write down their knowledge about *research methods*, the Control-Group gave 67 answers and the Project-Group gave 82 answers. During the analyzing process five categories were provided: Research methods, Positive Attitudes, Negative Attitudes, Operation/Resources and Applications.

Figure 4 shows that the students of the Project-Group had more positive Attitudes in research methods than the Control-Group. The Project-Group named more Research methods than the other group, so the students knew more research methods than the Control-group. This result showed the impact of the workshop at ACIN (see Sect. 2a) about the participative research and the using of methodology at the pensions club.

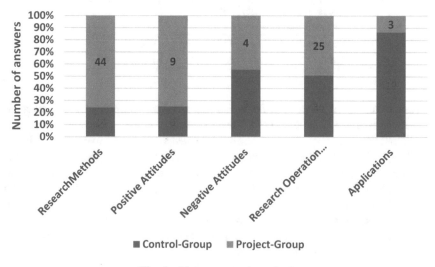

Fig. 4. Cluster research methods

When asked to write down their knowledge about robotics, the Control-Group gave 127 answers and the Project-Group gave 177 answers. During the analyzing process six categories were established: Positive Attitudes, Negative Attitudes, Technical Features, Term of Robots, Events and Applications. Technical Features includes technical components and features. The answers were clustered according to these six categories.

Figure 5 shows that the students of the Project-Group gave more positive replies than the Control-Group. The Project-Group learned more about the Technical Features of a robot than the Control-Group. The negative Attitude in Robotics is higher at the Control-Group. The data shows that the project had a positive impact to the attitude in robotics and knowledge of technical features and components in robotics. This gap could come from the topic robotic of the project and the introduction at the workshop at ACIN.

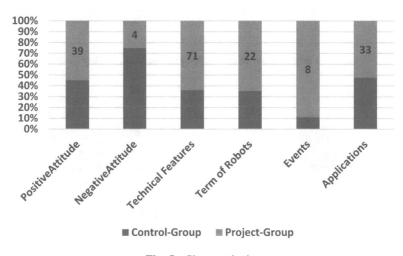

Fig. 5. Cluster robotics

(b) H2: The project engages adolescence students in technical content knowledge

Hypothesis 2 was that the project seeks to engage adolescence students in technical content knowledge. In this context, we analyzed the following statements from the POST-Questionnaires: A: I wish more exercises with the content of robotics at school, B: I am now more interested in Engineering Content and C: I am now more interested to study in the engineering field. Figure 6a shows that approx. 60% of all participates are interested in more exercises with robotics; Fig. 6b and 6c show that more students from the Project-Group are interested in Engineering Content and studying in an Engineering field.

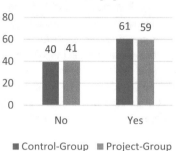

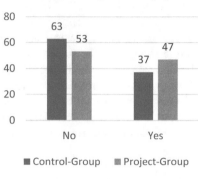

a

b

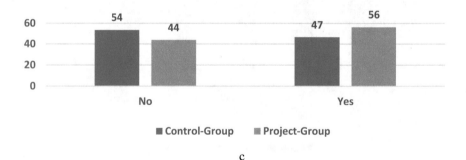

c

Fig. 6. .

(c) H3: The project fosters Softskills: Communication

The fostering of communication (H3) is measured through the following statements:

- S1: I improved my communication.
- S2: I improved my skill to connect better with other people.
- S3: I start a conversation with other people more easily.
- S4: I can be tuned better into the conversation partner.
- S5: I respond better to communication problems.
- S6: Out-of-school activities improved my communication skills.

As presented in Fig. 7, most of the students said "Yes" to the different statements of communication. The project fosters the perception of communication. As future work we plan interviews to understand the different problems existing in communication skills in more depth.

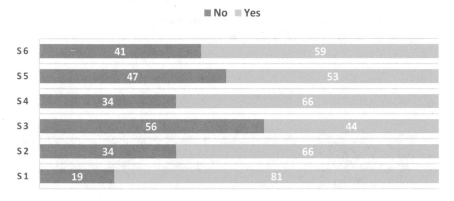

Fig. 7. Foster Communication Skills

(d) **H4: The project fosters Softskills: Collaboration**

The fostering of Collaboration (H4) is measured through the following statements:

- S1: I could support team members to involve their competences.
- S2: I could agree with others realistic and accurate work objectives.
- S3: I could comply common agreements.
- S4: I could develop my competence to working better together with others.
- S5: Out-of-school activities improve my competence in teamwork.

As presented in Fig. 8, most of the students had the perception that their collaboration skills improved during the project.

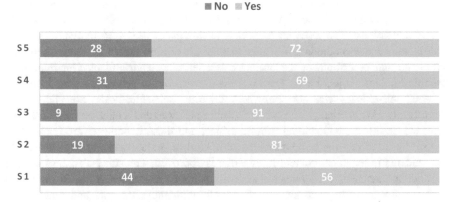

Fig. 8. Foster collaborative skills

4   Evaluation of the Out-of-School Learning Activities

The influence of out-of-school learning is evaluated through the following statements:

- S1: Out-of-school activities show me actual problems in life.
- S2: Out-of-school activities show me state-of-the-art of technology.

As presented in Fig. 9, the students showed that they are interested in out-of-school learning and prefer more out-of-school activities.

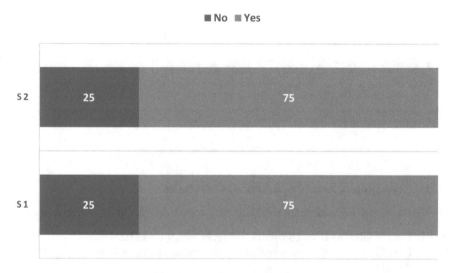

Fig. 9. Out-of-school learning

5   Conclusion

The presented project iBridge combines three approaches: project-based learning, out-of-school learning, and participant research. These three approaches have different purposes, but in combination, they promote self-efficacy, foster communication and collaboration skills, and motivate students to deliver responsible work efforts with robotic technology. The research presented in this paper used quantitative as well as qualitative research methods to measure these skills of high school students when they participate and contribute in the project-based learning environment. Our hypothese that this kind of project-based learning linked with out-of-school-learning increases students' self-efficacy was not confirmed, since the self-efficacy did not elicit a statistically significant change. However, communication and collaboration skills are increased during the educational robotic project and these hypotheses are completely confirmed. Similarly, the hypotheses that this kind of project seeks to engage adolescent students in technical content knowledge is approved. The study also showed that the students increased their interests in out-of-school learning activities. In the future work we are going to measure the same skills with more students and much more structured guideline interviews to get new information from the students to optimize the didactical designs of project-based learning. We also intend to evaluate additional skills such as critical thinking.

Acknowledgements. The authors acknowledge the financial support by the Sparkling Science program, an initiative of the Austrian Federal Ministry of Education, Science and Research, under grant agreement no. SPA 06/294.

References

1. Henshaw, R.: Desirable attributes for professional engineers", gehalten auf der Broadening horizons of engineering education. In: 3rd Annual conference of Australasian Association for Engineering Education, S. 15–18 (1991)
2. Bell, S.: Project-based learning for the 21st century: skills for the future. Clear. House **83**(2), 39–43 (2010)
3. Jones, B.F., Rasmussen, C.M., Moffitt, M.C.: Real-life problem solving: a collaborative approach to interdisciplinary learning. Am. Psychol. Assoc. (1997)
4. Blumenfeld, P.C., Soloway, E., Marx, R.W., Krajcik, J.S., Guzdial, M., Palincsar, A.: Motivating project-based learning: Sustaining the doing, supporting the learning. Educ. Psychol. **26**(3–4), 369–398 (1991)
5. Eshach, H.: Bridging in-school and out-of-school learning: Formal, non-formal, and informal education. J. Sci. Educ. Technol. **16**(2), 171–190 (2007)
6. Falk, J.H.: Field trips: a look at environmental effects on learning. J. Biol. Educ. **17**(2), 137–142 (1983)
7. Schwarzer, R., Jerusalem, M.: Skalen zur erfassung von Lehrer-und schülermerkmalen, Dok. Psychom. Verfahr. Im Rahm. Wiss. Begleit. Modellvers. Selbstwirksame Schulen Berl (1999)
8. Bandura, A., Wessels, S.: Self-Efficacy. (1994)
9. Städeli, C., Grassi, A., Rhiner, K., Obrist, W.: Kompetenzorientiert unterrichten: das AVIVA-Modell. hep, der Bildungsverl (2010)
10. McTaggart, R.: Participatory Action Research: International Contexts and Consequences. Suny Press, US (1997)
11. Jäggle, G., Vincze, M., Weiss, A., Koppensteiner, G., Lepuschitz, W., Merdan, M.: iBridge—Participative cross-generational approach with educational robotics. In: Lepuschitz, W., Merdan, M., Koppensteiner, G., Balogh, R., Obdržálek, D. (eds.) Robotics in Education—Methods and Applications for Teaching and Learning Volume of the series Advances in Intelligent Systems and Computing, Springer, Berlin Heidelberg (2018), in Press
12. Lepuschitz, W., Koppensteiner, G., Merdan, M.: Offering multiple entry-points into STEM for young people. In: Robotics in Education—Research and Practices for Robotics in STEM Education, Advances in Intelligent Systems and Computing, pp. 41–52. Springer, Cham. ISBN 978-3-319-42974-8

Practice Enterprise Concept Mimics Professional Project Work

Integration of on the Job Training in an Academic Environment

Phaedra Degreef(✉), Sophie Thiebaut, and Dirk Van Merode

Lecturer Thomas More University, Sint-Katelijne-Waver, Belgium
{Phaedra.degreef,Sophie.thiebaut,Dirk.vanmerode}
@thomasmore.be

Abstract. In the new "Practice Enterprise" concept, students from a professional bachelor in Electronics-ICT are confronted with managing a large project by themselves, as a team. They need to tackle both project management, soft skills as well as technical solutions to the proposed challenge.

Keywords: Practice enterprise · Project-based learning · Customer service · Communication skills · Project management · Scrum methodology

1 Analysis and Problem Definition

Recent research from Antwerp Management School shows that companies put increasing stress on so-called soft skills. They analyzed job offers and discovered a new "holy trinity". Companies want employees that can work autonomously, while still being able to function well in a team. But, above all, they need to be able to adapt quickly. Flexibility is the code-word, more than ever.[1]

Bachelor's and Master's degree education consists mainly of providing theoretical and some practical knowledge to students. But when they graduate, there still appears to be quite a **big gap with the expectations of the professional and business** world. Students are ill-prepared to the sheer size of real-world projects in the industry, and still lack the required skills and tools to operate in larger project teams.

Communication skills, problem solving skills, team working and ICT skills, are increasingly valued in modern economies and labour markets. This has clear implications for education, if young people are to be equipped with the types of skills they will need to succeed in the labour market of the mid-21st century. This is now a well-recognized and a highly visible feature of mainstream curricula in the developed world.[2]

---

[1] Source: Belgian newspaper, De Morgen (2018.05.12), *Wat moeten jongeren studeren nu de banen van vandaag bedreigd zijn* https://www.demorgen.be/economie/wat-moeten-jongeren-studeren-nu-de-banen-van-vandaag-bedreigd-zijn-b8b71a0b/.

[2] Wilson (2013).

© Springer Nature Switzerland AG 2019
M. E. Auer and T. Tsiatsos (Eds.): ICL 2018, AISC 917, pp. 372–380, 2019.
https://doi.org/10.1007/978-3-030-11935-5_36

- Students usually work in **very small teams** (1–2) on small-scale projects. Hence, documenting, planning and budgeting the project is considered (by the students) as pure overhead.
- The **scope of projects is often limited** to a part of the curriculum, requiring only limited analysis or correlation to other parts. In other words, the existing curriculum, as presented and offered by the teaching staff, forms the basis for the project.
- **Student motivation is driven by scores** or credits, i.e. personal success (not for the team or for a customer).
- **Students report only to teaching staff**, which is their customer as well as their manager and coach. Organization and planning is mostly driven by semester time constraints, and required effort is estimated by teaching staff to fit in those constraints. Students have no chance of learning how to estimate and plan in this system, or how to organize their work as a team.
- Challenges (impediments rather, like failing infrastructure for instance) are reported to the teaching staff; students expect the problems to be solved for them, i.e. they have very **limited responsibility**.
- They have **limited accountability**: when they fail at something, it will only impact themselves (their score).
- They are **not required to communicate with external people** (outside the own university or campus), let alone dealing with a language- or cultural barrier.

The purpose of this new concept is to close this knowledge and skills gap as much as possible.

2 Goal Setting

The main goal of this new concept is to **close the gap between the academic and the professional environment** by working on the following skills:

- **Increase the size and scope of the project,** requiring correlation between different parts of the curriculum and even beyond.
- Increase the **size of the project team**.
  - Grow and improve a team mindset.
  - Learn self-organizing team skills.
  - Functioning, operating in a larger team.
- **Autonomous learning**, as a preparation to continuous learning later in their careers.
- **Professional attitude**: learn to take ownership of a task or challenge.
- **On-the-job-training** is being pulled inside the academic world and educational system: students learn professional tools and skills.
- **Apply and improve technical knowledge** and skills.
- Students learn that **documenting** their work is essential.
- Being **constructively critical** about their own and other's work, using and providing feedback within the team.
- **Dealing with external resources** (international, preferably) such as customers, teams in other organizations, external advisors.

3   Approach

3.1   Phase 1: Practice Enterprise I, II, III

To improve on the **practical skills of the students**, a first version of the Practice Enterprise concept was implemented at Thomas More Campus De Nayer in 2015, in which students were asked to design and implement a **self-defined practical project**. Students were expected to submit a project proposal, and upon approval of the proposal, they were given around 12–13 weeks (given 3 h per week) to complete the project, either alone or in a group of maximum 2 students. At the end of the semester, they had to present the outcome of their project before a jury. For the academic year 2016/2017, **supplemental coaching and training** was added during the first semester: students participated at a design thinking and pitching workshop, received a **project management training** and learned more about **Intellectual Property** (IP) protection. During academic year 2017/2018, the design thinking workshop was dropped and the focus on project management was extended. **Communication skills** and self-reflection coaching were added to the programme. Insights in customer needs and **customer behavior** were stimulated (Fig. 1).

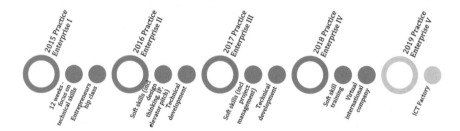

Fig. 1.   Evolution and history practice enterprise academic programme

Weaknesses

- Very **hard to follow-up for teaching staff**, hard to battle the procrastination of the students, resulting in less qualitative projects.
- Still **lack of good training on planning and reporting skills**, lack of realism compared to the project team work they will face in their later job environment.
- For **ICT students, the scope of these type of projects was too limitative** to give an added value to their training. The blueprint of the original concept of Practice Enterprise was designed with a more hardware and embedded systems education in mind, and doesn't quite fit the needs for ICT training:
  - ICT projects don't necessarily focus on innovative ideas, creating new devices or software, but rather on **implementing functional requirements** from a customer, using existing equipment.

- They often don't involve interaction or communication with the end-user, but rather with the **customer** (functional requirements of the system to be implemented).

3.2 Phase 2: Practice Enterprise IV—Virtual Multinational Company

An improved version of this concept is being prototyped now, in which the students will form one team that will be challenged in a way that is more realistic in terms of business context and given a task which they will have to tackle together as a team.

Not only the scale, but also **the scope of the project** will be larger than before, and they will be working in an environment that mimics the professional environment they can expect to be working in during their later career.

A **virtual multi-national company** is the basis of the project. The project team is asked to design and implement (including security) a multi-national ICT infrastructure, spread over several virtual geographical locations all over the world.

They will be allotted **4 h a week** during a whole semester (an average of 13 school-weeks) to work on this (they can continue working on the project outside of these allotted hours of required) (Fig. 2)[3].

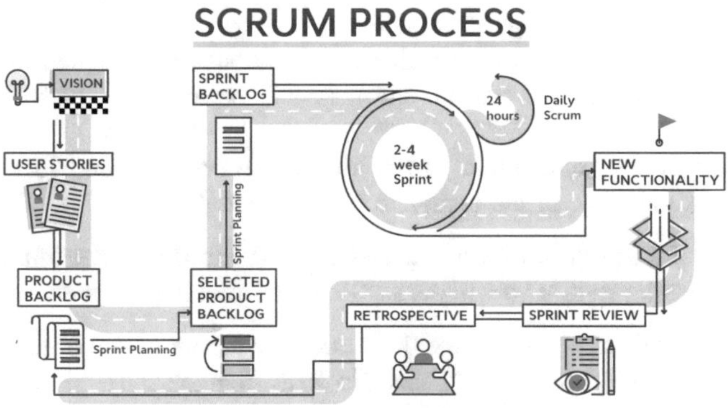

Fig. 2. Overview of scrum methodology

Scrum will be used as an **agile methodology**[4]:

- They will be fully **responsible** for organizing themselves, choosing the scrum parameters (sprint size, scrum master designation, etc.), updating of the burndown charts, organizing the scrum meetings, etc.
- The designated **scrum master** is responsible for tracking of all impediments of the team.

---

[3] https://www.linkedin.com/pulse/large-scale-scrum-simulation-lego-bricks-lego4scrum-less-krivitsky/

[4] Krivitsky (2017).

- To enable more **scrum sprint cycles**, the 4 h per week will be split in two 2-hour blocks, each with their own stand-up meeting and scrum board update.
- Scrum **retrospective meetings** will be used for regular evaluation and steering.
- The choice was made to use the "low-tech" version of a **scrum board**, instead of a digital solution (web-based). Practice has shown that this will improve on awareness of the status of the project, by actually seeing scrum board updates being done, and providing a single-view project status upon entering the room.

Teaching staff will be available as "external consultants" to help when the students get into trouble. The project team will be responsible to decide when to call upon these "external consultants". Virtual invoicing of this consulting is being considered to add to the budgeting process.

Some of the **networking theoretical skills** that they require for the project has not been covered yet in the curriculum by the time they need it. They will be given relevant course material:

- They will divide this subject matter amongst team members.
- Teaching staff will assist them in determining which part of the course material is required for which project task.
- They will process this assigned course material in self-study in the light of applying the acquired knowledge for the project tasks at hand, which will make them de facto subject experts in the assigned domain.
- During the semester, they will organize lectures about their assigned course material towards the other team members, and they will be evaluated on their theoretical knowledge.
- At the end of the semester, a written exam will be organized to evaluate their theoretical knowledge of the complete subject matter.

At the end of the project, they will **present and demonstrate the outcome of their work** to the teaching staff, who will act as "Board of directors" of the virtual company.

- They will be held accountable for the success or failure of the project, as a team.
- This will (hopefully) inspire a "Leave No Man Behind" spirit.

International partners (universities abroad) are being contacted to participate in this project. They could represent other sites with their own ICT infrastructure requirements, and our students will have to meet (via tele- or video-conference) on a regular basis to coordinate the complete project across borders.

During the first semester of the academic year, the students will be prepared for the project[5]:

- Students will have to apply for a position in the team. Teaching staff will provide some training and accompanying role playing on how to apply for a job.
- They will receive a training and workshop about the scrum methodology.
- Professional attitude, taking ownership, entrepreneurship attitude.

---

[5] The 2018–2019 programme is still in development, so the topics listed below are still not final and are subject for discussion in the coming months with the different stakeholders.

- Communication/reporting and documentation skills: job training and others.
- Presentation skills: elevator pitch, video, interactive presentations.
- Customer behavior: journey, needs, user stories.
- Real life cases from ICT companies: project management, teamwork, challenges (Fig. 3).

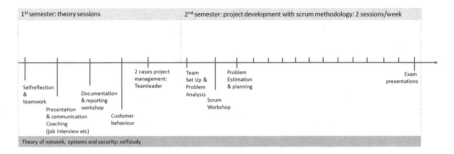

Fig. 3. Timeline academic year practice enterprise 2018–2019

Risks and challenges
Risks

- **Students taking advantage of other students' work**, "hiding in the group". This can be tackled by adhering to the scrum methodology, and leveraging the different scrum meetings (estimation, stand-up, retrospective...).
- **Managing the pressure** for ASD (Autism Spectrum Disorder) students, safeguarding their limitations. They will need a team position that copes with their needs (and added value).
- **Dependency** of the project team on the capabilities, motivation, work ethics, knowledge etc. of every team member (weakest link in the chain).
- **International partners / conferencing:** might impose extra stress on students. Perform poll and/or training during first semester.

Challenges

- **Evaluating** each student, both on hard skills as on soft skills and learning curve during the project. Peer evaluation is considered also.
- **Follow-up**.
- What in case of **blocking problems/impediments**?
  - Physical infrastructure limitations or problems;
  - Absence of team members;
  - Unforeseen problems.

Evaluation
For the evaluation of any course of the curriculum at Thomas More, a number of educational goals is defined, for which a set of levels has been defined.

For each course, a required level is mapped to each educational goal, and this will form the basis for the evaluation of the course.

For Practice Enterprise, the goals and levels have been determined as follows[6]:
Technical knowledge, analysis and research: level 2

- In cooperation with the stakeholders/clients/customers, the student is able to formulate a clear and concise problem description, goal setting and task description.
- He/she will substantiate the task at hand by means of analysis and research.

Design: level 2

- The student is capable of designing a system, prototype of system or service based on a set of requirements.
- He/she is capable of realizing a relevant implementation based on the set of requirements. He/she will substantiate the choices that were made in terms of tools, setup, configuration et cetera.
- He/she will consider additional requirements like safety, security, environmental, technical and economic life.

Implementation, realization: level 2

- The student will deliver a working system, prototype or service.
- The provided solution will be a complete and correct implementation of all the requirements set forth. He/she will verify and validate the delivered system, prototype or service versus the requirements. He/she will document the implementation process.

Managing: level 2

- The student will ensure optimal functioning of the system or service within its operational context.
- He/she will provide system management tools that will allow long-term and sustained correct functioning.

Communication: level 2

- The student will communicate purposefully, using the most common communication forms and techniques.
- He/she is capable of clearly summarizing the different aspects of the project, orally, in writing as well as visually presenting.

Advising: level 2

- The student can provide relevant advice to any third party about his/her design and the corresponding process.
- This advice can be provided to a great variety of stakeholders, target audiences or professional audience.

---

[6] Ing. P. Pelgrims, Ing. P. Degreef, Ing. J. Dieltiens; *ECTS Practice Enterprise ICT 2: Application Development & System Engineering* http://onderwijsaanbodmechelenantwerpen.thomasmore.be/2017/syllabi/n/YT0717N.htm#activetab=doelstellingen_idp545472.

Entrepreneurship: level 2

- The student has a broad knowledge about entrepreneurship and is able to apply this his/her own future professional position.
- He/she has improved upon the own enterprising attitude by taking risks, dealing with failure, user experience design, pitching ideas, negotiating, networking, etc.
- The student is able to detect chances and opportunities, and knows how to convert them into value creation via market-oriented thinking and acting.
- The student has obtained basic experience in the use of tools and instruments required to market products and services, like business model canvas, value proposition canvas, marketing strategies, etc.

Project and performance management: level 2

- The student masters the most important project- and performance management skills. He/she performs the won tasks independently and works in a solution-oriented fashion.
- He/she steers, evaluates and optimizes the own efficiency. He/she adjusts based on feedback. He/she handles tasks in a project-oriented fashion, respecting the planning and budgetary context, adjusting where required.
- He/she operates and functions in a team-oriented and enterprising fashion.

Cooperation/teamwork: level 2

- The student has a constructive and participative attitude as a knowledgeable (project) team member, considering his/her own (future) role and responsibility within a company or organization.
- He/she can manage a team with focus on optimal results. He/she will thus operate in an international and digital working environment.

The scrum retrospective meetings will be used as a fundament for follow-up and permanent evaluation. Students will develop a critical attitude towards their own functioning, as well as to the functioning of the other team members. This will provide a good basis for peer evaluation.

3.3   Phase 3: Practice Enterprise V: ICT Factory

The Practice Enterprise concept as described will cover only a part of the complete job content of a **Systems and Network Administrator**: design and setup of an ICT infrastructure.

In a later phase of the roadmap, the **ICT Factory** concept will be developed.

After implementing an ICT infrastructure, they will learn how to maintain and support it. This will not only require technical skills, but also organizational.

Instead of teaching them theoretically how to implement a helpdesk, they will be required to organize one themselves, and actively operate it.

ITIL[7] will be used as a basis to streamline the processes behind this.

---

[7] Orta and Ruiz (2018).

This will not remain a purely academic exercise as such: they will support and maintain the infrastructure that will be used by the first-year students for labs and exercises in the scope of the curriculum. Teaching staff will remain on standby to help them when things might go really wrong.

At this moment of writing, sponsors are being contacted for financing of this infrastructure.

4   Conclusion

The phased introduction of the Practice Enterprise concept as described above provides us with an iterative approach, closing the gap between an academic and professional environment. This project is in continuous improvement, so phase 3 will be adapted according to learnings from phase 2.

References

Krivitsky, A.: Teach agile thinking and demonstrate the scrum framework with Lego (2017). https://www.lego4scrum.com/

Orta, E., Ruiz, M.: Met4ITIL: a process management and simulation-based method for implementing ITIL. Comput. Stand. Interfaces 61 (2018). https://doi.org/10.1016/j.csi.2018.01.006

Wilson, R.: Skills anticipation—the future of work and education. Int. Journey Educ. Res. 61, 101–110 (2013)

Enhancing Understanding and Skills in Biopharmaceutical Process Development: Recombinant Protein Production and Purification

Abdellatif Elm'selmi[1(✉)], Guilhem Boeuf[1], Amina Ben Abla[1], Rabah Azouani[1], Rafik Absi[1], Ahmed Elmarjou[2], and Florence Dufour[1]

[1] Ecole de Biologie Industrielle, 49 avenue des Genottes, CS 90009, 95895 Cergy Cedex, France
a.elmselmi@hubebi.com
[2] Institut Curie, Centre de Recherche, 26, rue d'Ulm, 75248 Paris Cedex, France

Abstract. The aim of this pedagogical work is to enhance the understanding of fundamental recombinant protein production process and develop the required skills in pharmaceutical biotechnology of graduate students. In this study, the applied biotechnological module was reorganized by positioning specific scenarios of practical experiments in the central core of the course. We developed a training program by alternating lecture classes and practical lab experiments related to recombinant protein production process. Our results indicate that this new teaching method helps students in the understanding of different theoretical biotechnological concept and allows them to improve the abilities and skills required in recombinant protein production process.

Keywords: Recombinant protein · Production process development · Biotechnology · Practical lab teaching

1 Introduction

Experimental learning in pharmaceutical biotechnology provide to students, skills needed for process development. In this pedagogical work, we developed an innovative approach for teaching in this field to help students' understanding fundamental concepts required for the development process of recombinant protein production.

Recombinant protein production remains a central component of many biotechnological projects and emerging as important tool in bioproduction [1]. For development process in this field, biopharmaceutical industry requires engineers who have a good knowledge in chemistry, molecular biology, biochemistry, microbiology, process and strong core skills in recombinant protein production [2]. Students majoring in biotechnology should acquire solid skills and extensive experience in both drugs discovery strategy and therapeutic proteins production: upstream and downstream process [3, 4]

© Springer Nature Switzerland AG 2019
M. E. Auer and T. Tsiatsos (Eds.): ICL 2018, AISC 917, pp. 381–392, 2019.
https://doi.org/10.1007/978-3-030-11935-5_37

The production of recombinant protein needs the following steps: (1) identification and preparation of DNA sequence (gene or cDNA), coding for the protein of interest (2) cloning transgene in the expression vector (3) cell transformation (intro-duction of recombinant vector into a specific host cell) (4) production (biomass and induction) (5) extraction (6) purification (7) characterization and validation and (8) formulation of protein [5, 6]. Many expression systems dealing with the production of recombinant protein have been developed (bacteria, yeast, mammalian cells…) whose the ultimate objective is the obtention of proteins of interest in active form with lower production cost. The choice of expression system depends on the cost of production, the post-translational processing needs for the activity of target protein and process system simplicity.

In this context, the applied biotechnology program was developed in our school to provide graduate students fundamental concepts and skills required in the recombinant protein production field. The course is principally focused on the development of different strategies and processes of recombinant protein production and cover the following areas: (1) protein chemistry, (2) general method for recombinant protein engineering, (3) molecular cloning and expressions systems development: bacteria, yeast, mammalian cells…, (4) production development (fermentation/cell culture), (5) extraction and purification process development, (6) analysis and characterization of target protein, (7) large-scale production (principles of scaling up parameters) and (8) Recombinant protein formulation, stability and storage. In this field, the content of training pro-gram must evolve according to skills required in the bioindustry and teaching methodology must consider the evolution of student's profile [7, 8].

The aim of this pedagogical work is to enhance the understanding and develop the required skills in pharmaceutical biotechnology of graduate students in our institute by positioning specific scenario of practical experiments in the core of the course.

This work is motived by the difficulties encountered by students in understanding fundamental molecular biology and integration of biotechnology process design.

In the classical pedagogical method of this module, the program has been organized in theoretical class (master class) and ended by an experimental lab project in recombinant protein production.

This classical method shows low level students' understanding of the fundamental recombinant protein production concepts related to process development.

To overcome this limitation, we have designed a novel training method based on the alternation of lecture classes and practical lab experiments around recombinant protein production process.

This module is organized throughout one semester and concerns engineering students majoring in research and development, 5th year graduate level. Thirty students per year are registered to this teaching at School of Industrial Biology, Cergy Pontoise, France: www.ebi-edu.com.

The main research activity of the molecular biology department is centered on the biotechnology process: cloning and development of recombinant protein production and purification process.

The recombinant protein produced in the practical class experiments as a case study is the Tumor necrosis factor-alpha (TNF-alpha).

TNF-alpha is a proinflammatory cytokine produced by activated macrophages, T and B lymphocytes, natural killer cells, astrocytes, endothelial cells, smooth muscle cells, some tumor cells, and epithelial cells [9, 10]. This monocyte/macrophage derived protein has pleiotropic functions that include the inflammatory response and host resistance to pathogens [11]. Despite the protective role of TNF-alpha, it has been shown in many systems to induce human immunopathology. A wide range of human diseases of inflammatory nature such as Crohn's disease, Influenza virus and Alzheimer's disease, are mediated in part by TNF-alpha increasing [12]. Because of these various functions, it has been proposed as a therapeutic target for several diseases. Anti-TNF-alpha drugs are now licensed for treating certain inflammatory diseases. Therefore, production of recombinant TNF-alpha is a promising approach in the development of therapeutic agents that selectively inhibit the biologic activity of TNF-alpha. For this purpose, we investigated the production of recombinant human TNF-alpha (rhTNF-alpha) using DNA technology.

Fusion tags technology is used to achieve the purification and detection of recombinant proteins. It has become indispensable in production and purification process. Today, fusion technology is applied not only for purification but also for monitoring and solubilizing of the protein of interest (Table 1) [5]. Fusion tags are subsequently cleaved by several proteases such as TEV proteases [13], enterokinase and SUMO protease [14].

Table 1. Examples of fusion tags used in recombinant protein production

Tag	Size	Uses	Comments
His-tag	6, 8, or 10 aa	Purification	Most common purification tag used for immobilized metal affinity chromatography (IMAC) [15]
Thioredoxin	109 aa	Purification and enhanced expression	Affinity purification with phenylarsine oxide- modified (ThioBond) resin [16]
Streptavidin binding protein (SBP)/streptag II	38 aa/ 8 aa	Purification	Streptavidin affinity purification (strep-tag) [17]
Glutathione-S-tansferase (GST)	218 aa	Purification	Glutathione affinity or GST antibody purification. Enzymatic activity assay possible for quantitative analysis [18]
Maltose binding protein (MBP)	396 aa	Purification and secretion	Amylose affinity purification with maltose elution [15]
Green fluorescent protein (GFP)	220 aa	Detection	Used as reporter gene fusion for detection and monitoring [19]
ubiquitin	76 aa	Enhance solubility	Increase *E. coli* expressed recombinant protein solubility [14]
Small ubiquitin modifier (SUMO)	~100 aa	Enhance solubility	Used to increase *E. coli* expressed recombinant protein solubility. Can be cleaved by SUMO protease [14]

2   Methodology

2.1   Pedagogical Methodology

The objective of this work is to help students in the understanding of many theoretical biotechnological concepts necessary for the development of strategy and bioprocess production. The work proposed here is the reorganization of this module by positioning the practical class experiments in the core of training.

At the end of training, the students work on a project of recombinant protein for three days. They were evaluated based on the understanding of the concepts, process development and practical realization of the different required phases.

The main practical aspects of the program related to recombinant protein production process are summarized in the framework bellow (Fig. 1).

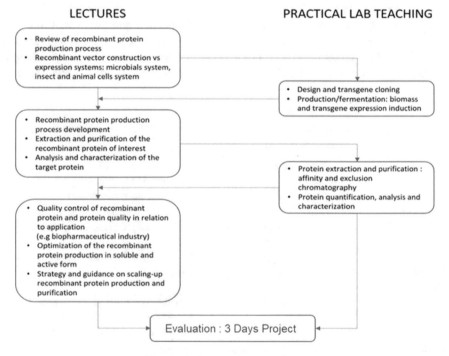

Fig. 1. Pedagogical methodology

2.2   Practical Lab Teaching

2.2.1   Materials and Methods

Materials:

Pipettes and micropipettes, 1.5 mL microtubes, thermocycler, incubator orbital shaker, laboratory centrifuges, chromatographic column.

Reagents

DNA polymerases, oligonucleotides, TA Cloning kit, T4 DNA ligase, molecular weight marker, Pierce Coomassie (Bradford) Protein Assay Kit, hTNF-alpha protein and gel electrophoresis were obtained from Thermo Fisher Scientific (Massachusetts, USA). Plasmid mini-preparation kit, NucleoSpin® Plasmid from Macherey-Nagel (Germany). His-select® Nickel affinity gel and Arabinose was purchased from Sigma-Aldrich (Missouri, USA). Other reagents including tryptone, yeast extract, chloramphenicol, kanamycin, agarose, IPTG were bought from VWR (Pennsylvania, USA).

Expression vector and strains

Expression vector Champion™ pET SUMO was obtained from Thermo Fisher Scientific (Massachusetts, USA). E. coli strain TOP 10 F' (Invitrogen, USA) was used as the recipient of the sub-cloning, strain BL21 DE3 Star (Thermo Fisher Scientific, Massachusetts, USA) transformed by pRARE vector, was used for the expression studies.

Methods:

Pratical lab 1

Construction of expression plasmid pET-6xHis-SUMO-rhTNF-alpha

The hTNF-alpha cDNA was amplified by Polymerase Chain Reaction (PCR). To generate the modified plasmid, the hTNF-alpha fragment coding sequence was PCR amplified by Pfu polymerase (for high-fidelity) with primers, forward (5'-GTGA GAAGCAGCAGCAGAAC-3') and reverse (5'-TCACAGGGCGATGATGCCGA-3'). PCR cycles were as following: denaturation at 94 °C for 1 min, annealing at 58 °C for 1 min and extension at 72 °C for 1 min for 30 cycles. To introduce the rhTNF-alpha amplicon, an incubation at 72 °C for 1 min with Taq polymerase for TA Cloning is performed then an incubation with pET-6xHis-SUMO vector and T4 DNA ligase at room temperature for 30 min. The construct was transformed into E. coli TOP 10 F' cloning host. Vector preparation and purification was performed by NucleoSpin® Plasmid from Macherey-Nagel (Germany). The recombinant expression vector pET-6xHis-SUMO-rhTNF-alpha was confirmed by both endonucleases digestion and DNA sequencing.

Expression of recombinant proteins

The recombinant expression vector was transformed into E. coli BL21 DE3 Star-RARE expression host. Pre-cultures, to test protein expression, from multiple colonies were grown overnight at 37 °C and used to inoculate 10 mL of fresh LB medium containing antibiotics: kanamycin (50 µg/mL) and chloramphenicol (34 µg/mL). Start OD 600 nm was standardized to 0.1 and shaking flask was incubated at 37 °C, 230 rpm until biomass reached the early log phase growth. At OD 600 nm 0.4–0.6, induction was performed by addition of IPTG to a final concentration of 0.5 mM. The induced culture was incubated for either 18 h at 20 °C. Induced and non-induced cultures were centrifuged at 4000 rpm for 15 min and cell pellets were stored at low temperature (−80 °C).

Practical lab 2

Protein extraction

Lysis was carried out by enzymatic lysis, cell pellets were re-suspended in 5 mL of phosphate-buffered saline (PBS) pH 7.4, 350 mM NaCl, 20 mM imidazole, 1 mg/mL lysozyme, 50 µg/mL DNase I supplemented with a protease inhibitor cocktail (Roche Diagnostics, Germany) in iced water. Supernatant and lysate pellet fractions were separated by centrifugation at 14,000 rpm for 30 min.

IMAC column purification

About 5 mL of the soluble fraction of 6xHis-SUMO-rhTNF-alpha extracted from induced *E. coli* BL21 DE3 Star-RARE were mixed with 500 µL equilibrated His-select® Nickel Affinity gel for 1 h at 4 °C. The charged resin was packed in a 15 mL column and washed with two column volumes of 20 mM Imidazol to remove contaminating *E. coli* proteins. 6xHis-SUMO-rhTNF-alpha was then eluted with phosphate buffered saline (PBS) pH 7.4, 50 mM NaCl containing 150 mM Imidazol.

Proteins quantification

The total protein amounts and the purified samples were determined by using a Coomassie (Bradford) protein assay kit.

Proteins analysis

Expression, solubility, and purification of 6xHis-SUMO-rhTNF-alpha ($\approx$30 kDa) were analyzed by SDS-PAGE (polyacrylamide gel electrophoresis) on 4–12% Bis-Tris gel and electrophoresis was performed according to the manufacturer's instruction. For soluble, insoluble and purified fractions, three volumes of samples, relevant to 20 µg of proteins, were mixed with one volume of SDS sample loading buffer and treated by heat at 95 °C for 5 min. Samples were loaded into gel and electrophoresis was run at 140 V for 1 h. Gels were stained with Coomassie Blue G (Sigma-Aldrich, Missouri, USA).

Final expression, extraction, purification, and analysis

A pre-culture from a single positive colony was grown overnight at 37 °C and used to inoculate 100 mL of fresh LB medium. The same conditions, previously validate, were performed to express, extract, and analyze 6xHis-SUMO-rhTNF-alpha protein. Purification was made by affinity chromatography on Nickel resin.

Activity assay

For activity efficiency test, a scratch test was performed. 6xHis-SUMO-rhTNF-alpha protein at 100 and 200 ng/mL and commercial TNF-alpha (Thermo Fisher Scientific) at 20 ng/mL was incubated with Fibroblasts into an incubator (CO_2 5%) for 18 h.

3　Results

3.1　Experimental Results

3.1.1　Construction of Expression Plasmid pET-6xHis-SUMO-rhTNF-alpha

The hTNF-alpha cDNA was PCR amplified with primers as explained in the "Material and methods" section. The 474 bp PCR amplicon was ligated into Champion™ pET SUMO by T/A Cloning to obtain the vector encoding 6xHis-SUMO-rhTNF-alpha (Fig. 2).

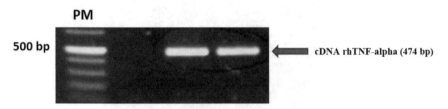

Fig. 2. cDNA-rhTNF-alpha PCR amplification

After cDNA amplification, the TNF-alpha sequence was validated by sanger sequencing, students then used bioinformatics tools to obtain predicted protein sequence (Fig. 3).

> VRSSSRTPSDKPVAHVVANPQAEGQLQWLNRRANALLANGVELRD-
> NQLVVPSEGLYLIYSQVLFKGQGCPSTHVLLTHTISRIAVSYQTKVNLLSAIKS
> PCQRETPEGAEAKPWYEPIYLGGVFQLEKGDRLSAEINRPDYLDFAESGQVYF
> GIIAL

Fig. 3. Predicted sequence of rhTNF-alpha (≈17 kDa)

The amplified and validated rhTNF-alpha fragment is then cloned into the Champion™ pET in fusion with His and SUMO tags (Fig. 4).

> MGSSHHHHHHGSGLVPRGSASMSDSEVNQEAKPEVKPEVKPETHINLKVS
> DGSSEIFFKIKKTTPLRRLMEAFAKRQGKEMDSLRFLYDGIRIQADQAPEDLD
> MEDNDIIEAHREQIGGVRSSSRTPSDKPVAHVVANPQAEGQLQWLNRRANAL
> LANGVELRDNQLVVPSEGLYLIYSQVLFKGQGCPSTHVLLTHTISRIAVSYQT
> KVNLLSAIKSPCQRETPEGAEAKPWYEPIYLGGVFQLEKGDRLSAEINRPDYL
> DFAESGQVYFGIIAL

Fig. 4. 6xHis-SUMO-rhTNF-alpha sequence (≈30 kDa)

3.1.2   Production and Characterization of Target Protein

After transformation, multiple colonies were studied by students to test the expression of the recombinant proteins.

The transformation of *E. coli* BL21 DE3 Star-RARE was carried out (Fig. 5). Positive colonies were selected for the production and characterization of recombinant human TNF-alpha.

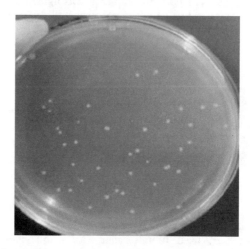

Fig. 5. Multiple colonies on LB-agar medium with kanamycin

As described in the materials and methods section, after the fermentation (bio-mass and induction), students performed the extraction of total proteins and the fused protein his-SUMO-rhTNF-alpha was subsequently purified by NI-NTA affinity and his-SUMO tag was removed by SUMO protease. Obtained fractions were analyzed by SDS PAGE (Fig. 6).

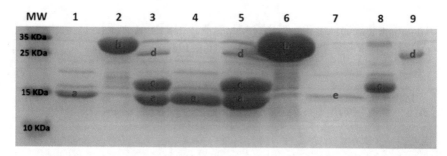

Fig. 6. Protein fraction analysis PM = Molecular weight, 1: rhTNF-alpha, 2: 6xHis-SUMO-rhTNF-alpha, 3: enzymatic digestion, 4: rhTNF-alpha, 5: enzymatic digestion, 6: 6xHis-SUMO-rhTNF-alpha, 7: laboratory reference rhTNF-alpha, 8: tag His-SUMO, 9: SUMO protease a: rhTNF-alpha (17,4 kDa), b: 6xHis-SUMO-rhTNF-alpha (30 kDa), c: tag SUMO (22 kDa), d: SUMO protease (28 kDa), e: labora-tory reference rhTNF-alpha

rhTNF-alpha resulted from cleavage test was then purified by size exclusion chromatography (Fig. 7). SDS PAGE analysis shows that the recombinant protein was successfully cleaved by SUMO-protease (Fig. 7a) and there is a clear, single band with a molecular weight about 17 kDa when eluted from the chromatography column (Fig. 7a).

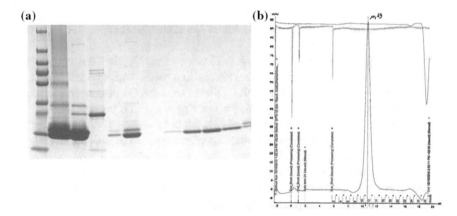

(a)   **(b)**

Fig. 7. Protein fraction analysis **a** SDS-PAGE analysis of the production and purification of rhTNF-alpha; a: 6his-sumo-TNF-alpha + SUMO protease, b: 6his-sumo-TNF-alpha + SU-MO protease, c: 6 his-sumoprotease, d: TNF-alpha native from column after digestion; e: TNF-alpha native from column, 9–14: Eluted fractions from size exclusion chromatography. **b** rhTNF-alpha monotrimer analyzed by size exclusion chromatography on superdex 75 column from GE. The elution volume of the tetramer is 10.69 ml correspond to the elution volume of the ovalbumin used as a caliber protein after digestion

The result of chromatogram presented in Fig. 7b shows a single pic of assembled 51 kDa monotrimer rhTNF-alpha. Those results indicate that students achieved efficient purification of protein of interest with a high degree purity and correctly folded.

3.1.3   Activity Test

To determinate whether the purified TNF-alpha proteins were biologically active, we carried out monolayer culture cell scratch assay. As expected, produced recombinant homotrimer TNF-alpha as validated (Fig. 7b), stimulate the migration of fibroblast in dose-dependent manner in concentration going from 100 to 200 ng/ml. The stimulation levels of rhTNF-alpha produced were comparable to that obtained with the native commercial hTNF-alpha.

3.2   Pedagogical Results

This pedagogical study describes a new teaching method developed in the module of applied biotechnology to help students in the understanding of fundamental molecular biotechnology and integrating different process concept required for the recombinant protein production.

In our work, we developed a program training enabling improvement of students' ability in recombinant protein production process. In biopharmaceutical industry this topic seems to become an important workforce in drug discovery and therapeutic proteins productions area: enzymes, cytokines, hormones, antibodies, antibody fragments, etc.

For each experimental lab section from cloning to purification and activity test, students collect data, analyse and interpret results in a report with the description and principle of used methods. In addition, the students are asked to discuss results and to propose strategies needed for the optimisation of the production process.

According to the results of the pedagogical work developed in this study, based on the integration of practical experiments with specific scenarios in the core of training, this new teaching method helps students in the understanding of several biotechnological concepts and allows them to acquire skills required in recombinant protein production process: expression system choice, process production and purification design, results analysing and process optimization.

At the end of the formation the students are able to design and perform a specific process required for the production of target protein, to optimise the production by varying different parameters, to develop chromatography methods needful for the obtention of high degree of purity of target recombinant protein and the development of the activity test.

Furthermore, we have seen a remarkable increase in the involvement and motivation of our students during this reorganised module.

This innovative teaching method allows a noticeable increasing of students' marks of applied biotechnology evaluation compared to those obtained in the classical method.

4   Conclusions

This study describes a new teaching method in recombinant protein production process adopted in our school to increase the quality of our students and to promote a good understanding of processes used in biopharmaceutical industry.

We designed and implemented bioproduction process courses using specific experimental lab works to positively influence students' understandings of strategies in recombinant protein production process.

According to this pedagogical experience, the use of TNF alpha protein production as a case-study, helps students to integrate the development process phases required for production of target proteins.

The content program developed in this course largely concerns the type of knowledge and skills that graduate engineering students need to acquire for process development in this field.

References

1. Expert Group on Future Skills Needs: Future Skills Requirements of the Biopharma-Pharmachem Sector (2010)
2. Elm'selmi, A., Boeuf, G., Elmarjou, A., Azouani, R.: Active pedagogy project to in-crease bio-industrial process skills. In: Auer, M., Guralnick, D., Uhomoibhi, J. (eds.) Interactive Collaborative Learning. ICL 2016. Advances in Intelligent Systems and Computing, vol. 544 (2016)
3. Gobert, J.D., Buckley, B.C.: Introduction to model-based teaching and learning in science education. Int. J. Sci. Educ. 22(9), 891–894 (2000)
4. Palomäki, E., Qvist, P., Natri, O., Joensuu, P., Närhi, M., Kähkönen, E., ... Nordström, K.: LabLife3D: Teaching biotechnology and chemistry to engineering students by using Second Life. Stanford University, H-STAR Institute, USA; to Associate Professor Jukka M. Laitamäki, from New York University, USA, and to Professor Yngve Troye Nordkvelle from Lillehammer University, 124 (2011)
5. Miladi, B., Dridi, C., El Marjou, A., Boeuf, G., Bouallagui, H., Dufour, F., Di Martino, P., Elm'selmi, A.: An improved strategy for easy process monitoring and advanced purification of recombinant proteins. Mol. Biotechnol. 55(3), 227–235 (2013)
6. Rosano, G.L., Ceccarelli, E.A.: Recombinant protein expression in *Escherichia coli*: advances and challenges. Front. Microbiol. 5, 172 (2014)
7. Absi, R., Nalpace, C., Dufour, F., Huet, D., Bennacer, R., Absi, T.: Teaching fluid mechanics for undergraduate students in applied industrial biology: from theory to atypical experiments. Int. J. Eng. Educ. (IJEE) 27(3), 550–558 (2011)
8. Absi, R., Lavarde, M., Jeannin L.: Towards more efficiency in tutorials: active teaching with modular classroom furniture and movie-making project. In: 2018 IEEE Global Engineering Education Conference (EDUCON), pp. 774–778, 17–20 April 2018, Santa Cruz de Tenerife, Tenerife, Islas Canarias, Spain, Spain (2018)
9. Abbas, A.K., Lichtman, A.H., Pober, J.S.: Cellular and Molecular Immunology, 4th edn, pp. 235–269. W. B. Saunders Company, Philadelphia, PA (1991)
10. Horiuchi, T., Mitoma, H., Harashima, S.I., Tsukamoto, H., Shimoda, T.: Trans-membrane TNF-α: structure, function and interaction with anti-TNF agents. Rheumatology 49(7), 1215–1228 (2010)
11. Brustolim, D., Ribeiro-dos-Santos, R., Kast, R.E., Altschuler, E.L., Soares, M.B.P.: A new chapter opens in anti-inflammatory treatments: the antidepressant bupropion lowers production of tumor necrosis factor-alpha and interferon-gamma in mice. Int. Immunopharmacol. 6(6), 903–907 (2006)
12. Tobinick, E., Gross, H., Weinberger, A., Cohen, H.: TNF-alpha modulation for treatment of Alzheimer's disease: a 6-month pilot study. Medscape Gen. Med. 8(2), 25 (2006)
13. Miladi, B., El Marjou, A., Boeuf, G., Bouallagui, H., Dufour, F., Di Martino, P., Elm'sel-mi, A.: Oriented immobilization of the tobacco etch virus protease for the cleavage of fusion proteins. J. Biotechnol. 158(3), 97–103 (2012)
14. Peroutka III, R.J., Orcutt, S.J., Strickler, J.E., Butt, T.R.: SUMO fusion technology for enhanced protein expression and purification in prokaryotes and eukaryotes. In: Heterologous Gene Expression in *E. coli*, pp. 15–30. Humana Press (2011)

15. Crowe, J., Döbeli, H., Gentz, R., Hochuli, E., Stüber, D. & Henco, K.: 6xHis-Ni-NTA chromatography as a superior technique in recombinant protein expression/purification. In: Harwood, A.J. (ed.) Methods in Molecular Biology, vol. 31, pp. 371–387. Humana Press, Inc., Totawa (1994)

16. Sherwood, R.: Protein fusions: bioseparations and application. Trends Biotechnol. **9**, 1–3 (1991)

17. Schmidt, T.G.M., Skerra, A.: The random peptide library-assisted engineering of a C-terminal affinity peptide, useful for the detection and purification of a functional Ig Fv fragment. Protein Eng. **6**, 109–122 (1993)

18. Abel, U., Koch, C., Speitling, M., Hansske, F.G.: Modern methods to produce natural-product libraries. Curr. Opin. Chem. Biol. **6**(4), 453–458 (2002)

19. Chalfie, M., Tu, Y., Euskirchen, G., Ward, W.W., Prasher, D.C.: Green fluorescent protein as a marker for gene expression. Science **263**, 802–805 (1994)

A Teaching Concept Towards Digitalization at the LEAD Factory of Graz University of Technology

Maria Hulla$^{(\boxtimes)}$, Hugo Karre, Markus Hammer, and Christian Ramsauer

Graz University of Technology, Graz, Austria
{maria.hulla, hugo.karre, markus.hammer, christian. ramsauer}@tugraz.at

Abstract. The ongoing digital transformation is offering significant potentials in the manufacturing industry. As a result, competences for digital technologies and their implementation are in great demand. In this context, a digitalization teaching concept has been introduced in the learning factory of Graz University of Technology. The aim of this paper is to identify competencies that will be required from the future workforce, to discuss if learning factories are suitable for transferring those competencies and to introduce a digitalization teaching concept. This teaching concept consists of alternating and complementary theoretical and practical training sessions. The training aims to enhance understanding of how digital technologies can improve the production process, how digital technologies and strategies can be implemented in a company, and how value is created with business models based on digitalization.

Keywords: Digitalization teaching concept · Learning factory · Digitalization projects · Digital learning environment

1 Introduction

Industry is currently undergoing a transformation towards digitalization. Manufacturing systems are becoming more intelligent and are steadily improving with high adaptability to changing environments, increasing resource efficiency hand in hand with the integration of technology and people. Future production systems are estimated to be characterized by small digitalized, decentralized elements which will be capable of acting autonomously and will thus be able to control their operations according to external inputs. In this production network products and materials will be uniquely identifiable and locatable along the entire life-cycle. Furthermore, they are customized while having costs similar to those of mass production [1]. This digital transformation also leads to an increased organizational and technological complexity of manufacturing [2].

As potentials, challenges and requirements of industrial companies change, so are competencies of the industrial workforce, which are necessary in a factory of the future [3]. Competencies of the workforce pose a key for success in highly innovative and

© Springer Nature Switzerland AG 2019
M. E. Auer and T. Tsiatsos (Eds.): ICL 2018, AISC 917, pp. 393–402, 2019.
https://doi.org/10.1007/978-3-030-11935-5_38

connected factories [4]. In order to capture the potentials of the factory of the future, it is essential to focus on the teaching of the (future) workforce [5]. Insights from industry and the findings from research suggest a change of focus and content of educational intuitions [6].

To meet the competencies required, the authors introduce a digitalization course for students and professionals of the industry in a learning factory, the LEAD Factory of Graz University of Technology. For the development of this concept, it was first necessary to evaluate the required competencies of the future workforce in the context of digitalization. A research of the recent literature and research studies was thus conducted. The results are summarized in this paper in Chap. 2. In Chap. 3 the concept of learning factories as an appropriate teaching and learning environment to gain relevant competencies will be illustrated. Chaps. 4 and 5 focus on the LEAD Factory of Graz University of Technology and the digitalization teaching concept, respectively. A summary and outlook is given in Chap. 6.

2   Competencies of the Future Workforce

Campion et al. defined the term "competence" as a *"collections of knowledge, skills, abilities and other characteristics that are needed for effective performance in the jobs in question"* [7]. The Fourth Industrial Revolution propagates the idea that the staff is focused on innovative, creative and communicative activities. Workers will have to do more complex and indirect tasks such as working with collaborating robots and machines [8]. Moreover, they will also need to deal with information generated in real-time as well as a higher quantity of data and communicate with machines [4]. However, repeating or routine tasks will mainly be taken over by machines [9].

In Fig. 1 the results of the review of the competency requirements are structured and summarized according to Erpenebeck and von Rosenstiel (2007) [10].

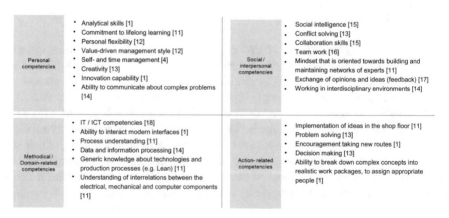

Fig. 1. Competencies of the future workforce

3 Concept of Learning Factories

The term "learning factory" first came up in 1994 when the National Science Foundation of the US announced a consortium led by the Penn State University to develop such a learning environment [11]. Over the past ten years a steadily increasing number of learning factories have been established in all parts the world, especially in Europe, imparting knowledge and competences for both industry and academics [11–13]. In a narrow sense learning factories are specified by authentic, including multiple stations and comprising technical as well as organizational aspects processes, a changeable setting, a manufactured physical product and a didactical concept which enables learning by own actions [14]. The learning factory itself represents a model of a production system offering trainees the opportunity to implement process improvements and enables to experience the results of these changes immediately [15]. Most of these learning environments are focused on process related learning, concerning lean manufacturing, logistics or energy and resource efficiency [11, 16, 17]. The main goal is to provide a close-to-industrial reality for the education environment by letting course participants experience hands-on activities though real-life projects [18]. It has been shown in preliminary studies that the concept of learning factories has a better knowledge performance as well as a better capability acquisition than traditional approaches [19]. Learning factories have become popular during not only in schools and universities but also in industry (e.g. BMW or Crysler) over the past few years [11].

Over the recent years an increasing number of learning factories addressing the emerging topic of digitalization were established [20, 21]. A review study conducted by Block et al. (2018) showed that learning factories with a digitalization focus are mainly showcases or pure research objects [11, 22]. Moreover, not the development of own creative solutions is the content of trainings but rather demonstration of new potentials as well as the use of information and communication technologies [22]. For instance, in the Application Center Industry 4.0 a teaching scenario enables students to experience the potentials of Internet of Things technologies [23]. Furthermore, Reuter et al. (2017) presented their didactical concept which allows participants to explore the capabilities of digital assistance systems [24].

The current literature shows that there is a general lack of trainings that focus on the creative solution finding and implementation of digitalization technologies as well as strategies [22]. Learning factories proofed to be an effective learning environment for teaching the introduction, effectiveness, tools and principles for lean management and resource efficient operations. Furthermore, it was shown that learning factories can make a significant contribution to understanding the general concept of digitalization and its related technologies [25]. The authors thus assume that the identification of potentials, the selection, the evaluation and the implementation of digital technologies together with the development of digitalization strategies can also be taught in this way.

4   LEAD Factory of Graz University of Technology

The LEAD (Lean, Energy Efficient, Agile, Digital) Factory of Graz University of Technology has been operated by the Institute of Innovation and Industrial Management since 2014. This learning factory enhances hands-on education, company trainings and research and enables practice oriented learning. More than 400 students and professionals from the industry have been trained in three-hour to three-day courses until now. During these courses, participants first learn about lean principles and energy efficiency in operations. Based on the theoretical knowledge gained, they experience the impact of lean principles directly through the process of assembling a functional market available scooter consisting of 60 separate parts. While 59 parts are bought, one part is produced by the course participants themselves using a 3D printer. The didactic approach in the LEAD Factory is performed in two setup states: the initial state and the optimized lean state (Fig. 2).

Fig. 2. Non-optimized initial state (left) and optimized lean state (right) of the LEAD Factory of Graz University of Technology

First, participants work in the initial (non-optimized) state where eight training participants need about 13 min and later in the lean state only four workers take 3.5 min for assembling a customized scooter. In 2016, a digitalization roadmap for the LEAD Factory was established [26]. During the recent years, many digital technologies such as the Augmented Reality glasses, gesture and mimic control, screens, RFID system etc. have been introduced in the LEAD Factory. These are essential elements for the teaching concept.

5   Teaching Concept

The teaching concept is based on the findings of the literature research regarding the future of work and the required competencies of the future workforce. The digitalization teaching concept is introduced in the LEAD Factory of Graz University of Technology in the summer semester 2018. It consists of eight major steps and is framed by a case study. The teaching concept strives to educate the future workforce, students and also professionals from the industry within the scope of a 6.5-h course. In this chapter the case study teaching method is introduced first. Thereafter an overview of the required steps is given and the competencies that are enhanced or acquired during the course are illustrated.

5.1 Case Study

The Harvard Business School established the case study teaching method 100 years ago. These cases are defined as *"stories with an educational message"* [27]. Case studies involve problem-based learning and encourage the development of analytical skills. There are several methods of case study teaching. For the digitalization teaching concept, the discussion method is used, where the whole classroom debates the problems stated in the case [28]. Teachers ask the right questions at the right time and provide feedback on answers, while not giving personal opinions. Students have to come well-prepared to the lectures and provide most of the content of a case discussion [29]. Case studies facilitate interdisciplinary learning and are capable of highlighting connections between theory and real-world issues [30]. Moreover, it has been shown that case study teaching motivates students to participate in class, which in turn enhances the learning performance [31]. Due to the facts outlined above the authors agreed on using the case study teaching method in the digitalization teaching concept.

The case study for the digitalization teaching concept was written by the authors of this paper and frames the concept. Starting with a short overview of the history, the employee structure, the products (variants of the scooter), and the supply chain of the LEAD Factory. Further, the assembly process of the scooter is described roughly and the strengths, problems and challenges of the production steps are illustrated. This part also contains the strengths and challenges of logistics, production planning and performance measurement. Furthermore, the position of the factory regarding safety and sustainability is stated. Based on this information and their experiences on the shopfloor of the LEAD Factory, the participants should be able to identify, evaluate and select appropriate digitalization potentials. Another part of the case study represents the current status of the company regarding external factors, including chances, weaknesses and some data for the PESTEL analysis. Moreover, a business model canvas of the LEAD Factory is illustrated to give further information.

The final part of the case study deals with the development of a digital business model for urban mobility. Students get information regarding a specific digital business model as an example and current challenges of urban mobility.

5.2 The Procedure of the Digitalization Teaching Concept

The digitalization teaching concept contains a theoretical part as the basis for the practical exercises which are performed in the LEAD Factory. The procedure of the course is illustrated in Fig. 3.

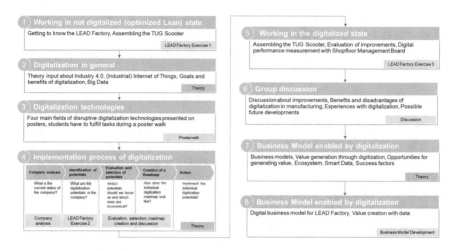

Fig. 3. Overview of the digitalization teaching concept

In step 1 of the digitalization course, the participants assemble the scooter in lean state of the LEAD Factory, getting to know the production process and the product in a hands-on way. Following on from this they are given a theory input focused on digitalization, industry 4.0, the (Industrial) Internet of Things and Big Data. Participants should get familiar with the basic terms, goals and benefits digitalization can offer.

A number of disruptive technologies enable digitization in the manufacturing sector which can be clustered in four main areas: data, digital power and connectivity, analytic and intelligence, human-machine interaction as well as digital-to-physical conversion [32]. These areas are introduced to the participants via a poster walk. The posters are already provided by the supervisors and the trainees must complete a number of specific tasks they are called on to perform and present the most important information regarding the technologies to the other colleagues.

The core of the digitalization teaching concept represents step 4, the implementation process, which is based on the procedure published by Seiter et al. (2016) [33]. In this step participants are introduced to the procedure by which digitalization potentials are identified, evaluated and selected and how a digitalization roadmap is created. They get alternating theoretical and practical inputs starting with a company analysis (Business model canvas, PESTEL, SWOT analysis) based on the case study. After that, participants need to identify digitalization potentials in the LEAD Factory. Based on an evaluation catalogue with predefined criteria and the case study they evaluate and select potentials. On the basis of this selection a digitalization roadmap is created and the implementation of the digitalization technologies performed.

In the fifth step of the digitalization teaching concept participants work in the digitalized state of the LEAD Factory. They are able to experience the improvements with the technology they implemented in step 4. In a group discussion in step 6 the participants talk about their experiences with digitalization in the LEAD Factory and their professional lives.

As a steadily increasing number of companies are shifting their business models to digital enabled ones this topic is also a part of the digitalization teaching concept. While in step 7 participants get a theoretical input regarding digital business models and value creation with digitalization, for instance based on Big Data, they are asked to develop a business model for the LEAD Factory in step 8. Furthermore, trainees should investigate how they can create value with the data created during the manufacturing process and while using the scooter in everyday life. The results are presented to the supervisors and colleagues.

5.3 Transferred Competencies

The digitalization teaching concept of the LEAD Factory strives to develop and build up the competencies that will be required from the future industrial workforce. Within the digitalization course personal, social, methodical and activity-oriented competences can be gained by the participants. The competency group addressed, the step of the course and the description of actions set in the course in order to acquire or enhance the competence are illustrated in Fig. 4.

	Competence	Step in the teaching concept	Desciption of actions how competentcies are aquired or enhanced
Personal competencies	Self- and time management	3, 4, 8	Participants are facing different challenges in terms of exercises which have to be tackled in a certain amount of time
	Creativity	3, 4, 8	Creative thinking is required when fulfilling tasks of the poster walk, the identification and selection of problems as well as in business model development
	Innovation capability	4, 8	Innovation inputs are asked in the implementation when participants should think about potential technologies and when they are developing an innovative business model
	Ability to communicate (about) complex problems	4, 6	Problems regarding the manufacturing process are stated in the case study and come up during the assembly process. Teams have discuss them and come to a solution
Social / Interpersonal competencies	Collaboration skills & team work	1, 3, 4, 5, 8	In every exercise participants are working in teams and have to collaborate with each other
	Conflict solving	1, 3, 4, 5, 8	In the course of group works several conflict potentials come up. The group members have to deal with the situation.
	Exchange of options and ideas (feedback)	3, 4, 8	When participants are discussing solutions regarding digitalization technologies and strategies, they have to give feedback on another
Methodical / Domain-related competencies	IT/ICT competency	2, 3, 5	Participants get ICT competency in the theory part and when should implement and experience the technologies on in the exercises
	Ability to interact with modern interfaces	5	Participants have the possibility to interact with human-machine interfaces (e.g. Google glass, gesture control, screens)
	Generic knowledge about technologies and production processes	1, 4, 5	During the theory part, participants get to know technologies, can then be implemented and tested by them.
	Understanding of interrelations between the electrical, mechanical and computer components	2, 3, 5	Information regarding electrical, mechanical and computer elements are given and can be experienced when working in the digitalized state in the learning factory
Action-related competencies	Implementations of ideas on the shopfloor	4	During the course ideas of the participants are implemented on the shopfloor
	Encouragement taking new routes	4, 8	Within the implementation project and business modelling, participants learn how they can encourage colleagues
	Problem solving	3, 4, 8	Participants are facing several problems in exercises and on the shopfloor which have to be solved in the group.
	Decision making and taking	3, 4, 8	Decisions should be made during the posterwalk tasks and particularly in the steps of implementation potentials and business modelling

Fig. 4. Competencies transferred during the digitalization training

6 Summary and Outlook

The ongoing digital transformation is entailing new potentials and challenges in manufacturing industry. This is leading to a shift in the competencies that will be required in the future workforce and further to the necessity of training in digitalization technologies and strategies for the workforce of today and the future. Classic lecture based teaching methods are not seen to be suitable in this case. Learning factories, such as the LEAD Factory of Graz University of Technology, are providing close-to-practice learning environments and have proven to be an appropriate learning environment to acquire the competencies required by manufacturing industry. A case study is framing the concept and serves as a source of information for the exercises.

The introduced teaching concept offers the possibility for participants to gain skills, knowledge and capability in the field of digitalization. It consists of theoretical input and exercises where the finding of a solution is in the center of attention. When comparing the required competencies of the future workforce in the current literature with the competencies that should be acquired or enhanced during the course of the teaching concept, many overlaps can be found.

Further research effort as part of this teaching concept will focus on the proof-of-concept in the summer semester 2018 and further developments of the digitalization course in the LEAD Factory will be made.

References

1. Erol, S., Jäger, A., Hold, P., Ott, K., Sihn, W.: Tangible industry 4.0: a scenario-based approach to learning for the future of production. Procedia CIRP **54**, 13–18 (2016)
2. Schuh, G., Potente, T., Varandani, R., Schmitz, T.: Global footprint design based on genetic algorithms—an "industry 4.0" perspective. CIRP Ann-Manufact. Technol. **63**, 433–436 (2014)
3. Hirsch-Kreinsen, H.: Wandel von Produktionsarbeit—"Industrie 4.0". WSI-Mitteilungen **67**, 421–429 (2014)
4. Gehrke, L., Kühn, A., Rule, D.: A discussion of qualifications and skills in the factory of the future: a german and american perspective. VDI Hannover Messe (2015)
5. Armstrong, M., Taylor, S.: Armstrong's Handbook of Human Resource Management Practice. Kogan Page Limited, London (2014)
6. Enke, J., Metternich, J., Bentz, D., Klaes, P.: Systematic learning factory improvement based on maturity level assessment. Procedia Manufact. **23**, 51–56 (2018)
7. Campion, M., Fink, A., Ruggeberg, B., Carr, L., Phillips, G.: Doing competencies well: best practices in competency modeling. Pers. Psychol. **64**, 225–262 (2011)
8. Levy, F., Murnane, R.: Dancing with robots: human skills for computerized work. https://www.thirdway.org/report/dancing-with-robots-human-skills-for-computerized-work (2013)
9. Lanza, G., Haefner, B., Kraemer, A.: Optimization of selective assembly and adaptive manufacturing by means of cyber-physical system based matching. CIRP Ann-Manufact. Technol. **1**, 399–402 (2015)
10. Erpenbeck, J.: Handbuch Kompetenzmessung: erkennen, verstehen und bewerten von Kompetenzen in der betrieblichen, pädagogischen und psychologischen Praxis. Schäffer-Poeschel (2007)
11. Abele, E., Chryssolouris, G., Sihn, W., Metternich, J., ElMaraghy, H., Seliger, G., Sivard, G., ElMaraghy, W., Hummel, V., Tisch, M., Seifermann, S.: Learning factories for future oriented research and education in manufacturing. CIRP Ann. **66**, 803–826 (2017)
12. Tisch, M., Ranz, F., Abele, E., Metternich J., Hummel V. (2015) Learning factory morphology—study of form and structure of an innovative learning approach in the manufacturing domain. Turk. Online J. Educ. Technol. (2), 356–363

13. Wagner, U., AlGeddawy, T., ElMaraghy, H., Müller, E.: The state-of-the-art and prospects of learning factories. Procedia CIRP **3**, 109–114 (2012)
14. Abele, E., Metternich, J., Tisch, M., Chryssolouris, G., Sihn, W., ElMaraghy, H., Hummel, V., Ranz, F.: Learning factories for research, education, and training. Procedia CIRP **32**, 1–6 (2015)
15. Lamancusa, J., Simpson, T.: The learning factory—10 years of impact at penn state (2018)
16. Rentzos, L., Doukas, M., Mavrikios, D., Mourtzis, D., Chryssolouris, G.: Integrating manufacturing education with industrial practice using teaching factory paradigm: a construction equipment application. Procedia CIRP **17**, 189–194 (2014)
17. Böhner, J., Weeber, M., Kuebler, F., Steinhilper, R.: Developing a learning factory to increase resource efficiency in composite manufacturing processes. Procedia CIRP **32**, 64–69 (2015)
18. Enke, J., Tisch, M., Metternich, J.: Learning factory requirements analysis—requirements of learning factory stakeholders on learning factories. Procedia CIRP **55**, 224–229 (2016)
19. Cachay, J., Abele, E.: Developing competencies for continuous improvement processes on the shop floor through learning factories-conceptual design and empirical validation. Procedia CIRP **3**, 638–643 (2012)
20. Wank, A., Adolph, S., Anokhin, O., Arndt, A., Anderl, R., Metternich, J.: Using a learning factory approach to transfer industry 4.0 approaches to small- and medium-sized enterprises. Procedia CIRP **54**, 89–94 (2016)
21. Schallock, B., Rybski, C., Jochem, R., Kohl, H.: Learning Factory for industry 4.0 to provide future skills beyond technical training. Procedia Manufact. **23**, 27–32 (2018)
22. Block, C., Kreimeier, D., Kuhlenkötter, B.: Holistic approach for teaching IT skills in a production environment. Procedia Manuf **23**, 57–62 (2018)
23. Gronau, N., Ullrich, A., Teichmann, M.: Development of the industrial IoT competences in the areas of organization, process, and interaction based on the learning factory concept. Procedia Manufact. **9**, 254–261 (2017)
24. Reuter, M., Oberc, H., Wannöffel, M., Kreimeier, D., Klippert, J., Pawlicki, P., Kuhlenkötter, B.: Learning factories' trainings as an enabler of proactive workers' participation regarding industry 4.0. Procedia Manufact. **9**, 354–360 (2017)
25. Prinz, C., Morlock, F., Freith, S., Kreggenfeld, N., Kreimeier, D., Kuhlenkötter, B.: Learning factory modules for smart factories in industry 4.0. Procedia CIRP **54**, 113–118 (2016)
26. Karre, H., Hammer, M., Kleindienst, M., Ramsauer, C.: Transition towards an industry 4.0 state of the LeanLab at Graz university of technology. Procedia Manufact. **9**, 206–213 (2017)
27. Herreid, C.: Start with a Story: The Case Study Method of Teaching College Science. NSTA Press, Arlington (2007)
28. Herreid, C.: Case study teaching. New Dir. Teach. Learn. **2011**, 31–40 (2011)
29. Ellet, W.: The Case Study Handbook: How to Read, Discuss, and Write Persuasively About Cases. Harvard Business Press, Boston (2007)
30. Bonney, K.: An argument and plan for promoting the teaching and learning of neglected tropical diseases. J. Microbiol. Biol. Educ. JMBE **14**, 183–188 (2013)
31. Flynn, A., Klein, J.: The influence of discussion groups in a case-based learning environment. Educ. Technol. Res. Dev. **49**, 71–86 (2001)
32. Manyika, J., Chui, M., Bughin, J., Dobbs, R., Bisson, P., Marrs, A.: Disruptive Technologies: Advances that will Transform Life, Business, and the Global Economy. McKinsey, New York (2013)
33. Seiter, M., Bayrle, C., Berlin, S., David, U., Rusch, M., Treusch, O.: Roadmap Industrie 4.0: Ihr Weg zur erfolgreichen Umsetzung von Industrie 4.0. tredition (2016)

34. Spath, D., Ganschar, O., Gerlach, S., Hämmerle, M., Krause, T., Schlund S.: Produktionsarbeit der Zukunft - Industrie 4.0. Fraunhofer Verlag (2013)
35. Cleary, M.: Management Skills in the future manufacturing sector, Precision Consultancy for the Department of Education and Early Childhood Development (2014)
36. Grzybowska, K., Anna, L.: Key competencies for industry 4.0. pp 250–253 (2017)
37. Dworschak, B., Zaiser, H., Martinetz, S., Windelband, L.: Zukünftige Qualifikationserfordernisse durch das Internet der Dinge in der Logistik. Freq - Früherkennung Von Qualif, pp 1–11 (2011)
38. Davies, A., Fidler, D., Grobis, D.: Future work skills 2020. Inst. Future Univ. Phoenix Res. Inst. (2011)
39. Grega, W., Kornecki, A.: Real-time Cyber-Physical Systems transatlantic engineering curricula framework. In: 2015 Federated Conference on Computer Science and Information Systems (FedCSIS). pp 755–762 (2015)
40. Uhrin, A., Moyano-Fuentes, J., Bruque Camara, S.: Lean production, workforce development and operational performance. Manage. Decis. **55**, 103 (2017)
41. Estrada, S., Cuevas-Vargas, H., Larios-Gómez, E.: The effects of ICTs as innovation facilitators for a greater business performance. Evid. Mex. Procedia Comput. Sci. **91**, 47–56 (2016)

One Graduate—Two Majors: Employers' Demands, Students' Interests

Maria Yurievna Chervach[1,2(✉)]

[1] National Research Tomsk Polytechnic University, Tomsk, Russian Federation
chervachm@tpu.ru
[2] Association for Engineering Education of Russia, Moscow, Russian Federation

Abstract. The article discloses a study on the identification of multidisciplinary master degree majors, which interest potential employers and students in Russia. Results of the initial phase of the research project conducted at Tomsk Polytechnic University are presented. The study focuses on collecting raw data from two categories of respondents: potential employers (industry representatives) and potential master students (4th year bachelor students of engineering majors, TPU). The respondents underline the demand for double major master programs and rate reasons for studying two majors. The study proposes particular pairs of master programs that should be harmonized on the latter stages of the project. The importance of the research lies at the root of communicating modern issues and requirements of industry and society with the responses provided by universities, i.e. the quality and relevance of students' professional training.

Keywords: Master programs · Engineering education · Double majors · Harmonized programs

1 Introduction

The trend on multidisciplinary education for engineers has gained popularity in early 2000s [1–3]. By now, the majority of the world leading universities have implemented the ideas of multidisciplinarity to their educational process and curricula by means of dual education, double-degree programs, multidisciplinary courses and projects, and other approaches [4–7].

The reasons for introducing multidisciplinarity have been studied widely [8–11] and, in most cases, come down to the modern companies' need to solve complex tasks that lie on the joint of several scientific and technological areas.

Highly motivated students show interest in applying for a second degree in a new area of knowledge at some stage of personal and professional development. An overview of the SESTAT statistics [12] that has been gained from over 1700 respondents gives an opportunity to identify students' reasons for pursuing two or three majors rather than one. The most popular reasons are practical ones, such as competitiveness in terms of employment, however personal interests are also in place.

Pitt and Tepper [13] in their report on double majors underlines not only the employers' demand in specialists with extra competences, but also argue that having

© Springer Nature Switzerland AG 2019
M. E. Auer and T. Tsiatsos (Eds.): ICL 2018, AISC 917, pp. 403–414, 2019.
https://doi.org/10.1007/978-3-030-11935-5_39

double majors is prosperous for students. However, the report provides only a brief overview of the possible combinations of majors that are of a bigger interest for students and employers, but does not indicate particular pairs of majors/specialties.

Multidisciplinarity is usually discussed in terms of project-based learning and considers modernization of particular courses. At the same time, students choosing to double major find and create opportunities for such education by themselves. But what if we upscale this principle of multidisciplinary education to the terms of providing educational programs that are not just cross-disciplinary in its content, but provide two different majors to one student simultaneously?

2   Problem Identification

Russian educational model is embedded in the European Higher Education Area, regulated by the Bologna Declaration [14]. However, Russian system of higher education has its peculiarities; among such is the fact that it is based on a strict curriculum with a certain number of ECTS and a corresponding set of disciplines (courses), internships and research projects. Students, after enrolling for a certain major, have to undertake a certain set of disciplines with a very low number of ECTS for personal election. This perplexes the process of educational trajectory (path) personification and limits students' opportunities to develop competitive advantages in terms of unique competences' formation for their future workplace.

According to the surveys and researches conducted by the Association for Engineering Education of Russia [15–18], there are several features of the Russian educational system that affect the considerably low level of satisfaction of the employers in terms of engineering graduates' competences. Among these restrictive features are:

- A low level of academic freedom, mandatory curricula at HEIs;
- A high percent of lectures in a curriculum;
- A low level of Project-based learning and Practice-oriented learning implementation;
- Exemplary projects (course work, individual tasks), that do not require breakthrough ideas and creative thinking;
- A focus on knowledge-based education instead of competence-based approach;
- A low interest of employers in providing real tasks for industrial internships;
- A low level of interaction between university management, faculty, students and industry.

These constraints result in insufficient level of competences' formation, specifically students' practical professional skills and multidisciplinary view of professional problems. Therefore, university education has to be modernized in order to provide a more relevant training of specialists, which, in its turn, should increase the level of graduates' competitiveness in the labor market.

3 Purpose of the Research

The research conducted at Tomsk Polytechnic University (TPU) under the author's Ph. D. research project sets several topical questions:

- Are students interested in receiving double-major master level education?
- Do employers seek for graduates with two majors?
- What combinations of competences and majors are in demand by industry/students?
- Which competences are the most important for engineering graduates in the opinion of students, professors, employers?

The questions raised allow assuring communication between industry, higher professional education and students by identifying employers' requirements towards specialists and creating the basis for curriculum adaptation to rapidly changing technological and societal issues and needs.

Finding out professional fields, which are needed for multifaceted specialists' training, gives an opportunity for any engineering university to enhance the quality of education and its relevance.

On the other hand, the collected data on students' interests in double majors allows designing such curriculum that is demanded by industry and, at the same time, is competitive among master programs' enrollees—new generation of engineers. This aspect of research provides universities with higher chances of attracting ambitious and prospective students.

This initial stage of the research and its results are to serve as a basis for developing a unique mechanism and algorithms for managing curriculum design modernization.

The second stage of the project implies the development of a specific mechanism for harmonization of two master level programs of different majors that will be carried out simultaneously and prepare students for multidisciplinary professional activity.

The harmonized master programs [19] are aimed at providing students with two master level educations, fostering multidisciplinary skills, and allowing gaining extra knowledge and skills within a shortened period of study. The features of the harmonized master programs are as follows:

- There will be an enrollment fee for the second master program (though it will be twice less than the normal fee for a master program);
- The amount of credits required by the second master program will be decreased by half (i.e. 60 ECTS instead of the regular 120 ECTS for a master program) due to the mutual recognition[1] of the multidisciplinary courses and project research work (60 ECTS).

The proposed modernization of master level engineering education gives any university a unique selling point, enhances its competitiveness, attracts motivated and prospective students and assures higher rates of graduates' professional employment.

---

[1] The credits to be transferred between programs include credits on similar courses and credits on courses and internships that develop similar professional competences and multidisciplinary skills (such courses are required to have a unified project to assure credits' transfer).

4 Approach

The research focuses on identifying multidisciplinary master degree majors that interest potential employers and students in Russia.

The initial phase of the research focuses on collecting raw data from two main categories of respondents: potential employers (industry representatives; TPU graduates, who currently work in the engineering field) and potential master students (4th year bachelor students of TPU). The current number of respondents is 205 TPU students of engineering bachelor programs, 37 employers. It is anticipated to collect questionnaires from another 120 employers by the end of 2018.

The respondents have been asked to fill out corresponding types of questionnaires.

Students' questionnaires have been spread out personally by the author during university lectures. The project has been presented to each student group, identifying the features of the harmonized master programs. Potential master students have been asked to state, whether they would like to receive second-cycle education and whether they would be interested in enrolling for two different fields of study (harmonized master programs). The respondents were asked to rate the additional (complimentary) fields of knowledge that could be harmonized with their engineering majors. Students were also asked to name desirable pairs of majors (for instance, a preferred major and an additional major, which, in their opinion, would give them competitive advantages; an open choice question).

Employers have been asked, whether they consider graduates with two degrees (i.e. two master degrees in different fields of study) to be more competitive than their potential runners-up with a single major degree. Then employers were to propose a list of pairs of professional majors, which, in their opinion, were in-demand by the industry and their enterprises (open choice questions). The research is on-going; the final results should be collected by the end of 2018.

The final part of this survey included identification of the in-demand groups of competences, required for all engineering students (a weight ratio of three groups of competences), as well as a weight ratio of students' competences within each group.

Research results are expected to change the educational practice at TPU and in Russia in terms of providing background for developing harmonized master programs.

5 Research Results

The outcomes of the initial phase of the research are based on the questionnaires for potential employers and future master students, and concern the topicality of providing master programs in more than one major.

Among the outcomes of the initial phase of the research are the following:

1. Lists of in-demand pairs of majors:
   (a) Determined by potential master students,
   (b) Determined by potential employers.
       These lists will serve as a scientific basis for choosing most relevant interacting fields of study for multidisciplinary master level education.

2. A list of in-demand competences of students (ratio). This list will be presented in a separate article and is to be used to initiate curriculum restructuring in order to put more emphasis on developing relevant and up-to-date skills and attitudes of students.

5.1   Assessment of Students' Demand Towards Harmonized Master Programs

The questionnaire for students had been disseminated among 205 bachelor students of the 4th year, who receive education in the field of engineering and technology in Tomsk Polytechnic University. Questionnaires have been concisely presented by the author during lecture classes and spread out to the students for individual completion.

TPU provides master programs in 16 integrated groups of engineering specialties [20] (each of them consists of 1–5 different majors grouped by an engineering area of knowledge). Under the research the assessments of students from 12 integrated groups of specialties have been collected, including 20 different majors, which provided for a holistic view on TPU students' opinion.

The results of the first part of students' questionnaire are presented in the Fig. 1 and indicate that among 205 students 178 respondents (86.8%) are planning to continue education at master's level, and only 27 students (13.2%) are going to finish their education, when graduating from bachelor programs.

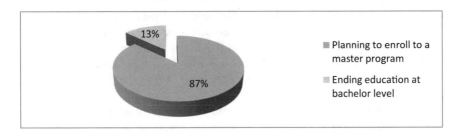

Fig. 1. Students' desire to continue education on master's level

Among those, who plan to enroll to master programs, 68.5% showed interest in applying for two harmonized master programs in order to receive multidisciplinary education, whereas only 31.5% rejected the proposal, mostly due to the fact that the second major program is to be non-budgetary (Fig. 2). Among those, who are not planning to continue their education, 30% of the respondents still perceive the project as worth attending.

The results of this part of questionnaire show that, overall, students are eager to receive multidisciplinary education, even though this would require extra payments and extra studying.

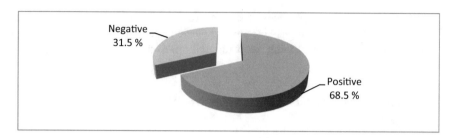

Fig. 2. Desire to enroll to harmonized master programs among students, who plan to continue education on master's level

Students rated 5 fields of knowledge to be paired with engineering majors in order to identify the most popular combination. According to their responds, the ratio is as follows (1—the most popular choice of pair):

1. Engineering majors;
2. IT majors;
3. Economic and/or managerial majors;
4. Majors in Natural Sciences;
5. Majors in Social Sciences and Humanities.

These results are quite predictable, since the overall trend of higher education and societal development focuses on the importance of IT and engineering specialists and provides better career development tracks for them. Students realize that having a good engineering major might not be enough to be highly competitive in the labor market, whereas having extra competences in a different area of engineering knowledge can give them a unique selling point. The same is true for complying engineering competences with IT skills, which might be an even bigger advantage for some engineering majors.

However, when the same respondents were asked to name the pairs of majors, which, in their opinion are the most relevant for employers in modern society, the results were slightly different. There were 257 pairs named by the respondents. They were later grouped according to their relation to the mentioned above fields of knowledge. The results are presented in the Fig. 3.

As can be seen from the mentioned above rating of the most popular fields of knowledge to pair with engineering majors and the Fig. 3, the latter evaluation sets the pair of "Engineering + Economic" majors on the first place, whereas the "Engineering + IT" exchanges its positions with the first pair and moves to the third place. This could be due to the fact that students conducted rating of coupling fields of knowledge with the view of their own majors, however, when providing examples of in-demand pairs they thought of overall pairs of major.

It should be noted that the pair of "Engineering + Economic" majors brings together examples of harmonizing the majority of engineering majors provided by TPU with the majors in Economy or Management. The top combinations include harmonization of economic and/or managerial majors with such engineering majors, as Electrical Engineering, Petroleum Engineering, Nuclear Physics, and others (Fig. 4).

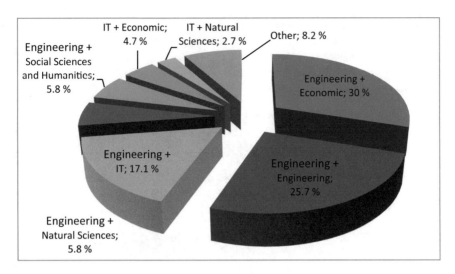

Fig. 3. In-demand pairs of fields of knowledge as seen by students

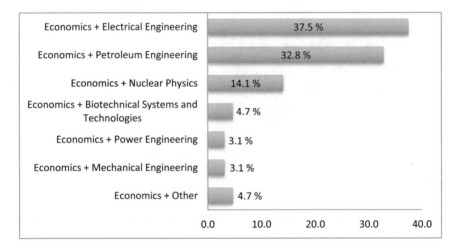

Fig. 4. The most demanded pairs of "Engineering + Economic" majors as seen by students

At the same time, the pair of "Engineering + Engineering" majors includes various combinations of areas of engineering knowledge, which are unified in Fig. 3 for the purpose of the research. Figure 5 provides evaluation of the pairs of engineering majors most commonly proposed by students as the demanded pairs of majors.

As of the pair "Engineering + IT", students see the following engineering majors to be the most demanded by industry if paired with IT majors (Fig. 6):

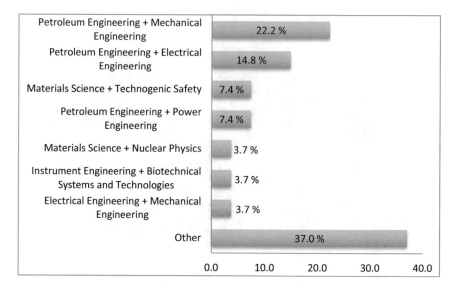

Fig. 5. The most demanded pairs of "Engineering + Engineering" majors as seen by students

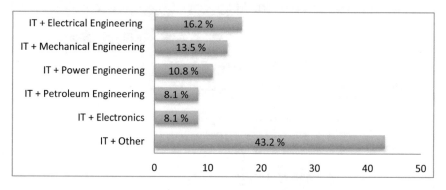

Fig. 6. The most demanded pairs of "Engineering + IT" majors as seen by students

Among the engineering majors proposed to be paired with majors in Natural Sciences are the following: Materials Science, Petroleum Engineering, Electrical Engineering, Mechatronics and Robotics, Nuclear Physics, Optical Engineering, etc. The Natural Sciences in the context of this study include such majors as Physics, Chemistry, Biology, Geology, Ecology, as well as Applied Mathematics.

IT, Electrical Engineering and Nuclear Physics are the most popular choice to be grouped with Social Sciences and Humanities. However, students, when providing popular pairs of majors, in the majority of cases implied Linguistics as Social Sciences.

Overall, the top pairs of majors for harmonization as indicated by students are:

1. Electrical Engineering and Economics/Management,
2. Petroleum Engineering and Economics/Management,

3. Mechanical Engineering and Petroleum Engineering,
4. IT and Economics/Management,
5. Electrical Engineering and Petroleum Engineering,
6. Nuclear Physics and Economics/Management.

The results of students' questionnaires are essential for the ongoing project, since students are one of the two main stakeholders of university education (the other stakeholder is industry, represented by employers). Both of them are the consumers of university's products: students are the consumers of the first level and consume the knowledge and skills provided; employers are the consumers of the second level and consume the knowledge and skills of universities' graduates. Therefore, students' opinions and desires have to be taken into account when developing educational products at the initial part of the products' marketing.

5.2 Assessment of Employers' Demand Towards Harmonized Master Programs

The survey includes responds from 37 employers. The survey is ongoing and will continue until September, 2018. It is expected to gain the responds from at least 150 employers. Among the respondents are heads of industrial companies, engineering officers, workshop supervisors from different regions of Russia and other employees, who are responsible for the work of industrial engineers, as well as their employment.

The results of the first part of the employers' questionnaire are presented in the Fig. 7 and indicate that among 37 employers 27 respondents (72%) are looking for university graduates with extra competences, preferably in two or more areas of knowledge, and 10 employers (28%) do not have a current need in specialists with two majors.

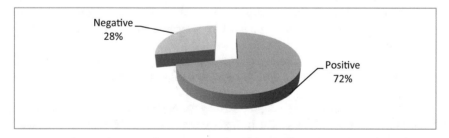

Fig. 7. Desire to employ graduates of harmonized master programs

The respondents were asked to propose pairs of majors that would give competitive advantages to graduates, when applying for an engineering job. Current results indicate that the most common choice of the employers is the pair of "Engineering + IT" majors (Fig. 8). The "Engineering + Engineering" pair received a slightly lower interest from the employers.

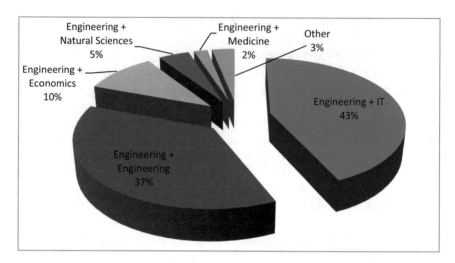

Fig. 8. In-demand pairs of fields of knowledge as seen by employers

In contrast to the responds of the students (Fig. 3), the pair of "Engineering + Economic" majors has moved to the third place with a considerable gap between itself and the above mentioned pairs. This inconsistency between the responds of the students and the employers can be due to the fact that employers, first of all, seek for graduates able to solve technical and technological problems, rather than for those, who can, for instance, provide economical evaluations or be able to manage personnel. The need for engineers with profound managerial skills or knowledge in economics comes on later stages of employment, but not right after graduation.

The number of pairs "Engineering + Natural Sciences" proposed by the employers is quite low, most likely, due to the fact that employers expect higher engineering education to provide profound training in terms of physics, chemistry, mathematics. Therefore, they believe that there is no need for a separate major in Natural Sciences to be undertaken.

In the rise of medical technological development, a part of the employers underlines the importance of training specialists with background in both Engineering and Medicine.

Particular examples of the pairs proposed by the employers, as well as the rating of the demanded pairs cannot be exhaustive, since the current number of the respondents is not yet sufficient. Once the final results of the questionnaires are gained, the top in-demand pairs of majors as outlined by the employers will be provided.

5.3   The Forthcoming Research

A rating of most important reasons for choosing to study two majors is to be conducted among graduating bachelor students. This stage of research will be executed through an on-line questionnaire for those participants of the initial research stage, who have shown interest in applying for harmonized master programs in their fields of interest.

This information will provide ideas for managing the organization and promotion of harmonized master programs.

Students' and employers' questionnaire results will serve as the basis for a data base of the harmonized major pairs; however, they will only provide the basic idea on the most demanded pairs. It is recommended for a university or an educational program developer to conduct a similar research with the use of the provided questionnaires among a particular range of bachelor students of the major under study and of the employers, who recruit graduates of this major, in order to gain a sufficient and narrow focused database of pairs of majors in demand.

The second phase of the research includes developing an algorithm for harmonization of master programs from two different fields of study in such a way that students receive second major in a shorter period of time and with less labor and financial inputs due to the transfer of credits between programs. The choice of majors to be harmonized will be based on the results of the first phase of the research.

6 Conclusions and Recommendations

Considering the fact that multidisciplinarity is not only a trend in higher professional education but is also a requirement for high quality education and in-time development of industry, the research conducted at TPU provides ideas and propositions on how to structure multidisciplinary education in a new and efficient way.

Through this research industry receives an opportunity to influence universities' choice of educational programs and to propose unique combinations of competences, skills and knowledge that are in high demand by industry, but have not been taken into consideration by HEIs yet.

The development of an algorithm for harmonization of in-demand master programs provides for training foremost specialists required by industry in a shorter period of time with cost-efficiency for students and for university due to the credit transfer system.

Such modernization of master level engineering education gives an HEI a unique selling point, enhances its competitiveness, attracts motivated and prospective students and assures higher rates of graduates' employment.

References

1. Irani, Z.: The university of the future will be interdisciplinary. The Guardian, Higher Education Network (2018)
2. Neeley, T.: Global teams that work. Harvard Bus. Rev. **93**(10), 74–81 (2015)
3. Woolf, D.: The future is interdisciplinary. Universities Canada (2017)
4. Puumalainen, K.: Interdisciplinarity & sustainability at Lappeenranta University of Technology, https://www.abis-global.org/content/documents/2016/11/abis-lut.pdf
5. University of Amsterdam. Institute for Interdisciplinary Studies. http://iis.uva.nl/
6. University of Glasgow. School of Interdisciplinary Studies. https://www.gla.ac.uk/schools/interdisciplinary/
7. University of Pennsylvania. Interdisciplinary. https://www.upenn.edu/programs/interschool

8. d'Hainaut, L.: Interdisciplinarity in General Education. UNESCO, Paris (1986)
9. Newell, W.H.: A theory of Interdisciplinary Studies. Issues Integr. Stud. **19**, 1–25 (2001)
10. Pokholkov, YuP: Engineers for interdisciplinary teams and projects: management of training process. Eng. Educ. **20**, 23–32 (2016)
11. Rogers, Y., Scaife, M., Rizzo, A.: Interdisciplinarity: an emergent or engineered process? In: Derry, S.J., Schunn, C.D., Gernsbacher, M.A. (eds.) Interdisciplinary Collaboration. LEA, Mahwah, New Jersey (2005)
12. Scientists and engineers statistical data system. surveys 2013. https://ncsesdata.nsf.gov/us-workforce/2013/
13. Pitt, R., Tepper, S.: Double Majors: Influences, Identities, and Impacts. Teagle Foundation, New York (2012)
14. European Higher Education Area and Bologna Process. http://www.ehea.info/
15. Ogorodova, L.M., Kress, V.M., Pokholkov, YuP: Engineering education and engineering in Russia: problems and solutions. Eng. Educ. **11**, 18–23 (2012)
16. Pokholkov, YuP, Gerasimov, S.I.: Training engineering workforce demanded by labor market (In Russian). Transportnaya Strategiya - XXI vek. **33**, 68–69 (2016)
17. Pokholkov, YuP: Quality of engineers' training as seen by academic community (In Russian). Eng. Educ. **15**, 18–25 (2014)
18. Pokholkov, YuP., Rozhkova, S.V., Tolkacheva K.K.: Practice-oriented educational technologies for training engineers. In: Interactive Collaborative Learning: Proceedings of 16th International Conference, pp. 619–620. KNSTU, Kazan (2013)
19. Chervach, M.Y., Pokholkov, YuP., Zaytseva, K.K.: Harmonized master programs for fostering extra in-demand competences. In: 11th International Technology, Education and Development Conference—INTED2017: Proceedings, pp. 7204–7208. IATED, Barcelona (2017)
20. National Research Tomsk Polytechnic University. http://tpu.ru

Multipurpose Urban Sensing
Equipment—An EPS@ISEP 2018 Project

Mostafa Farrag[1], Damien Marques[1], Maria Bagiami[1], Maarten van der Most[1],
Wouter Smit[1], Benedita Malheiro[1,2], Cristina Ribeiro[1,3], Jorge Justo[1],
Manuel F. Silva[1,2(✉)], Paulo Ferreira[1], and Pedro Guedes[1]

[1] ISEP/Porto—School of Engineering, Porto Polytechnic, Porto, Portugal
epsatisep@gmail.com
http://www.eps2018-wiki3.dee.isep.ipp.pt/
[2] INESC TEC, Porto, Portugal
[3] INEB – Instituto de Engenharia Biomédica, Porto, Portugal

Abstract. This paper describes the development of a Multi-purpose
Urban Sensing Equipment, named Billy, designed by a multinational
and multidisciplinary team enrolled in the European Project Semester
(EPS) at Instituto Superior de Engenharia do Porto (ISEP). The project
is set to design, develop and test an interactive billboard in compliance
with the relevant EU regulation and the allocated budget. The Team
benefited from the different background, multidisciplinary skills and the
newly acquired skills of the members, like marketing, sustainability and
design ethics, in activities both inside and outside of the University. The
challenge was to design a multi-purpose urban sensing and displaying
equipment to inform citizens of nearby environmental conditions. The
Team decided to design a system to monitor and display the temperature,
humidity, air pressure and air quality of leisure areas, featured with a
proximity detection sensor for energy saving. Billy will not only monitor
and display this local information, but also the air quality determined by
other billboards placed in other locations, creating a distributed urban
sensing network. The system has been successfully prototyped and tested
using the ESPduino Wi-Fi enabled micro-controller, different sensors and
displays (screen and map-based). The results show not only that the
prototype functions according to derived specifications and design, but
that the team members were able to learn, together and from each other,
how to solve this multidisciplinary problem.

1 Introduction

EPS challenges students from multiple educational backgrounds and nationalities to join their proficiencies [5] to solve multidisciplinary real-life problems
in close collaboration. On the 26th of February 2018, ISEP assembled its EPS
students into four groups, including the current team composed of Maarten van
der Most, from Netherlands, 22 years old, studying Industrial Engineering and
Management; Wouter Smit, also from Netherlands, 26 years old, studying Industrial Design; Damien Cordeiro Marques, from France, 20 years old, studying

© Springer Nature Switzerland AG 2019
M. E. Auer and T. Tsiatsos (Eds.): ICL 2018, AISC 917, pp. 415–427, 2019.
https://doi.org/10.1007/978-3-030-11935-5_40

Mechanical Engineering; Maria Bagiami, from Greece, 23, studying Environmental Engineering; and Mostafa Farrag, from Scotland, 20, studying Electrical Power Engineering. The Team chose to develop an interactive billboard to monitor and display the temperature, humidity, air pressure and air quality of leisure areas. By working together in a multidisciplinary team, the Team members had the opportunity to learn from each other and collectively, and achieve further than they would have individually.

Poor air quality has a negative impact on the quality of life. It causes many health issues, such as breathing or cardiovascular problems. These issues are even more critical in urban areas where there is often a poor air quality as a result of modern way of life [2]. The air in cities is polluted with small harmful particles. However, people are not fully aware of the actual level of air pollution. Although there are smart phone applications which provide related information, there is still a great lack of knowledge about this topic, as there is no system or object providing real and trusted local information. This is the Team's vision regarding the design of a smart billboard: a trusted equipment which informs and empowers people knowledge on how to improve air quality. This will not only benefit the public, but also governments by helping them to comply with the European Union (EU) rules [3].

In order to contribute to the minimisation of these problems, the team decided to build a connected billboard to provide citizens with real time air-related data concerning current and remote locations. The billboard was designed to measure local temperature, humidity and air pressure, estimate the local air quality, show the estimated air quality of the connected remote locations, display useful information and advice on how to improve air quality. The final objective of Billy is to raise the public awareness by offering as much information and advice as possible and, thus, contribute to the improvement of the air quality and the reduction of health related problems.

The team searched for similar equipments and performed several studies on marketing, ethics and sustainability to design Billy as a smart billboard. Billy not only collects and shares information about the current local time, weather and air pollution, but informs on how to reduce the carbon footprint and on upcoming local activities. To design and develop Billy, the team had to integrate what they learned in their individual field of study. Billy main potential clients will be governmental bodies and local public authorities.

This paper is organised in six sections, including this introductory section. Section 2 describes existing types of billboards; Sect. 3 details the background studies performed; Sect. 4 presents the design and development of the prototype; Sect. 5 reports the tests and results; and Sect. 6 draws the conclusions.

2   Billboards

A billboard is a large outdoor advertising structure (a billing board), initially found in high-traffic areas such as alongside busy roads and presenting large advertisements to pedestrians and drivers and distinctive visuals, billboards are highly visible in the top designated market areas. Typically showing witty slogans, billboard advertisements are designed to catch public attention and create a memorable impression very quickly, leaving a lasting impression on readers. They have to be short and easy to read because cars can drive by at high speed. This section presents the different purposes of billboards, leading to the team's decision to create Billy.

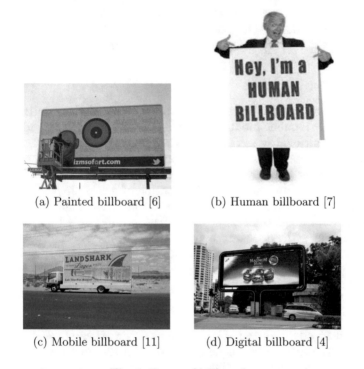

(a) Painted billboard [6]   (b) Human billboard [7]

(c) Mobile billboard [11]   (d) Digital billboard [4]

Fig. 1. Types of billboards

Figure 1 illustrates the most common types of billboards:

- Painted billboards: This traditional billboard displays a painted message or advertisement.
- Human billboards: The human billboard is a person who displays an advertisement. The person can just wear a T-shirt with a message, carry a small billboard or also "wear" the billboard. Frequently, the person will spin, dance or wear costumes with the promotional sign to attract attention.
- Mobile billboards: A mobile billboard, also known as "truck side advertising", is used for advertising on the side of a truck or trailer. Unlike a typical billboard, mobile billboards are able to go meet their target audience.
- Digital billboards: A digital billboard is a billboard that displays changing digital images. Imagery and text are created using computer programs and

software. Digital billboards are primarily used for advertising, but they can also be adopted by public services.

3   Background Studies

The team decided to create Billy, a smart billboard, to collect and share publicly information on the weather and pollution, using IoT cloud. For sustainability reasons, the team chose to use wood as building material and a solar panel as renewable energy source. This section will present different background studies conducted by the team to define the proposed solution.

3.1   Marketing

Since the European Union has set the maximum allowed air pollution, the team decided to create a public product to display the air pollution and advise on how to improve the air quality in their area. Consequently, the team chose to target city councils and offer a multipurpose urban sensing equipment. Billy stands out from its competitor because it provides: (i) information about local and remote air pollution, indicating the spatial distribution of urban air pollution; and (ii) advice on how an individual can improve the local air quality. The team has decided to use a differentiation strategy. The product will be promoted with direct marketing to the decision making entities of the city councils in Europe as a solution to measure air pollution, informing inhabitants with local information [10] on air pollution and how to mitigate it [1].

3.2   Eco-efficiency Measures for Sustainability

Engineers must adopt sustainable development practices when designing new products. Thus, the team took equally into account the environmental, economic and social aspects of sustainability in order to design a sustainable billboard. Based on this study, the team chose to use environmental friendly materials, *e.g.*, natural materials or materials which can easily be recycled and are not harmful to the environment. The team opted to display the air quality for social reasons and to use a power supply unit composed of a solar panel and battery for environmental and economic reasons. Moreover, by informing the public on how to protect against air pollution and on how to improve the air quality, the team increases the efficiency of billboard, making it even more sustainable [9].

3.3   Ethical and Deontological Concerns

In the ethical and deontological analysis, the team considered the production, servicing and recycling of Billy. The focus of the company will be on honesty, respect and high standards in every step of the production process, including choosing suppliers and components, selling and marketing the product and providing a warranty of two years, even if it increases costs. In the case of breakdown, the company will collect Billy from the customer to reuse or recycle the parts. Team Billy will not use suppliers for whom it is unsure if they use child labour. In terms of environmental ethics, the decision to adopt a solar panel and a battery reduces the usage of fossil fuels.

4   Project Development

4.1   Pre-defined Requirements

The project proposal specified the mandatory use of the International System of Units, sustainable materials, low-cost hardware solutions, open source software and technologies, the compliance with the European Union (EU) Machine Directive (2006/42/CE 2006-05-17), EU Electromagnetic Compatibility Directive (2004/108/EC 2004 12 15), EU Low Voltage Directive (2014/35/EU 2016-04-20), EU Radio Equipment Directive (2014/53/EU 2014-04-16), EU Restriction of Hazardous Substances (ROHS) in Electrical and EU Electronic Equipment Directive (2002/95/EC 2003-01-27), and a maximum budget of 100.00 €.

4.2   Functionalities

The main functionality of the billboard is to publicly display the temperature, humidity, air pressure and the air pollution. Billy is designed to sense these parameters, process the readings and display the results individually.

The map-based display is colour coded to indicate low, normal and high values of the air quality, using green and red light-emitting diodes (LED). The map of the enlarged urban area is dotted with green, amber and red LED over the locations of the billboards deployed in the urban area. It additionally displays complementary information about the air quality, health and safety, and public activities to assist citizens in keeping a high standard life style. Finally, Billy was designed to be echo-friendly as it is solar powered. It includes a photovoltaic system and a battery to sustain its operation day and night.

4.3   Design Structure

The structure of the product, which was drawn in SolidWorks, was divided into the top, middle and bottom part to make it easier to develop and assembly (Fig. 2).

The top part holds the sustainable energy source—the solar panels—which powers the system. These solar panels are wired to the power unit through the hollow stand of the billboard. This part has several requirements: (i) be waterproof; (ii) be robust and vandal proof; (iii) be easy to assemble and disassemble for maintenance; (iv) be easy to recycle; (v) protect the solar panels; and (vi) be easy to clean. To strengthen the structure, an additional piece was developed and attached to the top part and to the stand of the billboard. It makes the structure more robust and vandal proof and contributes to an attractive design.

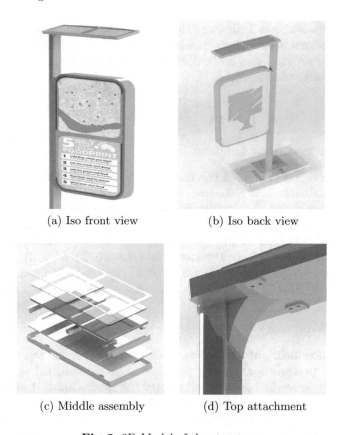

(a) Iso front view   (b) Iso back view

(c) Middle assembly   (d) Top attachment

Fig. 2. 3D Model of the structure

The middle part—the core of the billboard—has to: (*i*) be attached to the stand of the billboard; (*ii*) be watertight; (*iii*) have ventilation to make sure the warmed up air can escape; (*iv*) be dust proof (on the inside); (*v*) be easy to open, close and dissembled for maintenance; (*vi*) house the control system, the map-based display with the LED indicators and the text-based display to inform and advise people, e.g., on how to reduce their ecological footprint.

The bottom part or stand is intended to hold the other parts. The bottom part has a few important requirements: (*i*) be robust to hold the complete structure; (*ii*) be vandal proof; (*iii*) be adjustable to the deployment area (to position it as well as possible); (*iv*) be fitted as different parts to build the product up at the location where it should be installed.

4.4 Control System

The billboard is equipped with a control system to implement the identified functionalities. The power unit includes a photovoltaic solar panel for power supply and a battery for energy saving. The ESPduino micro-controller, which includes a Wi-Fi module, is responsible for reading, communicating and displaying the local sensor values in the organic light-emitting diode (OLED) display and on the map-based LED display as well as for obtaining and displaying the remote air quality on the map-based LED display. While the proximity detection sensor and gas sensor are directly connected to the ESPduino, the humidity, temperature and air pressure sensors are connected through a signal conditioner to adapt the sensors signals to the micro-controller input requirements. The billboard is connected to an Internet of Things (IoT) cloud platform (ThingSpeak) to save and share information through the public Wi-Fi network. The values are displayed on the LED display, using different colours to indicate the air quality status. Figure 3 displays Billy's conceptual diagram.

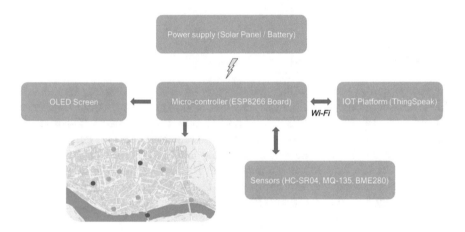

Fig. 3. Conceptual diagram

Table 1 holds the maximum expected voltage, current and power consumption per component. Considering that the prototype will display the air quality of up to two locations (2 sets of red and green LED), it will need a power source capable of providing a current higher than 3.885 A and a power of at least 14,076 W. Based on this analysis, the team selected of a power source with 4.0 A and 27 W.

The control algorithm starts by initialising the parameters and configuring the micro-controller and, then, implements the control loop. This continuous loop performs the following tasks: (*i*) sensor data acquisition and display (OLED); (*ii*) uploading of local air quality to ThingSpeak; (*iii*) downloading remote air quality (remote sites) from ThingSpeak; and (*iv*) map-based air quality display.

The MQ135 measures the concentration of CO_2, NH_3, NO_x, benzene and alcohol in parts per million (ppm). According to [8], if the concentration value is below 500 ppm, the air quality is good and, otherwise, the air quality is bad. Consequently, the green LED illuminates when the concentration value is below 500 ppm and, otherwise, the red light illuminates.

Table 1. Power budget

Component	Voltage (V)	Current (A)	Power (W)
Green LED	1.8–2.2	0.040	0.088
White LED	2.4–3.6	0.040	0.144
Red LED	1.8–2.7	0.040	0.108
MQ-135	5	0.150	0.750
BME280	1.2–3.6	0.000	0.000
Solar Panel	5	0.600	0.300
ESP8266 board	5	1.000	5.000
HC-SR04	5	0.015	0.075
OLED	3.3–5.0	0.020	0.100

5   Tests and Results

The experimental set-up, which was assembled in a breadboard, includes the ESPduino micro-controller, the gas sensor (MQ-135), the humidity, temperature and air pressure sensor (BME280), the ultrasonic range sensor (HC-SR04), the OLED screen and a set of two of each red, white and green LED. Figure 4 displays the experimental set-up. The sensors were connected to the ESPduino (Arduino) as follows: (*i*) the MQ-135 is connected to A0; (*ii*) the OLED, HC-SR04 and BME280 are connected to the Arduino via I2C interface; and (*iii*) the LED are directly plugged to the Arduino GPIO interface.

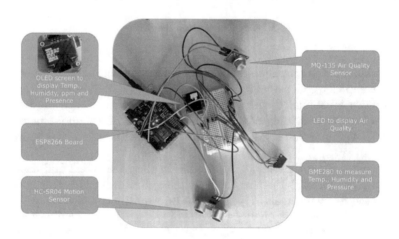

Fig. 4. Experimental set-up

Figure 5 displays the assembled prototype. It includes the main map-based LED display, the OLED and the complementary information panel bellow, advis-

Fig. 5. Prototype

ing people how to reduce their footprint. The different LED report the air quality of the different areas (green for good and red for bad). The OLED provides the following information: (*i*) the Internet Protocol (IP) address of the device connected to the Wi-Fi; (*ii*) the concentration of the measured gases in ppm; (*iii*) the temperature in °C; (*iv*) the relative humidity in %; (*v*) the pressure in hPa; and (*vi*) the output of the presence detection sensor.

5.1 Tests

The Team performed a set of functional tests to ascertain whether 'Billy' complied with the requirements and was ready to be transformed into a product for release in the market. The tests were performed as follows:

Temperature & Humidity Sensor: the sensor was placed near a cold source (cloth with ice) to lower the temperature and near a heat source (warm damp cloth) to raise the temperature and the humidity (warm damp cloth);

Air Quality Sensor: the sensor was placed near a gas source (kettle) to see if it detects gas and lights up the LED (green is low and red is high);

Proximity Detection Sensor: the sensor detected an object within the detection desired zone. First, in calibration mode, the sensor was adjusted to cover the desired distance and, then, was switched to regular mode;

Photovoltaic panel: the solar panel was placed outdoor and the voltage and current charging the battery monitored.

Table 2 summarises the functionalities tested. First, the OLED test involved sending and displaying "Hello World" on the screen. Second, the temperature, humidity and air pressure (BME280) test comprehended reading and displaying the three values on the screen. Last, the gas (MQ-135) test consisted of reading the values and lighting the four LED accordingly.

Table 2. Functionalities tested

Ref	Component	Test
DISP	OLED display	Draw strings and pictures
Wi-Fi	Wi-Fi on board	Connection to WLAN
BME	BME280 sensor	Temperature acquisition
SR04	HC-SR04 sensor	Motion detection
MQ135	MQ-135 sensor	Air quality measurement
LED	LED	Light the LED on
IOT	IoT cloud	Communication with ThingSpeak

5.2 Results

Table 3 displays the expected and achieved results of the functional tests. The expected results of each test were defined beforehand. The actual results matched the expected results, which means that the parts tested were working properly.

Table 3. Results of the functional tests

Part	Test	Expected result	Result
DISP.01	Draw a string	String displayed	Yes
DISP.02	Draw a picture	Picture displayed	Yes
Wi-Fi.01	Connect to Wi-Fi	@IP displayed	Yes
BME.01	Acquire temperature	Temperature displayed	Yes
BME.02	Warm the sensor	Temperature raise displayed	Yes
SR04.01	Acquire distance	Distance displayed	Yes
SR04.02	Establish proximity	System acti-vated/deactivated	Yes
MQ135.01	Acquire ppm values	ppm values displayed	Yes

continued

Table 1. continued

Part	Test	Expected result	Result
MQ135.02	Breath over	ppm values increase displayed	Yes
LED.01	Specify a colour code	Corresponding LED on	Yes
IOT.01	Send data toIoT Cloud	Value displayed on the dashboard	Yes
IOT.02	Get ppm values from IoT Cloud	Light LED accordingly	Yes
UAT.01	Connect to Wi-Fi	OLED displays @IP	Yes
UAT.02	Detect presence	OLED activation	Yes
UAT.03	Measure temperature	OLED displays temperature	Yes
UAT.04	Measure ppm	OLED displays ppm value	Yes
UAT.05	Send data to IoT Cloud	Data displayed on IoT dashboard	Yes
UAT.06	Display air quality of other bills	Download data and light LED	Yes

6   Conclusion

The goal of this project was twofold: (*i*) to design, develop and test a proof of concept prototype of Billy; and (*ii*) prepare engineering undergraduates for their future profession. The latter is achieved by providing a multicultural, multidisciplinary and collaborative learning experience where distinct visions of the problem need to be integrated to reach the final team solution.

Billy is a smart urban equipment intended to inform and raise the awareness of the public regarding the quality of urban air. In standalone mode, it displays the local temperature, humidity, air pressure and air quality as well as information and advice on how to improve the air quality and reduce related health problems. In addition, in network mode, it presents the air quality of connected locations. Currently, Billy meets most of the requirements set by the Team, and the members are on verge of concluding the development.

Collaborative learning is not always easy and, in a multicultural and multidisciplinary context, becomes even more demanding. The hardest challenge the Team faced was how to make the individual visions of the problem converge to a common solution since it was their first work and learning experience in a multicultural and multidisciplinary set-up. The Team members learned about themselves, from each other and together while participating in EPS@ISEP and staying in a new country, with its own culture and traditions. These are their testimonials:

- *"I'm grateful for having the opportunity to participate in the EPS@ISEP, it was an amazing experience to be in Porto learning many new things and meeting new people, while discovering Portugal and its culture. Within the EPS I have learned different information on different areas of studies. I'm glad working with team members who are willing to put 100% into their work to make our project to be standing out"*—Mostafa.
- *"EPS was for me a great opportunity to meet people from all over Europe and make friends and connections. EPS also enabled me to develop my skills and the best way to improve my English"*—Maria.
- *"The thing I liked the most about EPS was to learn to work with people from different nationalities and cultures. Before I did the project, I thought there were only minor differences between European cultures, but during the project I realised that every country has so many differences"*—Maarten.
- *"The European Project Semester was really good experience to learn about multiple cultures and their way of working, not only from Portugal, but also the different countries that your team members are from. Also, I loved to work on a project that is focussed on sustainable energy and materials and a solution for a problem that improves people's lives. It's a good structured project semester that has a lot to offer"*—Wouter.
- *"The European Project Semester at ISEP was a very rewarding experience. I'm used to work in team but it was the first time that I worked with people from different countries and backgrounds in a global project where we have not only to design an object but also think about marketing, management and sustainability. We are all coming from different cultures and working methods so we learned a lot from each other. I also learned different studies which I'm not specialised at such as sustainability and project management. This project also made me grow up, I'm now more tolerant and I trust more easily my team mates. As we used English to communicate, I'm now much more comfortable to speak this language than in the beginning. Even if I already knew some things about Portugal culture, history and language, working in Portugal give me the opportunity to increase my ability to speak this language and also to know new things thanks to some trips and visits"*—Damien.

Funding. This work was partially financed by the ERDF—European Regional Development Fund through the Operational Programme for Competitiveness and Internationalisation—COMPETE 2020 Programme within project POCI-01-0145-FEDER-006961, and by National Funds through the FCT—Fundação para a Ciência e a Tecnologia (Portuguese Foundation for Science and Technology) as part of project UID/EEA/50014/2013.

References

1. CleanTechnica. Best Air Pollution Startups—Breathing Easy, "State of Pollution" Series (2017). https://cleantechnica.com/2017/08/11/best-air-pollution-startups/. Accessed May 2018
2. European Council. The Clean Air Package: Improving Europe's Air Quality (2017). http://www.consilium.europa.eu/en/policies/clean-air/. Accessed May 2018
3. James Crisp: 23 Pays de l'UE Violent les Règles de Qualité de L'air (2017). https://www.euractiv.fr/section/climat/news/23-eu-countries-are-breaking-european-air-quality-laws/. Accessed May 2018
4. LEDtronics: LED Displays—Digital Billboards—Media Facade (2018). http://ledtronics.com.my/. Accessed May 2018
5. Malheiro, B., Silva, M., Ribeiro, M.C., Guedes, P., Ferreira, P.: The European Project Semester at ISEP: the challenge of educating global engineers. Eur. J. Eng. Educ. **40**(3), 328–346 (2015)
6. Izms of Art: Izms of Art Live Billboard Painting (2012). https://izmsofart.wordpress.com/2012/06/03/izms-of-art-live-billboard-painting/. Accessed May 2018
7. Pinterest: Human Billboard Advertising (2015). https://www.pinterest.co.uk/pin/571675746426128706/. Accessed May 2018
8. Shinde, P.E., Javeri, V.H.: Air Quality Monitoring System (2014). https://www.slideshare.net/Pravin1993/air-quality-monitoring-system. Accessed June 2018
9. PVthin—Thin Film PV Technology. Life Cycle Analysis (2018). http://pvthin.org/life-cycle-analysis. Accessed May 2018
10. Time: This Billboard Sucks Pollution from the Sky and Returns Purified Air (2014). http://time.com/84013/this-billboard-sucks-pollution-from-the-sky-and-returns-purified-air/. Accessed May 2018
11. Truckbillboards. Truck Advertising on Billboards Long Island NY (2018). https://truckbillboards.wordpress.com/tag/mobile-billboards/. Accessed May 2018

Vertical Farming—An EPS@ISEP 2018 Project

Anastasia Sevastiadou[1], Andres Luts[1], Audrey Pretot[1], Mile Trendafiloski[1],
Rodrigo Basurto[1], Szymon Błaszczyk[1], Benedita Malheiro[1,2],
Cristina Ribeiro[1,3], Jorge Justo[1], Manuel F. Silva[1,2(✉)], Paulo Ferreira[1],
and Pedro Guedes[1]

[1] ISEP/PPorto—School of Engineering, Porto Polytechnic, Porto, Portugal
epsatisep@gmail.com
http://www.eps2018-wiki1.dee.isep.ipp.pt/
[2] INESC TEC, Porto, Portugal
[3] INEB – Instituto de Engenharia Biomédica, Porto, Portugal

Abstract. This paper summarises the joint efforts of a multinational
group of six undergraduate students cooperating within the European
Project Semester (EPS) conducted at the Instituto Superior de Engenharia do Porto (ISEP). The EPS@ISEP initiative, made available as
a part of the Erasmus+ international students exchange programme,
employs the principles of problem-based learning, facing students with—
albeit downscaled—real-life scenarios and tasks they may encounter in
their future professional practice. Participation in the project initiative
outclasses most of the traditional courses through a wide spawn of its
learning outcomes. Participants acquire not only hard skills necessary for
an appropriate execution of the project, but also broaden their understanding of the approached problem through detailed scientific, management, marketing, sustainability, and ethics analysis—all in the atmosphere of multicultural and interdisciplinary collaboration. The team
under consideration, based on personal preferences and predispositions,
chose the topic of vertical farming and, in particular, to design a domestic indoor gardening solution, appropriate for space efficient incubation
of plants. The paper portrays the process, from research, analysis, formulation of the idea to the design, development and testing of a minimum
viable proof of concept prototype of the "Vereatable" solution.

Keywords: Collaborative learning · Project based learning ·
Technology · Education · European Project Semester · Vertical
farming · Aeroponics

1 Introduction

The European Project Semester (EPS) initiative is currently implemented at 19
universities, scattered across 12 different European countries [6]. The programme
is governed by the idea of facing the challenges of today's world economy and

© Springer Nature Switzerland AG 2019
M. E. Auer and T. Tsiatsos (Eds.): ICL 2018, AISC 917, pp. 428–438, 2019.
https://doi.org/10.1007/978-3-030-11935-5_41

job market, where engineers will often double as entrepreneurs and work in small teams of many specialists of various professions. While being tailored for undergraduate engineering students at the 3rd or the 4th year of their degree, the project is open to any student capable of a meaningful contribution to the work. The spring semester of the academic year 2017/2018 at the Instituto Superior de Engenharia do Porto (ISEP), Portugal, had the participation of 21 students. The teams were assembled according to the team worker profile (Belbin test), the nationality and field of study of the participants, aiming at the most optimal mix of nationalities, fields of studies, and predisposed teamwork functions. Following the guidelines of the initiative regarding the optimal team size, there were three groups of five and a single group of six students. This paper focuses on the so-called SAMARA team (acronym formed by the initials of the member's first names). Table 1 presents the composition of SAMARA, showing an appropriate mix of nationalities, with a slight bias towards electrical-oriented fields.

Table 1. Team SAMARA of EPS@ISEP 2018

Name	Country	Belbin team role	Field of studies
Anastasia Sevastiadou	Greece	Monitor evaluator	Env. & Geotechnical Eng.
Andres Luts	Estonia	Resource investigator	Electrical Eng.
Audrey Pretot	France	Complete finisher	Packaging Eng.
Mile Trendafiloski	Macedonia	Complete finisher	Comp. Science & Eng.
Rodrigo Basurto	Spain	Implementer	Mechanical Eng.
Szymon Błaszczyk	Poland	Implementer	Telec. & Comp. Science

In the first week of the programme, the teams were presented with a wide selection of possible topics to consider. Out of them, the SAMARA team, taking into consideration every member's preferences, motivation, skill set, and personal objectives, chose Vertical Farming. The SAMARA team recognised that there were multiple phenomena and social tendencies contributing to the importance of this subject. As societies become steadily more industrialised and people agglomerate in increasingly larger cities, there is a general will to reconnect with nature. The key aspect is an overall concern with the quality and purity of food—with even key fast-food market players introducing "healthy alternatives" to their core menu over the past decade. Many people are now actively avoiding ingredients and additions they believe unhealthy—and although the debate over some processes, *e.g.*, genetically modified organisms (GMO), is still ongoing without a definite conclusion [12], other practices, namely overuse of toxic pesticides in large field farming, is rightly perceived as alarming. Abuse of health standards is not the only problem faced by farming though. With the steady growth of human population, rises the demand for both food supply and the living space area. Conventional crop fields have however a tightly limited efficiency of the acreage use—and to provide more food, they require more space. Moreover, these fields are exposed to environmental threats, vermin and natural disasters alike. A single flood or drought can put at risk the well-being of a huge community. All above factors call for a transfer of our crops from—although considered beautiful by many—ineffective fields to a more controllable environment, where some risks can be eliminated and dedicated structures can be employed in order to utilise the third, vertical dimension, multiplying the spatial efficiency of farming. To top that, several sources claim that vertical indoor farms only use as little as 5–10% of water when compared with traditional means [2,11]. Yet still, simply moving the mass scale food production indoors and granting it one more dimension does not answer all concerns the SAMARA team has identified—nor it satisfies the team's set of goals for the EPS participation. Another aspect to be taken into consideration is the growing need and will to stay in touch with nature. By bringing a user-friendly farming solution directly to households, where herbs, minor fruits and vegetables like berries, lettuces, and tomatoes can be grown, not only daily contact with nature is guaranteed, but the aforementioned concerns regarding quality and healthiness of food are reassured, when each step of the food's growth can be observed and controlled in person. Hence, the SAMARA team decided to channel efforts into proposing a viable end-user consumer product, incorporating vertical farming solutions into households. Such a device would offer the user a steady supply of fresh and healthy food directly, supporting and encouraging good eating habits.

This paper presents the team's work and project outcomes organised in six sections: introduction, background analysis, complementary studies, prototype development, prototype testing and conclusions.

2   Background Analysis

Accessibility to fresh food is already a problem, expected to worsen with the growing population. It is vital for humankind to find more sustainable and environmental friendly solutions. With vertical farms, the required volume of water and land surface decreases dramatically. The following analysis helped the team designing a solution for common people and build a proof of concept prototype.

There are many vertical farming types and technologies in the market, ranging from simple soil based solutions to complex multi-level hydro-aeroponics. Typically, they provide basic seed pods, which the buyer can keep or substitute. Some products also provide a mobile application to help the user to tender for the crops. Examples of such products are:

Minigarden Vertical is a solution originating from Lisbon, Portugal [9]. The concept is an affordable, straightforward system for creation of green walls, big or small, outdoors or indoors [8]. It is a modular solution, allowing it to fit into different areas. Modules are made out of high strength polypropylene copolymer and contain additives to provide high life expectancy, so that the product will not be damaged by extreme weather conditions, such as solar radiation and changing temperatures [5]. Also the materials used are 100% recyclable. However, plant watering cannot be said to be fully automated due to the lack of an intelligent water distribution unit. It is up to the user to water the plants regularly or to create an automated system. The product is fully mechanical and contains no electrically powered elements.

Click&Grow is based in Tallinn, Estonia, and was founded in 2010. Their mission is to make healthy food available for all people. They offer different options meant for indoor only [3]. This product is socket based, but allows to choose from many sizes. Starting from three slots up to a 51 slot Wall Farm. Each capsule hosts a seed embedded in an advanced nanotechnology growing material, labelled Smart Soil. Everything is grown without any use of GMO or pesticides, leading to healthy naturally grown greens.

ZipGrow FarmWall is a Canadian company specialized in commercial scale wall-mounted, self-sustainable solutions. The product is designed to provide low maintenance, high yield hydroponic farming system, and is modular, automated and user-friendly [13]. The wall is made of food-safe polyvinyl chloride (PVC), holding the towers in place. The main base can contain five 152 cm towers. Plants are inserted into openings in the middle tower. There is no exact number of plants which can be put inside—the user can insert as many while there is room. The towers can be easily removed from the base, allowing to harvest and plant easily.

Although the most advanced vertical farming technologies found in the market are above the budget limits of this project (which was of 100€), the Team, after analysing the competitor devices and their market strategies, was able to identify which features to include in their own product.

3   Complementary Studies

To develop the project with adequate depth, SAMARA team has conducted research and analyses in three complementary fields. These analyses help to understand the impact and aim of the team's work.

3.1   Marketing

The marketing analysis helped the team to define the goals, target consumers and brand of the proposed solution. The "Vereatable" brand logo associates the *veritas* (truth), edible and table concepts to vertical garden, modularity, sustainability, smart control and smartphone connectivity. As nature-lovers, the SAMARA team members want to share this vertical garden way of life, where the goal is to bring the production of biological and healthy products to the household. Besides biological and healthy products being a current trend, according to the team, they are greener. The proposed Vereatable solution respects the environment, has low energy consumption, controls autonomously the water and light conditions and interfaces with the consumer via smartphone. Vereatable is created for urban, busy and connected people who want to eat more natural and healthier and improve their nature environmental consciousness. The market analysis revealed that Vereatable is competitive in the actual market thanks to the smart functionalities (smartphone connectivity and autonomous water and light control), the energy consumption and price (from 75€). This price covers the costs of production and remaining expenses. In terms of promotion, the ideal is to have the clients sharing this novel way of life and environmental consciousness.

3.2   Eco-efficiency Measures for Sustainability

The main purpose of sustainable development is to provide solutions for the preservation of natural resources, reduce the negative impact of people on the environment and promote a greener and healthier lifestyle. The team decided to use natural resources, such as wood and recyclable materials, to create low impact. By using aeroponics, on one hand, it does not use herbicides due to the absence of fungi and, on the other hand, it recycles the nutrient solution, which is re-used in fertilization. Vereatable distinguishes itself by the minimal use of water, as it manages to use its irrigation system in a reasonable and fully controlled manner and drastically reduces the unnecessary use of water. The SAMARA team decided to support and work through the guidelines of The Vertical Farming Association, a two-year, non-profit organization focussed on promoting the industry. Vertical farming allows people to produce crops throughout the year because all environmental factors are controlled. It produces healthier and higher yields faster than traditional agriculture and is resistant to climate change. In addition, as the world's population becomes more urbanized, vertical farms can help meet the growing demand for fresh, locally produced products.

3.3   Ethical and Deontological Concerns

Throughout the duration of this project there were five critical points related with both ethical and deontological concerns: Engineering, Marketing, Academics, Environmental and Liability ethics. The team decided to follow the National Society of Professional Engineers (NSPE) list of rules [10], which are a set of moral rules engineers should adopt, as well to support and work within the guidelines of the ICC/ESOMAR International Code on Market and Social Research [4]. Since the biggest priority are the consumers, the team will not provide misleading information about the general purpose and nature of the solution, preserving its reputation. Moreover, it will conform with the relevant national and international laws and ensure the project is designed, developed, reported and documented accurately, transparently and objectively. The goal is to design the most sustainable and efficient solution to grow healthier, chemical-free products indoors, giving consumers the opportunity of a greener lifestyle. The first concern revolves around ensuring the product works properly and that all materials used are from certified suppliers. Additionally, it shall be advertised using a strictly factual description and include detailed user-friendly instructions.

4   Project Development

The **Vereatable** indoor garden has been designed as a cost-efficient solution for automated household-scale indoor farming. This does not imply being the cheapest product on the market, although the aim would be to achieve lowest sustainable selling point. The leading idea is to respect the customer's investment through high durability of the product, efficient use of resources, and expandable nature of the product, allowing possible further module acquisitions.

4.1   Requirements

The project requirements were the following: a modular solution adaptable to different areas, a 100 € budget to prioritise the use of sustainable/reusable materials, open source tools and software as well as the mandatory adoption of several international regulations [1].

4.2   Functionalities

Working with living organisms—either animals or plans—is always sensitive. For a plant to grow, its environment has to support its development as a whole. Nevertheless, since some basic principles of operation for maintaining this environment can be predicted, the developed device should: provide water and nutrition in appropriate amounts and at appropriate intervals; provide lighting of appropriate intensity and at the appropriate periods of the day; offer space for the roots and the shoot of the plant to grow and develop.

In addition, the product should offer certain functionalities to its users: fall silent and turn off the lights at night-time, preferably defined by the user to their taste, not to disturb the owner; and notify about any maintenance operations necessary to keep it operational. These can be realised through the establishment of wireless connection and the development of a mobile application.

4.3   Structure

Figure 1 presents the computer-aided design (CAD) model of the prototype module. Figure 1a and b show the module as a whole, and Fig. 1c and d focus on the piping and plant socket chalice details. The overall size of the module is 30 cm by 20 cm by 80 cm (width, depth, height). The casing is made of plywood and the water distribution inside uses PVC piping elements.

4.4   Control

The core of the prototype control system is an ESP-12E micro-controller by Wemos, operating under NodeMCU firmware. It has been chosen over the Arduino family of micro-controllers, one very well established for use in similar scale applications, due to comparatively higher computational power, embedded wireless connectivity and lower average price [7].

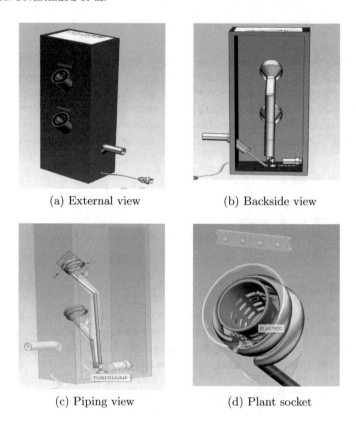

(a) External view   (b) Backside view

(c) Piping view   (d) Plant socket

Fig. 1. Design structure

A digital TSL2561 I2C luminosity sensor and a simple analogue liquid level sensor are connected to two ESP-12E inputs and a single red-green-blue (RGB) light-emitting diode (LED) and three transistor-based relay circuits, to control components with voltages higher than the operational voltage of the microcontroller, are connected to four ESP-12E outputs. The relays are managing two 12 V LED bars and a 9 V water pump. The circuit as a whole is powered by a 12 V direct current (DC) power supply unit, with a voltage step-down converter to 9 V DC. Appropriate limiting resistors and a fly-back diode are applied where needed.

The control software includes: a LUA program installed in the microcontroller, which controls the components directly connected to the microcontroller. A Microsoft .NET server installed on-line, providing endpoints to the prototype and the smartphone. Finally, a natively-developed Java mobile application for Android platforms, offering a friendly user interface. All three pieces are interconnected through the Internet.

The control flow is subdivided in two segments, one handling the start up procedures necessary to achieve full functionality and the other supervising the continuous, ongoing work of the device once it is enabled. Taking into consideration the limited capabilities of the micro-controller processor as well as its limited access to certain elements of the system, *e.g.*, the database, part of the decision-making process is being outsourced to the server, where a batch of sensor readouts is exchanged for a batch of control requests every one minute.

5   Tests and Results

The SAMARA team conducted a series of tests covering the different subsystems:

Water Distribution System: The water pump has to produce a water flow with the right pressure for the sprinklers to create water drops over the plants. The output pressure is controlled through pulse width modulation (PWM). This watering process occurs periodically during the day. The time interval is customisable via the user-device interface. The initial test identified the need to adjust the piping. Once fixed, the plant chalices were successfully irrigated drop by drop. While the duration and frequency of the irrigation was successfully controlled using the developed mobile application, it was not possible to control the intensity of the pump—the motor did not react to any PWM settings beside the full duty cycle of 100%, allowing only for a digital on/off control, rather than a gradual one.

Water Recollection System: The passive water collection system was designed collect any excess of water back to the water reservoir for the safe operation of the electrical components. The water recollection system conducted successfully any water excess back to the reservoir tank.

Reservoir Water Level Sensor System: The level sensor measures the water level inside the storage tank to inform the user, via the on-device feedback LED and the user-device interface, of the need to refill the tank. Once the proper relation between the sensor reading and the water level relation was found, the device worked as expected.

Lighting System: The LED bars, which are positioned above the plants, are intended to provide the luminosity required for growing plants. By default the bars are on, with the exception of user define curfew periods and when there is sufficient light. The LED bars offered full range of control via software, allowing for easy modulation of the light intensity. Due to the limited time available, the LED bars used in the prototype do not have adequate spectrum for growing plants, but they will be substituted in the final product.

Ambient Light Sensor System: The high-resolution luminosity sensor is expected to detect when the ambient lighting is sufficient to support the growth of plants. At such times, the lighting system is deactivated to increase the overall sustainability of the product. The digital sensor was placed on a range of different environments in order to check if its readings complied with the specification. The sensor worked as expected.

On-device Feedback LED: The RGB LED diode, located on the structure of module, provides a simple direct communication with the user. The diode in question is anticipated to react to the most important events, such as enabling wireless connection, depletion of resources or internal software errors. The diode should use different colours and uptime patterns for the various events-of-interest. While the diode was easy to connect and program, the quality of some colours was significantly poorer than others – while distinguishable shades of blue were easily achievable, yellows and oranges were mostly contaminated by their green component. Consequently, the choice of colours was adjusted. This motivates the need to find a different RGB LED.

User-device Interface: The smartphone mobile application is the main user-device interface. It is expected to offer the user insight into the operation of the device, allow the user to modify the user-dependent variables and notify the user about the status of the device. The mobile application and the micro-controller interact via a dedicated webserver, which offers endpoint methods for both client devices. The fine-tuning of these three software modules took the team more effort than anticipated. The remote webserver was deployed on an independent host, reachable by both the micro-controller and the mobile application through standard Hypertext Transfer Protocol (HTTP) messages. The webserver endpoint methods were successfully tested, first with artificial, mock-up requests from the Postman environment, and, then, through the counterpart code. In the end, a simple but stable communication system was established between the device and the smartphone.

6   Conclusions

The main objective of this project was to develop a modular and sustainable vertical farming solution for personal use. The project started with the analysis of the existing solutions, followed by complementary studies covering project management, marketing, sustainability and ethics. From this initial research, the SAMARA team set the requirements of the Vereatable indoor garden, a solution made for people who want fresh and healthy food, but without the time or the space required to grow it themselves. The proposed solution was designed to use recyclable materials and consume energy parsimoniously for sustainability reasons. Because of its specific features, Vereatable fosters sustainability and healthy eating in urban environments. The Team designed and assembled the prototype, including the control system, and, then, developed and debugged the different software modules. The research and development performed by team SAMARA provides a good base for new sustainable vertical farm products. The Team hopes to inspire new ideas to reduce the impact of today' s agriculture in the environment and provide a more greener and sustainable way of life.

Considering the EPS@ISEP process, the Team reports that, at the beginning of the semester, they were faced the problem of building a solid team with members from different cultures and using English as communication language. In the end, this experience was considered a preparation for future professional life, where similar situations are bound to occur. Regarding this collaborative learning experience, the team members reveal having learned to trust and help each other, to work in a multinational multidisciplinary team and to discover new things about themselves, as stated in the following opinions:

Anastasia Sevastiadou: *"I deeply believe that living abroad and participating in EPS was a crucial step in my life. Not only did it brought knowledge in various academic disciplines, but it also gave me the opportunity to work with people from different countries and different cultural backgrounds, to improve my English and learn some Portuguese. EPS taught me a lot about teamwork and how important communication skills are in a group project. The prospect of personal development is one more reason among the numerous others that made me participate in EPS, in that case, I conceder this experience very profitable."*

Andres Luts: *"EPS was an amazing experience for me since I was working with people from different fields of study. This greatly improved my research abilities and I also improved my abilities in my field of work. Additionally, living in a new country in an international environment was something new for me and I learned a good deal of life skills."*

Audrey Pretot: *"EPS was the most impactive experience of my short life. As French without a really good English, I felt my language skills improved along the semester. With some difficulties at the beginning of the semester, but, thank for goods team, mates, patience and motivation, I realized that I could overcome all difficulties. I learned a lot about project management, teamwork, marketing, communication and other fields which weren't mine. But I think that the most beautiful thing in EPS is learning about yourself, change your point of view about your own personality, become a little bit more mature and objective, improving yourself. Knowing your character even more in each difficult situation I needed to face at work with really different team mates from a different culture and in everyday life in a foreign country (a beautiful one) was really amazing. Even if sometimes this semester was frustrating thank you to my teachers, thank you to my classmates and especially thank you to my team mates."*

Mile Trendafiloski: *"EPS was a wonderful experience for me, having to work with people of different nationalities and different backgrounds. I learned a lot about teamwork, consistency in work, product development and I greatly improved my current field of knowledge. To add more, I believe that EPS did not only help me with my academic and practical knowledge, but I acquired some assets that can be useful in my life and made new friends that hopefully will last throughout my lifetime."*

Rodrigo Basurto: *"I have never worked like this before, I mean with engineers from different fields, making it possible to know and learn new things from them. Build a real project with its real issues that only appear in the real life made that project a real challenge to me. I really liked the semester and it has been profitable academically and also for growing as a person."*

Szymon Błaszczyk: *"Through participation in the EPS programme, I have shone a new light on the set of my skills and assets. Having a bit of a control-freak attitude towards the projects I get involved with, and having jack-of-all-trades interests, I usually try to pitch into every single aspect of the work done, be it for better or for worse. Faced with the mere size of this EPS assignment, throughout the semester I have learned how important it is to simply put trust in my team mates' qualifications and the quality of their work. The multicultural environment itself wasn't something new to me, as I have been active in a Europe-wide student organization in the past years. As far as the technical side of the project considered, I got to work with some new technologies that never before got my interest, but in the end proved to be interesting and may get more of my focus in my future work."*

Funding. This work was partially financed by the ERDF—European Regional Development Fund through the Operational Programme for Competitiveness and Internationalisation—COMPETE 2020 Programme within project POCI-01-0145-FEDER-006961, and by National Funds through the FCT—Fundação para a Ciência e a Tecnologia (Portuguese Foundation for Science and Technology) as part of project UID/EEA/50014/2013.

References

1. Sevastiadou, A., Luts, A., Pretot, A., Trendafiloski, M., Basurto, R., Błaszczyk, S.: Vereatable Indoor Garden Report, June 2018. http://www.eps2018-wiki1.dee.isep.ipp.pt/doku.php?id=report
2. BBC Future: How Vertical Farming Reinvents Agriculture (2017). http://www.verticalfarms.com.au/advantages-vertical-farming. Accessed 13 May 2018
3. Click and Grow: Click and Grow Indoor Herb Gardens (2018). https://eu.clickandgrow.com/. Accessed 21 May 2018
4. ESOMAR: ICC/ESOMAR International Code on Market and Social Research, December 2007. https://cdn.iccwbo.org/content/uploads/sites/3/2008/01/ESOMAR-INTERNATIONAL-CODE-ON-MARKET-AND-SOCIAL-RESEARCH.pdf
5. EU Commission: EU Machinery Legislation (2018). http://ec.europa.eu/growth/sectors/mechanical-engineering/machinery. Accessed 21 May 2018
6. European Project Semester: About EPS - Introduction (2014). http://europeanprojectsemester.eu/info/Introduction. Accessed 11 May 2018
7. James Lewis @ baldengineer.com. Arduino to ESP8266, 5 Reasons to Switch (2016). https://www.baldengineer.com/esp8266-5-reasons-to-use-one.html. Accessed 18 April 2018
8. Minigarden: Technical Specifications (2018). https://uk.minigarden.net/minigarden-vertical-technical-specifications/. Accessed 21 May 2018
9. Minigarden: Urban Green Revolution (2018). https://www.youtube.com/watch?v=gexVQ7XvELY. Accessed 21 May 2018
10. National Society of Professional Engineers. Code of Ethicsonline, July 2007. https://www.nspe.org/resources/ethics/code-ethics
11. Vertical Farm Systems. Advantages of Vertical Farming (2018). http://www.verticalfarms.com.au/advantages-vertical-farming. Accessed 11 May 2018
12. World Health Organization: Frequently Asked Questions on Genetically Modified Foods (2014). http://www.who.int/foodsafety/areas_work/food-technology/faq-genetically-modified-food/en/. Accessed 10 May 2018
13. ZipGrow: ZipGrow Farmwalls (2018). https://zipgrow.ca/collections/zipgrow-farmwalls/products/4-tower-farmwall-led-light-kit. Accessed 21 May 2018

Water Intellibuoy—An EPS@ISEP 2018 Project

Mireia Estruga Colen[1], Hervé Houard[1], Charlotte Imenkamp[1], Geert van Velthoven[1], Sten Pajula[1], Benedita Malheiro[1,2], Cristina Ribeiro[1,3], Jorge Justo[1], Manuel F. Silva[1,2(✉)], Paulo Ferreira[1], and Pedro Guedes[1]

[1] ISEP/PPorto—School of Engineering, Porto Polytechnic, Porto, Portugal
epsatisep@gmail.com
http://www.eps2018-wiki2.dee.isep.ipp.pt/
[2] INESC TEC, Porto, Portugal
[3] INEB – Instituto de Engenharia Biomédica, Porto, Portugal

Abstract. This paper reports the collaborative learning experience of a team of five Erasmus students who participated in EPS@ISEP—the European Project Semester (EPS) at Instituto Superior de Engenharia do Porto (ISEP)—during the spring of 2018. EPS@ISEP is a project-based learning capstone programme for third and fourth year engineering, product design and business students, focussing on teamwork and multidisciplinary problem solving as well as on the development of sustainable and ethical practices. In this context, the Team developed a drifting intelligent buoy to monitor the water quality of urban water spaces. Motivated by the desire to build an intelligent buoy for urban water bodies, the Team conducted several scientific, technical, sustainability, marketing, ethical and deontological analyses. Based on the findings, it has derived the requirements, designed the structure and functional system, selected the list of components and providers and assembled a proof of concept prototype. The result is Aquality, an intelligent drifting buoy prototype, designed for private sustainable pools. Aquality monitors the quality of the pool water by measuring its temperature and turbidity, while interfacing with the user through a mobile application. Considering the EPS@ISEP learning experience, the Team valued the knowledge and skills acquired, and, particularly, the collaborative learning and working component of the project, *i.e.*, working together towards one goal while maintaining high motivation and cohesion.

Keywords: Collaborative learning · Project based learning · Technology · Education

1 Introduction

The European Project Semester (EPS) is a one semester capstone project offered by 19 European engineering schools to engineering, product design and business undergraduates. EPS challenges students from multiple educational backgrounds

© Springer Nature Switzerland AG 2019
M. E. Auer and T. Tsiatsos (Eds.): ICL 2018, AISC 917, pp. 439–449, 2019.
https://doi.org/10.1007/978-3-030-11935-5_42

and nationalities to join their competencies to solve multidisciplinary real life problems in close collaboration with industrial partners and research institutes during the spring semester[1]. In 2018, Team 2 was composed of Sten Pajula, an Electrical Engineering student from Estonia; Geert van Velthoven, an Industrial Engineering and Management student from the Netherlands; Charlotte Imenkamp, a Biomedical Engineering student from Germany; Mireia Estruga Colen, a Mechanical Engineering student from Spain and Hervé Houard a Product Development student from Belgium. The Team chose to develop an intelligent drifting buoy to monitor water quality because this challenge matched the interests of the whole Team and allowed the Team members to individually contribute with their educational knowledge.

Water is one of the most important resources on Earth. When humans come in contact with water, in pools or urban lakes, it becomes necessary to monitor and maintain the water quality on a regular basis. The World Health Organisation states that the water quality is a "parameter of immediate operational health relevance (. . .) and should be monitored most frequently in all pool types" [7]. An unmaintained pool can lead to the growth of harmful infectious micro-organisms and bacteria. To efficiently maintain the water quality of a pool, it is mandatory to continuously monitor the condition of the water. This proves to be a time and money consuming activity [6]. With Aquality, this task is simplified since it autonomously monitors the water quality and informs the user of the current water quality status.

The Team's vision was to design, build and test a buoy equipped with sensors to collect data on the quality of the water in natural ponds or pools. The buoy should drift on the water, collect data and inform or alert the user according to the findings. To determine the quality of the water, two parameters were chosen: temperature and the turbidity level. Together, they provide information about the condition of the filtration system, which is directly connected to the oxygen saturation, a highly relevant water quality indicator for ponds or pools inhabited by fishes or plants.

The Team was able to design, create and test together the Aquality prototype, which autonomously monitors water quality and, with the help of the companion mobile application, informs and makes relevant suggestions to the user. This comfortable solution contributes to save time and money in terms of water pool quality monitoring and maintenance. By working together in a multidisciplinary team, the Team members had the opportunity to learn from each other and collectively, and achieve further than they would have individually.

This paper is divided in five sections. This Sect. 1 describes the context, team, problem, goals and contributions. The Sect. 2 describes the background of the project, namely the state of the art, marketing, sustainability and ethics studies. Based on this research, the Team designed and developed the product which is described in Sect. 3. Section 4 contains the functional test results. Finally, Sect. 5 presents the conclusions of the project and suggestions for further work.

---

[1] https://www.isep.ipp.pt/Course/Course/44.

2   Background

This section presents the different background studies conducted by the Team to define the proposed solution.

2.1   Related Work

After an initial brainstorm, the team decided to focus on natural swimming pools. A natural swimming pool uses biological filters, such as plants, to filter and maintain the quality of the water. Natural pools require less maintenance than conventional ones, the year-to-year costs are lower after construction is finished and they are chemical free. They only need to be kept well-skimmed and clean of debris. The filtration system works through the roots and sediments as in a natural lake. This causes turbidity depending on the presence of algae. Because of the filtration system it is impossible to completely remove sediment and living organisms and a high turbidity could indicate a problem in the filtration [3,10]. Even though the popularity of natural swimming pools is increasing [10], there are no dedicated products on the market for monitoring its water quality— only the companies who build and maintain these pools offer post-sale services. Nevertheless, this section presents four products of interest with a wide range of application both in terms of environments and sensors:

- DIY buoy: The Do It Yourself (DIY) buoy is a scientific product, whose purpose is to collect data for weather forecast. It's a start-up prototype made by a team based in North America. It runs on the Arduino Trinket Pro 5 V from Adafruit. A solar panel charges a small Lithium Polymer battery to provide additional power. The buoy collects data, including latitude and longitude, speed, instantaneous direction, water and internal board temperature and tilt angle. The data is transmitted via a satellite communication link [2].
- Bluetooth Pool Thermometer: The wireless pool thermometer shows the current temperature in an application. It includes a check for high and low temperatures and notifications, when the temperature exceeds the upper limit or the supported range and when the battery is low [1].
- Seneye Pond: The Seneye Pond measures the water temperature, Ammonia, pH, total light and water level. The sensors are connected via Wi-Fi or USB to a website, where additional advice is provided [9].
- pHin: pHin continuously measures the chemicals and temperature in a pool or hot tub. It is connected to a mobile phone app and a smart monitor via Wi-Fi and notifies whenever the water quality needs to be balanced. This product is sold as a product-service system where services, like repairs, upgrades and maintenance, are provided through the payment of a monthly fee. In the case of pHin, the monthly fee also includes the delivery of chemicals [8].

Taking into account the sensor comparison in [4], the team considered the use of pH, turbidity, temperature, oxygen and motion sensors. The comparison showed that pH, turbidity and temperature sensors were within the reach of the

100 € budget allocated to the project. Finally, the Team decided to measure the turbidity and temperature, estimate the oxygen value based on the correlation between oxygen saturation and water temperature, instead of using an oxygen sensor, and drop the pH sensor since it requires frequent calibration.

Based on this state of the art study, the conclusion was that the buoy should facilitate the maintenance of a natural pool. This rising business offers the perfect niche and the buoy will combine the best features of the related products. The main components will be a solar panel, a battery, a Wi-Fi module and turbidity and temperature sensors. The sensor data will be presented in a mobile application, which will also offer additional advice about natural pool maintenance and the state of the buoy. For a better customer service, the team will also adopt a product-service system.

2.2 Marketing

Based on the market analysis, the team decided to create an intelligent buoy for natural swimming pools. It will be sold as a Product Service System, meaning that customers will receive advice on how to maintain the water quality of their pool. Besides, the company is responsible for the maintenance of the product. The buoy should be sold through the Internet, targeting families with natural swimming pools, wishing to easily estimate water quality. These customers will pay a monthly fee around 20 €.

2.3 Sustainable Development

The team performed the life-cycle analysis of the product as well as for the future theoretical company involved in the manufacturing. The environmental, economic and social dimensions of sustainability were considered. The conclusions made the Team opt for polyvinyl chloride (PVC)—for recycling convenience—as the product's main material and offer client service, instead of simply a product. That means that the client will return the product sporadically, when updates on hardware are made. The company will reuse the exterior and upgrade only the interior, to reduce the waste produced, before returning the product to the client.

2.4 Ethical and Deontological Concerns

The ethical and deontological study helped the Team gain a better understanding of the issues which appear during the development of a product, namelly, in terms of environmental ethics. The Team believes the buoy could help making natural pools more attractive and the pool industry more sustainable. Therefore, the buoy must be sustainable and comply strictly with the environmental ethics. Combined with sustainability, the ethics analysis has again proved PVC to be the correct choice of building material. The sales and marketing ethics resulted in a clear vision of the advertising, price and employees policy.

3 Project Development

Aquality was developed to help monitor and maintain the water quality of natural pools. This water quality is ensured by a fragile ecosystem, filtering the water in an environmentally friendly way. Understanding the fluctuations in the quality of the water is thus of utmost importance. Different parameters are measured to allow the user to get a clear and understandable overview of the health of the pool's micro-biotope.

3.1 Requirements

The broad requirements for developing Aquality were defined in the project proposal. This included a limited budget of 100 €, the usage of open source software and technologies, opting for low-cost hardware solutions, complying with the applicable EU directives and using the International System of Units. Besides these general requirements, the Team derived the following requirements:

- Functional requirements: The buoy must be a self sufficient prototype, which floats on the surface and collects data.
- Usability requirements: The data must be read and presented in a user friendly way (mobile application/browser).
- Environmental requirements: The materials and manufacturing must comply with environmental requirements, according to environmental and sustainable development practices.

3.2 Functionalities

The user will communicate with the device through Wi-Fi via a smartphone application. The device will send out relevant data, as temperature and turbidity, and notify the user about ongoing changes in the pool. This gives the owner the possibility to be remotely notified.

3.3 Structure Design

The main architecture of the product was explored using hand-drawn sketches. With the help of the computer aided design (CAD) software SolidWorks, these sketches were translated into a 3D model. Multiple iterations were needed to obtain a model fit for a 3D printer. In each iteration the most relevant renders were made, using SolidWorks PhotoView 360, to allow the Team to assess the model in detail. During this design stage, all requirements were kept in mind.

The final result, depicted in Fig. 1, is a buoy consisting of two watertight compartments where the different components are housed. Between these compartments, the Team created a flooded area to place the sensor probes without compromising the aesthetics of the buoy. Very early, the Team realised that the prototype would have to be made of two separate shells. This meant that a reliable solution had to be found to join the two shells together tightly and ensure the water tightness of the buoy. The solution was to place a rubber seal between the two watertight compartments and fasten both halves together with five screws (two on the top and three on the bottom part of the buoy) plus o-rings. Due to the somewhat complex design of the buoy, the Team chose to 3D print the prototype. This meant that the design had to be adapted further. One of the major changes was related to the maximum printing dimensions of the available 3D printer. Each shell had to be split in two halves, so that they could be successfully printed, and, afterwards, be glued together. Figure 1 displays the details of the designed structure.

(a) Exploded view

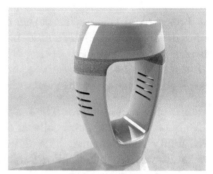

(b) Frontal view

(c) Lateral view

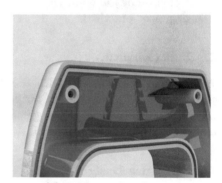

(d) Rubber seal and screws

Fig. 1. Aquality buoy structural design

3.4 Control System

The electronics inside Aquality features a Wi-Fi enabled controller, sensors and a power unit, according to the conceptual diagram of the prototype control system displayed in Fig. 2. Since Aquality is designed to be self-sustainable, the power unit includes a solar panel, battery and charger. The type of sensors installed was constrained by the budget.

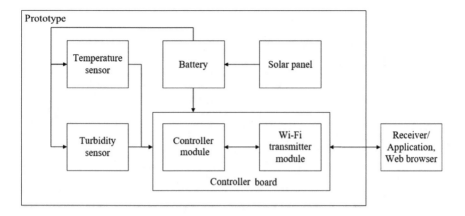

Fig. 2. Conceptual diagram

3.5 Mobile App Development and IoT Platform

The smartphone application is an important feature to offer a pleasant user experience and to drive competitive differentiation of the product. The first step of developing the application was to identify the use cases and define the main functionalities, which are to display the water temperature, air temperature, turbidity and water quality, offer support and recommend maintenance (animals or plants). Figure 3 presents the use case diagram, *i.e.*, the different ways the user may interact with the app. The on-board sensor data is communicated and stored in a cloud-based IoT platform named Easy Internet of Things (EasyIoT) [5]. Figure 4 shows the dashboard of the EasyIoT, displaying Aquality's sensor data.

4   Tests, Results and Discussion

To ascertain the correct functioning of Aquality, several software and hardware tests were planned, including the mobile application, consistent and correct data readings from sensors and water tightness to keep the electronics safe. The detailed planned functional tests were the following:

– Software tests: (*i*) reading sensor data; (*ii*) uploading sensor data to the cloud-based EasyIoT platform; and (*iii*) downloading from the EasyIoT platform and displaying sensor data on a web browser/mobile application.
– Hardware tests: (*i*) watertightness; (*ii*) floatability; (*iii*) consistency and credibility of the sensor readings; (*iv*) Wi-Fi reach; and (*v*) autonomy of the prototype.

4.1   Tests

In the current stage of the project, the software tests were performed, but the hardware tests were postponed due to the 3D printing delay.

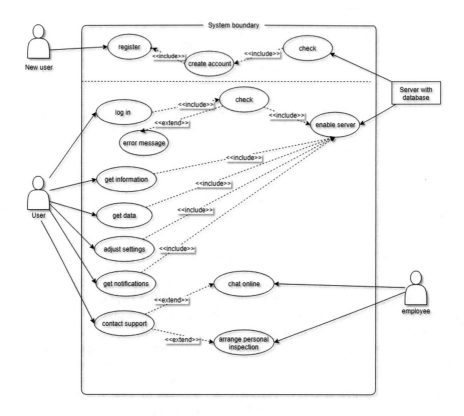

Fig. 3. Use case diagram

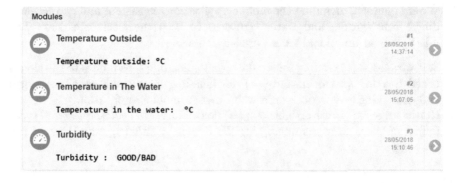

Fig. 4. Displaying the sensor data in the EasyIoT platform dashboard

The methodology behind the software testing was based on trial-and-error, which means that the code was tested, debugged and tuned until the desired results were achieved.

Concerning the planned hardware tests, *watertightness* will be tested by submerging the prototype underwater for six hours. Afterwards, the prototype interior will be examined for leakages. *Floatability* will be tested by putting the buoy in the water. To achieve the desired waterline on the product, the ballast shall be adjusted. The *sensors consistency and credibility* will be tested by submerging the sensors and logging the readings. These readings will be compared to calibrated scientific measurement devices and the sensor readings calibrated by software. The *Wi-Fi reach* will be tested by measuring the approximate distance between prototype and Wi-Fi router, from the spot where last signal from prototype was received. Finally, the *autonomy of the prototype* will be tested by measuring the time the battery takes to charge and achieve sufficient amount of energy to power the prototype.

4.2 Results

The software tests were performed by submerging one temperature sensor in a water cup and leaving the other outside. The turbidity was tested with two separate cups, one with hazy water and another with clear water. Figure 5 displays the results of both tests.

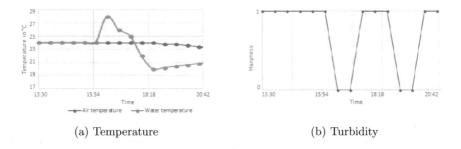

(a) Temperature   (b) Turbidity

Fig. 5. Sensor tests

Figure 5a depicts the temperature test: (i) before 15:54, both temperature sensors were unsubmerged; (ii) after 15:54, one sensor is held in palm of the hand; and (iii) after 16:00, the same sensor was kept submerged in the water. The temperature measured by this sensor slowly rose before decreasing and stabilizing near room temperature. The turbidity sensor was tested by changing the sensor between the clear and hazy water cups. Figure 5b shows the clear—zero (0)—and hazy—one (1)—turbidity readings. In this test, the placement of the sensor changed at 16:00, 17:00, 18:30 and 19:30.

Finally, Fig. 6 displays the tangible outcomes of this EPS@ISEP project: the EasyIoT mobile application displaying the sensor readings (Fig. 6a) and the final printed prototype (Fig. 6b).

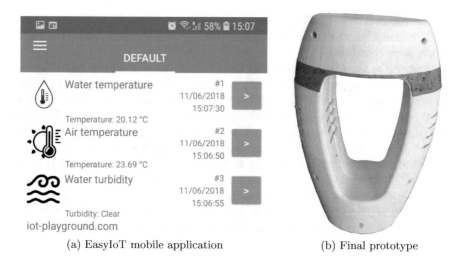

(a) EasyIoT mobile application   (b) Final prototype

Fig. 6. Final outcomes of the project

5   Conclusion

The main objective of the project was to motivate and unite the Team around the development of Aquality's proof of concept prototype, while improving the active learning and soft skills of each member. This paper reports this journey.

The product under development is user-friendly, features Internet connection and is self-sustainable in terms of power. The user-friendly application and the information it provides is unique as far as the team knows. Because of trends in pool management and worlds constant thrive for a greener future, the potential market is expanding.

Further development is necessary to transform the prototype into a product ready for release into the designated markets. The Aquality prototype was constrained by the time, budget, insufficient knowledge in some technical areas and lack of experience in product development. The following aspects should be considered for future development:

- Development of a professional mobile application;
- Upgrade the solar panel to fully meet the buoy's energy requirements;
- Reduction of the buoy dimensions, including the development of smaller dedicated sensors.

Regarding the collaborative learning and teamwork, maintaining the motivation and unity of the Team was not always an easy task. The differences in culture, study fields and personalities can lead to frictions within the teams. In the case of the Aquality's Team, none of the latter issues surfaced because the Team members were determined, but flexible. The acquired project management skills helped to allocate tasks, meet deadlines and achieve the goals of the project.

Quoting Sten Pajula, who was responsible for the prototype electronics and coding, "*The overall experience of the project has been positive. The weekly meetings with the supervisors and Team members made the semester pass seamlessly. The Team had the project deadlines always under control and constant progress was a result of well structured work, which would have been much harder without supportive supervisors and lectures. In this semester, I have learned the basics of product development, marketing and improved the knowledge in my study field*".

Funding. This work was partially financed by the ERDF—European Regional Development Fund through the Operational Programme for Competitiveness and Internationalisation—COMPETE 2020 Programme within project POCI-01-0145-FEDER-006961, and by National Funds through the FCT—Fundação para a Ciência e a Tecnologia (Portuguese Foundation for Science and Technology) as part of project UID/EEA/50014/2013.

References

1. CoolStuff. Bluetooth pool thermometer (2018). https://www.coolstuff.com/Bluetooth-Pool-Thermometer. Accessed 1 Mar 2018
2. DIYBuoy. DIYBuoy (2018). http://mdbuoyproject.wixsite.com/default. Accessed 2 Mar 2018
3. EcoHome. Natural ponds and natural swimming pools (2018). http://www.ecohome.net/guide/natural-ponds-natural-swimming-pools/. Accessed 19 Mar 2018
4. Herve Houard. Sensor comparision (2018). https://docs.google.com/spreadsheets/d/1dO_1b-xUnJIKaXS0Gknxx8hb4fXyWcNo6XjkdDLFAQQ/edit#gid=1719295485. Accessed Apr 2018
5. EasyIoT iot playground.com. EasyIoT framework (2018). http://iot-playground.com/. Accessed May 2018
6. US National Library of Medicine National Institutes of Health. A comprehensive review on water quality parameters estimation using remote sensing techniques (2016). https://www.ncbi.nlm.nih.gov/pmc/articles/PMC5017463/. Accessed 20 Mar 2018
7. World Health Organisation. Guidelines for safe recreational water environments (2006). http://www.who.int/water_sanitation_health/bathing/srwe2full.pdf. Accessed 15 Apr 2018
8. pHin. phin (2018). https://www.phin.co//. Accessed 19 Mar 2018
9. SENEYE. Seneye pond (2018). https://www.seneye.com/store/devices/seneye-pond.html. Accessed 1 Mar 2018
10. The spruce MAX VAN ZILE. The pros and cons of owning a natural pool (2018). https://www.thespruce.com/pros-and-cons-owning-natural-pool-2737100/. Accessed 19 Mar 2018

Implementing a Project-Based, Applied Deep Learning Course

A Case Study

James Wolfer[(✉)]

Department of Computer Science, Indiana University South Bend,
South Bend, USA
jwolfer@iusb.edu

Abstract. Deep Learning is having a significant impact on disciplines ranging from autonomous vehicles to medical imaging. As one component of a Data Science focus in a Masters degree in Applied Mathematics and Computer Science, as well as supporting the Computer Science B.S. degree at the senior level, the department created a topics course in Applied Deep Learning. This work describes the course, its institutional context, and highlights its first deployment. Particular emphasis is placed on the supporting infrastructure and lessons learned from its deployment.

Keywords: Deep learning · AI · Curriculum ·
High-performance infrastructure · Pedagogy

1 Introduction[1]

1.1 Overview

Deep Learning is having a significant impact on disciplines ranging from autonomous automobiles to medical imaging [1, 2]. As one component of a Data Science focus area for an M.S. in Applied Mathematics and Computer Science, as well as for a proposed undergraduate degree, the Computer Science department commissioned the development of a deep learning course. Since this course is embedded in an applied program where students are expected to acquire the skills necessary for real-life deployment of the knowledge and experience gained, the course was developed with student projects as a prominent feature. The balance of this report describes the unique challenges involved with the development and implementation of the course, including providing high-performance computing, addressing moving targets in both the course content and software infrastructure, and the necessity of providing custom support for student projects.

---

[1] This Research Was Supported in Part by Lilly Endowment, Inc., Through Its Support for the Indiana University Pervasive Technology Institute.

© Springer Nature Switzerland AG 2019
M. E. Auer and T. Tsiatsos (Eds.): ICL 2018, AISC 917, pp. 450–457, 2019.
https://doi.org/10.1007/978-3-030-11935-5_43

1.2 Deep Learning Overview

Deep learning can be viewed as the contemporary re-introduction of Artificial Neural Networks as a machine learning approach driven by the availability of very large, publicly available datasets, and very high-speed computing resources enabled, largely, by GPU technology. For supervised training, modeled loosely by the notion of Hebian learning, these networks form a nonlinear response to a given set of inputs, and through optimization, learn a set of weights to produce a desired response. Figure 1 shows an example of a single "neuron" or "unit." Here the response is formed by summing the product of the weights and the inputs, followed by a non-linear activation function.

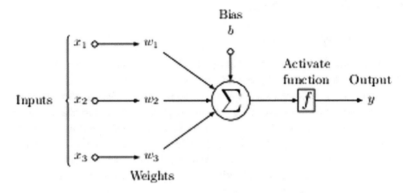

Fig. 1. Single "unit"

In current practice many of these neurons are arranged in layers as shown in Fig. 2, forming a powerful classifier. Often they are trained with convolutional layers that act as feature detectors trained along with the classification layers (Fig. 3). Characteristics of these deep networks include, for image processing, hundreds-of-thousands to millions of images to train, and often days of training time on high-performance computers.

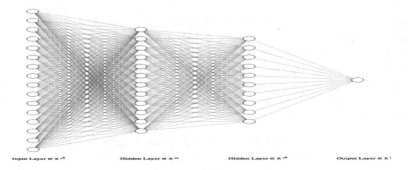

Fig. 2. Dense classifier

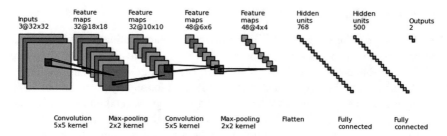

Fig. 3. Convolutional network with dense output layers

2   Instructional Framework

2.1   Project-Based Motivation

As a satellite campus of a major university the university serves a diverse student body ranging from students intending to work in the local community to those preparing for graduate studies. The campus serves approximately five-thousand students across over 65 undergraduate programs and 17 graduate programs. The computer science department serves approximately 200 undergraduate majors, and 25 graduate students. With an overt objective to serve both the local and wider communities, there is a stated value of "excellence in teaching, student-faculty interaction, research and creative activity…" Providing students with real-world skills to support the local community is often a priority. This makes a project-based course a natural fit to meet these objectives.

There has been a long history of Problem-based and Project-based pedagogy in a variety of disciplines. While this has lead to wide ranging discussion of what constitutes which, for our purposes we take the pragmatic view that there is enough overlap that, properly implemented, the approaches reach the same endpoint. The principles include [3]:

- Active Learning. The students take an active role in both the design and implementation of an appropriate project guided by the instructor.
- Real-life Problems. Learning is designed to a microcosm of real-life, with problems constrained by the limitations of class time and resources not by predetermined assignment.
- Self-directed learning. Students take an active role in the design, implementation, and evaluation of the project. This includes the expectation of prior knowledge appropriate to the course level and objectives.
- Learning for Transfer. The knowledge gained from the classroom experience will transfer to tangible product during the class, and presumably, beyond the classroom.
- Student-Student Interaction. Peer interaction and mentoring are an integral part of this learning process.

Recent meta-analysis supports the efficacy of these approaches in both medicine and engineering [4, 5]. For example, Strobel and van Barneveld [4] did a significant meta-analysis comparing problem-based pedagogy to conventional classroom approaches in science and medicine. Their analysis of the extant literature indicated that problem-based approaches were superior for long-term retention, skill development, and student satisfaction. They found that traditional approaches were slightly more effective for short-term retention such as standardized examinations.

These findings were reflected by Galand, Frenay, and Raucent [5] for engineering students. Based on a study of nearly 400 students they, in contrast to [4], found no negative effects for problem-based learning.

2.2 Project-Based Deep Learning Course

Given the advantages of a project-based approach for our students we formed a course that included initial lecture to establish a baseline of principles and practice then formed groups of students to propose and implement projects. Specific topics included:

- Supercomputer setup for Deep Learning.
- Application Areas (Vision, Natural Language, Textual Analysis, Robot Control, …)
- Background and history.
- Deep Learning Models:
  - Convolutional Neural Networks
  - Long Short-Term Memory models.
  - Gated Recurrent Units.
  - Others (GAN, Reinforcement, U-Net, Capsule)
- Data preparation
- Training, validation, measurement
- Visualization
- Workflow and project management

The class assumed that students code at a high-level, and that students are able to learn new programming languages "on-the-fly" as necessary to support their projects. The class was also predicated on the concept that, in addition to programming language proficiency, students are operating system agile. This proved to be optimistic, as will be described below.

The textbook for the class, Goodfellow, Bengio, and Courville, "Deep Learning" [6] was barely in print when the class was offered.

2.3 Hardware and Software Support

Deep Learning is often characterized by a propensity to need a large amount of compute time to learn sometimes millions of parameters. Currently GPU technology provides much of this compute power. Since most students don't have the resources to personally provide such capability, the university must provide the compute infrastructure.

One asset used is the university's Big Red II supercomputer. This computer registered within the Top-500 list of the worlds fastest computers from 2013-2017. Big Red II is a Cray XK7, with 31,288 cores and Nvidia K20 GPUs capable of a theoretical peak performance of one petaflop. Big Red II uses Cray Linux for its operating system.

Big Red II is designed to be a shared resource, with batch, as opposed to interactive, jobs performed on a scheduled basis. This is optimal for large runs taking days, but is less appropriate for smaller, interactive, experiments to test ideas and software prior to submitting a long run.

To supplement Big Red II, the Computer Science department provided a remotely accessible server including an Nvidia 1080Ti GPU. While the GPU could only support one project at a time, it provided the interactive infrastructure for setting up more extensive runs on the larger computer.

Software for this course centered on available Deep Learning fameworks. Python was chosen as the base language, and while it was expected that students could learn it on-the-fly, a "crash course" in Python and Numpy was provided.

At the time the course was offered many of the frameworks stable today were in a state of flux. Ultimately, while we elected not to prescribe a framework for any project group, instructional presentations centered around Tensorflow [7] and Keras [8, 9]. Theano [10] was also available as an alternative. Note that Theano is no longer being actively developed. The students used publically available datasets to support their projects.

2.4   Projects and Assignments

In addition to the project component, the course included a variety of assignments designed to gain experience with the DL frameworks, and to explore ancillary skills such as image processing and natural-language processing basics.

The class also included one examination. This was designed to be a practical exam appropriate to an applied course. Questions were included designed to test ideas such as appropriate activation functions, the nature of dropout for regularization, appropriate validation statistics, and recognizing overfitting during training.

Students were organized into groups of four or five. Each group formulated a plan for managing the project, including designating a single, or at their option, rotating managers. These managers were responsible for arranging meetings, collecting results, and for reporting progress on a weekly basis. From the onset students were informed that their grades were not based on the ultimate outcome of their trained networks, but on the totality of their design, implementation, and presentation. Classroom time was divided between lecture and in-class, instructor supported, group work. This ultimately resulted in five projects: leaf classification with convolutional networks, categorizing movies based on posters, identifying faces using transfer learning, classifying music genre from spectral images, and image colorization.

The final results were presented formally by each group to the class as a whole, with each student taking part in presenting.

3   Observations

3.1   Instructor Challenges

There were two prime challenges for the instructor to support this class. The first was the software infrastructure on Big Red II. The Deep Learning software environment was changing, literally, daily. When this project was initiated, for example, Theano was the only Keras tensor library. Big Red II did not support it so the instructor had to become system administrator as well and install a local (i.e. non-privledged) version of Python, Theano, Tensorflow, and the Nvidia CUDA libraries necessary to support the GPUs. By the time the class was offered, Tensorflow was the library of choice for Keras and the university IT office had installed a selection of the appropriate libraries. Unfortunately, not all the libraries necessary for supporting student projects were available so a significant amount of system administration was still required.

Related to the dynamically changing software environment was the rapidity of change in the discipline itself. Ten to twenty papers per day posted to the ArXiv pre-print repository, for example, related to the course. Topics, such as Generative Adversarial Networks gained attention over the course of the semester. Keeping the class both accessible and contemporary was challenging.

3.2   Student Challenges

While each project incurred individual challenges, the programming environment offered by Big Red II was uniformly frustrating for the students. While students in our department take a transition course introducing them to the Linux environment used on Big Red II, most students had forgotten many of the basic concepts. Reacquiring this expertise while trying to understand and implement new concepts proved to be too much for some students. For example, some supporting libraries needed to be installed locally in a student directory. This required adjusting path environment variables in a command-line interface, something some of the students struggled with.

The parallel programming aspects, while less directly used, presented a conceptual struggle for some students.

Finally, some of the more vulnerable students were the most reluctant to request help, significantly delaying progress. In the future the instructor is encouraged to pay more attention to some of the low-level details outside the direct course content to intervene earlier in these challenges.

3.3   Future Changes

Since the class was offered a variety of new, or maturing, alternative frameworks and libraries have emerged as viable alternatives to Tensorflow and Keras. Examples include PyTorch [11], CNTK [12], and even Knet [13] for the new scientific programming language Julia [14]. While Tensorflow and Keras still form a viable infrastructure, these tools change quickly and need to be reevaluated for each course instance.

In the future the instructor would give serious consideration to assigning project domains. While not wishing to excessively constrain student choice, a themed class in, for example, medical, social, and/or commercial applications, may allow for more inter-group collaboration.

As a pragmatic matter, a series of "warm-up" review exercises to reawaken knowledge of the operating system, scripting tools, and a light review of system administration will be provided very early in the class. Serious consideration will be given to making these online self-learning to minimize their impact on face-to-face class time.

3.4   Feedback

Aside from the Big Red II system administrative challenges, student feedback, while anecdotal, was generally positive. Seventy percent of anonymous respondents rated the course as excellent or good, and comments were constructive. Specific recommendations by students included having Artificial Intelligence as a prerequisite course, require more experience with Linux, and include more instructor-produced notes. Student observations about the class itself included good support material, very interesting subject matter, and an appreciation for the opportunity for guided, independent, "academic liberty" afforded students in the class. Or, in anonymous post-course feedback, "…the skills learned are hugely valuable to my future as a software developer. I am extremely grateful for the opportunity to learn cutting-edge techniques in Deep Learning".

4   Conclusion

Offering the Applied Deep Learning course was challenging for both instructor and students. Given the rapid change in the discipline, student expectations and preparation, and the challenges with infrastructure this can act as a caution as well as a template for reproducing the class. Recommendations include setting up an adequate support system, systematically and proactively assessing student understanding of off-topic, previously assumed, capabilities. Overall the class went well, and we look forward to its next iteration.

References

1. Huval, B., Wang, T., Tandon, S., Kiske, J., Song, W., Pazhayampallil, J., Andriluka, M., Rajpurkar, P., Migimatsu, T., Cheng-Yeu, R., Mujica, F., Coates, A., Ng, A.Y.: An empirical evaluation of deep learning on highway driving, arXiv. arXiv:1504.01716, http://arxiv.org/abs/1504.01716 (2015)
2. Litjens, G., Kooi, T., Bejnordi, B.E., Setio, A.A.A., Ciompi, F., Ghafoorian, M., van der Laak, J.A, van Ginneken, B., Sanchez, C.I.: A survey on deep learning in medical image analysis, arXiv. arXiv:1702.05747v2. (2017)
3. Mastascusa, E.J., Snyder, W.J., Hoyt, B.S., Effective Instruction for STEM Disciplines: From Learning Theory to College Teaching, Wiley (2011)
4. Strobel, J., van Barneveld, A.: When is PBL more effective? A meta-synthesis of meta-analyses comparing PBL to conventional classrooms. Interdisc. J. Prob-Based Learn. 3(1) (2009)
5. Galand, B., Frenay, M., Raucent, B.: Effectiveness of problem-based learning in engineering education: A comparative study on three levels of knowledge structure. Int. J. Eng. Educ. 28(4) (2012)
6. Goodfellow, I., Bengio, Y., Courville, A.: Deep Learning, MIT Press, Cambridge, MA (2016)
7. Tensorflow: Tensorflow Tensor Library. http://www.tensorflow.org
8. Keras: Keras Documentation. http://keras.io

9. Chollet, F.: Deep Learning with Python, Manning (2018)
10. Theano. http://github.com/Theano/Theano
11. PyTorch. https://pytorch.org
12. Microsoft: Microsoft Cognitive Toolkit (CNTK). http://github.com/Microsoft/CNTK
13. Denizyuret: Knet.jl. http://github.com/denizyuret/Knet.jl
14. Julia: Julia programming language. http://julialang.org

Real World Experiences

First Results of a Research Focused on Teachers' Didactic Technological Competences Development

Ján Záhorec[1]([⊠]), Alena Hašková[2], and Michal Munk[2]

[1] Comenius University in Bratislava, Bratislava, Slovak Republic
zahorec@fedu.uniba.sk
[2] Constantine the Philosopher University in Nitra, Nitra,
Slovak Republic
{ahaskova,mmunk}@ukf.sk

Abstract. The paper presents preliminary results of the research aim of which is to support modernization and optimation of teacher training study programs in their parts related to formation didactic technological professional competences of teacher trainees. The paper focuses on the research results of a screening in which practising teachers assessed significance of the use of various interactive educational activities and digital means in teaching process from the point of view of different aspects of education. Among these aspects were stated e.g. pupils motivation, easier understanding of the presented subject matter, skills to apply the acquired knowledge etc. The assessments given by teachers were tested in dependence on different factors, from which the most important was the factor of the teaching staff category the teachers belonged to. As the results show, there were identified some significant differences in the assessments of the contribution of the use of various interactive educational activities and digital means to the teaching process efficiency given by teachers of different levels of education.

Keywords: Teacher training · Teacher's professional profile ·
Didactic technological competences

1 Context of the Research and Its Goals

Multimedia technology based on new hardware and software technologies brings new opportunities for teaching different subjects and topics in more interesting ways and for obtaining difficult knowledge in easier ways. To use these means in teaching process assumes a teacher disposing of adequate didactic technological competences. With respect to the newest digital technologies, didactic technological competences of a teacher can be defined as teacher's professional competences to use digital teaching tools and their applications in real practice of education of the taught subject.

Main purpose of the presented research is to support modernization and optimation of teacher training study programs in their parts related to formation of didactic technological professional competences of teacher trainees. In particular, this means to find out an optimal structure of subjects, their curricula (content) and time allocation to give the Slovak higher education institutions offering teacher training study programs a model how to develop the relevant competences of students enrolled in these programs [1, 2].

In the first phase of the carried out research the current state and perspectives of primary and secondary school teachers' continuing education focusing on the teachers' continuous professional development in the area of their didactic technological competences has been surveyed. The survey has been based on a screening of the practicing teachers' opinions and attitudes. For this purpose a questionnaire was created and its reliability was verified in a pilot test [3]. Following the questionnaire reliability verification, screening of the practicing teachers' opinions and attitudes was started.

2   Methodology of the Screening of the Teachers' Opinions and Attitudes

In the period from September 2017 to February 2018 the final version of the questionnaire was administrated.

The proposed questionnaire consisted of 41 items structured in four parts A–D, from which 30 items were ordinary (C1–C13; D1–D17) and 11 items were nominal (A1–A4; B1–B7). Part A was focused on identification of the respondents' characteristics, part B observed how the respondents use interactive educational activities and digital means in their teaching practice, in part C respondents were asked to assess significance of the use of various interactive means in teaching process for selected specific aspects of education and part D was focussed on the respondents' self-assessment of their knowledge and skills to work with software applications and digital means within the scope of their own teaching practice.

More detailed information on the questionnaire parts B–D and the questionnaire reliability assessment is presented in [4]. All parts of the questionnaire were designed with the view of enabling to transfer teaching qualitative aspects, related to the use of selected software applications and digital teaching objects, into the quantitative one, what opens broader evaluation possibilities based on the use of different methodologies of the quantitative oriented research [5–8]. Presented results in this paper are connected with the research data gathered in the questionnaire part C.

The questionnaire was administrated to a research sample consisted of 210 primary and secondary school teachers representing primary and secondary schools in three of eight regions of Slovakia (Nitra region, Trnava region and Bratislava region). From this number of the addressed teachers, 173 fulfilled it, i.e. the questionnaire response rate was 83.3%, what also proves topicality and usefulness of the solved issue. From the final total number of the 173 respondents 68 of them were teachers of primary level of education (ISCED 1), 69 were teachers of lower level of secondary education (ISCED 2) and 19 of them were teachers of upper level of secondary education (ISCED 3).

Factual items of the questionnaire part A were focused on identification of the respondents' gender (A1), lengths of their teaching practice (A2) and identification of the official category (A3) and sub-category (A4) of the teaching staff to which the concerned respondent belongs according to the legislation rules. Slovak legislation [5] distinguishes 7 categories of the teaching staff, which are teacher, vocational education teacher (supervisor), governess, teacher assistant, foreign lecturer, sport school/classroom trainer, accompanist, and in relation to the regional schools (ISCED1–ISCED3) it categorizes teachers in three sub-categories, which are (a) *primary education teacher*, (b) *lower secondary education teacher* and (c) *upper secondary education teacher*. A detailed description of the research sample is summarized in Table 1.

Table 1. Structure of the research sample of the respondents

Factor	Factor category value	Absolute number	Relative number (%)
Gender (A1)	male	15	8.67
	female	158	91.33
Length of teaching practice (A2)	up to 5 years (including)	46	26.59
	from 5 up to 20 years (incl.)	87	50.29
	more than 20 years	40	23.12
Category of the teaching staff (A3)	teacher	156	90.17
	governess	17	9.83
Sub-category of the teaching staff (A4)	teacher of primary level of education (ISCED 1)	68	43.59
	teacher of lower level of secondary education (ISCED 2)	69	44.23
	teacher of upper level of secondary education (ISCED 3)	19	12.18

As it has been above-mentioned, at the ordinary items C1–C13 the respondents expressed their opinions and assessments to the use of various interactive educational activities and digital means in teaching processes taking into consideration different aspects of the teaching process. The assessment was done through a four-point scale, i.e. by assessments from 1 to 4 points (1 – *insignificant, unimportant, without any influence*, 2 – *rather insignificant, rather unimportant, rather without influence*; 3 – *rather significant, rather important, rather with influence*, 4 – *significant, important, with influence*). A choice of the neutral, emotionally indifferent attitude towards the given questions/statements was not included because we wanted to force the respondents to express themselves clearly and exactly. Each respondent's response to the particular ordinary items were recorded, i.e. we recorded the scale values by which the respondent evaluated impact of the interactive educational activities and digital means on the selected aspects of the teaching process (the level of his/her agreement or disagreement with the given statements on the observed phenomena, or the positive or negative assessment stated at the particular item). The selected aspects of the teaching process are presented in Table 2.

Table 2. Overview of the teaching process aspects in relation to which contribution of the use of the interactive education activities and digital means to increase teaching efficiency was observed

Aspect	Observed phenomenon
C1	*increase of pupils' motivation*
C2	*increase of pupils' interest in the taught subject*
C3	*increase of pupils' activity during the lesson*
C4	*development of pupils' creativity*
C5	*pupils' easier understanding of the presented new subject matter*
C6	*longer-term retention of the presented subject matter*
C7	*increase of the pupils' skills to apply the acquired knowledge in practical task solving*
C8	*increase of the taught subject popularity (favour)*
C9	*increase of pupils' mutual co-operation*
C10	*increase of pupils' "spirit of competitivity"*
C11	*positive influence on pupils' disciplined behaviour*
C12	*increase of the positive classroom climate*
C13	*development of pupils' digital literacy*

In the presented research results attention is not paid to the significance level of the use of different interactive educational activities and digital means in the teaching process for the given specific aspects of teaching (neither to the differences between the significance levels identified for the particular observed aspects). Attention is paid to the identification of the differences in the recorded assessments of the significance levels in dependence on the respondents' factor category values (in dependence on the respondents' affiliation to a certain sub-category of the teaching staff). This means that here are presented results of the analysis of the respondents' responses to the particular items in dependence on the segmentation factor SUB-CATEGORY OF THE TEACHING STAFF. Results of this analysis express significance of the use of the interactive educational activities and digital means in teaching in relation to the observed teaching aspect in dependence on the pupils/students' (recipients') age category (in frame of the ISCED 1–ISCED 3). The results provide an answer to the question whether there are any differences in the benefits, which these facilities bring into the teaching process, in dependence on the age category of the pupils/students.

Following the achieved results we were further interested what is the divergence of the means of the achieved scale scores of the respondents' responses to the observed items.

In frame of the statistical processing of the results, following null hypothesis, de facto presenting 13 particular null hypotheses (connected with the phenomena observed in the particular questionnaire items C1–C13, Table 2), was tested:

H0: Respondents' answers to the questionnaire item C do not depend on the level of the factor SUB-CATEGORY OF THE TEACHING STAFF.

Null hypotheses were tested on the 5% significance level through both parametric and nonparametric tests.

3 Results and Their Discussion

Following results of one-way ANOVA as well as its nonparametric alternative Kruskal-Wallis ANOVA null hypotheses are not rejected in case of the variables C1 (*increase of pupils' motivation*), C2 (*increase of pupils' interest in the taught subject*), C3 (*increase of pupils' activity during the lesson*), C4 (*development of pupils' creativity*), C7 (*increase of the pupils' skills to apply the acquired knowledge in practical task solving*), C8 (*increase of the taught subject popularity/favour*), C9 (*increase of pupils' mutual co-operation*), C10 (*increase of pupils' "spirit of competitivity"*) and C13 (*development of pupils' digital literacy*), i.e. these variables do not depend on the factor SUB-CATEGORY OF THE TEACHING STAFF. Statistical dependence was proved only for the items C5 (*pupils' easier understanding of the presented new subject matter*), C6 (*longer-term retention of the presented subject matter*), C11 (*positive influence on pupils' disciplined behaviour*) and C12 (*increase of the positive classroom climate*). Descriptive statistics of the final score of the responses to these items (C5, C6, C11, C12) are presented in Table 3 which comprises more detailed statistical view on the examined issues in dependence on the segmentation of the respondents—teachers into one of the four already above-mentioned categories (*primary education teacher* (a), *lower secondary education teacher* (b) and *upper secondary education teacher* (c)). Moreover in the table there are presented also descriptive statistics (values of the mean, standard deviation, standard error of the mean estimate and 95% confidence interval for the average value of the scale of the final score) of the given items overall, i.e. for the whole research sample, without any segmentation of the respondents on the factor SUB-CATEGORY OF THE TEACHING STAFF (TS-Cat).

Results of the dot estimation of the average scores of the assessments of the particular factors show that the group of the respondents – *primary education teachers* (a) in comparison to other two group of the respondents, *lower secondary education teachers* (b) and *upper secondary education teachers* (c), responded to all of the four tested ordinary items (C5, C6, C11 a C12) more positively. Average values of the scores of the respondents' responses to the items C5, C6, C11 and C12 are from the scale range 2 (*rather insignificant*) – 4 (definitely *significant*) from the maximal scale value 4, while majoritarian part of these items was evaluated by the respondents on the level *rather significant* (scale value 3). The tabulation (Table 3) of the results of the respondents' assessments of the level of the influence of the use of interactive educational activities and digital means on the specific aspects of education C5, C6, C11, C12 shows that the lowest average score was recorded in case of the item C11 (2.42) at which the respondents expressed their opinions on the positive influence of the use of interactive educational activities and digital means on pupils' disciplined behaviour. According to the group of the respondents – *upper secondary education teachers* (c) the intervention of the interactive educational activities and digital means into the education process has not any adequate influence on the positive behaviour affecting at teaching time (C11). The achieved results have been quite surprising as there were expected more positive opinions of the respondents in the context of the observed means influence on this aspect of education.

Table 3. Descriptive statistics of the items C5, C6, C11 and C12 of the questionnaire part C

Item **C5**	Factor value	N	Mean	Standard deviation	Standard error	Confidence Interval for the Mean	
						-95.00%	+95.00%
Total		173	3.329	0.611	0.046	3.238	3.421
"A4"	a	85	3.459	0.524	0.057	3.346	3.572
"A4"	b	69	3.203	0.632	0.076	3.051	3.355
"A4"	c	19	3.211	0.787	0.181	2.831	3.590

Item **C6**	Factor value	N	Mean	Standard deviation	Standard error	Confidence Interval for the Mean	
						-95.00%	+95.00%
Total		173	3.225	0.674	0.051	3.124	3.327
"A4"	a	85	3.388	0.599	0.065	3.259	3.518
"A4"	b	69	3.072	0.649	0.078	2.917	3.228
"A4"	c	19	3.053	0.911	0.209	2.613	3.492

Item **C11**	Factor value	N	Mean	Standard deviation	Standard error	Confidence Interval for the Mean	
						-95.00%	+95.00%
Total		173	2.699	0.910	0.069	2.563	2.836
"A4"	a	85	2.965	0.837	0.091	2.784	3.145
"A4"	b	69	2.449	0.916	0.110	2.230	2.670
"A4"	c	19	2.421	0.902	0.207	1.987	2.856

Item **C12**	Factor value	N	Mean	Standard deviation	Standard error	Confidence Interval for the Mean	
						-95.00%	+95,00%
Total		173	2.821	0.920	0.070	2.683	2.959
"A4"	a	85	3.000	0.873	0.096	2.812	3.188
"A4"	b	69	2.623	0.925	0.111	2.4009	2.845
"A4"	c	19	2.737	0.991	0.227	2.259	3.215

On the contrary, the highest average score was recorded at the items C5 (3.45) and C6 (6.38) in case of the group of the respondents—*primary education teachers.* The results indicate that the teaching has an object-lesson and attractive character for the pupils of the respective age category (based on the given possibility to enter actively into the object lesson teaching to both the teacher and the pupils).

In general quite satisfactory finding is the fact that the average score values achieved by the particular groups of the respondents for all the items did not occur bellow the scale value 2.

The final standard deviation values of the respondents' responses to the particular items C5, C6, C11 and C12 are not much different. The confidence interval estimation for the mean score values of the particular items ranged from the value 1.99 even to the value 3.59. In frame of the used scale this means evaluation of the significance of the intervention of the interactive educational activities and digital didactic means in the teaching process in range from *rather insignificant* up to definitely *significant.*

The most heterogeneous responses were recorded at the item C12 in case of the group of the respondents – *upper secondary education teachers* (variability index 0.99). In case of this sub-group of the respondents, the highest heterogeneousness of the stated assessments regarding the significance of the implementation of the interactive educational activities and attractive electronic teaching materials into the upper secondary education (ISCED 3) to increase the positive classroom climate during the lesson was found out. All the same a higher heterogeneousness of the responses occurred also in case of the assessment of the items C12 (variability index 0.93) and C11 (variability index 0.92) done by the *lower secondary education teachers*. Based on the interval estimation of the means, the score average values of the responses to these items ranged from the value 2.56 even to the value 3.21 (questionnaire item C12 assessed by the respondents – *upper secondary education teachers*), from 2.40 to 2.85 (item C12 assessed by the *lower secondary education teachers*), from 2.23 to 2.67 (item C11 assessed by *lower secondary education teachers*).

The lowest value of the standard deviation (0.52) was found out at the items C6 (range 3.26–3.52) and C5 (range 3.35–3.57) at the group of the respondents – *primary education teachers*. This means the lowest variability of the given statements to the specified teaching aspects given by the sub-category of the teaching staff *primary education teachers*. At the same time, at the items C5 and C6 assessments done by the *primary education teachers* recorded also the lowest value of the average score (C5–3.46; C6–3.39).

After the rejection of the null hypothesis, we were interested whether there are or there are not statistically significant differences among the responses to the items C5, C6, C11 and C12 given by the particular sub-groups of the respondents in dependence on the factor SUB-CATEGORY OF THE TEACHING STAFF, and if yes, then between which levels (particular sub-categories) of this factor they occur.

Identification of the homogeneous groups in dependence on the factor SUB-CATEGORY OF THE TEACHING STAFF was done by the means of the multiple comparison of the particular couples of the teaching staff sub-categories. In frame of each of the tested questionnaire items C5, C6, C11 a C12 two homogeneous groups were identified. Overview of the relevant results is presented in Table 4.

Table 4. Identification of the homogeneous groups

Item A4	Item **C5** Mean	1	2	Item A4	Item **C6** Mean	1	2
b	3.20	****		c	3.05	****	
c	3.21	****	****	b	3.07	****	
a	3.46		****	a	3.39		****

Item A4	Item **C11** Mean	1	2	Item A4	Item **C12** Mean	1	2
c	2.42	****		b	2.62	****	
b	2.45	****		c	2.74	****	****
a	2.96		****	a	3.00		****

Based on the multiple comparison (Table 4) statistically significant differences were identified between the categories of the teaching staff *primary education teacher* (a) and the rest of the categories, i.e. *lower secondary education teacher* (b), *upper secondary education teacher* (c) in case of the items C6 and C11. In case of the items C5 and C12 (Table 4) statistically significant differences were identified only between the categories of the teaching staff *primary education teacher* (a) and *lower secondary education teacher* (b) of the factor SUB-CATEGORY OF THE TEACHING STAFF. Statistically significant differences among the responses of the respondents to the questionnaire items in frame of the particular homogeneous groups were not proved. In frame of the identified homogeneous groups the respondents of the particular group (teaching staff sub-category), regardless of the factors GENDER—item A1, LENGTH OF THE TEACHING PRACTICE—item A2, CATEGORY OF THE TEACHING STAFF—*teacher* or *governess*—item A3, responded to each of the four observed questionnaire items more or less identically. At each of the four observed questionnaire items C5, C6, C11 and C12 the sub-group of the respondents of the SUB-CATEGORY OF THE TEACHING STAFF *lower secondary education teacher* (b) achieved the total mean even identical to the final mean of the sub-group of the respondents of the SUB-CATEGORY OF THE TEACHING STAFF *upper secondary education teacher* (c).

With respect to the identified normality variance, assumption of the variance equality for one-way ANOVA was tested by means of the non-parametric Levene's test. Results of this test did not proved failure of the assumption of the variance equality for any of the observed items C1–C13.

Not to decrease standard of the statistical test proofs in relation to the obtained research data, there was applied also the non-parametric alternative to one-way ANOVA, which is Kruskal-Wallis ANOVA. As the results were the same, they can be taken as robust.

Test results of the questionnaire items C5, C6, and C11, C12, i.e. of the relevant given aspects of the teaching process, according to the factor SUB-CATEGORY OF THE TEACHING STAFF are visualised at the graphs of the mean and interval estimation (Graphs 1 and 2). As the range of the interval estimation of the scale value mean for the item C5 (*pupils' easier understanding of the presented new subject matter*) shows (Graph 1), the most homogeneous responses of the respondents were recorded at the sub-group of the respondents—*primary education teachers* (a). On the contrary, *lower secondary education teachers* (b) and *upper secondary education teachers* (c) responded comparatively heterogeneously. Similar situation can be seen also in case of the teachers' reactions at the item C6 (*longer-term retention of the presented subject matter*).

Results of the repeated measure analysis (Table 4) confirmed the statistical significance of the response differences among the categories *primary education teacher* (a) and *lower secondary education teacher* (b) of the factor SUB-CATEGORY OF THE TEACHING STAFF for the ordinary item C5. This is proved also in Graph 1, in which the interval estimations of the scale value mean in case of the respondent sub-group—*primary education teachers* (a) and the respondent sub-group—*lower secondary education teachers* (b) do not overlap, or they overlap only partially. By contrast, it does overlap in case of the sub-group of the respondents—*upper secondary*

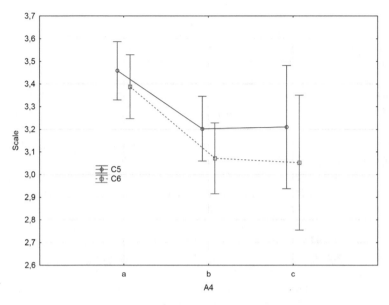

Graph 1. Dot and interval estimation of the assessments stated at the items C5 and C6 in dependence on the factor SUB-CATEGORY OF THE TEACHING STAFF

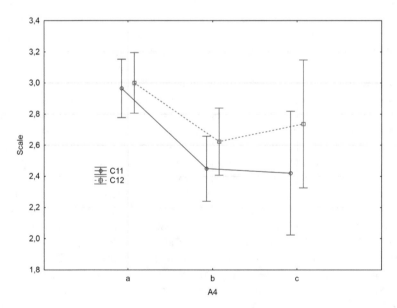

Graph 2. Dot and interval estimation of the assessments stated at the items C11 and C12 in dependence on the factor SUB-CATEGORY OF THE TEACHING STAFF

education teachers (c) with the two other sub-groups (*primary education teachers* (a) and *lower secondary education teachers* (b)), between which statistically significant differences in the respondents' responses were not proved.

Identification of the homogeneous groups in case of the specified teaching aspects C11 – *positive influence on pupils' disciplined behaviour* and C12 – *increase of the positive classroom climate* according to the SUB-CATEGORY OF THE TEACHING STAFF factor category value is visualised by the graph of the mean and interval estimation (Graph 2).

From the point of view of the respondent differentiation according to their affiliation to some of the SUB-CATEGORY OF THE TEACHING STAFF, the highest mean score value at the assessment of the both teaching aspects (item C11 and item C12) was recorded at the sub-group of the respondents – *primary education teachers* (a). At this teaching staff group the value of the total mean reaches significantly the highest value, what is 2.96 (C11) or 3.00 (C12) respectively. From the range of the interval estimation of the scale value mean at both of these questionnaire items it can be seen that the most homogeneous responses of the three tested variables were recorded just in case of this sub-group of the respondents. On the contrary, comparatively more heterogeneous responses were recorded in case of the sub-group of the respondents – *lower secondary education teachers* (b) as well as in case of the sub-group of the *upper secondary education teachers* (c).

Test results in case of the item C11 proved statistical significance of the response differences between the category *primary education teacher* (a), second homogeneous group, and the other two tested variables – *lower secondary education teacher* (b) and *upper secondary education teacher* (c), the first homogeneous group. This can be seen also in the graphical visualisation of the whole situation in Graph 2, where the interval estimation of the scale value mean in case of the sub-group of the respondents – *primary education teachers* (a) does not overlap, or it overlaps only partially, with the two other tested variables, in particular with the sub-group of the respondents – *lower secondary education teachers* (b) and the sub-group of the respondents – *upper secondary education teachers* (c). On the contrary, the graphs overlap in case of the respondent sub-groups *upper secondary education teachers* (c) and *lower secondary education teachers* (b), between which the statistically significant differences were not proved. In frame of the mentioned homogeneous sub-groups respondents responded to this questionnaire item almost in the same way.

As to the statistical significance of the differences between the responses, the same situation as it was recorded at the item C5 was identified also at the questionnaire item C12, i.e. significant differences between the responses given by the sub-group of the *primary education teachers* (a) and the sub-group of the *lower secondary education teachers* (b) were recognized also in case of the item C12 (Graph 1). This fact is visible also on the graphical visualisation presented in Graph 2, where the interval estimation of the scale value mean for the sub-group of the respondents *primary education teachers* (a) does not overlap just only with the value of the tested factor SUB-CATEGORY OF THE TEACHING STAFF *lower secondary education teacher* (b). On the contrary, in case of the sub-group of the respondents *upper secondary education teachers* (c) it overlaps with both other tested values of the factor – *lower secondary education teachers* (b) and *primary education teachers* (a), between which no statistically significant differences of the respondents' responses were proved.

4   Conclusion

Similarly to the values of the scale value means of the items C5, C6, C11 and C12 (Table 3) also the other items reached the scale value means above 2.0, what means that the practising teachers assess the influence of the use of interactive educational activities and digital means in teaching process on its efficiency at least as *rather significant* in relation to each of the considered aspects of the teaching process. As to the level of the significance of the use of interactive educational activities and digital means in teaching process for the considered aspects, there has been more or less consensus of the assessments given by the different groups of the respondents. Dependence of the assessments on the factor SUB-CATEGORY OF THE TEACHING STAFF was proved only for four of the assessed aspects, which were *pupils' easier understanding of the presented new subject matter* (C5), *longer-term retention of the presented subject matter* (C6), *positive influence on pupils' disciplined behaviour* (C11) and *increase of the positive classroom climate* (C12). The group of the respondents— *primary education teachers* assessed significance of the use of the interactive means in teaching on a higher level to all of the four tested aspects than the other two groups of the respondents (*lower secondary education teachers* and *upper secondary education teachers*). In our opinion this result follows the differences of the age categories of the pupils the particular sub-categories of the teachers are dealing with. Moreover the influence of the observed teaching aids on the youth can be seen also in other dimensions (e.g. [10]).

Acknowledgements. This work has been supported by the Cultural and Educational Grant Agency of the Ministry of Education, Science, Research and Sport of the Slovak Republic under the project No. KEGA 041UK-4/2017 and by the Slovak Research and Development Agency under the contract No. APVV-14-0446.

References

1. Brečka, P.: Teoretická príprava budúcich učiteľov na prácu so systémami interaktívnych tabúľ prostredníctvom LMS Moodle. In: Modernizace vysokoškolské výuky technických předmětů, pp. 35–38. Gaudeamus, Hradec Králové (2015)
2. Brečka, P., Olekšáková, M.: Implementation of interactive whiteboards into the educational systems at primary and secondary schools in the Slovak Republic. In: International Conference on Advanced Information and Communication Technology for Education (ICAICTE 2013), pp. 126–130. Atlantis Press, Hainan (2013)
3. Záhorec, J., Hašková, A., Munk, M.: Curricula design of teacher training in the area of didactic technological competences. In: Auer M., Guralnick D., Simonics I. (eds.) Teaching and Learning in a Digital World. ICL 2017. Advances in Intelligent Systems and Computing, vol. 716, pp. 383–393. Springer, Cham (2017). DOI:https://doi.org/10.1007/978-3-319-73204-6_43
4. Záhorec, J., Hašková, A., Munk, M.: Teachers' didactic technological competences: Results of the pilot research. In: 2017 IEEE 11th International Conference on Application of Information and Communication Technologies (AICT), pp. 345–349. Moscow (2017)

5. Aleandri, G., Refrigeri, L.: Lifelong education and training of teacher and development of human capital. In: Global Conference on Linguistics and Foreign Language Teaching (LINELT 2014), Procedia – Social and Behavioral Sciences. Vol. 136, pp. 1–564. Elsevier (2014)

6. Alt, D.: Science teachers' conceptions of teaching and learning, ICT efficacy, ICT professional development and ICT practices enacted in their classrooms. Teach. Teacher Educ. **73**, 141–150 (2018). https://doi.org/10.1016/j.tate

7. Bray, A., Tangney, B.: Technology usage in mathematics education research—a systematic review of recent trends. Comput. Educ. **114**, 255–273 (2017)

8. Brooks, E.P., Borum, N., Rosenørn, T.: Designing creative pedagogies through the use of ICT in secondary education. In: International Conference on Education & Educational Psychology 2013 (ICEEPSY 2013). Procedia – Social and Behavioral Sciences. Vol. 112, pp. 35–467, (2014)

9. Law No. 317/2009 on Teaching Staff and Specialists and its Amendments. (2009). Available from: https://www.minedu.sk/data/att/2918.pdf

10. Leskova, A., Valco, M.: Identity of adolescents and its dimensions in the relation to Mass media: philosophical-ethical reflections. XLinguae J. **10**(3), 324–332 (2017)

EPortfolio Maturity Framework

Igor Balaban[1(✉)], Serge Ravet[2], and Aleksandra Sobodić[1]

[1] University of Zagreb, Varaždin, Croatia
igor.balaban@foi.hr
[2] ADPIOS, Poitiers, France

Abstract. This paper presents activities and outcomes of a research project targeted to support individuals, organisations, communities and public authorities to reflect on and improve the use of technologies for learning with special focus on ePortfolio practices. With the support of the European ePortfolio community of practice, the 18 months long research resulted in the design of an ePortfolio Maturity Framework (EPMF) including five main components and forty-four indicators. The EPMF describes educational ecosystem with the support of ePortfolios and enables organizations to develop a tool to benchmark and measure their progress across time.

Keywords: Eportfolio · Eportfolio framework · Maturity framework
Eportfolio implementation

1 Introduction

A growing number of individuals and organisations is exploiting or planning to explore the benefits of ePortfolios as "selected collection of work presented electronically" [1]. The collection of work refers to "artefacts which provide evidence of learning, competencies and employability" [2].

Research done by [3] at the Association of American Universities (AAU) showed that in educational settings most people use ePortfolio at the program and course level, mostly experimenting with the integration of constructivist principles with the support of ePortfolio as a tool. This is in line with the research of [4] that reviewed a dozen of papers reporting on ePortfolio practices in which the majority of examples involve the use of ePortfolios at a course level, or even just case studies. Moreover, examples vary in terms of the extensiveness and comprehensiveness of ePortfolio use. Such scattered practice, often reflecting individual efforts in introducing ePortfolios to teaching and learning can be very confusing for organizations planning to use ePortfolios or wanting to further exploit the benefits of such use. Those facts were further supported by the results obtained during several workshops with members of the ePortfolio community and the stakeholders' consultations within this research project.

Therefore, the aim of this paper is to present the framework that can help organizations to reflect on and improve the use of ePortfolios as an institution-wide transformation taking into account not just ePortfolio as a tool for teaching and learning but also other key dimensions of teaching and learning transformations of the 21st century.

© Springer Nature Switzerland AG 2019
M. E. Auer and T. Tsiatsos (Eds.): ICL 2018, AISC 917, pp. 473–484, 2019.
https://doi.org/10.1007/978-3-030-11935-5_45

2 Background Research

We approached the problem by analysing the cases of ePortfolio implementations that go beyond single classroom use (i.e. testing user experience with ePortfolios for reflection, assessment etc.). It brought the opportunity to outline and describe ePortfolio as an institution-wide initiative which will present a solid ground for the creation of ePortfolio Maturity Framework (EPMF). Examples from [5–7] indicate that ePortfolios, as well as the 21st century education, bring learning as well as recognition of learning outside formal schooling system stressing the need to use top-down management approach to an institution-wide implementation of ePortfolios.

Analysis of a dozen papers on ePortfolio research and practice and showed that ePortfolio is an ecosystem supporting learning [4]. Educational ecosystem is formed of population inside (students, teachers, principals and other staff) and outside the school (parents, families, friends, entrepreneurs, institutions, etc.) alongside with the non-living elements (classrooms, electronic resources, buildings, etc.), and their mutual interaction [8, 9]. This is also consistent with findings from [10] who identified design principles for future iterations of an ePortfolio-based learning environment. Besides technology, they differ pedagogical approach that involves reflections and assessment and focus also on people as key players. In addition, a comprehensive review of literature from various disciplines and theoretical frameworks made by [5] showed that learning exists without technology (they differ other components besides technology), but that technology is indeed the driver for changes in the 21st century teaching and learning.

From these findings, it is evident that the ePortfolio should be observed within a complete educational ecosystem. Therefore, we did not focus only on the ePortfolio related frameworks, but also on the frameworks that consider the use of ICT in teaching and learning (see Table 1). It was concluded that the existing frameworks suffered from two major drawbacks:

1. Exclusiveness. Most of them are focused on the use of ePortfolio or ICT merely as a tool, without exploring or describing other intertwined areas of educational ecosystem;
2. Obsolescence. They fail to encompass most recent ePortfolio findings since they have been developed some time ago.

Based on the gaps mentioned above, this research aims to propose the ePortfolio Maturity Framework (EPMF) that describes the 21st century educational ecosystem supported by the ePortfolios by identifying the key areas (components) and the measurement indicators.

3 Research Design and Methodology

The research was divided into three phases during which the EPMF was developed and refined. The first phase resulted in the creation of an initial framework (EPMF) with five components and the potential indicators. The second phase led to the EPMF refinement through two additional rounds of experts' consultations. The third phase aimed at consolidating the work from the previous phase and ended with 44 indicators and their descriptors.

Table 1. An overview of the analysed frameworks

Name (Author)	Year	Purpose	Structure
The ePortfolio maturity matrix (European Institute for E-learning) [34]	2007	For professionals and organisations to identify their current practice in order to plan their future improvements	Six main components through five different maturity stages
The ePortfolio maturity model (SURF NL Portfolio) [15]	2007	To benchmark the state of HE ePortfolio practice in the Netherlands and to encourage continuous improvement	Five core elements through five phases
Self-review framework (Becta) [35, 12]	2008	To identify the maturity of ICT use in schools and prioritise the areas for improvement	Consists of six elements divided into strands, and various levels
Australian ePortfolio toolkit (Australian Learning and Teaching Council) [28]	2009	To inform diverse stakeholders in higher education about issues and opportunities associated with ePortfolio learning	Assessment of five core factors through five stages of maturity
The e-portfolio implementation toolkit (Joint Information Systems Committee) [14]	2013	To assist managers and practitioners in a large-scale e-portfolio implementation in a range of contexts	Case studies, video stories and guidance drawn from the experiences of the participating institutions
Updated self-review framework (Naace) [36]	2014	To provide a structure for reviewing the school's use of technology and its impact on school improvement	Six main elements through four stages of maturity

3.1 Phase 1: Development of the Initial Version of the EPortfolio Maturity Framework (EPMF)

The research approach taken for the development of EPMF was for the most part qualitative and included: (1) Comprehensive literature review on ePortfolio ecosystem, frameworks and toolkits; (2) Qualitative analysis of the selected frameworks and toolkits; and (3) Testing and feedback through a series of expert consultation, workshops and semi-structured interviews.

As a starting point for the development of EPMF, an extensive review of the academic and grey literature was conducted, as indicated in the previous section. The literature review was conducted with the following aims: (1) To identify, analyse and compare the existing frameworks and toolkits related to ICT in teaching and learning and ePortfolios and their maturity; (2) To get an insight into ePortfolio maturity including definitions, key elements, enablers and barriers; and (3) To identify recent findings that describe ePortfolio as an ecosystem in order to take into account its comprehensiveness. A qualitative analysis of the existing frameworks was performed in order to identify the commonalities, the points of divergence and gaps. Insights about their design, focus, methodology and implementation strategies were elicited. A brief overview of the identified frameworks is shown in Table 1.

Next, the findings from the analysed frameworks were used and combined with the findings drawn from relevant and up-to-date comprehensive ePortfolio related literature review. As a result, the first draft of the EPMF defining five main components and their indicators was proposed. Table 2 shows how five components of the EPMF are aligned with key domains of the existing frameworks.

Table 2. Mapping of the EPMF to existing frameworks

EPMF main components	Existing frameworks with their main components				
	EIfEL	Becta	Self-review framework	Australian ePortfolio toolkit	SURF maturity model
Learning	Learning	Learning and teaching Curriculum Extending opportunities for learning	Curriculum Planning Teaching and learning	Institutional factors	Curriculum embedding
Assessment	Assessment	Assessment	Assessment of digital capability	Teacher/tutor factors	
People	Vision	Professional development	Professional development	Teacher/tutor factors Student/learner factors	Freedom of choice regarding an educational programme
Technology	Implementation	Resources	Resources	Institutional factors EPortfolio system factors	ICT infrastructure
EPortfolios	Implementation	Leadership and management Impact on pupil outcomes	Leadership and management	Institutional factors	Freedom of choice regarding portfolio Consistency between policy/practice

The draft version of the EPMF was submitted during an online workshop to 16 experts from the Europortfolio community representing higher education institutes and experts from Croatia, Austria, France, Spain, Poland, UK, Denmark and the USA. The main task was to analyse the relevance and completeness of the five components and indicators then reach a consensus between all panellists. Following the workshop, the framework was uploaded to Europortfolio Portal to invite Europortfolio community members to reflect on the initial results. Their comments were analysed and incorporated into the EPMF as the last step in the first phase.

3.2   Phase 2: Experts Consultations

In order to further refine the EPMF, two workshops were organized during two international conferences dedicated to online learning that involved ePortfolios. The first workshop involved 19 panellists, ePortfolio experts, who were divided into groups and each group had to analyse one key component at a time with respect to its elements and descriptors.

Within each group, panellists wrote down their own comments, shared them with the group, got responses and worked towards establishing a consensus. After that, each group forwarded its comments along with key component to the other group and the procedure was repeated. During the workshop, each group analysed all key components with their indicators and descriptors. A workshop leader presented the results followed by a discussion. The same procedure was repeated during a second workshop that involved 14 panellists.

3.3   Phase 3: Consolidation of the Work

Based on the results from previous phases, the complete work was consolidated which resulted in the conceptual model of the EPMF (Fig. 1) including five key components, 44 indicators and their descriptors. The detailed description of the key components and their indicators can be found in Sect. 4.

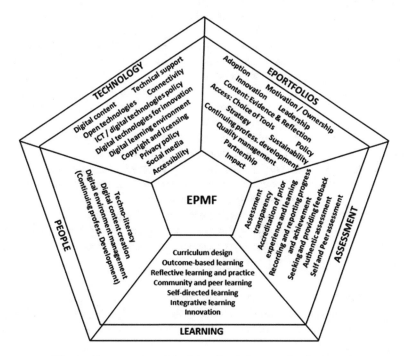

Fig. 1. Conceptual model of the ePortfolio maturity framework

4 The EPortfolio Maturity Framework (EPMF)

The EPMF (see Fig. 1) proposed below is a result of consolidated work as described in previous sections. It identifies and describes five key components that constitute the ePortfolio ecosystem. Three of them are not technology related but are essential for the maturity of "authentic learning": Learning, Assessment, and People. The fourth component is strictly related to technology, while the fifth component makes explicit reference to the ePortfolios. However, all the other four components are critical to the ePortfolio practice.

Learning component describes learning practices and processes, e.g. outcome-based learning. A variety of theoretical findings have been used. It is aligned with the work of [11] and [12], the work of [13] based on a report from 22 countries, exploring the advantages of recognizing non-formal and informal learning outcomes and recommending how to organize their recognition. It is also aligned with the work of [14] and [15] involving learning process and reflection, as well as the work from [16] analysing the use of reflection and self-directed learning with ePortfolios. The research from [17] also showed that the ePortfolio has a positive effect on students' self-regulated learning skills. Furthermore, the work from [18] was followed to incorporate elements of authentic learning context. In this context, up-to-date work from [19–23] was followed. As a result, seven indicators are proposed in Table 3.

Table 3. Indicators and descriptors for learning component

Indicators	Descriptors
Curriculum design	The local and/or regional community is actively involved, along with learners, staff and educational leaders in the design and review of the curriculum
Outcome-based learning	Staff members are actively involved with their community of practice in the definition and review of the learning outcomes within and across disciplines
Reflective learning and practice	Reflective practice is integrated within a global community of practice (e.g. professional body) and contributes to global innovation and change
Community and peer learning	Peer learning happens beyond institutional boundaries within cross-institutional networks of knowledge exchange
Self-directed learning	The organisation is actively involved in action research regarding self-directed learning in order to improve the policies, methods and tools supporting autonomous learning
Integrative learning	Integrative learning goes beyond disciplines and institutional walls to integrate outside resources, formal, informal and non-formal
Innovation	Innovation is the engine that drives the organisation. It is constantly re-inventing itself

Assessment component describes assessment practices and processes. Besides the previously mentioned frameworks and toolkits, the work of [14], as well as of [24] was followed. In this context, six indicators were identified also using recent work from [25–27] (see Table 4).

Table 4. Indicators and descriptors for assessment component

Indicators	Descriptors
Assessment transparency	Learners are actively involved in the definition of the policies regarding assessment process and criteria
Accreditation of prior experience and learning	The lessons learned through the practice of APEL are shared with the larger community, beyond the institutional boundaries
Recording and reporting progress and achievement	Individual and organisational progress and achievements are aggregated beyond the institution's boundaries to contribute to the improvement and/or transformation of policies
Seeking and providing feedback	The competencies involved in the provision of feedback are recognised and celebrated by the institution
Authentic assessment	Assessment is treated as "learning about learning" and is deeply intertwined with the learning process which is itself based on authentic learning experiences
Self and peer assessment	Teachers and learners are treated equally, i.e. learners' assessment of teachers is regarded as 'peer-assessment' within the learning community

People component describes teaching staff and learners, e.g. regarding techno-literacy and digital literacy. A work from [14] related to supporting transition was used in this context. Also, [28] following Becta model [12] describes students/learners and educators/tutors well. Such classification was also used in the proposed Framework. Although the existing models thoroughly describe the main elements of such context, findings were extended with the work from [29, 30]. This resulted in two sub-dimensions: (1) Teaching staff, and (2) Learners, presented in Table 5. Here, it needs to be noted that, in the context of reflective and authentic learning practice, teachers are also considered to be learners [4]. Therefore, the difference in assessment indicators is only in Continuing professional development.

Technology component explores the level of maturity in the exploitation of tech-nologies and particularly digital and social media. All identified frameworks and toolkits were focused on the use of technology, so this component was created by mapping elements from the existing frameworks and updated based on findings from [31]. Indicators and their descriptors are proposed in Table 6.

EPortfolios component is the only one in the EPMF strictly related to exploitation of ePortfolios. To be as inclusive as possible, beyond existing maturity frameworks focusing on ePortfolios, the work of [4] was reviewed by describing the ePortfolio ecosystem providing an extensive overview and elements of different contexts of ePortfolio use: it describes five different scenarios of ePortfolio use referring to edu-cational context, switching between educational institutions, job application, switching between employment institutions and part-time study/job retention. Furthermore, [24, 32] analysed portfolio as a workspace-process, [14] discussed learning process and reflection and [33] explained reflection portfolio and development portfolio. In total, thirteen indicators were identified and described for this component (see Table 7).

Table 5. Indicators and descriptors for people—teaching staff and learners components

	Teaching Staff	Learners
Indicators	Descriptors	
Techno-literacy	The organisation is experienced as a living laboratory where new technologies and practices emerge	The organisation is experienced as a living laboratory where new technologies and practices emerge
Digital content creation	Teachers work collaboratively beyond institutional borders for the creation/remix of learning resources and for seeking peer feedback	Learners from different levels of maturity and places actively collaborate in the creation/remix of digital contents
Digital environment management	Mastering the competencies required to create and update a tailored learning environment, allows the organisation to re-engineer its learning provision	The organisation has policies and systems to make learners the co-designers of the organisation's digital learning environment, according to their talents and maturity
Continuing professional development	CPD has a transformative effect on society. Learned and informed staff is fully empowered to influence and co-design the learning policies at local, regional and national levels	

Table 6. Indicators and descriptors for technology component

Indicators	Descriptors
Open technologies	The various Open "Things" (OER, Open Knowledge, Open Standards, Pên Trust, etc.) are clearly placed in the perspective of building an Open Society together
ICT/digital technologies policy	Organisation's leaders are actively involved and recognised in innovation networks, beyond the institutional boundaries
Digital technologies for innovation	The use of digital technologies is primarily sought to support innovation and organisational transformation
Copyright and licensing	Learners and teachers are active advocates and supporters of Open Knowledge and Open Educational Resource initiatives
Privacy policy	The organisation is part of a trust infrastructure, a federation facilitating the exchange of personal data under the control of the individuals
Social media	Social media has a transformative effect creating a participative and conducive culture
Accessibility	The creation of resources and tools accessible is not experienced as a constraint but as an opportunity to design for all
Connectivity	Using the institution's infrastructure, learners and teachers explore the potential of the Internet of Things
Digital content	Learners and teachers actively contribute to the curation of the Internet for the benefits of the learning community
Digital learning environment	Learners and teachers actively contribute to the reflection and the design of the future of personal learning environments
Technical support	The practice of technical support is valued and celebrated as a learning opportunity to develop a wide range of competencies at the service of the learning community

Table 7. Indicators and descriptors for ePortfolios component

Indicators	Descriptors
Adoption	Learners and staff engagement almost universally positive. EPortfolios used as a central tool for building both institutional and personal constructions of individuals' activity, achievements, life and identity
Motivation/Ownership	Portfolios are a central tool for building institutional and personal constructions of individuals' activity, achievements, life and identity
Innovation	EPortfolios are recognised as a key means to developing learners and teaching staff identities
Leadership	Senior staff and educational leaders contribute to the overall ePortfolio vision and strategy at local, regional and national levels
Content: evidence & reflection	The information collected for/provided to the institutional ePortfolios is seamlessly collected from the content of the individual ePortfolios
Access: choice of tools	Learners have access to ePortfolios with any device, from anywhere, for any purpose, supporting lifelong learning
Policy	EPortfolios data is used as a prime source of information to inform policies at local/regional/national levels
Strategy	EPortfolio initiatives are coordinated with external bodies beyond the institutional boundaries
Continuing professional development	Staff ePortfolios interact seamlessly with communities of practice, beyond the boundaries of the organisation
Quality management	The organisation's quality management ePortfolio is used for by external bodies for quality assurance (e.g. ISO 9000, TQM, etc.) or professional accreditation (e.g. AACSB, ABET, etc.)
Partnerships	The organisation is actively involved in many ePortfolio partnerships with local, regional, national and international partners
Impact	EPortfolio practice has a global impact on the organisation, changing the organisational culture
Sustainability	There is no need to earmark ePortfolio budget as ePortfolios are fully blended within institutional infrastructure and practice

5   Conclusion

In this paper, it was shown that the design of a maturity framework for the implementation of a specific educational technology, here ePortfolios, cannot be limited to that specific educational technology. It must consider the underpinning elements of any educational ecosystem that are Learning, Assessment, People, and Technologies.

It needs to be stressed also that EPMF has been designed to support organisations, not the individual authors of ePortfolios. With its five components and forty-four indicators, it can help organisations and communities willing to implement and develop ePortfolios to transform their learning environment and practice. It can be also used to review and/or plan the changes required to improve learning and more effective ePortfolio practice.

Based on the EPMF, as shown in the previous section, a series of matrices as a scoring rubric that allows organizations to assess their maturity in terms of ePortfolios can be very easily created. The authors are currently performing a new research with the aim to develop such instrument and to allow organizations to assess their maturity in terms of ePortfolios, to measure progress across time and to benchmark with others.

Acknowledgements. This study was carried out as a part of "Europortfolio: A European Network of Eportfolio Experts and Practitioners" (EPNET) project and was funded by the European Commission, Education, Audiovisual and Culture Executive Agency (EACEA) (grant number 2012-4138/001-001).

References

1. Australian ePortfolio Project.: Australian ePortfolio Toolkit. (2009) Retrieved from http://www.eportfoliopractice.qut.edu.au/information2/toolkit/index.jsp
2. Balaban, I., Divjak, B., Mu, E.: Meta-model of EPortfolio usage in different environments. Int. J. Emerg. Technol. Learn. **6**(3), 35–41 (2011)
3. Barab, S.A., Squire, K.D., Dueber, W.: A co-evolutionary model for supporting the emergence of authenticity. Educ. Tech. Res. Dev. **48**(2), 37–62 (2000)
4. Barrett, H.: Balancing the two faces of ePortfolios. Educação, Formação Tecnol. **3**(1), 6–14 (2010)
5. Baumgartner, P.: Educational Scenarios with E-portfolios—a taxonomy of application patterns. In: Sojka et al. (ed.), SCO. pp. 3–12 (2011)
6. Becta.: Measuring e-maturity in the FE Sector: Technical Report (2008)
7. Brown, S.: The impact of the ePortfolio tool on the process: functional decisions of a new genre. Theory Into Pract. **54**(4), 335–342 (2015)
8. Buyarski, C.A., Aaron, R.W., Hansen, M.J., Hollingsworth, C.D., Johnson, C.A., Kahn, S., Landis, C.M., Pedersen, J.S., Pedersen, J.S.: Purpose and Pedagogy: A Conceptual Model for an ePortfolio. Theory Into Pract. **54**(4), 283–291 (2015)
9. Carneiro, R., Draxler, A.: Education for the 21st Century: lessons and challenges. European J. Educ. **43**(2), 149–160 (2008)
10. Chu, R.J., Chu, A.Z., Weng, C., Tsai, C.-C., Lin, C.-C.: Transformation for adults in an Internet-based learning environment-is it necessary to be self-directed? Br. J. Educ. Technol. **43**(2), 205–216 (2012)
11. Hartnell-Young, E., Harrison, C., Crook, C., Joyes, G., Davies, L., Fisher, T., Pemberton, R., Underwood, J., Smallwood, A.: The impact of e-portfolios on learning (2007)
12. Devedžić, V., Jovanović, J.: Developing open badges: a comprehensive approach. Educ. Tech. Res. Dev. **63**(4), 603–620 (2015)
13. Himpsl, K., Baumgartner, P.: Evaluation of E-Portfolio software. Int. J. Emerg. Technol. Learn. **4**(1), 16–22 (2009)
14. Haddon, E.: Continuing professional development for the musician as teacher in a university context. In: Stakelum, M. (ed.) Developing the Musician: Contemporary Perspectives on Teaching and Learning, pp. 191–206. Ashgate Publishing Company, Aldershot (2013)

15. Ćorić Samardžija, A., Balaban, I.: From classroom to career development planning: Eportfolio use examples. Int. J. Emerg. Technol. Learn. **9**(6), 26–31 (2014)
16. Joint Information Systems Committee.: The ePortfolio implementation toolkit. (2013). Retrieved from https://epip.pbworks.com
17. Kearney, S.: Improving engagement: The use of "Authentic self-and peer-assessment for learning" to enhance the student learning experience. Assess. Eval. High. Educ. **38**(7), 875–891 (2013)
18. Lecourt, A.: Individual pathways of prior learning accreditation in France: From individual to collective responsibilities. Int. J. Manpower **34**(4), 362–381 (2013)
19. Lin, X., Hmelo, C., Kinzer, C.K., Secules, T.J.: Designing technology to support reflection. Educ. Tech. Res0 Dev. **47**(3), 43–62 (1999)
20. Mayowski, C., Golden, C.: Identifying e-portfolio practices at AAU universities. EDUCAUSE Research Bulletin, pp. 1–7 (2012)
21. Mueller, S., Toutain, O.: The Outward Looking School and its Ecosystem. Entrepreneurship360. (2015). Retrieved from http://www.oecd.org/cfe/leed/Outward-Looking-School-and-Ecosystem.pdf
22. Naace.: Self-review Framework Overview Six Questions. (2014). Retrieved from http://www.naace.co.uk/ictmark/srf
23. Nguyen, L.T., Ikeda, M.: The effects of ePortfolio-based learning model on student self-regulated learning. Act. Learn. High Educ. **16**(3), 197–209 (2015)
24. Osborne, R., Dunne, E., Farrand, P.: Integrating technologies into "authentic" assessment design: An affordances approach. Res. Learn. Technol. **21**(1) (2013). https://dx.doi.org/10.3402/rlt.v21i0.21986
25. Roberts, P., Maor, D., Herrington, J.: ePortfolio-based learning environments: recommendations for effective scaffolding of reflective thinking in higher education. Educ. Technol. Soc. **19**(4), 22–33 (2016)
26. Rock, K.Z.: Transferring learning from faculty development to the classroom. J. Nurs. Educ. **53**(12), 678–684 (2014)
27. Saskatchewan Learning.: A Journey of Self-Discovery: Facilitator's Guide to Reflection and Portfolio Development (2005)
28. EIfEL.: ePortfolio—a european perspective. A report on ePortfolio readiness and state of the art in technology and practice. (2009). Retrieved from https://formationdistance2.files.wordpress.com/2009/10/eportfolio-in-europe-v05.pdf
29. SURF.: Stimulating Lifelong Learning: The ePortfolio in Dutch Higher Education. In: Aalderink, W., Veugelers, M. (Eds.) SURF NL Portfolio (2007)
30. Watty, K., McKay, J.: ePortfolios: what employers think. Glob. Focus: EFMD Bus. Mag. **10**(3), 60–63 (2016)
31. Werquin, P.: Recognising Non-Formal and Informal Learning Outcomes, Policies and Practices: Outcomes, Policies and Practices (2010)
32. Wills, K.V., Rice, R.: ePortfolio Performance Support Systems: Constructing, Presenting, and Assessing Portfolios. WAC Clearinghouse, Fort Collins, CO (2013)
33. Yang, F., Li, F.W.B., Lau, R.W.H.: A Fine-Grained Outcome-based Learning Path Model. IEEE Trans. Syst. Man Cybern. Syst. **44**(2), 235–245 (2014)
34. Ćorić, A., Balaban, I., Bubaš, G.: Case studies of assessment ePortfolios. In: Proceedings of the 14th International Conference Interactive Collaborative Learning. pp. 89–94. Slovakia: IEEE Xplore, Piestany (2011)

35. Crosling, G., Nair, M., Vaithilingam, S.: A creative learning ecosystem, quality of education and innovative capacity: a perspective from higher education. Stud. High. Educ. **40**(7), 1147–1163 (2015)
36. Hall, P., Byszewski, A., Sutherland, S., Stodel, E.J.: Developing a sustainable electronic portfolio (ePortfolio) program that fosters reflective practice and incorporates CanMEDS competencies into the undergraduate medical curriculum. Acad. Med. **87**(6), 744–751 (2012)

Evaluation Automation of Achievement Tests Validity Based on Semantic Analysis of Training Texts

Olena Kovalenko, Tetiana Bondarenko$^{(\boxtimes)}$, and Denys Kovalenko

Ukrainian Engineering Pedagogics Academy, Kharkiv, Ukraine
bondarenko_tc@uipa.edu.ua

Abstract. The article describes the evaluation automation methodology of the performance validity tests based on the semantic analysis of training texts. The proposed methodology includes two stages. The semantic core of the training text and the semantic core of the tasks text is formed at the first stage (we use for this the semantic online analysis of the text—SEO-analysis). At the second stage, a comparative analysis of the semantic cores generated in the first stage is carried out using the Excel table processor. Based on the results of the analysis, a decision is made about the content validity of the test tasks. This method allows to significantly reduce the time and cost of the validity evaluation in comparison with traditional methods, and also allows teachers who independently develop test tasks to simply verify the validity of the achievement tests.

Keywords: Achievement test · Test validity · Content validity
Semantic text-analysis · SEO-analysis · Semantic core · Keyword
Keyword density

1 Problem Statement

At present, achievement tests, as a means of knowledge level control, are widely used in all spheres of education. However, the mass use of tests and their development by specialists that do not follow testing systems, but use specific disciplines, updates the issue of the quality of test assignments, since evaluation of the training quality is possible only if objective diagnostic tools are available.

Development of test tasks is a complex process of analysis, preparation and information processing. One of the main characteristics of the achievement test quality is the content validity, which characterizes the test according to the degree of its conformity to the subject area.

2 Analysis of Recent Research and Publications

The author of the work [1] notes that an essential condition for the tests validity is a clear and precise formulation of questions-assignments within the scope of knowledge that is learned by the students from the subject area of any educational discipline.

© Springer Nature Switzerland AG 2019
M. E. Auer and T. Tsiatsos (Eds.): ICL 2018, AISC 917, pp. 485–492, 2019.
https://doi.org/10.1007/978-3-030-11935-5_46

The test will not be effective for trainees if its questions are beyond the scope of knowledge acquired by the trainees, and also do not reach these frames or simply exceed the projected level of knowledge.

Content validity is an important characteristic of the test, without evaluation of which it cannot be considered a measuring tool. The article [2] describes a mixed method for test evaluation validity on the basis of specification sheets and adjustment of experts valuation.

The work [3] describes the technique of test content validity evaluation by independent experts, which did not participate in test development. Usually, the number of experts is at least three per each test. The content test validity examining procedure includes three stages. The first stage is the analysis of the contents of the individual test tasks. Analysis of the content quality of the entire test, which has several parallel options, is performed in the second stage. Preparation of general conclusions and recommendations for improving the content of the test completes the work of experts.

I.e., the traditional methodology for assessing the validity of a test is a complex and time-consuming procedure that requires the involvement of several experts.

Currently active work is carried out to automate the process of developing and evaluating the quality of test tasks in the field of educational achievements testing. For example, the article [4] describes a method for automated generation of test tasks from training texts and describes algorithms and the architecture of a generating system that uses the model of assessing the quality of received tasks on the basis of machine learning.

Authors of the article [5] address the challenge of automatically generating questions from reading materials for educational practice and assessment. Their approach is to overgenerate questions, then rank them. They use manually written rules to perform a sequence of general purpose syntactic transformations (e.g., subject-auxiliary inversion) to turn declarative sentences into questions. These questions are then ranked by a logistic regression model trained on a small, tailored dataset consisting of labeled output from our system. Experimental results show that ranking nearly doubles the percentage of questions rated as acceptable by annotators, from 27% of all questions to 52% of the top ranked 20% of questions.

However, a low percentage of automatic generation of acceptable questions and the need to write rules for sequence of general-purpose syntax conversions execution manually—all suggest that the problem of automatic generation of test tasks is still to be solved.

The goal of this article is to describe evaluation automation of achievement tests validity on the basis of the semantic analysis of training texts. This methodology will significantly reduce the time and costs spent on validity evaluation in comparison with the traditional methodology, and it will allow sufficient to simply verify the achievement tests validity for teachers who develop test tasks on their own.

3 Statement of Basic Material and the Substantiation of the Obtained Results

Before proceeding to the description of the technique, we will analyze the problem of highlighting keywords in the text of the scientific content and describe the main differential signs of keywords. On the basis of this analysis, we will justify the validity of the test when its semantic core coincides with the semantic core of the text.

Key words are lexical units that are basic from the point of view of conveying the meaning of a scientific text. In general scientific terms, the keyword is considered to be the most important semantic element that is essential for understanding. There are many different approaches to the interpretation of the keyword, but it is certain that the keyword determines the content of the text and is the bearer of its main meaning. Taking into account what has been said, it should be recognized that the keywords are words that are the most significant and essential for understanding the text content.

The word can belong to keywords is primarily due to its frequency, repeatability in this text. This criterion is determined by counting the repetitions of word forms throughout the text and highlighting as key words those, the frequency of which use in this text exceeds the frequency of their use in the language.

The author of the work [6] formulated the list of key words functions in the scientific text on the basis of the semantic structure research of scientific texts. The author notes that the keywords form the meaning of the text and provide that it is stored in memory. The keywords also mark the topic of the text (more than half of the core words of the thematic component consists of keywords).

However, the most important keywords function for us is following: when comparing primary and secondary texts (report, annotation, etc.), the set of keywords approaches the content variant and tends to reflect the semantic core of the text in its purest form. It should be clarified that the set of keywords of secondary texts will reflect the semantic core of the primary text in its purest form if these secondary texts reflect its content correctly.

In our case, the primary text is an instructional text, and the secondary text is test tasks. And it is important for us that the semantic core of the test task, in the case of its meaningful validity, should "reflect the semantic core of the training text in its purest form". Thus, taking into account the above function of keywords, we can propose a procedure for automated test tasks validation based on the comparison of the semantic cores of the text and test tasks.

Another feature of the keywords, which T. Moskvitin points out, is the ability of logically ordered keywords to convey the generalized content of the text. Key words allow the reader to quickly navigate across the text, choose exactly those texts that are needed for more thorough study, not wasting time for reading all the text. We want to draw the attention of test developers to this keywords feature. Even if you do not use evaluation automation of tests validity, simply create (and it's really easy) the two semantic cores of the text and test tasks and compare them. You can quickly navigate across the text and find gaps in the meaningful validity of the tests.

Semantic analysis of texts is widely used in search engines to determine the correspondence of sites' text content to incoming requests. The use of semantic text-analysis in our case has a similar goal: we need to determine how test assignments correspond to the text that was offered to the learner.

But if in the case of SEO analysis the text is optimized for the query, then in our case, it is on the contrary: we must optimize the test tasks for the text content.

Computerized semantic analysis has a deep history. Yorick Wilks in the distant 1968 describes the use of an on-line system to do word-sense ambiguity resolution and content analysis of English paragraphs, using a system of semantic analysis programmed in Q32 LISP 1.5 [7].

Now the market of software products presents a large number of applications for various types of text analysis. We went over text analysis software products that are described in Ann Smarty article, [8] and chose semantic and lexical text analyzer—**Textalyser.net**. The main application window is shown on Fig. 1. It offers an exhaustive lexical complexity and variety overview by looking into: different words count; characters count; complexity factor (lexical density); most frequent words analysis (keyword density) and other.

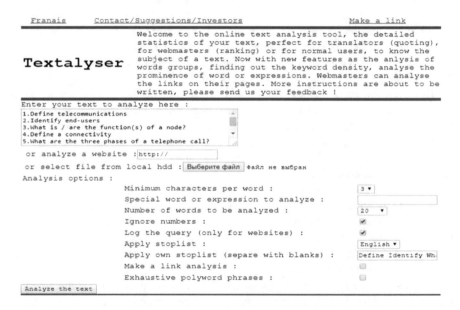

Fig. 1. Main window of semantic and lexical text analyzer Textalyser.net

Using additional analysis options, we can set minimum characters per word, special word or expression to analyze, number of words to be analyzed and other.

We enter the text language that is being analyzed (allowed languages are English and French) in *Apply stoplist* option.

Option *Apply own stoplist* is very useful for test validation analysis. This is due to the fact that there are often words in test tasks that are related to the formulation of the question and are in no way connected to the content of the training text. For example, these are words like: define, describe, find, etc. They must be included in the stoplist in order to not be the part of the semantic core of the test tasks and do not distort the results of the analysis.

Additional and very significant advantage for educational institutions is that this program is completely free and does not require registration.

The automating evaluation of test achievements validity technique, that we offer, includes two stages. Let's consider the use of this technique of test tasks validity evaluation on a concrete example.

For analysis purposes, we have chosen Chap. 1 *Introductory Concepts* from the book by Roger L. Freeman "Fundamentals of Telecommunications" [9]. This book is accessible online (the link is mentioned further in the References of the article) and anyone who wishes to do so can repeat our experiment.

At the first stage, the semantic core of the learning text and the semantic core of the task text is formed using the semantic and lexical text analyzer **Textalyser.net**. Here they preliminarily prepare for further processing. Figure 1 shows the test tasks semantic core formation. In *Apply own stoplist* option we set the words that are met only in test tasks: define, describe, what etc. Therefore, these words do not become the part of the semantic core do not distort the results of the analysis.

At the second stage, a comparative analysis of the semantic cores generated in the first stage is carried out, using the Excel table processor. Content validity evaluation of test tasks is being performed on the basis of comparative analysis.

Figure 2 demonstrates the difference between semantic cores of training text and test tasks.

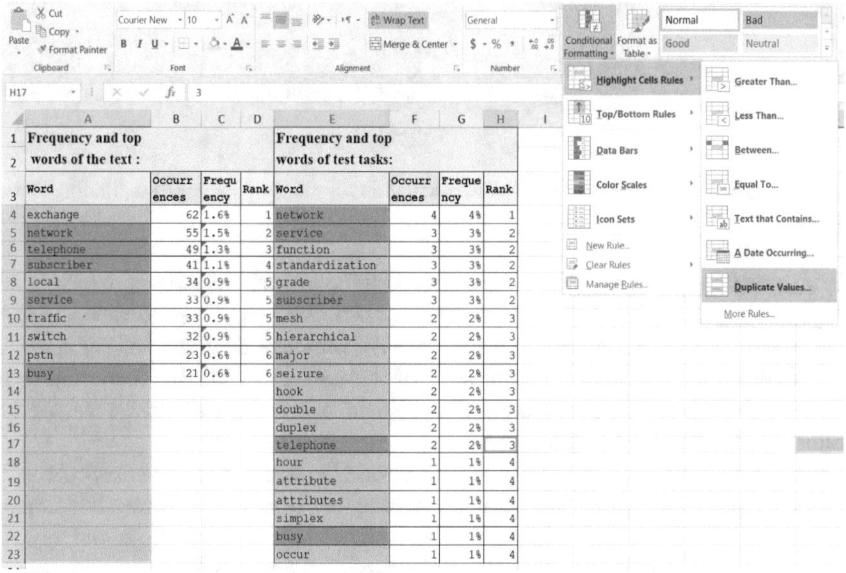

Fig. 2. Comparison of training text semantic core and test tasks

Note that the semantic core of training text is half size of the semantic core of the test tasks. The size of the semantic core of the text that we specify in the *Number of words to be analyzed* option depends on the size of the text. For our example, we selected a small introductory part of the textbook. Based on the analysis of the text, we received the following characteristics: *Total word count*—3916; *Number of different words*—1397. To assess the content validity, we recommend taking about 1% of number of different words (we have determined 10 key words for the analysis purposes).

The size of the semantic core of the test tasks should be twice as large, since this core is more blurred due to the frequent repetition of words with the same rank. For example, keywords with rank 3 are repeated 8 times in the semantic core of test tasks, and for the training text, the maximum number of repetitions is 4 (for rank 5).

To compare semantic cores we use the option of the Excel table processor *"Conditional formatting"* with *"Highlight Cells Rules"* mode and *"Duplicate Values"* parameter. On the basis of comparison results, the matching keywords are highlighted in a different color (see Fig. 3).

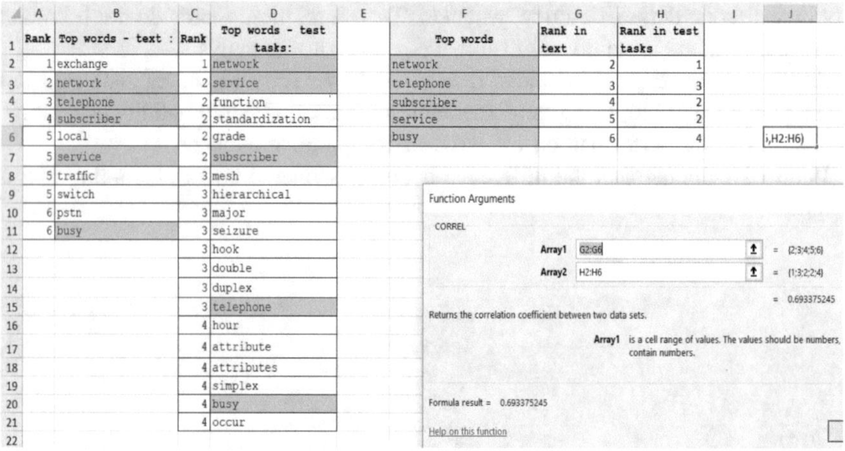

Fig. 3. The example correlation coefficient calculation for ranks of keywords in training text and test task

As a result of the comparison, we obtained five matching keywords, which is 50% for the analyzed training text. This is a good result, because a significant part of the material is included in the training texts from the previous sections, and the test tasks are formulated only within the framework of the new material. But according to our estimates, the minimum percentage of a coincidence for a sufficient level of tests validity should be at least 60%.

In order to get a full assessment of the test tasks validity, it is necessary to conduct additional studies, since the order of the matching key words in the core of the training text and the core of the test tasks can differ substantially. For example, the keyword *"service"* has rank 5 in the semantic core of the training text, but in the semantic core of test tasks—rank 2.

To assess the coincidence of the key words rankings in the semantic core of the text and test tasks, we use the Pearson correlation coefficient, which characterizes the presence of a linear dependence between two sets of numbers. If the ranks of the keywords in the semantic cores are the same, then the correlation coefficient is 1. If the key word rankings have the reverse order (the rank 1 keyword in one core has the last rank in the other core, etc.), then the correlation coefficient is—1. The formula of calculating the level of content test tasks validity is:

$$K_{CV} = \frac{n}{N} R, \qquad (1)$$

where
n the number of matching words in semantic cores of training texts and test tasks;
N total amount of keywords in the semantic core of training text;
R correlation coefficient of ranks of keywords in training text and test task.

Figure 3 shows the example correlation coefficient calculation for ranks of keywords in training text and test task. This calculation is performed in spreadsheet Microsoft Excel using CORREL function. In our case R = 0,69 and total evaluation of the content validity of test tasks equals

$$K_{CV} = 5/10 \times 0.69 = 0.345. \qquad (2)$$

According to our estimation, the value of K_{CV} coefficient must be not less than 0.5. A low estimate of the content validity of test tasks in this case is due to two factors.

First, only half of the keywords in the semantic core of training text coincide with the keywords in the semantic core of the test tasks.

Secondly, the order of the rankings of keywords in the semantic core of test tasks differs significantly from the order of the rankings of keywords in the semantic core of the training text. This suggests that in the test tasks more attention is paid to secondary issues.

4   Conclusions

Advantages of the proposed methodology for evaluation automation of achievement tests validity concluded in the following:

- the availability of solutions (free online text SEO-analysis tools are used to process training texts and tests);
- the ease of implementation (construction and processing of semantic cores of training texts and tests does not require special knowledge and significant time inputs).

Due to these advantages, the proposed approach to test validity evaluation can be used for frontal inspection of all tests in an education institution.

Now a front-line tests validity check in online courses has been started in the Ukrainian Engineering Pedagogical Academy, using this technique. The results of the analysis of the mass use of this technique will be the subject of the next publication.

References

1. Larin, S., Malkov, U.: Modern approaches to the modeling of tests: a system of requirements, advantages and disadvantages, the main stages of development. Int. J. World Sci. **4**(#1) (2016). Available at: http://mir-nauki.com/PDF/04PDMN116.pdf
2. Newman, I., Lim, J., Pineda F.: Content Validity using Mixed Methods Approach: Its application and development through the use of a Table of Specifications Methodology1 Paper presented at the 2011 Annual Meeting of the American Evaluation Association, Anaheim, California, Nov 2–5 (2011). Available at: http://comm.eval.org/HigherLogic/System/DownloadDocumentFile.ashx?DocumentFileKey=30758c14-2e59-4c94-af88-ecdf11b0dabc
3. Chelyshkova, M.: Theory and practice of designing of pedagogical tests. p. 432, Published by Logos (2002). Available at: http://booksshare.net/index.php?id1=4&category=pedagog&author=chelishkova-mb&book=2002
4. Kurtasov, A., Shvecov, A.: The method of automated generation of control-test tasks from the text of training materials. Bull. Cherepovets State Univ. (#7)7–13 (2014). Available at: https://cyberleninka.ru/article/n/metod-avtomatizirovannoy-generatsii-kontrolno-testovyh-zadaniy-iz-teksta-uchebnyh-materialov
5. Heilman, M., Smith, M.: Good question! statistical ranking for question generation. In: Human Language Technologies: The 2010 Annual Conference of NAACL. pp. 609–617, Los Angeles, California (2010). Available at: http://www.aclweb.org/anthology/N10-1086
6. Moskvitina, T.: Key words and their functions in the scientific text. Sci. J. Bull. Chelyabinsk State Pedagogical Univ. (#11), 277–285 (2009). Available at: https://cyberleninka.ru/article/n/klyuchevye-slova-i-ih-funktsii-v-nauchnom-tekste
7. Yorick Wilks.: On-Line Semantic Analysis of English Texts. Mech. Translat. Comp. Linguistics. **11**(#1, 2) (March and June) (1968) Available at: http://mt-archive.info/MT-1968-Wilks.pdf
8. Ann Smarty.: 5+ SEO Text analyzers for SEO diagnostics & copywriting. Search Eng. J. (2008). Available at: https://www.searchenginejournal.com/5-seo-text-analyzers-for-seo-diagnostics-copywriting/7597/
9. Freeman, R.L.: Fundamentals of Telecommunications. p. 676. Published by John Wiley & Sons, Inc. (2005). Available at: https://doc.lagout.org/network/4_Telecommunications/Telecommunications%20Fundamentals%2C%202nd%20Edition.pdf

Entrepreneurial Competency Development of the Engineering Students at the Research University

Mansur Galikhanov$^{(\boxtimes)}$, Sergey Yushko, Farida T. Shageeva,
and Alina Guzhova

Kazan National Research Technological University, Kazan, Russia
mgalikhanov@yandex.ru

Abstract. State-of-the-art science and industry changed professional activities of an engineer, e.g. he or she has to conduct entrepreneurial activity. One of the trends in university development is tending to University 3.0 (Education, Science, and Entrepreneurship). It determines inescapable changes in a future engineer education. Demand in such educational programs is obvious; they would combine technical aspect of the idea with its commercial implementation. Synergetic effect appears only when engineer's talent meets manager's talent. That's why development of entrepreneurial competencies of the students is set to become one of the learning objectives in engineering university. The paper describes system of the entrepreneurial competencies development in the context of a research university.

Keywords: Engineering student · Entrepreneurial competencies · Research university

1 Introduction

1.1 University 3.0 Concept and Entrepreneurship

Professional activities of a modern engineer are subjected to changes: he or she has to perform entrepreneurial activity along with operating and production ones. Both Russian and foreign researchers describe one of the trends in university development to be tendency to University 3.0 (Education, Science, Entrepreneurship) (see Table 1) [1, 2]. This development trend of the technical universities leads to unavoidable changes in engineer training structure aimed at formation of entrepreneurial competencies.

Strategic goals of modern university development include entrepreneurship development, education quality assurance, knowledge transfer and internationalization of universities [3]. For example, Lobachevsky State University of Nizhni Novgorod introduced project-based university management aimed at formation of entrepreneurial culture of the university employees and students [4]. Some researchers noting paradigm shift in higher education [5] suggest possible solution in a form of "social triangle of activity"—public-private partnership in implementation of investment projects; "triple innovative helix" model application in implementation of innovative projects; and staffing based on advanced professional and educational standards. On the other hand,

© Springer Nature Switzerland AG 2019
M. E. Auer and T. Tsiatsos (Eds.): ICL 2018, AISC 917, pp. 493–501, 2019.
https://doi.org/10.1007/978-3-030-11935-5_47

Table 1. University 3.0 distinctions

University 1.0	• Knowledge transfer • Staff training • Social elevator	– Educational standards – Approaches and educational materials
University 2.0	• New knowledge generation by research activity • Consulting-service center for market players	– Research and development for industry – Creating technologies on a by-order basis
University 3.0	• Technology commercialization • Entrepreneurship • Establishing companies (spin-out)	– Intellectual property management – Entrepreneurial ecosystem – Urban environment development

a number of authors considering the concepts of "the third mission" of a university and analyzing foreign practices of its implementation conclude that it is necessary to bear in mind national context and social mission of the Russian universities [6]. Researchers [7, 8] developed approach for ranking "the third mission" of universities and presented possible options of switching to a new type university giving considerable prominence to its social function. It is noted that the reason for "entrepreneurial culture" formation is not only external conditions (demands of innovation economics development, technology transfer, etc.), but internal needs of a university giving rise to curriculum updating [9].

Therefore, according to many researchers a new type of innovative university involves following indicators: horizontal communication in a management system; outrunning personnel trainings including online ones; interdisciplinarity; creating objects with a new value; developing technology transfer system; knowledge commercialization; contacts with the best world experience and expertise; collaboration with the leaders; developing international partnership; training of a new type specialists that are able to convert research into development and further into business; ability to manage the results of intellectual activity; incubator of startups and technostarters; creating entrepreneurial ecosystem (Fig. 1).

Knowledge commercialization   Start-up and technostarter incubator   Technology transfer system   Interdisciplinarity

Fig. 1. Features of the University 3.0

1.2 World Experience of Entrepreneurship Education

Papers [10–17] describe world experience in entrepreneurship education. It is based on common consent that the subject of entrepreneurship education is business itself. Authors of the papers analyzed challenges and problems of the Russian universities, and described approaches for entrepreneurship education in European and American universities. They specified necessity for the development of full-fledged educational programs, suggesting the applied Bachelor's program to be the basic form. At that, special attention should be given to both program content aimed at entrepreneurial competencies development of bachelor students and corresponding academicians training. Some researchers denoted low engagement of young people into entrepreneurship in the country and gave reasons for creating continuous multilevel educational system in this field that enables all interested students of all majors to get basic competencies in entrepreneurship. Implementation of this approach is possible within new educational policies, innovative educational technologies, personalizing of educational path, introduction of new forms and methods of interactive learning (university-wide issue-related elective courses; student competitions, social projects, etc.) [10–17].

Analysis shows that modern education is targeted at development of entrepreneurial ability of individuals as an important factor of production [2, 10]. For this purpose activities focused on revealing entrepreneurial abilities of the students by means of the entrepreneurial ideas they suggest are held since the first days at a university. Let's consider University of Arizona (USA) as an example, where we observed the work of teacher-trainers during our scientific internship [18]. University of Arizona is one of the US largest public educational and research university, it has about 72,000 students. The university ranks at the top of national and international ratings and is classified as «RU/VH» (Research University/Very High Research Activity) [19–21]. In class brainstorm showed that students were ready to organize own business, and more than that: business related to engineering solutions to the problems, most of which were innovative ones. For example, fantastic at first sight, small recycling company that recycles box-size wastes. However, this idea elicited response in the audience, there were students willing not only to support it, but to bring it to real project (technical aspect of the problem), implement it within a short time (management aspect of the problem) and even invest money (financial matter). It was a classical project-management triangle: cost, time and scope.

There are examples in Russian practice as well, e.g. the Republic of Tatarstan annually holds competition "50 Innovative Ideas" [22], where students take part and win as well. Their projects contribute to innovative development of the university and the region. However, students' projects at this competition fail in presentation and have a lack of knowledge and skills in business, organization and entrepreneurship. Besides, unfortunately the ideas and business are aimed at meeting the requirements of today's market or they are alternative to existing business.

2 Entrepreneurial Competency Development

Students of engineering occupation that by the graduation and diploma defense do not know the basics of entrepreneurship, do not see opportunities of their developments and perform thesis work according to supervisor's task, defending their technical decisions at cost level only. This practice shows that many engineering and scientific ideas are not commercialized. This is a big challenge that can be solved within university learning process. Doubtless, student returns at Master and PhD level to his/her scientific achievements obtained as a part of Bachelor's research activity, but time-factor should be considered. Ideas may be relevant at the moment of Bachelor's Thesis defense, however in two or three years there will be no need in the idea or it will be implemented by someone else.

It is obvious that there is a demand for implementation of the educational programs allowing combining technical aspect of the idea and its commercialization, similar to Apple Inc., where technical aspects were the duty of Stephan Wozniak, while Steve Jobs brought products to market [23].

We can provide many examples of such a successive collaboration, but all they are unified by one idea: synergetic effect appears only when engineer's talent meets manager's talent. It can be reached at least by organizing the learning process in a way that engineering students to be faced with the challenges empowering to think of the final result of the ideas aimed at meeting social needs. Entrepreneurial abilities development of the engineering students has to be top priority in their studying at a technical university.

2.1 Kazan National Research Technological University Experience

Possible ways to develop entrepreneurial competency of engineering students can be shown by an example of Kazan National Research Technological University. Entrepreneurial skills are developed there at all levels of education.

Bachelor's Degree Students and Entrepreneurship

Curriculum of the Bachelor program "Management" includes compulsory subject "Entrepreneurship organization" and the whole program is focused on formation of the following competencies: ability to evaluate economic and social conditions of entrepreneurship, reveal new market opportunities and form new business-models; skills of business-planning, starting-up and development of new organizations, coordination of entrepreneurship and preparation of the organizational and administrative documentation necessary for starting up new entrepreneurial structure. In other words, the educational program is intended to train graduates able to take a creative approach in problem solving in all spheres, including entrepreneurship.

Unfortunately, it is hard to find something related to entrepreneurship in Federal State Educational Standard of the "Economics and Company Management" training program. It is oriented towards bachelors' professional training in the spheres of banking, insurance, analytics, organization management, accounting and even pedagogics. There are no subjects covering entrepreneurial competencies development at least in basics in curriculum of engineering programs as well.

Nevertheless, most standards include such competencies as ability to find organizational and management solution in professional activity and readiness to be responsible for it; ability to organize activity of a small group that was created to implement specific economic project; ability to take stock of the suggested management solutions and draft and substantiate proposals on their improvement taking into account criteria of social and economic efficiency, risks and possible social and economic impacts. These competencies enable to focus on entrepreneurship within some subject of the curriculum. For example, subject "Management" is studied by all engineering students. One of the topics it comprises is "Trends of Management ideas" where management approaches review is followed by comparative analysis of a manager and entrepreneur management styles. Such case-study as "A. Sloan and F. Taylor" provides students an opportunity to see the difference in decision-making and distinguish manager and entrepreneur [24]. It lays the foundation of entrepreneurial competencies.

Developing Entrepreneurial Competency of the Postgraduate Students
Elements of project-based learning in seminars and practicals of the Master's programs are aimed at further development of entrepreneurial competencies of future engineers. Paper [13] describes an example of the Master's Program "Chemical Engineering of Plastics and Composite Materials Processing" at Kazan National Research Technological University. Teaching Staff and students recognize that the future of chemical engineering is related to innovative activity. To prepare students for innovative activity academicians incorporate methods and educational technologies intended to develop creativity, team-work and leadership, effective solution finding under uncertainty or risks and ability to work on multidisciplinary projects. Teams are formed at laboratory practicums to solve mini-projects oriented to entrepreneurship.

One of the projects is development and commercialization of production technology of high-efficiency filtering materials (HEFM). HEFM is a nonwoven polypropylene material manufactured using the spunbond technology that was treated in a specific way to be able to trap small (including nano-sized) particles [25]. This material can be used as individual (respirator) or general (industrial filters) protection at petrochemical enterprises of the Republic of Tatarstan and the Russian Federation.

Working on the projects Master's students are directed by the academician to do research on composite material formulation, production technology optimization and conduct experiments on physical and mechanical properties as well as filtering and separative power of the material. These experiments simulate operating conditions of the material. Students report on the results obtained at different conferences and publish scientific papers. Following this stage marketing research is performed, business plan is drawn up, investor search is made and grant applications are filed. (In 2016 project "Development and Implementation of the Project on High-Efficiency Filtering Material Manufacturing" won the competition "50 Best Innovative Ideas for the Republic of Tatarstan" held by Investment and Venture Fund of the Republic of Tatarstan, Science Academy of the Republic of Tatarstan and Ministry of Education and Science of the Republic of Tatarstan in nomination of "Start of Innovations". This is followed by the stage of product introduction process.

High-efficiency filtrating material is innovative for the Republic of Tatarstan, it is demanded by large petrochemical plants such as PJSC "Nizhnekamskneftekhim", JSC "Kazan Synthetic Rubber Plant", etc. It is driven by introduction of nanotechnology, development of new industrial safety laws and regulations and increased requirements to individual and general protection equipment. Consequently, establishing of small innovative manufacture of HEFM for petrochemical industry is promising and cost-effective.

Focus on innovation assumes that graduates have new competencies in perception and interpretation of the existing ideas and implementation of the new creative ones. In this case "standard" subjects of the curriculum lay the foundation of knowledge and skills for professional and career development, scientific research helps to improve out-of-the-box thinking, while work on mini-projects enhances the ability to handle with intellectual creations. Such system approach allows graduates of the technical university to have comprehensive strategy for innovations, implementation and commercialization of their ideas; creates learning and operational environment giving a boost to innovative performance of a person, and, eventually, creates the innovative university environment that is necessary for transition to new-style higher-education institution "University 3.0". Therefore, this stage of training forms advanced level of entrepreneurial competencies.

Entrepreneurship and Professional Retraining Programs

Advanced level can be mastered during additional training. Curriculums of some professional retraining programs that students study at Faculty of Additional Professional Education simultaneously with their major include subject "Basics of Entrepreneurship". Its features are focused on professional self-identification of students and reliance on system and personal-activity approaches. Content of the subject comprises three parts: Matter of entrepreneurship, Organizing entrepreneurial activities and Institutional operational environment of entrepreneurship. The first part considers the essence of entrepreneurship; inside and outside entrepreneurial environment; government secure of economic freedom of entrepreneurship. The second part describes standard organizational forms of entrepreneurship; self-financing mechanisms of daily operation and production development, corporate growth; development of complex business organizations based on sectoral and inter-sectoral competition; domestic and foreign practices of holding operation. The third one concerns regulatory affairs of entrepreneurship; responsibilities of business entities; contract relations in entrepreneurship. Academicians use business and role-playing games and problem-solving during class time.

Final assignment of the subject is a project "My business-plan". Work on the project promotes mastering elementary skills of entrepreneurship. It was worth mentioning that despite of conventional project title, students (of different faculties and majors) covered wide range of business areas and there were no theme duplication over the past several years. Students are concerned with business organization and view it through a current public needs lens. During project defense student not only listen attentively but are interested in the issue and discuss it with a speaker. It confirms students' interest to entrepreneurship-related subjects and demand for entrepreneurial trainings.

That's why we developed new project. It is a new training program "Entrepreneurship" that is mastered by engineering students simultaneously with their major. Its main goals include: formation of entrepreneurial mindset, respect and interest to entrepreneurship; development of self-identification and fulfilment in business; acquirement of project activities in economics and efficient playing the social and economic role of entrepreneur. Curriculum of the program comprises following modules: Introduction to the program, Marketing, Entrepreneurial Finances, Strategic Management, Risk Management in Entrepreneurship, Government-Business Interactions, Investment Design, Innovative Business Development, Intellectual Property in Company Management, Human Factor in Management Decision-Making, Franchising as a business strategy, Practical Psychology for Entrepreneur, Modeling Activities of a Company, Personal Financial Planning and Personal Efficiency Upgrading, Social Audit, Sales Management, Practical Aspects of Corporate Finances.

Additional professional program "Entrepreneurship" is organized on the principle of successive theory-to-practice transition. Since entrepreneur should have background knowledge of economics, then the first and the second units focus on basic principles of market economy and give insight into economic theory. In further units "Basics of Entrepreneurship" and "How to start a business?" amount of practical activity increases that enables to put knowledge into use during economic problem-solving, case-studies and projects (e.g. solving economic problems on taxes and subsidies, profit and costs; business game "Book Factory"; case-study "Entrepreneurship forms"; mini-research "Features of Small Business in the Republic"). The program is concluded by students' business-project defense at a scientific conference.

3 Conclusion

Thus, it is possible to organize and implement integral system of entrepreneurial competencies development of the future engineers under the unique conditions of the research university.

References

1. Verhovskaya, O.R., Dorohina, M.V., Sergeeva A.V.: National Report "Global Entrepreneurship Monitor, 64p. Nacional'nyj otchet «Global'nyj monitoring predprinimatel'stva». Graduate School of Management, Saint-Peterburg (in Russian)
2. Effects and Impact of Entrepreneurship Programs in Higher Education.: The Report of European Commission, Brussels (2012)
3. Grudzinskiy, A., Bedny, A.: Concept of university competitiveness: tetrahedron model. Economika obrazovania [Econ. Educ.] No. 1, 112–117 (2013) (In Russian, abstract in English)
4. Chuprunov, E.V., Strongin, R.G., Grudzinsky, A.O.: The concept and experience in designing the strategy for innovative development of university. Vysshee obrazovanie v Rossii [Higher Educ. Russia] No. 8–9, 11–18 (2013) (in Russia, abstract in English)

5. Shestak, V.P.: Triple helix model, novel state educational standards and educational programs at the University of Russia] Vysshee obrazovanie v Rossii [Higher Educ. Russia]. No. 2, 15–23 (2017) (in Russian, abstract in English)

6. Balmasova, T.A.: Third mission" of the university—a new vector of development. Vysshee obrazovanie v Rossii [Higher Educ. Russia]. No. 8–9, 48–55 (2016) (in Russian, abstract in English)

7. Marhl, M., Pausits, A.:Third mission indicators for new ranking metodologies. Neprerivnoye obrasovanie: XXI vek [Continuous Educ. XXI Century]. No. 1, 1–13 (2013) (in Russian, abstract in English)

8. Karpov, A.O.: University 3.0—social mission and realty. Sociologitcheskiye issledovaniya [Sociol. Res.]. No 9, 114–124 (2017) (in Russian, abstract in English)

9. Golovko, N.V., Zinevich, O.V., Ruzankina, E.A.: Third generation university: B. Clark and J. Wissema. Vysshee obrazovanie v Rossii [Higher Educ. Russia]. No. 8–9, 40–47 (2016) (in Russia, abstract in English)

10. Rubin, Yu.B.: Entrepreneurship education in Russia: diagnosis of the problem. Vysshee obrazovanie v Rossii [Higher Educ. Russia]. No. 11, 5–17 (2015) (in Russian, abstract in English)

11. Rubin, Yu.B.: Creation of graduates' entrepreneurial competencies within the educational area of Baccalaureate]. Vysshee obrazovanie v Rossii [Higher Educ. Russia]. No. 1 (197), 7–21 (2016) (in Russian, abstract in English)

12. Galikhanov, M.F., Ilyasova, A., Ivanov, V., Gorodetskaya, I.M., Shageeva, F.T.: Continuous professional education as an instrument for development of industry employees' innovational competences within regional territorial-production cluster. In: International Conference on Interactive Collaborative Learning (ICL 2015), Firenze, Italy, pp. 251–255, 20–24 Sept 2015

13. Ivanov, V.G., Galikhanov, M., Shageeva, F.T.: Research university as a center of internationally-focused training innovative-economy engineers. In: 2017 ASEE International Forum, Columbus, OH, United States, 28 June 2017

14. Ivanov, V.G., Barabanova, S.V., Galikhanov, M.F., Guzhova, A.A.: The role of the presidential program of training engineers in improvement of the research university educational activities. In: 2014 International Conference on Interactive Collaborative Learning (ICL 2014), Dubai, United Arab Emirates, pp. 420–423, 3–6 Dec 2014

15. Ilyasova, A., Galikhanov, M., Ivanov, V.G., Shageeva, F.T., Gorodetskaya, I.M.: Concept of implementing the programs of additional professional education within the cluster system. In: 2nd ASEE Annual Conference and Exposition: Making Value for Society, vol. 12, Seattle, United States, 14–17 June 2015

16. Galikhanov, M.F., Guzhova, A.A.: Complex approach for preparation and implementation of continuous professional education programs in technological university. In: 16th International Conference on Interactive Collaborative Learning (ICL 2013), Kazan, Russian Federation, pp. 54–55, 25–27 Sept 2013

17. Shageeva, F.T., Erova, D.R., Gorordetskaya, I.M., Kraysman, N.V., Prikhodko, L.V.: Training the achievement-oriented engineers for the global business environment. Advances in Intelligent Systems and Computing, vol. 716, pp. 343–348. 20th International Conference on Interactive Collaborative Learning (ICL 2017), Budapest, Hungary, 27–29 Sept 2017

18. Chepurenko, A.Y.: How and why entrepreneurship should be taught to students: polemical notes. Voprosy obrazovaniya [Educ. Stud. Moscow]. No. 3, 250–276 (2017) (in Russian, Abstract in English)

19. Kazin, F., Hagen, S., Prichislenko, A., Zlenko, A.: Developing the entrepreneurial university trough positive psychology and social entreprise. Voprosy obrazovaniya [Educ. Stud. Moscow]. No. 3, 110–131 (2017) (in Russian, Abstract in English)

20. Strekalova, G.R.: Motivational profile of the personnel in enterprise management decisions. Upravlenie ustoichivym razvitiem [Sustain. Develop. Manage.] No. 1(2), 48–52 (in Russian, Abstract in English)
21. https://www.educationindex.ru/university-search/arizona-state-university
22. http://ivf.tatarstan.ru/50ideas.htm
23. Elliot J., Simon W.L.: The Steve Jobs Way: iLeadership for a New Generation, 256p. Vanguard Press, New York (2011)
24. Strekalova, S.O., Strekalova, G.R.L: Computer business games in economic education enhancement. Uchenye zapiski Instituta sotsialnykh i gumanitarnykh znanii [Proc. Inst. Soci. Human. Knowl. No .1, 505–509 (in Russian, abstract in English)
25. Galikhanov, M.F.: Unipolar corona discharge effect on filtering capacity of polypropylene Non-Woven fabrics. Fibre Chem. **48**(6), 473–477 (2017)

Enhancing Lifelong Learning Skills Through Academic Advising

Abdelfattah Y. Soliman and Ali M. Al-Bahi[(⊠)]

Faculty of Engineering, King Abdulaziz University, 21589 Jeddah, Saudi Arabia
{ausoliman,abahi}@kau.edu.sa

Abstract. According to the National Academic Advising Association (NACADA), academic advising is a series of intentional interactions with a curriculum, a pedagogy, and a set of student learning outcomes. The curriculum represents the subject matter advising covers and ranges from academic and career educational planning, building campus community and social relationships, and developing lifelong learning strategies and capabilities. The present work addresses how the curriculum, pedagogy and learning outcomes of academic advising can contribute to develop and assess lifelong learning capabilities. The Key Performance Indicators for lifelong learning are identified and how they can be assessed, evaluated, and improved during a modified academic advising curriculum are discussed. The proposed approach shifts developing and assessing lifelong learning skills from selected program courses to an organized college-wise activity inside the framework of well-organized extra-curricular academic advising interactions.

Keywords: Lifelong learning · Academic advising

1 Introduction

Key competencies for the 21st century are based more on the ability to learn, or lifelong learning, rather than on the accumulation of knowledge [1, 2]. Curriculum and classroom teaching are not sufficient to instill the skills of lifelong learning in the students while academic advising, on the other hand, can play a fundamental role to enhance lifelong learning traits such as social inclusion, active citizenship, personal development, competitiveness, and employability [3].

According to the National Academic Advising Association (NACADA), academic advising is a series of intentional interactions with a curriculum, a pedagogy, and a set of student learning outcomes. The curriculum represents the subject matter advising covers and ranges from academic and career educational planning, building campus community and social relationships, and developing lifelong learning strategies and capabilities [4, 5].

Although an increasing number of universities in the United States offers a first-year experience class taught by academic advisors to their new students [6], the advisor/advisee relationship remains, in many schools, infrequent and personal and concentrates, more or less, on curricular aspects and educational planning. In fact, the history of this academic advising model is traced back by Kramer [7] to the year 1841

© Springer Nature Switzerland AG 2019
M. E. Auer and T. Tsiatsos (Eds.): ICL 2018, AISC 917, pp. 502–513, 2019.
https://doi.org/10.1007/978-3-030-11935-5_48

at Kenyon College. The college at that time required students to choose a faculty member as their advisor to help them determine what courses they need to take in order to graduate. The journey of academic advising is long but, according to Kook [8], it is only with the founding of NACADA in 1979 did academic advising begin the journey to professionalization.

Actually, the mission statement of academic advising in several high education institutions is articulated around two pillars [9]:

1. Support students in discovering their academic potential.
2. Foster personal development through lifelong learning.

As an example, the College of Humanities at the University of Arizona states that the goals of their academic advising include helping students to clarify their life and career goals, select appropriate courses, interpret institutional requirements, enhance awareness of available educational resources, evaluate their progress, and develop decision-making skills [10]. In this mission statement, the underlined goals are some of the expected key performance indicators (KPIs) of the students as lifelong learners.

The UN's Sustainable Development Goals [11] emphasize the need for creative lifelong thinkers to meet the current and future challenges. That is why the Association of American Colleges and Universities Creative Thinking VALUE Rubric [12] became a part of the KPIs used to assess lifelong learning in several institutions.

Lifelong learning has become a major consideration in different accreditation bodies including ABET in US and EUR-ACE in Europe. ABET, the accreditation body for engineering and technology, requires in their new criteria that engineering graduates should have *"an ability to acquire and apply new knowledge as needed, using appropriate learning strategies"* [13]. EUC-ACE, the European Accredited Engineer label, requires the learning process to *"enable Bachelor Degree graduates to demonstrate an ability to recognize the need for and to engage in independent life-long learning and an ability to follow developments in science and technology"* [14].

In several universities world-wide teaching and assessing lifelong learning is carried out, similar to other student outcomes, in some specific mandatory courses; while advising is infrequent and personal and concentrates, more or less, on curricular aspects and educational planning.

In the present work, it is proposed to shift development and assessment of lifelong learning skills from selected program courses to an organized college-wise activity inside the framework of well-organized extra-curricular academic advising interactions. Evaluation of the attainment of lifelong learning performance indicators is used to continuously improve the content and pedagogy of academic advising and to enhance lifelong learning resources on the college level.

2 Key Performance Indicators of the Lifelong Learning Outcome

The Association of American Colleges and Universities AAC&U introduced the lifelong learning VALUE rubric in 2009 [12]. The Key Performance Indicators in this rubric map the following aspects:

1. **Curiosity**: Explores a topic in depth, yielding a rich awareness and/or little-known information indicating intense interest in the subject.
2. **Initiative**: Completes required work, generates and pursues opportunities to expand knowledge, skills, and abilities.
3. **Independence**: Educational interests and pursuits exist and flourish outside classroom requirements. Knowledge and/or experiences are pursued independently.
4. **Transfer**: Makes explicit references to previous learning and applies, in an innovative (new and creative) way, that knowledge and those skills to demonstrate comprehension and performance in novel situations.
5. **Reflection**: Reviews prior learning and past experiences (inside and outside the classroom) in depth to reveal significantly changed perspectives about educational and life experiences, which provide foundation for expanded knowledge, growth, and maturity over time.

The ABET student outcome #7 requires the students to demonstrate their ability to acquire and apply new knowledge as needed, using appropriate learning strategies. As per ABET definition, student outcomes describe what students are expected to know and be able to do by the time of graduation and relate to the knowledge, skills, and behaviors that students acquire as they progress through the program [13]. Accordingly, should be assessed, evaluated and continuously improved before graduation.

In the faculty of Engineering at King Abdulaziz University (KAU), a rubric has been developed for each student outcome. Faculty members teaching courses identified by the program as the most likely to display convincing evidences from the students' work, use this rubric to demonstrate attainment of student outcomes. The rubric for outcome #7, in particular is based on five Key Performance Indicators, KPIs, which reflect curiosity, initiative, independence, transfer, reflection, and critical thinking (see Table 1). In each engineering program some courses are selected to build up the skills, knowledge, and attitudes specified for the outcome and carryout both formative and summative assessment. The major drawbacks of this strategy are as follows:

1. Individual courses inside one program can not reflect the curiosity of the student toward other engineering disciplines.
2. The link between different courses that address the same outcome can not reflect the student's progress, particularly if the courses belong to more than one program.
3. Evidences that reflect the attainment of all KPIs of the outcome remain limited within one course.
4. Attainment of the outcome is evaluated for the student cohort in one course and not for each and every student.

For these reasons, a new assessment strategy is proposed to track the student attainment of the outcome on a college level instead of courses level. The new KPIs can include a full track of the student's progress toward lifelong learning skill. Also, the assessment is based on direct judgment from the academic advisor using verified attainment evidences.

New KPIs need to be developed to assess the attainment of the lifelong learning skills in the new situation. These KPIs include but are not limited to the following:

Table 1. Key performance indicators for ABET Students Outcome #7

Outcome 7: Student work demonstrates an ability to acquire and apply new knowledge as needed, using appropriate learning strategies. For the following key performance indicators (KPIs), the student can

#	KPI	Excellent (3)	Good (2)	Needs improvement (1)	Unsatisfactory (0)
7.1	Recognition of the need, **(Initiative)**	Go beyond what is required in completing an assignment, by bringing credible value-adding information from outside sources	Go beyond what is required in completing an assignment, but the collected information may lack credibility, authenticity, or added values	Complete only what is required	Have trouble completing even the minimum required tasks
7.2	Accessing information **(Curiosity)**	Access information from a variety of sources and critically assess their quality, validity, and accuracy	Access information from a variety of sources and assess their quality, validity, and accuracy to some extent	Access information from a variety of sources without any attempt to assess their quality, validity or accuracy	Be unable to access information unless clearly guided to pending sources
7.3	Self-learning **(Independence)**	Analyze new content by breaking it down, comparing, contrasting, recognizing patterns, and/or interpreting information	Analyze new content with some difficulties	Need guidance to analyze new knowledge	Fail to analyze new knowledge
7.4	Reflection on learning **(Reflection)**	Regularly reflect on his/her learning process, evaluate personal performance and progress, and take required actions and improvements	Reflect on his/her learning process, evaluate personal performance and progress, but fail to take required actions	Occasionally reflect on his/her learning process if asked to do	Fail to recognize his/her own shortcomings or deficiencies
7.5	Apply new knowledge to perform tasks **(Transfer)**	Apply new knowledge successfully to solve real-life complex engineering problems	Apply new knowledge to solve complex engineering problems	Apply new knowledge to solve simple problems	Fail to apply new knowledge

1. Student has motivation toward lifelong learning.
2. Student attends seminars, meetings, and workshops.
3. Student engages in engineering voluntary work experience.
4. Student attends technical or skill oriented training.
5. Student participates in engineering competitions.
6. Student attends online lectures and has extra-curricular readings.

In order to measure these KPIs, three major assessment methods are used:

- Certificates
- Reflective reports on gained experience
- Oral discussion with academic advisor.

Depending on the targeted KPI, assessment can be carried out by the academic advisor, a faculty expert or an outside trusted entity in coordination with the college.

3 A New Approach for Developing and Assessing Lifelong Leaning

Developing and assessing lifelong learning is better carried out on a college level. This new approach depends mainly on the academic advisors who are assigned to create motivation, support motivation, track progress and assess performance for each student in addition to their classical role of providing guidance to the students on curricular aspects and educational planning. While the role of the college of engineering, under this approach, will be to:

- allocate resources for lifelong learning,
- provide guidance for outside resources, and
- coordinate with the academic advisors for evaluating the process, set improvement plans, and provide extra resources.

Figure 1 indicates the main elements of the new strategy. Students enter the college of engineering with different levels of motivation toward lifelong learning. Academic advisors organize regular meetings with the students to either create motivation or support motivated students through discussions and other active learning strategies. Academic advisors can easily track the student progress using the KPIs of the outcome and coordinate with the college administration for assessment, evaluation, and provision of required resources to support the students' attainment of the lifelong learning skills.

The college of engineering shall provide different resources. This may include the following:

- Guiding and providing data pool for online learning courses and/or training for different disciplines according to students' interests and recommendations from her/his academic advisor.
- Providing support to students' curiosity in building new technical skills through different technical training activities. These skills maybe more advanced than the skills introduced in the curriculum.

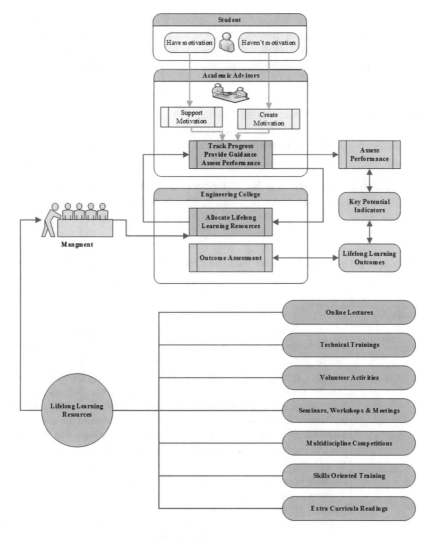

Fig. 1. The new strategy

- Announcing different technical volunteer activities by technical non-governmental organizations as well as domestic and international community.
- Coordinating with domestic and international industry for training.
- Promoting student chapters for professional non-governmental organizations and sponsoring their activities.
- Advertising local seminars, training, workshops, and meetings and allowing the students to reflect on their gained experience from these activities and present their work to their colleagues to encourage and motivate each other.
- Facilitating students' participation in multidisciplinary international and domestic competitions in order to go outside the boundaries of their specialization.

- Organizing specific training about the lifelong learning skills and competencies such as teamwork, project management, leadership, and communications.
- Encouraging classic and electronic libraries to participate in building lifelong learning skills through reading competitions and group discussions on multidisciplinary topics.
- Organizing job fairs to motivate students toward job market skills.
- Promoting multidisciplinary senior projects.
- Supporting curiosity via offering a large number of technical and free electives.
- Supporting creating and maintaining a lifelong learning electronic portfolio for each student.

The main differences between the course based and faculty based development and assessment approaches are shown in Table 2. The table indicates the major enhancements of different issues by implementing the new approach.

Table 2. Comparison between course based and college-based lifelong learning assessment

Topic	Course based assessment approach	College based assessment approach
Preforming assessment	Assessment is performed by course instructors without sufficient capabilities to map what is going outside the course	Assessment is performed by the academic advisor who can track how the student progresses inside and outside the discipline
Assessment tools	Different assessment tools are used in different courses that address the same outcome without an efficient tracking of the progress of the student from course to course or considering the student's activities outside these courses	Student lifelong learning portfolio efficiently track the student attainment of the outcome starting from admission to the college to graduation by including all lifelong learning activities inside and outside course work
KPIs	Lifelong learning KPIs are mapped in some courses and remain bounded by the content of these courses. Evaluation and continuous improvement require adjustments in several courses	Mapping of lifelong learning KPIs is simple and reliable and reflects the students' performance inside and outside their major. KPIs could be evaluated and improved in a strait forward way
Resources	Courses are not sufficient to provide the required pool of resources needed for lifelong learning	The students are able to identify required resource and the college is more capable of providing them
Student's motivation	One course is difficult to build and support student motivations toward different topics	Academic advisor can track and support student's motivations independent of the topics of a program

(continued)

Table 2. (*continued*)

Topic	Course based assessment approach	College based assessment approach
Lifelong learning activities	Activities are limited inside the boundary of each course; hindering the student's motivation to new topics	Activities outside course boundaries, such as volunteer activities and participations outside the campus, are included
Lifelong learning continuous improvement	Limited because of fixed course objectives which are more or less oriented to technical topics within the curricula	Efficient because it can address community-based activities to face the current and future challenges
Response of college management	Slow because it will depend on the performance of the program over successive assessment cycles	Fast because identification can be carried out within one cycle based on current needs and advisors recommendations

4 Curriculum and Pedagogy of Academic Advising for Lifelong Learning

The proposed approach relies on assigning one faculty member as an academic advisor for ten to twenty students. This is based on the fact that in major engineering institutes, student to faculty ratio is around 10–20. Advised students are from different engineering disciplines and expected year of graduation. The primary purpose of selecting the students from different disciplines and enrollment levels is to create a virtual lifelong learning environment disconnected entirely from engineering program courses. This promotes experience sharing between advised students and encourage them to work outside the specialization barriers while dealing with lifelong learning goals. Three full day sessions of six hours each are organized in the beginning, middle, and end of each semester for all the students enrolled in the college of engineering. Similar to the case of internship activity in KAU engineering curricula, the student is given "No-Grade-Pass" (NP) any time she/he scores Excellent or Good in all KPIs of the lifelong learning rubric before graduation. Otherwise the student is given "In Progress" (IP) and has to continue attending lifelong sessions for another semester.

During an academic advising session, the students are expected to meet their advisor and colleagues in active learning sessions to accomplish the following:

1. Presenting their progress toward achieving lifelong learning skills by discussing their curiosity about different topics, the initiative they take to achieve their goals, and their independent self learning activities outside their specializations.
2. Discussing the barriers and success stories to transfer their learning experience to their colleagues and reflect on their learning process, evaluate personal performance and progress, and take required actions and improvements.
3. Updating the academic advisor on availability of lifelong learning resources.
4. Creating and maintaining lifelong learning electronic portfolio.

5. Receiving support from the academic advisor and colleagues through practicing critical thinking and group discussions facilitated by the academic advisor.
6. Gaining motivation either from the academic advisor or colleagues.
7. Peer assessing teammates to help them achieve their goals.
8. Critically thinking for creative solutions during through group discussions.

Outcomes of these sessions are to create an active environment, which permits the academic advisor to assess the students' attainment of the lifelong learning skills according to the previously discussed KPIs. They also give the academic advisor a chance to motivate, support and assist the students' interests. Moreover and create a clear update for the required resources for lifelong learning. The list of required resources is delivered as recommendations to the college administration to update its lifelong learning pool of resources.

Students are assessed according to their degree of attainment of the KPIs as a direct measure of their success to attain the skills specified in the lifelong learning outcome. Their performance is cumulative over successive semesters, and the pass requirement is to scores Excellent or Good in each KPIs of the lifelong learning rubric as demonstrated in their lifelong learning portfolio.

Academic advising sessions in one semester are considered as one assessment and evaluation cycle archived in the student's portfolio. Scoring Excellent or Good in each KPIs of the lifelong learning rubric is required to obtain a "No Grade Pass" in lifelong learning independent of the student grades inside his department. Failing in archiving convincing activities will automatically assign the student for the further sessions as indicated in Fig. 2.

Student's electronic portfolio will contain all evidence of attaining the required skills and progress toward achieving the KPIs. Student's attendants after achieving the KPIs criterion will be voluntary until graduation.

The main advantages of this approach over the course-based one are the following:

1. Creating open boundary environment for the lifelong learning.
2. Creating and supporting students' motivation along with their engineering education.
3. Facilitating lifelong learning assessment and quality enhancement.
4. Helping the students through group discussions to express their interests and gain from the experience of other teammate.

5   Preparing Faculty as Lifelong Learning Academic Advisors

Orientation training should be provided for future academic advisors by lifelong learning and leadership experts. The purpose is to:

1. Develop awareness of the current status of available resources for lifelong learning,
2. Develop awareness of current and potential community needs,
3. Develop awareness of the status of the job market,
4. Emphasize the effect of passionate character on effective communication with the students

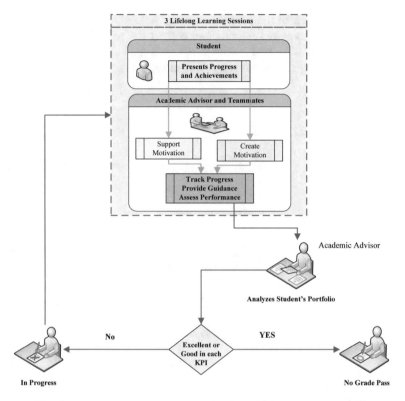

Fig. 2. Assessment cycle for the attainment of lifelong learning skills

5. Introduce the advisors to how to create, update and assess an electronic portfolio.
6. Emphasize that all interactions between the academic advisor and his students should be archived and become a part of the electronic portfolio.
7. Emphasize the importance of dealing seriously with the students' questions and setting rules and deadlines to answer them.
8. Introduce the advisors to how to transfer the students' needs for lifelong learning resources either from departments or out-of-campus entities to college administration.
9. Give examples of how to create motivation for lifelong learning purpose and community services.
10. How to help students to set plan to compensate for their progress in their disciplines and their curiosity about outside interests
11. Introduce advisors to support students in creating links between what they are doing in their disciplines and their lifelong learning interests.

The main objective of orientation training is to unify the efforts and the expertise of academic advisors to develop and assess lifelong learning skills. A college-level lifelong learning committee is set to assess the advisor's performance and recommend possible changes in the management of students' portfolio. Students' surveys and direct performance are indicators of the success of the advising process. Advisor are notified of the students' feedback to improve the advising process. The committee is respon-sible for preparing the required training and contact the administration regarding the institutional resources for lifelong learning. Figure 3 shows the management plan that integrates college management, advisory committee, and advisors.

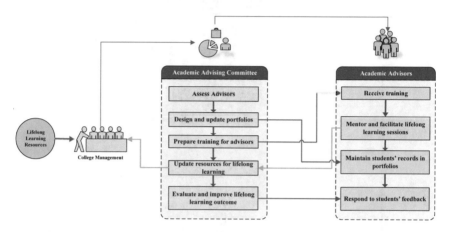

Fig. 3. Management plan that integrates faculty management, advisory committee and advisors

6   Conclusions

Academic advisors can play a pivotal role in enhancing the lifelong learning skills of the students. Curriculum and classroom teaching concentrates on the accumulation of knowledge and are not sufficient to instill the skills of lifelong learning in the students. A proposal is presented for a curriculum, pedagogy and learning outcomes of an academic advising system to contribute to develop, assess and continuously improve lifelong learning capabilities. The proposed approach shifts developing and assessing lifelong learning skills from selected program courses to an organized college-wise activity inside the framework of well-organized extra-curricular academic advising interactions. Modified academic advising is more capable of enhancing lifelong learning traits such as social inclusion, active citizenship, personal development, competitiveness, and employability. Academic advising sessions are carried out three times per semester using active learning strategies to replace the classical infrequent and personal academic advising interactions that concentrate, more or less, on curric-ular aspects and educational planning. The result of the new sessions is a lifelong learning electronic portfolio prepared and maintained by each student and contains all evidences of attaining the required skills as well as the progress toward achieving the KPIs of the lifelong learning outcome. The portfolio is used to assess the student's attainment of the outcome. Scoring Excellent or Good in each KPIs of the lifelong learning rubric is required to obtain a "No Grade Pass" in lifelong learning, indepen-dent of the student's grades inside his department. Failing in archiving convincing activities will automatically assign the student for the further sessions.

The process is complemented by a college-level committee to assess advisors, design portfolios, prepare training for advisors, and evaluate and improve the attainment of lifelong learning outcome. The committee also communicate with the college administration to update resources for lifelong learning.

References

1. Carneiro, R., Draxler, A.: Education for the 21st century: lessons and challenges. Eur. J. Educ. **43**(2), 149–160 (2008)
2. Scott, C.L.: The futures of learning 2: what kind of learning for the 21st Century? UNESCO Education Research and Foresight, Working Papers Series, No. 14, Paris, Nov 2015. Downloadable from: http://unesdoc.unesco.org/images/0024/002429/242996e.pdf. Retrieved 23 Mar 2018
3. Watson, L.: Lifelong learning in Australia. Australian Government, Department of Education, Science and Training, Canberra (2003). Downloadable from: http://www.forschungsnetzwerk.at/downloadpub/australia_lll_03_13.pdf. Retrieved 23 Mar 2018
4. Missouri Western State University: Concept of academic advising from NACADA (2014). Downloadable from: https://www.missouriwestern.edu/advising/wp.../Pillars-of-Academic-Advising1.pdf. Retrieved 10 Mar 2018
5. Erlich, R.J., Russ-Eft, D.: Applying social cognitive theory to academic advising to assess student learning outcomes. NACADA J. **31**(2) (2011)
6. Blanchot-Aboubi, G.: First-year seminars and advising: how advisors make a lifelong impact. Acad. Advising Today **38**(2). Retrieved 23 Mar 2018 from: http://www.nacada.ksu.edu/Resources/Academic-Advising-Today/View-Articles/First-Year-Seminars-and-Advising-How-Advisors-Make-a-Lifelong-Impact.aspx
7. Kramer, G.L.: Redefining faculty roles for academic advising. In: Kramer, G.L. (ed.) Reaffirming the Role of Faculty in Academic Advising, pp. 3–9. Brigham Young University Press, Provo (1995)
8. Cook, S.: important events in the development of academic advising in the United States. NACADA J. **29**(2). Retrieved 23 Mar 2018 from: http://www.nacadajournal.org/doi/pdf/10.12930/0271-9517-29.2.18
9. NACADA Resources (2010–2017): Examples of academic advising mission statements. Retrieved 23 Mar 2018 from: http://www.nacada.ksu.edu/Resources/Clearinghouse/View-Articles/Examples-of-academic-advising-mission-statements.aspx
10. The University of Arizona College of Humanities: Academic advising (2016). Retrieved 23 Mar 2018 from: http://advising.arizona.edu/content/mission-statement
11. UN Sustainable Development Goals: The 2030 agenda for sustainable development (2016). Retrieved 1 May 2018 from: http://www.un.org/sustainabledevelopment/sustainable-development-goals/
12. Association of American Colleges and Universities: Creative thinking VALUE rubric. Retrieved 1 May 2018 from: https://www.aacu.org/value
13. ABET Engineering Accreditation Commission: Criteria for accrediting engineering programs effective for reviews during the 2018–2019 accreditation cycle (2017). Retrieved 1 May 2018 from: http://www.abet.org/wp-content/uploads/2018/02/E001-18-19-EAC-Criteria-11-29-17.pdf
14. EUR-ACE Framework and Guidelines: Retrieved 1 May 2018 from: http://www.enaee.eu/wp-assets-enaee/uploads/2015/04/EUR-ACE-Framework-Standards-and-Guidelines-Mar-2015.pdf

University Teacher Professional Development in the Digital World

Julia Lopukhova[1] and Elena Makeeva[1,2(✉)]

[1] Samara State Technical University, Samara, Russian Federation
j.v.lopukhova@mail.ru, helenmckey2205@gmail.com
[2] Samara State University of Social Sciences and Education, Samara,
Russian Federation

Abstract. These are changing times in education systems around the world. With the start of the new millennium, many societies are engaging in serious educational reforms. One of the key elements in most of these reforms is the professional development of teachers. As universities become more autonomous, with open learning environments, university teachers assume greater responsibility for the content, organisation and monitoring of the learning process, as well as for their own personal career-long professional development. Systems of education and training for teachers need to provide them with the necessary opportunities. All this makes the field of teacher professional development a growing and challenging area, and one that has received major attention during the past few years. This is actual for high educational institutions in Russia as they are both eager and required to become a part of European educational system. This is why we should investigate the issue of professional development: what kind of instruments we can use for increasing the effectiveness of university teacher professional development.

Keywords: Professional development · Educational reforms
21st century competencies model

1 Introduction

The work of the university teacher has a great impact on each society. It is very demanding work that requires professional competences and ability in scientific research what is connected also with ability to transfer the science results in such a way to be understandable for students and inspiring them for their future development.

Over the past few decades, the Ministry of Education in Russia has put in place some important reforms as part of its Education Modernization Program, the goals of which are to improve education quality, access, and efficiency. Through the development of common standards of educational assessment both of students and of institutions, the Government has tried to ensure that decentralization needs not mean excessive disparity of educational outcomes. For that reason, comprehensive statements of Federal standards have been developed and widely disseminated. Now, flexible curricula are being piloted in high school to improve the relevance and career opportunities. In addition, the introduction of information communication technologies

© Springer Nature Switzerland AG 2019
M. E. Auer and T. Tsiatsos (Eds.): ICL 2018, AISC 917, pp. 514–524, 2019.
https://doi.org/10.1007/978-3-030-11935-5_49

in education is a positive step toward equipping the education system with the means which help to develop students' skill and adapt them to the modern and changeable world.

One of the key elements in most of these reforms is the professional development of teachers. As universities become more autonomous, with open learning environments, university teachers assume greater responsibility for the content, organisation and monitoring of the learning process, as well as for their own personal career-long professional development. Furthermore, as with any other modern profession, teachers have a responsibility to extend the boundaries of professional knowledge through a commitment to reflective practice, through research, and through systematic engagement in continuous professional development from the beginning to the end of their careers. Systems of education and training for teachers need to provide them with the necessary opportunities. All this makes the field of teacher professional development a growing and challenging area, and one that has received major attention during the past few years. This is actual for high educational institutions in Russia as they are both eager and required to become a part of European educational system.

2 Project Description

2.1 Purpose

Professional development is an essential element of teacher education and professional advancement. Such a continuous learning and training assures a high level of knowledge and enables teachers to keep their professional skills and knowledge up-to-date. At the moment, different models and experiences of teacher professional development are planned and implemented on a large scale in many countries. There are also many experiences that are being implemented on a much smaller scale, that is in one particular country or in one separate university. The paper focuses on determining the specifics, capabilities, and core tools which can be applied in university teacher professional development more effectively realizing 21st Century Competencies Model.

According to our model, ten most actual skills can be organised into four groups. They are: Ways of Thinking (among them we distinguish: 1. Creativity and innovation 2. Critical thinking, problem solving, decision making 3. Learning to learn, Metacognition); Ways of Working (4. Communication 5. Collaboration (teamwork)); Tools for Working (6. Information literacy 7. ICT literacy) and Living in the World (8. Citizenship—local and global 9. Life and career 10. Personal and social responsibility—including cultural awareness and competence). This is why we should investigate in much more detail the issue of professional development: more precisely, what kind of instruments can we use for increasing the effectiveness of university teacher professional development through developing these competencies.

2.2 Approach

The paper provides information on the impact of modern education systems on teachers' professional development, and reflects on the relation between teachers' professional development and effective university and education-system reforms. Professional development is defined as activities that develop an individual's skills, knowledge, expertise of a teacher. The literature review takes a performance-oriented perspective, with an emphasis on the meaning of professional development for the quality of education, in the sense of fostering educational performance and educational effectiveness. The authors believe that it is important to see teachers' professional development as a means of attaining the basic goals of the educational endeavour. They also acknowledge the relevance of other goals, such as enhancing teachers' job satisfaction. This performance-oriented perspective appears to encompass a number of aspects, while remaining targeted at the enhancement of educational quality.

The research further explores a complex 21st Century Competencies Model, in which it is hypothesised that the experienced impact of professional development is influenced by a set of interrelated university and teacher variables. The methodological part then deals with the results of the questionnaire survey that the authors were carrying out in the Samara State Technical University and Samara State University of Social Sciences and Education from 2014 till 2017. This paper presents a secondary analysis and an enlarged interpretation of the results. The examination findings in a sample of 90 teachers focus primarily on interdependence between 21st century competencies and the level of university teacher professional development. The survey also focused on the working conditions of teachers and the learning environment in universities which is the object of another paper. The survey results have shown as well how new education practices (online courses, MOOCs, shared online courses, etc.) evolve due to increased use of new digital technologies, especially among younger teachers. Such practices create reconceptions of key competencies and skills and help shape a new 21st Century Competencies Model. Digital technologies here serve as both a driver and lever for the transformation.

3 Literature Review

Competences and professional development of university teachers are of particular importance, mainly because teachers constitute the basis for the creation of new knowledge and new values beneficial to the university as well as to students. For that reason, numerous authors deal with the definition of competences of university teachers and their development [3–5, 7, 9].

A few initial studies on teaching competencies include those which point out six criteria for identifying effective teachers: in-service teaching, peer rating, pupil rating, pupil gain score, composite test score and practice teaching grade Barr [1]. Later, Gray and Gerrard [6] recommended sixteen competencies, stressing classroom behaviour and personal adjustment. Shulman [15] focussed on four such competencies that were related to subject matter knowledge, diagnosing students understanding levels, classroom monitoring and management and ability to differentiate curriculum.

The teachers professional and personal development (TPPD) is also understood as a process that is continuously formed, developed and changed by human interaction. It aims to provide for teachers authentic learning experiences in the sense that it builds on existing knowledge and encourages them to reflect on and to regulate their own learning in a context that stimulates the collaborative and ethical attitudes. As well, TPPD can be understood as a process of empowerment. The process of empowerment means the transfer from a state of helplessness to a state where there is greater control of life, fate and environment [14]. This process involves changes in three dimensions: people's feelings and abilities; the life of the collective to which they belong; and professional actions that intervene in the situation. Thus empowerment theory deals with three processes: individual empowerment, which is the personal, intimate change; community empowerment, which is the social change; and the empowering professional action which forms the functional and organisational change that encourages the fulfilment of both the processes [14].

Twenty years later, researchers presented a COACTIV model of teachers' professional competency based on professional knowledge, professional values and beliefs, motivation and self-regulation [8]. It is developed from a multidimensional perspective, closely referring to interplay of cognitive and self-regulatory characteristics needed to cope with work pressures. Selvi, in his turn, analysed the general framework of teachers' competencies and proposed the following nine aspects: field competence, research competence, life-long learning, communication competencies, curriculum, information technology usage, socio-cultural competencies, emotional competencies and environmental competencies [12]. Further studies also incorporate a set of digital competences and innovative teaching aids usage as key competencies: thus, a model of integration of digital competence into HE teachers professional development was developed and validated [11]. From the model, it was built a framework of 7 digital competences, that were found coherent in and for the modern knowledge society, with 78 digital competence units distributed in three competence development levels: Basic Knowledge, Knowledge Deepening and Knowledge Generation [11].

Other recent works on professional development also point to a number of teachers' key competencies. According to some authors, university teachers' competences can be divided into seven clusters: branch-specific; didactic and psycho-didactic; general educational; diagnostic and interventional; social, psycho-social and communicational; managerial and normative; professionally and personally cultivating [16].

One of the latest models developed is a competency model of university teachers based on student feedback [2]. Apart from the positive competencies, their study also explores the various negative characteristics that students do not desire in their teachers. Their competency model presents the positive and negative behaviour indicators that defined a particular desired competency. The major headings in their competency model are professional competence, educational competence, motivational competence, communicational, personal, science and research competence and publication competence [2]. Following that tendency, professional teacher development has been as well correlated with personal development [10].

Though different in their approaches, most researchers are unanimous in their conclusion that a competence model of university teachers should consider new, progressive and relevant educational strategies. They also agree, that university teachers today, in their turn, have to continually improve and adapt themselves according to changing requirements and never-ending reforms along with the resources offered by expanding technology [13, 17, 18].

4 Main Body

4.1 Questionnaire Survey Review

The questionnaire was developed based on an extensive review of the literature that provided the most important skills necessary for university teacher professional development in the digital world. The authors conducted this questionnaire survey in the Samara State Technical University and Samara State University of Social Sciences and Education from 2014 till 2017. Data collection was done while distributing questionnaires to 90 teachers working at these institutions who confirmed concernment in their own personal career-long professional development. In order to determine and evaluate the 21st Century Competencies Model structure, an exploratory factor analysis and a confirmatory factor analysis were carried out, respectively. After several exploratory factor analyses, certain skills were eliminated due to not reaching a minimum rotation. Then, a factor analysis was carried out for most actual skills, the results of which were ten skills organised into four enlarged groups. For choosing activities to be included into the model, the researchers used statistical evaluation methods appropriate for data processing: frequency and percentage-arithmetic mean-standard deviation. They identified 23 variables as essential activities for teacher professional development in the modern digital world.

4.2 21st Century Competencies Model Description

The analyses revealed the following:

Group 1. Ways of Thinking.

1. For creativity and innovation skills, the following is appropriate:

   - to role modelling creative habits. Nothing is more important than teachers exemplifying habits, behaviours and thinking they want students to demonstrate. Teachers need to exemplify such creative traits as curiosity and development of creative skills;
   - to appreciate the critical importance of questions, both their own and those asked by colleagues and students. Encouraging to ask questions can enable to look at the topic from different perspectives, to clarify a goal or plan for any investigation, to inspire yourself to find out the answer;
   - to treat mistakes as learning opportunities and encouraging learners also to take sensible risks in the classroom. Encouraging learners to take 'sensible risks' in their work is important for building up their creative confidence. It is important that this takes place in a supportive environment, and that teachers and learners have discussed what boundaries are acceptable in a certain context;
   - to give yourself sufficient time to complete an important project. Sometimes ideas need time to develop before becoming valuable. Delay judgement of ideas until working them out properly;
   - to participate in the discussion in professional forums, where creative discussions of professionals take place (theoreticians and practitioners).

2. Critical thinking, problem solving, decision making:

- to use the Six Steps to Effective Thinking and Problem Solving, or "IDEALS" (Facione, 2007), the problem-solver works through a case study or activity by responding to questions from the peer coach. The IDEALS are to Identify, Define, Enumerate, Analyze, List, and Self-Correct:

I	Identify the Problem: What is the real question we are facing?
D	Define the Context: What are the facts that frame this problem?
E	Enumerate the Choices: What are plausible options?
A	Analyze Options: What is the best course of action?
L	List Reasons Explicitly: Why is this the best course of action?
S	Self-Correct: Look at it again … What did we miss?

3. Learning to learn, Metacognition:

- to educate yourself (study of scientific and professional journals, either in printed or electronic form, and scientific books, textbooks of recognised authors, proceedings of the international scientific conferences, symposia and seminars, etc. bringing the latest knowledge in the field);
- to be an active member in scientific and/or academic communities, associations, chambers, groups, etc. (work on the defined challenges or international projects on the one hand is the source of new knowledge from other colleagues, and on the other hand it forces teachers to gain thorough knowledge in the scientific field);
- to publish in the scientific journals and proceedings (preparation of each article always means thorough analysis of the existing knowledge in the topic concerned and own research which renews and enriches existing knowledge of the teacher);
- to participate in the programs of professional development organised by firms and companies (training performed by the designers/producers of new equipment enables safe use of simulators and training under the supervision of practitioners);

Group 2. Ways of Working.

4. Communication:

- to participate in foreign academic and other internships and scholarships (international environment and foreign universities can strongly accelerate the training progress at university);
- to participate in scientific conferences (scientific discussions at conferences increase knowledge and enable to obtain experience of other professionals and teachers, confront one's own knowledge and research results with results of others);
- to replace strict criticism of students and colleagues by providing supportive and motivational feedback;
- to learn when it is appropriate to listen and when to speak.

5. Collaboration (teamwork):

- to work in interdisciplinary projects (teaching of specialised project subjects requires perfect preparation by teachers at a deeper level than the teaching of classical subjects, which forces teachers to keep in permanent touch with the latest knowledge), etc.
- to receive feedback from the colleagues at the department (survey of one's own teaching through the eyes of younger and older colleagues);
- to coach, mentor, consult experienced educational authorities (i.e. in advance planned and approved training in partnership with the recognised authority at the department, faculty or other institutions);
- to conduct collaborative research with different universities.

Group 3. Tools for Working.

6. Information literacy:

- to learn where to start looking for information and to develop awareness of a broad range of information sources (e.g., electronic and print periodicals, chapters in books, government documents, archival material, and microfilm), and to distinguish among the various types of resources (e.g., scholarly work, informed opinions of practitioners, and trade literature);
- to select key points from retrieved information and summarize them, rather than simply repeat material from research;
- to use high-quality content and reflect evaluative thinking in the context of students' academic level and discipline, as evidenced during classroom discussions, when writing papers, creating displays, or when speaking or performing publicly;
- to develop new insights or theories, or discover previously unknown facts, based on material teachers already knew;

7. ICT literacy:

- to apply technology effectively: use technology as a tool to research, organize, evaluate information;
- to use digital technologies (computers, PDAs, media players, GPS, etc.), communication/networking tools and social networks appropriately to access, manage, integrate, evaluate and create information to successfully function in the modern digital world;
- to create interactive e-course books and online courses for students;
- to work with MOOCs, shared online courses, etc.

Group 4. Living in the World.

8. Citizenship—local and global; 9. Life and career; 10. Personal and social responsibility—including cultural awareness and competence:

All these skills require to develop teachers' leadership and we consider them together. But leadership in this case is a process, not a position of authority. It has a moral purpose based on a deep sense of respect and responsibility for oneself and for the others. Involvement in leadership activities can improve teachers' own way of living as well as helping them to develop broader competencies needed for success beyond the classroom and university.

Teachers' leadership:

- starts with 'knowing yourself' and developing self-confidence, empathy, communication skills, resilience and resourcefulness;
- should see those who have positions of responsibility focusing on getting the best out of the team by encouraging others to lead at appropriate times;
- should be shared and distributed with collective responsibility and accountability;
- should respect and encourage quiet leadership—often the most effective leaders get things done without acknowledgement or recognition;
- should respect the culture and context of the high school—different cultures have important protocols and conventions that need to be understood and respected;
- involves teachers in meaningful high school development activities so that they are involved in getting the best out of the system.

Table 1 summarizes the research analysis.

The research made it possible to develop a 21st Century Competencies Model of professional development in detail, and to illustrate it with a description of experiences that have used that particular model. The factors that must be taken into account when designing and implementing models of professional development are also identified. One of the most striking findings is the relative importance of competencies

Table 1. 21st Century Competencies Model

Groups	Skills	Activities
Ways of thinking	Creativity and innovation	• modelling creative habit • appreciate the critical importance of questions • treat mistakes as learning opportunities • give yourself sufficient time to complete the project • participate in the discussion in professional forums
	Critical thinking, problem solving, decision making	• using the Six Steps to Effective Thinking and Problem Solving (IDEALS)
	Learning to learn, Metacognition	• self-educating • be an active member in scientific and/or academic communities • publishing in scientific journals • participating in training programs

(*continued*)

Table 1. (*continued*)

Groups	Skills	Activities
Ways of working	Communication	• participating in foreign academic and other internships and scholarships • participate in scientific conferences • replacing strict criticism by providing supportive feedback • learning when it is appropriate to listen and when to speak
	Collaboration (teamwork)	• working in interdisciplinary projects • receiving feedback from the colleagues at the department • coaching and consulting experienced educational authorities • conducting collaborative research
Tools for Working	Information literacy	• learning where to find information • selecting key points from retrieved information • using high-quality content and reflect evaluative thinking • developing new insights or theories
	ICT literacy	• applying technology effectively • using digital technologies • creating interactive e-course books and online courses • working with MOOCs, shared online courses
Living in the world	Citizenship—local and global	• developing teachers' leadership
	Life and career	
	Personal and social responsibility including cultural awareness and competence	

development and their impact on education quality and career perspectives. The more teachers have found that competencies have led to changes in aspects of their work, the greater their development needs are, the more they participate in different professional development activities, and the greater the experienced impact of professional development becomes. Further study may be conducted to explore the effectiveness of professional development on academic achievement of students.

5 Conclusion

As technologically advanced nations shift their economies from industrial to information-based, knowledge economies, a number of different systems have emerged across the world. The shift changed the way people lived and worked, it changed the way people thought, and it changed the kinds of tools they used for work. The new skills and ways of thinking, living and working, once recognised, demanded new forms of education systems to provide them. Similarly, as the products and the technology to develop them become more digitised, another set of management and production skills are needed, focusing on increased digital literacy and numeracy and new ways of thinking. These will increasingly be identified as essential, and pressure on education systems to teach these new competencies will intensify. Education faces a new challenge: to provide the populace with the information skills needed in an information society. Educational systems must adjust, emphasising information and technological skills, rather than production-based ones. The result of the study may lead to redesigning of professional development courses of teachers, at least in the two universities under consideration. In the end, well-designed and implemented professional development should be considered an essential component of a comprehensive system of teaching and learning that supports students to develop the knowledge, skills, and competencies they need to thrive in the 21st century.

References

1. Barr, A.: Teacher effectiveness and its correlates. In: Barr, A., et al. (eds.) Wisconsin Studies of the Measurement and Prediction of Teacher Effectiveness: A Summary of Investigation. Dembar Publication, Madison (1961)
2. Blašková, M., Blaško, R., Kucharþíkováa, F.: Competences and competence model of university teachers. In: Procedia—Social and Behavioral Sciences, vol. 159, pp. 457–467. (2014). [Electronic Resource]. URL: https://www.sciencedirect.com/science/article/pii/S1877042814065379. Accessed 25 April 2018
3. Blašková, M., Blaško, R., Matuska, E., Rosak-Szyrockac, J.: Development of key competences of university teachers and managers. In: Procedia—Social and Behavioral Sciences, vol. 182, pp. 187–196. (2015). [Electronic Resource]. URL: https://www.sciencedirect.com/science/article/pii/S187704281503030X. Accessed 25 April 2018
4. Creating Effective Teaching and Learning Environments: First Results from TALIS. Teaching and Learning International Survey. OECD (2009)
5. Darling-Hammond, L., Hyler, M.E., Gardner, M.: Effective Teacher Professional Development. Learning Policy Institute, Palo Alto, CA (2017)
6. Gray, W., Gerrand, B.: Learning by Doing: Developing Teaching Skills. Addison-Wesley Publishing Company, Sydney (1977)
7. Kraur, I., Shri, C.: Effective teaching competencies—a compilation of changing expectations from students and institutions. J. Contemp. Res. Manage. 10(1), 57–71 (2015). [Electronic Resource]. URL: http://psgim.ac.in/journals/index.php/jcrm/article/view/413/286. Accessed 29 April 2018

8. Kunter, M., Klusmann, U., Baumert, J.: Professionelle Kompetenz von Mathematik-lehrkra"ften: Das COACTIV-Modell [Teachers' professional competence: The COACTIV model]. In: Zlatkin-Troitschanskaia, O., Beck, K., Sembill, D., Nickolaus, R., Mulder, R. (eds.) Lehrprofessionalitat—Bedingungen, Genese, Wirkungen und ihre Messung, pp. 153–164. Beltz, Weinheim, Germany (2009)

9. Malik, S.K., Nasim, U., Tabassum F.: Perceived effectiveness of professional development programs of teachers at higher education level. J. Educ. Pract. **6**(13) (2015). [Electronic Resource]. URL: https://files.eric.ed.gov/fulltext/EJ1080484.pdf. Accessed 23 Feb 2018

10. Potolea, D., Toma, S.: The dynamic and multidimensional structure of the teachers professional development. In: Procedia—Social and Behavioral Sciences, vol. 180, pp. 113–118 (2015). [Electronic Resource]. URL: https://www.sciencedirect.com/science/article/pii/S1877042812013869. Accessed 20 April 2018

11. Pozos, K.V., Mas, O.: The digital competence as a cross-cutting axis of higher education teachers' pedagogical competences in the European higher education area. In: Procedia—Social and Behavioral Sciences, vol. 46, pp. 1112–1116 (2012). [Electronic Resource]. URL: https://www.sciencedirect.com/science/article/pii/S187704281501424X. Accessed 20 April 2018

12. Selvi, K.: Teachers' competencies culturass. Int. J. Philos. Cult. Axiol. **7**(1) (2010). [Electronic Resource] https://doi.org/10.5840/cultura20107133

13. Semradova, I., Hubackova, S.: Responsibilities and competencies of a university teacher. In: Procedia—Social and Behavioral Sciences, vol. 159, pp. 437–444 (2014). [Electronic Resource]. URL: https://www.sciencedirect.com/science/article/pii/S1877042814065331. Accessed 15 April 2018

14. Shor, I., Freire, P.: A Pedagogy for Liberation: Dialogues on Transforming Education. Bergin & Garvey Publishers, p. 203 (1987)

15. Shulman, L.S.: Those who understand: knowledge growth in teaching. Educ. Researcher **15** (2), 4–14 (1986)

16. Slavik, M., et al.: University Pedagogics, p. 253. Grada, Praha (2012)

17. Teachers' Professional Development: Europe in International Comparison. An Analysis of Teachers' Professional Development based on the OECD's Teaching and Learning International Survey (TALIS). OECD (2010)

18. Villegas-Reimers, E.: Teacher Professional Development: An International Review of the Literature. UNESCO: International Institute for Educational Planning (2003)

Project Based Learning a New Approach in Higher Education: A Case Study

Micaela Esteves[1(✉)], Rosa Matias[1], Eugénia Bernardino[1],
Vitor Távora[1], and Angela Pereira[2]

[1] CIIC - Computer Science and Communication Research - ESTG,
Polytechnic Institute of Leiria, Leiria, Portugal
{micaela.dinis, rosa.matias, eugenia.bernardino,
vitor.tavora}@ipleiria.pt
[2] CiTUR - Tourism Applied Research Centre - ESTM,
Polytechnic Institute of Leiria, Leiria, Portugal
angela.pereira@ipleiria.pt

Abstract. Project Based Learning (PBL) is a collaborative and teaching methodology which stimulates critical thinking, students' autonomy and creativity, among others. PBL has been implemented in higher education with success in many domains. In this context, the Polytechnic Institute of Leiria updated the curriculum of Web and Multimedia Development Technological course to a student-centred approach with PBL methodology. The purpose of this paper is to present the DWM curriculum design process and discuss the initials results. In contrast with previous scholar years, the results obtained with the new curriculum transformation are encouraging. Despite this, it was necessary to adapt the scenario for motivating the students with implementation of an enterprise simulator.

Keywords: Project based learning · Student-centred learning
Higher education · Professional higher technical course

1 Introduction

Since Bologna process, the student diversity at higher education increased with diverse academic orientations and commitments. In general, there are different students' profiles, which some of them are academically committed, bright, interested in studying and wanting to be successful. However, other students are academically uncommitted, plus are not being driven by curiosity in a particular subject and do not have ambitions to excel in a certain profession. Nevertheless, they want to be qualified for a different job. To this last range of students, the traditional teaching methods, mainly based on expositions classes is unsuitable. They are incapable of learning by conventional explanatory classes, leading to demotivation and consequently high dropout rates and low scores [1].

In Portugal, in 2015 were created several media higher superior courses designated by Professional Higher Technical Courses (TeSP) with the purpose to bring more students to the higher educational system. In the first editions of these courses, namely

M. E. Auer and T. Tsiatsos (Eds.): ICL 2018, AISC 917, pp. 525–535, 2019.
https://doi.org/10.1007/978-3-030-11935-5_50

in Information and Communication Technology (ICT) areas, there were low success achievements and high dropout rates (more than 50%). The reality in Polytechnic Institute of Leiria was not different, in the first edition of the TeSP course, named by Web and Multimedia Development (DWM), in the school year of 2015/2016, attended 49 students and only 41% finished the course. In the second edition, in 2016/2017, attended 44 students and only 43% finished it. One reason for these results is associated with the students' low profile for higher education. Some of them had low self-esteem, were not motivated to learn and presented fundamental flaws, namely, in maths and logical assumption. Special the maths abilities are very important for technical courses.

Furthermore, in the Portuguese labour market there is a lack of professionals with digital skills. Regarding this, Portuguese Educational Ministry recommends the use of PBL methodology in TeSP courses. In line with this, in 2017 the Polytechnic Institute of Leiria changed the DWM TeSP curriculum with the aim to adopt the Project Based Learning (PBL) methodology to improve the digital skills and students results.

In literature the PBL methodology has been for a long time considered an example of an active teaching method suitable for that kind of students [2]. It requires students to question, speculate, and generate solutions. In this way, they use the higher order cognitive activities.

PBL is based on a collaborative teaching and learning approach, by changing the focus from the teacher to the learner. The learning process begins with a real-life problem based on identified learning needs, which will require searching for information by students to solve. This process promotes the integration of different subjects, improves students' autonomy and prepare them for lifelong learning.

The aim of this paper is to present a case study concerning the implementation of PBL methodology in the context of a Portuguese higher education TeSP course.

This paper is organised as follows: first, our motivation and related work is outlined; then the DWM Curriculum Design is presented and the implementation process is described. Finally, the results, discussions and conclusions are exposed.

2   Related Work

Project Based Learning (PBL) has been employed in different countries and domains for more than 20 years. PBL appeared during the sixties associated to medical science courses. Since then, its applicability has been reported across all education levels [3]. In higher education, a practical implementation has been reported across a range of educational fields such as medicine [4, 5], architecture [6], management [7], engineering [8, 9] among others [10, 11].

PBL is a collaborative teaching-learning methodology which provides an active and coherent construction of mental models as a result of knowledge, rather than the simple process of subjects that happens in traditional teaching. Furthermore, it is a contextualised way of teaching-learning. Since its principles, ideas and mechanisms are not studied in abstract, instead, they are studied in context of a concrete situation which should be recognised by learners as relevant as interesting. Ideally, the scenario should be like the future professional environment in which the knowledge will be used [12]. This methodology focuses on solving complex real-life problems as a stimulus for learning, integration and organisation of acquired knowledge, to ensure its use in future problems [13].

In addition, PBL enhances the self-reflection, cooperation and responsibility which are integrated in the learning process [14].

To summarise, one characteristic that makes PBL an interesting educational methodology is that it allows to achieve broad educational goals. What is more, not only it grants the construction of knowledge, but also the development of skills and behaviours that will be useful in the students' future career.

The PBL learning model has two main features: the problem nature, which is the learning booster, and the learning is student-centred.

The problem is the start point of students learning process, so the quality of problem is the main concern (Fig. 1). The problem' complexity should be adapted to students' profiles. The problem should be clear with an open-end solution, in this way the students could find out different solutions for the same problem. During the problem resolution, students should find the theoretical concepts for their learning needs. This research promotes the integration of concepts to reach a final solution [3].

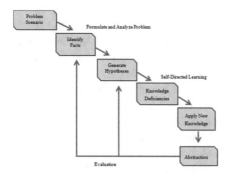

Fig. 1. The project base learning cycle (adapted from [4])

Students work in groups of five to six elements and start by organising their ideas using the information and the knowledge they have. Through the group discussion, students will define and consider the different problem issues, make assumptions and raise questions about what they do not understand. In this way, students analyse and evaluate their prior knowledge and acquire a deeper understanding of the problem. After this, students engage in a self-directed study and research for information. Students meet again, explore previous learning issues, integrate their new knowledge into the problem context and propose a solution. This process repeats when a solution is not found. In the end, students evaluate the learning process, their performance and their peers. This way, they develop self-assessment habits and constructive assessment of their peers, essential for autonomous and effective learning [4].

3   PBL—DWM Curriculum Design

In general, the TeSP courses have four semesters, each one with twenty weeks. The first three semesters consist in several subjects, practical classes and lectures, and the last one is the professional internship. At the internship, students put into practice the knowledge acquired during the whole course.

In the new PBL DWM curriculum design it was decided to keep the same structure, i.e. three semesters of classes and lectures and a fourth dedicated to the professional internship.

The new curriculum design process began with a three-month training for all teachers that were assigned to the course, with a PBL expert.

Initial training sessions were dedicated to introduce the PBL methodology. Although some teachers knew the concept, they were not familiarised how to implement it. The teachers' feelings about this methodology ranged from enthusiasm to scepticism. A lot of doubts were raised concerning the process feasibility, namely about the students' profile and their adaptation. Some teachers doubted that students were prepared to the PBL methodology. Moreover, there were concern about the lack of time to accomplish the curriculum. On the other hand, experienced teachers were not very enthusiastic to switch from teacher-centred learning to student-centred learning.

To prepare the PBL implementation, three teachers' teams were created, one for each semester (1st, 2nd and 3rd). Each team started to establish the learning outcomes that students should achieve in a semester. The next step was to establish each learning outcome assessment. With these in mind, each team defined a project for the semester, as well as the aggregated curricular units. Each curricular unit corresponds to an important theme to be addressed by the project. The curricular units associated with the project became concrete thematic models named by Active Learning Modules (ALM). The idea behind the ALM is to give a theoretical support to the project development (Table 1). This structure follows the Aalborg PBL model [15] that was implemented in engineering programmes at the Higher Education Polytechnic School of Agueda from Aveiro University [16].

Table 1.   Active learning module example

Active Learning Modules (ALM)—Introduction to Web Programming and Multimedia	
• Web contents production	Curricular units aggregated to the project
• Image contents production	
• Project	

The Aalborg model does not consider time assigned to the project in the class schedule. However, all teachers' teams agreed that it was important to have some hours assigned in the class schedule to the project. This is due to the students' profile that do not have enough maturity to work outside the classes. Along the course the difficulty of the projects increases and, consequently, it also increases the ALM ECTS's and the time per week assigned to the project.

In addition to Active Learning Modules the course also has autonomous subject units responsible for some essential subjects' themes, such as maths and programming. Those subject units are also taught in PBL format. Either the ALM and the autonomous units are practical (PL), i.e., there are not theoretical classes (Tables 2 and 3).

Table 2. The DWM-PBL Course Structure (1st Year)

1st year					
1st semester			2nd semester		
Subject	PL	ECTS	Subject	PL	ECTS
Applied mathematics	45	3	English language	30	2
ICT	30	2	Web server programming	75	5
Web programming	75	6	AML: Client/Server Web Programming and Audio-visual	240	23
AML: Introduction to Web Programming and Multimedia	165	19	• User Interaction	30	3
• Web Content Production	45	6	• Audio-Visual Content Production	45	3
• Image Content Production	30	4	• Data Base for Web Application	30	3
• **Project**	90	9	• **Project**	135	14

Table 3. The DWM-PBL Course Structure (2nd Year)

2nd year					
1st semester			2nd semester		
Subject	PL	ECTS	Subject	PL	ECTS
Digital Marketing and Social Web	45	3	Internship	640	30
Video Edition and Pro-Production	45	3			
Active Learning Module: Multimedia Web Portal Development	240	24			
• Multimedia Project Methodology	30	3			
• Web Administration and Publication	30	3			
• CMS Systems	45	4			
• **Project**	135	14			

3.1 Project Structure

Every student should develop three big projects along the course, one in each semester. These projects stimulate knowledge integration and are the start point of the learning process. In this way, projects outlines are clear enough but without a possible solution consideration. Each group should provide its own solution. The projects are defined to aggregate themes from both ALM unit and autonomous units.

The project level of difficulty increases each semester to improve the students' knowledge and independence. The project's problems are presented in the ALM first class. The students work in groups of 5–6 elements, and each group is assigned to a different project. Within each group, students share information and work together to solve the problem. Each group has a teacher assigned to them, whose main role is to facilitate students' progression. Students in group should discuss and search for a solution. This way, they should find out what they need to learn in order to solve the problem.

The subjects covered in the aggregated curricular units will be the topics that students pointed out as important for resolving the problem. The project involves an interdisciplinary approach in both the analysis and the solving phases [15].

The project assessment was designed to have several public presentations with public discussions. The final assessment would result through a written report and a public presentation/discussion for a jury formed by three teachers. The presentation is followed by a period of discussion with directed questions for each student. In the project assessment it would be also considered the students' peer evaluation. The student project mark depends of the project quality and the individual performance during the final discussion as well as the peer evaluation.

In the 2nd and 3rd project it is planned to invite external entities to the jury, including companies' elements. In the 1st semester the project scope does not justify the invitation of the external entities.

3.2   Active Learning Modules Assessment

The assessment model is presented in Fig. 2. The module assessment results from the weighted average of the students' grades obtained in the project and the aggregated curricular units (Eq. 1).

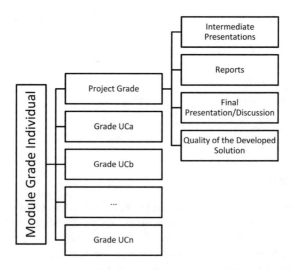

Fig. 2. Module assessment components

$$Module\ Grade = Grade(Project) \times ECTS(Project) + Grade(UC_A)$$
$$\times ECTS(UC_A) + Grade(UC_B)...+ Grade(UC_n) \tag{1}$$

The project only has one assessment moment although the aggregated curricular units could have several assessment periods, like continuous assessment or exam. This means that the module assessment period is only one.

After the design phase, in the beginning of the 1st semester in the 2017/2018 scholar year all teachers met with the aim to prepare the classes. Teachers defined one open-end project outline without any considerations about the possible solutions. It was also defined that the project should have four intermediate presentations and a final discussion/presentation at the end of the semester. It was also discussed the way it would be done the articulation of Curricular Units aggregated with the project. It was decided that each group should have two supervisors one from computer science and another from graphical designer, since the project covered these two main areas.

In the ALM first class it was presented to students the project outline. Then, for a week, students worked in groups and searched for solutions that they would present and discuss with their tutors. After this, students developed the project on the project classes with the tutors' supervision.

4 Results and Discussions

In this study it was used a qualitative methodology in which the data collected was obtained by the researchers through direct observation in the field and through written narratives. Also, teachers were interviewed with the aim to know their own opinions, experiences and to identify difficulties. Moreover, students were interviewed to know their opinions, the perceptions of their engagement in the course and to identify areas for further development.

4.1 1st Semester

In the first year, entered in the course forty-seven students which were divided into two classes: practical laboratory 1 (PL1) and practical laboratory 2 (PL2) with, respectively, twenty-six students in the first and twenty-one students in the second. One student with disabilities attended PL2, requiring differentiated treatment and some careful in his integration on class, especially in the group. The students were assigned to groups in the first class. Since students did not know each other, neither teachers, so students randomly organised into teams. This resulted in nine groups of five to six elements. PL1 had one group with six elements and three groups with five elements and PL2 had one group with six elements and four groups with five elements.

In first semester, the students developed an academic project that consisted of an implementation of a static web site using technologies such as HTML, JavaScript and CSS.

From students' interviews researcher observed that there was a diversity of students' profiles. The majority of them confirms that they never had studied outside the classroom, what was a big issue, since they needed to work and study for learning. The situation was particularly worrying in maths, due to students had a very low background in the area. To overcome the situation, students started to attend a psychologist to guide them on how to organize their study and on how to work in teams.

During the semester several problems arise, namely inside the groups with strong disagreement between their elements, despite the teachers' help. The lack of maths knowledge bases was an evidence, since only nine students get approval. In contrast, at the ALM module only five students did not get approval and six students dropped out.

Confronting these results with the last two editions (2015/2016 and 2016/2017), that adopted the traditional teacher-centred methodology, we can notice that there was an improvement in students' dropout rates (Fig. 3). These results are in line with authors from literature [14, 17, 18], which refer that with the PBL methodology the students' dropout rate decreases since students are more motivated and engagement.

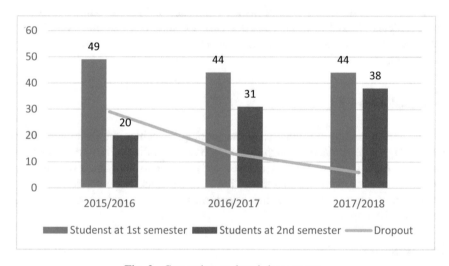

Fig. 3. Comparing students' dropout rates

4.2 2nd Semester

To overcome some difficulties from the 1st semester, in the 2nd semester the project was restructured. It was created an enterprise simulator where each teacher plays a role. The headmaster was the CEO, the teacher responsible for the project subject. The Human Resources department, with the aim to monitor the teams and their interaction, was associated to a teacher from Communication area. The Design department, responsible for accomplishing the design of all projects, had two teachers from Design and Video areas. Finally, the Technical department, in charge of the project development with two chiefs, one for each PL class, had three teachers from Computer Science area.

In the hope that teams work will be less problematic than in 1st semester, the new teams were organised with a Psychologist help. Since, this Psychologist accompanied the students along the semester, she knew then very well. This way resulted in eighteen students in PL1, and twenty-two students in PL2 (Table 4).

Table 4. Division of Students by Groups, 2nd semester

Number of elements	PL1	PL2
6	3 groups	2 groups
5	–	2 groups

Four teams were developing real projects for local companies and the others were developing academic self-proposed projects with the goal to put them in the market. The project had seven milestones, in each one teachers rewarded the team which presented the best work. The rewarded team, each element, received a certificate of wining and a small gift, such as a cup with sweets.

A student who dropped out in the 1st semester has returned to the course after speaking with his its colleagues. This is another good signal of the improvement that have been made in the course.

In the teachers' opinion, during the 2nd semester, students' presentations were better both in oral expression and in support materials' quality. Furthermore, students became more active during their colleagues' presentations, making suggestions and detecting small gaps. Moreover, have improved the students' performance and assessment. Teachers' also highlight the students' motivation and the constructive discussions about their project solutions. However, besides these positive aspects teachers referred that was important to have more time available to support students work and more full-time teachers.

From students' interviews it was notorious that the relationship between students inside the groups have improved since they were more balance. Furthermore, students highlighted the importance of the reward in each project milestones, as it can be observed in this student reaction:

> I gave the reward to my father, because it was the first time I have received one and he stayed very proud of me.

The researcher considers from the observation that the rewards had a big impact in students' self-esteem and motivated them to be more productive due to this feeling of pride and achievement (Fig. 4). Some of these students had never been acclaimed along their academic path. Several authors mention the importance of reward for students' motivation and commitment [19, 20].

In general, students mention that they liked learning in a collaborative environment, since they saw a practical application of what they had been learning and they considered this methodology will help them in the future career.

Fig. 4. Students' rewards

5   Conclusion

All in all, in this study, researchers concluded that PBL implementation in Higher Education is a hard process, due to the changes of traditional methodologies, not only for students, but also for teachers. Besides that, this approach has critical factors to succeed. For students, it is necessary a psychological monitoring in order to accomplish dynamic groups, since students have difficulties to work as a team, and alone they cannot overcome their difficulties. For the students' motivation it is important to work in real life projects because they have more responsibility and are proud of make a product to be used by a company. Other critical factor for students' commitment is to learn in an enterprise environment, where the reward is made by the performance. The public recognition of their work also has a strong psychological effect, since through their learning process they never had this type of encouragement.

For teachers, the process requires a hard work as it is necessary to prepare all the process before the beginning of the semester, and to spend more time to accompany the students. It is also required to have experienced teachers to evaluate the students' solutions proposals and to deal with their doubts.

The important result to be taken out of this experience is the success in reducing dropouts' rates with this methodology. Since, this is a problem in higher education courses special in computer science areas.

Based on our experience in teaching at higher education, we believe, that PBL methodology could increase the academic results, decreasing the students' dropouts, and could improve their motivation and their skills.

References

1. Biggs, J.B.: Teaching for Quality Learning at University: What the Student Does. McGraw-Hill Education (UK)
2. Esteves, M., Fonseca, B., Morgado, L., Martins, P.: Improving teaching and learning of computer programming through the use of the Second Life virtual world. Br. J. Educ. Technol. **42**(4), 624–637 (2011)
3. Barrows, H.S.: A taxonomy of problem-based learning methods. Med. Educ. **20**(6), 481–486 (1986)
4. Barrows, H.S., Tamblyn, R.M.: Problem-based Learning: An Approach to Medical Education. Springer Publishing Company (1980)
5. Idowu, Y., Muir, E., Easton, G.: Problem-based learning case writing by students based on early years clinical attachments: a focus group evaluation. JRSM Open **7**(3), 2054270415622776 (2016)
6. Cabodevilla-Artieda, I., Torres, T.L., Muniesa, A.V.: PBL. Problem-based learning cross-application to the first year graphical courses of the degree in architecture. In: Congreso Internacional de Expresión Gráfica Arquitectónica, pp. 265–276. Springer, Cham (2016)
7. Maxwell, N.L., Bellisimo, Y., Mergendoller, J.: Problem-based learning: modifying the medical school model for teaching high school economics. Soc. Stud. **92**(2), 73–78 (2001)
8. Mills, J.E., Treagust, D.F.: Engineering education—Is problem-based or project-based learning the answer. Australas. J. Eng. Educ. **3**(2), 2–16 (2003)
9. Perrenet, J.C., Bouhuijs, P.A.J., Smits, J.G.M.M.: The suitability of problem-based learning for engineering education: theory and practice. Teach. High. Educ. **5**(3), 345–358 (2000)
10. Raine, D., Symons, S.: Experiences of PBL in physics in UK higher education. In: PBL in Context–Bridging Work and Education, p. 67 (2005)
11. O'Kelly, J., Gibson, J.P.: RoboCode & problem-based learning: a non-prescriptive approach to teaching programming. ACM SIGCSE Bull. **38**(3), 217–221 (2006) (ACM)
12. Duch, B.J., Groh, S.E., Allen, D.E.: The Power of Problem-based Learning: A Practical "How To" for Teaching Undergraduate Courses in Any Discipline. Stylus Publishing, LLC (2001)
13. Tsai, C.W., Shen, P.D.: Applying web-enabled self-regulated learning and problem-based learning with initiation to involve low-achieving students in learning. Comput. Hum. Behav. **25**(6), 1189–1194 (2009)
14. Hmelo-Silver, C.E.: Problem-based learning: what and how do students learn? Educ. Psychol. Rev. **16**(3), 235–266 (2004)
15. Kolmos, A., Fink, F.K.: The Aalborg PBL Model: Progress, Diversity and Challenges. L. Krogh (Ed.). Aalborg University Press, Aalborg (2004)
16. Oliveira, J.M.N.: Nine years of project-based learning in engineering. Revista De Docencia Universitaria, Redu (2011)
17. Freeman, S., Eddy, S.L., McDonough, M., Smith, M.K., Okoroafor, N., Jordt, H., Wenderoth, M.P.: Active learning increases student performance in science, engineering, and mathematics. Proc. Natl. Acad. Sci. **111**(23), 8410–8415 (2014)
18. Bell, S.: Project-based learning for the 21st century: skills for the future. The Clearing House **83**(2), 39–43 (2010)
19. Lopes, D., Esteves, M., Mesquita, C.: Video game interaction and reward mechanisms applied to business applications: a comparative review. In: 2012 7th Iberian Conference on Information Systems and Technologies (CISTI), pp. 1–6. IEEE (2012)
20. Esteves, M., Pereira, A., Veiga, N., Vasco, R., Veiga, A.: The use of new learning technologies in higher education classroom: a case study. In: International Conference on Interactive Collaborative Learning, pp. 499–506. Springer, Cham (2017)

Framework and Ontology for Modeling and Querying Algorithms

Baboucar Diatta[1]([⊠]), Adrien Basse[1], and Ndeye Massata Ndiaye[2]

[1] University Alioune Diop of Bambey, Bambey, Senegal
{baboucar.diatta, adrien.basse}@uadb.edu.sn
[2] Virtual University of Senegal, Dakar, Senegal
massata.ndiaye@uvs.sn

Abstract. To improve algorithmic problem solving skills and thus coding skills, we need to practice algorithms. In that regard, databases like algoBank and websites provide algorithmic exercises with solutions but with contents that may or may not be semantically structured to be queried efficiently. To better exploit the potential of exercise databases, research suggests to represent their content using ontologies. Nevertheless, such research focuses mainly on representing and querying exercises as well as algorithmic skills or specific programming languages. In this work, we choose to represent, through the ontology CodOnto, the solutions of exercises in pseudocode to enhance understanding and querying of algorithms for learners and facilitate translation into the programming languages.

Keywords: Pseudocode · Algorithm · Ontologies · Distance education

1 Introduction

Learning programming languages provide multiple benefits. Besides the possibility to start a technology business, the benefits include the capacity to develop mobile or desktop applications and websites. Overall, learning programming languages provide a better understanding of how technologies (used daily) work and how to break down any problem into easier and logical steps.

Improving programming skills requires learning (i) rules of a particular programming language, (ii) basics and core concepts of any programming language (variable, selective structure, loop structure, array, ...), (iii) algorithms (step by step method to accomplish a task). It is also necessary to practice both programming and algorithm. In that regard, databases like algoBank,[1] EDBA,[2] some platforms like codederbyte[3] and hackerRank[4] provide algorithmic exercises with solutions but with contents that may or may not be semantically structured to be queried efficiently.

---

[1] https://moodle.utc.fr/course/view.php?id=503.
[2] https://edba.imag.fr/index_EDBA_Full.html.
[3] https://coderbyte.com/.
[4] https://www.hackerrank.com/.

© Springer Nature Switzerland AG 2019
M. E. Auer and T. Tsiatsos (Eds.): ICL 2018, AISC 917, pp. 536–544, 2019.
https://doi.org/10.1007/978-3-030-11935-5_51

To better exploit the potential of exercise databases, research suggests to represent their content using ontologies. Indeed, ontologies defined as "explicit formal specifications of the terms in a domain and relations among them" [11], provide not only semantic description of domain concepts but also new knowledge. Nevertheless, research about exercise databases focuses mainly on representing and querying exercises as well as algorithmic skills [4], programming domain knowledge [20] or specific programming languages [2, 7, 13, 14, 15, 19].

In this work, we choose to represent, through the ontology CodOnto, the solutions of exercises in pseudocode to enhance understanding and querying of algorithms for learners. Choosing pseudocode instead of a particular programming language helps to focus more on algorithmic skills and less on the syntax of a programming language with the possibility to automatically translate the code into the programming languages.

The rest of the paper is organized as follows: Sect. 2 surveys related works. Section 3 describes the development process of ontologies and ontology designed for pseudocode. Section 4 presents an example of using this ontology. Section 5 is the conclusion.

2   Related Work

The development and deployment of ontologies have many benefits including "sharing common understanding of the structure of information among people or software agents and, enabling reuse of domain knowledge" [17]. Therefore, research has proposed many ontologies to enhance understanding and querying of algorithms and programming domain.

Kouneli et al. [14] proposed a java ontology to capture the semantics of java concept with the reserved words of the language, the general object-oriented concepts (class, constructor, interface, ...), the different types of statements, ... As [14, 15] proposed a java ontology and added a C programming language ontology. The two ontologies focus on the C and Java language elements in an aim to help teachers in designing and organizing the Learning Objects (smaller learning unit to combine in order to build course) and help learners to better interact with the computer programming courses. In the same way to model object-oriented concepts and features, [1, 2] have proposed two ontologies with a component to automatically populate them from Java code. Because SCRO [1] ontology uses elements strongly depending on java language like java.lang.String, [7] decided to build the new ontology CSCRO to represent object-oriented concepts but dedicated to C#. Ganapathi et al. [9] uses an ontology to describe some tools used for java programming and the java language elements. Ivanova [13], to facilitate the learning of java programming to young Bulgarian programmers, capture in a bilingual (English and Bulgarian) ontology the semantic of Java and object-oriented concepts. As [15, 19] proposed an ontology for teaching/learning C programming. This ontology represents the basic syntactic elements of C language and other knowledge including helpful programming techniques.

Underlining the importance of pseudocode, [4] proposed to represent, through an ontology, algorithmic skills, exercises description and organization. Nevertheless, the proposed ontology does not represent detailed exercise solutions.

The above ontologies focus mainly on object-oriented programming and essentially on a particular language syntax. If the syntax is important, it is fundamental to acquire the algorithmic skills to be able to propose the set of primitive steps to solve a problem. For this reason, this work proposes codOnto ontology to describe essentially in pseudocode the exercise solutions. This is all the more important as a well-designed pseudocode solution is easily translated into any procedural languages (C, Pascal ...).

3   CodOnto Ontology

Our main task is to construct CodOnto, an ontology of algorithms. In this section we describe the adopted methodology to construct our ontology and the resultant ontology.

3.1   Methodology

In the literature, there are many methodologies [3, 8, 17, 21] that address the issue of ontologies development. In order to achieve CodOnto, we mainly adopt the methodology proposed by [17] and used to construct, for instance, AlgoSkills [4] and Java ontology [14]. We follow mainly five steps:

1. Determine the scope of the ontology: This first step helps to limit the scope of the model by clarifying its purpose and the type of expected queries and answers;
2. Enumerate significant terms: In this step we write down all significant domain terms which appear in queries or answers defined in step 1;
3. Define the classes and class hierarchy: From the list of terms created in step 2, we pick up the classes and organize them into a taxonomic hierarchy;
4. Define properties of classes with their facets: At first, this step consists in identifying the properties (datatype properties) of each class and the interactions (object properties) among classes. Then, for each of this two kinds of properties, we define their domain, range and if necessary property restrictions (value constraints and cardinality constraints);
5. Create instances: In this step, we populate the ontology with individual instances of classes created in step 3.

3.2   CodOnto Ontology

We construct our ontology by following the five steps listed above. We begin with determining the main goals and intended use of CodOnto.

Scope of codOnto.
CodOnto has been proposed to conceptualize pseudocode exercise solutions in order to enhance learners' algorithmic understanding and querying. Thus, CodOndo is expected to contain exercise descriptions and all pseudocode constructs so as to describe an algorithm.

Significant terms.
Since programming languages and pseudocode have a set of concepts in common, we particularly rely on programming language ontologies during this step. We also consult

some pseudo code courses,[5,6] latex packages[7] for producing pseudocode listing, books [5, 6, 18], pseudocode and flowchart editors,[8,9,10] and research papers like [12].

Figure 1 shows a part of the concept map developed in this step. Concept map and mind map are very useful in this brainstorming step especially with the available conversion algorithm from concept map to ontology [10, 22]. The represented concepts in the concept map include:

- *pseudocode* and *exercise* with their authors;
- *input*, *output* and *intermediate* data of pseudocode;
- *main function* and *user function* with their lists of statements;
- *defined functions*, a kind of statement, that we can call in our pseudocode.

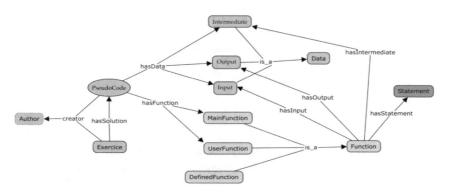

Fig. 1. Part of concept map of significant terms

Classes and class hierarchy.
From this step, we use the Protégé[11] editor to build codOnto ontology in particular our class hierarchy from the significant terms identified in the previous step. The open source ontology editor Protégé provides complete OWL 2 [16] editing capabilities. Figure 2 shows a part of class hierarchy obtained with mainly:

- *Data* class and its subclasses *input*, *output* and *intermediate*;
- *foaf:Person* class, from FOAF[12] vocabulary, to describe the *exercise* or *solution* authors;

---

[5] http://www.cs.ucc.ie/~dgb/courses/toc/lectures.html.

[6] https://en.wikibooks.org/wiki/GCSE_Computer_Science/Pseudocode.

[7] https://www.tex.ac.uk/FAQ-algorithms.html.

[8] http://www.flowgorithm.org/.

[9] https://www.rapidqualitysystems.com/TechnicalDocumentation/CodeRocketDesigner.

[10] https://algobuild.com/en/index.html.

[11] https://protege.stanford.edu/.

[12] http://xmlns.com/foaf/spec/.

Fig. 2. Part of the class hierarchy

– *Statement* class with two subclasses: *CompoundStatement (loopStatement* or *SelectionStatement)* which contain statements and *SimpleStatement* like *assignmentStatement* and *inOutStatement*.

Properties of classes with their facets.

This step consists in identifying data properties for each class as defined previously and the object properties between the individual members of such classes. Thus, for each property, we define its domain (class it describes) and its types of values. Figure 3 shows part of the data properties and object properties obtained.

Fig. 3. Part of the data and object properties

This figure includes object properties such as:

- *block* to assign a set of statements to a compound statement like *forStatement*, *whileStatement* or *repeatStatement*;
- *hasStatement* to link a main, user or defined function to statements;
- *dc:creator* to link exercises and solutions to their author;
- *hasData* and its inverse *isDataOf* to assign to each pseudocode its input, output and intermediate data. The *has* properties generally have corresponding *is...of* properties to represent bi-directional relationships;
- *foaf:name, foaf:firstname, foaf:lastname* to describe authors.

Instances.
The last step consists in creating individual instances of classes in a hierarchy with their properties. Thus, we have created many instances of exercises with their complete solutions using protégé. Figure 4 shows the individual pseudocode1 with two input data, one output data and a main function. Each of this has, in turn, property assertions.

Fig. 4. View of individual pseudocode assertions

4 Implementation

CodOnto ontology can be used in many scopes of application. At the end of its design, we developed a web application to facilitate its usage through an information querying module.

On the one hand, our information querying module helps to query our pseudocode data using directly SPARQL[13] (SPARQL Protocol And RDF Query Language). SPARQL is a standard query language and protocol to facilitate access, querying and manipulating RDF[14] graph in triple stores.

On the other hand, our module provides a user-friendly interface to retrieve available pseudocodes and to highlight pseudocode keywords in displayed algorithms. Figure 5 shows the list of pseudocode exercises with their title, author, difficulty level and the possibilities to display other properties or solutions. The solution of the first exercise (see Fig. 6) have two input data, one output data and eight statements.

---

[13] https://www.w3.org/TR/sparql11-query/.
[14] https://www.w3.org/RDF/.

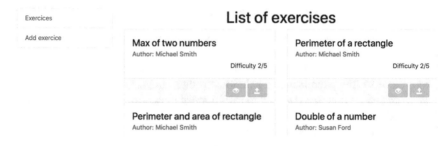

Exercices

Add exercice

List of exercises

Max of two numbers
Author: Michael Smith

Difficulty 2/5

Perimeter of a rectangle
Author: Michael Smith

Difficulty 2/5

Perimeter and area of rectangle
Author: Michael Smith

Double of a number
Author: Susan Ford

Fig. 5. List of pseudocode exercises

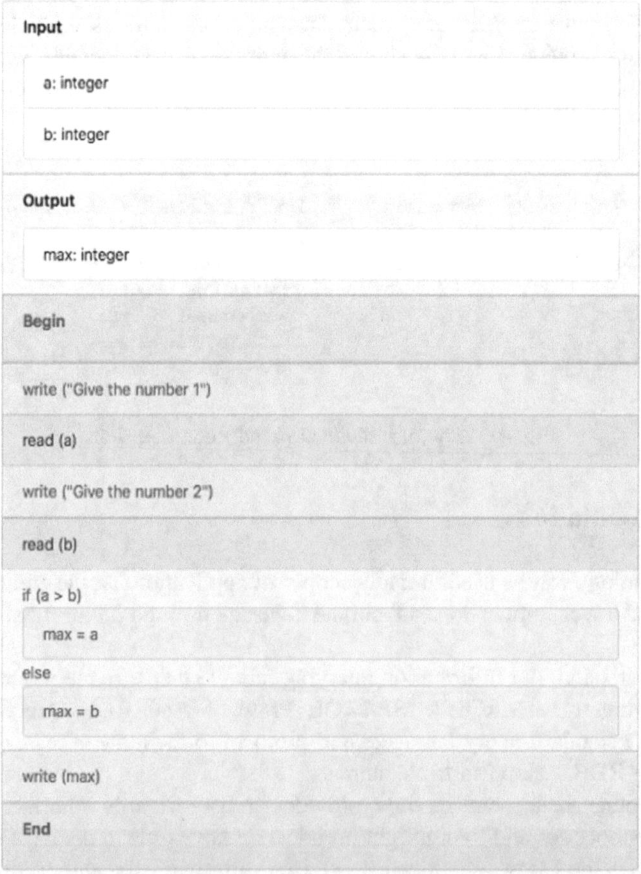

Input

a: integer

b: integer

Output

max: integer

Begin

write ("Give the number 1")

read (a)

write ("Give the number 2")

read (b)

if (a > b)

   max = a

else

   max = b

write (max)

End

Fig. 6. Display of the solution of exercise "Max of two numbers"

To build web application for storing and querying RDF data, we used as SPARQL end-point Fuseki and as programming language PHP. Fuseki is a web server that implements SPARQL Protocol to expose triple stores over HTTP.

5 Conclusion and Future Work

Practicing algorithms is an essential step to gaining skills in programming. However, in the literature, much research relies on directly structuring databases for a specific programming language (C, Java ...). Then programming skills upstage problem solving skills. For this reason, this work proposes to build an ontology to structure and describe algorithms in pseudocode. We also propose to develop a platform around this ontology so as to query it and highlight pseudocode keywords in displayed algorithms.

Moving forward, we plan to add modules:

- to populate our ontology from platforms like codeByte which already provides exercises with their solutions in many programming languages;
- to automate translation into the programming languages.

References

1. Alnusair, A., Zhao, T.: Component search and reuse: an ontology-based approach. In: 2010 IEEE International Conference on Information Reuse and Integration (IRI), pp. 258–261. IEEE (2010)
2. Atzeni, M., Atzori, M.: CodeOntology: RDF-ization of source code. In: International Semantic Web Conference, pp. 20–28. Springer, Cham (2017)
3. Bachimont, B., Isaac, A., Troncy, R.: Semantic commitment for designing ontologies: a proposal. In: International Conference on Knowledge Engineering and Knowledge Management, pp. 114–121. Springer, Heidelberg (2002)
4. Belhaoues, T., Bensebaa, T., Abdessemed, M., Bey, A.: AlgoSkills: an ontology of algorithmic skills for exercises description and organization. J. e-Learning Knowl. Soc. **12** (1) (2016)
5. Chaudhuri, A.B.: The Art of Programming Through Flowcharts & Algorithms. Firewall Media (2005)
6. Cormen, T.H.: Introduction to Algorithms. MIT press (2009)
7. Epure, C., Iftene, A.: Semantic analysis of source code in object oriented programming. A case study for C. Rom. J. Human-Computer Interact. **9**(2), 103 (2016)
8. Fernández-López, M., Gómez-Pérez, A., Juristo, N.: Methontology: From Ontological Art Towards Ontological Engineering (1997)
9. Ganapathi, G., Lourdusamy, R., Rajaram, V.: Towards ontology development for teaching programming language. In: World Congress on Engineering (2011)
10. Graudina, V., Grundspenkis, J.: Algorithm of concept map transformation to ontology for usage in intelligent knowledge assessment system. In: Proceedings of the 12th International Conference on Computer Systems and Technologies, pp. 109–114. ACM (2011)
11. Gruber, T.R.: Toward principles for the design of ontologies used for knowledge sharing? Int. J. Hum. Comput. Stud. **43**(5–6), 907–928 (1995)
12. Haowen, L., Wei, L., Yin, L.: An XML-based pseudo-code online editing and conversion system. In: Conference Anthology, IEEE , pp. 1–5. (2013)
13. Ivanova, T.: Bilingual ontologies for teaching programming in java. In: Romansky, R. (ed.) Proceedings of the International Conference on Information Technologies, pp. 182–195 (2014)

14. Kouneli, A., Solomou, G., Pierrakeas, C., Kameas, A.: Modeling the knowledge domain of the java programming language as an ontology. In: International Conference on Web-Based Learning, pp. 152–159. Springer, Heidelberg (2012)
15. Pierrakeas, C., Solomou, G., Kameas, A.: An ontology-based approach in learning programming languages. In: 2012 16th Panhellenic Conference on Informatics (PCI), pp. 393–398. IEEE (2012)
16. Motik, B., Patel-Schneider, P. F., Parsia, B., Bock, C., Fokoue, A., Haase, P., Smith, M.: OWL 2 web ontology language: structural specification and functional-style syntax. W3C Recommendation 27(65), 159 (2009)
17. Noy, N.F., McGuinness, D.L.: Ontology Development 101: A Guide to Creating Your First Ontology (2001)
18. Skiena, S.S.: The Algorithm Design Manual: Text, vol. 1. Springer Science & Business Media (1998)
19. Sosnovsky, S., Gavrilova, T.: Development of Educational Ontology for C-programming (2006)
20. Su, X., Zhu, G., Liu, X., Yuan, W.: Presentation of programming domain knowledge with ontology. In: First International Conference on Semantics, Knowledge and Grid, SKG'05, pp. 131–131. IEEE (2005)
21. Uschold, M., King, M.: Towards a Methodology for Building Ontologies (1995)
22. Yao, J., Gu, M.: Conceptology: using concept map for knowledge representation and ontology construction. JNW 8(8), 1708–1712 (2013)

Digital Entrepreneurship: MOOCs in Entrepreneurship Education the Case of Graz University of Technology

Stefan Vorbach[(✉)], Elisabeth Poandl, and Ines Korajman

Institute of General Management and Organisation,
Graz University of Technology, Graz, Austria
{stefan.vorbach,elisabeth.poandl,ines.korajman}
@tugraz.at

Abstract. Digital Entrepreneurship has gained substantial attention in theory and practice in the last years. In addition, new technological as well as pedagogical approaches have emerged. A special focus is set on massive open online courses (MOOCs) in entrepreneurship education. MOOCs are one of the strongest trends in online education and influence the content and procedure of teaching and learning. The paper contributes to a better understanding of necessary capabilities, chances and risks of using MOOCs as a new way of teaching entrepreneurship for engineers at Graz University of Technology. It empirically investigates challenges and drivers for using MOOCs as a novel pedagogical concept. Results show that a lack of self-discipline to finish a MOOC as well as missing interaction with others are main hurdles compared to lectures with compulsory attendance at University. However, findings also reveal that MOOCs are flexible in terms of time and location, thus can add convenience in reaching education, particularly entrepreneurial education.

Keywords: Entrepreneurship education · Digital entrepreneurship · MOOC

1 Introduction

The relevance of entrepreneurship education (EE) to foster entrepreneurship culture and activity is widely recognized. Entrepreneurial education provides key skills to identify a winning business model. EE also improves the managerial expertise to run the business competitively by creating core competencies, to build a loyal customer base through brand visibility, and to sustain growth [1].

Although more and more engineering students are being exposed to EE, minimal research has examined engineering student attitudes toward it, its impact on their learning, or professional competence. This is not surprising given the fact that the integration of entrepreneurship in engineering is a relatively new effort, where definitions of what it means to be entrepreneurial within an engineering program as well as program models vary greatly [2].

Several studies proposed that EE cannot be taught with traditional methods [3]. Traditional education teaches students to obey, duplicate, and be employed while

M. E. Auer and T. Tsiatsos (Eds.): ICL 2018, AISC 917, pp. 545–555, 2019.
https://doi.org/10.1007/978-3-030-11935-5_52

entrepreneurship tells students to make their own judgements and create their own jobs and these cannot be taught using traditional teaching [4].

Massive open online courses (MOOCs) are changing the way in which people can access digital knowledge, thus creating new opportunities for learning and competence development. MOOCs leverage the free and open use of digitized material through supportive online systems. Many education providers have started to offer courses in different domains such as entrepreneurship tackling recent demands for better self-employability [5].

2   Digital Learning

In recent decades, digital technologies have seen widespread use across global society and adoption at all levels of education. Today's connected classrooms provide both teachers and students easier, faster, and more affordable access to information, learning resources, experts, peers, and a wider community of educators. Class-delivered lectures can be successfully replaced by rich media formats including videos, podcasts of lectures, online presentations or interactive content or online tutorials, and are effective in instructing large amounts of conceptual content [6]. Table 1 gives an overview of modern digital learning methodologies, tools and contexts.

Table 1. Digital learning methodologies, tools and contexts [7]

Methodologies, tools and contexts	Examples
Digital learning methodologies	Project based learning; problem based learning; digital stories; online learning environments; digital moments; technology integrated teaching methods; digital storytelling; educational games; authentic learning
Digital learning contexts	Collaborative communities; cooperative learning; digital combinational system; collaborative learning; flipped classroom using digital media; moving from fixing to online space; experiential online development; open educational practice; network participation
Tools and simulators	Web-based video; computerised environments; spatial science technology; slowmation: narrated stop-motion animation; generic modelling language; digital video; augmented reality; design based research; gamification; learning manager; simulation; computer based teaching; library webinars
Support system for digital learning	eLearning; mobile learning; learning object repository; blended learning; blackboard; moodle learning manager; twitter; videoconferencing; MOOC—massive open online courses

Video lectures can allow self-paced learning and provide effective overviews or illustrative examples showcasing diverse situations and cases. They can be aimed at particular knowledge points or known problematic concepts and real-world problems [8].

Students reported the following benefits of videos over live lectures: taking notes from video lectures were convenient, as they need not to miss content while writing; some students found it easier to focus attention on videos than in a live class. Most students watched videos at suitable times and re-watched them to prepare for exams, eventually spending more time on the subject than otherwise. Being able to view the video multiple times, as well as pause and rewind, helped students to understand content, particularly those who struggled with the language [9], while others preferred listening to the subject being "talked". Video also provides alternative representations of information, accommodating diverse learners who prefer visuals, audio or text. Distributed flipped classes incorporating massive online open courses (MOOCs) open new possibilities for providing online content resources like videos and assignments, if they run concurrently with the campus courses [8, 10].

While traditional lecturing may be more effective at transmitting information than supporting development of skills, values or personal development, the value of digital learning is contested if it represents a teacher-centred pedagogy [9]. Teacher-centred pedagogies which lack engagement or interaction, often result in students' adopting a passive attitude to learning whilst encountering focus difficulties, and not taking responsibility for their own learning. Moreover, the limited interaction causes difficulties for lecturers to differentiate pacing and instruction that adapt to the different progress levels of students [8].

Flipped classrooms (FC), however, may fail due to monotonous and impersonal video lectures that can inevitably lead to loss of interest and poor class attendance. Videos that represent technology-delivered lectures that merely aim at transmitting content, compare poorly with well-planned interactive lectures. Videos can also be overrated as a teaching tool if students are unable to view videos due to unavailability of computers and internet [11]. Thus, the format of the wide variety of self-paced preclass materials and activities employed in different contexts was less important than making sure students really accessed those materials. In addition, teachers also find it challenging to source suitable videos, in spite of copious online offering [8].

However, the most important component in an FC was engaging students in the face-to-face component. Teachers of FC agree that instructional videos on their own do not improve teaching; it depends on how they are integrated into an overall approach [9]. A well-integrated approach may disseminate students' attitudes towards flipping which are sometimes negative due to perceptions of a higher workload or a lack of cohesion between in-class and out-of-class work.

3   Digital Entrepreneurship

Entrepreneurship is recognized as one among the so-called 21st century skills [12], namely those skills that are required to succeed in learning, working, and living in the knowledge society [13]. Entrepreneurship and entrepreneurs have become increasingly important worldwide considering the positive impact on employment, productivity, innovation and economic growth, by analysts, economic theoreticians and researchers as well as by policymakers and international organizations [7].

To become a successful entrepreneur requires more than an identified set up of competencies or skills. Even the combination of opportunity, capabilities and resources may not necessarily be sufficient to lead to entrepreneurship if opportunity and start-up costs outweigh the potential benefits [7].

Digital entrepreneurship is claimed as not just a context but an entirely new field of research, given the unique characteristics of the Internet [14].

The use of technology in EE is increasing and information technology skills are viewed as a critical factor in determining new business success in a global market [7]. A wide network of leading universities such as Harvard, MIT and Stanford have launched MOOC platforms such as Udacity, Coursera, edX, MIT Open Courseware, and Stanford eCorner. Courses have been realized using many different kinds of technological support for self-learning (papers, short videos on well-focused contents, flash animation) and for synchronous and asynchronous interaction. Asynchronous mode delivery brings to reality the idea that anybody can approach education at any time and from anywhere; synchronous delivery demands the learner to synchronize his or her learning agenda with anyone else [5].

To prepare students for this new reality, universities are increasingly aware that they must graduate engineers who not only understand science and technology, but who are also able to identify opportunities, understand market forces, commercialize new products, and have the leadership and communication skills to advocate for them. This has prompted a significant increase in the delivery of EE to engineering students through new courses, programs, and experiential learning opportunities [2].

4 Massive Open Online Courses (MOOCs) in Entrepreneurship Education

4.1 MOOCs in General

Massive open online course (MOOC) has appeared as a disruptive innovation that permits to engage a large number of persons in an open online course available through internet to anyone aiming to enroll [15].

MOOCs are one of the strongest trends in online education [5]. They offer the possibility to learn online to a massive number of students, and part of their features is free of charge for the participants. Over the past few years, MOOCs have achieved a widespread, global profile. Enabled by technology, they have arisen from a mixture of experimentation with educational technology and pedagogic approaches. [16].

MOOCs have a high potential to allow the massive development of knowledge and certain competences among adult learners' showing enough motivation, self-regulation and cognitive quality time to engage, and succeed in this online courses. For this reason, MOOCs could be considered as an excellent opportunity to achieve education objectives among massive number of participants in informal contexts, such the development of an entrepreneurship culture [15].

MOOCs have four main characteristics. First of all, they are open to everyone, meaning that there are no entry requirements. Secondly, there is no participant restriction regarding the number of participants. Thirdly, the courses are offered free of charge. Fourthly, the courses are conducted completely online. Therefore, there are no technical laboratory phases [17].

MOOCs can be thus considered forerunners of course exemplars—early prototypes of improved learning environments which frequently recover flexible educational practice. The online courses are made for various target groups e.g. for school students, individuals or university students. Teachers are allowed to make selected learning materials available as Open Educational Resources (OER). Without violating copyrights, they can use the materials for the purpose of teaching [18].

However, there are also critical aspects when it comes to MOOCs. Insights of media didactics regarding the structuring of the subject matter, the depth and speed with which content is conveyed and the design of performance reviews for learners are not yet sufficiently taken into account [17]. Moreover, MOOCs have high abort rates. Depending on the course, only 2–10% of those who have registered for a course take the final examination [19]. The development, implementation and support of a MOOC involves considerable effort. This concerns the universities or university lecturers who are developing the courses. However, this also applies to the platform providers who make the courses available to participants [20].

4.2   The MOOC: Start-Up Journey

Students have plenty of ideas, from which many could be developed further, transferred to interesting business ideas and become the basis to set up a new company or startup. The path from the initiation of an idea to its implementation into a business model raises many questions and requires entrepreneurial competences as an obligatory prerequisite for founding a company.

Encouraging entrepreneurship has become a topic of high priority in the university policy of Graz University of Technology. Several entrepreneurial activities and lectures are offered, encompassing a MOOC in digital entrepreneurship education, the "Startup-Journey: Business Model Generation".

The MOOC "Startup-Journey" consists of four units with eight videos in total (two videos each unit), where one unit is offered weekly. The course doesn't require any special prerequisites and aims to be used by students for their own business idea. Therefore, within the course basic knowledge and methods as well as their handling are imparted to gain the competence for generating a business model by the end of the course. Additionally, elements and methods are explained step-by-step, with a focus on the customer value as the reason why customers want to consume or purchase the product or service.

An important step in the foundation process is the creation of a business model. For this reason, students learn in the MOOC what a business model is and how to create one. Figure 1 shows an overview of the units of the MOOC "Startup-Journey: Business Model Generation". Unit 1 starts with an introduction of the topic with definitions, basics of the business model framework and patterns. Unit 2 explains, why the USP (unique selling point) is important, how customer value is identified and which helpful methods and tools can be used to fill in a business model canvas. Building up on this, the generation of a business model canvas is demonstrated in Unit 3, using Tesla as an example. Optionally a task is given to create Lego's business model canvas and a link to a possible solution is provided. To underline the practicability of the concept, two founders are interviewed in another video, giving an insight into their experiences with startups. To top off the "Startup-Journey" Unit 4 provides experience reports with dos and don'ts when creating a business model canvas and how to successfully communicate the business idea with a pitch. Completing the "Startup-Journey", participants get an overview of helpful methods for their project and are able to create a business model for their own start-up idea.

Fig. 1. Overview of the units of the MOOC "Startup-Journey: Business Model Generation"

For actively participating in the course students receive an automatic confirmation of participation (certificate) which confirms that the user answered at least 75% of the self-assessment questions correctly. Furthermore, the MOOC can be completed by TU Graz students as part of a corresponding course and is therefore provided with 1 ECTS as an elective subject.

4.3 Experiences with the MOOC Start-Up Journey

To get an insight into student's experiences regarding MOOCs in EE, an empirical study was conducted. Information was derived from surveying 40 students of Graz University of Technology in total who got in touch with MOOCs. The questionnaire used for this purpose consisted of 10 questions, whereby three of them were open ones.

Of the 40 respondents who were surveyed, 34 claimed that they have never attended a MOOC before participating at the lectures Entrepreneurship, Gruendungsgarage or Process Management held by the Institute of General Management and Organization of TU Graz. 7,5% (k = 3, n = 40) respondents maintained that they have joined a MOOC several times before and 7.5% only once before the three listed lectures.

The pie chart (see Fig. 2) indicates that 35% (k = 14, n = 40) take the online courses on their own, while 20% (k = 8) prefer conducting it in groups. Moreover, 5% (k = 2) claimed that they absolved the MOOC both, alone and in groups. According to the chart, 25% (k = 10) rated that they have watched the MOOC in addition to a face-to-face course and 6 students watched the MOOC directly in a lecture.

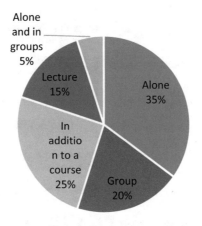

Fig. 2. The setting for absolving MOOCs

With reference to the graph illustrated in Fig. 3, it was indicated that most respondents (k = 25, n = 40) use their PCs or notebooks to watch MOOCs. It is interesting that 10 students watched MOOCs on a beamer although only 6 students were shown them directly in a lecture. Moreover, 4 students in total claimed that they used more than one hardware device to watch MOOCs. Surprisingly, only one person made use of a tablet and no one solely used the smartphone.

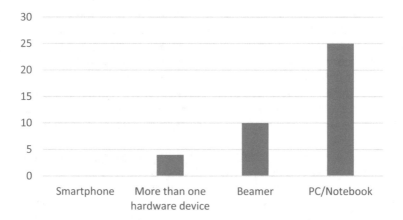

Fig. 3. Used hardware devices for conducting MOOCs

In addition, students were asked to evaluate the statements shown in Fig. 4. With reference to the data, it was stressed that 25 out of 40 students fully agreed to regularly use digital media for learning. However, only a small number of respondents expressed they would prefer a pure online course. Overall, students claimed that they would recommend the MOOC.

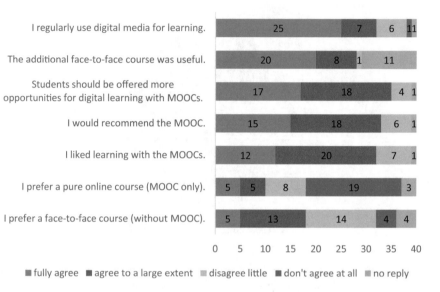

Fig. 4. Statements on MOOCs

Finally, students were asked about advantages and disadvantages in correlation with MOOCs. According to the ratings in Fig. 5, 39 out of 40 respondents claimed that time flexibility is a main advantage and 38 fully agreed on the local flexibility of MOOCs. Moreover, students appreciate the fact that there are no course fees, only 4 students disagreed little and 1 student fully disagreed. Overall, respondents also agreed on the fact that someone can adapt the online courses to the individual learning pace. Surprisingly, three respondents see it differently and disagreed little.

Several disadvantages in accordance with MOOCs have been stated in literature. The results of the survey show that 16 students (n = 40) fully agree that "less exchange with fellow students" is a disadvantage. It is interesting that 10 students claimed that they don't agree at all on the disadvantage that more motivation is required to complete a MOOC (Fig. 6).

In addition, participants were asked to answer open questions about the MOOC to gain further insight and input for possible developments in future.

As successful and well done in particular, participants perceived the MOOC as very understandable, professional and vividly explained with examples. The MOOC gives a good overview of the topic briefly, is available at any time, easy to use and a valuable supplement to lectures with physical presence. The quiz at the end of each unit of the MOOC was mentioned positively e.g. as a helpful tool for self-monitoring in the learning process.

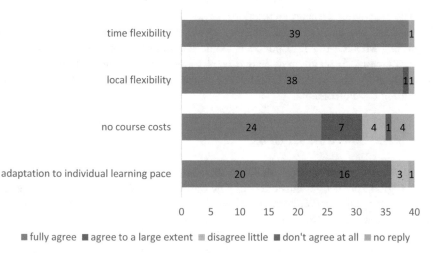

Fig. 5. Advantages of MOOCs

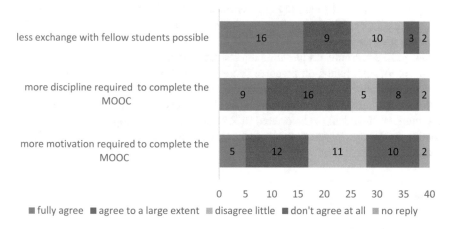

Fig. 6. Disadvantages of MOOCs

Challenges of digital EE using MOOCs were identified by the respondents too. The lack of self-discipline to finish the MOOC is mentioned as a hurdle compared to lectures with compulsory attendance at university. Challenges for lecturers were identified (e.g. costly to create to contents and develop the videos, need of special equipment and infrastructure for the video recording and production, focus on a small area of content) as well as for students who are confronted with theoretical input only online. Taking a course only online, respondents would miss the chance to ask the lecturer for rephrasing, discuss the content in real time and learning in interaction with others.

The respondents made suggestions as to where a MOOC could also be used: mechanical engineering, mathematics, software development and programming. Regarding MOOCs in EE respondents would like to learn more about a business plan, success factors, financial analysis and legal framework conditions. In general, participants identify the potential of MOOCs to be used in all courses in some ways, depending on the contents.

5   Conclusion and Outlook

Massive open online courses (MOOCs) are changing the way in which people use and share digital knowledge, thus creating new opportunities for learning and competence development in market-relevant areas such as innovation management and entrepreneurship. MOOCs are predestined to reach a large audience that can enjoy the autonomy of self-paced instruction with the assistance of the network of online peers. MOOCs can be considered as an appropriate tool to teach entrepreneurship-related courses, because they can increase personal entrepreneurial attitudes and inclinations, and improve problem solving capabilities and multiple tasks execution. With its easy scalability, operational flexibility and cost advantage, MOOCs can add economy and convenience in reaching education, particularly entrepreneurial education, to a large heterogeneous audience.

But unlike learning traditional subjects, entrepreneurial learning can best be disseminated through hands-on practical process, though the available pedagogy is deficient on this aspect. Further research therefore is required to evolve methods for improving cognitive skill, maintaining regularity and reducing dropouts. Given the large number of students to be catered to, it is essential to develop meaningful and cost-effective evaluation process, innovate new meta-tutoring for effective learning, and make teaching increasingly practical.

References

1. Mondal, M.K., Kumar, A., Bose, B.P.: Entrepreneurship education through MOOCs for accelerated economic growth. In: Proceedings of the 2015 IEEE 3rd International Conference on MOOCs, Innovation and Technology in Education (MITE) (2015). https://doi.org/10.1109/MITE.2015.7375354
2. Duval-Couetil, N., Reed-Rhoads, T., Haghighi, S.: Engineering students and entrepreneurship education: involvement, attitudes and outcomes. Int. J. Eng. Educ. **28**(2), 425–435 (2012)
3. Löbler, H.: Learning entrepreneurship from a constructivist perspective. Technol. Anal. Strat. Manage. **18**(1), 19–38 (2006). https://doi.org/10.1080/09537320500520460
4. Chen, S.C., Hsiao, H.C., Chang, J.C., Chou, C.M., Chen, C.P., Shen, C.H.: Can the entrepreneurship course improve the entrepreneurial intentions of students? Int. Entrepren. Manage. J. (2015). https://doi.org/10.1007/s11365-013-0293-0
5. Cirulli, F., Elia, G., Lorenzo, G., Margherita, A., Solazzo, G.: The use of MOOCs to support personalized learning: an application in the technology entrepreneurship field. Knowl. Manage. E-Learning (2016). https://doi.org/10.1007/978-3-642-40790-1_21

6. O'Flaherty, J., Phillips, C.: The use of flipped classrooms in higher education: a scoping review. Internet Higher Educ. (2015). https://doi.org/10.1016/j.iheduc.2015.02.002

7. Sousa, M.J., Carmo, M., Gonçalves, A.C., Cruz, R., Martins, J.M.: Creating knowledge and entrepreneurial capacity for HE students with digital education methodologies: differences in the perceptions of students and entrepreneurs. J. Bus. Res. 1–6. https://doi.org/10.1016/j.jbusres.2018.02.005

8. Le Roux, I., Nagel, L.: Seeking the best blend for deep learning in a flipped classroom—viewing student perceptions through the Community of Inquiry lens. Int. J. Educ. Technol. High Educ. 15(16), 1–25 (2018). https://doi.org/10.1186/s41239-018-0098-x

9. Sams, A., Bergmann, J.: Flip your students' learning the best use of face time. Technol. Rich Learn.0 Pages 70(6), 16–20 (2013)

10. Israel, M.J.: Effectiveness of integrating MOOCs in traditional classrooms for undergraduate students. Int. Rev. Res. Open Distrib. Learn. 16(5), 102–115 (2015)

11. Roehl, A., Linga, S., Gayla, R., Shannon, J.: the flipped classroom: an opportunity to engage millennial students through active learning strategies. J. Family and Consum. Sci. 105(2), 44–48 (2013)

12. Voogt, J., Roblin, N.P.: 21st century skills, Discussienota, Universiteit Twente (2010)

13. Antonaci, A., Dagnino, F.M., Ott, M., Bellotti, F., Berta, R., De Gloria, A., Lavagnino, E., Romero, M., Usart, M., Mayer, I.: A gamified collaborative course in entrepreneurship: focus on objectives and tools. Comput. Hum. Behav. 51, 1276–1283 (2015). https://doi.org/10.1016/j.chb.2014.11.082

14. Zaheer, H., Breyer, Y., Dumay, J., Enjeti, M.: Straight from the horse's mouth: founders' perspectives on achieving 'traction' in digital start-ups. Comput. Hum. Behav. 2 (2018). https://doi.org/10.1016/j.chb.2018.03.002

15. Milian, R.P., Gurrisi, M: Educ. + Train. 59(9), 990–1006 (2017). https://doi.org/10.1108/ET-12-2016-0183

16. Porter, S.: InTo MOOC or Not to MOOC, pp. 3–6. Chandos Publishing, Sawston (2015)

17. Schulmeister, R.: Massive Open Online Courses, pp. 22-23, 32. Waxmann, Münster (2013)

18. Ebner, M., Kopp, M., Wittke, A., Schön, S.: Das O in MOOCs – über die Bedeutung freier Bildungsressourcen in frei zugänglichen Online-Kursen. HMD 4–7 (2014)

19. Clark, D.: MOOCs: who's using MOOCs? 10 different target audiences (2013). URL: http://donaldclarkplanb.blogspot.de/2013/04/moocs-whos-using-moocs-10-different.html

20. van Treeck, T., Himpsl-Gutermann, K., Robes, J.: Offene und partizipative Lerkonzepte. E-Portfolios, MOOCs und Flipped Classrooms. peDOCS 9–10 (2013)

Practical Work in the Digital Age

Brigitte Koliander[(✉)]

Pädagogische Hochschule Niederösterreich, Baden, Austria
b.koliander@ph-noe.ac.at

Abstract. Why should students design and carry out real life experiments when they could use applets or learning games to see what may happen and to check which factors affect the outcome of a chemical reaction? The integration of digital media seems to be the preferred method of improving learning. For the development of some competences, this may be true. Other competences may need students to confront the real world. What can students learn by direct interaction with the physical world? This question came up in science teaching a hundred years ago, when only books and lectures rivaled practical work in science. I will sum up some answers to this question in the theoretical part of this article. The answers found in the literature lead to the study I will present in this paper: How do tasks given to students in chemistry lab work in Austria fit in with theoretical considerations about the importance of practical work in school?

Keywords: Practical work · Inquiry · Chemistry

1 Introduction

All of us interact with the physical world—even in a world full of digital information and media. We walk, breathe, eat and drink, wear clothes, heat rooms and drive vehicles, use electricity, take medication and so on. Some of us build houses, others grow wheat. During such interactions, we collect and interpret data; we use and construct ideas and concepts concerning the world around us.

Interacting with the physical world is part of practical work in school. What can students learn by direct interaction with the physical world that they could not learn otherwise?

The theoretical part of this article gives an overview of 100 years of discussion about what students can learn during practical work in science teaching. The empirical part shows how tasks for practical work in chemistry teaching at schools in Austria fit in with these theoretical considerations.

2 Theoretical Part

As Hodson [1] states, practical work as part of science teaching is expensive, time-consuming, and there is no clear evidence that students learn more about science content through practical work than through teacher lecture or the use of other media. Beyond that, practical work may impart a distorted picture of scientific inquiry.

© Springer Nature Switzerland AG 2019
M. E. Auer and T. Tsiatsos (Eds.): ICL 2018, AISC 917, pp. 556–565, 2019.
https://doi.org/10.1007/978-3-030-11935-5_53

On the other hand, there are some arguments why lab work is an indispensable part of science teaching and learning. In this chapter, I will present some of these arguments.

For comprehensibility, in the following section I will give some definitions and clarify the use of the terms "practical work" and "inquiry" in this article.

2.1 Definitions

I use the terms **lab work, practical work** and **laboratory experiences** synonymously. "Laboratory experiences provide opportunities for students to interact directly with the material world (or with data drawn from the material world), using the tools, data collection techniques, models, and theories of science" [2, p. 3]. Practical work does not just mean performing a predetermined experiment, but includes every interaction with the material world. These interactions can be various activities: producing and observing a phenomenon, training practical skills in interacting with materials and tools, or performing an investigation (which includes defining a question, designing and performing data collection, interpreting data and discussing the results).

I use the concept of inquiry for modeling scientific investigations. **Inquiry** is (following Dewey [3]) "the controlled or directed transformation of an indeterminate situation into one that is as determinate in its constituent distinctions and relations as to convert the elements of the original situation into a unified whole". "Scientific inquiry" includes all activities done by scientists in performing research. Inquiry in school means student activities "echoing some subset of the practices of authentic science around the answering of questions or the solving of problems" [4, p. xxxiv].

The chosen definitions indicate that I follow a particular epistemological approach. Therefore, within the next part I will present some epistemological considerations.

2.2 Epistemological Considerations

Dewey [3] represents an epistemological approach called pragmatism. Many current publications cite his ideas about inquiry [5, 6]. Official documents take up Dewey's formulations and descriptions of inquiry [7].

Pragmatism represents a constructivist approach. "Knowledge" in pragmatism does not mean an objective representation of the world but refers to conceptual structures as result of proper inquiry. Nevertheless, Dewey acknowledges an external world we can interact with. During interaction, the reaction of the external world can demonstrate our conceptions to be inaccurate. The concepts must prove to be viable in dealing with the unyielding world.

Table 1 shows the correspondence between the thoughts of Dewey [3] about inquiry and the "Scientific Practices" [7] published in 2012 as a guideline for learning in science and engineering.

Table 1. Inquiry and scientific practices

Inquiry [3, pp. 101–120] (summarized by the author)	Scientific practices [7, p. 3]
Inquiry starts with an indeterminate (unsettled, uncertain, disturbing) situation. There has to be a self (epistemic agent, person, group of persons) and an environment to establish such a situation. The next step is for the person to realize there is a problem, to view the situation as problematic. This includes the motivation to search for a solution	1. Asking questions (for science) and defining problems (for engineering)
Then the individual determines factual conditions and imagines ideas for a possible solution. The ideas are anticipations of something that may happen	2. Developing and using models 3. Planning and carrying out investigations
The person has to check facts and ideas using their capacity to work together to introduce a resolved unified situation	4. Analyzing and interpreting data 5. Using mathematics and computational thinking 6. Constructing explanations (for science) and designing solutions (for engineering)
He/she or the group leads a rational discourse to find the most appropriate solution to the problem. Facts are not self-sufficient and complete in themselves, the person has to select and describe them and give them meaning relevant for the solution to the problem. In the same way, ideas and concepts are not true or false but they have to demonstrate their ability to solve the problem	7. Engaging in argument from evidence

Dewey, and other authors discussing inquiry and practical work [1, 8], explicitly states, that during practical work students may construct some common misconceptions:

Misconception 1: The result of an experiment can prove a concept (or a theory).
Misconception 2: If you collect data accurately, the theoretical explanation will emerge out of the data.

Herron [8] harshly criticizes a reduced form of inquiry and a distorted understanding of scientific inquiry. This distorted understanding evolves from practical work in school science during which students should come to a scientific theory solely by observing phenomena. Herron, like Wellington [9], argues that theory does not emerge from an experiment or observation. Students may realize that a solution of sodium chloride in water can conduct electricity and find some hints about which changes have an effect – but they will not construct knowledge about ionic particles by observing this phenomenon.

Practical work seems to be dangerous, not only in respect to toxic substances, fire, and other possibilities of injury. Practical work may also result in serious misconceptions. Tasks given to students to be solved during practical work should be resistant to the possible construction of such misconceptions. Are the tasks given in the lab lessons suitable in this regard?

Before discussing tasks taken from chemistry lab lessons at Austrian schools, I will give arguments why practical work is an indispensable part of science teaching and learning.

2.3   The Need for Practical Work

In the discussion about the need for practical work, the classification of tasks for practical work by Woolnough and Allsop [10] may be useful (see Table 2).

Table 2. Aims and tasks for practical work [10]

Aims for practical work	Classes of tasks
Developing practical scientific skills and techniques	Exercises
Getting a feel for phenomena	Experiences
Being a problem-solving scientist	Investigations

Exercises. During practical work, students can develop their practical skills in interacting with materials and tools. During a chemistry lab lesson, for example, students can learn how to use a burette and perform a titration. It will not be the same as performing an animated titration. Compliance with safety regulations is important when handling acidic solutions. Craftsmanship is necessary to operate the burette. Students have to watch carefully to recognize the ongoing change of color.

When a teacher selects exercises that students have to do, he should reflect upon the meaning of the exercise to the students. Often this decision is not trivial. Even if they become chemists, with state-of-the-art automation being common in their future profession, they may never have to perform a titration by hand. Nevertheless, it could make sense to have done a titration in school, especially when students reflect afterward about their experiences. They have learned what is important in performing a chemical analysis and that there may be restrictions in the applicability.

Experiences. During practical work, students can get a feel for phenomena. Do you know how it feels to immerse a big balloon filled with air down into a tub of water? You feel a force that pushes the balloon up; it is hard to press the balloon down. Have you ever watched magnesium burning? The grey metal burns with a white flame and changes into a white substance you can grind with your fingers. Is it important for students to get a feel for this or that phenomenon? It may be important for understanding a term (buoyancy) or the usefulness of a chemical concept (oxidation). Sometimes the observation of an unknown phenomenon can help students with testing their preconceptions in interaction with the physical world.

During practical work, students can not only get a feel for phenomena, they also can learn how to produce these phenomena and learn to demonstrate mastery of the physical world. Some tasks will specifically demonstrate this mastery: Look at what we can do with these substances, how we can change their color, their smell! Look how we can produce light and fire and electricity and master it!

Investigations. During practical work, students may investigate a problem. Why should students perform an investigation in which they have to interact with the physical world?

Practical work in school can provide the opportunity to test scientific key-concepts in predicting, planning and interpreting interactions with the physical world. If the concepts prove their worth in dealing with the physical world, students will trust them and accept them.

Beyond that, students should know about the reliability of scientific knowledge. About the way scientific knowledge is constructed and tested, for example. This includes knowledge about scientific procedures for the collection of data as well as ways scientists use data for building up evidence through argumentation and discussion within the scientific community.

As Dewey [11, p. 395] states:

> Surely if there is any knowledge which is of most worth it is knowledge of the ways by which anything is entitled to be called knowledge instead of being mere opinion or guess-work or dogma. Such knowledge never can be learned by itself; it is not information, but a mode of intelligent practice, a habitual disposition of mind. Only by taking a hand in the making of knowledge, by transferring guess and opinion into belief authorized by inquiry, does one ever get a knowledge of the method of knowing.

Through this interaction with these practices students have during their science lessons and in lab work at school, they implicitly construct a picture **about how** scientists warrant knowledge [12]. They may develop the competence called the "empirical attitude" [13]: To be able to coordinate theory, phenomena, data, and the data collection events that yield the data. Students also may learn to coordinate theory and data in using textbook-tasks or computer animations. Textbooks or animations present data that are not equal to data drawn from the natural world. Textbook data is not plagued with measurement errors and ambiguity [2]. Textbook data and data used in animations often lead too easily to a relationship or a law. They are made for showing a certain relationship. Data drawn from the physical world is never as clear. The complexity of the material world and the variety of possible influencing factors does not allow any data "to speak for itself". Scientists have to develop the competence of collecting reliable data and the knowledge of how to prevent undesired influences. In interactions with the physical world during investigations students may experience the ambiguity of data. Then they can develop an understanding why even experts can come to different results in interpreting data with regard to complex phenomena in the material world.

Practical work offers opportunities to learn about epistemic and procedural approaches. In dealing with questions, data and evidence during laboratory lessons, students implicitly learn how scientific knowledge is constructed [14]—but are the tasks given in lab work at school adequate for this purpose?

3 Approach

I analyzed experimental tasks from 27 different Austrian schools with laboratory work in chemistry. The descriptions of the tasks were taken from reports teachers wrote about lab work under the initiative IMST.[1] In these reports, teachers described an innovation they had implemented in their classes. In the reports chosen for this study, the teachers presented selected tasks they gave to their students during lab work. To identify tasks that allowed students to coordinate theory, phenomena, data, and the data collection events that yielded the data, a model for the openness of tasks from the literature was used for analysis. Tasks were analyzed with content analysis [15] using the four levels of inquiry [4] as deductive main categories (Table 3).

In sharpening the four categories, I looked at the single steps of the tasks and compared them with the demands for practical work taken from the literature.

Table 3. The four levels of inquiry [4]

	Source of the question	Data collection methods	Interpretation of results
Level 0	Given by the teacher	Given by the teacher	Given by the teacher
Level 1	Given by the teacher	Given by the teacher	Open to student
Level 2	Given by the teacher	Open to student	Open to student
Level 3	Open to student	Open to student	Open to student

4 Outcomes

4.1 Other Tasks

Some of the tasks described within the reports did not fit into any of the four categories of inquiry. That was not surprising. As shown in the theoretical part, there are different classes of tasks for practical work and not all of the tasks have to lead to inquiry. Tasks aimed at training skills in using tools and materials or tasks aimed at producing phenomena need a proper description about what has to be done. There is no need to formulate a question or a problem to be solved; it is not necessary to construct an interpretation. For these classes of tasks a new category was introduced to Table 3: "Other tasks" (Table 4).

About 30% of the tasks found in the reports were not inquiry tasks but exercises for using materials or equipment correctly, or illustrations of phenomena. Such other tasks are justifiable part of lab work as discussed in the theoretical part. But the aim of the task, either an exercise or an illustration (and not an investigation) is often not clearly formulated. The following task taken from one of the reports illustrates that a simple exercise is introduced as an investigation.

---

[1] The initiative IMST (Innovations in Mathematics and Science Teaching) is a project of the Austrian ministry of education and serves the further development of science education.

Table 4. The category "other tasks"

	Source of the question	Data collection methods or description of how to use the tools and materials	Interpretation of results
Other tasks	There is no question	There is a clear description given by the teacher to follow	There is no interpretation of the data, neither given by the teacher nor demanded from the students

Example 1: T8/645-660,[2] Translated by the Author

Investigation

Aim of the investigation: You will investigate how a flip style pipette filler works.

Tools: A flip style pipette filler, a 10 mL graduated pipette, a 5 mL volumetric pipette, test tubes.

Chemicals: A bottle with colored water

Procedure: You use the flip style pipette filler. First, you pipette with the graduated pipette 2 mL of water into one test tube. Then you pipette with the graduated pipette 3 mL water into a second test tube.

You pipette with the volumetric pipette 5 mL water into a third test tube.

Pour the content of the first test tube into the second test tube and compare the water level of the second and the third test tube. If you worked accurately, the water level in both test tubes should be the same.

A flip style pipette filler is an orange rubber bulb that serves as a vacuum source for filling reagents through a pipette. The flip style pipette filler has a two-valve system and it will take some practice to use it properly. It would be a nice investigation to figure out how the valves are used to draw the water. Surely, this is not the aim of this task. If you look at the procedure given, students learn to use the pipette filler as a tool for transporting a defined amount of liquid substance. So the term "investigation" and the named goal are misleading.

4.2 The Absence of a Question

An unexpected result was the absence of a question or a problem in many of the tasks guiding students with precise instructions through an investigation. Students had to find some answer without knowing the question.

---

[2] The 27 reports were imported into the program MAXQDA12. The first number shows the number of the report, the second numbers present the lines within that report

Example 2: T7/302-318, Translated by the Author

Examination of ink
Material: 1 beaker (50 mL), 1 piece of blackboard chalk, syringe (10 mL), black ink, water
Preparation:
In the beaker mix four drops of black ink (syringe) with 4 mL of water. Place the chalk in the beaker and watch what happens.
Observation:
Attempt an interpretation.
My explanation:

The students are asked to interpret their observations. What will you observe in this experiment? With high probability you will notice that the black color changes and other colors become visible. You will notice that the colors in the chalk move upwards. You may also observe that the colors become paler; they spread and need more space. You may see, smell, hear, or measure something else when trying to perceive everything accurately. Which of the observations or measurements should you interpret?

Which concepts should you use to interpret the observation? Should the black ink be recognized as a mixture of several substances? What concept would still be possible to explain the change in color? Could the color change be interpreted as a chemical reaction? That would not be scientifically appropriate, but could be logically derived from the observations. Should you explain the varying speed of the migration of different colors?

In this task, a focus on a question would be helpful. This could be: Is the black ink a mixture or a pure substance? The focus would thus already be on the concept of pure substances versus mixtures. If a separation into several colors becomes visible, then you may interpret this with regard to the question as a separation of a mixture.

Without the focus, this task could be used as an example how manifold and complex the handling of observations and evidence is. In this case, the teacher would have to be prepared to value all observations, and to evaluate the interpretation of the observations only on the basis of logical reasoning and not with regard to the well-known scientific knowledge about the analysis of ink. However, the analysis of goals for laboratory practice shows that teachers do not pursue the discussion of complexity and ambiguity in dealing with the material world as an aim of lab work [16]. Therefore, it can be assumed that observations and interpretations that do not fit to the interpretation goal by the teacher are either ignored or classified as false. Thus, the complexity that can be experienced in dealing with the unyielding world is not explicated and not visible to the students.

In some other tasks there was a question given by the teacher—but the practical work, and the data collected could not help answer the question.

4.3   Inquiry Level 2 and 3

All tasks described in the 27 reports were classified with one of the five categories according to the four levels of inquiry and the one class of "other tasks". In nine of the reports at least one task was identified that allowed students to design data collection by themselves (inquiry level 2). Seven teachers documented in their reports how they guided students to find their own research question (inquiry level 3). Inquiry level 2 and level 3 require that students coordinate theory, phenomena, data, and the data collection events that yield the data.

5   Conclusion

During laboratory work, by watching, listening, smelling, and feeling, students get acquainted with phenomena of the physical world. Students may train competencies in using methods to handle these phenomena and to collect data about the phenomenon. By designing and performing procedures for the collection of data suitable for constructing evidence, students may realize the complexity of the physical world and the ambiguity of data interpretation.

Tasks for practical work from reports written by teachers who participated in the initiative IMST have been analyzed for their openness (level of inquiry). About 30% of the tasks are exercises or experiences. These may help students to become familiar with phenomena of the physical world and to become skilled in the use of tools and substances. More than half of the teachers list examples for more open tasks where students have to plan an investigation. These tasks may support students in developing the competence called the "empirical attitude" [13]: To be able to coordinate theory, phenomena, data, and the data collection events that yield the data.

Nevertheless, the in-depth analysis of tasks with predetermined procedures (level 0, level 1, other tasks) gives some evidence, that teachers are not sensible in dealing with data and its use as evidence for finding an answer to a question through inquiry. One way to improve this sensibility may be the accurate use of terms not only for chemical and technical devices but also for different tasks given in practical work.

References

1. Hodson, D.: Re-thinking old ways: towards a more critical approach to practical work in school science. Stud. Sci. Educ. **22**(1), 85–142 (1993)
2. Singer, S.R., Hilton, M.L., Schweingruber, H.A.: America's Lab Report. Investigations in High School Science. The National Academies Press, Washington, DC (2005)
3. Dewey, J.: Logic. The Theory of Inquiry. Henry Holt and Company, New York (1938)
4. Abrams, E., Southerland, S.A., Evans, C.: Introduction. Inquiry in the classroom: identifying necessary components of a useful definition. In: Abrams, E., Southerland, S.A., Peggy, S. (eds.) Inquiry in the Classroom: Realities and Opportunities, pp. xi–xlii. Information Age Publishing, Charlotte (2008)

5. Ødegaard, M., Haug, B., Mork, S.M., Sørvik, G.O.: Challenges and support when teaching science through an integrated inquiry and literacy approach. Int. J. Sci. Educ. **36**(18), 2997–3020 (2014)
6. Chu, S.K.W., Reynolds, R.B., Tavares, N.J., Notari, M., Lee, C.W.Y.: 21st Century Skills Development through Inquiry-Based Learning. Springer Science + Business Media Singapore, Singapore (2017)
7. National Research Council.: A framework for K-12 science education: practices, crosscutting concepts, and core ideas. committee on a conceptual framework for new K-12 science education standards. Board on Science Education, Division of Behavioral and Social Sciences and Education. The National Academies Press, Washington, DC (2012)
8. Herron, M.D.: The nature of scientific inquiry. School Rev. **79**(2), 171–212 (1971)
9. Wellington, J.: Practical work in science—time for a re-appraisal. In: Wellington, J. (ed.) Practical Work in School Science—Which Way Now? pp. 3–15. Routledge, Oxon (1998)
10. Woolnough, B., Allsop, T.: Practical Work in Science. Cambridge University Press, Cambridge (1985)
11. Dewey, J.: Science as Subject-Matter and as Method. Science (New Series) **31**(787), 121–127 (1910)
12. Anderson, C.W.: Perspectives on science learning. In: Abell, S.K., Ledermann, N.G. (eds.) Handbook of Research on Science Education, pp. 3–30. Routledge, New York (2007)
13. Apedoe, X., Ford, M.: The empirical attitude, material practice and design activities. Sci. Educ. **19**(2), 165–186 (2010)
14. Matthews, M.R.: The nature of science and science teaching. In: Fraser, B.J., Tobin, K.G. (ed.) International Handbook of Science Education, vol. 2, pp. 981–999. Kluwer Academic Publishers, Dordrecht (1998)
15. Mayring, P.: Qualitative Inhaltsanalyse, Grundlagen und Techniken, 10th edn. Beltz, Weinheim (2008)
16. Koliander, B.: Laborpraxis im Chemieunterricht: Ziele und Wege österreichischer Lehrpersonen. Dissertation (2017)

Remote and Virtual Laboratories

Work In Progress: CoCo—Cluster of Computers in Remote Laboratories

C. Madritsch[(✉)], T. Klinger, and A. Pester

Carinthia University of Applied Sciences—CUAS, Villach, Austria
c.madritsch@cuas.at

Abstract. This paper gives an overview about the design, implementation, and practical use in Engineering Education of "CoCo—Cluster of Computers," a massive parallel IoT-computing cluster developed by bachelor degree students. First, the development goal and its expected application is explained. Next, hardware and software design and implementation are described in detail. Furthermore, the practical use of the cluster including the first test and demo applications are shown. Here, current super-computing methods like Message Passing Interface (MPI) and Open Multi Processing (OpenMP) are used to depict the ability and functionality of the cluster. Finally, the integration of CoCo into the curriculum of several lectures like Computer Science and Computer Engineering, as well the integration into the CUAS Remote Laboratories are described.

1 Introduction

The trend towards multi core computing systems is increasing. Due to the rising number of processing nodes, new parallel software architectures and development methodologies are becoming relevant. In engineering education, this trend needs to be reflected, since tomorrows engineers will have to face this challenge. CUAS decided to develop a multi core computing cluster using low cost IoT nodes in order to facilitate several areas of education.

Open Multiprocessing (OpenMP) is used to increase the computing performance by automatically distributing sequential repeating code (like For-loops) into parallel threads within the same process and thereby within the same memory area.

Using the Message Passing Interface (MPI), processes can be distributed among several computing nodes via the network in order to increase the level of parallelism. MPI uses a distributes memory architecture. Both, OpenMP and MPI are used in Supercomputing architectures and can be combined in order to get the best possible system performance (see Fig. 1).

Our Computing Cluster allows students to apply these advanced concepts first hand and thereby getting experienced with the design of parallel software architectures using real hardware.

© Springer Nature Switzerland AG 2019
M. E. Auer and T. Tsiatsos (Eds.): ICL 2018, AISC 917, pp. 569–576, 2019.
https://doi.org/10.1007/978-3-030-11935-5_54

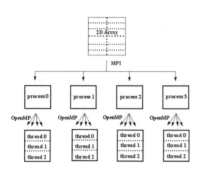

Fig. 1. MPI and OpenMP combined

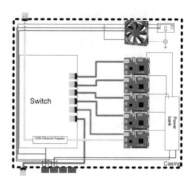

Fig. 2. Computing cluster block diagram

2 Design

The first design of the computing cluster was done during a master thesis project [4]. The Linux-based computing cluster uses five Raspberry Pi 3 modules, one as a master and the other four as computing slaves. Linux is an advantage, since software like Ansible, which automates software provisioning, configuration management, and application deployment, is available. Additionally, a managed Ethernet switch as well as power-supply and cooling concepts were developed (see Fig. 2).

The developed cluster is still functional (see Fig. 3). It is used as a test-system for student projects as well as the basis for the large-scale cluster.

Fig. 3. First computing cluster

Fig. 4. Rack

2.1   Hardware Design and Implementation

The large-scale cluster shall consist of 96 nodes, thus providing 384 processor cores. Therefore, the following actions must be taken to ensure a proper and reliable functionality:

- A mounting solution had to be found to group the Raspberry Pi single-board computers (SBC). We decided for a 19" rack system (see Fig. 4). Each SBC is mounted on a 3D-printed carrier, so that it can be easily removed and changed (Fig. 5);
- A power supply concept had to be established. The power lines to the SBCs are switched using relays on a Uniboard Pi (Fig. 6). For the power connection to each SBC, it is not advisable to use the built-in micro-USB connector, as they are not very reliable. Therefore, power is connected to each SBC via the GPIO port, thus requiring additional safety circuitry;
- Additionally, a cooling system had to be designed and integrated into CoCo, which can also be seen in Figs. 4 and 5.

Fig. 5. 19 inch slot containing the computing nodes

2.2   Software Design and Implementation

Due to the large number of computing slaves, a concept for the distribution, configuration and maintenance of software components needs to be applied. After careful considerations, Ansible has been chosen to accomplish this task. Ansible [5] is a command line based tool, which uses a scripting language (playbooks) to store configuration information. With little effort, software components can be distributed among all nodes in an automated fashion (see Fig. 7).

In order to monitor the status of the computing slaves, a web-based interface called Dashboard was developed. The dashboard provides a simple way to turn

Fig. 6. Uniboard Pi

Fig. 7. Ansible software distribution

on and off individual slaves or groups of slaves. Furthermore, it indicates the current operating conditions like temperature and power consumption. Using the dashboard, the user can start a parallel computation and monitor its progress. Finally, the dashboard also displays the current load of the individual slaves in terms of CPU load and memory consumption (see Fig. 8).

Fig. 8. Dashboard to indicate several performance parameters

3   First Demo

During the next phase of Software implementation, several demo applications have been developed. As a first performance test, a distributed sorting algorithm, based on Bubble Sort and Merge Sort, has been implemented. On the master node (a Laptop Computer), the user enters the array size. Next, the array is filled with random numbers.

The array is split into several chunks according to the MPI distribution algorithm. The dataset is sent to the computing slaves where the actual sorting with Bubble Sort takes place. Within a computing slave, OpenMP is used to distribute the load among the individual processor cores (four per node).

The last step is to merge the individual sorting results together and the sorted data set is sent back to the master (see Fig. 9).

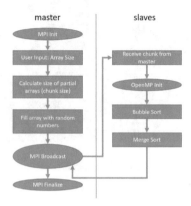

Fig. 9. MPI/OpenMP Bubble Sort/Merge Sort Algorithm

As can be seen in Fig. 10, the overall performance scales from one to three to five nodes quite evenly.

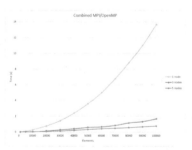

Fig. 10. Performance comparison

To be able to show the available computing power more graphically, a second demo application calculating the Mandelbrot set (see Fig. 11) has been developed. Here, the user can experience firsthand the time duration differences between the computations on one slave versus using many or all slaves.

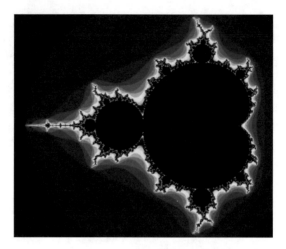

Fig. 11. Mandelbrot set

4   Curriculum Integration of CoCo

The lecture on Computer Engineering focuses on parallel software architectures like Multithreading, Open Multiprocessing (OpenMP), and Message Passing Interface (MPI) [1]. Multithreading deals with the problem of how to increase software maintainability and decrease software complexity by splitting an application into several parallel hand-coded threads.

The lecture on Digital Image Processing (DIP) provides students with the basics of image processing and analysis algorithms, especially the detection of image features. Simple image features are edges, corners, or areas of equally distributed information. Also, color information of images such as color statistics or color histograms can be used as features. Finally, methods like Histogram of Oriented Gradients (HOG) [2] or Hu Moments [3] are part of the group of more complex image features.

The extracted features can then be used to build image descriptors out of feature vectors, which themselves can be used for machine learning algorithms. For instance, object detection or image classification are typical machine learning tasks. In case of multiple object classes, this task is suitable for parallel computation.

5   Remote Labs Integration

It is planned to integrate the CoCo in a remote access environment. In our case we plan to use our Remote Lab for integration. In general, the integration of simulations or experiments can be done in different ways, Remote Lab Management Systems (RLMS) line WebLab Deusto [8] or a direct connection of their client interface with the laboratory machine like in Smart Device specification [9].

In our case we will use a method, developed by Danilo Garbi-Zutin, named Experiment Dispatcher. It is a framework, developed special for cloud based integration of online experiments, "that provides Online Laboratory server infrastructure as a service (LIaaS)... to enable and/or facilitate the deployment and development" [10] of such kind of integration architectures (see Fig. 12).

Fig. 12. Lab Infrastructure as a Service [11]

The easiest way will be the integration via a web-site. The CoCo will be combined with an experiment engine, which is listening to requests for computation, process them (run the computation) and return the results. The operation mode can be interactive or batched. Depending from the computation task just the computation result, the image, a video or a screen streaming from the virtual screen will be exchanged with the client. In the case of the Mandelbrot example a screen streaming will be the most appropriate method. Another method would be the conversion of the screen output to a video, which can be downloaded. The experiment dispatcher will manage different experiment engines and balance the load. In any case all runs, and requests are running in a highly controlled environment, which follows a very specific sequence (see Fig. 13).

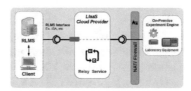

Fig. 13. LIaaS relay service [11]

The implementation of that service will be finished in several month. The user group will be open, but moderated.

6   Conclusions and Outlook

CUAS has accumulated a lot of experience in the field of Remote- and Pocket Labs [6] over time. Some previous projects have even led the way for other universities and institutions. The use of IoT Technologies in education is a logical next step [7]. Since several projects have been based on the Raspberry Pi platform, we felt confident to begin the development of our Computing Cluster. The Cluster does not only allow CUAS to teach concepts of Supercomputing, Multithreading and related topics, it also enables us to use its computing power for Digital Image Processing applications.

The next step will be to implement Machine Learning algorithms—based on different forms of Neural Networks—on our cluster. Students will have the chance to test and try out advanced concepts and by doing so, students will have a definite advantage in understanding and mastering these emerging technologies.

Concluding, the use of small, low-priced single-board computers like the Raspberry Pi or similar, provides low- to mid-budget solutions also for the emerging field of parallel computing.

References

1. Barlas, G.: Multicore and GPU Programming: An Integrated Approach. Morgan Kaufmann (2014)
2. Dalal, N., Triggs, B.: Histograms of oriented gradients for human detection. In: IEEE Computer Science Conference on Computer Vision and Pattern Recognition (CVPR). San Diego, CA (2005)
3. Hu, M.: Visual pattern recognition by moment invariants. IRE Trans. Inf. Theor. **8**, 179–187 (1962)
4. Vrisk, S.: Implementation of a Single-Board Computer Cluster, Mater Thesis, Fachhochschule Kärnten (2017)
5. https://www.ansible.com/ . Accessed 11 Nov 2017
6. Klinger, T., Madritsch, C.: Use of virtual and pocket labs in education. In: International Conference on Remote Engineering and Virtual Instrumentation, Technologies and Learning, Spain, February 2016
7. Madritsch, C., Klinger, T.: Work In Progress: Pocket Labs in IoT Education, vol. 26–29 (2017)
8. http://weblab.deusto.es/website/
9. Salzmann, C., Govaerts, S., Halimi, W., Gillet, D.: The smart device specification for remote labs. In: 2015 12th International Conference on Remote Engineering and Virtual Instrumentation (REV), pp. 199–208, February 2015
10. Zutin, D.G., Auer, M., Orduña, P., Kreiter, C.: Online lab infrastructure as a service: a new paradigm to simplify the development and deployment of online labs. In: 2016 13th International Conference on Remote Engineering and Virtual Instrumentation (REV), pp. 208–214, February 2016
11. Zutin, D.G., Kreiter, C.: A software architecture to support an ubiquitous delivery of online laboratories. In: 2016 International Conference on Interactive Mobile Communication, Technologies and Learning (IMCL), pp. 93–97. IEEE, San Diego, 17–19 October 2016

Virtual Security Labs Supporting Distance Education in ReSeLa Framework

Anders Carlsson[1]([✉])[iD], Ievgeniia Kuzminykh[1,2][iD], and Rune Gustavsson[1][iD]

[1] Blekinge Institute of Technology, Karlskrona, Sweden
Anders.Carlsson@bth.se, http://www.bth.se
[2] Kharkov National University of Radio Electronics, Kharkov, Ukraine
http://nure.ua/en/

Abstract. To meet the high demand of educating the next generation of MSc students in Cyber security, we propose a well-composed curriculum and a configurable cloud based learning support environment ReSeLa. The proposed system is a result of the EU TEMPUS project ENGENSEC and has been extensively validated and tested.

Keywords: Training lab · Remote lab · OpenStack · Virtualization · CDIO · Distance learning · Cyber security

1 Introduction

A challenge of practical Engineering education is to decide on and implement environments allowing students to conduct and evaluate experiments on equivalents of real equipment, software or technologies. Acquiring practical skills is an important part of training. Complementing theoretical knowledge and giving seminars the student must perform exercises with selected equipment, special software, applications and/or configuring networks. To that end, the student must either use a specific prepared laboratory, or download and install the necessary software on his/her computer.

The instructor or teacher can deploy the laboratory in mainly two ways:

– prepare a physical laboratory with equipment or simulation laboratory, install programs on all machines that will be used by students;
– create a directory with shared access where all material for the practical exercises will be placed.

But both ways have their drawbacks. In the first case, the teacher should devote time to set up and copy the material to each machine, and check that everything is started properly. It is also necessary to find a place to place all equipment. In the second case, the time for copying from the shared directory and the installation time is spent by the students but, anyway, it is time consuming. Moreover, it is a possibility that one of the students run the software directly from the shared folder, and thus might modify the initial version.

© Springer Nature Switzerland AG 2019
M. E. Auer and T. Tsiatsos (Eds.): ICL 2018, AISC 917, pp. 577–587, 2019.
https://doi.org/10.1007/978-3-030-11935-5_55

To mitigate those drawbacks we propose a virtual cloud-based laboratory ReSeLa as a framework. In this case, the deployment of experiments is limited to only one machine, and each student receives an individual clone of the virtual equipment and software that can also be remotely accessed. Remote access allows virtual labs to be used in distance education process which are in high demand nowadays. The instructor needs access to the administrative tools to create a lab and evaluate the students work. The student needs access to the lab, to execute (complete) it and submit the assignment. Such cloud-based laboratories will consume minimal time and minimal physical space to set up.

The purpose of ReSeLa was to provide a tool supporting education of future Master students in cyber security in safe environments [1].

2   Purpose and Goals

The challenge of practical education is to give students opportunities to conduct experiments within environment where real equipment, software and technology are faithfully modeled. In such laboratory the configuration and installation of environment as well as running of experiments should be time efficient and require minimal physical space.

A successful education environment should also include a well-defined purpose and pedagogical supporting material for students and teachers.

At present we are witnessing a trend of moving IT infrastructures into the cloud. As a result, clouds need to be protected against cyber attacks similarly to on-premises network environments. This requirement also applies to all types of cloud environments: private, public and hybrid.

The Man-in-the-Cloud attacks (MitC) have been coined in [2] describing possible scenarios of MitC.

The increased number of cyber attacks, theft of personal data, ransomware, credit cards frauds, and hacker intrusions are increasing threats to our ever-increasing networking societies. We are also witnessing examples of digital based cyber warfare [3]. In short, there is a high demand of well-trained professionals in the fight against cyber crime and for protection of our infrastructures.

To that end, EU funded the 4-years TEMPUS project ENGENSEC Educating future Experts in Cyber Security, Coordinated by BTH [5]. A result of that project is Remote Security Laboratory (ReSeLa) and supporting teaching and learning materials.

The design, development, implementation, and validation of ReSeLa had to take into account the following requirements:

1. Rapid deployment of an experiment supporting many students.
2. Flexibility of the platform that allows to install different software on different OS.
3. Software and machines must be isolated from external networks.
4. The laboratory should grant the rights to the administrator of the VM.
5. Storage and backup of current experiments should be stored, allowing students the opportunity to suspend the execution and then continue with the task.

3   Approach

To meet the five requirements above, the following architecture for design and implementation was chosen:

- The requirements 1 and 5 were met by a cloud approach. The cloud approach allows easy deployment, quick access to experiments, both for the teacher and students, locally as well as remotely. To provide redundancy, in case of hard disk failure, and without interruption of the experiment, data storage virtualization technology, such as RAID, was used.
- The virtualization approach for software and applications meets requirements 2, 3, 4, and 5. Allowing reduction of software and hardware costs of importance for education institutions. Creating a VM allows to host on a single physical machine several operating systems with different software. Each student gets a clone of the experiment with necessary set of images, with administrative rights to manage these images. By using the snapshot function the student has access to stored state of the experiment and can continue the experiment later.
- The network approach allows the user to connect physical elements of the virtual laboratory and provide access to the Internet. The network combines the components such as the controller node, the compute node, and the router.

On the router, the firewall policy and VPN management can be configured as specified in requirement 3 above. Isolation is very important in the experiments related to ethical hacking and malware analysis. Only machines inside the laboratory should be able to attack the authorized machine within the laboratory, and not be able to communicate with or attack machines outside the laboratory.

Following those guidelines for the ReSeLa framework, the following components were selected:

- OpenStack as a cloud structure and backend.
- SDN router Mikrotik as a firewall and a VPN backend.
- Python based applications as a frontend that connected to OpenStack data bases provides a graphical interface. ReSeLa frontend interacts with the network via the open source Flask server. A more detailed architecture is described in work [1].

As a pedagogical framework, CDIO was selected by ENGENSEC. The CDIO INITIATIVE is an innovative educational framework for producing the next generation of engineers. The framework provides students with an education stressing engineering fundamentals set in the context of Conceiving Designing Implementing Operating (CDIO) real-world systems and products [4].

4   Results from ENGENSEC

At the start of the ENGENSEC project a relevant curriculum for a new Masters program in cyber security was selected. The selection was based on answers to a survey included 52 IT security companies from different countries that were involved in the ENGENSEC project. The outcome was the following set of seven courses:

- Advanced network & Cloud security.
- Wireless and Mobile security.
- Secure software development.
- Advanced malware analysis.
- Web security.
- Pentest and Ethical hacking.
- Digital forensic.

The second step was to identify a suitable framework for future MSc programs in Cyber security at level A2N or A2F, and to identify and implement suitable course developer teams for each course. The framework should be instantiated during MSc thesis work, supervision and examination.

Mixing selected teachers from different member partners composed the teams. This to ensure that the correct mix of competences and skills was met. And, not the least, also to enable an "out of the box" innovative approach of the work at hand.

The final MSC in Cyber Security should have:

- a minimum of 30 credits from above courses at level A2;
- 7 credits of Research Methodology;
- 30 credits for a MSc thesis/project at advanced level, A2E.

The joint proposed framework also recommends a process of writing thesis/projects as described in the project book [6].

To train and test necessary skills and knowledge (CDIO) of students suitable course material and a cloud based configurable raining environment ReSeLa was developed.

During the last project year the focus was on tendering processes of equipment purchase, teacher training, Summer schools and validation activities.

At the end of the project suitable ReSeLa equipment has been installed at partner's HEI in Russia and Ukraine. Furthermore, agreements of double diploma programs has been signed by BTH in Karlskrona, Sweden and NURE in Kharkiv, Ukraine, as well as by BTH and SpbSUT in St Petersburg, Russia.

Those agreements also describe the processes for supervision and examination to obtain double diploma examination.

To ensure sustainability of the ENGENSEC results the following three categories of activities have been agreed upon:

- 'In real' meetings.
- Off-line meetings.
- Maintenance and further developments of the ReSeLa environments.

To support universities, the ENGENSEC project has also delivered the following two books:

1. A. Carlsson, I. Sokolianska, A. Adamov: *Educating the Next Generation MSc in Cyber Security.* ISBN: 978-91-7295-962-0 (2018).

   Illustrating the ENGENSEC MSc Roadmap on MSc program in Cyber security. Including short descriptions of the framework, courses, lab instructions and additional material. The book gives guidelines on how individual universities can compose and implement a MSc in Cyber security.
2. T. Surmacz and A. Carlsson: *Cyber Security for Next Generation Experts.* ISBN: 978-91-7295-962-0 (2018).

The book is a study guide and includes the fundamentals in cyber security.

5   Training and Validation

Throughout the ENGENSEC project, extensive mobility and training activities have taken place to enable knowledge and skill sharing among partners. Those activities included workshops for course developers, teacher training and summer Schools.

During first and second year thee one-week workshops were arranged. Each were attended by 25 teachers/trainers from partners and additional 10 trainers from BKA (Federal Police of Germany). The workshops were deployed at the BKA Cyber Crime training center at Schloss Waldthausen, Budenheim outside Wiesbaden. During these workshops over 100 teachers and trainers were trained.

The first workshop targeted primary on training of course's managers and consortium steering committees. The first workshop included education on CDIO framework and the preliminary NICE framework and combined those with rules and demands from Bologna Master Level A2N/A2F. It also included education using modern tools supporting distance education, such as Adobe Connect.

During second workshop the target audience was course's developer teams, the purpose was to coach them with tools that can create interactive presentations, infographics, and scribing.

Third workshop supported training in course development and 'intercoaching' between teams. An outcome was content thread flows through the different courses, avoiding, e.g. duplication in parts that could be overlapping or missing links between courses.

Three summer schools, each for 2 weeks, were planned and given at;

- Lviv, Ukraine, June - July 2016
- St Petersburg, Russia, July 2016
- Lviv Ukraine, June 2017

In total, 90 students were attending the summer schools. During each summer school minor workshops were implemented with invited specialists. During he summer schools the course material was evaluated and validated by students and teachers.

The results of the ENGENSEC projects have also had (very) favorable evaluations by internal as well as external international evaluators [6].

6   Results and Education Lab Examples

To illustrate the ENGENSEC framework we first give a summary of advantages
we have found by using the ReSeLa platform compared with classical platforms
supporting distance learning (Table 1). The advantage of using a controlled virtual system is threefold:

1. Time saving related to installation and setting up and deployment of labs
   serving 12—16 students in a class.
2. Quality of experiments and teaching.
3. Proper access rights for students.

In summary, the main advantages of executing labs in the framework of
ReSeLa are:

- Short deployment time.
- Ability to perform laboratory work remotely at a convenient time.
- No interference between hardware, operating systems and software.
- Minimizing the cost of proprietary software and its effective use.
- Simple scaling (adding compute nodes).

As a result, the virtual laboratory allows the teacher and student to plan the
work efficiently. Time for implementation of the experiment is not limited by the
time of the session, but also by the abilities and needs of each individual student. Reducing the speed with which the laboratory works are developed, allows
saving the teacher's time for routine operations and allowing concentrating on
the development of new exercises, improving existing, and other methodological
work. In addition, the above differences are also related to scientific experiments.

Requirements for downloadable distributions:

- Minimum size. Since the same image can be simultaneously launched many
  times by different students, a minimal size allows to save significantly memory
  and network resources this increase the speed of loading and work.
- Preloaded software. Using proprietary software such as compiler, you should
  limit the number of copies run simultaneously, which corresponds to number
  of available licenses.
- Minimizing the number of software available to the student reduces the risk
  of errors and unnecessary complications while the student performs the laboratory task.
- Restricting access to the Internet (in case of the laboratory work, especially
  in important courses using malware).
- Restricting access from Internet is especially important in courses that test
  vulnerable systems, because external hacker can hack the system.
- Creation of laboratory works with gradually increasing complexity that
  increases the interest of the student.

We include in the paper three typical lab sessions from the Book [6].

Table 1. Comparison of Classical and Virtual laboratories

Parameter	Training laboratory	
	Classical	Virtual
Deployment time	The preparation of the laboratory takes time comparable in time to the performance of the work or even more	One-time preparation of working images of operating systems
Break down of the system	Requires re-installation of the laboratory environment, comparable over time with the initial roll-out	It is sufficient to restart the laboratory environment
The order of implementation	The equipment is prepared for the performance of specific laboratory set up. The order of tasks is fixed	Each student can perform any laboratory assignment, (s)he is not tied to other students in the execution of work
Remote work	A rigid link to the laboratory room and the scheduled time of training	Ability to perform tasks remotely and only restricted by the deadline for the delivery of task
Repeated execution	Limited by the load of laboratory premises	Ability to re-run the lab at any time
Quality of experiment	Its hard to repeat mistakes in the students actions	Can be repeated to clarify measurements
Heterogeneity of operating systems	The unfolding of laboratory tasks can be complicated	Any operating system can be deployed at approximately the same speed
"Rubbish" in the operating system	When preparing a cycle of laboratory work it is easier to create one distribution with all the necessary software for a series of laboratory works	For each job its easy to create your own "lightweight" distribution with only the necessary software
Number of equipment	In the case of inadequate equipment, students are forced to perform laboratory work in shifts	The number of instances started is limited only by the performance of the server software
Obsolete operating systems	Need to maintain the old hardware in working order	Running obsolete operating systems and software without restrictions
Equipment safety	Critical failure of equipment is possible	In an isolated environment the risk of failure is minimal

6.1   Lab Session I: Hijacking Using DNS Spoofing

Each student or each team need there own network, four computers and a physical switch. This lab is not possible to run on a shared network due to the potential disturbing interactions between students. A virtual environment such as ReSeLa is needed.

The configuration scenario have a PC of a victim with browser and Internet access; a PC of an attacker in the same LAN; a web server for hosting a fake web-page and fake web-page template. Topology of experiment is shown on Fig. 1.

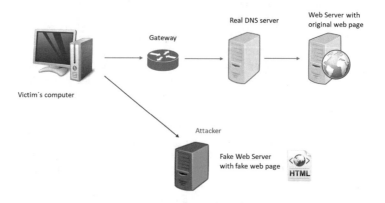

Fig. 1. The topology for Hijacking using DNS spoofing scenario

Malicious attacker redirects the traffic for original web-page to its own 'fake' version of the page through the spoofing.

Experiment set up:

For teachers: Create a VM for victims computer with any OS and web browser, create VM for attacker with OS where we install web server and fake web-page. Also, on attacker VM we need install tool that can implement ARP and DNS spoofing. Include these images to a lab in the course, make description of lab and study guide for experiment with tasks.

For students: Launch lab, follow study guide.

Running experiment: After all VM's are running student should implement tree steps.

1. From attacker VM scan network to discover victims IP address and gateway IP address.
2. Launch APR spoofing against gateway IP address using suggested tools.
3. Launch DNS spoofing against domain name of original web site. Now domain that we spoofed will be assigned to attacker PI address where fake web server installed. Victim will be redirected to fake web page when enter original URL in the browser line.

6.2   Lab Session II: Exercises with Malware

To work with malware is a security challenge. For instance, the malware should no be allowed to reach Internet, students should not (unintentionally) be able to copy it, or any other possibility for the malware to sneak out from the lab environment. Furthermore, it is quite tricky to set up debugger and network switches with wiretapping capabilities for each lab group.

The configuration scenario has a victim PC, with browser using and internet access. Within this PC analyze the attack of a spyware infected by a video launched from the browser. Study how the malware download new executable part from Internet, while the victim watch a funny movie, it continue even if the movie is stopped and deleted.

Screen capture of the experiment is shown on Fig. 2.

Fig. 2. Screen capture showing the Flashplayer running the Duke spyware and monitoring tools

Experiment set up:

For teachers: Install MS Win 7 client, with analysis tool Wireshark, SysInternals tool, Internet browser and the "CozyDuke APT" inside a "Office Monkey - Flash movie".

For Students: Start the Lab set "CozyDuke" in ReSeLa, login as student with password student. Prepair the lab and launch: Wireshark.exe (network analysis tool), Procmon.exe (Process Explorer from Sysinternal tools), Regedit.com (standard windows tool to monitor Windows registry changes), then launch "Office Monkey flash video" with FlashPlayer from the Explorer browser. Student should analyze network communication and processes.

Running experiment: After the VM is launched the student should:

1. Analyze processes: using Sysinternal software Procmon.exe to monitor the FlashPlayer.exe.

2. Wiretapping and analyzing network communication with Wireshark.exe
3. Collecting the HTTP traffic and analyze witch files that are downloaded.
4. Continue to watch and analyze the new downloaded files and there behavior.

6.3   Lab Session III: Android Ransomware, Analysis and Mitigation

The lab is based on a virtual machine of an Android Phone running in a Windows environment. Management of reverse engineering is a quite complex task. In our set up the mix of CPU architectures (Intel X86 and ARM) is potentially unstable and might need a re-launch as a fresh system. This task is very easy and fast in a cloud based system such as ReSeLa.

The task is that the student shall debug and analyze the ransomware inside the virtual Android phone.

Using Whireshark and debug tools from "Android SDK" to monitoring processes and files in a virtual Android smartphone that is pre-infected with a ransomware.

For teachers: To create the Lab set "Android Ransom", create a VM as a Windows Client with Wireshark, Hexedit.exe, TotalCommander and Android SDK. Inside the Windows Client install a Android VM. Inside the VM "Android Phone" install the "Sex_xonic service".

For students: Start the Lab set "Android_Ransom" in ReSeLa, launch Wireshark, Hexedit.exe, TotalCommander and inside the virtual Android phone "adb_shell (cmd.exe)".

Running experiment: After the VM is running student should:

1. Prepare to analyze and debug the Android.
2. Manually start "Sex_xonic_service" .
3. Monitor how files are encrypted, to find the start point address where ransomware makes the choice - start encrypt / -start decrypting, and what CPU registry content is used to make the choice.
4. Manually change this CPU register or force the PC (program counter) so the malware start to decrypt on restart of the phone.
5. This lab requires multiple reboots from initial configure, which is time consuming in a normal computer class room but in a virtual lab a simple and fast procedure.

Screen capture of the experiment is shown on Fig. 3.

7   Conclusions and Further Work

The requirements of future education environments for Master Science in Cyber Security are outlined in Sects. 1 and 2. The requirements are related to:

– Education level and education support.
– Education environment.
– Education topic.

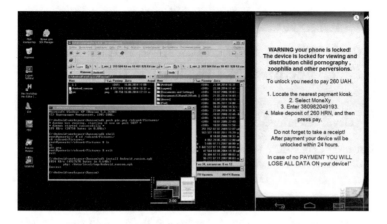

Fig. 3. A virtual Android phone inside a Windows client

In the EU supported TEMPUS project ENGENSEC Educating Next Generation of MSc students in Cyber security we have the following refinements of those requirements:

1. Education level Next generation of MSc in Cyber Security.
2. Education environment A suitable curricula meeting industrial requirements and the configurable Cloud based environment ReSeLa supporting the CDIO model.

The organization and work of the project has resulted in a high-quality product that has been extensively tested and validated by different target groups such as teachers and students (Sects. 4 and 5). Findings and conclusion of the ENGENSEC project are given in Sect. 6. We also include here three illustrative lab sessions supported by ReSeLa. Planned future work includes upgrading of supporting teaching and learning material as well as the Learning system ReSeLa [6].

References

1. Carlsson, A., Gustavsson, R., Truksans, L., Balodis, M.: Remote security labs in the cloud ReSeLa. In: IEEE Global Engineering Education Conference (EDUCON 2015), pp. 199–206. IEEE, Tallin (2015)
2. IMPERVA. http://www.imperva.com/docs/HII-Man-In-The-Cloud-Attacks.pdf. Accessed 18 July 2018
3. Carlsson, A., Gustavsson, R.: The art of war in the cyber world. In: IEEE PIC S&T 2017, pp. 42–44. IEEE, Kharkiv (2017)
4. CDIO Homepage. http://www.cdio.org/. Accessed 18 July 2018
5. ENGENSEC Homepage. http://www.engensec.eu/. Accessed 18 July 2018
6. Carlsson, A., Sokolianska, I., Adamov, A.: Educating the Next Generation MSc in Cyber Security (2018). ISBN: 978-91-7295-962-0

Effectiveness and Student Perception of Learning Resources in a Large Course "Experimental Physics for Engineers"

David Boehringer$^{(\boxtimes)}$, Elena Rinklef, and Jan Vanvinkenroye

Computer Center (ICT Services), University of Stuttgart, Stuttgart, Germany
{david.boehringer, elena.rinklef, jan.vanvinkenroye}
@tik.uni-stuttgart.de

Abstract. The 1st year course "Experimental Physics for Engineers" at the University of Stuttgart was analyzed to find evidence for restructuring the module. Despite a cautious approach to the statistical data, justified statements about the effectiveness of the learning resources of this course can be made. The analysis provides a basis for the restructuring of the module and its further evaluation.

Keywords: Experimental physics · Effectiveness of learning resources

1 Introduction and Context

At the University of Stuttgart "Experimental Physics for Engineers" is a mandatory class for almost all engineering students in their first semester. About 1500 participate in each winter term. Due to the high number of participants and limited room capacity, the lecture is held twice a day, giving all students a chance to attend. During a lecture the lecturer is performing the experiments in the lecture hall. Most students are following the experiments via the live video on the beamer projection in the lecture hall. The curriculum does not include any mandatory exercises for this lecture, nor is there an accompanying lab course during which the students could get hands on experience. The hands on lab has to be taken after the exam of the theoretical part of the course.

Students experience the class as difficult and about two thirds of those taking the exam for the first time fail. The number of students taking the exam two or three or more times is quite high. A lot of those who fail quit their engineering degree program completely. Experimental physics is not the only reason for this; other classes with high failure rates are e.g. higher mathematics, technical mechanics, or constructional design.

The appreciation and importance of experimental physics for engineers has decreased tremendously in the engineering programs over recent years, starting with the Bologna process. Currently the whole module is reduced to just three credit points (ECTS, valued with a work load of 30 h for one credit point, i.e. 90 h for the whole module), with two credit points for the lecture (theoretical part) and one for the lab sessions (practical part). The module is not graded, and so has no importance for the final grade of the students. They only have to pass the theoretical exam and attend the

M. E. Auer and T. Tsiatsos (Eds.): ICL 2018, AISC 917, pp. 588–599, 2019.
https://doi.org/10.1007/978-3-030-11935-5_56

lab sessions. The change from grading the course to not grading it increased the failure rate from about half to two thirds.

To activate the students during the lecture the lecturer introduced an Audience Response System, used in a variation of Eric Mazur's peer instruction. Flipping the whole classroom seemed impossible with the large number of students and their often inadequate learning attitude. To alleviate the problems a little, at least, and to help those students who are willing to make use of them, the lecturer started to accumulate all kinds of optional learning resources such as an online script, slides of the lecture, lecture recordings, the questions of past examinations, and last, but not least online laboratories [1]. All learning resources are available via the central Learning Management System (LMS) of the university where they are arranged according to media. Despite these efforts the failure rate remained constant.

2   Restructuring of the Course and Goal of the Inquiry

Since the arrangement of the class with 1500 students remaining passive throughout the lecture, the plan is to re-arrange the set-up of the whole module. The separation of theory and practice and "experimental physics" in a passive group of students has always been somewhat absurd, and a contradiction in terms. Digital innovations such as online laboratories had been introduced in the past, but now a comprehensive digital transformation of the module is in preparation. For this the topics of the lecture will be represented as self-contained modules in the LMS. These modules will include all available learning resources of the respective topics.

In particular the online laboratories are to be redesigned. They are the only activating method of this class leading to experimentation outside the hands-on lab. Learning objectives of the lab work are being defined right now and suitable kinds of labs (remote, virtual, simulated or hands-on) [2] will be chosen for the different topics. At least one of the experiments, preferably a remote lab, will include keeping a laboratory notebook during the phase of the theoretical lecture. The definition of learning objectives of the lab work will ultimately lead to different examinations of student performance.

An important part of planning the re-arrangement of the class was the analysis of the effectiveness of the different learning resources as they are used in this class [3]. We wanted to find out the students' opinions about the learning resources, the way they worked with them and the effectiveness of the resources for a good result in the exam. Which resources are good, which should be reworked, and which can be omitted? And which need a different context or explanation?

3   Methodology and Approach

3.1   The Data Sample

The data sample for this study comes from the course "Experimental Physics for Engineers" of the winter term 2016/2017. 1266 students took the exam, 869 for the first time, 397 in a repeated attempt. We have four different sources of data: student questionnaires (one in context of the midterm test from December 2016, one after the final exam from February 2017), tracking data of the online-experiments, and the grade of the final exam.

The *student questionnaire* asked about intensity of usage and opinions (e.g. "The online experiments increased my motivation to deal with the topics of the lecture more intensively" in a bipolar scale with six options) about the learning resources as well as the group context of the learning. The midterm test is intended as feedback for the students to show where they stand and as indicator how much they have to learn for the final exam. It is designed as an online assessment with immediate feedback about correct answers after completion of the test. The participation in the online questionnaire was optional (n = 210). The questionnaire after the final (pencil and paper) exam was optional as well. But it was put as mandatory task in the LMS, if students wanted to know their grade shortly after the exam and did not want to wait for the official notification which takes a couple of weeks. 903 students started the questionnaire, 830 finished it; 811 of these actually took the exam, 19 participated in the questionnaire for unknown reasons. The midterm-questionnaire was biased by answers of students who worked as they should, i.e. who constantly worked on the topics of the lecture. But that was a minority. The sample of the questionnaire after the final exam was fairly representative for the whole group. The average score in points was 28.35 (out of 50) for the whole group, 29.14 points for our sample. With a probability of 95% this difference is by chance; therefore our sample seems representative.

For the *online experiments* we have tracking data. With this it is possible to compare students' statements about their usage of online experiments with their actual usage. The comparison of data will be discussed in Sect. 3.3.

32% of the 1266 students who took the *final exam* passed. For this it was necessary to reach 31 points.

3.2   "Effectiveness" of Learning Resources

Making valid assumptions about the effectiveness of learning resources is a tricky business [4]. "Media doesn't matter" is a well-known assumption in pedagogical research [5, pp. 11–14 with further references, 6]. It is certainly true for the fact that results of one investigation cannot be generalized easily or not generalized at all. Circumstances and influences differ much too much from course to course, all the more from field to field, institution to institution or country to country. In our case we have a much smaller focus. We are looking at one year of one course and trying to find indicators of the learning effect of different media as they are used the students. We want to compare these results to the results of coming years. Step by step we want to change the content and the structure of the course and observe the effect of those changes. The investigation is meant to indicate what changes should be made.

Claims about learning effect in our investigation are reduced to success in the exam [7]. This assumes that the exam actually tests the understanding of physics. We cannot be sure if this is true, but we have no other indicators for the learning success nor any means to investigate the long term development of the students' understanding of the field. There is the possibility that we are detecting the learning resources that prepare best for the test and not for the understanding of physics, but calling it "effectiveness" or "learning success".

The significance of the effectiveness of learning resources sometimes changed surprisingly depending how the data was viewed. Divisions of our sample by learning context or success in the exam, or number of attempts to pass the exam showed surprising differences that cannot be easily explained. Results for which we have no reasonable answers (yet) lead us to be careful in drawing conclusions or making changes in our pedagogical setup, even if results should suggest a significant importance. Such uncertainties can only eliminated after we repeat our investigation in following years. In the winter term 2017/18 we had the first repetition of our investigation. The multivariate regression of the whole group showed that results are comparable; we have not completed the detailed analysis for 2017/18 [8].

Comparing students' claims about their working behavior and what was actually observed in the tracking data shows that students' statements about their learning have to be judged carefully [9–11]. Students who claimed to have completed all tasks around the online experiments in fact completed only half of them on average. We exchanged the data of the online experiments in the multivariate regression with the ones of the actual usage and observed that the effect of online experiments on the exam is much higher than thought. According to students' claims of usage, online experiments had a significant positive effect of 1.56 points on the exam score. According to their actual usage there is a significant effect of 2.79 points (see Table 1). In this case, students' statements gave us a correct indication of direction, but not of intensity.

Table 1. Effectiveness of learning materials

	Model 1				Model 2		
	b	SE	Sig.	b	b	SE	Sig.
(Constant)	25.28	0.43	***	25.29	0.40	***	
Use of lecture recordings	−0.45	0.41	n.s.	−0.40	0.40	n.s.	
Use of slides	0.96	0.42	**	0.86	0.41	**	
Use of online script	−0.26	0.43	n.s.	−0.18	0.42	n.s.	
Use of online experiments	1.56	0.39	***				
Observed use of online experiments				2.79	0.40	***	
Use of exams from previous years	2.79	0.45	***	2.85	0.44	***	
Use of students' lecture notes	−0.24	0.41	n.s.	−0.08	0.40	n.s.	
Use of STEM course	1.01	0.44	**	0.96	0.43	**	
Use of learning groups	1.51	0.44	***	1.42	0.43	***	
No. of observations	811			811			
Adjusted R^2	0.13			0.16			
Significance of model	***			***			

Note *, **, *** indicates significance at the 90, 95 and 99% level, respectively. The dependent variable (points reached in the final exam) has a range from 0 to 50; independent variables are dichotomous

3.3   Methods of Analyzing the Data Sample

Because of the variety of learning resources that were offered to the students and the different ways they could work, it is difficult to isolate the effectiveness of a single resource. We did not interfere in the pedagogical setting of the lecture, but designed our investigation in connection with its current procedure.

Among the learning resources we analyzed are: lecture recordings, the slides of the lecture, an online script (delivered as an online-module), online experiments, the exams of previous years, and students' own lecture notes. Accompanying the course the university offers a physics-course as part of the "MINT College" that offers propaedeutic and accompanying courses in STEM subjects. Another context for learning are self-organized learning groups of students. All learning resources and both, STEM course and learning groups, are optional.

For all learning resources and learning contexts we made bivariate analyses to exam scores. We added a multivariate regression to discriminate effects of a single learning resource or context. We also divided the sample according to gender, number of attempts to pass the exam, students who passed and those who failed as well as dividing the group by the median. We also looked at the participants of the STEM course and of learning groups and what effect the learning resources had for them.

4   Analysis

4.1   Introduction and General Remarks

The sources and analyses described in the previous chapter generated significant quantities of data.

- We observed how intensely the learning resources were used (at least according to students' claims); see Fig. 1;

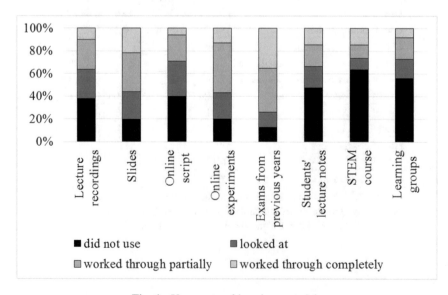

Fig. 1. Usage rate of learning materials

- what the students' perception of the learning resources was in different aspects, e.g. learning outcome; see Fig. 2;

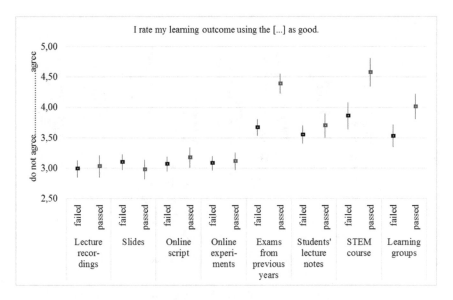

Fig. 2. Perception of learning outcome of learning materials, split by "passed final exam" or "failed final exam"; Note: Variable ranges from 1 (do not agree) to 6 (agree)

- how the use of the learning resources influenced the exam score; see the multi-variate regression of all students in Table 1;
- how the intensity of use of the learning resources influenced the pass rate; see Fig. 3.

The results will be discussed mainly according to type of learning resource in Sect. 4.2. In this introduction to the analysis we will concentrate on a few general remarks.

- The results of the mid-term questionnaire are hardly taken into account because of the bias of participants.
- Students who passed the exam drew conclusions from their engagement with all learning resources and learning contexts which they still had to learn (which they did, obviously) to a significant larger extent than those who failed. They also learnt with more kinds of learning resources and learnt more intensively. On the other hand students who just had a look at the learning resources without really using them scored worse than those who did not look at the respective learning resource at all (with the exception of lecture recordings); see Fig. 3.

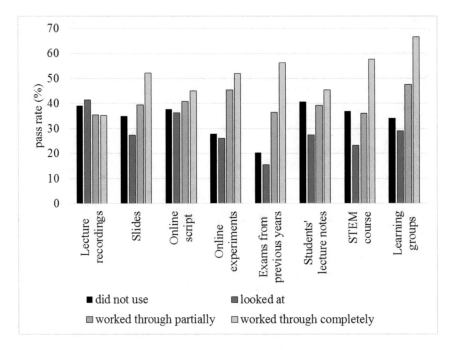

Fig. 3. Pass rate of the final exam by learning resource usage rate

- We have observed almost no effect of gender on the results. Male students of our sample scored a little (but significantly) better than female students (29.49 compared to 27.62 points). Females worked with their own lecture notes a bit more often, had less fun and motivation working with the online script and exams of previous years and liked learning in a group context a little more than males. All mentioned effects seem to be significant, but are weak.

4.2   Effects and Effectiveness of Learning Resources and Contexts

Lecture recordings and *slides* are discussed together because they have much in common and show significant relations in their usage. Lecture recordings are more or less slides with spoken explanations of the lecturer. Additionally they show videos of the experiments that were demonstrated during the lecture.

Slides were used more often than lecture recordings. Only 20% of the students did not work with slides, whereas twice as many did not use the recordings; see Fig. 1. This is somewhat surprising since lecture recordings are passionately and often requested by student representatives. One explanation for this contradiction may be that representatives are second year students or older.

50% of the students who did not use the recordings (20% of all students) considered them as useless for learning; with slides this value is 66% (but only 13% of all students), the highest for all learning materials.

The students' perceptions of lecture recordings and slides are comparable, but with slight differences: slides appear more demanding, they seem to indicate better what is still to be learnt and it is less fun to work with them. Slides have the lowest "fun-factor" of all learning resources. Dividing the students who passed the exam and those who didn't shows that the better students estimated the usefulness of slides as preparation for the exam lower than those who failed. They also thought that their learning success was rather low. The evidence does not support this view. Working with slides had a significant positive effect (0.86 points according to the multivariate regression; see Table 1) and, dividing the whole group according to exam results, this was significant only for the students who passed. Vice versa, for lecture recordings only the students of the lower half of the median of the exam-score had a significant benefit.

Lecture recordings differ remarkably in usage from the other learning resources. Only with lecture recordings students who had a brief look at them have better results than those who did not use them at all or worked with them a lot; see Fig. 3. Overall the results according to usage intensity are quite the same. This seems to indicate that there are differing reasons for using lecture recordings and it is not the case that more diligent students score better. Possibly students who visit the lecture and work selectively with the lecture recordings have the best learning practice. This assumption needs to be confirmed by a qualitative exploration.

A remarkable difference in usage of slides and lecture recordings can be seen between first year students and those who tried to pass the exam for the third time (or more often). First year students had a significant positive effect for slides (1.40 points) and a significant negative one for lecture recordings (−1.33 points). For students in their third attempt it was exactly the other way round: a positive effect for lecture recordings (2.43 points) and a negative one for slides (−2.69 points). Possibly a few students did not attend the lecture, but learnt with the learning resources only. In this case it was smarter to take the lecture recordings instead of the slides. For first year students we assume that they had inferior learning methods, maybe even taking the lecture recordings as an excuse to procrastinate. This has to be substantiated by a qualitative study.

The *online script* is a (non-interactive) online module in the LMS explaining the central phenomena and includes tasks for calculations. The lecturer considers it as "deprecated" and wants to remove it. What do the students think?

First of all, it is the least used learning resource. 26% of total students did not know about its existence. This finding is somewhat surprising since the online script was well visible in the LMS. Those who used it found it relatively demanding and not well suited to prepare for the exam, or to understand the lecture topics. Neither did it foster their motivation.

The usage of the online script had no significant effect on exam scores. There is one exception: in the group of learners of the STEM course it had a significant negative effect of −1.4 points according to the multivariate regression of all STEM course participants (see Table 2). We have no explanation for this. Perhaps none is needed since the lecturer's assumption that the online script should be removed seems justified. A new online module as replacement, preferably interactively connected with the online-experiments could be helpful for the students, though.

Table 2. Effectiveness of learning materials, split by participation in STEM course, learning group or neither

	Neither STEM course nor learning group			Participants of STEM course			Participants of learning groups		
	b	SE	Sig	b	SE	Sig	b	SE	Sig
(Constant)	25.30	0.44	***	26.78	0.99	***	26.67	1.40	***
Use of lecture recordings	−0.67	0.52	n.s.	−0.31	0.82	n.s.	0.16	0.85	n.s.
Use of slides	1.16	0.52	**	0.62	0.81	n.s.	0.01	0.90	n.s.
Use of online script	0.64	0.56	n.s.	−1.40	0.76	*	−0.12	0.81	n.s.
Observed use of online experiments	2.15	0.53	***	4.62	0.77	***	3.08	0.81	***
Use of exams from previous years	2.59	0.51	***	3.62	1.05	***	3.58	1.29	***
Use of students' lecture notes	0.04	0.54	n.s.	−1.64	0.78	**	−0.12	0.81	n.s.
No. of observations	465			213			218		
Adjusted R^2	0.12			0.20			0.08		
Significance of model	***			***			***		

Note *, **, *** indicates significance at the 90, 95 and 99% level, respectively. Dependent variable (points reached in the final exam) has a range from 0 to 50; independent variables are dichotomous

Online-Experiments were introduced in this course in 2010 [1]. They are embedded in an online assessment. An experiment (simulation or remote) has to be conducted in order to answer some questions. A detailed long term study on the effectiveness of the different kinds of online experiments is set up [12], but the large amount of collected data is yet to be analyzed.

There is considerable difference between asserted and actual usage of the online experiments. 20% of the students claimed not to have used them whereas in fact it is 50%. This limits the value of some of the following statements, because we have added the data of the actual usage only in the multivariate regressions up to now.

41% of the students did not know about the online experiments (which are easily visible in the course in the LMS!). Increasing interest in the topic and motivation to deal more with physics, as well as the perceived learning success were considered quite low. This is a remarkable difference to the view of the participants of the midterm questionnaire, and even more to the students we interviewed in 2010 [1]. Since using the online experiments was, and is optional, it seems that in earlier studies we asked the more motivated and better students. Our latest sample of students seems to be more representative.

The value for perceived comprehension of the subject matters of the lecture is especially low, which indicates that the online experiments should be more closely integrated in the lecture. However, working with online experiments has a significant positive effect on the exam results (2.79 points). Two-thirds of the first-year-students used the experiments, but the effect was stronger for students who repeatedly tried to pass the exam (2.98 points). The biggest positive effect was for STEM course participants who were also engaged in learning groups. The positive effect for this group was 5.03 points. Online experiments are significantly effective for good students as well as weak ones, but for the good students the positive effect is higher.

Exams of previous years are the most popular of all learning resources and the most appreciated. They make clear what is still to learn and are considered the best preparation resource for the exam. Students who passed the exam evaluate the old exams significantly better in all aspects than those who failed. Almost 90% of them had worked with exams of previous years, but only 65% of those who failed.

The positive effect of learning with old exams is higher for 1st-year students (3.64 points) and declines with the number of attempts students make to succeed (1.90 points for those with the third attempt). It is also higher for good students than for weak ones, although it is significantly positive for both. The positive effect according to the multivariate regression for all students is 2.85 points, the highest of all learning resources. Students in learning groups (3.58) and the STEM course (3.62) had higher positive effects than those who learnt on their own (2.59).

Up to now old exams were delivered as pdf. All questions have now been collected in a question pool in the LMS combined with an item analysis. Therefore in the future learning behavior and strategies as well as connected effects with other learning resources in the LMS are going to be detected, and in more detail.

All students are asked to take their *own lecture notes*. We do not know how may did, but only 50% said they used their notes to learn. Those who did not use them considered their own lecture notes (if they exist) as useless; i.e. 25% of the whole group. The student that took their notes to learn were quite confident with it. They felt motivated, thought them to be adequate to prepare the exam and thought they understand the topics of the lecture well. There is a stronger tendency for good students (i.e. who did well in the exam) to think so.

This confidence in the lecture notes is not justified by the results of the exams. For the whole group there is no significant effect in the multivariate regression. But there is for the 1st-year students who made their first attempt at the exam. For them, their own notes had a significant negative effect of -0.85 points. For the participants of the STEM course the negative effect is even more, at -1.64 points. For students who took the exam for the third time a positive effect of 1.62 points is observed. All other divisions of the whole group brought no significant effects.

It can be summarized that students overestimate the quality of their own notes and that less experienced students need better learning competency, either because of the quality of their notes, or their tendency to rely on them too much.

Everybody can learn on their own, but learning in groups is usually more fun and more effective. Two different kinds of groups can be compared in our study: the *STEM course*, a formal offer of the university, and informal *learning groups* of students. The lecturer encouraged the students to meet in groups and learn together, but more than half of them did not follow this suggestion; and two-thirds did not visit the STEM course. Why not? Around half of those who worked on their own claimed not to have known about the STEM course or the possibility to create learning groups. 40% said they had not had enough time to join the STEM course; 30% considered learning groups as not helpful for learning.

Those who participated in the STEM course had remarkably different experiences and opinions. The STEM course ranks best in student opinions about motivation, fun, increased interest in the topic, understanding the topic and learning success as well as usefulness to prepare for the exam. The same is true for students' perceptions of learning groups, but to a lesser degree in all aspects. The positive view of learning in the STEM course or in learning groups was especially strongly expressed by the students who passed the exam (see e.g. Fig. 2).

There is a significant positive correlation between participation intensity and score in the exam, especially for the STEM course. Moreover, the STEM course is the only category (learning resource or learning context) for which it appears that the more students use it the better their perception is of increased motivation and learning success. For usefulness to prepare for the exam this is true for learning groups as well.

Indeed: the more engaged the students were in the groups, the higher the pass rate; see Fig. 3. The multivariate regression shows a significant positive effect for the STEM course (0.96 points) as well as for learning groups (1.42 points). It is interesting to see that this significant effect can be seen for 1st-year students in both group types (1.15 points in STEM course and 1.11 points in learning groups), but only in learning groups for those who took the exam for the third time or more often (3.15 points). There is hardly any effect to be observed between more and less successful students in the exam. There is a slight significant indication that students above the median of the exam score profited from learning groups.

For the participants of the STEM course, online experiments had the highest positive effect (4.62 points), even higher than for old exams (3.62 points); see Table 2. For learning groups the effect of old exams was slightly higher: 3.58 points compared to 3.08 points for online experiments. There were no other positive effects. We cannot explain the negative effects within the STEM course for the online script (-1.40 points) and students' own notes (-1.64 points).

All positive effects were also to be found for students who worked on their own, but to a lesser degree (2.59 points for old exams and 2.15 points for online experiments); it seems that the more formally the learning is organized the higher is the positive effect. Only for students who learnt on their own lecture slides had a significant positive effect (1.16 points) as well. The results clearly encourage use of online experiments in the STEM course or a supervised conduction of online experiments for all students.

5   Summary and Conclusions

The detailed analysis gave many indications where and how the set-up of the course and its representation in the LMS could be improved. Despite the methodological limitations of statistical representations of reality and the effectiveness of learning resources, the analysis met our expectations to build a basis for the evaluation of the current set-up of the course and its coming successive restructuring.

We still need further qualitative information and discussions with the students. The analysis raised questions we cannot answer with the data we have. Moreover, there is one detailed analysis of the whole data set still to be made: for each question of the exam we want to know with which learning resources the students with correct answers had prepared the exam. Are there patterns? And deviations from patterns? And how can this be combined with a long-term item analysis? We are going to carry out this detailed analysis especially for online experiments.

In the winter term 2017/18 the structure of the exam had to be changed for legal reasons. The weight of multiple choice questions is lower, the weight of calculations higher. This change led the failure rate to rise from 68 to 76%. We mentioned the rising failure rate after the change of the curriculum in 2010 from 50 to 66% in the introduction. The disillusioning result of the study is that pedagogical improvements to the course will always have a much weaker effect than such curricular and structural changes. However, hopefully the comprehensive restructuring of the whole module might have a comparably strong effect as well; this time for the better.

References

1. Tetour, Y., Boehringer, D., Richter, T.: Integration of virtual and remote experiments into undergraduate engineering courses. In: Global Online Laboratory Consortium Remote Laboratories Workshop (GOLC), Rapid City, SD, pp 1–6. IEEE, Oct 2011
2. Orduña, P., et al.: Classifying online laboratories: reality, simulation, user perception and potential overlaps. In: 13th International Conference on Remote Engineering and Virtual Instrumentation (REV), pp. 218–224. IEEE (2016)
3. Rinklef, E., Vanvinkenroye, J.: Die Nutzung von freiwilligen Lernangeboten. Gesammelte Erfahrungen aus einer Befragung von Ingenieursstudenten. In: Ullrich, C., Wessner, M. (eds.) Proceedings of DeLFI and GMW Workshops 2017 Chemnitz, Germany, 5 Sept 2017. http://ceur-ws.org/Vol-2092/paper2.pdf
4. Bernard, R.M., et al.: How does distance education compare with classroom instruction? A meta-analysis of the empirical literature. Rev. Educ. Res. **74**(3), 379–439 (2004)
5. Clark, R., Mayer, R.E.: e-Learning and the science of instruction. Proven Guidelines for Consumers and Designers of Multimedia Learning (2016)
6. Means, B., et al.: The effectiveness of online and blended learning: a meta-analysis of the empirical literature. Teachers Coll. Rec. **115**(13), 1–47 (2013)
7. Gijbels, D., et al.: The relationship between students' approaches to learning and the assessment of learning outcomes. Eur. J. Psychol. Educ. **20**(4), 327–341 (2005)
8. Pituch, K.A., Stevens, J.P.: Applied multivariate statistics for the social sciences. In: Analyses with SAS and IBM's SPSS (2016)
9. Bowman, N.A.: Can 1st-year college students accurately report their learning and development? Am. Educ. Res. J. **47**(2), 466–496 (2010)
10. Fredricks, J.A., McColskey, W.: The measurement of student engagement: a comparative analysis of various methods and student self-report instruments. In: Handbook of Research on Student Engagement, pp. 763–782. Springer, Boston
11. Gonyea, R.M.: Self-reported data in institutional research: review and recommendations. New Direct. Inst. Res. 73–89 (2005)
12. Boehringer, D., Vanvinkenroye, J.: The effectiveness of online-laboratories for understanding physics. In: Auer, M., Zutin, D. (eds.) Online Engineering & Internet of Things. Proceedings of the 14th International Conference on Remote Engineering and Virtual Instrumentation REV 2017, Columbia University, New York, USA, pp 459–465, 15–17 Mar 2017. Springer, Berlin

Recommendation System as a User-Oriented Service for the Remote and Virtual Labs Selecting

Anzhelika Parkhomenko[(✉)], Olga Gladkova,
and Andriy Parkhomenko

Zaporizhzhia National Technical University, Zaporizhzhia, Ukraine
parhom@zntu.edu.ua

Abstract. The remote and virtual laboratories are actively used for the educational process all over the world. They give a lot of possibilities for inclusive education, for working with unique equipment, for safety performance of various experiments. Internet space offers a huge amount of online laboratories and projects. The number of variants is increasing but the time for their searching is usually limited. The variety of remote and virtual laboratories for different fields of study makes their identification, classification and selection problematic. Therefore, the task of user-oriented web service development for making recommendations for potential users of remote and virtual laboratories is relevant. This recommendation system will help teachers, students and developers of new labs with the analysis of existing remote and virtual laboratories, it will also provide information about their possibilities and features of their realization as well as simplify the task of laboratories selection according to the users' criteria (subject domain, category of laboratory, type of experiment, used equipment, method of experiment control, accessibility, reusable components, languages, etc.). The developed recommendation system is realized on knowledge-based method and it gives recommendations to users even if their requirements to the desired online laboratories are contradictory.

Keywords: Remote and virtual laboratories · Recommendation system
Knowledge-based method · User oriented service

1 Introduction

Modern Remote and Virtual Labs (R&VL) are the tools of online engineering that provide access to virtual and remotely controlled devices and installations, as well as to the study of various control and design technologies, effects, lows, etc. Thus, they provide the efficient organization of the joint usage of various software and engineering equipment, taking into account financial and economic constraints.

The International On-Line Engineering Association and the Global Consortium of On-Line Laboratories promote the development, dissemination and application of online design technologies and their impact on society [1]. They also encourage the development of new laboratories in the common infrastructure and their application in education and industry.

© Springer Nature Switzerland AG 2019
M. E. Auer and T. Tsiatsos (Eds.): ICL 2018, AISC 917, pp. 600–610, 2019.
https://doi.org/10.1007/978-3-030-11935-5_57

The investigations have shown that today, the most significant number of online laboratories is in the field of electronics and robotics. These laboratories can be successfully used not only by students but also by design engineers to prototype created systems based on reusable solutions [2].

The main problem of remote labs is the limitation of users' number who can simultaneously work with the equipment. Another problem of already developed R&VLs is the lack of a common infrastructure. Existing diversity of such laboratories, as well as of their components (interfaces, hardware/software parts, implementations of algorithms, etc.), the problems of standardization and unification complicate their application for solving the certain tasks.

Therefore, today in the world experience of R&VL development the clear trends are information aggregation of existing Internet resources with access to laboratory experiments, as well as unification of labs architecture and hardware/software solutions [3–7].

The variety of the R&VL for different fields of study makes it difficult to search and choose necessary tool for usage in educational process or research. The number of possible variants is increasing, but the time for decision adoption is usually limited.

There are several projects that try to streamline information about remote and virtual labs, as well as to simplify the search for the desired one [8, 9]. They offer an extensive database of laboratories and a large number of filters to work with them. However, the user has to make the final decision on the choice of a particular laboratory, based on the study of the possibilities that are offered after the filtering options. Thus, the stage of existing labs analysis and selection needs additional user-oriented support.

Today recommendation systems (RS) are widely used in World Wide Web for various web-sites of electronic commerce, the services with music, videos, books and others. These systems become more and more popular as they can help users to choose necessary products or services in accordance with their demands.

Thus, the task of user-oriented web service development for making recommendations for teachers, students and developers of new R&VLs is relevant. The goal of this work is the development of RS that will help potential users with selecting of necessary labs according to their criteria (subject domain, category of laboratory, type of experiment, used equipment, method of experiment control, accessibility, reusable components, languages, etc.).

2 The Application and Selection of Online Labs for Engineering Education

As known, the usage of R&VL makes it possible to improve the educational process in the different areas of study.

The most popular online labs for engineering education today are: iLab, WebLab Deusto, VISIR, GOLDi, RELLE, WEBENCH, Labshare and others. For example, one of the well-known laboratories that offers both remote and virtual experiments is GOLDi (Grid of Online Lab Devices), developed at the Department of Integrated Communications Systems at the Ilmenau University of Technology, Germany [10].

It provides a set of tools that supports all stages of the complex control tasks solving in engineering, robotics and telecommunications. Today the equipment of this laboratory is installed in several universities of Ukraine, Georgia and Armenia. At Zaporizhzhia National Technical University at the Laboratory of Embedded Systems and Remote Engineering of Software Tools Department, GOLDi software and hardware have been installed for four experiments to study the principles of design and control of digital systems. In addition, the department has its own remote REIoT complex, which unites the Remote Laboratory for Embedded Systems Design (RELDES) [11] as well as the Remote Laboratory Smart House & Internet of Things [12, 13].

Unfortunately, Ukrainian universities have only recently begun to create and apply online labs for engineering education, as well as to interact with known educational resources created at universities all around the world. The teachers and students encountered with the problems of relevant online laboratories' search in accordance with the goals and objectives of training. Another problem concerns the search of information on ready-made hardware and software components for developing their own online laboratories based on reusable solutions.

Therefore, we have analyzed the possibilities of several web services that allow to solve these problems. The first one is the Go-Lab Ecosystem [8]. It is intended for teachers from primary and secondary schools and it is also oriented towards STEM education (Fig. 1).

Fig. 1. Go-Lab interface form

The system includes several subsystems or platforms. The Go-Lab Sharing platform allows to select an appropriative to the learning objectives R&VL based on the extensive labs database and provided filters. Developers of the service offer a sufficient number of filters for laboratories search:

- Sort (Most Viewed, Alphabetically, Newest, Updated).
- Subject Domains (Astronomy, Biology, Chemistry, Engineering, Environmental Education, Geography and Earth Science, Mathematics, Physics, Technology).

- Big Ideas of Science (Energy Transformation, Fundamental Forces, Our Universe, Structure of Matter, Microcosm, Evolution and Biodiversity, Organisms and Life Forms, Planet Earth).
- Lab Types (Remote Lab, Virtual Lab, Data Set).
- Age Range (before 7, 7–8, 9–10, 11–12, 13–14, 15–16, above 16).
- Languages (64 languages).

The system uses hard filtering method and there are filters for each criteria. However, the usage of proposed set of filters requests systematization and it gives sometimes unexpected results due to the fixing of the area of search.

The Go-Lab Authoring Platform allows teachers to create Inquiry Learning Spaces by combining online laboratories with applications-specialized software tools to support all stages of the educational process (Orientation, Conceptualization, Research, Conclusion and Discussion), as well as to communicate teachers and students. Also ready-to-use scenarios for different subject domains and languages are available on the platform.

Nevertheless, the system doesn't provide users with the information about reusable solutions for R&VL development.

The goal of the LabsLand project is also to streamline the information and help with the selection of the necessary online laboratory [9]. LabsLand developers offer two categories of remote labs—Electronics and Physics (Fig. 2). For Electronics category, you can choose such labs as Electronics, Robotics, AC Electronics. For the Physics category, you can select two subcategories. The first one is Electronics for physics (Electronics, Robotics, AC Electronics), the second one is Forces and Motion (Archimedes, Kinematics, Electronics, Robotics, AC Electronics). The additional types of online labs are Biology, Radioactivity, Optics, Pendulum. All the experiments are hosted by the laboratories of the University of Deusto (Bilbao, Spain), the University of Queensland (Brisbane, Australia), UFSC (Florianópolis, Brazil), BIFI (Zaragoza, Spain). The proposed amount of labs is limited and the filters for their search are absent.

Fig. 2. LabsLand interface form

However, the important advantages of this service are the useful information about Open Source and the proprietary tools (such as weblablib) for online laboratories' development, provided by developers of WebLab Deusto. The developers of the service emphasize that the creation of an online laboratory is a laborious process involving the solution of many tasks to create a user registration subsystem, a queuing subsystem, a laboratory administration subsystem, a security subsystem as well as the main subsystem for conducting remote and virtual experiments. Therefore, they propose Open Source Remote Laboratory Management System (RLMS) for creation the common layers of online laboratories, that gives multiple ways for R&VL development (in different programming languages, etc.).

Thus, we can conclude, that the basic requirements for the user-oriented web service that is being developed for selecting online laboratories are: an extensive knowledge base of online laboratories for various subject domains, the effective search system, the availability of information for both—users and laboratory developers, as well as an intuitive interface.

3   The Recommendation System Implementation

The research has shown that a lot of different recommendation methods exist [14]. Depending on the principle of operation with data, the recommendation approaches can be divided into three common groups: collaborative filtering, content filtering, and knowledge-based filtering. All recommendation methods have their features, as well as advantages and disadvantages. Knowledge-based (KB) technique uses the domain knowledge about how objects fit to users' preferences [15, 16]. The system can be based on KB technique if the information about users' behavior in the past is absent, in the case when the user uses the recommendation service not often, or when he needs recommendation on specific task every time, and when the facts about recommended objects are known in details. The RS for R&VL selection is a one-time system and as a rule there is no information about user's activity in the past. In this case the collaborative filtering and content filtering are not useful.

Thus, KB techniques can be used for the recommendation algorithm development and the core of the RS is a knowledge base. To recommend the laboratory, it is necessary to take into account a set of criteria, so in this case the task of multicriteria analysis is realized. To structure and classify the criteria, the hierarchy diagram was created and the fragment for main criteria is presented in Fig. 3. The additional criteria are: used equipment (standard (for example, Arduino, Raspberry Pi, etc.) or original), method of experiment control (remotely controlled or visualized experiment). The additional search of online lab also can be done by the lab name and home country.

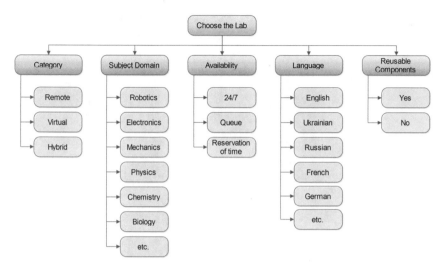

Fig. 3. Hierarchy diagram of the main criteria

To formalize knowledge about the R&VL, different knowledge representation models can be used [17–19]. One of them is the semantic networks, that gives the next advantages: great opportunities of network models; convenience and logical transparency; the set of relationship between concepts and events' forms is small and well-formalized [19]. After the analysis of expert opinions and R&VL specifications, the main attributes (concepts) that can be used for recommendations search were highlighted. As the result, the knowledge based model was proposed which represents as the semantic network scheme and it reflects the information about R&VL.

The RS is oriented to different types of users – students, lecturers and developers of new online labs. The activity diagram (Fig. 4) illustrates the activity of the registered in the RS users. When a user logs into the system, he needs to fill out his profile. The user can modify and supplement his profile.

The developed recommendation algorithm consists of the following steps.

Step 1. Formation of the set of criteria. At this step the user selects the user category to which he belongs. In accordance with the chosen category, a set of criteria oriented for defined user is formed.

The RS for choosing a laboratory includes two main categories of users:

- User (a student or a lecturer) looks for online laboratories for the purpose of performing experiments in the area of interest.
- Developer looks for online laboratories in the areas of interest for the purpose of own lab development as well as borrowing existing reusable solutions for development.

The set of main criteria for each user is shown in Fig. 5. The list of criteria for the developers is more extended than for students or lecturers.

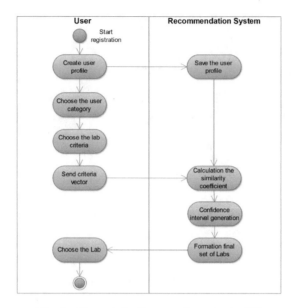

Fig. 4. The activity diagram of user in the RS

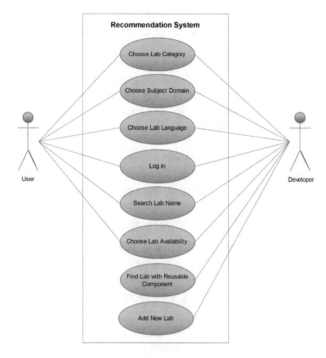

Fig. 5. UML use case diagram

Step 2. The formation of criteria vector for user/developer. The recommendation for laboratory selection is based on the user criteria filled in the RS web-page. It is possible to fill the criteria partly. The vector of criteria is formalized as (Eq. 1):

$$X = [f_1(x), \ldots, f_n(x)]^T, \tag{1}$$

where, $f_1, \ldots, f_n$—developer's criteria, $f: X \rightarrow D_f$, where D_f—the set of all possible criteria.

Step 3. The formation of the recommendation. The similarity measurement is used for the recommendation formation. Since the laboratory's criteria relate to non-numerical data, the set of existing numerical metrics and methods for comparing similarity cannot be used. There are a lot of coefficients to solve the task of comparison or finding the similarity of non-numerical data [20]. For our RS, the Dice coefficient [14] is used.

When the set of user's criteria X_1 is given, the similarity coefficient to each set of laboratory attributes X_2 from the knowledge base is calculated by the Eq. 2:

$$dice(X_1, X_2) = \frac{|criteria(X_1) \cap attributes(X_2)|}{|criteria(X_1)| + |attributes(X_2)|}, \tag{2}$$

The similarity coefficient compares two sets of words and takes a value from the interval [0, 1]. It shows how two sets of words are similar to each other: 1—if sets are totally identically, 0—if they don't.

Step 4. Confidence interval generation. The set of similarity values, that were found, leads to the task of working with a discrete random distribution. As a recommendation, the first 10 laboratories can be given.

The result of RS application is the list of recommended online labs. RS selects the nearest criterion of labs from the knowledge base according to the criteria entered by the user to the system.

The architecture of the RS is shown in Fig. 6. The user interacts with the RS through the web interface (Fig. 7). In general, RS includes several modules. The Web page formation Module and the Recommendation search Module interact directly with the Web interface and the Knowledge base. Besides the formation of the main web page with recommended online labs, the module of web page creation also generates additional pages with expanded information about R&VL. The Recommendation search Module performs the formation of recommendations for the selection of R&VL. The Dialog box formation Module forms a static area for work with user's criteria. Knowledge base is formed on the basis of online labs specifications, information from R&VL web sites, as well as expert conclusions.

As a development tool, the popular high-level Open Python framework Django was used. It has a lot of advantages: high level of security, internationalization, library for working with forms, imitation and construction of forms under the existing database model, managed to best satisfy the requirements of the established recommendation system. The knowledge base was implemented in a Django-compatible format. It gives the end-users the flexibility in selecting of the desired R&VL and allows to choose the necessary solution among the large number of existing online laboratories.

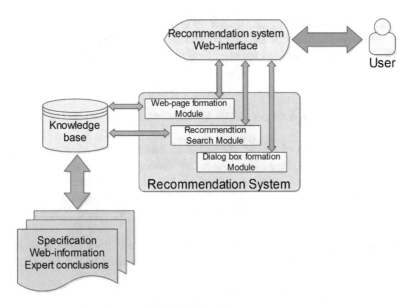

Fig. 6. RS architecture

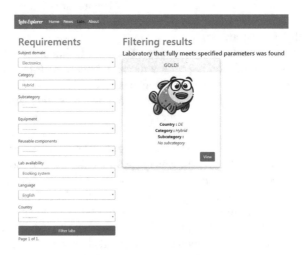

Fig. 7. RS web interface

4   Conclusion

The development of efficient RS is an actual scientific and practical task, which requires a comprehensive solution based on modern methods and tools.

The RS can be effectively used as a tool for selecting the necessary online lab by students and lecturers according to the set of their criteria. The RS can be useful for developers as it can give them the information about reusable software and hardware components for new labs development. The usage of RS for R&VL effective search will give a lot of possibilities for potential users and it will open the ways for R&VL prevalence.

In the future work another similarity coefficient can be used for recommendation algorithm, as well as weight coefficients to rank the users' criteria. The comparison of several coefficients will be done for the finding the best decision of the solving task.

Acknowledgements. The authors are grateful to Dr. Karsten Henke from TU Ilmenau, Germany for his help as an expert in the field of online laboratories during this research work.

References

1. The International Association of Online Engineering: The Global Online Laboratory Consortium. http://www.online-lab.org/
2. Parkhomenko, A., Gladkova, O., Sokolyanskii, A., Shepelenko V., Zalyubovskiy Y.: Reusable solutions for embedded systems design. In: The 13th International Conference on Remote Engineering and Virtual Instrumentation, Madrid, Spain, pp. 313–317 (2016)
3. Tawfik, M., Salzmann, C., Gillet, D., Lowe, D., Saliah-Hassane, H., Sancristobal, E., Castro M.: Laboratory as a service (LaaS): a novel paradigm for developing and implementing modular remote laboratories. iJOE 10(4), 13–21 (2014)
4. G.Zutin, D., Auer, M., Orduna, P., Kreiter, Ch.: Online lab infrastructure as a service: a new paradigm to simplify the development and deployment of online labs. In: The 13th International Conference on Remote Engineering and Virtual Instrumentation, Madrid, Spain, pp. 202–208 (2016)
5. Orduna, P., Rodriguez-Gil, L., Angulo, I., Dziabenko O.: Towards a microRLMS approach for shared development of remote laboratories. In: The 11th International Conference on Remote Engineering and Virtual Instrumentation, Porto, Portugal, pp. 375–381 (2014)
6. Niederstaetter, M., Klinger, Th., Zutin, D.: An image processing online laboratory within the iLab shared architecture. iJOE 6(2), 37–40 (2010)
7. Parkhomenko, A., Gladkova, O., Sokolyanskii, A., Shepelenko, V., Zalyubovskiy Y.: Implementation of reusable solutions for remote laboratory development. iJOE, 2(07), 24–29 (2016)
8. Go-Labz repository for online labs, apps and inquiry spaces. https://www.golabz.eu/
9. LabsLand network for remote experimentation. https://labsland.com
10. Henke, K., Vietzke, T., Hutschenreuter, R., Wuttke H.-D.: The remote lab cloud "goldi-labs. net". In: The 13th International Conference on Remote Engineering and Virtual Instrumentation, Madrid, Spain, pp. 31–36 (2016)
11. Parkhomenko, A., Gladkova, O., Ivanov, E., Sokolyanskii, A., Kurson, S.: Development and application of remote laboratory for embedded systems design. iJOE, 11(3), 27–31 (2015)
12. Parkhomenko, A., Tulenkov, A., Sokolyanskii, A., Zalyubovskiy, Y., Parkhomenko, A.: Integrated complex for IoT technologies study. In: Online Engineering & Internet of Things. Lecture Notes in Network and Systems, vol. 22(31), pp. 322–330 (2017)
13. Parkhomenko, A., Tulenkov, A., Sokolyanskii, A., Zalyubovskiy, Y., Parkhomenko, A. Stepanenko A.: The application of the remote lab for studying the issues of Smart House systems power efficiency, safety and cybersecurity. In: The 15th International Conference on Remote Engineering and Virtual Instrumentation, Dusseldorf, Germany, pp. 632–639 (2018)
14. Gomzin, A.G., Korshunov, A.V.: Recommendations systems: an overview of modern approaches. Proc. Inst. Syst. Program. Russ. Acad. Sci. 22, 401–418 (2012). (in Russian)

15. Jannach, D., Zanker, M., Felfernig, A., Friedrich, G.: Recommender Systems. An Introduction, 352p. Cambridge University Press, Cambridge (2011)
16. Tarus, J.K., Niu, Z., Mustafa, G.: Knowledge-based recommendation: a review of ontology-based recommender systems for e-learning. Artif. Intell. Rev. **50**(1), 21–48 (2018)
17. Dopico, J.R.R., de la Calle, J.D., Sierra, A. P.: Commonsense Knowledge Representation. Encyclopedia of Artificial Intelligence, pp. 327–333. Information Science Reference, Hershey (2009)
18. Brachman, R., Levesque, H.: Knowledge Representation and Reasoning, 381p. Morgan Kaufmann, San Francisco (2004)
19. Subbotin, S.O.: Representation and processing of knowledge in the artificial intelligence and decision support systems. In: ZNTU, 341p (2008) (in Ukrainian)
20. Choi, S.S., Cha, S.H., Tappert, C.C.: A survey of binary similarity and distance measures. J. Syst. Cybern. Inform. **8**(1), 43–48 (2010)

Widening the Number of Applications for Online and Pocket Labs by Providing Exercises for Measurement of DC Motor Characteristics

T. Klinger[1(✉)], C. Madritsch[1], C. Kreiter[1], A. Pester[1],
S. Baltayan[1], and V. Zaleskyi[2]

[1] Carinthia University of Applied Sciences, Villach, Austria
t.klinger@cuas.at
[2] National Technical University Kharkiv Polytechnic Institute,
Kharkiv, Ukraine

Abstract. Even though modern laboratory concepts like Online Labs and Pocket Labs are implemented in a number of engineering education programs, the available exercises are still limited regarding the use of standard com-ponents (in case of Pocket Labs) and standard experiments (Online Labs). The goal of the work described here is to increase the number of available exercises for the said modern laboratory concepts. In this case, we chose exercises that are measuring the characteristics of small DC motors, as they are easily available at a reasonable price and thus can be classified as — more or less — standard components. The project of designing these exercises focuses on the feasibility of creating appropriate hardware setups both for Pocket and Online Labs, and, in a next step, on providing a suitable software frame for students. The accompanying didactic material should not only give a step-by-step walkthrough, but also provide deeper understanding.

1 Introduction

Remote Labs are already existing for several years. In this concept, laboratory hardware is located at a place, where storage and maintenance of the experiments is possible, which mostly is a University or School campus. Students access the lab exercises via the Internet using a service broker; an example is the iLab Shared Architecture (ISA) [1], a Web services based distributed software framework to manage heterogeneous remote labs. The federation model of ISA allows for an easier sharing of remote labs across different institutions assuming the remote labs implement the ISA Web services API [2].

Pocket Labs became considerable when the prices of hardware reached such a low level that it was possible to provide each student with his or her own piece of laboratory equipment. This paper discusses the combination of both principles, which was first time carried out in Fall 2016 at CUAS [2–4].

© Springer Nature Switzerland AG 2019
M. E. Auer and T. Tsiatsos (Eds.): ICL 2018, AISC 917, pp. 611–619, 2019.
https://doi.org/10.1007/978-3-030-11935-5_58

2   Lab Concepts

Online or Remote Labs provide a solution for students to perform laboratory exercises at a self-chosen time and also from remote places, mostly at home. As personal interaction with the measurement and experimentation object is a critical issue for engineering students, especially Remote Labs contain real hardware. Access to this hardware is provided via web interfaces and very often also via video showing the exercise itself. Nevertheless, students are not in physical contact with the experiment.

As Pocket Labs provide actual and physical contact of students with real hardware, they can be used especially for basic exercises and for students without much experience. If electric or electronic components are used, a limitation will be for standard and cheap components, as any other solution will not be feasible due to financial reasons.

Figure 1 shows different lab infrastructure concepts, starting with the classic lab, where students have to come to the University campus and perform their exercises in a predefined time. Remote Labs have the advantage that exclusive and expensive hardware has to be installed and maintained only once and can be shared among students. Finally, Pocket Labs bring students again together with lab hardware, combined with the advantage of more or less free-chosen location and time [2].

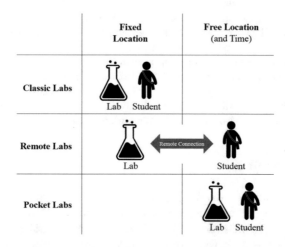

Fig. 1. Laboratory infrastructure concepts and methods [2]

3   Measuring DC Motor Characteristics

Conducting a test of a DC-motor may bring a significant practical benefit to the education in the fields of mechatronics, automation and electrical engineering. Moreover, implementing a lab experiment with already developed online lab technologies, such as a Remote Lab Management System (RLMS), enables students to interact with the real hardware using a modern Web browser and practically without any requirements, like their physical presence in a classical lab. Nevertheless, students obtain important practical knowledge, about the real behavior and performance of the tested equipment.

In order to evaluate an electrical motor, certain parameters are needed, such as energy consumption, torque and speed in idle and load state of the system, oscillations (vibrations), temperature drifting and noise exposure. Furthermore, such a DC-motor test bench has to monitor additional parameters regarding the functional durability of the tested hardware, which has to be recorded and analyzed [5].

The main purpose of the engine test bench is to check the performance of the engine. If basic values such as torque and driving speed can be checked, then it is a power testing plant. For a complete engine test values, such as energy consumption, system load, oscillation/vibration, temperature drifting and noise exposure are needed. Additionally, an engine test bench should provide an exact insight into the functionality and durability of an engine [5].

4   The Remote Lab

4.1   Experiment Setup

The system consists of the below-described hardware components, as shown in Fig. 2. The experiment setup consists of the following hardware:

- A programmable power supply unit with serial link;
- Raspberry Pi 3 Model B; single-board computer;
- Arduino Uno; single-board microcontroller;
- 2 brushless DC-motors; ESCAP 28D2R with encoder and a PCB to improve the encoder signal;
- 2 PT1000 temperature sensors;
- A webcam.

The core of the online lab are two coupled brushless DC-motors ESCAP 28D2R, which can perform in a drive and load mode. Attached to the motors are encoders which generate 144 impulses per rotation (2 pulses dislocated at 90°). The signal of this encoder is improved by a self-made PCB, which additionally detects the direction of the rotation and provides a digital signal of it. Additionally, a PT1000 temperature sensor is mounted on each motor to avoid malfunction due to long-lasting high load.

The Raspberry Pi has the task to communicate with all the other system units, as well as tracking, recording and analyzing all data from the tested electrical motors and controlling them in accordance with the received controlling commands. On top of it is an add-on board that provides a serial communication to the programmable DC power supply.

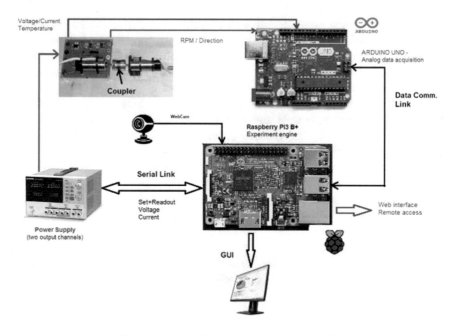

Fig. 2. Remote lab experimental setup [5]

The Arduino Uno is needed to complement the Raspberry Pi. First, the pulses of the DC motor encoder and the direction of rotation are counted. On the other hand, it reads the analog signals of the temperature sensors and the voltage and current measured values of the two DC motors. All this information is continuously sent to the Raspberry Pi via a serial-over-USB connection. The USB port also serves as power supply and for programming the Arduino.

4.2   Management System Setup

The online lab utilizes software frameworks, which were created to ease the development of online laboratories. The first is the iLab Shared Architecture (ISA) [1] which is a Remote Lab Management System (RLMS) and second is the Experiment Dispatcher [6] which implements the API of ISA by abstracting the setup of a lab server, allowing the lab owner to deploy several instances of virtual lab servers. This approach is defined as Laboratory Infrastructure as a Service (LIaaS) [7].

Figure 3 demonstrates the components of an online lab and the connection between them. A lab developer has to create a web client, as well as a so-called Experiment Engine (or Subscriber Engine), which serves as the link between the Experiment Dispatcher and the hardware-controlling software or simulation.

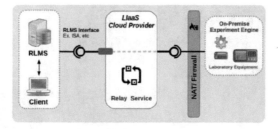

Fig. 3. LIaaS relay service [5]

4.3   User Interface

The user interface front panel mock-up is based on the design shown on Fig. 4. It
has a number of features to create a rich and interactive experience in order to
fulfill the requirements of a complete DC-motor test bench and to minimize drawbacks, which are naturally caused through the distance between experimenter
and lab. Via the web client interface the user can get to know the theoretical
background of the PMDC machines and their important parameters and properties. In addition, the user can put this theory into practice through the extensive
experiments supplied and gain valuable experience through visual observation
of the experiment itself and a synchronized data result.

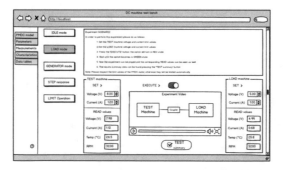

Fig. 4. Web user interface [5]

5   The Pocket Lab

5.1   Experiment Setup

The system consists of the following components, as shown in Fig. 5. The experiment setup consists of the following hardware:

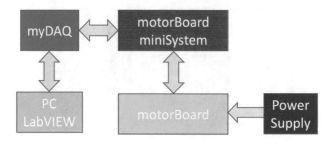

Fig. 5. Pocket lab overview

- NI myDAQ connected to a PC;
- motorBoard miniSystem;
- Motor Board (encoder, motor, and load generator);
- Power supply.

The main elements of the pocket lab are the motorBoard miniSystem as well as the Motor Board itself. The motorBoard miniSystem consists of a LMD18200 H-bridge, designed for motion control applications, several LEDs and connectors. Figure 6 shows the schematic diagram of the miniSystem and Fig. 7 the developed miniSystem in combination with myDAQ.

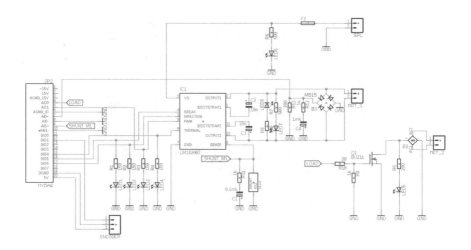

Fig. 6. motorBoard miniSystem schematic

The Motor Board (Fig. 8) consists of the motor under test, an incremental encoder as well as a second motor used as an adjustable load. The Motor Board is connected to the motorBoard miniSystem as well as to the Power Supply.

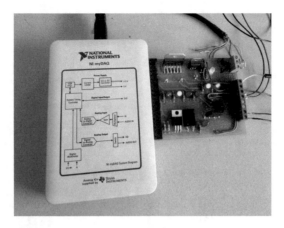

Fig. 7. NI myDAQ with motorBoard miniSystem

Fig. 8. Motor board

5.2   LabVIEW Software

The software, controlling the Pocket Lab, was developed using LabVIEW. The main components can be seen in Fig. 9. A producer/consumer design pattern was used as the basic architecture. Several parallel loops are controlling the different aspects of the experiment.

5.3   User Interface

The current provisional user interface (Fig. 10) allows the user to control all aspects of the Pocket Lab experiment. First, the Sample Clock Rate, Frequency, Offset and Amplitude are adjusted. The Duty-cycle setting controls the motors RPM. Several indicators, like the output waveform, motor voltage, motor current and motor RPM give feedback to the user about the progress of the experiment.

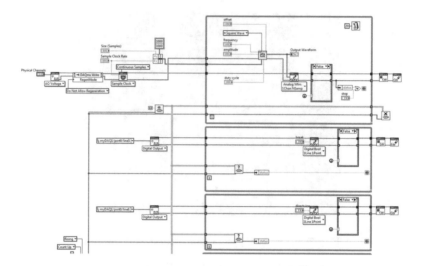

Fig. 9. LabVIEW block diagram

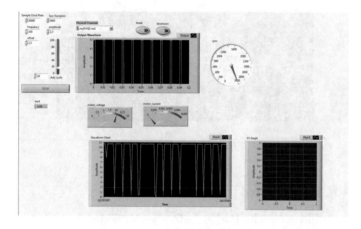

Fig. 10. LabVIEW front panel

6   Conclusion

It can be seen from the previous chapters, that the topic of measuring DC motor characteristics is both suitable for Remote and Pocket Labs. However, the measurement possibilities with Pocket Labs are somewhat limited due to the capabilities of the usually cost-efficient hardware.

It is therefore advisable to realize more sophisticated exercises, like load balances, four-quadrant operation, and others, with a Remote Lab. In this case, costs are not the big issue as with Pocket Labs, because the lab equipment does not have to be provided for every student.

References

1. Harward, V.J., et al.: The iLab shared architecture: a web services infrastructure to build communities of internet accessible laboratories. In: Proceedings of the IEEE 2008, pp. 931–950 (2008). https://doi.org/10.1109/JPROC.2008.921607
2. Klinger, T., et al.: Parallel use of remote labs and pocket labs. In: 14th International Conference on Remote Engineering and Virtual Instrumentation (REV2017), New York City, NY (2017)
3. Klinger, T., Madritsch, C.: Use of virtual and pocket labs in education. In: 13th International Conference on Remote Engineering and Virtual Instrumentation (REV2016), Madrid, Spain (2016)
4. Klinger, T., Madritsch, C.: Collaborative learning using pocket labs. In: International Conference on Interactive Mobile Communication, Technologies and Learning (IMCL2015), Thessaloniki, Greece (2015)
5. Baltayan, S., et al.: An online DC-motor test bench for engineering education. In: Global Engineering Education Conference (EDUCON2018), Santa Cruz de Tenerife, Spain (2018)
6. Zutin, D.G., Kreiter, C.: A software architecture to support an ubiquitous delivery of online laboratories. In: International Conference on Interactive Mobile Communication, Technologies and Learning (IMCL), San Diego, CA (2016)
7. Zutin, D.G., et al.: Online lab infrastructure as a service: A new paradigm to simplify the development and deployment of online labs. In: 13th International Conference on Remote Engineering and Virtual Instrumentation (REV2016), Madrid, Spain (2016)

The Timeless Controversy Between Virtual and Real Laboratories in Science Education—"And the Winner Is…"

Charilaos Tsihouridis[1](✉), Dennis Vavougios[1], Marianthi Batsila[2], and George S. Ioannidis[3]

[1] University of Thessaly, Volos, Greece
{hatsihour, dvavou}@uth.gr
[2] Directorate of Secondary Education of Thessaly,
Ministry of Education, Thessaly, Greece
marbatsila@gmail.com
[3] University of Patras, Patras, Greece
gsioanni@upatras.gr

Abstract. The present research investigates the virtual and real laboratories in science education. Adopting one method over the other is a matter of interest but also too often a matter of controversy regarding their effectiveness in the educational practice. With this being kept in mind, a number of 106 research papers (articles, doctoral dissertations, and reviews) related to the comparison of real and virtual laboratories across science teaching research were identified and reviewed. The research questions focused on the most effective environment, whether findings vary over time and level of education in which the labs are used. The basic search tools were the Internet, e-libraries, e-journals, databases, thematic guides and Portals, Catalogues of other libraries, offered to the researchers from collaborating universities. Key words and critical analysis were employed for the analysis and interesting conclusions are made as regards the effectiveness of the two laboratories.

Keywords: Virtual · Real · Laboratory · Science · Controversy

1 Introduction

Natural sciences are considered an important part of human culture, with their most significant roles being their ability to enhance analytical and cooperative thinking and the provision of explanations for the behavior of the material and natural world and their properties. The importance of the role of natural sciences in the 21st century is particularly highlighted by the European Commission for Science Education [1], which states that the well-being of a nation depends not only on how well students are educated in general, but also on how well this is done in the study of natural sciences. The importance of natural sciences is reflected both on the daily routine of the world and its need to have a better understanding of the phenomena surrounding it as well as on the field of education, as being the dominant and essential pillar of knowledge. Experimental teaching is considered to be the most basic means of achieving the above

© Springer Nature Switzerland AG 2019
M. E. Auer and T. Tsiatsos (Eds.): ICL 2018, AISC 917, pp. 620–631, 2019.
https://doi.org/10.1007/978-3-030-11935-5_59

goals as it is known to promote scientific thinking, inquiry based learning and provision of documented answers for the acquisition of scientific knowledge [2]. Research indicates that experiments display a significant pedagogical value, and it has been found that, compared to traditional lecturing lessons, experiments are able to enhance the transmission of knowledge and the understanding of difficult and especially abstract concepts, phenomena and laws of natural sciences [3]. Educational benefits discussed do not seem to have differentiated since they were first recorded: provision of exploratory, organizational and communication skills, ability to process concepts—creation of hypotheses—use of scientific models, promoting cognitive skills, critical thinking, problem solving, application, analysis and synthesis of ideas, understanding of the nature of science, knowledge and understanding of the scientific method, knowledge of the relationship between science and technology, motivation and interest enhancement, initiative taking, objectivity of thinking, accuracy of actions, persistence, assumption of responsibilities, consensus, collaboration, communication, interaction [4] (Travers, 1973 as reported in Blosser, 1990).

Experiments demonstrate a significant pedagogical value in science teaching because they involve students actively in the educational process as opposed to traditional teacher-centred lecturing approaches [5]. It is believed that they promote learning, contribute to the detection of learners' alternative perceptions and examine the scientific appropriateness of their ideas. They develop life skills such as persistence, patience, responsibility, and equipment operation skills. They enhance creativity, objective thinking, and scientific reasoning [6]. It is also claimed that even simple demonstration experiments may have positive effects on students' thinking provided they are properly designed and followed by relevant questions and activities [7]. Research claims that virtual laboratories display advantages such as attractive environment, simulation of concepts and phenomena studied in less time than the time needed for real experiments, minimizing the need for costly or specialized apparatuses, and offering an easy and with minimum risk frequent repetition of the experiments [8]. It is also argued that virtual workshops contribute to building knowledge through a pleasant, interactive environment that encourages students to participate in the experimental process in order to develop a scientific approach towards knowledge [9]. On the other hand, real experiments foster learners' acquaintance with scientific methodology and help them become familiar with the use of real objects or instruments of everyday life [10]. They increase their motivation for learning and lead to positive thinking towards science education. Real experiments offer learners the possibility to experience phenomena as these evolve in real-life authentic situations, to become aware of the experimental errors, and familiarize themselves with the use of real objects, equipment and their operation [11].

2 Purpose of the Research

The importance of laboratory practice in science studies at all levels of education is acknowledged by the educational research community, while the pedagogical value of the experiments in science teaching has been acknowledged for years. An attempt to detect what type of lab (real or virtual) has better learning outcomes and/or is preferred to be used in science teaching has been reflected in research around the world. Adopting one over another method has been a matter of controversy as, throughout the years, a great interest has been shown by researchers in the world regarding the effectiveness of real or virtual laboratories in different age groups, with sometimes or vague conclusions. Focusing on this issue and on the need to detect the effectiveness of laboratory work, the authors of this study decided to investigate this controversy by reviewing comparative studies between virtual and real laboratory experiments.

3   Methodology

Our main research interest was to investigate the "long-standing" controversy regarding the effectiveness of the learning outcomes of laboratory practice in real and virtually-simulated environments by reviewing international literature. In this context, scientific papers (articles, doctoral dissertations, reviews) related to the comparison of real and virtual laboratories across science teaching research were identified and reviewed. The basic search tools were the Internet (e.g. Google Scholar Search Engine), e-libraries, e-journals, databases [EBSCO host, SCOPUS, Web of Science (Citation Indexes)], e-books (HEAL- Link, EBSCO), thematic guides and Portals, Catalogues of other libraries, offered to the researchers from collaborating universities.

The research, which lasted for quite a few months and employed key words and critical analysis, finally resulted in a number of 106 international research papers [2-107] on comparative studies between real and virtual workshops focusing on the most effective laboratory environment for the teaching and learning process. It should be noted that a particular effort has been made to ensure that the papers meet the criteria of comparison between groups with the same characteristics, the same methodological teaching approach and the same duration of the teaching interventions.

Thus, the present research includes a two-targeted meta-analysis of 106 studies: encoding the findings of each survey and classifying them to be grouped together and be interpreted for comparisons between them. Our research questions focused on the following: (a) Which laboratory environment is more effective for students in the educational process, real or virtual-simulations? (b) Do research findings vary over time? (c) Do research findings vary according to the level of education in which they are used? For the purposes of the aforementioned research questions, the authors of this paper analyzed the results of 106 papers carried out from the year 1978 to April 2018. In the research we will focus on three key classifications:

1. Classification relating to the time of the research, keeping into consideration the rapid development of educational technology, especially in the last decade, during which an extensive number of virtual educational environments and simulations have appeared. These environments tend to simulate the real environment with particular respect to three main categories: (i) before 2006 (ii) between 2006 and 2012; and (iii) after 2012. More specifically, out of the total of 106 papers found and analyzed 29 (27.4%) were carried out before 2006, 36 papers were implemented between 2006 and 2012 (34.0%) and finally 41 papers (38.7%) were implemented from 2012 onwards (Table 1):

Table 1. Classification of papers relating to the time of the research implementation

Year of study		
	Frequency	Valid percent
<2006	29	27.36
≥ 2006 & <2012	36	33.96
≥ 2012	41	38.68
Total	106	100.0

2. Classification of surveys by level of education of the target group that participated in the surveys. In particular, 14 studies (13.2%) concerned primary education, 29 papers (27.4%) were about the secondary education, 47 studies (44.3%) dealt with higher education and 16 studies (15.1%) aimed at secondary education with emphasis on teachers or other educational parties. It can be seen that the target groups are mainly secondary and tertiary education groups that actually constitute the core of the educational process. We consider this categorization and the number of studies per target group to be sufficient enough to give us possible answers. Meanwhile it allows us to use statistical methods to draw useful conclusions on the issue of "conflict" between real and virtual laboratory environments at all levels of primary, secondary, tertiary education over the last 20 years (Table 2).

Table 2. Classification of studies per target groups

Target group		
	Frequency	Valid percent
Primary Education	14	13.2
Secondary Education	29	27.4
Higher Education	47	44.3
Other	16	15.1
Total	106	100.0

3. Classification of the studies relating to the most effective educational approach in terms of the best learning outcomes (i) the real (ii) the virtual (iii) similar results (Fig. 1).

It can be observed that the majority of researchers have resulted in 51.9% with similar findings after comparing virtual to real environments in their studies, while equally important is the percentage regarding virtual labs (32.1%) as the most effective learning tool, followed by real laboratories (16%). Right below, there will be an attempt to investigate whether this differentiation is statistically significant and whether it is affected by the time of the survey and the level of education drawn by the surveys.

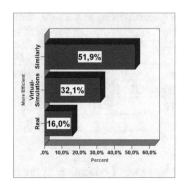

Fig. 1. Results of the experimental practices of a real and virtual environment as regards the effectiveness of learning outcomes

3.1 Investigating the Frequencies of the Categories of the Factor "Most Effective Experimental Environment"

The criterion that can be used to interpret the frequency of categories derived from a qualitative variable is Chi-square (Criterion Chi-square as a "goodness-of-fit" test). In this case, we formulate our hypothesis:

Alternative hypothesis: The frequencies of the three levels of the Variable "Most effective experimental environment" (Real, Virtual-Simulations, Similar) are different between them. The result with respect to the value of χ^2 derived from the statistical package is: $\chi^2(2) = 20.509$, $p < 0.001$. The result of χ^2 is statistically significant and the alternative hypothesis is accepted. Consequently, the researcher's findings are sufficient to support (not to prove) the assumption that there is a level of the variable "Most effective learning environment" that is more frequent than the other levels of the variable and this is "Similar" results.

3.2 Effect of the Factor "Time of Publication of the Research" on the Factor "Experimental Approach Effectiveness"

Based on the papers meta-analysis and the grouping—classification of the research results the two above variables have qualitative characteristics (categorical measurement scale) and the appropriate criterion of independence control is χ^2. Before proceeding with calculating the χ^2, let us formulate the hypothesis:

Alternative hypothesis: The two variables are dependent (related to one another), i.e. the effectiveness of the two different experimental approaches is influenced by the time of the survey (will be different). The table that follows (Table 3) presents the frequencies of the dual input table depicting the numbers of the most effective experimental environment per time period of the studies.

The results as derived from the SPSS statistical package are given on the following table (Table 4).

Table 3. The crosscheck table of the variables "Time" and "Most effective experimental approach"

		More efficient		
		Real	Virtual-simulations	Similarly
Year of study	<2006	7	10	12
	≥ 2006 & <2012	4	11	21
	≥ 2012	6	13	22

Table 4. Effect of the factor "Year of Study" on the factor "More Efficient"

			More efficient		
			Real	Virtual-simulations	Similarly
Year of study	<2006	% within Year of Study	**24.1**	**34.5**	**41.4**
		% within more efficient	**41.2**	**29.4**	**21.8**
		% of total	6.6	9.4	11.3
	≥ 2006 & <2012	% within year of study	**11.1**	**30.6**	**58.3**
		% within more efficient	**23.5**	**32.4**	**38.2**
		% of total	3.8	10.4	19.8
	≥ 2012	% within year of study	**14.6**	**31.7**	**53.7**
		% within more efficient	**35.3**	**38.2**	**40.0**
		% of total	5.7	12.3	20.8

The result relating to the value χ^2 is: $\chi^2(4) = 5.314, p = 0.256$. The above result of χ^2 is not statistically significant, therefore the alternative hypothesis is rejected and the zero is accepted. In conclusion, the findings are not sufficient to convince that there is a relationship between the variables "Time of the research implementation" and "Experimental environment effectiveness" (the variables are independent between them).

3.3 Effect of the Factor "Research Educational Level" (Target Group) on the Factor "Experimental Approach Efficiency"

Based on the meta-analysis of the papers and the grouping-classification of the research results the two above variables have qualitative characteristics (categorical measurement scale) and the appropriate criterion to control their independence is χ^2. Before proceeding with the calculation of χ^2, let us formulate the hypothesis: Alternative hypothesis: The two variables are dependent (interrelated), i.e. the effectiveness of the two different experimental approaches are influenced by the educational level of the research implementation (will be different). The table that follows (Table 5) presents the frequencies of the dual input table depicting the numbers of the most effective experimental environment per educational level of the target groups of the surveys.

Table 5. The crosscheck table of the variables "Educational level of the research groups" and "Most effective experimental approach"

		More efficient		
		Real	Virtual-simulations	Similarly
Target group	Primary education	5	2	7
	Secondary education	2	9	18
	Higher education	7	21	19
	Other (teachers, special groups et)	3	2	11

The results as derived from the SPSS statistical package are given on the following table (Table 6).

Table 6. Effect of the factor "Educational level" on the factor "Most Efficient"

			More efficient		
			Real	Virtual-simulations	Similarly
Year of study	Primary education	% within year of study	35.7	14.3	50.0
		% within more efficient	**29.4**	**5.9**	12.7
		% of total	4.7	1.9	6.6
	Secondary education	% within year of study	6.9	31.0	62.1
		% within more efficient	**11.8**	**26.5**	32.7
		% of total	1.9	8.5	17.0
	Higher education	% within year of study	14.9	44.7	40.4
		% within more efficient	**41.2**	**61.8**	34.5
		% of total	6.6	19.8	17.9
	Other (teachers, special groups et)	% within year of study	18.8	12.5	68.8
		% within more efficient	**17.6**	**5.9**	20.0
		% of total	2.8	1.9	10.4

The result relating to the value χ^2 is: $\chi^2(4) = 13.289$, $p = 0.039$. The result of χ^2 is statistically significant and the alternative hypothesis is accepted. In conclusion, the findings are sufficient to convince that there is a relationship between the variables "Educational level of the research groups" and "Experimental environment effectiveness" (the variables are dependent on each other). The graphs below (Fig. 2) present the efficiency rates of the experimental environments per educational level, as these result from the table:

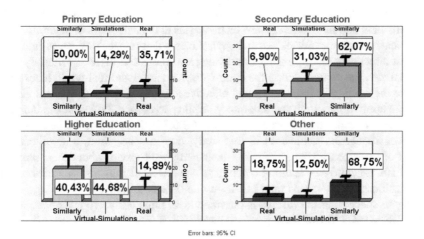

Error bars: 95% CI

Fig. 2. Graphical representation of the dual port table for the factor "educational level" and "most efficient"

From the above figures, it can be seen that real labs display better learning outcomes with primary school children at 35.7%, followed by teachers and other groups of trainers (18.8%), higher education students (14.9%) and finally, students of secondary education at 6.9%. Virtual workshops—simulations display better learning outcomes with tertiary level students (44.7%), followed by secondary level (31%), primary (14.3%) and teachers at 12.5%. Finally, it can be seen that the majority of the two educational environments display similar learning outcomes with groups of teachers being at 68.8%, secondary school students at 62.1%, primary level pupils at 50%, and tertiary level learners at 40.4%.

4 Conclusions

This study referred to the grouping of research results and to the meta-analysis of papers relating to the comparison of two experimental environments (real and virtual) and as regards their effectiveness in the learning process. Based on the results we wish to highlight the fact that constituted the primary purpose and concern of this research: there is no final winner in this timeless controversy between the two experimental lab environments according to our research. The recording and classification of the research results of 106 surveys conducted over the last 40 years by majority present similar learning outcomes. If we attempt to emphasize specific features resulting from the aforementioned grouping it can be argued (not proven) that the real labs still fascinate young learners over time whereas their charm gradually fades in secondary education giving way to virtual experiments and simulations. The virtual experiments and especially the specialized simulations impress and stimulate the interest of students resulting to better learning outcomes mainly in higher education. In conclusion, we can observe an increasing "trend" of learning outcomes with virtual environments, from primary to higher education. On one hand, this trend derives, from the gradual familiarization and integration of new technologies in the daily lives of students. On the other hand, this trend "creates" virtual environments that replace difficult to implement and cost-effectively environments simulating them more and more as faithfully as can be. Finally, while contemplating the controversy of virtual and real-world laboratories, we note the decline in the interest and effectiveness of real laboratories from 24.1 to 14%, the approximately virtual laboratory stability to around 32%, and the increasing trend towards similar results from 41 to about 58%. Finally, we would like to point out that, even though not addressed in this study, due to space limit, a significant proportion of researchers who have asked their participant groups about their preferable type of lab, regardless of learning outcomes, have revealed students' preference on the combination of the two labs rather than on one or another only.

Due to space limit the full list is provided here: https://drive.google.com/open?id= 1n8kCDEFdTkU758wm5JBerLfcN5uKskwW.

References

1. Abreu, P., Barbosa, R., Lopes, A.M.: Experiments with a virtual lab for industrial robots programming. iJoe **11**(5), 10–16 (2015)
2. Addair, E.: Ceci n' est pas une Mouse-Trap Car: Physical Versus Virtual Materials in the Classroom. Bachelor of Science in Psychology, Carnegie Mello University Pittsburgh, Pennsylvania, U.S. (2012)
3. Ajredini, F., Izairi, N., Zajkov, O.: Real Experiments versus Phet simulations for better high-school students' understanding of electrostatic charging, EJPE. Eur. J. Phys. Educ. **5**(1), 59–70 (2013)
4. Akai, C.: Depth Perception in Real and Virtual Environments: An Exploration of Individual Differences. Master of Science in the School of Interactive Arts & Technology, Simon Fraser University, Canada (2007)
5. Alqahtani, A.S., Daghestani, L.F., Ibrahim, L.F.: Environments and system types of virtual reality technology in STEM: a survey. (IJACSA) Int. J. Adv. Comput. Sci. Appl. **8**(6), 77–89 (2017)
6. Amgen Foundation and Change the Equation Students on STEM: More Hands-on, Real World Experiences. AMGEN Foundation, Inspiring the Scientists of Tomorrow, California, U.S.A. (2015)
7. Baser, M.: Effects of conceptual change and traditional confirmatory simulations on pre-service teachers' understanding of direct current circuits. J. Sci. Educ. Technol. **15**(5), 367–381 (2006)

8. Baxter G.P.: Using computer simulations to assess hands-on science learning. J. Sci. Educ. Technol. **4**, 21–27 (1995)
9. Booth, C., Cheluvappa, R., Bellinson, Z., Maguire, D., Zimitat, C., Abraham, J., Eri, R.: Empirical evaluation of a virtual laboratory approach to teach lactate dehydrogenase enzyme kinetics. Ann. Med. Surg. **8**, 6–13 (2016)
10. Choi, C.A., Gennaro, E.: The effectiveness of using computer simulated experiments on junior high students' understanding of the volume displacement concept. J. Res. Science Teach. **24**(6), 539–552 (1987)
11. Corter, J.E., Nickerson, J.V., Esche, S., Chassapis, C., et al.: Remote versus hands-on labs: a comparative study. In: 34th ASEE/IEEE Frontiers in Education Conference, pp. 17–21, Savannah, GA (2004)
12. Edward, N.S.: Evaluation of computer based laboratory simulation. Comput. Educ. **26**(1–3), 123–130 (1996)
13. Evaggelou, F.V., Kotsis, KTh: Comparative study of the impact of real and virtual experiments on learning for the phenomenon of water boiling to 5th and 6th grades elementary learners. Sci. Technol. Issues Educ. **7**(1–2), 5–24 (2014)
14. Evaggelou, F.V.: The Impact of Real and Virtual Physics Labs in Learning. PhD Thesis, University of Ioannina, School of Educational Studies, Pedagogical Department of Primary Education, Vo. A., Ioannina (2012)
15. Faour, M.A., Ayoubi, Z.: The effect of using virtual laboratory on grade 10 students' conceptual understanding and their attitudes towards physics. J. Educ. Sci. Environ. Health (JESEH) **4**(1), 54–68 (2018)
16. Fielder, M., Haruvy, E.: The lab versus the virtual lab and virtual field—an experimental investigation of trust games with communication. J. Econ. Behav. Organ. **72**(1), 716–724 (2009)
17. Finkelstein, N.D., Perkins, K.K., Adams, W., Kohl, P., Podolefsky, N.: Can computer simulations replace real equipment in undergraduate laboratories? In: Physics Education Research Conference, Sacramento, CA, U.S.A. (2004)
18. Gandole, Y.B., Khandewale, S.S., Mishra, R.A.: A comparison of students' attitudes between computer software support and traditional laboratory practical learning environments in undergraduate electronics science. e-J. Instr. Sci. Technol. **9**(1), 1–13 (2006)
19. Gibbins, L., Perkin, G.: Laboratories for the 21st Century in STEM Higher Education. A Compendium of Current UK Practice and an Insight into Future Directions for Laboratory-based Teaching And Learning. The Centre for Engineering and Design Education: Loughborough University, U.K. (2013)
20. Herga, N.R., Grmek, M.I., Dinevski, D.: Virtual laboratory element of visualization when teaching chemical contents in science class. Turk. Online J. Educ. Technol. (TOJET) **13**(4), 157–165 (2014)
21. Jaakkola, T., Nurmi, S.: Academic impact of learning objects: the case of electric circuits. In: British Educational Research Association Annual Conference (BERA), University of Manchester (2004)
22. Jeschke, S., Richter, T., Scheel, H., Thomsen, C.: On remote and virtual experiments in eLearning. J. Softw. **2**(6), 76–85 (2007)
23. Keller, C.J., Finkelstein, N.D., Perkins, K.K., Pollock, S.J.: Assessing the effectiveness of a computer simulation in conjunction with tutorials in introductory physics in undergraduate physics recitations. In: Proceedings of the 2005 Physics Education Research Conference, AIP Press, Melville, N.Y. (2005)
24. Keller, C.: Substituting Traditional Hands-On Laboratories with Computer Simulations: What's Gained and What's Lost? Department of Physics, University of Colorado at Boulder (2004)

25. Keller, H.E., Keller, E.E.: Making real of virtual labs. Sci. Educ. Rev. **4**(1), 2–11 (2005)
26. Kennepohl, D.: Using computer simulations to supplement teaching lab in chemistry for distance delivery. J. Distance Educ. **16**(2), 58–65 (2001)
27. Klahr, D.: Hands on what? The relative effectiveness of physical vs virtual materials in an engineering design project by middle school children. J. Res. Sci. Teach. **44**(1), 183–203 (2007)
28. Konstandinou, N., Zacharia, Z.H.: The impact of real and virtual learning environments on 1st grade primary school learners to achieve conceptual change about temperature and heat. In: Cyprus 10th Conference of Pedagogics-Quality in Education: Research and Teaching, Nicosia, pp. 294–316, 6–7 June 2008
29. Kotsis, K.: Primary school learners' attitudes about the experiment at science teaching. In: Papageorgiou, G., Kountouriotis (eds.) Proceedings of the 7th Panhellenic Conference of Teaching Science and ICT in Education-Interactions of Educational Research and Actions in Sciences, vol. A., Alexandropolis, pp. 238–247, 15–17 April 2011
30. Kotsis, KTh, Evaggelou, F.V.: Learning outcomes after using real and virtual physics experiments with 5th and 6th grades of primary school children about the concepts of electric circuits. Issues Sci. ICT Educ. **3**(3), 141–158 (2010)
31. Kotsis, K.T., Evaggelou, F.V.: Virtual or Real Experiment in Teaching Physics for Changing Students' Alternative Ideas? Literature Review. Scientific Workshop of the Pedagogical Department of Ioannina University, (20), pp. 57–90 (2007)
32. Mutlu, A., Şeşen, B.A.: Impact of virtual chemistry laboratory instruction on pre-service science teachers' scientific skills. SHS Web of Conferences, 26 (01088) 2016, ERPA International Congresses on Education 2015 (ERPA 2015) (2016)
33. Nancheva, N., Stoyanov, S.: Simulations laboratory in physics distance education. In: Proceedings of the 10th Workshop on Multimedia in Physics Teaching and Learning (EPS—MPTL 10), pp. 1–7. Berlin (2005)
34. Nguen H.D.K., Le Cong N.: Distinguishing advantages of virtual experiments on solving multiple choice questions in physics test. In: Proceedings of the 62nd ISERD International Conference, pp. 25–27. Boston, USA, 14th–15th Jan 2017
35. Nickerson, J.V., Corter, J.E., Esche, S.K., Chassapis, C.: A model for evaluating the effectiveness of remote engineering laboratories and simulations in education. Comput. Educ. (49), 708–725 (2007)
36. Olimpiou, G., Zacharia, Z.: A comparative study for the effectiveness of the experimentation in a real or virtual lab to achieve conceptual understanding in physics. In: Kariotoglou, P., Spirtou, A., Zoupidis, A. (eds.) Proceedings of the 6th Panhellenic Conference on Science and ICT in Education-Multiple Approaches to Science Teaching and Learning, Florina, pp. 621–629, 7–10 May 2009
37. Oral, I., Bozkurt, E., Guzel, H.: The effect of combining real experimentation with virtual experimentation on students' success. World Acad. Sci. Eng. Technol. **30**, 1599–1604 (2009)
38. Paxinou, E., Zafeiropoulos, V., Sypsas, A., Kiourt, C., Kalles, D.: Assessing the impact of virtualizing physical labs. Comput. Sci. Hum. Comput. Interact. 1–5 (2017)
39. Pyatt, K., Sims, R.: Learner performance and attitudes in traditional versus simulated laboratory experiences. In: Proceedings Ascille Singapore, pp. 870–879 (2007)
40. Ranjan, A.: Effect of virtual laboratory on development of concepts and skills in physics. Int. J. Tech. Res. Sci. (IJTRS) **2**(1), 15–21 (2017)
41. Ronen, M., Eliahu, M.: Simulation as a home learning environment—students' views. J. Comput. Assist. Learn. **15**, 258–268 (1999)
42. Sari Ay, O., Yilmaz, S.: Effects of virtual experiments oriented science instruction on students' achievement and attitude. Elementary Educ. Online **14**(2), 609–620 (2015)

43. Sauter, M., Uttal, D.H., Rapp, D.N., Downing, M., Jona, K.: Getting real: the authenticity of remote labs and simulations for science learning. Distance Educ. **34**(1), 37–47 (2013)
44. Scanlon, E., Colwell, C., Cooper, M., Di Paolo, T.: Remote experiments, re-versioning and re-thinking science learning. Comput. Educ. **43**, 153–163 (2004)
45. Sonnenwald, D.H., Whitton, M.C., Maglaughlin, K.L.: Evaluating a scientific collaboratory: results of a controlled experiment. ACM Trans. Comput. Hum. Interact. **10**(2), 150–176 (2003)
46. Srinivasan, S., Perez, L.C., Palmer, R.D., Brooks, D.W., Wilson, K., Fowler, D.: Reality versus simulation. J. Sci. Educ. Technol. **15**(2), 137–141 (2006)
47. Taramopoulos, A., Psillos, D., Hatzikrianiotis, E.: Teaching electric circuits by guided inquiry in virtual and real laboratory environments. In: Jimoyiannis, A. (ed.) Research on e-Learning and ICT in Education, pp. 211–224 (2012)
48. Taramopoulos, A., Psilos, D., Hatzikraniotis, E.: Can open virtual environments be used instead of real ones? AMAP experience in electricity, In: Papageorgiou, G., Kountouriotis (eds.) Proceedings of the 7th Panhellenic Conference of Teaching Science and ICT in Education-Interactions of Educational Research and Actions in Sciences, (A), Alexandropolis, pp. 679–686, 15–17 April 2011
49. Tarekegn, G.: Can computer simulations substitute real laboratory apparatus? Lat. Am. J. Phys. Educ. **3**(3), 506–517 (2009)
50. Tatli, Z., Ayas, A.: Effect of a virtual chemistry laboratory on students' achievement. Educ. Technol. Soc. **16**(1), 159–170 (2013)
51. Triona, L.M., Klahr, D.: Point and click or grab and heft: comparing the influence of physical and virtual instructional materials on elementary school students' ability to design experiments. Cogn. Instr. **21**(2), 149–173 (2003)
52. Tsihouridis, C., Vavougios, D., Ioannidis, G.: The effectiveness of virtual laboratories as a contemporary teaching tool in the teaching of electric circuits in Upper High School as compared to that of real labs. In: Proceedings of 2013 International Conference on Interactive Collaborative Learning (ICL), 25–27 September 2013, Kazan National Research Technological University, Kazan, Russia ISBN: 978-1-4799-0152-4/13©2013 IEEE, pp. 816–820 (2013)
53. Tsihouridis, C., Vavougios, D., Papalexopoulos, P.F.: The use of calculating packages for electrical circuit problem solving by Secondary Education students: a comparative educational evaluation. In: Proceedings of 15th International Conference on Interactive Collaborative Learning ICL2012 11th International Conference Virtual University (vu'11), 26–28 Sept 2012, Villach, Austria, International Association of Online Engineering, IEEE Catalog Number: CFP1223R-USB, ISBN:978-1-4673-2426-7, pp. 815–825 (2012)
54. Tsihouridis, C., Vavougios, D., Ioannidis, G.S.: Evaluation of educational software regarding its suitability to assist the laboratory teaching of electrical circuits. In: Auer, M. (eds.) Proceedings of ICL2007 Workshop: Interactive Computer Aided Learning, Villach, Austria, pp. 1–15, Kassel University Press, ISBN: 978-3-89958-279-6 (2007)
55. Tsihouridis, C., Vavougios, D., Ioannidis, G.S.: Students designing their own experiments on heat transfer phenomena using sensors and ICT: an educational trial to consolidate related scientific concepts. Int. J. Emerg. Technol. Learn. iJET (4), 74–82 (2009)
56. Tsihouridis, C., Vavougios, D., Ioannidis, G., Alexias, A., Argiropoulos, H., Poulios, S.: Mobile physics labs as contemporary study tools in electric circuits teaching: PCB301-a case study. In: Proceedings of the Panhellenic Science Teaching Conference and ICT in Education, Thessaloniki, 8–10 May 2015

Integrating Cyber Range Technologies and Certification Programs to Improve Cybersecurity Training Programs

Demitrius Fenton[1], Terry Traylor[2(✉)], Guy Hokanson[1],
and Jeremy Straub[1]

[1] North Dakota State University, Fargo, ND, USA
{demitrius.fenton,guy.hokanson,jeremy.straub}@ndsu.
edu
[2] United States Marine Corps, Fargo, ND, USA
terry.o.traylor4.mil@mail.mil

Abstract. Today more than ever, computer science students with actionable security skills are in high demand and this requirement is not expected to change in the near future [1]. Schools that provide traditional computer science education programs do a satisfactory job of exposing students to cybersecurity principles and some schools small labs that provide training. Industry essential cyber skills aren't traditionally taught inside most undergraduate and graduate programs [2]. Recent college graduates are forced to enroll in numerous certification programs in order to develop critical industry useful hands-on skills that are not taught in college [1]. What is needed today is a comprehensive training environment with extensible configurations for deeper learning and adaptive teaching. This will enable students to gain firsthand experience with the tools for the attack and defense of systems. We present the results from an Action Research project focused on improving cybersecurity training by introducing elements of gamification and a cyber range.

Keywords: Cyber range · Game · Virtual training laboratories · Extensible structures · Knowledge transference · Blended education and training · Action research in education · Mixed-methods

1 Introduction

During the 2017–2018 academic school year, our team integrated a cybersecurity training range for Computer Science student at North Dakota State University in order to address the talent development gap mentioned in the abstract. The lab provide a blended education and training environment for up to 24 students who needed to learn cybersecurity principles while simultaneously training for industry certifications. We propose that building labs with extensible configurations designed around creating an experience and delivers an immediate return on investment when developing a network lab. We will discuss how we used accepted research methods to improve the training and education experience. In this paper we also discuss how our team integrated gaming scenarios into the training platform to improve cybersecurity education and

M. E. Auer and T. Tsiatsos (Eds.): ICL 2018, AISC 917, pp. 632–643, 2019.
https://doi.org/10.1007/978-3-030-11935-5_60

training. Finally, we will expose our basic extensible lab design and outline future work on our cybersecurity laboratory environment.

The Research Questions (RQ) for this paper is How can we develop and integrate extensible cyber range technologies and gamification techniques into a blended education and training program that improves cybersecurity training. We have three supporting research questions that will be used to guide data collection, analysis and reporting. The supporting research questions are how can we: (1) improve cybersecurity teaching and knowledge transference, (2) improve industry cybersecurity testing performance, and (3) increase positive self-reported classroom atmospherics or quality of training experience for the student?

2 Background

2.1 Bloom's Taxonomy and Learning Domains

Between 1949 and 1953, Benjamin Bloom an Educational Psychologist, chaired a committee that developed three hierarchical models to classify and describe different student learning levels [3, 4]. The three hierarchical models addressed how students learning information in the cognitive or "knowledge-based" domain, affective or "learning attitudes" domain, and psychomotor or "skills-based" domain. Inside each hierarchy students progressed through different learning levels and mastered tasks or learning objectives in each level to advance from a lower level of learning to a higher level of learning. While in Blooms Committee's original construct (recreated in Fig. 1), the three domains were depicted as mutual exclusive; later work demonstrated that learning in one domain can trigger learning in other domains [5, 6]. For example, learning how to physically play one person in a basketball game will also trigger learning about basketball game rules.

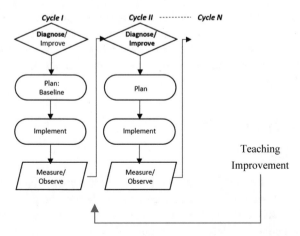

Fig. 1. Action research cycle implemented to improve training environment

In the case of cyber educational research, there have been multiple models developed to describe how to map Information Security Objectives into a learning taxonomy [7–10] and effective methodologies to integrate Cybersecurity curricula into IT Education programs [11]. The most successful cyber application of Blooms Taxonomy was put forth by Harris and Patten [12] and depicted the learning domains across the range of cyber activities or learning objectives.

For this research project we used Bloom's original descriptions of the learning domains and definitions for Cognitive and Affective domains. We used Simpson's [13] and Dave's [14] definitions for Psychomotor domain learning.

2.2   Learning and Transfer of Learning

Transfer of Learning is an educational psychology concept that describes how learning in one environment with specific materials affects performance or use of similar materials in a similar environment [15]. As an illustration a transfer of learning occurs when a student learns a skill in a classroom environment and that same student is able to apply that classroom learned skill in a similar but different environment. When a student applies that new skill in the different environment or context, the learning is said to have been transferred [16, 17]. Learning and Transfer of learning has been hypothesized and shown [5, 18, 19] as optimized when learning is relevant to the student, experience based, and easy to decode or process [16, 17].

2.3   Gamification and Scenario-Based Learning Integration

Drawing largely on the work originally done by Deterding et al. [20] to define gamification and the Gamification in Education Survey done by Dicheva et al. [21]; gamification is, "the use of games and game elements in non-game contexts." When an activity such as a course, business process, or other activity is "gamified" elements of gaming (which will be described in the following subsection) such as badges, scoring systems, detailed interfaces, and social engagement through group work are incorporated into the activities required to complete a task or achieve a goal [21, 22]. When applied to learning, training or education; gamifying a course means to use game thinking and game design elements to motivate the learner and increase student engagement [20, 21].

3   Methodology and Data

3.1   Methodology

For this study and the supporting technology development, our team adopted a mixed-methods approach with Action Research (AR) [23] as the primary methodology that drove initial RQ development and data collection. We also used traditional quantitative and qualitative data management and analysis techniques to collect, analyze, and report results. With Action Research as our primary research method we implemented a cyclical process model where we diagnosed, planned, implemented, measured/observed, and improved our training program and learning technologies over multiple cycles. Between each cycle we used the data collected during the measure/observe research step to plan the successive learning and teaching improvement.

Using the RQs we collected both quantitative and qualitative data. The quantitative data included test data results from two examinations where the time between each examination represented a cycle. The first examination test results represent that baseline student performance data with no technology support and the second examination results represent the second cycle data with the technology integration as the cycle 1 to cycle 2 improvement. Qualitative data were collected through face-to-face after-action review sessions.

Finally, we also collected and reported technology performance data and technical design issues encountered in order to expose lessons learned associated with building an extensible learning and education cyber range. Quality of Service (QoS) data was collected to study how well the system worked. Quality of Training Experience (QoTE) data was collected to assess student attitudes to how well the system performed.

There were 22 participants in the study. All participants self-selected to participate in the course. The group of students that participated in the course represented a mixed-cyber technical and experience group. The 22 participants were between the ages of 18 and 38 and were all students at North Dakota State University. The students in the group had between 1 and 11 years of IT experience.

4 Gamification, Cyber Range, and Education System Engineering Design

Faced with a small development budget, the research team used elements of gamification design, systems approach to training and education techniques [24] quote, and traditional systems engineering principles to develop an optimal system design. We were forced to implement a minimalist system design in order to engineer a system that met our cyber training objectives at a fraction of the cost of a complete physical system with similar capabilities. We called, this blended process of designing minimized systems to gamify an education or training program; Gamification, Education, and System Engineering Design or GESED. This section describes our design approach, resulting technical design, and basic GESED system capabilities.

4.1 Gamification, Education, System Engineering Design

Designing a cyber range that incorporates gamification elements and provides students with enabling learning objective (ELO) exposure requires a focused approach to ELO reinforcement design and training tool selection. Our team developed labs or games based on practical application ELOs such as systems administration commands and knowledge items that would be difficult to memorize in order to pass an industry certification test. The ELOs sub-divided into two groups. Group 1 contained ELOs that required open-source research or active online tools. Group 2 contained ELOs that required ethical hacking practice in a safe closed-network environment. We designed games, labs, and challenges to provide the student with integrated scenarios to facilitate learning in a fun environment.

Group 1 ELO-based challenges were designed to provide students that completed the challenges with information that would lead to or help students complete lab tasks or challenges in a closed-network environment. For example, our first lab (depicted in Fig. 2) was an online Capture the Flag (CTF) exercise that once completed provided students with exposure to 8 industry specific practical application ELOs.

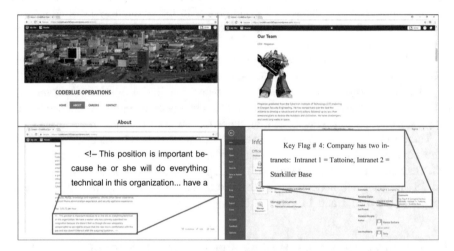

Fig. 2. The first gamified lab based on Group 1 ELOs. Students could navigate through the online website, finding flags or interesting open-source ethical hacking research facts like the accidentally exposed comments on the web page (lower-left) or the explicit closed-network intranet flags inside downloadable web documents (lower-right)

Group 2 ELOs were designed as challenges to be completed on the closed network cyber range. All cyber range ELOs and challenges were designed to provide training that simulated experience gained within the first two years of work in the IT industry or knowledge that would be difficult to memorize without proper context or practical exposure. In general, we tried to select obscure but highly testable industry certification commands and integrate the commands into games, challenges, and labs. The tasks or challenges were largely psychomotor tasks designed to expose students to unfamiliar commands/processes and force them to recall cognitive domain cyber data. Designing challenges around highly testable commands or tasks allowed our team to implement a minimalist design when actually developing the GESED.

4.2   GESED and Cyber Range Design and Development

Configuration of the training technology emphasized low cost and high return on investment for the most effective cyber range possible for a given budget. The only hardware acquired to build the cyber range was one Cisco server (depicted in Fig. 3) to the host virtual machines (VM). The server has 28 CPUs, 256 GB of RAM, and 15 TB of storage. VMware ESXi 6.5 was used as an operating system because it enabled VM development using an open source tool.

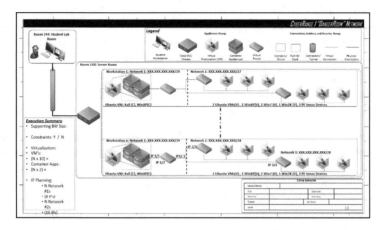

Fig. 3. GESED environment diagram. The GESED environment became known as the "Danger Room" named after the famous XMEN from Marvel Comics training area. Every student account was named after a XMAN in order to highlight the "gamelike" aspects of the course

All student environments were identical and designed to maximize ELO exposure using multiple virtual machines on one flat network. Advantages of a fully virtualized environment included adding or subtracting machines from the environment or and adding network complexity without adding additional cost to the course. Operating systems on the virtual machines are also completely interchangeable. The initial gamified training environment included Windows XP, Windows ME, Kali Linux, Ubuntu Linux, the metasploitable framework, the Damn Vulnerable Web Application (DVWA) [25] and a Smartphone Pentest Framework [26]. Each student had 7 virtual machines in their environment. All 7 machines were connected by a single virtual port to a shared virtual switch, that was not connected to the outside network. Isolating student virtual machines from the external network was critical to the success of the cyber range, because it allows a completely sandboxed environment where the students are able to play with cyber security tools without the risk of damaging or interfering with any other infrastructure. Students interact with their environment with either the VMware Remote Console application, or using the ESXi web console. Each student had access to their own private environment which enabled command reinforcement practice and tailored training. Isolating student environments also enabled the course instructor to add virtual machines to the student environment throughout the class, which the student has to identify and interact with over the network using real industry tools.

5   Preliminary Results and Findings

This section describes our preliminary Action Research implementation results and is divided into subsections. Each of the subsections describe either preliminary answers to one of our research vectors or a positive unintended result from our work. Since we only conducted two cycles during the semester, the results shown and lessons learned correspond to the cycles between January 2018 and May 2018.

5.1   Gamification May Positively Impact Transfer of Learning and Certification Testing Attitudes

Data collected from the two cycles indicate that using the cyber range to integrate gamification elements into the training may have improved student performance on skill-based test items, increased transfer of learning, and improved attitudes towards registering for and taking industry certification tests. As mentioned in previous sections, we completed and deployed the cyber range design between cycles one and two. Hands-on labs that used the cyber range were designed and deployed immediately after cycle one. All hands-on labs were developed as games to be completed as homework assignments. For example, we developed a series of penetration test labs as capture the flag challenges that encouraged ethical hacking skill development.

We compared student performance on skill-based test items on test 2 to student performance based on skill-based questions on test 1 to measure potential changes on skills-based test items. Table 1 shows the results from that comparison.

Table 1. Student performance on tests 1 and 2

Metric	Cycle 1	Cycle 2	Performance difference
# of skill-based questions	14	13	
Average class performance on skill-based questions (%)	61	71.3	10.3
# of analysis based questions	11	18	
Average class performance on analysis questions (%)	68.7	76.2	7.5
# of knowledge-based questions	39	41	
Average class performance on knowledge-based questions (%)	49	53	4
# of students who would have passed the industry examination at test time	52.2% or 12 students	73.9% or 17 students	21.7% or 5 additional students

Average Class performance on skill-based questions is the average of the student performance on all skill-based test items per cycle so for example, the 61% shown for cycle one indicates that the average student performance on all 14 questions in cycle one for all 22 students taking the exam is 61%. Said differently, on average each student got at least 8 skill-based questions correct.

Table 1 illustrates a potential increase on skill-based, analysis, and knowledge-based elements between tests one and two. Direct comments from the students during post-test in class after-action reviews indicate that students viewed skill-based test items as "easier" after executing the gamified labs. Student comments during the after-action reviews also indicated that the gamification aspects of the course such as the cyber range and CTF ethical hacking challenges "added meaning" and "made the course more realistic." Student's also commented during after-action reviews that "doing the labs made it easier to remember [knowledge-based] test information." Because the cyber range and online challenges exercised specific industry skills, out team hypothesized that the improved performance on the skill-based, knowledge, and overall pass rate could have been affected by the added game elements exercised using the cyber range.

The data highlighted in Table 1 and student after action comments also reveal that the cyber range may have introduced a transfer of learning effect between the cycles. As can be seen in Table 1, performance on analysis questions increased by 7.5%. As course designers, we introduced no additional training aspects designed to improve student analytical skills. However, several qualitative indicators during the after-action reviews and during in-class lab/game briefs indicate that analytical skills were improved through gaming on the cyber range. During in-class labs/games when students would ask questions regarding how to complete a specific lab element, we used this time to ask them analytical questions. As an example, during the third in-class lab/game several students could not determine how to deploy a penetration testing application and gain access to a vulnerable machine. We asked the students leading questions to help them determine how to complete the game task and then asked follow-on IT security examination analysis questions, not inside their game. The majority of the students were able to correctly answer the analysis questions in-class and most of the students usually expanded the discussion into other key IT security areas indicating that they were able to transfer the knowledge obtained mastering penetration testing systems to IT security planning practical knowledge. The ability to apply learning in one context—ethical hacking—and apply the skills mastered to other IT Security-related domains such as good system coding; may demonstrate that knowledge has been learned and transferred. This ability to relate ethical hacking practical skills to good system design is a potential transfer of learning effect that we did not have time to completely study; but annotated as a future gamification research topic that could be used to confirm the impacts of hands-on technology-based training such like those described in [6].

As a final note on the positive impact of the cyber range and gamification, we noted that gaming to prepare for industry certification may have improved attitudes towards taking industry certification tests. At the completion of the course four students were signed up to take the examination in less than 30 days, 10 students planned to take the examination in less than 60 days, and three students planned to take the examination in less than 12 months, and six students did not plan on taking the examination. While there is no baseline college student industry test data to compare our student attitude test data to, out team found it promising that 17 students planned to take the examination within the next year. We also noted that the school received more than 100 requests that the course be taught in the following semester.

5.2 Good GESED, Configuration Management, and Optimization Delivered a High Return on Technology Investment

Cyber range system performance data collected and analyzed during the project indicates that applying good GESED to the server delivered a high return on investment and acceptable student training experience. The optimal and extensible GESED that we implemented provided our students with a cost-effective and flexible system at a fraction of the estimated cost for a physical cybersecurity lab or training range. As described in Sect. 4.2 to this paper, the labs were designed to reinforce hard to memorize knowledge that would normally be mastered during the first two years in the IT/Cybersecurity industry. Focusing lab and game elements around the fun and essential training tasks (ELOs) let us eliminate unnecessary services and optimize the training game network design on the server. Iterating the virtual machine configurations on our GESED using AR feedback between cycles also let us customize our environment and deliver an improved experience to our students.

We studied system performance and used traditional cost-effect analysis techniques to assess system value. The cyber range/gaming platform was cost-effect compared to quotes for equivalent medium and large size physical systems that would deliver an equivalent training capability and network quality of service. Cost-effect analysis data is displayed in Table 2. Table 2 shows that our team cost avoided between $60,000 and $130,000 building a minimized GESED that focused on essential and fun ELOs. Minimizing costs and maximizing flexibility allowed our team to provide training and education benefits such as improved student test performance, increased projected student industry certification test registrations, and increased program prestige (measured as increased course media coverage [27–30] and increased collaboration requests), and improved program outcomes.

Table 2. Cost-avoidance Comparison Table

	Our cyber range	Medium size physical cybersecurity lab	Large size physical cybersecurity lab
Cost	$14,970	$75,366	$145,961
Costs avoided by implementing our GESED	–	$60,396	$130,991

Finally, GESED QoS and QoTE were assessed as functions of network performance. Between AR cycles, we benchmarked GESED performance to study how well the cost-effective system worked. GESED QoS benchmarking was conducted using 22 student accounts with 7 active virtual machines per student during a lab exercise/game. The lab consisted of 22 students conducting password cracking a CPU intensive task. During the lab all students were logged on using at least one of their machines,

Memory utilization was at 155.29 GB average, the CPU peaked at a maximum of 14.66% utilization, and average disk latency was at 9.68 ms. The network was an average of 1.75 mbps reaching a maximum of 10.14 mbps.

QoTE was initially assessed using system response data during high CPU cycle operations and in face to face in class after-action responses. Average ping response time during high CPU cycle operations was 1.2 s and average nmap scan of 7 virtual machines during the same peak operation evaluation period was 1.58 min. Verbal feedback from the students included statements that, " the system [worked] fine," and "[the students] didn't see any real delays." The research team assessed that the minimized and ELO focused virtual machine footprint provided the students with an acceptable and quality training experience. Moreover, the minimal and extensible system design allowed our team to customize our environment and deliver a much more tailored learning experience.

5.3 Integrating Career Exploration/Development Focused Training Improves Student Attitudes Towards Learning and Cybersecurity Program Focus

Good cyber security education not only prepares students to further their education, but prepares them for future jobs. Course offerings like the Certified Ethical Hacker certification preparation course have been some of the most attractive to students out of all offered cyber security courses. Courses that give students actionable skills and/or certifications that employers seek are especially interesting to students. As demonstrated by feedback from the students involved in this very class, usable skills attract students. A direct quote from one of the students featured in a news article summarizes the increased interest in industry testing, "Since I'm going into computer engineering and the computer science fields there's definitely going to be many places where I can apply this," said Isaac Burton, a junior from Two Harbors, Minnesota, who's majoring in computer engineering. "If this is going to be my primary job title, like as an ethical hacker, that's unclear right now but it's definitely going to play a big role in the future."

Interest in the application of this class toward jobs drives students to enroll, but media attention [27–30] surrounding the class, as demonstrated by the above article [28] also builds interest in the program and classes. Media coverage showing students excelling in their studies and expressing their interest in the field of cyber security appeals to other students who had some interest.

6   Future Work/Continued Action

Future work on the cyber range and developing education environments using our GESED approach shows much potential. The paper has provided initial AR research results from an iterative research project designed to incorporate games and industry certification training into cybersecurity training program. The extensible nature of the cyber range implemented improved student performance on tests and may have even triggered learning across Bloom's taxonomic learning domains.

References

1. Cybersecurity Training Online, Combat Threats with Cybersecurity Training. www.villanovau.com/resources/iss/cyber-security-training/#.Wrm05IjwbIU. 20 July 2018
2. While, S.: Cybersecurity Skills Aren't Taught in College. IDG Communications, Inc., www.cio.com, 13 Dec 2016
3. Bloom, B.S., Engelhart, M.D., Furst, E.J., Hill, W.H., Krathwohl, D.R.: Taxonomy of Educational Objectives: The Classification of Educational Goals. Handbook I: Cognitive Domain. David McKay Company, New York (1956)
4. Krathwohl, D.R.: A revision of Bloom's taxonomy: an overview. In: Theory Into Practice, vol. 41, issue 4, Special Issue: Revising Bloom's Taxonomy, pp. 212–218. Routledge, London (2002)
5. Leaman, C.: Improving learning and transfer: using brain science to drive successful learning transfer. In: Training: The Source for Professional Development. http://trainingmag.com/improving-learning-transfer. (2014)
6. Jones, R., Korwin, A.: Do hands-on, technology-based activities enhance learning by reinforcing cognitive knowledge and retention? J. Technol. Educ. **1**(2) (1990)
7. Andel, T., McDonald, J.: A systems approach to cyber assurance education. In: Proceedings of infoSecCD'13: Information Security Curriculum Development Conference (2013)
8. Bicak, A., Liu, X., Murphy, D.: Cybersecurity curriculum development: introducing specialties in a graduate program. Inf. Syst. Educ. J. **13**(3), 99 (2015)
9. Dark, M., Ekstrom, J., Lunt, B.: Integration of information assurance and security into education: a look at the model curriculum and emerging practice. J. Inf. Technol. Educ. **5**(5), 389–403 (2006)
10. Futcher, L., Schroder, C., Von Solms, R.: Information security education in South Africa. Inf. Manage. Comput. Secur. **18**(5), 366–374 (2010)
11. Rowe, D., Lunt, B., Ekstrom, J.: The role of cyber-security in information technology education. In: Proceedings of the 2011 Conference on Information Technology Education, pp. 113–122 (2011)
12. Harris, M., Patten, K.: Using Bloom's and Webb's taxonomies to integrate emergeing cybersecurity topics into a computing curriculum. J. Inf. Syst. Educ. **26**(3), 219–234 (2015)
13. Simpson, E.: The Classification of Educational Objectives in the Psychomotor Domain. Gryphon House, Washington, DC (1972)
14. Dave, R.: Psychomotor Levels in Developing and Writing Behavioral Objectives. In: Armstrong, R.J. (ed.) pp. 20–21. Educational Innovators Press. Tuscon, AZ (1970)
15. Wikipedia: Definition of Learning. https://en.wikipedia.org/wiki/Learning
16. Transfer of Learning Slide deck
17. Kolb, D.A., Kolb, A.Y.: Learning Styles and Learning Spaces: Enhancing Experiential Learning in Higher Education. In: Academy of Management Learning and Education, vol. 4, issue 2, pp. 198–212. Academy of Management, Briarcliff Manor (2005)
18. Salas, E., Wildman, J.L., Piccolo, R.F.: Using simulation-based training to enhance management education. In: Academy of Management Learning & Education, vol. 8, issue. 4, pp. 559–573, Academy of Management, Briarcliff Manor (2009)
19. Stanford University: Speaking of Teaching: Teaching with Cases. In: Stanford University Newsletter on Teaching, Winter Edition, vol. 5, issue 2, Stanford (1994)
20. Deterding, S., Dixon, D., Khaled, R., Nacke, L.: From game design elements to gamefulness: defining "Gamification." In: Laugmayr, A., Franssilia, H., Safran, C., Hammouda, I. (eds.) MindTrek 2011: Proceedings of the 15th International Academic

MindTrek Conference: Envisioning Future Media Environments, pp. 9–15. Association for Computing Machinery, New York (2011)

21. Dicheva, D., Dichev, C., Agre, G., Angelova, G.: Gamification in education: asystematic mapping study. In: Educational Technology and Society, vol. 18, issue 3, pp. 75–88. Athabasca, Canada (2015)

22. Kapp, K.: The Gamification of Learning and Instruction: Game-Based Methods and Strategies for Training and Education. Pfeiffer, San Francisco (2012)

23. Lewin, K.: Action research and minority problems. J. Soc. Issues **2**, 34–46 (1946)

24. Buchanan, L., Wolanczyk, F., Zinghini, F.: Blending Bloom's taxonomy and serious game design. In: Proceedings of 2011 International Conference on Security and Management. Las Vegas, NV (2011)

25. Damn Vulnerable Web Application. www.dvwa.co.uk

26. Smartphone Pentest Framework. www.bulbsecurity.com

27. https://www.wday.com/news/4424170-ethical-hackers-expose-businesses-security-problems

28. https://www.mprnews.org/story/2018/04/09/class-combats-cyber-crime-by-promoting-ethical-hacking

29. https://www.usnews.com/news/best-states/north-dakota/articles/2018-04-21/north-dakota-class-combats-cybercrime-with-hacking-skills

30. https://www.indianagazette.com/leisure/class-members-combat-cybercrime-by-learning-advanced-hacking-skills/article_197c1790-caa5-58ba-9e35-792938ac5594.html

31. Webster's Dictionary for Learning (2015)

32. Stanford University: Speaking of Teaching: Problem-Based Learning. In: Stanford University Newsletter on Teaching, Winter Edition, vol. 11, issue 1, Stanford (2001)

33. Armstrong, E.: Case Based Teaching: Overview: Advantages of the Case Based Approach. http://pedicases.org/teaching/overview/approach.html

34. Christensen, C.R.: Teaching with cases at the Harvard Business School. In: Barnes, L., Christensen, C.R., Hansen, A. (eds.) Teaching and the Case Method, 3rd Edition, p. 34. Harvard Business School Press, Boston (1994)

35. USMC Systems Approach to Training and Education

The Role of an Experimental Laboratory in Engineering Education

Maria Teresa Restivo[1(✉)], Maria de Fátima Chouzal[1], Paulo Abreu[1], and Susan Zvacek[2]

[1] LAETA-INEGI, Faculty of Engineering, University of Porto, Porto, Portugal
`trestivo@gcloud.fe.up.pt`, {`fchouzal,pabreu`}`@fe.up.pt`
[2] Consultant for Teaching and Learning in Higher Education,
SMZTeaching.com, Castle Rock, USA
`susan@smzteaching.com`

Abstract. The Laboratory of Instrumentation for Measurement started its activities in September of 2000 and initially was intended to support the Instrumentation for Measurement course of the Mechanical Engineering Degree of Faculty of Engineering, University of Porto, Portugal. Since that time its multidisciplinary team has promoted and enlarged the scope of responsibility and interests. During these 18 years, the lab has been involved in experimental engineering education, mainly in the field of mechanical engineering but also in other engineering disciplines and fields such as medicine, nutrition, rehabilitation, sports, and multimedia, for example. This intense involvement attracted and helped to support students for master's theses and final projects, Ph.D. students, and younger learners within non-formal learning programs, all focused on experimental tasks, project-based learning, and mentoring and coaching approaches. The lab has also been involved in informal learning initiatives and has supported regular R&D activities of a research group (funded by the National Science Foundation), while maintaining high quality standards for its students. This work describes the work of the lab and examines results, not in an analytical perspective, but based on real education-related outputs.

Keywords: Experimental engineering education · Formal · Non-formal and informal learning

1 Introduction

Experimentation is an important component of engineering coursework and as such is required by accrediting bodies to enable significant, essential student outcomes, including, "Design and conduct experiments as well as analyze and interpret data" [1] and, "Investigations of technical issues in the execution of experiments, the interpretation of data, and computer simulations" [2]. Laboratory experimentation, however, can support a much broader range of desirable outcomes, such as the ability to work in teams, learn from failure, and apply creative thought to practical problems. These boundary-crossing skills involve the use of higher-order thinking that can take learning beyond necessary foundational knowledge (i.e., the ability to memorize and

© Springer Nature Switzerland AG 2019
M. E. Auer and T. Tsiatsos (Eds.): ICL 2018, AISC 917, pp. 644–652, 2019.
https://doi.org/10.1007/978-3-030-11935-5_61

comprehend key ideas) into the realm of application, design, analysis, synthesis, and evaluation, for example [3, 4].

This article is intended to heighten awareness of how laboratories are integrated into many academic programs. Multiple examples, across institutions, will not only help us to learn from others but develop consistency in what we expect lab work to include. Such a consensus could provide testbeds for multi-institutional research on engineering education and encourage the development of products to facilitate innovative teaching practices, such as games, simulations, and shared resources for remotely-operated technologies.

Section 2 describes the laboratory and its main activities. Section 3 presents the lab's formal, non-formal, and informal educational contributions, while Sect. 4 looks at how R&D activities are shared with the engineering education mission. A sample of the lab's significant outputs are described and comments on the work carried out related to contributions to engineering education, as well as to society, are presented.

2   The Laboratory of Instrumentation for Measurement

The Laboratory of Instrumentation for Measurement (LIM) started its activities in September of 2000, and supported the undergraduate course, "Instrumentation for Measurement" within the Mechanical Engineering Degree of Faculty of Engineering (FEUP), University of Porto (UPorto), Portugal, up to the academic year 2005–06. From the beginning its leader—supported by a multidisciplinary team—has been promoting and enlarging the lab's scope of interests to multidisciplinary projects involving students.

During these past 18 years, the Lab has been involved in experimental engineering education by providing modern online resources to support coursework and considerably enlarging and intensifying experimental activities in multidisciplinary areas such as engineering, medicine, nutrition, rehabilitation, sports, and multimedia, involving students from these varied disciplines. This intense involvement attracted and helped to support students for master's theses and final projects, Ph.D. students, and younger learners within non-formal learning programs, all focused on experimental tasks, project-based learning, and mentoring and coaching approaches. LIM has also been involved in informal learning initiatives and has supported regular R&D activities of a research group (funded by the National Science Foundation). This helped to keep quality standards for its students.

Linking experimental engineering education and R&D activities, LIM has been focused in two main areas: online experimentation (OE) and instrumented devices (ID). These developments allowed LIM to participate in national and international events, to integrate international networks, and to be involved in national and international organizations and societies, enabling its students and collaborators (recently graduated) to contribute by fostering their experience and developing their soft skills.

The LIM team has promoted project-based teaching/learning and mentoring and coaching approaches, as well as encouraging the use of many online-emergent teaching tools. The activities developed at LIM, involving the experimental mechanical engineering education as well as other areas, can be characterized by the:

- development of OE tools based on the use of different emerging technologies, leading to the use of modern online experimental teaching methods for all students, and freely available to anyone else;
- use of project-based learning in formal and non-formal education;
- promotion of multidisciplinary educational approaches at the project level and among team members to facilitate students' involvement;
- contribution to the development of students' soft skills;
- promotion of mentoring and coaching teaching approaches: (i) mentoring students and recent graduates to act as mentors with young students in practical aspects or as mentors in non-formal and informal approaches of LIM intervention; and (ii) coaching teachers from different areas and levels on the use of OE for teaching purposes at the national and international level;
- assessment of how teachers use OE resources, whether students find them motivating, and how these resources influence learning; and
- involvement in successful national/international projects, networks and events, where students were often invited to participate.

3 LIM and Formal, Non-formal, and Informal Education

3.1 Formal Education Activities

As previously mentioned, the main goal of LIM during its first six years was to support the teaching of "Instrumentation for Measurement" for the Mechanical Engineering degree students. This course, focused on concepts, principles and methodologies of instrumentation measurement for laboratory and industrial purposes, used more than 50% of the class time in lab activities. The Lab also supported initiatives like b-learning courses on experimental measurements, having been involved in the first b-learning course of FEUP, in 2000–2001. This experience enabled the production of experimental and multimedia content for different books with national and international publishers.

The applicability to other areas where instrumentation plays an important role led the laboratory to naturally attract and support students for final course projects and seminars on a regular base in the period 2000–2006 as well as for the M.Sc. degree in Automation, Instrumentation and Control. Since the Bologna Curriculum implementation (2006), it has also supported integrated M.Sc. theses, mainly from the Mechanical Engineering course, but also from other courses within FEUP, and academic programs across UPorto.

LIM has also provided experimental support to many funded projects, some of them oriented to the development and use of emerging technologies in engineering education. These projects have brought opportunities to include students and later to guide them as future young granted collaborators.

3.2 Non-formal Education Collaboration

LIM has been participating in many non-formal learning activities, all of them based on experimental tasks using project-based learning with mentoring and coaching approaches, such as the UPORTO initiatives, "Annual Public Showroom" (since 2003), the "Younger Research Projects Program"-iJUP- (2004–2012), the "Junior University"-UJr- (since 2005), and the European Researchers' Night (since 2007), as well as many other FEUP initiatives, where students are asked to meet with visitors, provide explanations, and respond to questions.

Regarding the cooperation with UJr, the LIM team members integrated its steering committee (2005) and scientific committee (2005–2016). Team members have also organized activities related to instrumentation and measurements and mechatronics topics every July since 2005 with the support of students as monitors. In the period 2005–2017 the estimated number of participants within the LIM environment was around 2000 youngsters and over 50 monitors.

LIM has been involved in 12 iJUP projects gathering students in strong multidisciplinary teams with staff from engineering areas as mechanical, materials, electrical, informatics, and civil, as well as the field of nutrition, multimedia, physics, sports, architecture, and psychology (http://ieeexplore.ieee.org/document/6402056/). The team members supported around 50 undergraduate students in writing and submitting papers and presentations to iJUP Annual Conferences, an annual UPorto event of around 1000 students. Outcomes of these small projects resulted in 12 prototypes, 7 software applications, and students shared two different awards.

In 2018, two LIM team members have been within the national jury of 35 elements cooperating with the National Young Scientist Contest, organized by the Young Portuguese Foundation and by Ciência Viva Agency. This Contest involved 109 projects at National level, enfolding 289 high school students, 66 teachers, and 44 schools.

3.3 LIM Collaboration with Informal Education and Dissemination

Concerning informal education and dissemination, LIM has been co-organizing and/or participating in exhibitions where students or recent graduates are representing the team. As a few examples: Geneva Inventions, 40th Salon International des Inventions de Genève, 2012; Porto4Ageing (since 2017); Porto Innovation Hub (2016); HCP (Health Cluster Portugal) (www.healthportugal-directory.com); Demo Sessions on many different International Conferences, some with IEEE technical support (since 2006).

4   R&D Activities Side by Side with Engineering Education

As already mentioned, LIM Team links experimental engineering education and R&D activities, by focusing them in two main activities: online experimentation and instrumented devices.

4.1   Online Experimentation

OE is based on emerging technologies, like remote access to sensorized set-ups, virtual 2D and 3D experiments, and augmented and virtual reality apps. OE uses sensing devices for remote or virtual reality interaction, enriched by additional tools such as sound and live videos, available through web collaborative platforms for exploring and increasing user immersion in real systems or in their virtual replicas. OE is available through web collaborative platforms and is an unquestionably valuable tool for STEM students, particularly in helping students develop skills to be part of the driving force behind IoT. LIM has offered these IoT resources to students since 2003.

In the OE area, LIM has been promoting the *online exp@FEUP for all* project, a repository of 40 Lab resources (https://remotelab.fe.up.pt/), based on many technologies, 25 of them being of direct application and use in the mechanical engineering area. The majority of these resources (designed, developed and built within LIM) also support colleagues from other areas who are interested in sharing its experiments or in integrating their own resources. Presently, LIM supplies all the maintenance support to *online exp@feup for all*.

Topics of OE are also useful in continuing education, as well as for STEM at secondary and higher education. These resources received the GOLC Online Laboratory Award 2015, in the Simulation Laboratory category, and Honorable Mention of 2016 International E-Learning Awards, within Academic Division.

4.2   Instrumented Devices

The instrumented devices area integrates early developments of transducer prototypes designed at LIM, some of which originated patents (Relative acceleration transducer—2005; Displacement transducer based in magnetic flux sensors—2007; Device for acquiring and processing data for body mass estimation—2009; etc.). Since 2010, many of the instrumented devices were developed and oriented for health, rehabilitation, training and occupational therapy, following EC eHealth Policy and one of its goals "to make eHealth tools more effective, user-friendly and widely accepted by involving professionals and patients in strategy, design and implementation" [5]. Some of the developments are described at (https://remotelab.fe.up.pt/#health). In this perspective, the cooperation with health care areas and institutions (hospitals and clinics) has been fundamental and opened the opportunity to establish contacts and provide experiences for students during their final year projects, their M.Sc. theses, and other learning activities at LIM, contributing to the continued improvement of their soft skills.

Other developments of Instrumented devices are focused on environmental monitoring and control systems where mechanical engineers are traditionally involved. Different works were also described at (https://remotelab.fe.up.pt/#wireless_sensors).

5   LIM Outputs: Facts and Numbers

Usually results are presented in analytic form, based on inquiries, students' marks, any possible measured values, etc. In the present work the evaluation should be performed by the reader based on the presented outputs. As an indicator of LIM's foundational mission regarding experimental support to undergraduate students, an average of 200 students per year are in contact with online resources offered by LIM. In the last three academic years, more than 800 were able to use the OE resources supported by LIM.

At the time of its creation, the main goal of this Lab was the teaching support of one particular course and then, on a regular base, the support of some Final Course Projects in which instrumentation plays an important role. Later on, LIM also supported M.Sc. theses, mainly from the Mechanical Engineering course, but also from other M.Sc. courses at FEUP. Figure 1 shows the number of theses by area and course, totaling more than 40 M.Sc. theses since 2006.

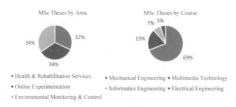

Fig. 1. LIM M.Sc. Theses since 2006

Moreover, from its activities in a multidisciplinary basis, the Lab holds patents at national and international levels, has commercialized technologies under option contracts between UPORTO and private companies, has supported content for books and articles with national and international publishers, and received national and international awards. In all these outputs many students are included as co-authors, co-inventors, and co-winners. These initiatives have allowed fundamental synergies with Lab's foundational mission to support engineering education (Table 1).

Table 1. LIM outputs

LIM outputs (Most involving students)	
Topic	Number
Awards: national and international	12 + 10
Main funded projects (national and international)	>35
Organization of Special Tracks/Conferences/Workshops	>30
Books: National and international	3 + 4
Publications (proceedings and journals)	>30
Involved students co-authors	>55
Patents: national and international (PT + USA) granted	4 + 1
Plus 3 pending (past students as co-inventors)	3

These numbers explain the previously mentioned students' systematic involvement in many different activities of formal, non-formal, and informal learning approaches. The following examples will describe several activities involving students, including research, developments, publications, public presentations, awards, and patents.

Displacement transducer based on magnetic sensors was an iJUP project that started in 2004. A multidisciplinary team of students and staff of two Faculties of Engineering from two Universities, involving areas as mechanical, electrical and civil engineering, ended with a full concept proof of the transducer communicating in a CAN bus. A few papers were co-written with students and they assumed the respective public presentations during 2005 and 2006. This project received the best project award on this projects edition in 2005. Later, based on the Lab's continuing efforts a final prototype was the basis of a national patent granted in 2008. At that point the mechanical engineering students decided to use this example as a case study of a spin-off in order to accomplish the work for a discipline in the Mechanical Engineering and Industrial Management area. A final prototype was tested and used by the Civil Engineering building lab.

LipoTool: a system created in a M.Sc. student's thesis project named "Virtual instrument for monitoring, digitally recording and assessing body composition" originated a system able to perform the body composition measurement and to communicate with a computer software app. A similar title was used for a paper presented by the student during an international conference REV'07, in 2007. In 2008 a M.Sc. thesis introduced a new version of the prototype using wireless communication with a new software app. This system was tested in nutrition and sports areas by students from those respective areas. A final work was written including all the students and presented by them, getting the best written communication award in X Congresso Anual da Associação Portuguesa de Nutrição Entérica e Parentérica (APNEP). In 2009 a graduate student received funding to carry out this project with the LIM team and a totally new prototype was built. From all of these efforts two national patents were granted (2009 and 2013), and a US patent in 2015 with a student included as co-inventor. This system has been used in areas such as Nutrition, Medicine, and Sports and articles were co-authored by students and team members from those areas, as well. In 2010 two awards were presented to the areas of mechanical engineering and nutrition and were received by the team and past students. In 2012, a Silver Medal (Medical Devices category) was given to the prototype in Salon International des Inventions Genève, Geneva, Swiss, where LipoTool was presented by the student, being in 2012 a graduate student team member. Currently the system is under limited production for commercialization in a spin-off company in which the previous student is presently working.

BodyGrip: an instrumented device with wireless communication with a software application for body grip evaluation was launched in 2008 as an M.Sc. thesis and the first concept proof of this device was successfully created. On 2012 a graduate student was funded to work on this system with LIM and a prototype with totally new characteristics was used in a Ph.D. thesis in the area of medicine for performing studies on handgrip strength. A considerable number of publications were written and co-authored by all the participants working on this particular system, including students or staff. The system is under patent process, and has been intensively used in Sports, Rehabilitation,

Health and Nutrition Schools, involving staff and students sharing the prototypes and participating in discussions intended to result in new features in the near future. Among these discussions is the inclusion of data analysis techniques as a present goal of a Ph.D. student.

Many other examples could be given above but these intend to explain how the lab continues to fulfill its mission in Engineering Education, looking at the progress of knowledge without forgetting societal commitments.

6 Final Comments

Apart from the numbers just presented, the work carried out at LIM has resulted in contributions to engineering education as well as to society. As an example, the collaboration with colleagues from the health area led to the development of specific instrumented devices such as the LipoTool. M.Sc. students were involved in the development of this tool, M.Sc. and Ph.D. theses and other scientific publications were written and published about it, presentations were given at different events, awards were received, and patents (national and international) were granted. As a societal benefit, a commercial agreement between the University and a spin-off company was established and a prototype short series is under development. In other examples, students have participated in the development of instrumented devices for rehabilitation and evaluation such as the SHARe system for hand rehabilitation in dexterous manipulation of daily objects, the WEST wrist evaluation system, and the pulse trial system, for example. These activities have been extending the role of the laboratory and fostering collaboration with fields outside engineering providing a rich platform for formal and non-formal education.

As further evidence of student engagement, LIM has supported two of its past students and later young researchers in the international Young Scientist Award, the annual competition for the International Society for Engineering Education (IGIP), in 2012 and 2016.

The Laboratory of Instrumentation for Measurement has served as an essential resource for learners as they gain skills required by successful engineering professionals. With the integration of practical tasks, creativity, and problem-solving into lab activities, these students have engaged in higher-order thinking that benefits them, their peers, their instructors, and society.

Acknowledgements. This work has been also funded by the Project LAETA - UID/EMS/50022/2013 from the Portuguese Science Foundation.

References

1. ABET. Criteria for Accrediting Engineering Programs, 2016–2017. http://www.abet.org/accreditation/accreditation-criteria/criteria-for-accrediting-engineering-programs-2016-2017/#outcomes. Accessed 14 Nov 2017
2. Eur-ACE. EUR-ACE Framework Standards and Guidelines. 31st edn., 3-15-2015

3. Simon, N.: Improving higher-order learning and critical thinking skills using virtual and simulated science laboratory experiments. In: Elleithy, K., Sobh, T. (eds.) New Trends in Networking, Computing, E-learning, Systems Sciences, and Engineering. Lecture Notes in Electrical Engineering, vol. 312. Springer, Cham (2015)
4. Madhuri, G.V., Kantamreddi, V.S.S.N., Prakash Goteti, L.N.S.: Promoting higher order thinking skills using inquiry-based learning. Eur. J. Eng. Educ. **37**(2), 117–123 (2012). https://doi.org/10.1080/03043797.2012.661701
5. ISC. Intelligence in Science website. http://www.iscintelligence.com/tema.php?id=4. Accessed May 2018

An Authoring Tool for Educators to Make Virtual Labs

Dimitrios Ververidis[1(✉)], Giannis Chantas[1], Panagiotis Migkotzidis[1], Eleftherios Anastasovitis[1], Anastasios Papazoglou-Chalikias[1], Efstathios Nikolaidis[1], Spiros Nikolopoulos[1], Ioannis Kompatsiaris[1], Georgios Mavromanolakis[2], Line Ebdrup Thomsen[3], Antonios Liapis[4], Georgios Yannakakis[4], Marc Müller[5], and Fabian Hadiji[5]

[1] Centre of Research and Technology, Hellas, Greece
ververid@iti.gr
[2] Ellinogermaniki Agogi, Pallini, Greece
[3] Aalborg University, Aalborg, Denmark
[4] University of Malta, Msida, Malta
[5] goedle.io Gmbh, Cologne, Germany

Abstract. This paper focuses on the design and implementation of a tool that allows educators to author 3D virtual labs. The methodology followed is based on web 3D frameworks such as three.js and WordPress that allowed us to develop simplified interfaces for modifying Unity3D templates. Two types of templates namely one for Chemistry and one for Wind Energy labs were developed that allow to test the generalization, user-friendliness and usefulness of such an approach. Results have shown that educators are much interested on the general concept, but several improvements should be made towards the user-friendliness and the intuitiveness of the interfaces in order to allow the inexperienced educators in 3D gaming to make such an attempt.

1 Introduction

Educational organizations often use electronic games in order to facilitate learning. These games simulate real-life situations, and allow the learner to be trained in a controlled environment. 3D games have been significantly improved in the last years with the advent of Virtual Reality (VR) technologies. Their maturity has raised the interest of several organizations in order to be exploited in education and training [9,12]. However, several barriers such as the high cost and the luck of proper design prevent their expansion. In our work, we seek into surpassing these obstacles with a proper designed authoring tool that allows to make virtual labs with a low cost.

Games in general cost a lot to develop [13]. The process of making a game starts with the scenario writing, proceeds to the game-play, the artistic content, and finally the programming. It is difficult for educational organizations to accomplish these tasks as expert knowledge required. There is a need to automate this process by providing templates that allow several these tasks to be

© Springer Nature Switzerland AG 2019
M. E. Auer and T. Tsiatsos (Eds.): ICL 2018, AISC 917, pp. 653–666, 2019.
https://doi.org/10.1007/978-3-030-11935-5_62

auto-filled. Our approach is based on game project templates that incorporate high level organization to allow object behavior inheritance, i.e. by selecting a category for each item, its behavior in the virtual lab is concretely defined. Thus, unnecessary details are hidden from the educators by allowing items to inherit a pre-programmed behavior, e.g., all items on a table should have a collider so that they stay on the table and do not pass through it. Our methodology relies on the development of a user-friendly platform that is used to create and design virtual labs by using high quality game engines and web interfaces. More specifically, we employ WordPress [7] web content management system in order to develop an editor for Unity3D game engine [15]. This editor is actually a web portal for educators that allows them to build educational game projects. The game projects can be compiled by Unity3D game engine and the game output is therefore of a high quality lab.

Analytics are an essential part of improving software products. In the developed templates and the authoring tool, analytics and visualizations of them were implemented, enabling educators to receive a feedback about the effectiveness of their labs and make changes accordingly. To achieve this, we have borrowed technology from game analytics [11], i.e., tracking infrastructure and game data analysis, and have apply it to the educational context of the authoring tool.

The outline of this paper is organized as follows. In Sect. 2, the existing tools for making Virtual Labs are outlined and discussed. The methodology for making our Virtual Labs authoring tool is provided in Sect. 3. The user evaluation of the resulted implementation is provided in Sect. 4. Finally, conclusions and future work are discussed in Sect. 5.

2   Existing Tools

Few authoring tools for making Virtual Labs exist, due to the fact that it is not financially viable to make tools with the strict limitation to create only games for learning. Thus, in order to create educational games, developers and designers resort to one of two solutions: use desktop based game design engines (desktop game-makers) or cloud based game design engines (web game-makers).

Desktop game-makers are game editors and engines that are developed using C code and should be downloaded and installed locally in a PC. The advantages of desktop game-makers are (a) the realistic high-quality graphics, (b) the large user community that develops new components and functionalities (Leap, Kinect, Oculus Rift, etc.) and (c) the support of multiple output formats, e.g., Android, Windows, iOS, Playstation, Xbox, WebGL and others. On the other hand, they require programming skills, and they are computational heavy requiring local installation in high-end PCs. A list of desktop game-makers can be found in Table 1. Unity3D and Unreal 3D engines are the most popular solutions for research projects as they are open source and can be distributed for free assuming non-commercial use. Another option is the Torque3D, being a completely open software with MIT license and quite mature. However, it has a very small community and it does not export the games into mobiles and consoles. Open source solutions such as Blender, Godot, and Copperlicht are face similar problems. For our developments we have exploited Unity3D as it is the game engine that exports to most of the operating systems in desktops and mobile devices.

Table 1. Desktop based runtime technologies for making Virtual Labs

Desktop game-makers

Name	Compiler - GUI	License	Exported for	Features
Unity3D	Unity - Unity	Open or Proprietary	Desktop, Mobile, Web, Consoles	Leap, Kinect, Oculus++
Unreal Engine	Unreal - Unreal	Open or Proprietary	Desktop, Mobile, Web, Consoles	Leap, Kinect, Oculus++
Godot	Godot - Godot	MIT	Desktop, Mobile, Web	Lightweight
Torque3D	GFX - Torque3D	MIT	Desktop, Web	Leap, Oculus, RazerHydra
Blender	Blender - Blender	GPL	Desktop	Incomplete
Blend4Web	Blender - Blender	Proprietary	Web	Plug-in for blender
Copperlicht	Copperlicht - Copperlicht	GPL	Web	No plug-ins for third parties

Web game-makers is an emerging area of the game industry due to the relatively new WebGL feature of HTML5. The benefit of web game-makers is that they are easily accessible through Internet and they are easier to learn. However, as a new software area, web game-makers have strong weaknesses. More specifically, they lack of several features of desktop game-makers such as exporting into several formats, supporting various hardware peripherals, and having a robust physics engine (Physijs was tested and found inadequate for games as regards speed and robustness[1]). Although WebGL is exploiting the graphics card, it is not as fast as a standalone game made with desktop game-makers. A list of the

---

[1] https://github.com/chandlerprall/Physijs.

existing web game-makers can be found in Table 2. Three.js is by far the most popular open web framework for 3D games. However, it hasn't any reliable GUI for making games. Babylon, Superpowers, XeoEngine, PlayCanvas, Goo, and Cyberix3D have a GUI but it is targeting for programmers as it shows many generic details. In our approach, we want to make a GUI that hides programming details and it leads the educator into making a game with certain steps under certain templates. From the aforementioned review, we have concluded that only desktop game-makers can really offer high quality games. Web technologies can be used though for 3D scene editing only.

Table 2. Web based technologies for making games

Web-gamemakers					
Name	Compiler	GUI	License	Export game	Features
Three.js	Three.js	–	MIT	Web	–
Babylon	Babylon	Babylon Editor	ASL 2	Web	–
Superpowers	Three.js	Superpowers HTML5	ISC (GPL like)	Web, Desktop	Real-time collab.
XeoEngine	XeoEngine (SceneJS)	–	MIT	Web	–
Turbulenz	Turbulenz	–	MIT	Web	–
PlayCanvas	PlayCanvas	PlayCanvas	Proprietary	Web, iOS	Real-time collab.
Goo	Goo	Goo Create	Proprietary	Web	–
Cyberix3D	Cyberix3D	Cyberix3D	Proprietary	Web, Android	–

3 Methodology

The authoring tool supports the authoring of Virtual Labs through the configuration of certain templates, namely the Chemistry and the Wind-Energy templates. In order to make a template several steps are followed. First, the game is designed and it is implemented into Unity3D. Analytics tracking functions are also embedded during game implementation. In this way, any generated game has already the tracking functionality incorporated. Next, the YAML code of the Unity3D game project is split into pieces of code that are inserted into WordPress data structures (taxonomies metadata).

The authoring tool web front-end interface is built according to the requirements for modifications on Virtual Labs templates. The Virtual Lab template can be modified by an educator using the front-end and compiled on the server side. If the game is compiled for a standalone format, e.g. Windows or Mac, a link of the game is provided for downloading the binary of the Virtual Lab. If the Virtual Lab is compiled for WebGL, a link is provided for directly playing the game. When a Virtual Lab is played, any game data are send to an analytics server that performs all the analysis required. The game analytics are led back in the authoring tool through proper visualizations in order to be inspected by the educator and to make edits in the lab accordingly.

The overall architecture of the authoring tool is shown in detail in Fig. 1. The backbone of the system is the Master Server, which contains the creation, editing and design functionalities for authoring virtual labs. Another server called as "Analytics Server", collects, stores, and process raw game data from each deployed game in order to provide meaningful analytics visualizations back to the Master Server. Initially, the Educator access the platform front-end via a web browser as shown in Fig. 2. The front-end has also a 3D editor interface that allows to position the 3D objects correctly. The platform has already uploaded 3D Assets of certain behavioral categories so that the Educator can easily use them. The Educator makes the necessary configurations and saves them into a new Game Project. After configuring the Game Project, and setting up the Scenes with the required Assets, the Educator defines the export format such as Windows, Mac, or WebGL in order for the game to be compiled and waits to receive the binary (Windows or Mac), or the link for WebGL compiled games. Next, the Educator provides the game to the learners for playing. When the game is played it sends game data and the game id to the Analytics server through an API. The data are aggregated and augmented with time statistics, and various features extracted with machine learning methodologies which can be found in a technical reports [4–6]. The analytics are served back to the Master server through an API in order to be visualized and thus provide a feedback to the Educator.

The software to make an instance of the Master Server is open, namely it consists the developed plug-in for WordPress [8], WordPress itself, and the Unity3D game engine. The aforementioned technologies are open and free to install. As regards Analytics server, it is a commercial system and it is not provided for free.[2] An instance of the Master Server connected with the Analytics server is accessible through the following link: https://envisagelabs.iti.gr

The 3D editor (based on three.js framework) of the lab scene consists of several widgets that help into authoring a scene. Briefly, the upper part consists of buttons that allow to view the lab in 2D, 3D, first person, or in third person. The left bar has an hierarchy view widget that allows to manage all the objects in the scene whereas the right-side bar contains all the available assets that can be dragged-n-dropped in the scene. Ray-casting allows drag-n-drop to be efficient, i.e. the assets are placed where the mouse pointer intersects an object. By right clicking on objects of the scene, their properties popup. For example,

---

[2] http://Goedle.io.

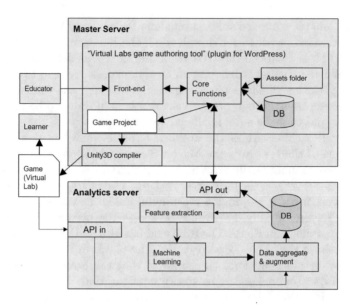

Fig. 1. Overall architecture of the system

some objects that serve as gates to other scenes, and therefore their properties is the scene to load.

Below the 3D editor, the scenes of the game project can be managed. In the specific lab, i.e. the Chemistry Lab, there are two other scenes to be edited, namely the "Exam 2D naming of molecules" and the "Exam 3D construction of molecules" scenes. In the "Exam 2D naming of molecules", the educator defines molecules names and formulas in order to examine the ability of the students to find the formula of a molecule out of its name, e.g. find that water formula is H20. In the "Exam 3D construction of molecules", the educator uploads some 3D structures of molecules that are posed to the students without the position of the atoms in the molecules and asks students to put the atoms in the correct position in 3D space. More details about the Chemistry Lab and the Wind Energy Lab can be found in [2,14]. In the lower part, the Game Settings are defined. It is mainly configurations about the texts and images in the standard scenes of the games such as the Main Menu, Credits, and Help scenes.

4   User Evaluation

In this section, we present the results of the evaluation of the authoring tool and the virtual labs created using it. Two methods are applied to evaluate the authoring tool: heuristic evaluation and user testing. Next, we describe both methods along with the evaluation results.

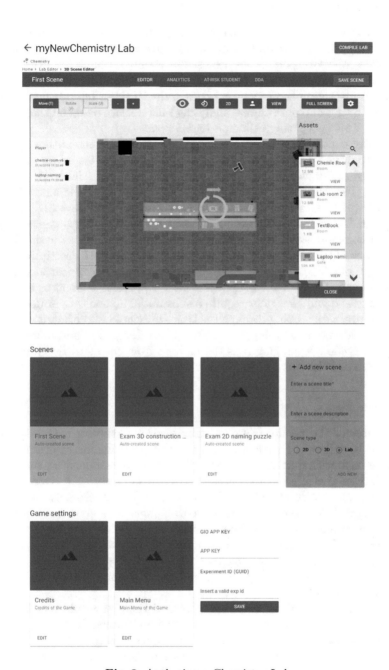

Fig. 2. Authoring a Chemistry Lab

4.1   Heuristic Evaluation

Heuristics are broad design guidelines, which can be used either for creating a user-friendly design or evaluating an existing solution in order to increase its usability. Moreover, heuristics are used as a rule of thumb for either making decisions for new designs or for pinpointing weak points when evaluating existing ones. Their inspection helps identify issues in the UI (User Interface) and is often performed by a group of reviewers analysing the interface based on the heuristics principles. This is typically called an *expert review*.

Usability experts performed a heuristic evaluation on the authoring tool in an effort to improve the user-friendliness of its user interface. More specifically, they utilized the 10 usability heuristics originated by Jakob Nielsen [16] as shown in Table 3. The evaluation focuses on the functionalities of the authoring tool, which are: (a) Create a new game project, (b) create/edit/save 3D scenes, (c) create/edit 3D assets, (d) insert assets into scenes, (e) delete game projects. The most common issues found during the analysis were related to missing previews, a lack of tool tips and help functions and inadequate descriptions of the authoring tool's functionalities. Such issues are natural to appear at this point of development. Many of the issues are related to missing content or ambiguous terminology and will hence not require a substantial amount of resources to correct. In future, we are going through the issues presented in the heuristics reports and will then rename functionalities, change the design and add more tool tips where it is needed, leveraging thus the findings from the heuristic evaluation to improve the next iteration of the authoring tool.

4.2   User Test of Authoring Tool

In this section, the results from the user tests are presented. Questionnaires were used to collect the feedback from the test participants for evaluating the authoring tool, with questions based on [10]. In the questionnaires, the teachers were asked to give a score to statements based on Likert-like scale with rating options ranging from unlikely (1) to likely (7) and "strongly disagree" to "strongly agree". For the tests, the test participants were given a set of tasks in the form of scenarios that led them through the different functionalities of the authoring tool for creating a new lab, adding 3D objects, and finally compiling a lab. The tests of the authoring tool were conducted with 15 educators. In Tables 4 and 5, the different questions and the frequency of the ratings by the testers are presented.

One of the participants consistently ranked the statements relating to the usability of the tool in a positive manner, while all others had a more neutral response to the authoring tool's usability. The questions that ask directly about the user friendliness of the authoring tool's interface are among the questions that get the most negative feedback. Several of the comments described the need for extended help functionalities such as tool tips on the different elements of the interface, a user manual, tutorial or step-by-step guide for using the tool.

Table 3. Distribution of usability issues across Nielsen's 10 heuristics in the authoring tool, as found by the inspectors

No.	Heuristic name	Issue count	No.	Heuristic name	Issue count
1	Visibility of system status	1	6	Recognition rather than recall	0
2	Emphasis on realism	7	7	Flexibility and efficiency of use	1
3	User control and freedom	3	8	Aesthetic & minimalist design	6
4	Consistency and standards	4	9	Help recovery from errors	0
5	Error prevention	2	10	Help and documenta-tion	15

Some comments specifically mention that it was difficult to interact with the 3D scenes. Rotating, scaling and placing objects as well as moving the avatar around in the 3D view is described as confusing and difficult. Some testers found the 3D view hard to understand and were unsure what the students would see in the compiled version of the created lab. Since then, we have increased significantly the user-friendliness of the tool.

The questions related to the usefulness of the tool show that most teachers have a positive attitude, most being above the mid-point and one being very sure about the usefulness. While there were quite a few comments that remarked negatively on the tools user-friendliness in the tested iteration there were also several comments that expressed a confidence that the authoring tool could offer benefits for the teachers going forward once more iterations are completed.

In general, what the answers reveal is that the test participants encountered difficulties working with the authoring tool but at the same time they expressed an interest in the tool and enjoyed the possibility being able to create a 3D experience for their classroom.

4.3   Virtual Labs and Learning Content

To evaluate the type of virtual labs and learning content that can be created using the authoring tool the teachers were asked to assess a demo version of such a lab, i.e., the Wind Energy Lab [1]. The test participants evaluating the virtual lab were the same as for the user test for the authoring tool, see Sect. 4.2. They

Table 4. Summation of answers in relation to the perceived usefulness of the authoring tool. The score range is 1 (unlikely) to 7 (likely), representing degrees of agreement

Question	Rating						
	7	6	5	4	3	2	1
Using the system in my job would enable me to accomplish tasks more quickly	1	2	6	4	6	1	0
Using the system would improve my job performance	2	1	6	6	2	2	0
Using the system in my job would increase my productivity	2	2	9	4	3	0	0
Using the system would enhance my effectiveness on the job	2	2	5	6	3	0	0
Using the system would make it easier to do my job	2	2	8	4	3	0	0
I would find the system useful in my job	3	3	6	5	1	1	0
The system would enable me to accomplish tasks more quickly	1	3	7	1	4	4	0
I would find it easy to get the system to do what I want it to do	2	1	4	5	3	5	0
My interaction with the system was clear and understandable	1	1	4	6	6	3	0
I would find the system to be flexible to interact with	1	3	4	5	5	2	0
It would be easy for me to become skillful at using the system	2	3	7	3	4	1	0
I would find the system easy to use	3	1	4	4	5	3	0

were asked to consider how much they (dis)agreed with statements regarding the students' engagement with the lab, quality of educational contents, the fit in terms of the students' abilities, and the teachers' expectations.

Regarding the quality of the learning content and the fit with the curriculum, the teachers agreed that the virtual lab's can be integrated into a learning context. The interface however was rated as needs improvement, thus, we have improved dramatically since then the interface of the authoring tool according to the evaluator rating and comments. The statements relating to the students' engagement and enjoyment of the virtual lab, are concentrated on the neutral midpoint. Though the teachers were satisfied with the content in the lab, they all proposed to further develop the content of the labs so as to better support their teaching.

Table 5. The table summarizes the testers' responses to the questionnaire on system usability of the authoring tool. The score range is 1 (min) to 7 (max), representing degrees of agreement

Question	Rating						
	7	6	5	4	3	2	1
Overall, I am satisfied with this system	2	2	2	3	3	0	1
It was simple to use this system	0	3	3	1	5	1	0
I can effectively complete my work using this system	1	2	2	1	5	0	2
I am able to complete my work quickly using this system	1	2	2	3	3	0	2
I am able to efficiently complete my work using this system	1	1	5	2	3	0	1
I feel comfortable using this system	2	2	3	4	2	0	0
It was easy to learn to use this system	2	1	2	6	1	1	0
I believe I became productive quickly using this system	1	0	5	4	1	0	1
The system gives error messages clearly tell me how to fix problems	1	1	2	1	0	4	4
When making a mistake, I recover easily and quickly	1	3	0	4	1	2	2
The provided information is clear	2	0	4	4	2	1	0
Easy to find the needed information	2	1	1	3	4	1	0
The information provided for the system is understandable	1	2	4	2	2	1	0
The information is effective in helping me	1	2	3	4	2	1	0
Organization of the information is clearly presented	1	2	3	5	2	0	0
The system interface is pleasant	2	0	6	1	2	2	0
I like using the system interface	2	0	5	3	0	3	0
This system has all the expected functionalities	1	1	3	6	1	1	0
Overall, I am satisfied with this system	0	2	5	3	1	1	0

4.4   Game Analytics

Game analytics is something that educators are not familiarized with, and therefore we have conducted a research-evaluation in order to find which types of game analytics visualizations would be useful to them. The visualizations shown to the participants were a dashboard, bar charts, force-directed graphs, chord diagrams and an absolute time-line. For example, a dashboard as shown in Fig. 3, is used as an overview of KPI's (Key Performance Indicators) connected and often also customized to fit a particular objective of its user. Linked to a database, the dashboards can be updated constantly and are frequently used for websites to tack user retention, daily users, revenue, page views etc.

For evaluating the visualizations, the test participants were given a questionnaire with three metrics, each being visualized in two to three different ways. Participants were asked to rank the visualizations internally with the metric and in relation to, e.g., best overview and most informative hereof. The participants were also encouraged to add more in-depth descriptions of why they had ranked the visualizations in the order they did. This evaluation therefore helped us narrow down which visualization is more useful.

Most of the testers had experience in reading and extracting the information from a visualization, which helped them understood them better in our case.

Fig. 3. Visualizations of a dashboard

Thus, a central conclusion was that providing data analytics efficiently depends on not only the data being visualized but also on the receiver. Another conclusion is that simpler visualizations were more understandable than complex ones. More details can be found in the respective technical report [3].

5   Conclusions

The authoring tool that we presented has been proven to be a complete solution for authoring Wind Energy and Chemistry labs. However, the development of a tool that allows inexperienced users to author 3D games is a great challenge. We saw by the answers to the questionnaires that the tool was accepted by the educational experts as a potential tool to be used in the class on the condition of several improvements in its user friendliness. Thus, we can conclude that the authoring tool architecture by the combination of the web based authoring capabilities with the compiling mechanism of Unity3D was proven successful, but more have to be done in order to use them in real life educational context.

As regards game analytics, it was shown that analytics is feasible to be embedded automatically in a lab, and their use is understandable and meaningful. In the future, we aim to include more data analytics in a context of a larger scale, i.e., many schools, even belonging in different countries. Lastly, we plan to develop new virtual lab templates for our tool, designed for different scientific fields, such as physics, maths, etc.

Acknowledgements. The research leading to these results has received funding from the European Union H2020 Horizon Programme (2014–2020) under grant agreement 731900, project ENVISAGE (Enhance virtual learning spaces using applied gaming in education).

References

1. ENVISAGE, Wind Energy Virtual Lab (2017). http://www.envisage-h2020.eu/games/energy/v1_1_3/
2. Final version of the "Virtual labs authoring tool", project Deliverable (2018). http://www.envisage-h2020.eu/wp-content/uploads/2017/10/D4.4Final.pdf
3. Implementation of the educational scenarios and evaluation report, project Deliverable (2018). http://www.envisage-h2020.eu/wp-content/uploads/2017/10/D5.2-Implementation-of-the-educational-scenarios-and-evaluation-report_Final-version_V3.pdf
4. Preliminary predictive analytics and course adaptation methods, project Deliverable (2018). http://www.envisage-h2020.eu/wp-content/uploads/2017/12/D3.1.pdf
5. User profiling and behavioral modeling based on shallow analytics, project Deliverable (2018). http://www.envisage-h2020.eu/wp-content/uploads/2017/09/D2.2Final.pdf
6. Visualization strategies for course progress reports, project Deliverable (2018). http://www.envisage-h2020.eu/wp-content/uploads/2017/07/D2.3-Visualization-strategies-for-course-progress-reports.pdf

7. Wikipedia official site for WordPress. https://en.wikipedia.org/wiki/WordPress. Accessed 2017
8. WordPressUnity3DEditor plugin for WordPress. https://github.com/Envisage-H2020/Virtual-labs-authoring-tool
9. Bavelier, D.: Your brain in video games. https://www.ted.com/talks/daphne_bavelier_your_brain_on_video_games/transcript?language=en
10. Davis, F.D.: Perceived usefulness, perceived ease of use, and user acceptance of information technology. MIS Q. **13**, 319–340 (1989)
11. El-Nasr, M.S., Drachen, A., Canossa, A.: Game Analytics. Springer (2016)
12. Labster: Empower and engage your STEM students. https://www.labster.com
13. Lovato, N.: What is the budget breakdown of AAA games? https://www.quora.com/What-is-the-budget-breakdown-of-AAA-games
14. Migkotzidis, P., Ververidis, D., Anastasovitis, E., Nikolopoulos, S., Kompatsiaris, I., Mavromanolakis, G., Thomsen, L.E., Müller, M., Hadiji, F.: Enhanced virtual learning spaces using applied gaming. In: Proceedings of the International Conference on Interactive Collaborative Learning (2018)
15. Miles, J.: Unity 3D and PlayMaker Essentials: Game Development from Concept to Publishing. CRC Press, Boca Raton (2016)
16. Nielsen, J.: Usability Engineering. Elsevier, Amsterdam (1994)

Work-in-Progress: Development of a LEGO Mindstorms EV3 Simulation for Programming in C

Valentin Haak[✉], Joerg Abke, and Kai Borgeest

University of Applied Sciences Aschaffenburg, Wuerzburger Strasse 45, 63743
Aschaffenburg, Germany
{valentin.haak, joerg.abke, kai.borgeest}@h-ab.de

Abstract. Embedded Systems are finding their way in more and more
areas of our daily life. Therefore, the ability to program such systems will
be more and more important for engineering topics. However, this is only
one good reason why students in mechatronics Bachelor Degree Program
learn a programming language at our University of Applied Sciences.
This programming language is C. Students learned this programming
language only by programming local console applications for a windows
computer. Since 2016, the Students learn it not only by programming
these applications but also by programming a LEGO Mindstorms EV3
robot. Therefore, they are using a toolchain that was also developed
at our University. This increases the student's interest and fun [1] in
programming, but it has also a big drawback. The LEGO Mindstorms
EV3 can only be used in the laboratory so there is no possibility for the
students to practice programming them at home. To provide a solution
for this, a developing process for a LEGO Mindstorms EV3 simulation
for programming in C is in progress.

Keywords: Computer science · Programming · C · Lego Mindstorms
EV3 · Simulation

1 Introduction

The students in the Bachelor Degree Program (B.Eng.) in Mechatronics at the
University of Applied Sciences Aschaffenburg learn the programming language
C as part of the module Informatics. These lessons are split in two parts. One
theoretical lecture per week (90 min) in which the students learn the basics
of computer science and the programming language. The second part is one
practical lesson per week (90 min) in which the learned skills can be practised.
Until 2016, the students only programmed console application exercises in this
practical part. Programming an embedded system was only part of the further
study program in the late fourth semester. Since the students learn C already in
the second and third semester this is way too late. Therefore, a special Toolchain
has been developed to program LEGO Mindstorms EV3 robots in ANSI C via

© Springer Nature Switzerland AG 2019
M. E. Auer and T. Tsiatsos (Eds.): ICL 2018, AISC 917, pp. 667–674, 2019.
https://doi.org/10.1007/978-3-030-11935-5_63

an API and an Eclipse plug-in [2]. Now it was also possible to create exercises for programming an embedded system the LEGO Mindstorm EV3 in the practical lessons.

Since 2016, the students learn C programming in both ways. Some exercises are still console applications and others are EV3 applications. The evaluation of the usage of LEGO Mindstorms EV3s showed that the students have more fun during the lessons and the exercises are closer to their later professional life as engineers in Mechatronics [1,3]. But it also showed a big disadvantage. There is no possibility of using the EV3 outside of the laboratory [3]. The time to access, the laboratory is not limited to the practical lessons, but it is limited by other lectures in this room and also the student's timetables. So there is nearly no time to prepare exercises for the practical lessons or the exam. This collaborates to the newest evaluation in which the students were asked how much time they invest for exercise preparation at home. In this evaluation the students also were asked some questions about how a simulation could help them in their learning process. The results of this questions can be found in Fig. 1. The following four questions were asked:

1. Do you consider a simulation for the preparation and the postprocessing could be useful?
2. Would a simulation support the self-responsible learning process?
3. Do you think a simulator helps better internalize the learned content?
4. Do you consider a simulation to be useful for the preparation of the exam?

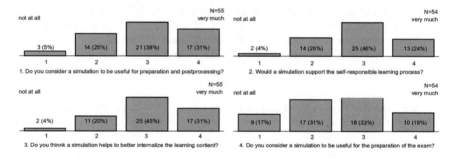

Fig. 1. Results of the evaluation from the winter term 2017/18 concerning to the EV3 simulation (4 point Likert Scale)

Due to this and the fact that the LEGO Mindstorms EV3 robots are very expensive, a simulation for this seems to be an effective solution. Since there is no useful simulation that can be used with ANSI C or the toolchain [2], the development of such a tool is now in progress. This paper will sum up the basic information of the development of the simulation and will give an overview of the status quo. In Sects. 2 and 3 the concepts for this development will be explained. Afterwards the used environment and its architecture will be explained. Section 6 will sum up the current progress and the next steps.

2 Concepts

A simulation shall be used to help the students exercise independent of the laboratory and its opening hours. Before the development started, a concept was developed [4]. Therefore, different existing simulation environments have been researched. The results showed that there are many simulation environments that simulate LEGO Mindstorms EV3 robots, but none of them could be programmed in C. This means that this would be something new. Subsequently, two different concepts were considered closer.

Chip Simulation. The first one was a complete simulation of the EV3 intelligent brick, which means a simulation of the ARM9 core [5] and the peripherals around it. In addition, the actors and sensors need to be simulated in a proper way. A graphical environment that works with data of such a simulation also needs to be developed. This only sums up the major things to do for this concept. Due to this, the concept was discarded after some evaluation, because it would take too much time.

Existing simulation environment. The first one was a complete simulation of the EV3 intelligent brick, which means a simulation of the ARM9 core [5] and the peripherals around it. In addition, the actors and sensors need to be simulated in a proper way. A graphical environment that works with data of such a simulation also needs to be developed. This only sums up the major things to do for this concept. Due to this, the concept was discarded after some evaluation, because it would take too much time.

A closer inspection of these two concepts showed, that it is more promising to use an existing environment [4]. Section 3 will explain the Selection of a fitting environment.

3 Selection of an Existing Environment

After the preselection, an evaluation matrix was used to find the most fitting of the remaining five environments. The five main criteria that have been used for this can be taken from the following Table 1. The Table also shows the results of the evaluation.

Open Roberta is an open source project of the Fraunhofer-Institut IAIS [9]. The goal of Open Roberta is to bring programming into primary schools. After the evaluation, the decision was made to use this concept for the development of a simulation environment that can be used with our toolchain. This marks the starting point of the current development (Table 2).

4 Environment

The standard build of the LEGO Mindstorms EV3 that is used for the exercises in the practical computer science lessons is displayed in Fig. 2. It shows that three

Table 1. Evaluation matrix for the already existing simulators with rating scheme

Criteria	Weighting	Environments					Mark	Value
		(1)	(2)	(3)	(4)	(5)		
Time effort	5	1	1	2	2	1	Very good	4
Completeness	4	4	1	3	3	2	Good	3
Documentation	3	1	3	3	4	3	Satisfying	2
Support	2	1	2	3	4	2	Sufficient	1
Easy handling	1	4	0	2	4	2	Not satisfying	0
Total	–	**30**	**22**	**39**	**46**	**28**	Insufficient information	0

Table 2. Simulation environments of the evaluation matrix

(1)	(2)	(3)	(4)	(5)
Trikstudio [6]	Greenfoot [7]	EV3 Jlib [8]	Open Roberta [9]	TigerJython [10]

sensors are used. A colour sensor that is adjusted to the ground to check the floor colour. The next sensor is the touch sensor that is used to touch obstacles in front of the robot. The third one is the ultrasonic sensor that can detect obstacles in a range up to 255 cm. The robot also has two medium servomotors that make it possible to drive. A small servomotor can be used to pull up or down the fork in front of the robot. However, an evaluation of the exercises showed that this part is never used [4]. Last but not least, the environment should be able to display the Mindstorms Brick functionality i.e. the LCD, the LED and the Buttons. This chapter will show how open Roberta displays these functionalities.

Fig. 2. Standard build of the LEGO Mindstorms EV3

Since Open Roberta is a program dedicated to unexperienced programmers and beginners it is not possible to program the robots in a higher programming language. Instead the robots can be programmed in NEPO, a simple but powerful functionblock programming language that is based on Blockly [11]. After building a program with these functionblocks, the Open Roberta-user can decide whether he wants to send his program to a real robot or see what the robot would do in a top view 2-dimensional (2D) simulation environment.

This environment specified for the LEGO Mindstorms EV3 is represented as a 2D top view terrain in two layers. The first layer is the background layer, which builds the floor of the environment. Every object in this layer has a fix position. The objects can have different colours which can be scanned by the colour sensor. The colour sensor in this build is adjusted to the floor. This is the only relevance that this layer has to the robot.

The robot itself is part of the robot layer. This layer is above the background layer and the robot can drive "over" every part of the background layer. In addition, obstacles can be initialised in the robot layer. Different to the objects in the background layer these objects can be moved during runtime. It is also possible to scan these obstacles with the ultrasonic sensor of the robot or touch it with the touch sensor. This functionality is again similar to the functionality of the standard build of the robot.

The LED is displayed at the top of the robot as a grey point as it is off. In addition, the colours green, red, and orange can be displayed like on the real EV3. The LCD and the buttons can be displayed in an extra window. This window looks like the top view of the EV3 intelligent brick. In this view the display can be used for writing text on it and pushing the virtual buttons will also be realized by the simulation.

This shows that this 2D environment completely fulfills the requirements of a LEGO Mindstorms EV3 simulation for our university of applied sciences. It covers all the basic functionalities of the standard build used for the exercises.

5 Architecture

Open Roberta is a server based application which means that it can be opened by accessing it via web browser. After the code in form of the Blocks is "written" on the client side as explained in Sect. 4, this blockly program is transferred into an Abstract Syntax Tree (AST). This happens back on the server side of the application. The AST can be transferred into code that can be used by the robot. Open Roberta provides this for different types of educational robot platforms. Since we develop a simulation for the LEGO Mindstorms EV3, only this robot will be assumed in this chapter.

For the EV3 the AST will be transformed into Java, based on the lejos [12] firmware for the EV3. The drawback here is that this code cannot be used on a robot running on the original LEGO Mindstorms firmware. In addition, it is not C so this part cannot be adapted to our toolchain in any way. However, the AST can also be transformed into JavaScript. This JavaScript will be sent back to the client side where an interpreter can display the code in form of a simulation in the environment that was explained in Sect. 4. Figure 3 shows a block diagram of this procedure.

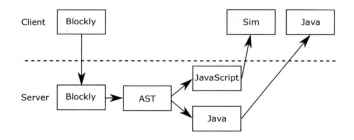

Fig. 3. Block diagram of Open Roberta programming procedure

The actual work in this development is the find te best possibility to integrate the API C-code into this simulation. Therefore different concepts to bring the C-code into the Open Roberta application will be explained in the following:

Client Side. The first possibility is to transform a valid C-code directly into JavaScript. Since our toolchain plug-in for ECLIPSE [13] checks the C-code, the result will be a valid code. Otherwise there will be a warning or an error. However, the actual compiler is transforming the Code into an ELF file that can be handled by the original LEGO Mindstorms EV3 firmware. So this plug-in must be extended with the possibility to create the JavaScript.

A matching of the API functions shows that nearly all functions can be matched to the JavaScript-commands that can be interpreted by the simulation. Some others can be displayed by the combination of two or more commands. This means that one possibility could be to use an existing open source compiler that compiles C-code into JavaScript like Emscripten [14] for example. This compiler could be expand in a way that the API functions can be transformed into the matching JavaScript. Another way is writing an own transformer that transforms the C-code into the matching JavaScript.

By starting the simulation process on the client side the process runs as described before. Nevertheless, the interpreter will not get the JavaScript from the Blockly program but the program that is transformed of the C-Code. The disadvantage here is that every local PC needs to host the server local so that it can access the local code and push it into the simulation.

Server Side. This alternative is similar to the first one. The difference here is that the created JavaScript will not be pushed in on the client side. After starting the simulation the Blockly program will be transformed into the AST and out of this into JavaScript on the server side. However, the server will not send this code back. Instead, it will send the JavaScript back that has been created out of the C-code the same way as in alternative one.

Server Side AST. The third alternative is again similar to the second one. However, here the transformation is not done from C to JavaScript, but form C to the AST used in the Open Roberta application. The advantage here is that the AST can be used in the complete Open Roberta Application. Out of this, not only the simulation can be interpreted but also the Java-code for the

Lejos firmware can be created. More interesting is that this is universal in this application and it is easier to match it with updates or new features of Open Roberta. Another advantage of the last both options is that the application can run on a server. The disadvantages of these two last options is that a file need to be sent to the server. This requires some strong safety-restrictions.

6 Current Progress

The current work in this development is to evaluate the possibilities explained in Sect. 5. Which one of them is the best fitting for the actual goal of this development. This means which one of them is feasible. Therefore, different things will be playing a role. The following criteria need to be considered:

Time effort: The time that it needs to get a usable prototype and after this a fully functional program.

Usability: How easy is the handling of the program and the simulation of the different alternatives.

Performance: How fast can the C-code be transformed int JavaScript or into the AST.

Recyclability: How good can the project be used to combine it with other projects at our university of applied sciences.

After the best alternative is evaluated, it will be implemented. If this part is working correctly, it can be integrated into the Toolchain Eclipse plug-in. This means that the user should have the possibility to decide if he wants to load his program up to the real robot or into the simulation. This should be realized by an extra button in the IDE plug-in. This makes it possible to check the C-code whether it is valid or not. Only if it is valid it should be possible to start compiling or transform the code so that the simulation interpreter can run it. If this is possible the prototype will be accessible for the student to test it and use it. If the simulation works well it can be used in every semester and it can be evaluated if the students still have as much fun as with using the real robots or maybe more and if the students use the simulation for home exercise preparation.

7 Summary

This work shows a way to counteract the problem that students can only access the teaching aids in the laboratory. It shows the actual state of the development of a tool that will help them doing this. With such a simulation, the students can use a virtual LEGO Mindstorms EV3 outside of the laboratory. Since Open Roberta provides not only a nearly perfect environment for this purpose but also the fact that it is a server application it could be combined with another project that is nearly at the end of its development. A Server based programming tool that makes it possible to write C-Code in a browser window, send it to the server, where it is compiled and executed. The command-output will be displayed in an output-window at the browser. With a combination, it could be possible that not the command output will be sent back but also the simulation interpretation.

Acknowledgements. The present work as part of the EVELIN project was funded by the German Federal Ministry of Education and Research (Bundesministerium für Bildung und Forschung) under grant number 01PL17022B. The authors are responsible for the content of this publication.

References

1. Perez, S.R., Gold-Veerkamp, C., Abke, J., Borgeest, K.: A new didactic method for programming in c for freshmen students using lego mindstorms ev3. In: 2015 International Conference on Interactive Collaborative Learning (ICL), pp. 911–914, September 2015
2. Simón Rodriguez Perez, B., Fatoum, A.: c4ev3 plug-in (2016). https://github.com/c4ev3. Accessed 22 Jan 2018
3. Simón Rodriguez Perez, B.: Erarbeitung und experimentelle Erprobung einer neuen Lehr- und Lernform für die Programmierung mit C mittels Robotereinsatz, Masterarbeit, Hoschule Aschaffenburg, Aschaffenburg, 03 March 2016
4. Haak, V., Abke, J., Borgeest, K.: Conception of a lego mindstorms ev3 simulation for teaching c in computer science courses. In: 2018 IEEE Global Engineering Education Conference (EDUCON), pp. 478–483. IEEE, 17–20 April 2018
5. The LEGO Group: Lego mindstorms ev3 firmware developer (2017). https://education.lego.com/de-de/support/mindstorms-ev3/developer-kits. Accessed 15 Nov 2017
6. CyberTech Co.Ltd.: Trikstudio. http://blog.trikset.com/p/trik-studio.html. Accessed 20 Oct 2017
7. University of Kent, La Trobe University: Greenfoot. http://www.greenfoot.org/. Accessed 20 Oct 2017
8. Plüss, A.: Ev3jlib + jgamegrid. http://www.aplu.ch/home/apluhomex.jsp?site=145. Accessed 20 Oct 2017
9. Fraunhofer Institut IAIS: Open roberta lab. https://lab.open-roberta.org/. Accessed 20 Oct 2017
10. Arnold, J., Kohn, T., Plüss, A.: Tigerjython. http://www.tigerjython.ch/index.php?inhalt_links=navigation.inc.php&inhalt_mitte=robotik/roboter.inc.php. Accessed 20 Oct 2017
11. Google LLC: Blockly. https://developers.google.com/blockly/. Accessed 20 May 2018
12. Sun Microsystems: leJOS. http://www.lejos.org/. Accessed 20 May 2018
13. The Eclipse Foundation: Eclipse (2017). http://www.eclipse.org/. Accessed 15 Nov 2017
14. Zakai, A.: Kripten/emscripten (2018). https://github.com/kripken/emscripten. Accessed 25 May 2018

Educational Data Mining from Action LOG Files of Intelligent Remote Laboratory with Embedded Simulations in Physics Teaching I

Sayan Das, Franz Schauer$^{(\boxtimes)}$, and Miroslava Ozvoldova

Faculty of Applied Informatics, Tomas Bata University in Zlin,
Zlin, Czech Republic
fschauer@fai.utb.cz

Abstract. Remote laboratories enter the teaching process, especially in research based education. In connection of this, new demands on both teacher and student occur. Especially acute is the need for a fast feedback about the process of measurements and its correctness. All these is enabled by the static LOG file, providing a time record of all steps, executed during measurements. We present the detailed case study of remote laboratory "Transient phenomena in electric oscillations" with about 170 university students of bachelor studies. We discovered the "knowledge barrier" at the beginning of process of measurements, leading to the excessive time losses and discovering the sources of it in simple mathematical relation of formula—graph and inability to evaluate data accordingly. We also found the ill effect of this inability on the total time of laboratory exercise. As a solution we suggest the use of components of virtual and augmented reality and artificial intelligence for reasonable influence of students' activities and advice together with cooperating experienced teacher.

Keywords: Remote laboratories · LOG file building · LOG file analysis · Case study bottle neck in measurement

1 Introduction

In connection with computer oriented activities, including controlling and measurements of various processes in general [1–3], especially connected with great hazards, request is raised to take time record of taken steps and collect it into so called LOG files. In parallel evolved a branch in computer science, dealing with the development, use and evaluation of LOG files [4].

With the invasion of computers in teaching, especially teaching laboratories, there arose a possibility to evaluate both the effectiveness of the teaching process and activities of individual students and pupils. Interestingly enough, there are to be found only very limited number of attempts in this direction [5, 6]. We realized the strength of the detailed stored information in LOG file especially of remote laboratories, information content of individual steps taken by students, their sequence and corresponding time

© Springer Nature Switzerland AG 2019
M. E. Auer and T. Tsiatsos (Eds.): ICL 2018, AISC 917, pp. 675–686, 2019.
https://doi.org/10.1007/978-3-030-11935-5_64

consumed by individual steps during measurements. For the purpose of the case study of evaluating the LOG file of students, we used one of the real intelligent remote laboratory, supplemented with several free-parameters synchronized simulations and diagnostics. We are starting the research process towards better understanding and easier evaluating of Remote Laboratory (RL) data by disclosing the reasons of misunderstandings and failures via the exploitation of stored information in LOG files. For the purpose we undertake the analysis of our state of art RLs, intend to discover the bottleneck of each RL with the goal to remove it. Our general intention is to help in next steps students using virtual reality and artificial intelligence to smooth the measuring process.

At present we start with the analysis of data like which and for how long time the step in question was used, the sequence of individual steps, if the student was successful, where he/she found the difficult step and how long it took to surmount it. All these results are rich in information, where are the problems both in pre-laboratory time of preparation, where can teacher (instructor) help and what can be improved on a specific RL. As we want to stress, all our activities are oriented on research—based education with the help of Integrated e-Learning (INTe-L) [7]. Such items and the present state of our RLs is to be found in our recent monograph [8].

In the present paper we try to understand to what extend these resources are useful for improving learning, taking one of our remote laboratories "Transient phenomena in electric oscillations" that has been subjected to deeper investigation both from the applicational and pedagogical views.

Illustration photo: Students of Engineering Informatics of the Faculty of Applied Informatics, Tomas Bata University in Zlin, Czech Republic in the research—based physics laboratory with hands on, virtual and remote experiments

2 Activity Description

2.1 Teaching with Remote Laboratory in General

The answer to better level of results in education are **research-based courses** [9, 10].

The clue to solving the situation is the outcome of educational and cognitive research, which can be reduced to the following basic principle: People learn by creating their own understanding by deliberate practice. However, that does not mean they must, or even can do it without assistance. Effective teaching facilitates that by getting students engaged in thinking deeply about the subject at an appropriate level and then monitoring that thinking and guiding it to be more expert-like [11].

We came on the long teaching trajectory to the opinion that laboratories and simulations can deeply change the teaching of physics, but new strategies, including these new teaching tools, are needed. For that reason, we suggest the method of e-Learning, including RL experimentation, and call it Integrated e-Learning (INTe-L), which is e-learning strategy enhanced by the missing element—experiment [7].

On the other hand, the RL, as a rather new technique of experimentation, needs new approaches and taking into consideration the rules of man-machine interaction. For this purpose, experimenting teacher should have means for RL optimization with respect to new way of cognition and communication with experimental hardware. An excellent tool for this is static LOG file giving, if properly organized, detailed and unexpected possibilities for RL optimization.

2.2 Description of the Remote Laboratory "Transient Phenomena in Electric Oscillations"

The remote laboratory "Transient phenomena in electric oscillations" is used in the curricula Electromagnetic theory and paragraph Electronic passive circuits and delivered via MOODLE and INTe-L for home work, lecture, seminar and laboratory. Before laboratory exercise students should study accompanying material in advance to understand theoretical and experimental background. For the purpose serves the embedded simulation with 4 variable circuit components (C, L, R_1 and R_2 see diagram in Fig. 1) to learn their influence on the final damped time response. On top of this, at the beginning of every laboratory work the students undergo a simple test for basic knowledge of the remote laboratory in question (quantities, their units, what to be measured, etc.). Then are students allowed to start the measurements using RL itself.

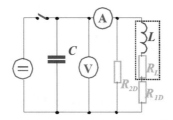

Fig. 1. *RLC* circuit with artificial damping due to resistors R_{1D} and R_{2D}

In RL, behavior of *RLC* circuits is examined in the time domain as a response to the applied voltage step. The circuit used in Fig. 1 contains a source of DC voltage and relay and measuring modules ISES (V-meter and A-meter) for pick-up of the time responses as well as constituent *RLC* elements of the circuit. The goal of the experiment is to determine the individual parameters of the *RLC* circuit from the time domain responses for varying damping due to the variable resistors R_{1D} and R_{2D}. The circuit in Fig. 1 may be described by the system of Kirchhoff Eqs. 1 and 2, leading to the differential equation for the instantaneous voltage u

$$\frac{d^2u}{d_t^2} + 2b\frac{du}{d_t} + \omega_0^2 u = 0,\tag{1}$$

with the general solution

$$u(t) = u(0)e^{-bt}\sin(\omega_1 t + \varphi),\tag{2}$$

with the natural frequency ω_1

$$\omega_1^2 = \omega_0^2 - b^2\tag{3}$$

and the damping coefficient b

$$2b = \frac{1}{R_2 C} + \frac{R_1}{L}.\tag{4}$$

It is $R_1 = R_{1D} + R_L$, $R_2 = R_{2D}$, φ is the initial phase. The damping coefficient b $[b] = s^{-1}$, can be thus expressed by the values of all constituent parts of the *RLC* circuit. The procedure of the experiment rests in the change of the damping coefficient b by varying the resistors R_{1D} and R_{2D}. From these dependences we may determine all the values of constituent elements of *RLC* circuit in question. The example of the measured response by Internet School Experimental System (ISES) system is shown in Fig. 2. As it is obvious from Eq. (4) and Figs. 3 and 4 both dependences $b = f(R_1)$ and $b = F(1/R_2)$ give straight lines with the slopes $1/2L$ and $1/2C$, respectively, giving the

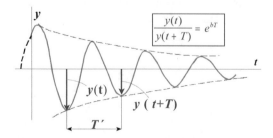

Fig. 2. Oscillations in damped *RLC* circuit

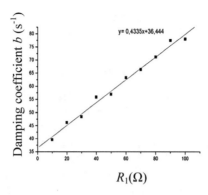

Fig. 3. The damping coefficient b dependence (Eq. 4) on the series resistance R_1

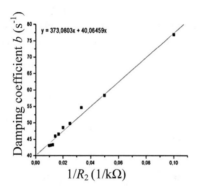

Fig. 4. The damping coefficient b dependence (Eq. 4) on the parallel resistance R_2

possibility to get the values of circuit components C and L together with the internal resistance of the inductor R_L.

3 Methodology—LOG File Design and Data Analysis

In general approach to the LOG file organization, we set up an assignment to store information about all the steps by the client in preparatory and measurement phases of the measurements to be able to follow later his/her measurement procedure on time scale. So every controlling step of the client (with the detailed values of chosen quantities), including start and stop, data storing of both the real measuring process and embedded simulation are recorded with the execution time for subsequent analysis. We adhered to the general recommendations given in a review paper [5]. For the pedagogical data mining we used the standard method of parsing with procedure with Python.

4   Case Study: Remote Laboratory "Transient Phenomena in Electric Oscillations"

4.1   Basic Exploitation of LOG File Data Mining

Educational data mining and analysis of LOG files was first aimed at standard evaluation of distribution of connections with respect to month in the calendar (not school) year and their corresponding hour of logins distribution (Figs. 5 and 6). We may reasonably expect the majority of connections come from the students active in the subject Electromagnetic theory depicting both summer and winter terms. More surprising and puzzling is the distribution according the login hours, as our (obligatory) laboratory start at 8.00 and finish at 17.00. The proportion of logins from outside the laboratory is substantial and so we came to the conclusion that students repeat their measurements and collect or refresh their new/renewed data before their evaluation. So we analysed the distribution of logins in greater detail (Fig. 7a, b) distributed in both laboratory and out of laboratory hours. The students logins are distributed, starting from evening time (18.00–22.00), in individual events even through the night and early morning hours.

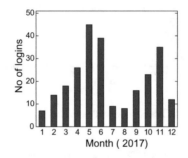

Fig. 5. Distribution of total clients' logins to the RL "Transient phenomena in electric oscillations" with respect to login months (2017)

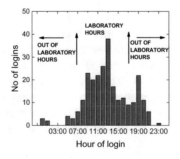

Fig. 6. Distribution of total clients' to the same RL by the login hour (2017)

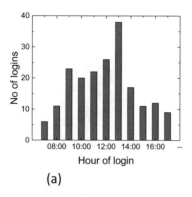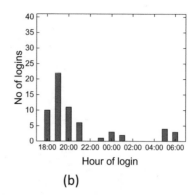

(a)   (b)

Fig. 7. Detailed distribution of clients' logins to the RL "Transient phenomena in electric oscillations"with respect to login hours **a** laboratory hours, **b** out of laboratory hours

4.2   Students' Working Attitude Pedagogical Data Mining

When analysing the measurements and processing steps in RL "Transient phenomena in electric oscillations" we could distinguish:

1. Preparation for measurements—study of accompanying material and simulation,
2. Login to RL,
3. Adjustment of variables,
4. Graph and first signal,
5. Adjustment of variables for required tables,
6. Required graphical dependencies and their evaluation (outside RL—most probably in ORIGIN environment).

In Fig. 8 is the corresponding pictorial representation of individual cognitive, decision and processing steps in the RL "Transient phenomena in electric oscillations" and in Fig. 9 is the corresponding flow chart diagram of the measurement process. We realized, from the very beginning, the low starting level of mathematics knowledge of students of the 2nd term. This was the reason, of not relying of the knowledge of differential equations and provided for students the simplified approach (see Eqs. 1–4). Then, when designing the RL and accompanying material, we naively supposed the straightforward use of Eq. (4) for evaluation of RLC parameters of the real resonant circuit.

5   Results

Analysis of the LOG file of 160 students (the final analysis was executed only for about 100 students, the rest withdrew from studies, did not finish the RL or were given later terms for measurements) discovered following facts depicted in Fig. 10. Students encountered problems to use the simple Eq. (4) with respect to organizing dependences

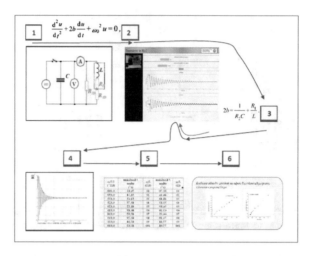

Fig. 8. Individual cognitive, decision and processing steps in the RL "Transient phenomena in electric oscillations"

$$2b = \frac{1}{R_2C} + \frac{R_1}{L}$$

(a) how to choose the values of resistors R_1 and R_2, (b) how to organize tables to get dependencies $b = f(R_1)$ and $b = F(1/R_2)$ in Figs. 3 and 4, (c) with respect to get corresponding parameters of the circuit. This is in our pictogram depicted as an "knowledge barrier "No 3–4(5) and results in erratic movements on the RL controls, not needed waiting times, switching - off, etc.

Looking at Fig. 10 we can distinguish several regions of the time schedule of the measurements. First, there is the time of the efficient students (about 50%), who master the steps 3–4(5) within less then 10 min. The time distribution for overcoming the "knowledge barrier" for the rest of students is enormously wide, spanning from 11 min to the record 60 min. The distribution seems to be exponential

$$N_{exp} = N_{o\,exp}e^{-\frac{t(3-4)}{\tau_{exp}}}, \tag{5}$$

with the time constant $\tau_{exp} = 22$ min.

6   Discussion

In the present study we intend to show the peculiarities of using RL both in laboratory and outside it. Predominantly we want to concentrate on pedagogical data mining, using LOG files. To our knowledge we, for the first time, use LOG files of REAL experiments for the purpose of tracking every step of the client (in our case university student in his/her bachelor course of physics) in active measurements with the goal of discovering problems and bottle necks students may encounter. We were spurred to

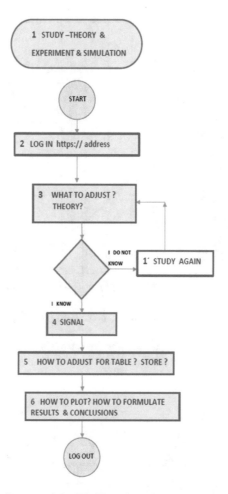

Fig. 9. Flow chart diagram of the RL "Transient phenomena in electric oscillations"

such research by the fact of large spans of time, which individual students need for the measurements. When looking into total time distribution for executing the experiment with the RL "Transient phenomena in electric oscillations" our teachers found a wide, seemingly exponential distribution of measurements time, spanning up to 60 min of only data collection

$$N = N_o e^{-\frac{t}{\tau}}, \tag{6}$$

with the time constant $\tau = 25$ min. The data with preliminary results of about 170 students are collected in Fig. 11. Striking for us was the wide distribution of the whole ensemble into two groups, one, where the execution times for RE spans to about 10 min (about 50%), and the second part with a widely distributed execution time of RE that span from 15 to 60 min. We decided to study the reason, what is the output of

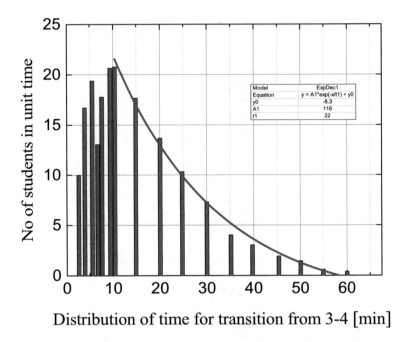

Distribution of time for transition from 3-4 [min]

Fig. 10. Distribution of time needed for transfer from step 3–4 (5) of RL "Transient phenomena in electric oscillations"(see Figs. 8 and 9)

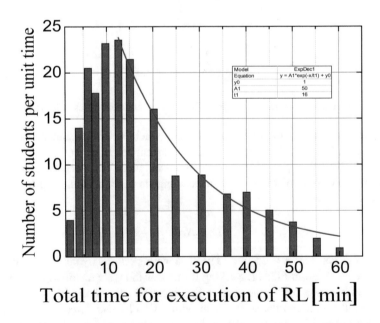

Total time for execution of RL [min]

Fig. 11. Distribution of total execution time for the RL "Transient phenomena in electric oscillations"

the present study. For us it was even more interesting and beneficial, as we constantly approach the teaching process as research-based education [8].

We examined in detail the "timetable "of each student on a group of about 160 students. The whole procedure of RL was distributed into 6 subsequent steps, binding on each other, described in detail in 4.2. starting from login, first movement of control elements of experiment till the data storing and plotting with data evaluation. LOG file was then designed for straightforward parsing process without excessive loss on information and we analysed the measuring process first from the collection of LOG files, then individually with arbitrary choice. It is worth noticing that in spite of the average LOG file was about 1–2 MB, the whole process was fully automatic and relatively fast giving rich and condensed information.

We choose for the examination mentioned goals the RL "Transient phenomena in electric oscillations" giving insight of electric oscillations in passive RLC circuits with the possibility to study effects like tuning, damping, energy and signal transfer, etc. Also contributing was the fact we have built embedded synchronized simulation of the identical effects as measured and might be available before measurements on internet.

When examining in detail the "time table" measurements of students, we found surprising fact the most of them met the "knowledge barrier" at organizing experiment and aligning experiment organization with the theory simple general results (here Eq. 4), on more detailed insight we found even more surprising fact – lack of abilities to setup graphs and evaluate them in other words steps 3 and 4 (even 5) in 4.2 according our methodology. When we plotted the time spent for steps 3–4 (5) only we found striking similarity of Fig. 10 with results in Fig. 11 meaning the main reason for time spent in whole experiments comes from the simplest decision at the beginning of the RL measurements.

As we discovered from discussions with students the students admit their inferior preparation from secondary schools, neglecting preparation for the measurement from for the purpose available sources (assignments) and embedded simulations, and during measurements itself they then relay on random help from schoolmates, sometimes afraid of asking teacher for help (!) and trying find the help on available during measurements internet sources. This is the reason we start massive support for RL measurements first by virtual and amended reality, later by artificial intelligence, where the LOG file and pattern recognition will be the main data, whose analysis will lead to the advice or even help in the situation, where student loses time in vain.

7 Conclusions

With emergence of RL in research based education [8], where the experiment is in focus of the whole teaching process new demands occur. All involved "actors" may contribute:

- Teacher (creator) of the RL and accompanying material should realize the weak points of the RL, on the level of man-machine cooperation, clarity of accompanying materials, accessibility of help and advice,

- Teacher in laboratory, even in situation of research base teaching, he/she should know its trickery and follow either by net means or by inspection which activities is student undertaking and give advice,
- Students are till now not ready for research based education and only mentoring by teacher, enabling the development of creative skills and teamwork,
- The components of virtual and augmented reality and artificial intelligence are to be used for reasonable influence of students' activities and advice.

Acknowledgements. The authors acknowledge the support of the Swiss National Science Foundation (SNSF) – "SCOPES". Also, the support of the Internal Agency Grant of the Tomas Bata University in Zlin, Czech Republic is highly appreciated.

References

1. Pasler, M., Kaas, J., Perik, T., Geuze,J., Dreindl, Künzler, R., Wittkamper, F., Dietmar, G.: Linking LOG files with dosimetric accuracy—a multi-institutional study on quality assurance of volumetric modulated arc therapy. Radiother. Oncology, 11 407–411 (2015)
2. Barringer, H., Groce, A., Havelund, K., et al.: Formal analysis of LOG files. J. Aerosp. Comput. Inf. Commun. **7**(11), 365–390 (2010)
3. Urhahne, D., Jeschke, J., Krombass, A., et al.: The valudation of questionnaire data: on interest in animals and plants with LOG files. Zeitschrift fur Pedagogische Psychologie. **18** (3–4), 213–219 (2004)
4. Athanasios, M.I.: LOG file formats for parallel applications: a review. Int. J. Parallel Prog. **37**(2), 195–222 (2009)
5. Lustigová, Z., Brom, P.: Educational datamining in virtual learning environments. Int. J. Adv. Corp. Learn. (iJAC) **7**(1) 39–42 (2014)
6. Cuadrosa, J., Artigasa, C., Guitartb, F., Martoria, F.: Analyzing a virtual-lab based contextualized activity from action LOGs. In: 4th World Conference on Educational Technology Researches, WCETR-2014, Procedia—Social and Behavioral Sciences. 182 441–447 (2015)
7. Schauer, F., Ožvoldová, M., Lustig, F.: Integrated e-learning—new strategy of cognition of real world, in teaching physics. In: W. Aung et al. (ed.) Innovation, pp. 119–136. World Innovations in Engineering Education and Research, iNEER Spec. 2009, USA, (2009)
8. Ozvoldova, M., Schauer, F.: Remote laboratories in research-based education of real world, Peter Lang, Int. Acad. Publ. Frankfurt, Germany, 157, ISBN 978-80-224-1435-7
9. Wieman, C.E.: Applying new research to improve science education. Issue Sci. Technol. **1**, 1–7 (2012)
10. Wieman, C.E.: Why not try a scientific approach to Science education? Change. Mag. Higher Learn. **5** 9 (2007)
11. Bransford, J., Brown, A., Cocking, R. (eds.): How People Learn: Brain, Mind, Experience, and School. National Academy Press, Washington DC (2002)

Research in Engineering Pedagogy

Computational Thinking, Engineering Epistemology and STEM Epistemology: A Primary Approach to Computational Pedagogy

Sarantos Psycharis[(⊠)]

ASPETE, Marousi, Greece
spsycharis@gmail.com

Abstract. This article presents a case study in which Computational STEM pedagogy is applied to improve STEM epistemology in pre-service engineering educator students. We report findings of a teaching course for pre-service teachers in a Greek Higher Education Institute. The findings indicate that: (a) students reported gains in their STEM content knowledge, (b) they also showed an improvement in their engagement in the STEM content epistemology by combining Math and Science concepts with Technology and Engineering, (c) students considered the model as the fundamental instruction unit used in the computational experiment approach, and (d) confidence to develop Computational STEM pedagogical inquiry based scenario as future teachers was enhanced by engagement in the Computational Pedagogy approach.

Keywords: STEM epistemology · Computational pedagogy · Engineering pedagogy · Arduino · Mathematics education · Science education

1 Introduction

1.1 Computational Thinking

Jeanette Wing [21] introduced the concept of "computational thinking" (CT) and considered that it includes solving problems, designing systems, and understanding human behaviour, by drawing on the concepts fundamental to computer science. She also argued [22] that CT is a universal skill and attitude that complements thinking in mathematics and engineering with a focus on designing systems that help to solve complex problems. In the absence of a consensus about its definition, a set of core concepts/dimensions and skills is continuously emerging from the literature to lead to a more complete definition for CT. These include: abstraction (considering a problem at different levels of detail, usually using the inductive process), modelling (developing of models with variables and the relation between the variables), algorithmic thinking, automation, decomposition of a problem in a set of smaller problems, debugging, pattern recognition(seeing a new problem as related to problems previously encountered) and generalization [6]. The concept that CT is a universal skill, attitude, competency practice and problem solving approach that impacts nearly all disciplines was

© Springer Nature Switzerland AG 2019
M. E. Auer and T. Tsiatsos (Eds.): ICL 2018, AISC 917, pp. 689–698, 2019.
https://doi.org/10.1007/978-3-030-11935-5_65

also suggested by many researchers in the field [15]. Bundy [5] also stated that "the ability to think computationally is essential to conceptual understanding in every field, through the processes of problem solving and algorithmic thinking". Many researchers put an emphasis on the fact that while there are "connections" between CT and computer science, CT needs to be taught in disciplines outside of computer science [1, 23]. Weintrop et al. [20] also support that Science and Mathematics are becoming computational endeavours and this fact is reflected in the recently released Next Generation Science Standards [13] and the decision to include "computational thinking" as a core scientific practice. We should remark here that there is a lot of ambiguity for the terms "Computing", "Computation" and "Computational", and there is no consensus for their meaning. A detailed analysis about the different meaning of these terms in the literature is presented in [15]. In this paper we adopt the view presented in [25] which states that "Computational pedagogy is an inherent outcome of computing, math, science and technology integration". In the same article computing is related to algorithmic and programming. They also suggest that computational modeling and simulation technology (CMST) can be used to improve technological pedagogical content knowledge (TPACK) of teachers. Our perspective is also close to that of [1], where Computer Science is related to computational processes and "scientists can promote understanding of how to bring computational processes to bear on problems in other fields and on problems that lie at the intersection of disciplines" and to that of [3], who state that "Projects with an orientation to computational science tend to emphasize data, modeling, and systems thinking". In this article a strong link between Computational Science and Computational Thinking is presented. The Computational Science in Education (CSE) is the integration of Mathematics, Computer Science and any other discipline to explore authentic-complex problems and is considered to have its own core knowledge area subjects [9]. Applying CSE, we accept that we can conduct experiments and iterations that could never be conducted in the real world with results equivalent to the classical experiments [15].

CSE "can be an effective methodology to support learners to solve a STEM problem using computer simulations and this includes diverse tasks, such as: formulating the problem in a way suitable for simulations using models (connection with CT); choosing an efficient computational algorithm (connection with CT); running the simulations and collecting numerical data) (connection with CT); analyzing the data obtained a (connection with CT); finding patterns in order to generalize the method to other problems (connection with CT), extracting the solution of the problem in a form that can lead to the creation of artefacts (connection with the engineering education epistemology" [15]. According to [14–16], "One of the crucial components of (CSE) is the abstraction of a physical phenomenon to a conceptual model and its translation into a computational model that can be validated against the experimental data using an inductive approach.

Modelling is closely related to the three dimensions of CT as suggested by [4], namely the computational concepts, the computational practices and the computational perspectives. According to the authors, CT involves concepts (e.g. loops), practices (e.g., abstraction and debugging) and perspectives, some of which are shared with other subject areas taught in schools such as science and mathematics. In our approach for the Computational Pedagogy we consider the CSE as the theoretical framework and the

CE experiment as the space where computational practices can be applied and students start with the model and follow a process and select an evidence based on data acquisition and analysis.

1.2 The Engineering Epistemology Education(EEE)

According to Shirey [18], the discipline of engineering can be divided into engineering content and engineering design. Engineering content arises from the integration of science, mathematics, and embraces a collection of tools, which engineers can use to design solutions to specific problems based on criteria and constraints. Rugarcia et al. [17] described engineering education "as the development of engineering knowledge (facts and concepts), skills (design, computation, and analysis), and attitudes (values, concerns and preferences)".

According to Katehi et al. [8] "perhaps the most important for engineering is design, the basic engineering approach to solving problems and when students are engaged in the design process, they can integrate various skills and types of thinking—analytical and synthetic thinking and detailed understanding". We can recognize the fundamental aspects of CT included in engineering design, while we consider that engineering content contributes also in CT dimensions, like pattern recognition and abstraction.

In [11–13], is stated that "while in the past an engineering design-based activity was viewed as a design or model construction activity, the new science education standards in USA require students to engage in engineering design and engineering practices as they learn and apply crosscutting concepts and disciplinary core ideas". We notice that emphasis is given to crosscutting effects that later will be connected to the transdisciplinary approach for STEM education. Katehi et al. [8] also report that one fundamental question that should be answered is "How does engineering education "interact" with science, technology, and mathematics?" This report also discusses another very fundamental issues, about the description of the ways in which K–12 engineering content has incorporated science, technology, and mathematics concepts, as context, to explore engineering concepts, or—reversely-how engineering is used as context to explore science, technology and mathematics concepts. This dual relationship is of fundamental importance in order to define the STEM epistemology [15].

1.3 The STEM Epistemology

There are two approaches for STEM education integration: the content integration and the context integration. STEM content approach [10] follows the so-called transdisciplinary approach, which focuses on the "integrated" approach to teach the four disciplines included in the STEM cognitive areas and "focuses on the merging of the content fields into a single curricular activity or unit to highlight "big ideas" from multiple content areas". In the context STEM integration approach, the focus is on the content of one discipline and next contexts from other disciplines are "used to make the content more relevant." We claim that STEM epistemology is closely related to Mode-2 system as it faces problems that emerge from different disciplines with loose organizational structures, flat hierarchies, and open-ended chains of command are dominant [15]. As a conclusion, STEM epistemology should follow the Mode-2 Transdisciplinarity and STEM integration in the classroom is a type of curriculum integration that combines the concepts of STEM in a transdisciplinary teaching approach which ends in the development of an artefact during the teaching sequence [7].

1.4   The Computational Pedagogy Approach

The term Computational Pedagogy was introduced by Yasar et al. [25] as an extension of TPACK, and was called Computational Pedagogical Content Knowledge. In this article we adopt the model of Yasar et al. [25] and we add the computational spaces practices related to the engineering design.

In Fig. 1 we present our model (Computational Pedagogy), where we have expanded the model of [25], in order to include the engineering epistemology and the STEM content epistemology.

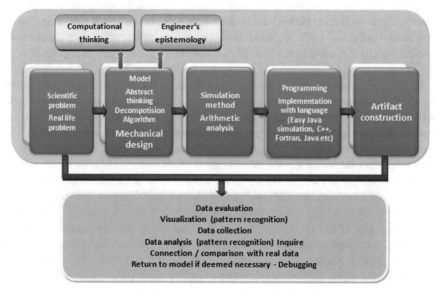

Fig. 1. The Computational Science Experiment (CE experiment)—Computational Pedagogy

In our approach, the model is the basic instructional unit, and it is used in the inductive process of teaching while it is closely connected to abstraction [24]. To use the model and simulation in the inductive process of teaching, we need proper environments that support the use of mathematics and algorithms, so the computational experiment will be "equivalent" to the physical experiment.

2 Methodology

The aim of this research is to examine the impact of the Computational STEM pedagogy (modeling and simulation, computational experiment, engineering design practices) on pre-service students' willingness to be engaged in computational STEM and on their self-efficacy in the creation of STEM artefacts and didactic scenario related to the content epistemology of STEM.

2.1 Research Model

The present study is an attempt to investigate students':

1. Intrinsic Values; engagement in the integration (STEM content) in STEM subjects, questions 1–8 of the questionnaire,
2. Attainment Value, questions 9–17 of the questionnaire: How much importance pre-service teachers place on working with the computational experiment approach integrated with the engineering design process,
3. Utility Value/Self efficacy; questions 18–24 of the questionnaire: How likely pre-service teachers feel that being successful in the creation of the STEM model will lead to success integrating STEM content in their future classrooms. Within this framework, students' "STEM content" perceptions have been attempted to detect.

2.2 Study Group

The study group of the research consists of collectively 45 prospective engineering education students from a Higher Education Institute. The 3rd year students had to attend the course "Education Technology", which consists of 3 h per week, one hour devoted in the theoretical aspects and 2 h in lab. The STEM epistemology followed was the STEM content.

2.3 Measurements

The data of this research have been collected using a tool, based on the questionnaire proposed by Yasar et al. [25]. Some questions from the Intrinsic Values part of the questionnaire were changed to be in alignment with the Computational Pedagogy model. Questions refereed to Physics were transformed to questions about Engineering for three reasons. The first one came from the fact that Physics is included in the Engineering content, at least for the purposes of the course. The second reason raised from the fact that we wanted to examine the engagement in the STEM content epistemology and not in Science and Mathematics engagement. The third reason was justified by the fact that students study to be engineering educators and engineering is their main subject of study. Questions about Attainment Value were not changed, but one question for the importance of the collection of data was added.

Two questions about Utility Value/Self efficacy were added after suggestions of experienced STEM teachers, in the category "Utility Value" in order to include self-efficacy questions. The alpha reliabilities of the factors were: (1) Intrinsic values a = 0.659, (2) Attainment Value, a = 0.733, and (3) Utility Value/Self efficacy, a = = 0.805. T-test methodology was adopted and results were recorded before and after the instruction. The course lasted thirteen weeks and after the course students had to develop a didactic scenario which included concepts from Engineering, Science and Mathematics. Students used software like Tracker, Easy Java Simulations, Arduino, Scratch and App inventor. Data were collected from engineering educator students at one Higher Education Institute in Greece, and were analysed using SPSS. T-test was applied for every question and for the three categories (Intrinsic Values, Attainment Value, and Utility/Self efficacy Value).

2.4   Example of Applications

Students were engaged in physical computing using Arduino and one example was about the algorithms and how these can be implemented in Scratch and Python. They were also engaged in the development of an artefact that "implements" an algorithm for the mystery triangles (http://www.ehrhard-behrends.de/pdf_zaubern/behrends_humble. pdf) in a physical artefact that was developed in Arduino (Fig. 2). Students discovered the algorithm presented at the left of Fig. 2 and they created the Arduino artefact (right hand side of Fig. 2). For the example presented in Fig. 2, the corner of the triangle should be programmed so that the yellow led should be switched on.

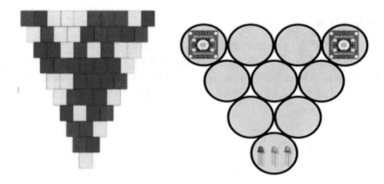

Fig. 2. The mystery triangles algorithm and the implementation of artifact using Arduino

3   Results

Results show a significant change of students' responses before and after the instruction. Results also show a significant relation between the three categories. The normality tests for the mean values before and after the intervention, for every category (MO_1, MO_2, MO_3 stand for Intrinsic Values, Attainment Value, and Utility/Self efficacy Value respectively), show that the results follow the normal distribution and t-test was used. The differences in the mean values for each category are presented below (Table 1).

Table 1. Differences for the mean values for each category (Paired Samples Test)

		Paired differences					t	df	Sig. (2-tailed)
		Mean	Std. deviation	Std. error mean	95% Confidence Interval of Difference				
					Lower	Upper			
Pair 1	MO_1 initial MO_1 final	−0.9853	0.0409	0.007	−0.9996	−0.9710	−140.5	33	0.000
Pair 2	MO_2 initial MO_2 final	−0.9935	0.0265	0.0046	−1.003	−0.9842	−218.3	33	0.000
Pair 3	MO_3 initial MO_3 final	−0.9622	0.1129	0.0194	−1.002	−0.9228	−49.69	33	0.000

MO_1 stands for the category "Intrinsic Values", questions 1–8, MO_2 stands for the category "Attainment Value", questions, 9–17 and MO_3 stands for the category "Utility/Self efficacy Value", questions 18–24

4   Discussion/Conclusions

The Computational STEM content pedagogy, integrated with engineering design, seems to be effective in teaching and learning as well as on students' capacity to implement this in a form of didactic scenario. Students developed inquiry based scenario and they suggested ways to collect and analyze data and to decompose the problem. They were also engaged in fundamental concepts of CT, like abstraction, algorithmic thinking and decomposition of the problem and pattern recognition. In addition they suggested the creation of artefacts based on engineering design using Arduino, they engaged in the creation of prototype and they understood that prototype should be tested according to the outcomes of the data they collected from their model, according to the CE experiment approach. This article adds to the literature as an example of application of Computational Pedagogy when the engineering education is added in the computational experiment approach. Research is in progress for investigation of the impact of this approach to large scale as wells as on the impact of this approach on higher school students. Results show that the Computational Pedagogy model has a significant impact on students' "Intrinsic Values", "Attainment Value", and "Utility/Self efficacy Value" respectively. Findings will be of special interest to individuals, Engineering School and tertiary Education educators and stakeholders working towards the STEM integration in the curriculum, the quality of STEM education and it will trigger discussions of what STEM education should be. Results can be useful also for Science and Mathematics Education, with focus on "big" ideas of Science and on algorithms like the one used in the mystery triangles. According to the literature, self-efficacy has been suggested as a measure for computational thinking [2] and our results is a first indication that when Computational Pedagogy model is implemented, it could enhance the self-efficacy for STEM content epistemology. Literature states that Computational thinking education allows the cultivation of student's creativity [19], and they could be empowered to become producers of technology with the creation of programming artifacts. Our preliminary results support this, as was indicated by students' willingness to develop artifacts, but further research is necessary to verify this statement.

Acknowledgements. Author acknowledges financial support for the dissemination of this work from the Special Account for Research of ASPETE through the funding program "Strengthening ASPETE's research".

References

1. Barr, V., Stephenson, C.: Bringing computational thinking to K-12: what is involved and what is the role of the computer science education community? Acm Inroads **2**(1), 48–54 (2011)
2. Bean, N., Weese, J.L., Feldhausen, R., Bell, R.: Starting from scratch: developing a preservice teacher program in computational thinking. Frontiers in Education (2015)

3. Bienkowski, M., Snow, E., Rutstein, D.W., Grover, S.: Assessment design patterns for computation-al thinking practices in secondary computer science: a first look (SRI technical report). SRI International, Menlo Park, CA (2015)
4. Brennan, K., Resnick, M.: New frameworks for studying and assessing the development of computational thinking. Presented at the American Education Researcher Association, Vancouver, Canada (2012)
5. Bundy, A.: Computational thinking is pervasive. J. Sci. Practical Comput. **1**(2), 67–69 (2007)
6. Hoyles, C., Noss, R.: Revisiting programming to enhance mathematics learning. Paper presented at the Math + Coding Symposium. Western University, (2015)
7. Honderich, T.: The Oxford Companion to Philosophy. Oxford University Press, NY (1995)
8. Katehi, L., Pearson, G., Feder, M.: Engineering in K-12 education: Understanding the status and improving the prospects. National Academy of Engineering and National Research Council, Washington, DC (2009)
9. Landau, R.H., Páez, J., Bordeianu, C.: A Survey of Computational Physics: Introductory Computational Science. Princeton University Press, Princeton and Oxford (2008)
10. Moore, T.J.: STEM integration: crossing disciplinary borders to promote learning and engagement. In: Invited Presentation to the Faculty and Graduate Students of the UTeachEngineering, UTeachNatural Sciences, and STEM Education Program Area at University of Texas at Austin, 15 Dec 2008
11. National Research Council: A framework for K-12 Science Education: Practices, Crosscutting Concepts, and Core Ideas. National Academies Press, Washington, DC (2012)
12. National Research Council: Discipline-based Education Research: Understanding and Improving Learning in Undergraduate Science and Engineering. National Academies Press, Washington, DC (2012)
13. NGSS Lead States: Next Generation Science Standards: for States, by States. The National Academies Press, Washington, DC (2013)
14. Psycharis, S.: The Impact of Computational Experiment and Formative Assessment in Inquiry Based Teaching and Learning Approach in STEM Education. J. Sci. Educ. Technol. **25**(2), 316–326 (2015) (JOST) https://doi.org/10.1007/s10956-015-9595-z
15. Psycharis, S.: STEAM in Education: A literature review on the role of computational thinking, engineering epistemology and computational science. Computational STEAM pedagogy (CSP). Sci. Cult. **4**(2), 51–72 (2018)
16. Psycharis, S., Kotzampasaki, E.: A didactic scenario for implementation of computational thinking using inquiry game learning. In: Proceedings of the 2017 International Conference on Education and E-Learning, pp. 26–29. ACM, Nov 2017
17. Rugarcia, A., Felder, R.M., Woods, D.R., Stice, J.E.: The future of engineering education: I. A vision for a new century. Chem. Eng. Educ. **34**, 16–25 (2000)
18. Shirey, K.: Teacher productive resources for engineering design integration in high school physics instruction (Fundamental). In: Proceedings of the 2017 ASEE Annual Conference, Columbus, OH, June 2017
19. Voogt, J., Fisser, P., Good, J., Mishra, P., Yadav, A.: Computational thinking in compulsory education: towards an agenda for research and practice. Educ. Inf. Technol. 1–14 (2015)
20. Weintrop, D., Beheshti, E., Horn, M., Orton, K., Jona, K., Trouille, L., Wilensky, U.: Defining computational thinking for mathematics and science classrooms. J. Sci. Educ. Technol. **25**(1), 127–147 (2016). https://doi.org/10.1007/s10956-015-9581-5
21. Wing, J.M.: Computational thinking and thinking about computing. Commun. ACM **49**, 33–35 (2006)
22. Wing, J.M.: Computational thinking and thinking about computing. Philos. Trans. Royal Soc. London Math. Phys Eng. Sci. **366**(1881), 3717–3725 (2008)

23. Yadav, A., Zhou, N., Mayfield, C., Hambrusch, S., Korb, J.T.: Introducing computational thinking in education courses. In: Proceedings of the 42nd ACM Technical Symposium on Computer Science Education, pp. 465–470. ACM (2011)
24. Yaşar, O.: Teaching science through computation. Int. J. Sci. Technol. Soc. 1(1), 9–18 (2013) https://doi.org/10.11648/j.ijsts.20130101.12
25. Yasar O., Veronesi P., Maliekal J., LittleL. J., Vattana S.E., Yeter I.H.: Presented at: ASEE Annual Conference and Exposition. Presented: June 2016. Project: SCOLLARCIT (2016)

Economics' Digitalization—New Challenges to Engineering Education

Irina Makarova$^{(\boxtimes)}$ ⓘ, Ksenia Shubenkova ⓘ, and Vadim Mavrin ⓘ

Kazan Federal University, Naberezhnye Chelny, Russia
kamIVM@mail.ru, ksenia.shubenkova@gmail.com

Abstract. Digital Era and the growing competition require higher education graduates have skills relevant for competing under conditions of rapid technology development and economical growth. The system of engineering education should ensure the quality of engineers that are needed for the business and society. For these purposes, there are new opportunities, which envisages the use of virtual reality, simulations and other IT. The requirements for the automotive industry have risen dramatically in recent years: cheap vehicles, a wider range of models and higher quality are needed to reach the global market. The intelligent functions of the vehicles' on-board systems are expanded through driver assistance systems and IT integration with various devices. The peculiarity of the proposed method is to use in the educational system such digital environment, that is implemented in industry. We suggest using Blended learning, which can become one of key solutions to existing problems in an educational sphere. Interaction with business will allow educators to learn new technologies and apply them in teaching students. This is especially important in transition to Digital Era.

Keywords: Industry 4.0 · Digitalization · Engineering education · Blended learning

1 Introduction

We are at the dawn of the Fourth Industrial Revolution (also known as 4IR), which will bring together digital, biological and physical technologies in new and powerful combinations. When compared with previous industrial revolutions, the Fourth is evolving at an exponential rather than a linear pace. Moreover, it is disrupting almost every industry in every country. Fourth Industrial Revolution has changed the way we need to measure some aspects of competitiveness, particularly in relation to innovation and ideas. the value of ideas and collaboration within companies; the values of open-mindedness, of connectivity, and the value of an entrepreneurial spirit is growing. Now we need a new kind of education, that is more conducive to students' creativity, their ability to observe and generate ideas. The 4IR forces us to pay more attention to all these aspects of a nation's innovation ecosystem.

The digitalization of the economy and society materially change the requirements as well as to the content and the organization of engineers training. Using the capabilities of the industrial Internet of things, universities can also go in the category of

M. E. Auer and T. Tsiatsos (Eds.): ICL 2018, AISC 917, pp. 699–709, 2019.
https://doi.org/10.1007/978-3-030-11935-5_66

"intelligent enterprises" and become "Smart University". The University needs to stay ahead of the situation for 5–10 years, to be a platform where technology of the future are not just used, but "live". The University, as the production, must be "digital" that is digital textbooks and teaching materials, electronic documents, digital online courses, digital logistics of the educational process, etc. It does not replace the formation of professional thinking, but can significantly reduce costs of the educational process. The transition to the model of SMART University is necessary, which uses digital technology as a means of process improvement and is positioned as an entrepreneurial University and innovation ecosystem. The main thing for it becomes the ability to prepare the graduate able to formulate a problem in technical language and solve it using the new intelligent technology. In order to provide conditions for continuous adaptation of the University to the needs of the business it is necessary to create a single information environment, allowing to predict the required competencies at the end of a University student. This is possible due to a flexible educational standards modular and individual educational trajectory. It is necessary to develop a mechanism for the harmonization of professional and educational standards and rapid adaptation to changing needs of the market. This will allow the emergence of new professions and professional standards to identify those educational standards that are more easily adapted to the requirements of the labour market.

2 Economy Digitalization as Systemic Challenges for Engineering Education System

2.1 Changing the Educational Paradigm in the Digitalization Era

It was noted in the report "The Future of Employment" that the drastic changes will affect more than 35% of workers' skills available to modern people within 5 years. Such rates of development of technique and technologies require quite inert educational system and radical change in paradigm. In the WEF report on global competitiveness, it is indicated that global competitive-ness will increasingly determine innovation potential of the country. A new ranking of the digital competitiveness introduces several new criteria to assess the ability of countries to introduce and explore digital technologies, leading to transformation in governmental practices, business models, and society as a whole, which includes Knowledge, Technology, and Future Readiness. To enhance digital competitiveness, the work on the three fronts is required. The central place must be taken by the inter-action of education and business. In order to stimulate the development of the regions, the EU actively encourages the development of regional strategies of smart specialization, despite the fact that the strategy "Europe-2020" is aimed at solving structural problems through progress in three mutually reinforcing priorities:

- "smart" economic growth based on knowledge and innovation;
- sustainable economic growth through a more resource-efficient, green and competitive economy;
- inclusive economic growth through increasing employment and ensuring economic, social and territorial integration.

To solve these problems, we need engineers able to generate ideas, develop innovative projects, create new products and services. To provide opportunity for training of an engineer "for the future" we need to create conditions to change people's consciousness, because the need for engineers with creative thinking is becoming more acute. There is a connection between lack of talent and training with a lack of business activity. In this situation, the main priority is the production and production of knowledge. There is a connection between lack of talent and training with a lack of business agility. In this situation, education and knowledge production are the key. For countries where an innovative smart society will be formed, the main idea of which will be the development of human potential, the World Bank has determined the national wealth forecast structure: natural resources amount to only 5, 18%—material, production capital, while 77%—personnel knowledge and skills which will determine their possessor's future.

The new digital competitiveness rating introduces several new criteria for assessing the country's ability to introduce and study digital technologies that lead to transformation in government practices, business models and the society as a whole that includes Knowledge, Technology and Future Readiness. To increase digital competitiveness, work on these three areas is necessary. The main direction should be to improve the interaction between education and business.

2.2 World Trends in Personnel Management and Demands of Changes in the Educational System

Given the technologies' accelerating development, the 4IR will pay special attention to the workers' ability to continually adapt and acquire new skills and approaches in a variety of contexts [1]. Undoubtedly, digital technologies allow to reduce time for communication, accelerate all processes of economic activity. However, the result of accelerating these processes (progress or economic degradation) depends on the human capital development vector.

The 2017 Deloitte Global Human Capital Trends report [2], drawing on a survey of more than 10,000 h and business leaders globally, takes stock of the challenges ahead for business and HR leaders in a dramatically changing digital, economic, demographic, and social landscape. Today, a new set of digital business and working skills is needed. companies should focus more heavily on career strategies, talent mobility, and organizational ecosystems and networks to facilitate both individual and organizational reinvention. The problem is not simply one of "reskilling" or planning new and better careers. Instead, organizations must look at leadership, structures, diversity, technology, and the overall employee experience in new and exciting ways. The trends in this report identify 10 areas in which organizations will need to close the gap between the pace of change and the challenges of work and talent management.

One of the main trends is the change in core "career" concept that driving companies toward "always-on" learning experiences that allow employees to build skills quickly, easily, and on their own terms. Robotics, artificial intelligence (AI), sensors, and cognitive computing have gone mainstream, along with the open talent economy. These on- and off-balance-sheet workers are being augmented with machines and software. Together, these trends will result in the redesign of almost every job, as well as a new way of thinking about workforce planning and the nature of work. In his 2016 book "Thank You for Being Late", Thomas Friedman refers to a graph created by Eric "Astro" Teller, CEO of Alphabet's Google X division, which suggests that technology is increasing at an ever-faster rate while human adaptability rises only at a slower, linear rate [3].

Therefore, the main production factors in the future economy are human and information capital, while the key factor's role is enshrined in human capital. The founder and president of the Davos Economic Forum K. Schwab is convinced that the main production factor will not be capital, but human resources [4]. As a result of the analysis of employment, K Schwab comes to the conclusion that: "The fourth industrial revolution seems to be creating fewer jobs in new industries than previous revolutions… only 0.5% of the US workforce is employed in industries that did not exist at the turn of the century, a far lower percentage than the approximately 8% of new jobs created in new industries during the 1980s and the 4.5% of new jobs created during the 1990s … innovations in information and other disruptive technologies tend to raise productivity by replacing existing workers, rather than creating new products needing more labour to produce them".

Two researchers from the Oxford-Martin school—economist Carl Benedict Frey and computer training expert Michael Osborne—quantified the potential impact of technological innovation on unemployment by distributing 702 occupations in terms of the likelihood of their automation. According to the results of this study, about 47% of jobs in the US are at risk of automation, most likely for the next two decades. This will cause a more rapid destruction of a much larger range of occupations than changes in the labor market as a result of previous industrial revolutions. In addition, the labor market tends to increase the polarization. Employment will grow in high-yielding cognitive and creative professions and in low-income manual labor, but it will drop significantly in middle-income monotonous standard occupations [5]. Consequently, in industries with high automation risks, employees must have the skills to self-improvement and adapt to new needs in their industry. This task should be solved with the help of an innovative educational system.

2.3   The Transition to a SMART Education System in the Digitalization Era

Digitalization covers all activity areas, resulting in the emergence of the "digital city" concept. With the development of technologies, the new philosophy of "smart education" replaces the "traditional" e-learning. Despite the fact that Smart Education was first discussed in 2009, in many respects due to the so-called second digital gap, it is currently impossible to talk about a single formed paradigm, but rather the description of individual solutions and technologies in the field of education. Thus, the authors of the article [6] based on the questionnaire of 875 respondents and analysis of the results come to the conclusion that task characteristics and social interaction improve media richness, media experience, and media technostress, which in turn enhance ICT adoption behavior. The authors emphasize that in rural areas it is difficult for users to gain access to information technologies and overcome the so-called "digital disruption". However, the authors found that increasing the digital literacy will help to change the behavior of the community in the introduction of ICT.

The main distinguishing feature "Digital City" and "Smart City" is "Smart people". Smart Education is the implementation of educational activities in the Internet on the base of common standards, technologies and agreements established between the network of educational institutions and scientific and pedagogical staff. It is a fundamentally new educational environment, uniting the efforts of teachers, professionals and students to implement global knowledge and to shift from passive content to active. The factors that characterize the level of development of education in Smart City include:

1. Education and training: the number of people with higher education and conformity of the educational programs' basic directions to the real needs of economy.
2. E-Learning: plans of the digital technologies' development in classrooms; the use of ICT in education; implementation of E-learning programs; lifelong learning.
3. Human resources: cooperation of companies with scientific and training centers; training specialists able to solve the real problems of production, taking into account the factors of sustainable urban development [7].

The concept of Smart City assumes a special kind of "Smart Citizen", an active citizen, who must be educated enough to participate in urban processes in order to optimize them. The concept of Smart Education is aimed at the creation of Smart People. Therefore, the quality of training, the correspondence of educational system to the current needs of economics and the ways of how people implement their knowledge and skills are very important. The extent to which the city's population is "smart" can be assessed in terms of the level of qualifications and education of residents, as well as the degree of their social activity and the ability to be open to the outside world.

3 Results and Discussions

3.1 Possible Changes of the Educational System

There is no doubt that higher education transformation is already underway, with every university leader indicating they are at least part way through their digital journey. However, university leaders, edtech founders and students have divergent views when it comes top priorities for change, and how imminent disruption of the traditional university model will be. Overall, universities see digital transformation as a way to enhance their current model, rather than change it.

The university must outstrip the current situation for 5–10 years, be a platform where the future's technologies not only are used, but they "live" already. The university, like production, should be "digital"—it is digital textbooks and methodological materials, electronic document management, digital online courses, digital logistics of the educational process, etc. This will not replace the formation of professional thinking, but can significantly reduce the costs of the educational process. We need a transition to the SMART University model, which uses digital technologies as a means of improving processes and is positioned as an entrepreneurial university and an innovation ecosystem. Universities and research centers play a key role in the innovative ecosystem (Fig. 1). They are not only a source to create new enterprises, but they are also responsible for the development of leaders who will be able to organize the interaction of educational institutions, business and government. Transition to the University 4.0 model, which should contain all characteristics of the SMART University, is necessary.

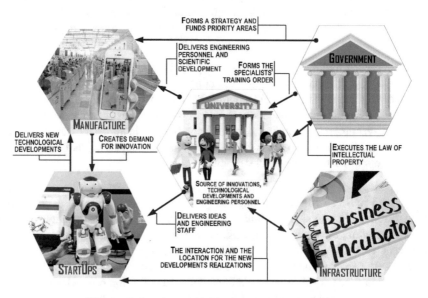

Fig. 1. University role in the innovations' ecosystem

Thus, from the point of view of the processes taking place at the university, digitization will mean a gradual transition from traditional business models to innovative ones. The university, just as the production, must be "digital". For such a university the main goal is to prepare an engineer who can formulate a task in a professional language and solve it with the help of new intelligent technologies.

An important task is the creation of a single digital platform that will enhance the university's sustainability and ensure its fast adaptation to changes. Standardizing, optimizing and integrating systems more efficiently means that valuable technical resources can be used to undertake activities that add real value to the university, for example developing innovative solutions that enhance a university's strategy. IT must be considered a strategic asset across the entire institution, with common standards and principles in place and solution reuse broadly promoted, because IT can analyses systems and map these to business functionality, which means opportunities for simplification and rationalization can be identified and in turn, frees up expert resources for more innovative solution development. It is important to ensure access to any digital resources. This means the use of digital textbooks and resource materials, e-document flow, digital online courses, digital logistics of the educational process, etc.

Now there came free or cheap digital learning platforms like Coursera and the University of Phoenix, offering distance learning to people who would not have otherwise gone to a traditional college. This will not replace the shaping of professional thinking, but it can significantly reduce the costs of the educational process. In article [8] authors say, that open data repository that manages and shares information of different catalogues and an evaluation tool to OER will allow teachers and students to increase educational contents and to improve relationship between open data initiatives and education context. Authors provide new modality of digital educational resources access as a personalized way to meet the knowledge and interests needs of different citizens. This work covers how open data creates value and can have a positive impact in education and research context. In addition to exploring the background, the work also provides concrete information on how to produce open data.

3.2 Methods and Technologies of Teaching at the Digital University

Today, neither traditional education nor E-learning methods do not allow the full use of advanced educational technologies to improve the education quality. This is due to the fact that the applied teaching methods (lectures, practical exercises, seminars, tests) do not allow preparing specialists for production activities at the concrete enterprises. In addition, there is the problem of objectivity in assessing the education quality. An effective solution is the use of the Blended Learning (BL), which unites advantages of both methods, and also allows using progressive educational technologies (modeling, virtual reality application, gaming education, etc.) intensifying the process of studying.

The fundamental difference between BL [9] and the traditional education system is the combination of organizational learning forms in the real and virtual campus of the university and a combination of traditional teaching methods with E-Learning technologies. At the same time, the process of combining technologies can take place both at the level of a separate course, discipline, and at the level of the educational program as a whole. More detailed comparison of existing learning types was presented in our previous study [10].

The use of BL can help to solve problems existing in engineering education, because it allows: (1) more efficient use of time in classrooms, focusing on the problems faced by students, (2) identification of students experiencing difficulties, (3) picking up materials and assignments that are optimal to a particular group and a particular student, taking into account individual characteristics, (4) students' knowledge assessment is based on the use of objective criteria, (5) improvement of motivation and education quality by implementation of progressive educational technologies. Feedback is the most effective tool of the complex systems' reactive management. The educational system has considerable inertia due to the duration of specialist training process. That is why, the effective feedback is also a tool to overcome this inertia.

With the rapid development of network technology and information technology, the information literacy situation students have attracted more and more attention of the society. In urgent need of society, it attaches great importance to the cultivation of information literacy, which pays attention to the research on College Students' information literacy. In paper [11], research is to further explore the situation under the background of big data challenges and corresponding solutions for large, it can provide a solution for reference of the training and development of students' information literacy.

The authors of [12] analyzed how higher education students engage with technology during self-studies and how they in particular utilize different semiotic affordances of information and communication technologies in order to learn course content. Consequently, focus is put on how university students design their learning during self-studies through exploiting multimodal literacy and by constructing knowledge through different modes and media. Currently, expansion of opportunities of this interaction is accompanied with the expansion of the range of methods of educational process organization (E-Learning, co-learning, BL, flipped classroom [13], etc.) as well as learning technologies (modeling and simulation experiments [14], gamification [15, 16], virtual and augmented reality [17]). These methods contribute the students' motivation increasing and developing the necessary competencies.

3.3 Cooperation Between University and Automotive Company in the Era of Digitalization

To ensure that the competence of engineers meet the expectations of business, we conducted a survey of managers of various companies. The aim of the survey was to study necessary competencies required for future engineers, and what software and technical solutions applied in practice. The survey was conducted among companies involved in designing and development of intelligent vehicles, companies operating the vehicles and engaged in logistics, as well as among organizations responsible for the security of transport systems. The aim of the survey was to identify the most important competencies among graduates for their future successful work and career. The survey results have been processed and are grouped into appropriate categories. Learning courses that form necessary competencies were included in the curriculum areas of engineer training for different companies and professional activities. Details of the conducted survey are presented in our previous paper [18].

By participating in real projects, the students got a better idea of requirements imposed on modern vehicles and developed them by employing new design technologies.

The ITS area most in need of a use engineers in different activity lines is that of transport systems control because it is here that the greatest scope of tasks is solved. The engineers, creating programs for intelligent onboard systems, study the computer vision systems and pattern recognition techniques. The issues of logistics, transportation planning and simulation are dealt with in different companies. There are companies which creating the infrastructure, building and reconstructing the street-road networks, determining the location of sites for public transport stops, petrol filling and service centers. Others companies are involved in planning of traffic along the street-road networks and of traffic control with the purpose of enhancing its safety and effectiveness.

Engineers in these companies must know how to obtain adequate information and operate with great data volumes, analyze them and make strategic and operative decisions. Training for these positions envisages instruction in the systems theory, intelligent control systems, statistical data analysis, the theory of experiment planning, and methods of optimization.

Constructing of public transport routes requires exploring the consumer demand for transportation both within and outside the city. The engineers involved in logistics processes and cargo delivery must know the means and methods of cargo shipment planning and managing of motor vehicle fleets, methods of building logistical chains accounting for interaction of different kinds of transport. They must also be trained in how to select the optimal routes with account for different factors and risks. These aspects are studied by doing such courses as statistical data analysis, methods of planning and forecasting, theory of constraint, systems analysis methods, theory of experiment planning, management in logistical systems, and the theory of transport flows.

The students solve real problems which help them to understand the problems of the industry and professional tasks already during their study time. This approach facilitates professional adaptation. Survey results of the employers after the introduction of above described method shows that the degree of satisfaction with the quality of students training increases (Fig. 2).

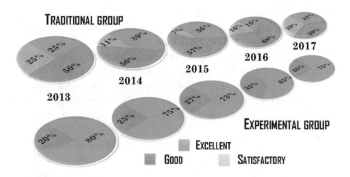

Fig. 2. Dynamics of satisfaction with the quality of students training

4   Conclusions

Digital Education concept implies, first of all, the creation of an educational system that will form not only professional and digital competencies by future engineers, but also develop their environmental consciousness and social responsibility for future generations. Integration of progressive teaching methods will give a synergistic effect by improving the quality of processes in the industrial, transport, educational and other Digital City systems. The gained experience shows that the use of new methods of teaching in engineering education contributes to the development of engineering competencies that are required for the high-tech industries. Students will be prepared to implement complex solutions when designing, modernizing and managing production systems that will increase their efficiency, sustainability and ensure the quality and competitiveness of their products. An integrated platform for implementation of such solutions will improve the quality of both operational and strategic management in both production and educational systems. In addition, the intellectualization of managerial processes will increase the sustainability of these systems and will facilitate their adaptation to the changing needs of the market.

References

1. If Your Employees Aren't Learning: You're not leading—forbes. https://www.forbes.com/sites/markmurphy/2018/01/21/if-your-employees-arent-learning-youre-not-leading/#3713f9d29478
2. Rewriting the rules for the digital age : Deloitte Global Human Capital Trends (2017) https://www.legal-island.com/globalassets/pdf-documents/digital-hr.pdf
3. Friedman, T.L.: Thank You for Being Late. Farrar, Straus & Gioux, pp. 213–219 (2016)
4. Schwab, K.: The fourth industrial revolution. World Economic Forum 91–93 route de la Capite CH-1223 Cologny/Geneva Switzerland www.weforum.org 172 p. https://luminariaz.files.wordpress.com/2017/11/the-fourth-industrial-revolution-2016-21.pdf
5. Frey, C.B., Osborne, M.A.: The Future of Employment: How Susceptible are Jobs to Computerisation? (2013) https://www.oxfordmartin.ox.ac.uk/downloads/academic/The_Future_of_Employment.pdf
6. Yu, T.-K., Lin, M.-L., Liao, Y.-K.: Understanding factors influencing information communication technology adoption behavior: the moderators of information literacy and digital skills. Comput. Hum. Behav. **71**, 196–208 (2017)
7. Uskov, V.L., Bakken, J.P., Karri, S., et al.: Smart university: conceptual modeling and systems' design. Smart Innovation Syst. Technol. **70**, 49–86 (2018)
8. Piedra, N., Chicaiza, J., López, J., Caro, E.T.: A rating system that open-data repositories must satisfy to be considered OER. reusing open data resources in teaching. In: Global Engineering Education Conference (EDUCON). 2017 IEEE, pp. 1768–1777. Athens, Greece (2017)
9. Baytiyeh, H., Naja, M.K.: Students' perceptions of the flipped classroom model in an engineering course: a case study. Eur. J. Eng. Educ. **42**(6), 1048–1061 (2017)
10. Makarova, I., Shubenkova, K., Tikhonov, D., Pashkevich, A.: An Integrated Platform for Blended Learning in Engineering Education. In: CSEDU 2017, vol. 2, pp. 171–176
11. Ying, Y.: Research on college students' information literacy based on big data. Cluster Comput. https://doi.org/10.1007/s10586-018-2193-0
12. Nouri, J.: Students Multimodal Literacy and design of learning during self studies in higher education. Technol. Knowl. Learn. https://doi.org/10.1007/s10758-018-9360-5
13. Tsai, T.P., Lin, J., Lin, L.C.: A flip blended learning approach for ePUB3 eBook-based course design and implementation. EURASIA J. Math. Sci. Technol. Educ. **14**, 123–144 (2018)
14. Makarova, I., Shubenkova, K., Tikhonov, D., Buyvol, P.: Improving the quality of engineering education by developing the system of increasing students' motivation. In: ICL 2017. Advances in Intelligent Systems and Computing, vol. 716. Springer, Cham (2018)
15. Çakıroglu, U., Basıbüyük, B., et al.: Gamifying an ICT course: influences on engagement and academic performance. Comput. Hum. Behav. **69**, 98–107 (2017)
16. Yáñez-Gómez, R., et al.: Academic methods for usability evaluation of serious games: a systematic review. Multimed Tools Appl **76**, 5755–5784 (2017)

17. Ucán Pech, J.P., Aguilar Vera, R.A., Gómez, O.S.: Software testing education through a collaborative virtual approach. In: Advances in Intelligent Systems and Computing, vol. 688, pp. 231–240 (2018)
18. Makarova I., Shubenkova K., Buyvol P., Mavrin V., Mukhametdinov E.: Interaction between education and business in digital era. In: 1st International Proceedings on 2018 IEEE Industrial Cyber-Physical Systems (ICPS), pp. 503–508. IEEE, St. Petersburg, Russia (2018)

Interrelation of Enthusiasm for Work and Professional Burning Out at Teachers of Engineering Higher Education Institution

Khatsrinova Olga, Mansur Galikhanov$^{(\boxtimes)}$, and Khatsrinova Julia

Kazan National Research Technological University, Kazan, Russia
{khatsrinovao,khatsrinoval2}@mail.ru,
mgalikhanov@yandex.ru

Abstract. The teacher of the higher school needs to create a training environment now. It assumes not only knowledge of a subject, but an opportunity to build up the personal relationship with students. It is possible under a condition when in educational process participants feel satisfaction, but not a stress. But professional burning out also is also characteristic of teachers. It leads to a negative side of activity. It is possible to prevent development of burning out, if constantly to hold the training seminars on development of the emotional sphere and development of positive motivation of professional and pedagogical activity with teachers.

Keywords: Professional burning out · Enthusiasm for work · Mental phenomena

1 Introduction

Recently interest in studying of favorable states and strengths of the person increases—it a referral of researches was got by the name of positive psychology. It is caused by the fact that knowledge of factors of development and consequences of negative phenomena does not give a full picture about mental activity of the person. Prevention of adverse states and development of measures for optimization of level of operability of subjects of activity are possible if systematically to create conditions for development of strengths and abilities of people and communities, but not just to correct shortcomings and frustration. Especially it is important for teachers of the higher school who are in interrelation with students of different age and the directions of preparation. Has this appearance of professional activity as negative qualities: for example, surplus of communication in the course of work, physical and psychoemotional activities. On the other hand, the considered type of activity contains also positive characteristics (dynamism, creativity, possibilities of career and professional development, gratitude from students) which can contribute to the development of enthusiasm of workers. The matter was studied in works about professional burning out (S. by Maslach, M. Leiter, M. Burisch, A. Pines, V.E. Orel, T.I. Ronginskaya, N.E. Vodopyanova, E.S. Starchenkova, V.V. Boyko, A.A. Rukavishnikov) and enthusiasm for work (W. Schaufeli, A. Bakker, M. Salanova, V. Gonzalez-Roma). It is represented to us that the

© Springer Nature Switzerland AG 2019
M. E. Auer and T. Tsiatsos (Eds.): ICL 2018, AISC 917, pp. 710–719, 2019.
https://doi.org/10.1007/978-3-030-11935-5_67

combination of expressiveness of enthusiasm and burning out can be differentiated that will expand ideas of interrelation of these phenomena. However not all researchers divide a point to sight that the interrelation of enthusiasm for work and professional burning out has unambiguously negative sign. Some scientists [1–3] point that the "burnt-out" workers are initially keen on work, show activity and enthusiasm in professional activity. Excessive positive installation in the work relation limits the interests and requirements which are not connected with work that can lead to decrease in vigor of workers, development of fatigue and burning out over time. But representatives of this approach do not provide empirical proofs of the point of view and consider enthusiasm as the initial stage of development of burning out, but not as an independent psychological phenomenon. Valuable information the empirical typology of a combination can provide them to expressiveness for clearing of features of interrelation of enthusiasm and burning out. As a rule, researchers assume existence of 2 types—a combination of high degree of enthusiasm to the low level of burning out and, on the contrary, the high level of burning out with low degree of enthusiasm. The first type is considered as favorable, the second—as adverse (for health of the worker and efficiency of activity). It is represented to us that the combination of expressiveness of enthusiasm and burning out can be differentiated that will expand ideas of interrelation of these phenomena.

1.1 The Purpose

The research objective consisted in studying of features of interrelation of enthusiasm for work and professional burning out at teachers of engineering higher education institution.

There is no uniform point of view on interrelation of enthusiasm and burning out—the phenomena reflecting degree of wellbeing of mental activity of the subject of work and its professional motivation. For studying of features of this interrelation it is necessary to carry out the substantial analysis of the studied phenomena. Identification of empirical types of a combination of expressiveness of enthusiasm and burning out and comparison of communications of the considered phenomena with other variables allows to expand ideas of a ratio of enthusiasm and burning out.

For achievement of this purpose we solved the following problems: to carry out the theoretical analysis of researches of enthusiasm for work and professional burning out, their interrelations, development factors; to analyse and compare the content of manifestations of enthusiasm and burning out; to establish interrelations of enthusiasm and burning out with personal properties and value judgment of characteristics of activity; to reveal empirical types of a combination of expressiveness of enthusiasm and burning out. We consider that in pedagogical activity there are manifestations of enthusiasm and burning out which are not reduced to opposite. Also the enthusiasm for work is connected with positive, and burning out—with negative estimates of activity. The enthusiasm is connected negatively with a neyrotizm and is positive—with an ekstraversiya, openness, tendency to consent, conscientiousness; communications of burning out with these personal properties have opposite signs. Different empirical types of a combination of expressiveness of enthusiasm and burning out meet.

Representatives of empirical types differ in personal properties and value judgment of characteristics of activity.

2 Approach

The Teoretiko-metodologichesky basis of a research was made: provisions of system and subject and activity approaches in psychology (B.F. Lomov, C.JI. Rubenstein, V.D. Shadrikov, A.N. Leontev, K.A. Abulkhanov-Slavskaya, L.G. Dikaya, V.A. Bodrov, D.N. Zavalishina, A.A. Oboznov, E.A. Klimov), concepts of professional burning out (S. of Maslach, M. of Leiter, M. of Burisch, A. Pines, V.E. Orel, T.I. Ronginskaya,

N.E. Vodopyanova, E.S. Starchenkova, V.V. Boyko, A.A. Rukavishnikov) and enthusiasm for work (W. Schaufeli, A. Bakker, M. of Salanova, V. Gonzalez-Roma), provisions of positive psychology (M. of Seligman, M. Csikszentmihalyi), modern paradigms of studying of a psychological stress in professional activity (V.A. Bodrov, JI.A. Kitayev-Smyk, A.B. Leonova, R. Lasarus, S. Folkman), concepts of mental self-control (O.A. Konopkin, A.A. Oboznov, L.G. Dikaya, V.I. Morosanova), concepts of resistance to stress and resources of overcoming a stress (V.A. Bodrov, S.K. Nartova-Bochaver, R. Lasarus, S. Folkman, S. Hobfall, A. Bandura), scientific approach to a psychological research of activity in the sphere of professional education.

Collecting empirical data was carried out with use of the following techniques: "The questionnaire on burning out" (the author K. Maslak, N.E. Vodopyanova's adaptation), "The Utrecht scale of enthusiasm for work" (authors of U. Skhaufeli and the colleague, D.A. Kutuzova's adaptation), "the Five-factorial questionnaire of the personality" (authors P. Costa and R. Makkrey, M.V. Bodunov and S.D. Biryukov's adaptation), the author's questionnaire for estimation of characteristics of activity in the sphere of pedagogical activity.

For designation of the phenomenon opposite to burning out, S. to Maslach and M. Leiter [4] offered the term "enthusiasm for work" (work engagement). The Dutch researcher of W. Schaufeli with colleagues [5] suggested to consider enthusiasm for work as the independent phenomenon, but not as the complete antithesis of burning out. Also defined enthusiasm as the positive, bringing satisfaction and the phenomenon connected with work which is characterized by vigor, enthusiasm and preoccupation. To add a picture of a ratio of enthusiasm and burning out results about communications of these phenomena with other variables can. In our work personal properties (model "Big Five" P. Costa and R. Makkrey) and the characteristics of activity estimated by participants of a research act as such variables. Manifestations of enthusiasm, and manifestation of burning out belong to such categories of the mental phenomena as states, the relations, properties. But the maintenance of the relations and properties reflecting manifestations of enthusiasm for work is not opposite to the content of the same categories reflecting burning out manifestations. From what the conclusion was drawn on substantial groundlessness of the data of enthusiasm and burning out to the polar phenomena. Results of substantial comparison of enthusiasm and burning out have the theoretical importance as it will allow to provide training of teachers taking into account these factors.

3 Results

As burning out arises when resources of the person on overcoming a stressful situation are exceeded, measures for decrease in level of burning out first of all have to be directed to restoration of deficiency of internal energy and restriction of their loss. It is possible to prevent development of burning out if constantly to accumulate resources and in due time to restore their expenses and also to use koping-strategy adequate to situations.

Object of our research are representatives of the sphere of engineering education. 38 examinees participated in a flight research (teachers of departments of the Kazan National University of Science and Technology, from them 25 men, 13 women). The volume of selection of the main research was 107 examinees (representatives of four departments, of everyone on 24, 26, 26 and 31 workers), of them 33 women, 74 men. Age of examinees—from 28 to 67 years (with average value 46,3 years). Length of service at department—from 2 to 35 years (with average value of 25 years).

Table 1 summarizes all the investigated psychodiagnostic indicators

Table 1. System of psychodiagnostic indicators

Variables	Techniques	Indicators (scales)
Enthusiasm for work	Utrecht scale of enthusiasm for work	1. vigor; 2. enthusiasm; 3. preoccupation
Professional burning out	The questionnaire on burning out	1. exhaustion; 2.cynicism; 3. Reduction of personal achievements
Personal properties	Five-factorial questionnaire of the personality	1.neyrotizm; 2.ekstraversiya; 3. openness to experience; 4. Tendency to consent; 5. conscientiousness
Characteristics of activity in the sphere of engineering education	The author's questionnaire for estimation of characteristics of activity	estimates of 46 characteristics and total score

The characteristics of the states that make up the phenomena can be considered as opposites. Enthusiasm is described by the flow of forces, the experience of positive emotions, loss of sense of time (time flies by unnoticed) [6]. Burnout, by contrast, is characterized by a lack of emotions, constant fatigue, depression, a sense of the ultimate tension of the internal forces [7] Table 2.

The relations with students are allocated by us in a separate subcategory since cover only aspect of interaction with them, unlike a subcategory of the relations to the work including both aspect of interaction, and scientific, methodical, organizing and other kinds of activity characteristic of teachers of the higher school. Professional abilities are referred to category of mental properties as reflect the steady ways of actions based on correct use of the acquired knowledge.

Table 2. Manifestations of enthusiasm for work and professional burnout by categories of mental phenomena

Manifestations of enthusiasm for work	Mental categories	Manifestations of professional burning out
vigor, positive emotions, absorption in work "with the head"	states	emotional devastation, constant fatigue and desire to have a rest, apathy
relations		
enthusiasm, pride, interest	in work	–
–	to students	indifference, cynicism
–	to life	disappointment, loss of belief in implementation of plans, dissatisfaction
properties		
persistence	personal properties	–
–	professional abilities	low level of abilities of work with students and colleagues

Manifestations of burning out cover features of the attitudes of teachers towards students: indifference, callousness, lack of interest in them, aspiration to communicate is as little as possible also without emotions. And manifestations of enthusiasm describe features of the relations to work (enthusiasm, inspiration, pride of work, job evaluation as purposeful and interesting). At the same time it is not concretized to what parties of professional activity (scientific, methodical and others) the relations of keen workers are directed. Possibly, the "burning-out" worker is unaffected also by other types of the professional activity. However with confidence we cannot claim that the relations of the carried-away and "burning out" subjects to the work have unambiguously opposite contents. Also manifestations of burning out reflect specifics of the relation of workers to life: dissatisfaction with achievements, loss of belief in implementation of plans, loss of interest in what pleased earlier. However the teacher who is carried away by a profession can be not satisfied with other aspects of the life.

Therefore by consideration of features of the relations of the teachers who are a part of phenomena we do not find their full opposition.

Is not mutually exclusive the manifestations of enthusiasm and burning out reflecting features of personal properties of teachers. The enthusiasm includes properties which reflect such features of workers as: persistence, aspiration to be engaged in scientific or methodical work. And manifestations of burning out reflect features of properties which disclose professional abilities of workers: ability to find out needs of students; ability to avoid conflict situations with students and colleagues, to create the atmosphere of goodwill in collective.

Therefore, the assumption that manifestations of enthusiasm and burning out are not completely opposite according to contents was confirmed.

Based on results of the analysis of content of manifestations of enthusiasm and burning out and on representations of other researchers, our understanding of maintenance of these mental phenomena is stated [8]. The enthusiasm for work is a motivational phenomenon and reflects aspiration of the subject to immersion in activity, to achievement of good results and receiving satisfaction from work. Authors [9–11] note that the enthusiasm is rather motivational aspect, but not reaction to external conditions. Though maintenance of high enthusiasm requires existence of balance between external requirements and internal opportunities. A burning out phenomenon more than the enthusiasm, is reaction to external conditions. In spite of the fact that the predisposition to burning out development (as well as enthusiasm) is caused by individual properties. Burning out is formed gradually and in development has similar lines with professional deformation [12]. Similar to deformations, the created burning out is shown also out of work, and the enthusiasm is found generally at work. Burning out is the demotivating factor in work [13–16] as includes states and installations, adverse for the subject of activity.

In selection all 3 levels of expressiveness of scales of burning out and enthusiasm are presented, but at most of examinees the low level of scales of professional burning out and the average level of scales of enthusiasm for work are shown. Personal properties of examinees are characterized by prevalence of very high values on a scale of a neyrotizm, high values on an openness scale to experience, average values on scales of an ekstraversiya, tendency to consent, conscientiousness. Selection is described by prevalence of averages and appreciation of activity (judging by a total score of characteristics) that speaks about the general satisfaction of examinees with various aspects of the work. For synthesis of data on estimates of characteristics of activity the factorial analysis is carried out (a method main a component, a way of rotation Varimaks rated). The 7-factorial structure of characteristics of work in the sphere of professional education explaining 53.4% of the general dispersion and covering 30 of 46 estimated aspects is received. Let's list the allocated factors: 1. satisfaction with work in general; 2. free time and rest; 3. professional growth and relations with the management; 4. stability; 5. relations with students and colleagues; earnings; 7. predictability of work. The further analysis considers both indicators of separate characteristics, and their factorial structure generalizing most of them.

Based on results of the analysis of content of manifestations of enthusiasm and burning out and on representations of other researchers, our understanding of maintenance of these mental phenomena is stated [7]. The enthusiasm for work is a motivational phenomenon and reflects aspiration of the subject to immersion in activity, to achievement of good results and receiving satisfaction from work. Authors [8–11] note that the enthusiasm is rather motivational aspect, but not reaction to external conditions. Though maintenance of high enthusiasm requires existence of balance between external requirements and internal opportunities. A burning out phenomenon more than the enthusiasm, is reaction to external conditions. In spite of the fact that the predisposition to burning out development (as well as enthusiasm) is caused by individual properties. Burning out is formed gradually and in development has similar lines with professional deformation [12]. Similar to deformations, the created burning out is shown also out of work, and the enthusiasm is found generally at work. Burning out is the demotivating factor in work [13–17] as includes states and installations, adverse for the subject of activity.

In selection all 3 levels of expressiveness of scales of burning out and enthusiasm are presented, but at most of examinees the low level of scales of professional burning out and the average level of scales of enthusiasm for work are shown. Personal properties of examinees are characterized by prevalence of very high values on a scale of a neyrotizm, high values on an openness scale to experience, average values on scales of an ekstraversiya, tendency to consent, conscientiousness. Selection is described by prevalence of averages and appreciation of activity (judging by a total score of characteristics) that speaks about the general satisfaction of examinees with various aspects of the work. For synthesis of data on estimates of characteristics of activity the factorial analysis is carried out (a method main a component, a way of rotation Varimaks rated). The 7-factorial structure of characteristics of work in the sphere of professional education explaining 53.4% of the general dispersion and covering 30 of 46 estimated aspects is received. Let's list the allocated factors: 1. satisfaction with work in general; 2. free time and rest; 3. professional growth and relations with the management; 4. stability; 5. relations with students and colleagues; 6. earnings; 7. predictability of work. The further analysis considers both indicators of separate characteristics, and their factorial structure generalizing most of them.

Results of the correlation analysis showed that scales of enthusiasm are not connected with estimates of the relations of workers with colleagues and students and also with assessment of degree of risk to lose work and satisfaction with earnings in general. With indicators of burning out communications with estimates of uniformity of distribution of working loading and degree of independence, independence of the teacher are not revealed. Neither with burning out indicators, nor with indicators of enthusiasm correlations with estimates of predictability of work, a possibility of "discharge" of emotions at work are not revealed. For drawing up complete idea of interrelations between the studied variables the correlation analysis between estimates of activity and indicators of personal properties of examinees is carried out. It is received that favorable estimates of activity positively correlate with scales of an ekstraversiya, openness, tendency to consent, conscientiousness, it is negative—with a neyrotizm scale. And adverse estimates of characteristics, on the contrary, are connected directly with a scale of a neyrotizm and back—with scales of an ekstraversiya, openness, tendency to consent, conscientiousness. As high extent of burning out and adverse estimates of activity speak about discrepancy between the personality and work, the personal profile combining the high level of a neyrotizm and low level of an ekstraversiya, openness, tendency to consent, conscientiousness is designated as "undesirable" to representatives of a pedagogical profession. And the return personal profile (low level of a neyrotizm and high level of an ekstraversiya, openness, tendency to consent, conscientiousness) is designated as "desirable" for this profession—workers with such profile estimate characteristics of activity favorably and are keen on the carried-out activity. Differences on expressiveness of scales of enthusiasm and burning out in groups of different age are revealed: cynicism level (a burning out scale) accepts the highest values at examinees at the age of 35–47 years (middle-aged group), in comparison with younger (up to 35 years) and seniors (of 48 years); at examinees with the greatest age (of 60 years) it is significant above value of enthusiasm (an enthusiasm scale), in comparison with group of middle age (45 years). Therefore, the level of cynicism is most expressed in middle-aged group of examinees (35–47 years), and enthusiasm level—at the most senior workers (of 60 years). Differences of expressiveness of components of enthusiasm and burning out in groups of examinees of different length of service are found. In group with the smallest length of service (to 2 years) the level of all three scales of enthusiasm is higher, and burning out scales—below, than in group with an average experience (25 years). Also in group with the smallest experience values of scales of enthusiasm (vigor and enthusiasm) in comparison with group of the greatest experience (more than 20 years) are higher. Means, the highest degree of enthusiasm for work is characteristic of workers with an experience to 2 years (the smallest on selection), and the highest level of burning out—of workers with an experience of 25 years (average on selection). The discriminant analysis showed high reliability of division of examinees into 4 clusters (accuracy of prediction of the proposed cluster solution is 98.13%). The revealed empirical types of a combination of expressiveness of enthusiasm and burning out are designated as "keen" (25 examinees), "moderate" (40 examinees), "not keen" (17 examinees) and "burning out" (25 examinees). At examinees of the specified types distinctions on expressiveness of personal properties and estimates of characteristics of activity (a total indicator and separately taken characteristics) are revealed. These results confirmed the assumptions of existence of several types of a combination of expressiveness of variables and of distinctions of personal properties and estimates of characteristics of activity at examinees of different empirical types. We revealed 4 empirical types of a combination of expressiveness of enthusiasm and burning out. Differences in values of personal properties, estimates of characteristics of activity and length of service are found in representatives of types. Detection of empirical types and their description taking into account properties of the personality and characteristics of activity is the new result expanding idea of features of interrelations between the phenomena of enthusiasm and burning out of teachers of the higher school.

Carried away" the type is described by the highest level of enthusiasm and the lowest extent of burning out. Teachers of "keen" type are characterized by small length of service in the pedagogical sphere, low values of a neyrotizm, high values of an ekstraversiya, openness, tendency to consent, conscientiousness. The "moderate" type is described by the raised indicators of enthusiasm and the lowered burning out indicators. Bigger length of service, than at teachers of "keen" type, average expressiveness of values of a neyrotizm, ekstraversiya and openness is characteristic of teachers of this type. Neuvlechenny type groups workers with the lowest level of enthusiasm and average extent of burning out. Representatives of "not keen" type are described by the low level of an ekstraversiya and openness to experience. For the fourth, "burning out", type the highest values of burning out and average values of enthusiasm are characteristic. Teachers of this type are described by the highest values of a neyrotizm and low values of an ekstraversiya.

The result about distinctions of examinees of "moderate" and "keen" types of length of service says that at examinees at the beginning of a professional way the level of enthusiasm and below—extent of burning out is higher. Such result was confirmed by results of comparison of scales of phenomena in groups of different length of service: the highest level of enthusiasm is observed at workers with the smallest length of service (up to three years), in comparison with examinees of an average and most long standing. Extent of burning out higher at workers with an average experience (10–

15 years), in comparison with examinees with the smallest experience. Examinees of "keen" and "moderate" types estimate characteristics of the activity as favorable (but workers of "keen" type estimate some aspects more positively, than workers of "moderate" type). Representatives of two other types ("not keen" and "burning out") negatively estimate characteristics of the work (but some aspects of activity more favorable by estimates of workers of ("not keen", but not "burning out" types).

4   Conclutions

Among personal properties most closely the expressiveness of an ekstraversiya and openness to experience is connected with the level of enthusiasm. With extent of burning out the expressiveness of a neyrotizm and introversion most closely correlates. Personal properties of tendency to consent and conscientiousness have less noticeable communication with the studied phenomena. It is revealed that the more favorably teachers estimate characteristics of the activity, the level of their enthusiasm is higher, and, on the contrary, the activities for estimates of workers are less favorable, the extent of their burning out is higher. Therefore, the enthusiasm is connected with positive estimates of characteristics of activity, and burning out—with negative. At the same time communications of estimates of activity with burning out components closer, than with enthusiasm scales. For each empirical type of a combination of expressiveness of enthusiasm and burning out practical recommendations about maintenance of high degree of satisfaction with work and working capacity, possible measures for improvement of the health, increase in labor motivation and professionalism are offered.

Estimates of characteristics of activity (in general) have more close connection with burning out components, but not enthusiasm for work. The expressiveness of an ekstraversiya and openness to new experience are the main personal determinants of high enthusiasm for work, and expressiveness of a neyrotizm and introversion are the main personal determinants of burning out. The received result about positive communication between some components of enthusiasm and burning out sets thinking on how manifestations of enthusiasm for work can affect mental characteristics of the subject of work. To us it is obviously possible that under certain conditions (when there are no sources of replenishment of the resources spent by the worker), readiness of the worker for considerable efforts at work, immersion in activity and absence desire to interrupt it can contribute to the development of adverse states (exhaustion, exhaustion) and installations of the worker, to reduce his professional motivation. As a result there is probable an emergence of signs of professional burning out which level will increase in the absence of energy restoration.

As a result of interpretation of results of a research we came to a conclusion that under certain conditions for health of the subject and result of activity not the high, but average level of enthusiasm for work can be optimum. Also some manifestations of burning out can be natural in similar conditions, and if their level does not exceed average values, they can promote economy of resources throughout the long period of work.

Conclusions that personal determinants of enthusiasm and burning out have differences can be considered when developing concrete practical recommendations, for example, at creation of psychological selection and training of teachers of the higher school.

Referenses

1. Berezovskaya, R.A.: Psychology of professional deformation. In: Nikiforov, G.S (ed.) Psihologichesky Ensuring Professional Activity: Theory and Practice. Speech, pp. 598–632. St. Petersburg State University, St. Petersburg (2010)
2. Bondarenko, I.N.: Personal determinants procedural moti-work vation. Avtoref. yew. ... edging. психол. sciences. – M, YIP RAHN, p. 26 (2010)
3. Bondarenko, I.N.: Adaptation of the questionnaire "diagnostics of the worker motivation" R. Hakmana and G. Oldkhem on Russian-language selection. Psychologistsc Mag. **31**(3) 109–124 (2010)
4. Maslach, C.: Understanding burnout: Definitional issues in analyzing a complex phenomenon. In: Paine, W.S. (ed.) Job Stress and Burnout. Sage, Beverly Hills (1982)
5. Schaufeli, W.B., Leiter, M.P., Maslach, C.: Burnout: 35 years of research and practice. Career Dev. Int. **14**(3), 204–220 (2009)
6. Vodopyanova, N.E.: Counteraction to a burning out syndrome in a context the resource concept of the person. The Messenger St. Petersburg University -sitet. It is gray. **12**(1), 75–86 (2009)
7. Garanyan, N.G.: Perfectionism and mental disturbances (the review for—rubezhny empirical researches). Therapy of mental races -stroystvo, vol. 1, pp. 23–31. (2006)
8. Glauberman, D.: Joy of combustion. As doomsday can become new beginning. – M.: Kind book, p. 368 (2004)
9. Gordeeva, T.O.: Theory of self-determination E. Desya and R. Ryan. Psihologiya of motivation of achievement: studies. grant. – M.: Sense: Academy, pp. 201–245 (2006)
10. Grishina, N.V.: The helping relations: professional and ekzistentsialny problems. In: Krylov, A.A., Korostyleva, L.A. Psychological problems of self-realization persons. pp. 143–156. Publishing house of St. Petersburg State University, St. Petersburg (1997)
11. Dimova, V.N.: Personal determinants and organizational factors development of mental burning out of the personality in professions "the subject—object" type, Yaroslavl, p. 25 (2010)
12. Koshelev, A.N.: Syndrome of "a white collar" or prevention "professional burning out". – M.: ,2008. p. 240
13. Eagle, V.E.: Issledovaniye of a phenomenon of mental burning out in otechestvenny and foreign psychology. Problem of the general and organizational psychology. Yaroslavl pp. 76–97 (1999)
14. Formanyuk, T.V.: Syndrome of "emotional combustion" as indicatorprofessional disadaptation of the teacher. Psychology Questions. vol. 6. pp. 54–67 (1994)
15. Kuznetsova, Y.M.: Psychosemantic approach to the description of influence of emotional burning out on a valuable context pedagogical deya-telnost. Psychol. Diagn. **5** 36–58 (2007)
16. Lengle, A.: Emotional burning out from positions existential ro the analysis. Psychol. Questions. **2** 3–16 (2008)
17. A stress, burning out, mastering in a modern context. Zhuravleva, A.L., Sergienko, E.A. (eds.) – M.: Ying t of psychology of RAS, p.512 (2011)

Development of Digital Competences of Teachers of Social Sciences at Secondary Vocational Schools

Pavel Andres[✉] and Petr Svoboda

Masaryk Institute of Advanced Studies, Czech Technical University in Prague, Prague, Czech Republic
{pavel.andres,petr.svoboda}@cvut.cz

Abstract. The paper points out to digital competence as an essential part of the competence model of a teacher in education. It points out to the fact that existing competence models need to be further explored, decompiled, and illustrated by the digital competences extensions. This is related to the importance of applying digital technologies to teaching and management.

Keywords: Digital competences · Teachers of humanities and social sciences
Continuing education of pedagogical staff

1 Introduction

Digital technologies have undergone rapid development in recent years, with implications for education as well. Many of us are sure to ask a few questions. Why is it necessary to deal with digital competencies? What are the reasons for disposing of these competences? What are the preconditions for using digital technologies? Are there barriers affecting the spread of new technologies in education sector? What is the look into future? This is related to the possibilities of using mobile devices and m-technologies in the education process and effective school management through digital technologies.

The reflection of the modern didactical trends takes place on the boundaries of the pedagogical, psychological and also sociological disciplines. The development is considerably accelerating in consequence of the ongoing technological changes and innovations. What is the real impact on the key and professional competences of the secondary school graduates? What does these competences follow? Do the education programmes reflect the demand of competences required by the industry 4.0 and the practice generally? Wherein is the present education reality determined in the sense of a broader conception of education technologies?

Present-day time brings new technologies called as e-technologies. Communication in the traditional school was and is concentrated on the direct verbal and nonverbal contact of the communicating persons. Currently enter into the education space the electronic communications. To the best known belong for example: E-mail, Chat, ICQ, Skype, WhatsApp, Viber, LinkedIn, Facebook, Cloud, LMS systems, Webinars,

M. E. Auer and T. Tsiatsos (Eds.): ICL 2018, AISC 917, pp. 720–731, 2019.
https://doi.org/10.1007/978-3-030-11935-5_68

Educasting, Podcasting. They are an effective and prospective support of education and also a positive supportive mean of upbringing.

In looking for this causes and effects it is necessary to accept the present situation but also take a look further ahead on the development of new technologies that are suitable and meaningful for application into the education reality. From the report Innovating Pedagogy 2013–2016 it is possible to estimate that expansion concerns primarily personal learning environment (PLE), m-learning, massive online open courses (MOOC), new objects in the distance education, wiki, blogs, RSS, use of the licence Creativecommons, sharing of electronic study supports in the cloud, u-learning, t-learning, educasting, seamlesslearning, social webs, omnipresent clever telephones and tablets, extended mobile reality, and in general a shift towards mobile education technologies. To the fore come new skills, frequently called as skills for the 21st century.

The paper points out to digital competence as an essential part of the competence model of a teacher in education. It points out to the fact that existing competence models need to be further explored, decompiled, and illustrated by the digital competences extensions. This is related to the importance of applying digital technologies to teaching and management.

The aim is to contribute to the solution of the actual problem of insufficiently developed digital competences of teachers of humanities and social sciences at the secondary vocational education. If these teachers are not digitally competent, they will not be able to develop pupils' digital competence, which will negatively impact their employability on the labor market in terms of transition to industry 4.0, readiness for further professional education and the ability to participate in lifelong learning. Respondents will increase digital competencies to be able to handle the demands of digital education paradigma. In cooperation with NÚV as an application guarantor no1., the results will be transformed so that they can be applied to the system of continuing education of pedagogical staff.

2 Using Digital Technologies—Assumptions and Reasons

As a result of technological change and innovations, developments have been greatly accelerating. Management activity involves the promotion of the formation of a large number of competencies [1] and becomes more flexible and fulfills the needs of effective management of school organizations. Digital technologies are included in the pedagogical practice management model and the role of the director is crucial when introducing new technologies into the educational process. It is necessary to use digital technology throughout the school year in organizing and managing the school.

It is a prerequisite that school staff are already inadvertently using current technologies [2] to prepare for meetings, to manage pedagogical conferences remotely, to implement electronic conferences, to self-study, to present a school, to prepare a school agenda, to organize and manage a school through appropriate software products (e.g., school information systems), use cloud services, interactive technologies, advanced didactic resources in teaching, e-learning, m-learning and Internet telephony.

Without the inclusion of digital technologies in education, the school loses credit of modern educational institutions, one of the reasons for the use of digital technologies in education [3]. Without digital technologies, students cannot be prepared for further education and application in the knowledge society and industry 4.0. Digital technologies help to educate talented and disabled students, evaluate students, enable parents to interact with the school, get acquainted with school documents/school rules, school annual report, ICT plans, school council, admissions, student activities, projects, competitions, offers of courses, tenders, school organizations. Students and parents have the opportunity to watch online classifications on the school website, results of competitions, educational counseling, participation in workshops, etc. In some schools, students use their own mobile devices with an Internet connection (BYOD[1]).

Digital competences correspond to key lifelong skills and must be considered in the field of teacher education [4]. A school worker who has a digital competence effectively works with information and data using modern information and communication technologies. It is orientated in current new trends in education and is able to apply it to practice.

The reasons for the manager to have (also developed) digital competences [5]:

- innovative teaching practices, examples of good practice, motivation,
- importance of digital technologies for management activity, the contribution of change and school development,
- Management skills—e.g. change management, knowledge management, implementation management, time management.

The expansion of new technologies in education affects barriers
According to research in the project *Professionalization of key competences of school and school management* [6], barriers to the spread of new technologies in education are: lack of school equipment (30%), lack of student interest (20%), distrust of new and untested practices 35%). Only 15% of respondents believe that the extension of new technologies to education is barrier-free.

Barrier 1—Insufficient school equipment
Teachers do not include new technologies in their education due to the lack of material facilities of schools, limiting the transmission of data over the network, finding suitable software for teaching, teachers' reluctance to use these resources, and financial restrictions on acquisition and operation.

Barrier 2—Disinterest of students, educators
Teachers do not incorporate new technologies into their teaching, because some students prefer to search for information in books rather than using digital technology. In the case of teachers, it is mainly about the older generation, the lack of interest because of the fact that technology is complicated for them, they need a longer period of learning, and digital technologies are becoming more complicated. There are also opinions that in some fields technology is not suitable for use in teaching. The barrier is

---

[1] BYOD = Bring your own device.

also students who prefer to use these technologies for entertainment rather than studying. Their attention from learning is taken away.

Barrier 3—Distrust in new and untested practices
Teachers are reluctant to incorporate new technology into their teaching because they have a lack of confidence in new things, excessive fears of untested, and financial risks. Also, ignorance and no experience are reasons for not being included in educational programs.

These statements indicate that the main obstacles to the use and subsequent expansion of new technologies and learning practices are both financial problems and some distrust of new things and the lack of information. It is clear that the personality of the teacher, his/her interest and willingness to adopt digital technologies, as a part of increasing the efficiency and attractiveness of teaching, plays a significant role here.

Looking to the future of developing new technologies that are appropriate and meaningful to apply to educational reality, [7] the m-learning, personal learning environments, MOOC, new distant learning objects, wikis, blogs, RSS, use of Creative Commons, sharing of cloud electronic learning support, u-learning, t-learning, educasting, seamless learning, social networks, omnipresent smart phones and tablets, mobile widespread reality, mobile technologies play an important role. New skills, often referred to as skills for the 21st century, are coming to the fore. This can also be deduced from the Innovating Pedagogue report [8].

3 Digital Competence

If we want to talk about digital competences, first of all we need to realize the fact that modern times bring new technologies called e-technology. Communication in traditional school was and is focused on direct verbal and non-verbal contact of communicators. At present, electronic communications penetrate the educational space. Some of the most popular ones include: Email, Chat, ICQ, Skype, WhatsApp, Viber, LinkedIn, Facebook, Messenger, MOOC, Cloud, LMS, Webinars, Educasting, Podcasting, other social networks and IP telephony. They are effective and prospective support for education as well as a positive supportive means of education. It is up to the teacher and supervisor of today's school to focus on digital technologies and to use these options to make the teaching process more effective and manage their activities.

At present, it is more than desirable to address selected **aspects of the educational environment** [9, 10]:

- comparison of traditional communication in class with current ways of electronic communication (e.g.: E-mail, Chat, social networks, IP telephony, LMS systems, MOOC, Webinars, Educasting);
- modern didactic means (e.g.: multimedia classrooms, interactive whiteboard, visualizer, hypermedia);
- new objects in distance education (remote laboratories, virtual laboratories, e-technology park with remote and virtual laboratories);
- new technologies, called as e-technologies, digital technologies, m-technologies (with high motivation value).

According to [9], appropriate supplements as a support will allow increase the efficiency of education, to extend the teaching methods for students, teachers and managers. These aspects will also make it possible to appeal to the necessity of lifelong learning for all school staff to obtain immediate, up-to-date information. We are also talking about creating conditions for a flexible, more accessible and individual learning process. Improving the work of teachers and enhancing their competencies in the removal of barriers to equal access to education is certainly a place, including the provision to each individual to effectively realize all their potential. A stimulating environment for stand-alone and combined study is created.

More specifically, we can focus, for example, on m-learning on specific schools, see [11, 12] or on the use of tablets in teaching generally, see [13]. Young educators and managers mostly have a very positive relationship with digital technologies and like to try out non-traditional forms of work [2, 10]. Managing activity involves the promotion of the formation of a large number of competencies [1]. With new technologies, the way of teaching and management can be changed. The assumption is that e-learning and m-learning, on-line and off-line courses (e.g. Blended learning, C-learning) can be said to be the next generation of "correspondent" courses.

Benefits of using advanced technologies by educators and school leaders [9]:

- space for gifted and disabled students,
- immediate availability of educational materials,
- school materials updated by means of useful information and new case studies, derived from concrete real situations,
- interactivity and the possibility of continuous innovations of textbooks,
- choice of individual learning paths and goals,
- acceptance of responsibility for own learning and decision-making encourage self-control and self-assessment,
- finding activating methods and forms of learning, new learning opportunities recognition,
- applicability to lifelong learning, obtaining information quickly,
- wider range of learners without age limitation,
- appropriate supplement, support and increase of the efficiency of education, extended way of teaching,
- the ability to learn anywhere, anytime, shared learning.

Effective school management using digital technologies implies the realization of these selected conditions: the need to use digital technologies throughout the school year, m-technology included into pedagogical practice management, digital technology for collective education, digital technology in school organization and management.

Teachers and school leaders inadvertently use these technologies, for example, to prepare for meetings, to manage the pedagogical college remotely, to implement electronic conferences, to access the shared space of materials in electronic form (e.g. digital teaching materials, documents, web links, scenarios, methodical guides, etc.), communication, self-education, school presentation, economic analyzes, statistical calculations, school agenda processing, organization and management of the school through appropriate software products (e.g. Open Source, freeware).

The dynamic development of digital technologies in the area of m-technologies leads to their greater expansion, not only in the commercial sphere but also in the educational field and school management. M-technologies are very popular mainly due to their properties (accessibility, modernity, practicality, interest, etc.) [2]. Their use in traditional teaching corresponds to the needs of improving the quality of education and strategies of digital education by 2020 of the Ministry of Education, Youth and Sports of the Czech Republic [14, 15]. The management activity will become more flexible and will meet the needs of effective management of school organizations.

As has been said, using new technology can change the way of teaching and management. The role of the director for the introduction of new technologies is crucial. Let's say it is about any reform within the school, see [16]. This is confirmed in the research plan for the introduction of the school information system [17]. Digital competence includes [18] the ability to use ICT, especially LMS or other systems applicable to this form of education.

Depending on the thematic structure of the issues addressed by the European ICT cluster and the overview of the main recommendations outlined, [4, 19] correspond to digital competences with lifelong skills and must be considered key in the field of teacher training. In addition, it is necessary to place emphasis on the application of new technologies to teaching related to the changing needs of the labor market and industry. Support for research into the impact and impact of digital technologies on the educational process is also essential.

"The digital competence of teachers must include the ability to critically access educational technologies, i.e. the ability to recognize their learning potential" [4].

Teachers and managers need to have **digital competences** for other reasons [20, 21]:

- to motivate co-workers to implement change and develop the school,
- to persuade the importance and benefits of change,
- to promote further education in digital technologies,
- to take responsibility for making changes,
- to facilitate the integration of digital technologies into the education and administrative processes of the school.

Without developed digital competencies, it will be difficult to flexibly respond to the dynamic development of digital technologies and their application to practice (for future industry development 4.0).

4 Relationships Between School Processes

The dynamic development of digital technologies is linked to the demand for systematic changes in the use of digital technologies in teaching and in connection with school processes. There is a change in the requirements for the managerial role of school heads [16]. The decision-making of school management is increasingly taking up digital technology due to the allocation of time and resources available. There is a certain mismatch between the penetration of technology into various aspects of society (critical aspects of education) and the considerable uncertainty among teachers about how best to use digital technology. Relations between new technologies and management issues in education are addressed [22].

From the point of view of the effective use of digital technologies, educational courses for school and teacher management plays an important role in enabling the continuous dissemination of skills in digital technology within training courses for teachers [16]. It should also be noted that new technologies encourage school principals and teachers to change ways to plan, acquire and evaluate continuing education [22]. It is important to address the relationship between new technologies and leadership in the field of education [22]: How do they change management prerequisites? How do technologies decentralize leadership support? Why does the education sector become more democratic? How much does resource allocation affect? How do they support the development of new forms of leadership?

According to the OECD publication [23], the manager must be able to:

- manage the real change, where necessary,
- manage the allocation of material resources,
- managing knowledge (managing knowledge systems, managing their efficiency) and ideas,
- planning time (time management),
- keep together an efficient team of co-workers.

5   Digital Competence—A Prerequisite for Success in All Areas of Human Activity

According to the DigComp 2.0 [21] Joint Research Center, the Digital Competence penetrates all areas of human activity. The most significant are:

Working with information
Digital content is searched for and processed. This also includes information evaluation, critical assessment, analyzing, organizing and storing.

Communication and cooperation
Digital media is a means of communication. Communication and collaboration requires effective interactions and sharing capabilities through digital technologies. It also makes it possible to engage in civic activities. For these activities carried out in the digital environment, it is necessary to know and respect information ethics, and to be able to take care of their digital identity.

Creating digital content
It is important to create new, but also rework or remix existing digital content. It is necessary to understand copyright and licenses. In order to solve some problems or perform certain tasks, the student should focus on the basics of algorithmizing and programming.

Security
Security includes multiple sub-areas ranging from protection of computer equipment through privacy, to health protection, the maintenance of quality of life and environmental protection.

Problem solving
Solving technical problems arising from digital devices, as well as selecting and using adequate digital tools and appropriate technology solutions. Creative use of technologies, innovation of traditional practices and cooperation with others in this area are increasingly important. It is important to improve our digital competencies in relation to the dynamically evolving digital technologies following industry 4.0.

6 Conclusion

Digital education responds to the needs of improving the quality of education and lifelong learning. It contributes to the possibilities of meaningful use of new technologies in teaching and is desirable in the current conditions of Czech education. Digital technology has become a significant helper in education and management.

It is clear that digital competences are an essential part of the competence model of a worker in education. Consequently, existing competence models need to be broadened by digital competences, to explore, decompile and propose alternatives to their appropriate expansion for life in the information society and industry 4.0.

The objective of the project is based on the development and experimental verification of the proposed educational program development of digital competences of neglected professional group of teachers of social sciences in secondary vocational education. The process of solution is based on Phase 1 on the research of the level of digital competences of teachers of the target group, in the 2nd phase of the creation, pilot verification and evaluation of the training program, in the 3rd phase of the experience sharing workshops, in the 4th stage of the transformation of results for use in practice. The project exploits exploration and educational methods using action research. Innovativeness lies in the choice of a neglected target group, the utility in transforming the results into a system of further teacher education and in supporting regional action planning in education.

For the project, the National Institute for Education, a directly managed organization of the Ministry of Education, and the Secondary Electro-Technical School of Ječná, have been acquired as application guarantors, with whom the research team has long been cooperating in the field of polytechnic education and teacher training. The project will be implemented in the context of national curriculum revisions and in support of regional action planning in education. It will also be used as a basis for assessing the European Standard of Digital Competencies for Teachers. It will also benefit from a target group of secondary school teachers of social science subjects who are insufficiently prepared to integrate digital technologies into education in terms of the requirements of the forthcoming standard of digital competences for teachers. The result will also be applied within the DVPP (further education of pedagogical staff).

Main goals

1. *Research of digital competences of teachers of social sciences in secondary vocational education*

Knowledge from empirical research to determine the level of digital competences achieved by teachers of social sciences will be implemented through collaboration with the Application Guarantor (NÚV) into national curriculum revisions (framework education programs for secondary vocational education) that respond to technological innovations and the needs of society. Furthermore, they will be used in the innovations of educational programs of specialization in pedagogy realized at CTU in Prague.

2. *The program of lifelong learning of digital skills development of teachers of social sciences in secondary vocational education*

The result is the proposal and pilot verification of a modular program for the development of digital competences of teachers of social sciences in secondary vocational education. It will focus not only on the ability of teachers to work with digital technologies but also on identifying their teaching potential in the field of classroom management, methods and forms that support students' co-operation, the development of creative thinking and individual learning strategies and motivation to learn. Educational modules will be accredited by the Ministry of Education, Youth and Sports.

3. *Workshop to share experiences with the development of digital competences for teachers of social sciences*

The workshop aims to enable social science teachers to share experience and best practice in the development of digital competences, the output will be implemented within the framework of the lifelong learning system of teachers and implemented through NUV as an application guarantor to support regional action planning in education.

4. *Didactic materials*

The result will be didactic materials supporting the use of digital technologies by a teacher of social sciences in teaching. The outputs will take the form of methodological sheets with the structure according to the three-phase model of learning process E-U-R.

This study is a part of project funded by TACR, Technology Agency of the Czech Republic, TL01000192, Developing digital competence of teachers of social sciences and secondary vocational schools.

The project responds to the needs of society, whose competitiveness requires the readiness of the population to the demands of industry, 4.0 (National Industry's initiatives 4.0). It also responds to the Strategy of Czech Educational Policy (Resolution of the Government of the Czech Republic, 927/2014), which includes the challenge to open education to new methods and ways of learning through digital technologies and to ensure the conditions for the development of digital literacy and informative thinking of teachers. The benefit consists in the creation of an educational program for the development of digital competences of teachers of social sciences at secondary vocational schools. As a result, technology will become more integrated into vocational education and thus also the development of competences necessary for lifelong learning.

References

1. Lhotková, I., Trojan, V., Kitzberger, J.: Kompetence řídících pracovníků ve školství. Wolters Kluwers ČR, a. s, Praha (2012). ISBN 978-80-7357-899-2
2. Svoboda, P.: M-learning—use of mobile technologies in teaching. In: Caha, Z., Stellner, F. (eds.) Littera Scripta 3/2016. Recenzovaný časopis. 1. vyd. České Budějovice: VŠTE, 2016. Dostupný z WWW: http://journals.vstecb.cz/category/littera-scripta/9-rocnik/3_2016/. ISSN 1805-9112 (Online)
3. Dixon, B.: Anytime anywhere learning. [online]. [2013] [cit. 2013-4-4]. Dostupný z WWW: http://www.aalf.org/
4. Brdička, B.: Doporučení evropského ICT clusteru. [online]. [2012] [cit. 2012-15-1]. Dostupný z WWW: http://www.spomocnik.cz/index.php?id_document=2460
5. Vymětal, J., Diačiková, A., Váchová, M.: Informační a znalostní management v praxi. LexisNexis CZ s.r.o, Praha (2005). ISBN 80-86920-01-1
6. Svoboda, P.: Profesionalizace klíčových kompetencí řídících pracovníků škol a školských zařízení, modul ICT, 2011–2012, projekt CŠM PedF UK, CZ.1.07/1.3.00/08.0235
7. GARTNER.COM.: Leading in a digital world. [online]. [2015] [cit. 2015-12-6]. Dostupný z WWW: http://www.gartner.com
8. INNOVATING PEDAGOGY 2013–2016: Open Univerzity inovation report. [online]. [2016] [cit. 2016-15-7]. Dostupné z WWW http://www.open.ac.uk/blogs/innovating/
9. Svoboda, P.: M-learning ve výuce technických předmětů. Sborník příspěvků z mezinárodní konference – Modernizace vysokoškolské výuky technických předmětů. 1. vyd. Hradec Králové: Gaudeamus, 2008. s. 172–175. ISBN 978-80-7041-154-4
10. Svoboda, P., Andres, P.: Multimedia as a modern didactic tool—windows EDU proof of concept project at CTU in Prague. In: Advances in Intelligent Systems and Computing. London: Springer, 2017. pp. 29–40. ISSN 2194-5365. ISBN 978-3-319-50336-3. Dostupný z WWW: http://www.springer.com/gp/book/9783319503394?wt_mc=Internal.Event.1. SEM.ChapterAuthorCongrat
11. VIDEOUKÁZKA - ČESKÝ JAZYK, 3. ročník ZŠ praktická. Využití m-technologií iPad. [online]. [2012] [cit. 2012-5-10]. Dostupný z WWW: http://www.youtube.com/watch?v= 7BaRulsbLJ0&feature=youtu.be
12. VIDEOUKÁZKA - MATEMATIKA, 5. ročník ZŠ speciální. Využití m-technologií iPad. [online]. [2012] [cit. 2012-5-10]. Dostupný z WWW: http://www.youtube.com/watch?v= eGYNjHWtEFA&feature=related
13. VIDEOUKÁZKA - využití ipad. [online]. [2012] [cit. 2012- 5-10]. Dostupný z WWW: http://numerato.posterous.com/tablety-ve-vyuce-vyber-vzdelavacich-aplikaci
14. MŠMT ČR: Strategie digitálního vzdělávání do roku 2020. [online]. [2016] [cit. 2016-19-11]. Dostupný z WWW: http://www.msmt.cz/uploads/DigiStrategie.pdf
15. JEDNOTA ŠKOLSKÝCH INFORMATIKU: Jak české vzdělávání využívá současné technologie? Sledujte s námi realizaci Strategie digitálního vzdělávání! [online]. [2017] [cit. 2017-20-8]. Dostupný z WWW: http://digivzdelavani.jsi.cz/

16. Flanagan, L., Jacobsen, M.: Technology leadership of the twenty-first century principal. J. Educational Adm. **41**(2), 124–142 (2003). ISSN 0957-8234

17. Yuen, A.H.K., et al.: ICT implementation and school leadership: case studies of ICT integration in teaching and learning. J. Educ. Adm. **41**(2), 158–170 (2003). ISSN 0957-8234

18. Mesarošová, M., Cápay, M.: Competences for teaching in modern society. Technológia vzdelávania. Vědecko- pedagogický časopis. XIX. ročník. [online]. [2014] [cit. 2014-2-10]. Dostupný z WWW: http://www.technologiavudelavania.ukf.sk/

19. KSLLL.NET.: Dokuments. ICT cluster. [online]. 2012 [cit. 2012-01-27]. Dostupný z WWW: http://www.kslll.net/Documents/Key%20Lessons%20ICT%20cluster%20final%20version.pdf

20. Schiller, J.: Working with ICT: perceptions of Australian principals. J. Educ. Adm. **41**(2), 171–185. ISSN 0957-8234

21. Vuorikari, R., Punie, Y., Carretero gomez, S., Van Den Brande, G.: DigComp 2.0: The Digital Competence Framework for Citizens. Update Phase 1: The Conceptual Reference Model. Luxembourg Publication (2016). ISBN 978-92-79-58876-1 Dostupné z WWW: https://ec.europa.eu/jrc/en/publication/eur-scientific-and-technical-research-reports/digcomp-20-digital-competence-framework-citizens-update-phase-1-conceptual-reference-model

22. Webber, C.F.: New technologies and educative leadership. J. Educ. Adm. **41**(2), 119–123 (2003). ISSN 0957-8234

23. OECD: New School Management Approaches. Paris: OECD

24. Cochard, G.-M.: La nécessité de la certification des compétences numériques les Certificats Informatique et Internet. Université de Picardie Jules Verne, Université de Versailles. [online]. 2012 [cit. 2012-01-01]. Dostupný z WWW: http://www.elearningeuropa.info/et/download/file/fid/19416

25. Beneš, P., Rambousek, V.: kolektiv autorů. Vzdělávání pro život v informační společnosti I. a II. 1. vyd. Praha: Vydavatelství ČVUT, Praha 2005. 500s. ISBN 80–7290-202-4

26. Vašutová, J.: Kvalifikační předpoklady pro nové role učitelů. In: Učitelé jako profesní skupina, jejich vzdělávání a podpůrný systém. 1. vyd. Praha: Ped. fak. UK, Praha 2001. 1. díl, str. 19–46

27. Vališová, A. Application of Electronics in the Formative Educational Process. In: Studia z teorii wychowania. TOM III 2012 nr. 2. Warszawa: Widawnictvo naukowe CHAT, 2011. pp. 114–122 (2012). ISSN 2083-0998

28. Andres, P., Svoboda, P.: Vybrané aspekty celoživotního vzdělávání učitelů - techniků. In: Danielova, L., Schmied, J. (eds.) Sborník příspěvků ze 7. mezinárodní vědecké konference celoživotního vzdělávání Icolle 2015. 1. vyd. Křtiny: Mendelova univerzita v Brně, 2015. s. 17–34. ISBN 978-80-7509-287-8

29. Semrád, J., Vališová, A., Andres, P., Škrabal, M.: a kol. Výchova, vzdělávání a výzvy nové doby. Brno: Paido, 2015. ISBN 978-80-7315-258-1

30. Snelling, J.: New ISTE standards aim to develop lifelong learners. Dostupné z WWW: https://www.iste.org/explore/articleDetail?articleid=751

31. Neumajer, O.: Být digitálně gramotný už neznamená jen ovládat počítač. Řízení školy. Praha: Wolters Kluwer, 2017, roč. 14, č. 3, s. 28–31. ISSN 1214-8679

32. Miština, J., Jurinová, J., Hrmo, R., Krištofiaková, L.: Design, development and implementation of E-learning course for secondary technical and vocational schools of electrical engineering in Slovakia. In: Auer, M., Guralnick, D., Simonics, I. (eds.) Teaching and Learning in a Digital World. ICL 2017. Advances in Intelligent Systems and Computing, 2017. ISBN print 978-3-319-73209-1, vol. 715, pp. 915–925 [online] (2017). https://doi.org/10.1007/978-3-319-73210-7_104

33. Hrmo, R., Miština, J., Krištofiaková, L.: Improving the quality of technical and vocational education in Slovakia for European labour market needs. Int. J. Eng. Pedagog. (iJEP), **6**(2), 14–22 (2016). [online]. ISSN 2192-4880
34. Hrmo, R., Krištofiaková, L.: Quality assurance in higher education with a focus on the quality of university teachers. In: Miština, J., Hrmo, R. (eds.) Key Competences and the Labour Market. Warsaw Management University Publishing House, Poland, Warsaw (2016). 298 s. ISBN: 978-83-7520-223-6

Psycho-Pedagogical Support of Students Project Activities in Multi-functional Production Laboratories (Fab Lab) on the Basis of Technical University

Marina V. Olennikova$^{(\boxtimes)}$ and Anastasia V. Tabolina$^{(\boxtimes)}$

Department of Engineering Education and Psychology, Institute of Humanities,
Peter the Great St. Petersburg Polytechnic University, Saint Petersburg, Russia
mariole@mail.ru, stasy335k@yandex.ru

Abstract. The article deals with the organization of psychological and pedagogical support of project activities of students on the example of fab lab. The authors describe the model of psychological and pedagogical support of project activities of students, define the psychological and pedagogical support of this activity, and consider the principles of its organization. The article presents a study of communicative competence, social intelligence and motivational orientations in interpersonal communications.

Keywords: Project activity · Creativity · Fabrication laboratory · The lab of digital production · Engineering education

1 Context

In the modern state concept of modernization of Russian education for the period up to 2020, much attention is paid to the intellectual and moral development of the individual, the formation of creative, ideological, behavioral qualities of students necessary for independent productive activities in the information society. In this regard, the concept of "psychological and pedagogical support of the project activities of the participants of the educational process" is gaining popularity in the national system of higher education. The problems of psychological and pedagogical support are reflected in the studies of A. G. Asmolova, K. A. Abulkhanovoi, A. G. Leders, etc. [1–3].

T. Yanicheva described psychological support as a system of organizational, diagnostic, educational and developmental activities for teachers, students, administration and parents, aimed at creating optimal conditions for the development of the individual [7].

L. M. Mitina, U. A. Korelyakov consider psychological and pedagogical support in terms of systemic activities aimed at providing social and psychological care of the individual social institution [7].

L. V. Bayborodova, M. I. Rozhkov, T. M. Alexandrova, G. V. Kupriyanova consider psychological and pedagogical support as professional activity of the educator-psychologist capability to give help and support in individual education of the

M. E. Auer and T. Tsiatsos (Eds.): ICL 2018, AISC 917, pp. 732–740, 2019.
https://doi.org/10.1007/978-3-030-11935-5_69

student as the technology including a number of consecutive stages of activity of the educator, the psychologist and other experts in providing educational achievements to students [6].

M. I. Rozhkov notes that effective psychological and pedagogical support is a complex of purposeful pedagogical actions that help students to prepare for project activities, to carry out multiple elections at different stages of activity and successfully bring the selected activity to the result [3].

Liisychenkova S. A. notes that productive psychological and pedagogical interaction in the educational process is possible if the educator is passionate about his work, capable to inspire to work, is able to maintain the students' faith in their own strength, gives a chance to correct the unsuccessful work, ready for cooperation [4].

A. G. Maklakov notes that in the organization of the system of psychological and pedagogical support of students it is important to take into account the criteria of the age stage of personality development associated with the transition from adolescence to early adulthood. Student hood—is a period of fundamental changes in the usual activities of the individual and his lifestyle, the time to achieve a new social status and the transition to a new age stage. In this regard, the process of psychological support cannot be considered, organized and provided without taking into account the essence of the process of adaptation of students to new conditions of life. Psychological and pedagogical support of project activities will be effective in compliance with such principles as regularity, consistency and complexity [5].

2 Purpose or Goal

The modern state concept of modernization of Russian education for the period up to 2020 implies the introduction of a competence-based approach, characterized by increased attention to the quality of education, which requires the training of qualified, highly educated, creative-minded, mobile, competent professionals. This model of specialist training will make it possible to focus on the intellectual and moral development of the individual, the formation of creative, ideological and behavioral qualities of students needed for independent productive activities in the information society. One of its manifestations is the formation of the project activities of students in the information and educational environment through the synthesis of activity and personal development with the dominant value of the last meaning.

Issues related to the development of project activities were highlighted in many works of domestic and foreign scientists in the field of philosophy, pedagogy, psychology—V. P. Bespalko, N. V. Bordovskii, L. S. Vygotsky, V. S. Hersonskaja, V. I. Zagvyazinsky, M. S. Kagan, A. N. Leonteva, I. P. Podlesovo, L. S. Rubinstein, V. A. Slastenina, L. V. Fridman, A. V. Khutorskoy [1–3].

Goals:

1. To carry out psychological diagnostics of personality development and small groups in the conditions of joint scientific activity

2. To provide psychological and pedagogical assistance to the participants of fab lab in the development of a common technical language, the formation of intra-group cohesion, assistance in solving group and individual personal and professional problems.

3 Approach

The relevance of the topic and the formulation of the research problem. In recent decades, qualitatively changed the target and value of the domestic system of higher education, which is characterized by an increase in professional and social mobility of graduates. In the context of the information society and the formation of a global information and communication environment requires new approaches to the organization of the educational process at the University, aimed at satisfying requests for educational services of a certain level and quality. In this case, it is important that the training of specialists correspond to the dynamically changing conditions in the field of information technologies and the social environment in obtaining and improving previously acquired knowledge and skills.

Meanwhile, the traditional educational system cannot yet adapt to the continuous growth of information, the introduction of new technologies in various fields of science and technology, lead to the rapid updating of knowledge systems, in need for more flexible skills and knowledge and the development of creative intellectual initiative. Undoubtedly, desirable properties, according To M. N. Akhmetova, should become analytical mind, systemic thinking, drive to experiment, and ability to cooperation. At the same time, the main tool is search and design, the style of communication—cooperation, and the distinctive feature is the skill, competence and own position of the future specialist.

The analysis of scientific research allows us to state that the question of how to support a student in his/her professional development in the conditions of higher education remains unclear, what mechanisms of psychological support can be used, and what conditions, effectiveness of the support, the role of services of practical psychology in solving these problems.

Objectives:

1. To determine the main communicative orientations and their harmony in the process of formal communication among fab lab students. The method of "Diagnostics of motivational orientations in interpersonal communication" (I. D. Ladanov, A. V. Urazaeva) [1, 2];
2. To study the features of social intelligence by the method of J. Guildford and M. Sullivan, adapted by E. S. Mikhailova [1, 2];
3. To study the indicators of communicative social competence (CSC) using the CSC technique [1, 2];
4. To carry out the program of psychological and pedagogical support "Together in a happy future";
5. Re-test;
6. Formulate conclusions and recommendations.

The study was of a longitudinal nature, was conducted on a group of technical students, 62 people, of whom (54 boys and 8 girls) aged 19–21 years, engaged in active project activities. The start of the study was 2nd semester of the 2nd course and the end was the 2nd semester of the 3rd course. At the end of the annual program of psychological and pedagogical support, a repeated diagnostic section of the presented group of respondents was carried out.

4   Actual or Anticipated Outcomes

To determine the main communicative orientations and their harmony in the process of formal communication among fab lab students we used the method "diagnostics of motivational orientations in interpersonal communications" (I. D. Ladanov, V. A. Urazaev). Value attitude to interpersonal, intercultural, professional communication, expressed in the desire to maintain the activity of communicative activities, we analyzed through a comparative analysis of the severity of the three scales: focus on the adoption of the partner, focus on the adequacy of perception and understanding of the partner, focus on compromise. The technique also allowed to analyze the level of overall harmony of communicative orientations.

The results of diagnostics of the initial level of formation of communicative competence is shown below (Fig. 1).

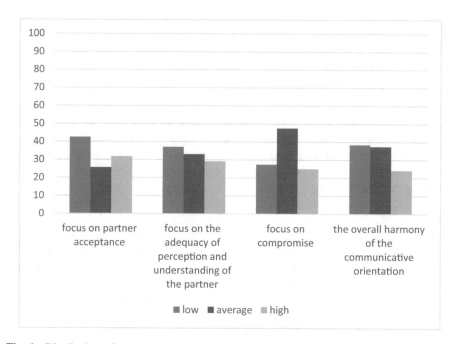

Fig. 1. Distribution of students according to the levels of formed communicative competence before the program of psychological and pedagogical support

The analysis of the results among students of technical specialties engaged in project activities, allow to conclude that in the context of orientation on the adoption of the partner and the adequacy of perception and understanding of the partner in the experimental group is dominated by a low level (42.5%) and (37.1%) students. The orientation towards compromise is dominated by the average level (47.6%).

In determining the level of overall harmony of communication orientations, the data were distributed as follows: low (38, 4%) and medium (37.4%) level is represented in the majority of respondents. The data obtained indicate the lack of formation of communicative competence. Such conclusions can be explained by the specifics of the technical profile of training, which is the predominance of technical and special disciplines, the dominance of laboratory work and production practices, which does not allow, as experience and research in this area, to keep a high level of motivational orientations in interpersonal communications.

Research has shown the following result: more than half of the students do not know the norms and methods of productive communication at a sufficient level. Specified contingent characteristic impulsiveness, irritability in communication, avoidance of discussing serious topics, the superficiality in judgment.

Indicators of social intelligence were identified using the technique of J. Guildford and M. Sullivan, adapted by E. S. Mikhailova (Aleshina).

The engineering students show high results in subtests "the History end" and "the history of addition". This suggests that they are more able to anticipate further actions of people based on the analysis of real situations of communication; to recognize the structure of interpersonal situations in dynamics; by logical reasoning, they can complete unknown, missing links in the chain of these interactions, to predict how a person will behave in the future; to find the causes of certain behavior; they know how to navigate the rules and regulations governing behavior in society.

The students showed low rates in terms of "group expression", "verbal intelligence" and the overall level of social intelligence. This suggests that students of technical specialties have difficulties in assessing the states, feelings, intentions of people by their facial expressions, poses, gestures; they have a low sensitivity to the nature and shades of human relationships, which makes it difficult for them to quickly and correctly understand what people say to each other in the context of a certain situation, as well as their ability to extract maximum information about people's behavior, which has a negative impact on work in collective project activities.

The indicators of communicative competence were identified by the method of diagnostics of communicative social competence (CSC).

Based on the data it can be concluded that the students of technical specialties show lower results in terms of communicative competence as the level of sociability, emotional stability, cheerfulness and self-control, and high values on indicators of logical thinking, sensitivity and independence, focus on their own solution.

To improve social intelligence, communication skills and adaptability of students we have developed a model of psychological and pedagogical support of project activities of fab lab students, it is aimed at:

1. Formation of professional technical orientation of students of engineering University;
2. Ensuring the normal development of the students;
3. Activation of motivational and intellectual potential;
4. Formation of interest in cognitive activity and own capabilities;
5. Assistance in overcoming educational difficulties, in interpersonal interaction with peers and teachers;
6. Assistance in choosing the educational route of study (Fig. 2).

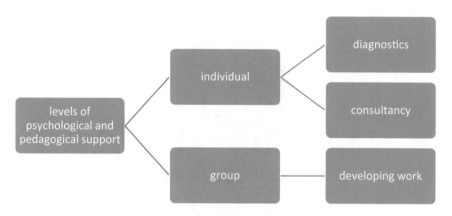

Fig. 2. Model of psychological and pedagogical support of project activity of fab lab students

At the individual level of psychological and pedagogical support, diagnostics is carried out of individual psychological features of students and consultation on personal difficulties arising in the process of project activities.

At the group, level of psychological and pedagogical support developing work is carried out using modern interactive technologies, training, master classes, analysis of real cases, business games, projects.

On the basis of the St. Petersburg Polytechnic University Peter the Great the work on psychological and pedagogical support of the participants of the educational process is carried out by the division "Higher school of engineering pedagogy, psychology and applied linguistics" together with "Psychological club SPBPU".

To help students of technical specialties created a team of children (adapters), who can at any time come to the aid of participants of fab lab.

Who is the adapter? It is an exemplary student, a carrier of the University traditions, which will help to solve the problems associated with learning, self-realization and adaptation to new conditions. First, future assistants of first-year students are trained in the "school of adapters". There is a specially organized a course of lectures, trainings and master classes for them aimed at the development of public speaking, psychology, team work, as well as the study of the legal framework and structure of the home University. Coordinators at the St. Petersburg Polytechnic University competently monitor and track the results of the adapters.

The duration of the program of psychological and pedagogical support is a year, during which students were provided with comprehensive assistance (Fig. 3):

1. Psychological consultation
2. Short-term therapy
3. Psychological correction
4. Psychoprophylaxis
5. Psychological rehabilitation
6. Psychological development.

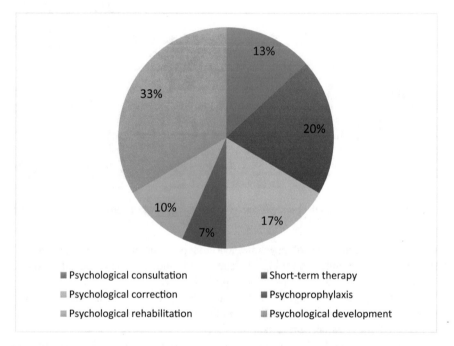

Fig. 3. Comprehensive assistance to fab lab students under the program of psychological and pedagogical support

After the program, there is a general positive dynamic among fab lab students on such parameters as "orientation to a partner" (48.4%), orientation to compromise (50.4%), the level of overall harmony of communicative orientation (40.1%) there is an increase in the average group indicators (Fig. 4).

In addition, there is an increase in indicators on the scale of "verbal expression". Students who have passed the program of psychological and pedagogical support feel free to communicate with others, can freely and competently express themselves, and learn the techniques of verbal presentation.

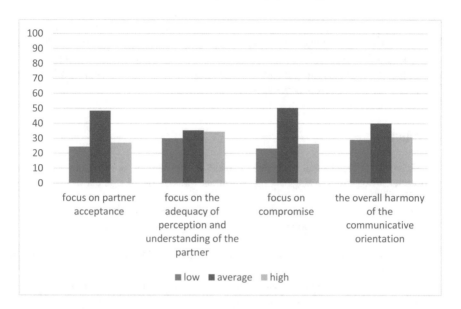

Fig. 4. Distribution of students according to the levels of formation of communicative competence after the program of psychological and pedagogical support

4.1   Conclusions/Recommendations/Summary

While we were conducting the program of psychological and pedagogical support of technical students, we have implemented the following pedagogical ideas:

1. Development and implementation of the dynamic psychological and pedagogical program "together in a happy future" in the educational process of a technical University
2. The choice and use of personality-oriented activity technologies (creative research projects, case-study technology, scenario-specific technologies, portfolio) and the use of pedagogical interaction between teachers of humanitarian and special departments in the practical implementation of project activities.
3. The use of person-oriented activity technologies through the development and demonstration of real projects and cases demonstrated that it is in the activity of conversation/communication that arises, enriches and improves.
4. Creation, content and actual saturation of the communicative educational environment on the basis of system integration of humanitarian and professional contexts of the educational process.
5. Organization of students involvement in the communicative educational environment and support of the whole process. The educational platform chosen by us, on the basis of which a communicative environment was created, contributed to the implementation of productive professional and educational communication in the spirit of cooperation and mutual assistance.
6. Organized systematic reflexive activity of students, aimed at assessing their own communicative actions and awareness of ways of communicative improvement.

In general, pedagogical observations, analysis of the products of design and reflexive activity of students, positive dynamics of formation of communicative competence of students of technical areas of training and positive expert evaluation of the level of formation of employers and representatives of departments confirmed the effectiveness of the developed pedagogical conditions.

References

1. Alisultanova, A.D.: Kompetentnostnii podhod v injenernom obrazovanii: monografia. [Competence Approach in Engineering Education: Monograph.] Academia Estestvoznaniya, Moscow (2010)
2. Borisova, K.V.: Model formirovaniya professionalnoi kulturi budushih injenerov. Rossiiskii nauchnii jurnal. [Model of formation of professional culture of future engineers. Russ. Sci. J.] **4**(29), 240 (2012)
3. International Standard Classification of Education: ISCE 2011, 88p. UNESCO Institute of Statistics, Montreal (2013)
4. Istufeev A.V.: Krizis gumanizma v usloviah sovremennoi tehnogennoi tsivilizatsii. Vestnik OGU. [Crisis of humanism in the conditions of modern technogenic civilization. Bulletin of OGU] **7**, 58–63 (2007)
5. Maklakov A.G.: Lichnostnyy adaptatsionnyy potentsial: ego mobilizatsiya i prognozirovaniye v ekstremal'nykh usloviyakh [Personal adaptational potential: its mobilization and forecasting in extreme conditions]. Psikhologicheskiy zhurnal. Psychol. J. **22**(1), 16–24 (2001)
6. Mitina L.M., Korelyakov Yu.A., Shavyrina G.V., et al.: Lichnost' i professiya: psikhologicheskaya podderzhka i soprovozhdeniye: ucheb. posobiye dlya studentov vyssh. ped. ucheb. zavedeniy [Personality and occupation: psychological support and maintenance: Textbook for students of higher pedagogical educational institutions], 336p. Akademiya Publ., Moscow (2005)
7. Yanicheva, T.G.: Psychological support of school activities. An approach. An experience. Finds. Zhurn. Pract. psychologist. **3**, 101–119 (1999)

Needs-Oriented Engineering Pedagogy Research Projects in Chilean Universities

Hanno Hortsch[1], Diego Gormaz-Lobos[2]($\boxtimes$),
Claudia Galarce-Miranda[2], and Steffen Kersten[2]

[1] IGIP, International Society for Engineering Pedagogy, Technische Universität
Dresden, Dresden, Germany
hanno.hortsch@tu-dresden.de
[2] Technische Universität Dresden, Dresden, Germany
{Diego_Osvaldo.Gormaz_Lobos,Claudia.Galarce_Miranda,
Steffen.kersten}@tu-dresden.de

Abstract. This paper presents the results of two surveys focused on the interests and needs of academic training related to the different pedagogical and technological aspects that influence the training of engineers in the engineering faculties of Chilean universities, seeking to improve the quality of the academic training and teaching-learning process. The first research was organized in the frame of the "Pedagogy in Engineering in Chilean Universities" project, led by the Technische Universität Dresden (TUD) of Germany, and consisting of a group of academics from engineering faculties of three Chilean universities: Universidad Autónoma de Chile (UA), Universidad de Magallanes (UMAG) and Universidad de Talca (UTAL). The objective of the project is the development of tools to determine the training needs of trainers of engineers in the pedagogical field. The second research led by TU Dresden in cooperation with the Universidad de Santiago de Chile (USACH) is oriented to the development and testing of training modules for students and teaching qualifications of teaching staff in academic engineering education. Methodologically, two instruments were developed for the needs assessment, based on specific theoretical concepts of the engineering training as well as the experience in research projects of the TUD in this area. Booth projects are financially supported through of German Academic Exchange Service (DAAD).

Keywords: Engineering pedagogy · Needs-oriented engineering education
Engineering education · PEDING and STING projects

1 An Approach of Engineering Pedagogy

Prof. Hans Lohmann, in his quest to systematize and to professionalize at an institutional level the teaching and research in engineering, founded in 1951 the Engineering Pedagogy Institute (Institut für Ingenieurpädagogik) at the Dresden Technical University (Technische Universität Dresden, TU Dresden). With his work in the relationship between the "Technique" and "the teaching of the Technique",

M. E. Auer and T. Tsiatsos (Eds.): ICL 2018, AISC 917, pp. 741–753, 2019.
https://doi.org/10.1007/978-3-030-11935-5_70

the foundations of Engineering Pedagogy are established, whose aim is no other than the conformation of teaching and learning processes that are specific for the technical and technological spheres.

A central role in Lohmann's approach of engineering didactics was played by the concept of technology. He defined technology by its function "to transform the natural world" (see Lohmann 1954). The task of an engineer is to develop this technology. Engineers are therefore to become qualified in such a manner that they are able to solve technical design problems. In contrast to this, the activity of natural scientists is focused on the discovery of relationships in the world and, thus, solving scientific knowledge problems. Invention and discovery require different ways of thinking and, thus, different methods of academic training.

The requirement to gear engineering education to the demands of the economy, which is determined by the specifics of the engineering labour, is meant when speaking about demand-oriented and employment-based engineering education, respectively. Requirements are understood as necessary personal dispositions for successfully managing the profession-specific work activities. They are thus determined by the prevailing structures of production and service. For example, the change from Tayloristic production structures to structures of lean production in the past 40 years has considerably changed the engineering activities and with them the requirements on engineers. In addition to these dynamic requirements, a variety of stable long-term requirements related to the personality dispositions of engineers result from typical engineering activities. An example is the typical way engineers think/reason. For instance, in the analysis or the design of technical systems, the thinking in the categories of "part-whole" in the relationship between structure and function plays a vital role.

Closely related to the term "technology" is the society as a factor influencing engineering education. Technique not only arises from the application of natural laws and theories in engineering sciences but is also part of the technical possibilities and the socially desirable aims (see Heidegger and Rauner 1991, p. 20). In this respect, the development of technique and technology is also driven by social needs.

In addition, a society also has an idealised image of its members. Maturity, ability to be democratic, and willingness to active shaping are just a few personality traits that are included in this ideal. From this follows the educational mission of our universities as well. Nearly to this influences factor of engineering education is the individual needs of those who study engineering and their capacities, abilities, predispositions, etc.

Another factor influencing the training of engineers is the field of engineering sciences itself. A scientific discipline is defined by particular matters and methods of research.

Regarding the matters of engineering sciences, the terms technique and technology play a key role. Technique and technology contain processes of change (form and structure), transport, and storage of material, energy, and information (see Wolffgramm 1994). The views on what technique is and which function it has in relation to nature and society is also subject to changes. A change of the matter of a scientific discipline has an impact on teaching in this discipline.

Kersten (2015) proposes a scheme (see Fig. 1) that describes the factors that influence and condition (demands) engineering education: (i) the economic and production sectors of a country, (requirements determined by the prevailing structures of production and service), (ii) engineering sciences (regarding the matters of engineering sciences; the terms technique and technology play a key role), (iii) society and culture of the country, (the development of technique and technology is also driven by social needs), and (iv) the individuals who study engineering (as person) (see Kersten et al. 2015).

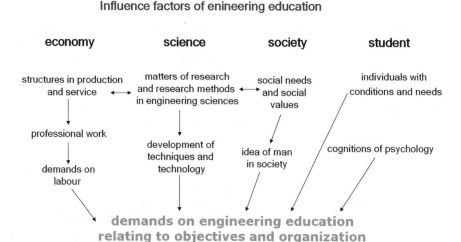

Fig. 1. Factors influencing on demand-oriented design of engineering education, Kersten (2015)

In this context, the design of a needs-oriented education in engineering sciences (planning, implementation, and evaluation of teaching and learning in engineering education) should consider the complexity of these influence factors.

2 Engineering Pedagogy at Chilean Universities

2.1 Needs-Oriented Engineering Pedagogy

Research in engineering education and didactic leaded by the Institute for Vocational Education at TU Dresden, in charge of Prof. Dr. Hanno Hortsch (President of IGIP, General Director PEDING and STING Projects in Chile) show numerous contributions that can deliver systematic training offerings in engineering pedagogy for faculty and instructors of engineering schools at universities in different countries. Financed by the DAAD we had the opportunity to develop and test a needs-oriented continuing education course for academic teachers in the field of engineering science in cooperation with Chilean universities since 2014.

The aim of the first project "Engineering Didactics at Chilean Universities" (PEDING-Project) is the development and testing of training modules for teaching qualifications of teaching staff in academic engineering education based in the IGIP

Curriculum offered for the TU Dresdens' Institut of Vocational Education. Through an analytical adaptation of the results of one study about the training requirements of lecturers at engineering faculties from Saxony (Germany) (see Köhler et al. 2013, pp. 17–18), Gormaz (2014) systematized in clusters categories and indicators/aspects, which later were used in the recollection instrument on teaching needs of the engineering faculty of the three Chilean universities (see Gormaz et al. 2014). In general the instrument and indicators seek to obtain information about: (i) characteristics of lecturers (years of experience, subject matter, etc.), (ii) experience and needs related to engineering didactic fundamentals, (iii) requirements for the structuration of Teaching-Learning forms in a university context, and the setting of objectives and contents of an engineering degree, and, (iv) identification of strengths and weaknesses, together with the conditions to enrol in a training course (see Table 1). The results of this research about the needs in engineering education were used in the development of the training modules and created the bases of a training course offered in 2018, modelled from the learning module structure according to IGIP (International Society for Engineering Education) and the Technische Universität Dresden, Faculty of Education.

The aim of the second Project "Strengthening engineering training at Chilean universities through practice partnerships" (STING-Project) is the development and testing of training modules for students (either for electrical or mechanical engineers) and teaching qualifications of teaching staff in academic engineering education based in demands and employment-requirements of German and Chilean companies. For this reason was developed a questionnaire by the TU Dresden and the USACH as part of a stage of information gathering to obtain the strategic positioning and future development of the participant enterprises. The goal of the application of this questionnaire is to know the opinion of strategic staff of the enterprises regarding the actually needs and the projected future scenario for engineers, and the type of "technology transferences" between university-company (see Table 2). The results of this survey were used in the development of two training modules for students at the USACH.

2.2 Methodology

In the first research study was applied a "cross—focus" strategy (Lincoln and Guba 2000), with the propose to integrate the opinions of the participants with the assessment of engineering education needs and interests, that are most required for the education of engineers. Thus, in this way, it was called upon a mixed research design for integrating the obtained data and "the convergence of the findings as a way to strengthen the knowledge claims oft he study" (Creswell 2003).

The second survey was designed using a mixed model of qualitative and quantitative methods. Through a concurrent triangulation strategy, Creswell (2003) states that quantitative and qualitative data can be collected simultaneously. The aim is to use two different survey methods to confirm, supplement or validate the research results.

2.3 Population and Available Sample

The sample of the first study (PEDING-Project) was composed by 144 academics of the Faculties of Engineering of the three Chilean universities, considering the

Table 1. Instruments categories and indicators for needs analyse PEDING-project (Gormaz 2014)

I. Engineering didactics fundamentals	
Category	Indicator/aspect* (*simplified version for this publication)
I.1. Design of teaching-learning processes in engineering sciences	I.1.1. Psychological foundations of teaching and learning I.1.2. Theoretical and practical bases of eng. didactics I.1.3. Didactic principles I.1.4. Organisation of the teaching-learning processes I.1.5. Structuring of the teaching-learning processes
I.2. Didactic media for teaching in engineering	I.2.1. Concepts and classification of didactics media I.2.2. Functions of didactic media and technological tools I.2.3. Field of action of didactic media I.2.4. Elaboration of didactic media
I.3. Communication	I.3.1. Design of communication processes I.3.2. Monologic and dialogic communication procedure I.3.3. Conflict identification and resolution
I.4. Control and Evaluation of the learning outcomes in engineering education	I.4.1. Registration and evaluation of the learning outcomes I.4.2. Operalisation of learning outcomes I.4.3. Procedures for the registration of learning outcomes I.4.4. Evaluation of the learning outcomes
II. Forms of structuring the teaching-learning processes in university contexts	
II.5. Lectures (theoretical courses)	II.5.1. General structure of a university course planning II.5.2. Preparation of a university course II.5.3. Execution of a university course II.5.4. Feedback in a university course
II.6. Laboratory practical training/self-study	II.6.1. Laboratory training II.6.2. Experiment functions in the teaching-learning processes II.6.3. Exercises and self-study planning
II.7. Engineering internships, written reports, research colloquium	II.7.1 Engineering Internship preparation and research preparation II.7.2 Support systems for internships and for autonomous research II.7.3. Internship analysis and research activities analysis
III. Determining objectives and contents of engineering studies	
	III.8.1. Analysis of the activities in engineering

(continued)

Table 1. (*continued*)

I. Engineering didactics fundamentals	
Category	Indicator/aspect* (*simplified version for this publication)
III.8. Determination of the study programme objectives	III.8.2. Analysis of the activities related to an university engineering study programme III.8.3. Analysis of social aspects in engineering III.8.4. Analysis of personal aspects in engineering
III.9. Determination of the engineering study programme contents	III.9.1. Fundamentals for the determination of contents III.9.2. Contents determination of an university study programs with regard to the academic activities III.9.3. Contents determination of an university study programs with regard to the societal activities III.9.4. Contents determination of an university study programs with regard to the personal activities

Table 2. Instruments categories and indicators for needs analyse STING-project

Engineering education needs	
Category	Indicator/aspect* (*simplified version for this publication)
I.1 Vision about needs for engineers	I.1.1. Competencies I.1.2. Knowledge and technological tendencies
I.2. Open questions	I.2.1. Aspects about university-company technology transference I.2.2. Aspects about the professional internships I.2.3. Aspects about professionals skills for engineers

indications of a minimum sample of 100 individuals (Hulland et al. 1996). The final sample consisted of 54% of lecturers belonging to UA, 26% to UMAG, and 20% to UTAL.

From a mixed approach of a quantitative and qualitative research (see Flick 2015), the sample of the second survey (STING-Project) was composed by 14 engineers on strategic positions of seven different Chilean and German companies from Santiago de Chile.

2.4   Instruments

With a view to identify the training needs and interests in the pedagogical and didactical aspects requirements of major importance for the formation of engineers, was applied an opinion poll type instrument with open and closed questions. The goals of this instrument are oriented to identify the perceptions about the teaching needs of different pedagogical aspects related to the engineering subjects at universities.

The Instrument two (STING-Project) for a needs-Analyse was applied with the goal to identify the opinion of the enterprises (experts) regarding the actually needs for their engineers and the projected future scenario, and the actually type of "technology transferences" between university-company. Table 2 presents the conceptual categories and the indicators of the instrument. The results of this survey were used in the development of two training modules for students at the USACH.

2.5   Procedure

Both instruments were individually applied, considering the ethical aspects according to the Chilean social sciences research criteria. The research process had two phases for both surveys. The first phase corresponded to information collection of the closed questions carried out by the research teams of the four Chilean universities. The statistical analysis applied was exploratory-descriptive with the aim to raise problems. The second phase of the study examined the open questions of the sample through a textual content analysis by codifying the discourse of each of the academics and experts, based on the item generating conceptual categories.

2.6   Characterization and Obtained Results

The characterization and the results obtained with the surveys applied to the academics of three Faculties of Engineering and the strategic Staff (experts) of seven companies are presented below. These results were analysed in three dimensions: (1) Characterization of the surveyed group, (2) Perception and needs in engineering pedagogy and education, and (3) Open questions.

2.7   Results of PEDING'S Survey

Characterization of the sample. The selected sample of academics that participated in the survey was in total 117 academics were gathered (62 UA, 33 UMAG, 22 UTAL) with 23% women (15 UA, 11 UMAG, 1 UTAL) and 77% men, that's represent approximately 30% of the total number of academics attached to each of the Engineering Faculties. More than 40% of survey participants are between 30 and 40 years old and approximately 11% are over 60 years old.

Of the total respondents, 64% were engineers by profession (56% UA, 67% UMAG, 81% UTAL), the rest had similar professions that help to complement the total training of the future engineers. In relation to years of teaching experience, over 75% is between 1 and 20 years (79% UA, 76% UMAG, 72% UTAL). Of the total number of participants, 70% have been trained in university teaching (74% UA, 55% UMAG, 81% UTAL) and approximately 39% (32% UA, 36% UMAG, 64% UTAL) have participated on graduates/magister programs in the area of university teaching.

Perception and needs in engineering pedagogy and education. In this section are presented the results about the perception of the respondents regarding the need for different skills and pedagogical tools for university teaching in engineering careers. It was asked *"How necessary do you consider the following aspects of engineering pedagogy in relation to your teaching experience?"* For this section, 28 aspects were considered based on the indicators of Table 1, being the most relevant those related to the evaluation methods, among which stand out with more than 90% of the preferences aspects such as: *"Evaluation and assessment of achieved learning"* and *"Knowledge about design for effective measurement of achieved learning"*. Then with more than 85% of the preferences are *"Structuring of teaching-learning processes in the scientific training of engineers"*, *"Use of didactic resources and information and communication technologies"* (ICTs).

Among the aspects considered less relevant (less than 70% of the preferences) were found: *"Psychological foundations for teaching and learning"* and *"Dialogic and monological communicative processes for teaching"*. It is important to note that all aspects had at least 60% relevance for the respondents.

The results by university do not suffer major modifications (Fig. 2). But it is observed that for the participants of the UTAL, there are three aspects that obtain preferences less than 60%: *"Dialogic and monological communicative processes for teaching"*, *"Knowledge about strategies to support professional practices and independent research activities"* and *"Analysis of the personal scope of engineering in Chile"*. Some discussed aspects present a great difference between the institutions. In 3 aspects, the UA has preferences above 85%, while UMAG and UTAL are under 66%: *"Recognition and resolution of conflicts within the classroom"*, *"Planning of activities for individual study"* and *"Analysis of the personal scope of engineering in Chile"*. Another aspect where there is a marked difference is *"Knowledge about strategies to support professional practices and independent research activities"* where the UA and UMAG have preferences over 81% while UTAL does not reach 55%. These differences may be due to the different programs given at each University, as well as to the institutional and social context and to the training given to the participants.

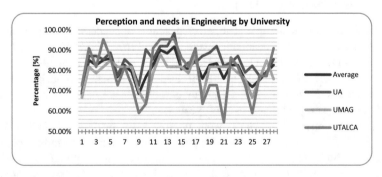

Fig. 2. Relevance of the different aspects consulted about perception and needs in engineering by university

By grouping the participants by gender (Fig. 3), the female participants (27) have 100% preferences on aspects "*Structuring of teaching-learning processes in the scientific training of engineers*" and "*Evaluation and assessment of achieved learning*". In the case of men (90), the aspects "*Knowledge about the design for effective measurement of learning achieved*" and "*Evaluation and assessment of learning achieved*" have preferences of 92 and 91% respectively.

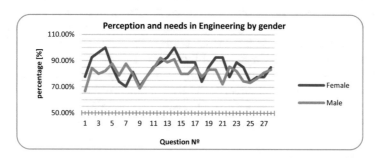

Fig. 3. Relevance of the different aspects consulted about perception and needs in engineering by gender

The worst evaluated aspects by the female gender correspond to the "*Knowledge about the design of didactic means for the teaching-learning processes*" and "*Dialogic and monological communicative processes for teaching*", both with 70% of preferences, while for males the worst evaluated aspects correspond to "*Psychological foundations for teaching and learning*" and "*Communicative dialogic and monological processes for teaching*" with 66 and 69% of preferences.

With respect to the results obtained in the 11 questions about strengthening of teaching methods respondents considered all aspects with relevance over 70%. Among the aspects considered, the most relevant are: "*Use and development of new didactic means in the training of engineers*", "*Design, choice and use of didactic means*" and "*Planning and structuring of teaching-learning processes at university level*", all of them with more than 80% of preferences. The aspects with the lowest relevance were the aspect "*Realization of communicative processes for teaching at university level*", "*Planning and materialization of evaluation and evaluative processes*" and "*Curriculum development for academic training at the university level*".

Open Questions. In this part of the survey we asked about 3 aspects: (1) strengths in engineering pedagogy; (2) aspects to be improved in the teaching task; and (3) interest and availability to train in the engineering pedagogy area.

Regarding the strengths of the teachers in the sample, the five most relevant results are grouped in strengths associated with: The *fundamentals for the determination of technical contents within the engineering area* (20.9%); the *Organization of teaching-learning processes* (15.69%); the *Knowledge for the determination of contents of teaching in Engineering in relation to personal, technical and social fields of the work of Engineers* (13.77%); the *Structuring of teaching-learning processes* (12.68%) and the *Analysis of specific subjects of the specific engineering activity* (11.30%).

With respect to the aspects to be improved in teaching, the five most relevant categories are grouped based on: *Fundamentals for the determination of technical contents within the area of engineering* (35.03%), *Evaluation and assessment of achieved learning* (18.77%), *Didactic principles for teaching-learning in Engineering* (12.70%) and *Knowledge and skills for the preparation, execution and feedback of teaching* (12.22%). In this review emerge two relevant categories associated to the improvement of the infrastructure and the time for preparing the teaching.

Finally, in response to the question related to interest in engineering pedagogical training, 83.44% would be willing to improve and only 8.06% would not. These results are mainly associated with lack of time, however they are available to review associated material, without having to attend formal courses. In addition, the results about the necessary conditions to attend for a engineering pedagogical training showed that the academic staff considered important to have a practical training course (13.96%), dictated by a specialist with expertise on the contents (10.95%) and ideally be a course dictated by engineers, with content only in engineering (9.31%).

2.8 Results of STING'S Survey

Characterization of the sample. The selected sample of engineers that participated in the survey was in total 14 experts of seven Chilean and German companies from Santiago de Chile. 28% of surveys participants are women and 72% men.

Of the total respondents, all were engineers by profession. About the position of the participants at the companies: 36% are mechanical engineering heads, 29% are heads of electrical engineering department, 21% are project managers and 14% product manager.

In relation to the "type of organization" of the companies, all are "private enterprises". Additionally, the participants were asked about the organization size, 40% of the enterprises are big, 40% are medium and only 20% correspond to small enterprises.

Perception and needs in engineering pedagogy and education. In this section are presented the results about the perception of the respondents regarding the need for different knowledge, skills and technological tools for the teaching in engineering careers. It was asked *"What are the most important competences for engineers?"* The results are presented in Fig. 4 and show many different competences like "leadership", "team working" and "autonomy" are the most valuable skills for companies, whit 87.5%, 87.5% and 75% of the preferences, respectively.

Another question was oriented to the importance of *innovation and research*. The companies were asked *"How relevant is for you that engineers students have experience in innovation and research project through their university time?"* The 37.5% gave 5 out of 5 points (most relevant) to these characteristics, while a 37.5% gave out of 4 points and 25% gave 3 out of 3 points. Therefore, the tendency to appreciate the experience of students in innovation and research projects is noticeable.

In relation to needs about *technical software for electrical and mechanical engineers*, the most popular option was Microsoft Office, which includes Excel, PowerPoint, Word, and Outlook, with five preferences (93%). Then, AutoCAD was the second option (86%), and finally "Project" comes in the third place.

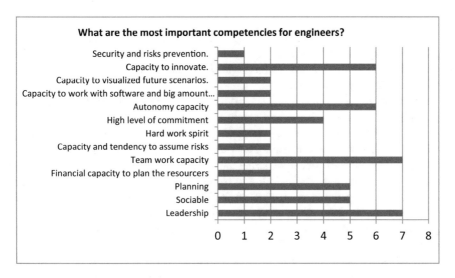

Fig. 4. Most important competencies for engineers by companies

In the same line of thought, the following question is *"Choose the most important technical subjects in electrical and mechanical engineering"*. Results are depicted in Fig. 5. The preferred option was PLC programming, followed by Instrumentation, Electrical Drives, and Renewable Energies.

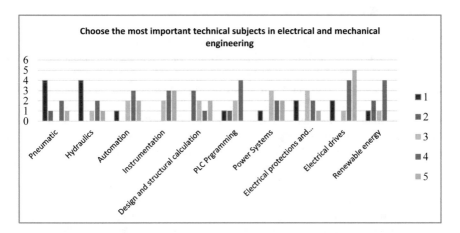

Fig. 5. Most important technical subjects for elec. and mecha. Engineers by companies

Open Questions. In this part of the survey we asked about 3 aspects: (1) about university-company technology transference; (2) aspects about the professional internships; and (3) aspects about professionals skills for engineers. The results will be presented as qualitative contents analyse (Mayring 2008).

According to *university-company technology transference* the companies staff recommend: (1) to renew high-tech equipment to improve laboratories and thesis projects, (2) to increase the link company-university through the development of joint projects, (3) to incorporate new technologies into classical engineering education, and (4) to promote applied research.

About the *professional internships*, all the participants were agree about the many drawbacks of internships. The main identified problems to resolve between university and companies are listed below: (1) internships do not have clear goals, (2) the length of internships is low, should last at least two months, (3) internships should be oriented to solve a problem, not just studies, (4) annual plan to define a pool of "internships projects", and (5) to increase flexibility to be able to start the internship all year round.

In relation to aspects about *professionals skills for engineers at companies in Chile,* all the participants were agree about the wide range of professional positions that engineers may perform. These positions go from Project Assistant, Sales Manager, Senior Engineer to head or CEO.

Though the summary of the participant´s answers, the following task were identified either for electrical or mechanical engineers: (1) engineers must have the analytic capability, and they must be able to learn on their own, (2) team working and leadership are fundamental either for junior or senior engineers, (3) the communication company-client is fundamental and must be considered in engineering education, and (4) because of the characteristics of Chile, engineers must know how mining industry works. Usually, they don't know the process or the security regulations.

3   Conclusions

From the results obtained it is possible to conclude, that the academic communities of the studied engineering faculties, tend to converge on the pedagogical capacities that are required to train the future engineers. Chilean academics from different engineering faculties are willing to train and incorporate systematic knowledge and skills, based on the tools of Engineering Pedagogy, to enhance the skills they already possess and thus improve the strategies and methods of teaching directed to its students. From the results of the survey of the PEDING project were identified many different needs in the field of engineering didactics: *(i) "Evaluation and assessment of the students' learning achievements", (ii) "Organisation of teaching and learning processes for the scientific formation of engineers", (iii) "Theoretical and practical knowledge about the didactics for the teaching and learning process in engineering", (iv) "Knowledge about how to design effective measurements of the learning accomplishments", and (v) "Use of didactic resources and of information and communication technologies (ICTs)"*.

As shown on the results of the STING survey, the companies identified many needs for the education process of engineers in relation to current and future industrial requirements. In general, the main needs in engineering education for the companies were: (i) to increase team working during student's careers, (ii) to renew high-tech equipment to improve laboratories and thesis projects, (iii) to increase the link company-university through the development of joint projects, (iv) to incorporate new technologies into classical engineering education, and (v) to promote applied research.

Additionally were identified for example that: (vi) engineers must have the analytic capability, and they must be able to learn on their own, (vii) the communication company-client is fundamental and must be considered in engineering education, (viii) team working and leadership are fundamental either for junior or senior engineers, and (ix) because of the characteristics of Chile, engineers must know how mining industry works and the security regulations.

To meet these and other needs, student-training and teacher-training modules were developed and subsequently implemented. With these actions it is expected to increase the academic success of the engineering students and their successfully employment at the companies, as well as to develop an improvement line in the area of engineering pedagogy for teachers of engineering faculties in Chilean universities.

References

Creswell, John W.: Research Design: Qualitative, Quantitative, and Mixed Method Approaches, pp. 217–219. Sage Publications University of Nebraska, Lincoln NE (2003)

Flick, U.: Qualitative Forschung: Ein Handbuch, pp. 15–36. Rowohlt Taschenbuch Verlag, Reinbeck bei Hamburg (2015)

Gormaz, D., Kersten, S.: Zur Analyse der ingenieurpädagogischen Weiterbildungsbedarfe von Lehrende. In: Jahresbericht DAAD Projekt 2014. TU Dresden, Institut für Berufspädagogik, Dresden (2014)

Heidegger, G., Rauner, F.: Berufe 2000: Berufliche Bildung für die industrielle Produktion der Zukunft. Düsseldorf., 20 (1991)

Hulland, J., Chow, Y.H., Lam, S.: Use of causal models in marketing. Int. J. Res. Mark. 13(2), 181–197 (1996)

Kersten, S., Simmert, H., Gormaz, D.: Engineering Pedagogy at Universities in Chile—A Research and Further Education Project of TU Dresden and Universidad Autónoma de Chile. In: Expanding Learning Scenarios. EDEN Conference Barcelona 2015

Köhler, M., Umlauft, T., Kersten, S., Simmert, H.: Projekt Ingenieurdidaktik an Sächsischen Hochschulen - e-didact. Projektabschlussbericht. Dresdner Beiträge zur Berufspädagogik Heft 33, Dresden (2013)

Lincoln, Y.S., Guba, E.G.: The only generalization is: there is no generalization. Case Study Method, pp. 27–44 (2000)

Lohmann, H.: Die Technik und ihre Lehre- Ein Forschungsteilprogramm für eine wissenschaftliche Ingenieurpädagogik, Wissenschaftliche Zeitschriften der TH Dresden, 602–621 (1954)

Mayring, P.: Qualitative Inhaltsanalyse: Grundlagen und Techniken, p. 21. Beltz, Weinheim, Basel (2008)

Wolffgramm, H.: Technische Systeme und Allgemeine Technologie. Bad Salzdetfurth (1994)

Modern Trends of the Russian Educational System Amid Global Changes

Natalya Ran[✉], Kseniya Kuzovenkova[✉],
and Marianna Kashirina[✉]

Samara State Technical University, Samara, Russia
natalirahn@mail.ru, {kuzovenkova.ks,mvkvv}@yandex.ru

Abstract. World globalization is gradually penetrating into all spheres of social life and, in particular, into education. Differences between Russia's traditional institutional social norms and market rules don't make it possible to fully benefit from globalization, while Russia's social and economic underdevelopment draws it back to ever-overtaking modernization, even in the educational sphere. Thus, this problem is still relevant.

Keywords: Globalization of education · Internationalization of education · Humanization of education · Democratization of education · Integrativity of education · Informatization of education · Humanitarization of education · Educational mobility and variability

1 Introduction

Fundamental items of the study include the analysis of modern trends of the Russian educational system amid world globalization, experience of global educational changes in the Samara region of the Russian Federation, and the need to introduce new organizational forms of combining science, education and production, which determine the development of global integration processes in education and science.

Globalization has several definitions. We consider globalization of education as a process of creating a universal unified system in education erasing the differences between its own educational systems [18]. The world educational system features the following global trends: access to education by the entire population of the country and the continuity of its levels and levels; granting autonomy and independence to educational institutions; ensuring the right to education for everyone; an increasing number of educational and organizational activities; the growth of the educational services market; expanding the network of higher education and changing the social composition of students; education becomes a financial priority in the developed countries; constant updating and adjustment of school and university educational programs; a shift away from the focus on the 'average student', an increased interest in gifted children and young people, specifics of the disclosure and development of their abilities in the educational process; search for additional resources for education of special needs children or disabled children [7].

M. E. Auer and T. Tsiatsos (Eds.): ICL 2018, AISC 917, pp. 754–761, 2019.
https://doi.org/10.1007/978-3-030-11935-5_71

Foreign scholars Bell [2], Toffler [23], Brzezinski [3], etc. believe that the planetary unity of humanity has immeasurably increased in recent times. It represents a fundamentally new super-system bound by common destiny and shared responsibility. Therefore, despite the striking socio-cultural, economic, political differences of regions, states and peoples, it is significant to talk about the formation of the single civilization and needs for a new global planetary thinking style.

Global changes in pedagogical activity were studied most extensively by American (Becker, Darling-Hammond, Hanvey, Evans, McLaunghlin, Talbert) and Russian scientists (V. Spasskaya, B. Vulfson, Z. Malkova, I. Tagunova, A. Lifferov, etc.), but the scientific approaches are different in terms of historical and social conditions, and the definition of global education has not yet been sufficiently studied. This problem is still relevant.

Education in the era of globalization is the area where a young specialist is attached to the global values, broadens his horizons and his knowledge of not only professional competencies, but also the working conditions that can be provided to him in various countries. Through the development of professional self-awareness, the specialist focuses on individual values and the search for better conditions for his own creative activity without taking into account state borders and the interests of his country. We believe that Russia's global education system is a consequence of world globalization.

2 Discussion

A review of the above-mentioned studies allowed us to show the main trends of modern education as a scheme (Fig. 1).

Let's study main changes in modern Russian education in detail.

The first trend is internationalization of education. It is a consequence of globalization of the whole world. A new approach to education provides the integration of social and economic life bringing together different nationalities and cultures. The world ultimately benefits from this in all directions. Russian universities endeavour to enter international ratings (for example, QS) by establishing cooperation with representatives of the foreign educational community, opening joint programs with foreign universities, attracting international staff. That's why the problem of internationalization is relevant for Russian universities [11].

Internationalization of education in European universities initially implied only academic mobility (training and internship in foreign countries), and the exchange of knowledge, ideas, etc. At present, internationalization of education is considered as measures to create the international academic community that will prepare students for life in globalizing environment [19].

Academic mobility continues to be seen as one of the main tools for improving the quality of higher education. The official website of the Russian Ministry of Education and Science declares that 'the Russian Federation promotes the development of cooperation between Russian and foreign educational and scientific organizations, international academic mobility of students'. One of the goals of the development of academic mobility is 'enhancing the efficiency and competitiveness of the Russian educational system' [16].

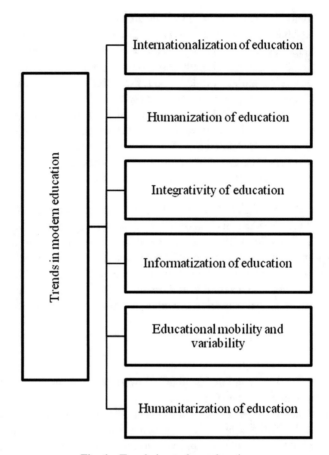

Fig. 1. Trends in modern education

Support for the development of academic mobility is carried out through various programs (ERASMUS, COMMETT, LINGUA, TEMPUS). Such purposeful development of students' academic mobility serves as a means of expanding the international market for the training of professionals and highly qualified specialists [6].

Humanization of education now plays an important role in the Russian educational system. Humanization of education is the maximum available attention to the student, respect for his human dignity, overcoming the aloofness of students and teachers from the educational process, eliminating the orientation toward the average student, creating conditions for the development of his social activity and the disclosure of creative potentials. Humanization of education also consists of renewing the content of educational programs [13].

Humanization of education is a system of measures aimed at the priority development of general cultural educational components and the formation of the students' personal maturity. Humanitarization includes an increase in the number of humanitarian disciplines, an increase in time spent studying humanitarian disciplines, and the development of new programs on normative humanitarian subjects.

Concerning humanization and humanitarization of higher professional education, we must keep in mind that technical education of the 21st century must necessarily take into account relationships between engineering (project) activity and environment, society, and mankind. In other words, activities of specialists must be humanistic. The problem of educational humanitarization in technical universities should be solved using the following items:

- expansion of the number of humanitarian disciplines;
- interconnection of humanitarian and non-humanitarian (science and technical) disciplines, while the humanitarian ones include the sciences of human, society, human-social interaction;
- the prognostication of social processes and the development of human nature;
- educational interdisciplinarity;
- training to solve scientific and technical problems at the interface of technical and humanitarian disciplines;
- providing students of technical universities with the possibility of obtaining a second humanitarian or socioeconomic specialty;
- strengthen the training of specialists in legal, linguistic, environmental and ergonomic areas;
- creating a humanitarian environment at the university;
- promoting the formation of students' worldview based on the dependence of socioeconomic, scientific and technological progress on the personal and moral qualities of a person, his creative abilities;
- personality-oriented training [22].

Educational integration plays an essential role and is manifested in various forms and at various levels (content, organizational and pedagogical, etc.). Considering the education system as an integrative phenomenon, it is worth analyzing all integrative connections of its separate parts. Pedagogic integration is usually considered 'horizontally' and 'vertically' [25].

As for informatization of education, an open educational system is gradually being implemented in Russia. Such a system allows each person to choose his own learning path, and also makes fundamental changes in the technology of obtaining new knowledge. Introduction of IT with their didactic property of individualizing the learning process while preserving integrity increases the effectiveness of education [10].

Variability is one of the main ways of humanizing not only the content, but also the learning process itself. It is manifested in ways of obtaining education, types of educational institutions, types of training courses (mandatory, elective and optional), in methods used by a teacher and organizational forms of training. Variability is the main trend of innovative changes in present-day education. At the same time, it is important to ensure each student has not only the right, but also a real possibility of choice [24].

In addition to the above-mentioned globalization trends, informatization is also singled out in the Russian education. Russian federal and non-governmental programs on informatization of education are widely distributed, each region has its own programs on computerization and telecommunications. It's worth mentioning that information technologies have recently become a long-awaited reality in Russia's education. We didn't therefore include informatization in the main current trends of Russian education.

Russia has developed a state policy reflecting national interests in education, simultaneously taking into account general trends of global development, which necessitate significant changes in the education system. Russia has also defined the principle of democratization as one of the main ones realized both at the level of organization and management of the whole education system and at the level of the direct organization of the educational process.

An analysis of world educational experiences testifies to the growing need of modern society for democratization of education at all its levels. Democratization of education is the creation of conditions for socialization of the country's future citizens and specialists capable of self-expression in democratic conditions. As for Russia, the education democratization principle was fixed in the National Doctrine of Education in the Russian Federation until 2025 and is regarded as 'education of Russian patriots, citizens of a legal, democratic, social state respecting human rights and freedoms and possessing high morality' [25].

We believe that global changes make it expedient to create new organizational forms of combining science, education and production, similar to those in technoparks and technopolises, which will determine the development of global integration educational and scientific processes in the future. Reinforcing the reality of interaction in education, the intensive expansion of telecommunications, the Internet and other means reflect modern emerging trends and needs [12].

Thereby, a number of educational events were held in the Samara region.

Firstly, it is university merging processes. Thus, the Samara National Research University named after academician S. P. Korolev, one of the region's largest universities, entered the 2017 QS World University Rankings annually compiled by UK-based Quacquarelli Symonds. It was included in a group of universities that took places from 800 to 1000. Being the supporting university of the Samara region, the Samara State Technical University (SamSTU) is developing its new educational technology: implementing the educational process within interdisciplinary project teams (IPTs)—a personnel training technology with a unique set of interdisciplinary competencies. The traditional educational process has been transformed to ensure the implementation of individual educational programs for teaching gifted students within IPTs. The educational process is formed according to the principle of designing target competencies for solving specific scientific and technical or engineering tasks within the real project implementation. Selection of projects and IPT leaders is conducted on a competitive basis. An expert council including international specialists has been established to evaluate and select the projects. The project may result in a competitive innovative product ready for implementation or a project team consisting of specialists with unique interdisciplinary competences able to solve not only technological tasks, but to create new jobs by implementing managerial functions and marketizating innovative products. The new educational technology aroused great interest among students and employers—10 interdisciplinary teams were formed in 2016, and two of them were created on request of the University's industrial partners—RKTs-Progress, Russia's space-rocket enterprise, and Novokuibyshevsk Refinery [5].

Secondly, the centre for gifted children—an educational institution of an innovative type—has been established. The centre is to identify, develop, and support gifted children by using modern scientific methods and technologies of teaching, education and personal development. Based on the main educational activities, the children's technopark includes:

- a VR laboratory equipped with augmented reality tools;
- robotic laboratories with modern educational kits, resource sets, electronic components, microcontrollers;
- a space laboratory equipped with spacecraft simulators and microsatellite sets on a special test bench that simulates the satellite's motion along the Earth's orbit;
- IT-laboratories equipped with modern computing devices, peripheral and mobile devices and other technical means of informatization (laboratories can be transformed into a lecture hall);
- an aircraft modelling workshop equipped with educational sets of aircraft models, unmanned aerial vehicles, quadrocopters;
- a laboratory for the implementation of nanotech industry's educational programs and research;
- a workshop equipped with educational sets for model-car construction classes;
- a laboratory for the implementation of industrial design educational programs;
- a special workshop equipped with modern drilling, milling, CNC machines, 3D printers, 3D scanners and other direct digital production and prototyping equipment.

As the children's technopark develops, the following facilities are scheduled to be established on its base:

- a co-working center to provide an additional zone for the implementation of students' technology start-ups;
- an interactive quantum-museum as a scientific and entertainment center to give an opportunity to participate in experiments, tests and other research actions;
- a media library with a rest area as an additional zone for individual work [20].

Thirdly, SamSTU is the initiator of the development, the regional operator and coordinator of the Unified Regional System of Measures for the Identification and Development of Creatively Gifted Youth in Science, Technology and Innovation Development of the Samara Region. This system includes a program for involving talented students, VZLET, and a program for involving students, undergraduates and post-graduate students of higher educational institutions, POLET. In 2017, 691 students were included in the Governor's Register of the creatively gifted youth in science and technology of the Samara Region. The implementation of the Unified Regional System of Measures gives the university the right to introduce innovations in the educational process by attracting gifted young people to the university and ensuring their further comprehensive development within its interdisciplinary educational projects [17]. The University thus acts as a system integrator of students' individual learning paths.

3 Conclusion

Having regard to the above said, we can conclude that globalization is an objective reality of the development of society and the education system. This phenomenon features the large-scale application of information and telecommunication technologies, including the Internet in all spheres of human activity, in particular education.

Globalization is comprehensive; it covers all aspects of human activity and affects all spheres of social and individual existence, including education.

Global changes provided Russian higher education institutions with new opportunities of opening new private universities, new specialties and implementing disciplines that are included in a high school component. A large number of new textbooks and teaching aids are being published. There is a wide choice in the specialties; the curriculum includes elective disciplines.

Thus, globalization of education in our country is entering a new phase, merging with transformation of the educational process. Considering separate technical, organizational, pedagogical aspects is now insufficient. The integrated globalization process calls for its rethinking and changing. The main current trends of globalization described in the research, as well as the global educational changes exemplified above through the Samara region, can serve as a positive experience for other regions of the country. Further studies of globalization in Russian education should provide a comparative analysis of global changes in the regions and across the country.

Acknowledgements. The authors express deep appreciation to colleagues for their assistance in conducting the research and to the University's administration for financial support of the research.

References

1. Becker, H.J.: Pedagogical motivations for student computer use that lead to student engagement. Educ. Technol. **40**(5), 5–17 (2000)
2. Bell, D., Inozemtsev, V.: An Epoch of Disunity. Reflections on the World of the 21st Century. Center for Studies of Post-Industrial Society, Moscow (2007)
3. Brzezinski, Z.: Choice: World Domination or Global Leadership. International Relations, Moscow (2004)
4. Darling-Hammond, L., Richardson, N.: Research Review/Teacher Learning: What Matters? pp. 46–53 (2018)
5. Development program until 2020 of the federal state budget education institution of higher education Samara State Technical University. (2018, May 17). Retrieved from the website: https://www.samgtu.ru/sites/default/files/2016/programma_samgtu_na_sayt.pdf/
6. Devyatova, I.E.: Problems of academy mobility of students. High. Educ. Russ. **6**, 112–116 (2012)
7. Esaulova, M.B., Kravchenko, N.N.: General and Professional Pedagogy. FGBOUVPO SPGUTD, St. Petersburg (2011)
8. Evans, C.: Exploring the use of a deep approach to learning with students in the process of learning to teach. In: Gijbels, D., Donche, V., Richardson, J.T.E., Vermunt, J. (eds.)

Learning Patterns in Higher Education. Dimensions and Research Perspectives, pp. 187–213. Routledge, London and New York (2015)

9. Hanvey, R.: Possibilities for International/Global Education. A Report (1979)
10. Ilidzhev, A.A.: Actual problems of computerization of education. Yurist-Pravoved **1**(74), 16–22 (2016)
11. Imperatives of internationalization. (2018, May 17). Retrieved from the website: http://window.edu.ru/resource/843/79843/files/imperativyiinternacionalizaciiitog.pdf
12. Isaeva, O.N.: World Education as a System: Extended Abstract of Cand. Ped. Dissertation. Ryazan, p. 20 (2009)
13. Ivanova, S.V.: Humanization of Education. Purposes, Objects and Terms. Values and Meanings, pp. 91–108 (2010)
14. Kamashev, S.V., Kosenko, T.S.: Globalization of education and global education in the modern world. Philos. Educ. **6**(45), 124–132 (2012)
15. McLaughlin, M.W., Talbert, J.E.: Building School-Based Teacher Learning Communities: Professional Strategies to Improve Student Achievement. Teachers College Press, New York (2006)
16. Ministry of Education and Science of the Russian Federation. (2018, May 17). International Academic Mobility. Retrieved from the website: http://im.interphysica.su/
17. On the formation of the Unified Regional System of Measures for the Identification and Development of Creatively Gifted Youth in Science, Technology and Innovation Development of the Samara Region in the 2016/17 academic year. (2018, May 17). Retrieved from the website: http://creative-youth.com/media/1111/materialswebmeeting_08_09_2016.pdf
18. Pedagogical terminological dictionary. (2018, May 17). Retrieved from the website: https://pedagogical_dictionary.academic.ru/773
19. Prokhorov, A.V. Higher education internationalization: state, problems, prospects. Alm. Theor. Appl. Res. Advert. pp. 8–17 (2012)
20. Samara regional center for gifted children. (2018, May 17). Retrieved from the website: http://codsamara.ru/
21. Samara University entered the global QS rating for the first time. (2018, May 17). Retrieved from the website: http://www.samru.ru/society/novosti_samara/99311.html
22. Shitikova, I.B.: Problems of humanization and humanitarization of education in Russian Technical University during professional training of designers. Mod. high Technol. **1**, 46–47 (2007)
23. Toffler, A.: Third Wave. AST, Moscow (2010)
24. Tulchinsky, G.L.: Digital transformation of education: challenges for higher education. Philos. Sci. **6**, 121–136 (2017)
25. Tyurina, Yu.A.: Democratization of education in the national educational policy of Russia (based on the results of a sociological survey in the Russian Far East). Political sociology: theoretical and applied problems: a collection of scientific articles dedicated to the 70th anniversary of the Honored Worker of the Higher School, Academician of the Russian Academy of Social Sciences, Professor Vinogradov V.D., p. 397 (2007)

An Empirical Study on Pair Performance and Perception in Distributed Pair Programming

Despina Tsompanoudi[1]([⊠]), Maya Satratzemi[1], Stelios Xinogalos[1], and Leonidas Karamitopoulos[2]

[1] Department of Applied Informatics, University of Macedonia, Thessaloniki, Greece
{despinats,maya,stelios}@uom.edu.gr
[2] Alexander TEI of Thessaloniki, Thessaloniki, Greece
lkaramit@otenet.gr

Abstract. This paper reports students' perceptions and experiences attending an object-oriented programming course in which they developed software using the Distributed Pair Programming (DPP) technique. Pair programming (PP) is typically performed on one computer, involving two programmers working collaboratively on the same code or algorithm. DPP on the other hand is performed remotely allowing programmers to collaborate from separate locations. PP started in the software industry as a powerful way to train programmers and to improve software quality. Research has shown that PP (and DPP) is also a successful approach to teach programming in academic programming courses. The main focus of PP and DPP research was PP's effectiveness with respect to student performance and code quality, the investigation of best team formation strategies and studies of students' attitudes. There are still limited studies concerning relationships between performance, attitudes and other critical factors. We have selected some of the most common factors which can be found in the literature: academic performance, programming experience, student confidence, "feel-good" factor, partner compatibility and implementation time. The main goal of this study was to investigate correlations between these attributes, while DPP was used as the main programming technique.

Keywords: Pair programming · Distributed pair programming

1 Introduction

Distributed Pair Programming (DPP) is a computer programming technique in which programmers develop software remotely using a specialized infrastructure. The aim of this technique is not only to make remote collaboration feasible, but also to gain the advantages of Pair Programming (PP). PP has its origins in the software industry as a part of Extreme Programming and is intended to improve software quality [1]. Typically it is performed on one computer, involving two programmers working collaboratively on the same code or algorithm. One programmer acts as the "driver" and the other one as the "navigator" (also called "observer"). The driver has possession of

© Springer Nature Switzerland AG 2019
M. E. Auer and T. Tsiatsos (Eds.): ICL 2018, AISC 917, pp. 762–771, 2019.
https://doi.org/10.1007/978-3-030-11935-5_72

keyboard and mouse and types the program code. The navigator reviews the inserted code and gives guidelines to the driver. DPP on the other hand is performed remotely, allowing programmers to collaborate from separate locations. Anecdotal positive feedback from professional programmers inspired researchers to perform educational studies and to incorporate PP in academic courses. As a result, many researchers followed this paradigm and numerous studies were published the following years [2, 3]. The main focus of them was PP's effectiveness with respect to student performance and code quality. Attention was also given to the investigation of best pair formation strategies and studies on students' attitudes [4].

Performance is one of the most investigated factors regarding the effectiveness of PP, indicating that PP has a positive effect on students' grades [4]. Another well-studied factor is pair *compatibility*. Research suggests that pairing students with similar skill levels has positive results on motivation and participation [5, 6]. Therefore, in order to achieve greater pair compatibility students should be paired according to their *programming experience*. Similarly, studies showed that pairs' performance is correlated with how comfortable students feel during a PP session (the so-called *"feel-good" factor*) [7]. *Implementation time* is also a common measure used in PP studies to evaluate PP's effectiveness [4]. Most of them report that PP requires less time to complete assignments. Finally, students' self-rated *confidence* has been used to evaluate PP satisfaction, although with contradictory results [8, 9]. Concluding, the most common factors which can be found in the literature are academic performance, programming experience, student confidence, "feel-good" factor, partner's compatibility and implementation time. The objective of this research was to investigate relationships between these factors that could affect the performance of Computer Science students working on their homework assignments in a DPP environment in the context of an object-oriented programming course.

The remaining article is organized as follows. In the next section follows a presentation of related work in the field. Then, the research questions and the context of the study are presented (Sect. 3). Section 4 contains the results of the statistical tests. A discussion and conclusions follow in the last section (Sect. 5).

2   Related Work

Most research studies conducted by academic and industry researchers conclude that PP has positive effects on programmers' performance and software quality. Williams et al. [2] studied several years the application of PP in the classroom. They found that the collaborative nature of PP helps students to achieve advanced learning outcomes, to be more confident and to receive better grades in programming assignments. Other studies indicate that PP leads in higher program quality, continuous knowledge transfer and more student enjoyment [3]. Group formation is considered to be a very important factor that affects the effectiveness of PP, and consequently DPP as well. Pairs can be defined by the instructor or students themselves. In the former case, group formation is based on students' programming skill level, their personality, or even randomly. Relevant studies suggest that pairing students with similar skill levels has positive results on motivation and participation. Toll et al. [10] concluded that the outcomes of

PP are better when the skills of the one partner are slightly better or worse than those of the other partner. Williams et al. [6] also concluded that pairs are more compatible if students with similar perceived skill level are grouped together.

Students' self-rated confidence in programming ability has been used to evaluate PP satisfaction, although with contradictory results. Thomas et al. [8] report that students with less self-confidence seem to enjoy pair programming the most. On the other hand, Hanks [9] states that in his study the most confident students liked pair programming the most, while the least confident students liked it the least. Muller and Padberg [7] define the "feel-good" factor of a pair as how comfortably the developers feel in a pair session. They study correlations between the "feel-good" factor and pair performance, as well as between programming experience and performance. They found that pair performance is uncorrelated with a pair's programming experience and that the "feel-good" factor is a candidate driver for the performance of a pair.

In our study we investigate correlations between performance and other critical factors as performed in the previous studies. The results are presented in the following sections.

3   Research Questions and Methodology

3.1   Context of the Study

The study presented in this paper was carried out in the context of an undergraduate course on Object-Oriented Programming (OOP) during the academic year 2016–17. The course offers an introduction to the Java programming language and is part of the second-year curriculum. It runs over thirteen weeks with a 3-hour lab class per week. As homework, students were assigned five Java projects to be solved in pairs using a DPP system. Eighty-eight students chose a partner and formed 44 groups. In order to solve the assignments students had to utilize a DPP system (SCEPPSys) which provides all means for remote collaboration and some logging capabilities [11]. An overview of the course is provided in Table 1.

Table 1. Course outline

Course	Object-Oriented Programming
Programming Language	Java
Academic year	2016-17
Semester	3rd
Duration	13 weeks, 3 h per week
Participants	88 (44 groups)
Assignments	5 Java projects as homework
Programming approach	Distributed Pair Programming

At the end of the course a questionnaire was delivered to the participating students in order to obtain their feedback on the DPP assignments. Data from the log files and students' responses were analyzed in order to run the statistical tests for this study. To measure correlations between the variables we used the Pearson correlation and ANOVA with Bonferroni post hoc tests.

3.2   Research Questions

The study aimed to answer the following research questions:

Q1: Does pair's performance correlate with pair's programming experience?
Q2: Does pair's performance correlate with pair's perception (confidence) about the ability on programming?
Q3: Does pair's "feel-good" factor correlate with pair's performance, pair's prior experience and pair's perception (confidence) about the ability on programming?
Q4: Does pair's implementation time correlate with pair's perception (confidence) about the ability on programming?
Q5: Does pair's performance correlate with pair's perceived compatibility?

The data used in the study was gathered from three different sources: the log files of the DPP system, the questionnaire and students' grades. The variables are summarized in Table 2.

Table 2.   Variables

Name	Description
Performance	Grades received in current course and Java assignments
Experience	Grades received in previous programming courses
Confidence	Self-perception on programming interest and ability
Feel-good	Pair programming evaluation
Implementation time	Total time spent in assignments
Compatibility	Pair's compatibility degree based on perceived compatibility

The variable *performance* is based on the mean grade that pairs had received in the OOP-course and the overall grade that they had received for their homework assignments. In order to evaluate *programming experience* student's grades in previous programming courses were considered. More specifically, the average grade from two courses, "Algorithms" and "Procedural Programming", was calculated in order to specify programming experience for each student. Grades in all courses are measured on a scale from 0 to 10, where 10 is "excellent".

Confidence is typically estimated using students' self-assessment in a survey. Just like in the study of Thomas et al. [8] we asked students to place themselves on a scale from 1 ("code-a-phobe") to 9 ("code-warrior"). This scale ranges between students who dislike programming and face difficulties while coding, and students who like and find challenging the programming process.

The *"feel-good"* factor was determined using students' responses on the question of how they evaluate the overall pair programming experience on a scale from 1 (very bad) to 5 (very good). The average value for each pair was then calculated. This approach is in accordance with the study of Muller and Padberg [7] where the term "feel-good" factor first appeared. The *implementation time* was calculated by the system, and it indicates the total time a pair spent on completing the assignments. In order to evaluate the impact of pair *compatibility* three subgroups were studied: (1) high compatibility pairs, (2) moderate compatibility pairs, (3) low compatibility pairs. The compatibility degree was based on the pair's perceived compatibility. Each student evaluated on a scale from 1 (non-compatible) to 3 (very compatible) how compatible he felt with his partner regarding his programming ability. For instance, pairs with a high degree of compatibility represented pairs with the same or similar programming skills.

The type of each one of the above variables is either continuous or ordinal. To measure correlations between the variables we used the Pearson correlation coefficient. To compare means among various groups we applied ANOVA with Bonferroni post hoc tests.

The research questions of the current study investigate meaningful relationships between aforementioned factors. The results are provided in the following section.

4   Results

In this section we present the results of our research questions. First we provide some general results of the studied factors. As presented in Fig. 1, the majority of students (79%) evaluated the overall experience in distributed and collaborative solution of assignments as a good (52%) or a very good (27%) experience ("*feel-good*" variable). Only 2 students reported a negative experience and both of them were in the same group.

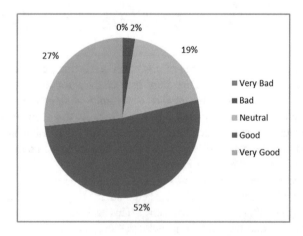

Fig. 1. Overall experience with DPP

Figure 2 depicts the distribution of students' self-perception on their programming skills (*confidence* variable). Half of the students rated themselves as "code-warriors" (scale 7–9), 40% of them rated themselves as 4–6 and only 10% of the students placed themselves as 1–3 ("code-phobes"). Finally, Fig. 3 depicts the distribution of perceived pair compatibility. As shown, students' vast majority report that they were very compatible (49%) or satisfactorily compatible (50%) with their partner.

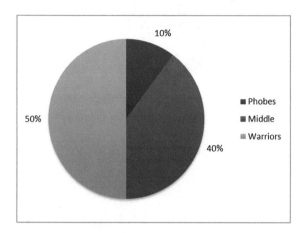

Fig. 2. Distribution of variable confidence

Q1: *Does pair's performance correlate with pair's programming experience?*
The first research question studies the correlation between the variables of *performance* and *programming experience* (Fig. 4). Although the result might seem self-evident, we decided to include it in the study as a sort of sanity check to test whether prior programming skills are reflected in the performance of future courses. For each pair the mean values of prior programming experience, exam grade and projects grade were calculated in order to run the tests. Grades in programming experience and exams were measured on a scale from 0 to 10 (where 10 is "excellent"), while grades in Java projects were measured on a scale from 0 to 1.5.
The correlation analysis revealed that:

- "Pair's mean programming experience" is correlated positively with "pair's mean Java exam grade" ($r = 0.766$, $p < 0.001$).
- "Pair's mean programming experience" is correlated positively with "pair's mean Java assignments grade" ($r = 0.511$, $p = 0.013$).

We can conclude that students with prior programming experience performed better in the OOP-course and the Java projects.

Q2: *Does pair's performance correlate with pair's perception (confidence) about the ability on programming?*
This research question studies the relationship between pair's confidence level and performance. We found that confidence and performance are positively correlated

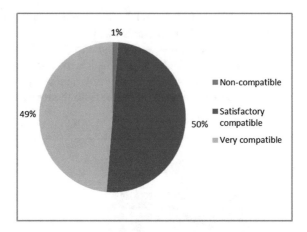

Fig. 3. Perceived Pair Compatibility

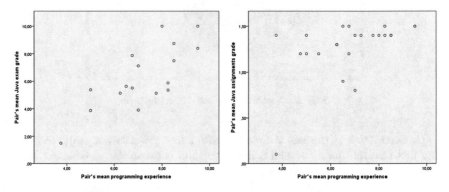

Fig. 4. Correlations between performance and experience

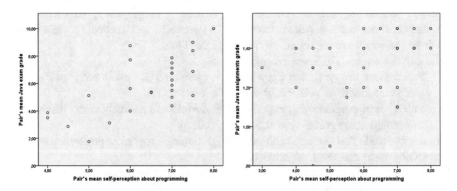

Fig. 5. Correlations between performance and confidence

(Fig. 5). The most confident students perform better in the course and the programming assignments.

More specifically, the statistical tests showed that:

- "Pair's mean self-perception about programming" is correlated positively with "pair's mean Java exam grade" (r = 0.715, $p < 0.001$)
- "Pair's mean self-perception about programming" is correlated positively with "pair's mean Java assignment grade" (r = 0.355, $p = 0.031$).

Q3: Does pair's "feel-good" factor correlate with pair's performance, pair's prior experience and pair's perception (confidence) about the ability on programming?
As presented in the previous section, Thomas et al. [8] found that students who have considerable self-confidence do not enjoy the experience of PP. Hanks [9] measured in a different way the level of confidence and his results contradict the findings of Thomas et al. [8]. However, in our study no correlation was found between "pair's mean feel-good factor" and any of the above variables. Students seem to feel comfortable during PP sessions regardless of their self-perception in programming competence.

Q4: Does pair's implementation time correlate with pair's perception (confidence) about the ability on programming?
We found that implementation time and "pair's mean self-perception about programming" are correlated negatively (r = −0.359, $p = 0.029$). This means that pairs with a high level of confidence on their programming skills needed less time to complete the assignments (Fig. 6).

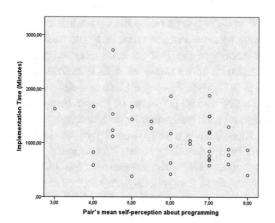

Fig. 6. Correlation between implementation time and confidence

Q5: Does pair's performance correlate with pair's perceived compatibility?
The statistical test showed that the performance in Java final examination was statistically different with regard to pairs' compatibility (F[2, 25] = 3.912, $p = 0.033$), meaning that pairs that perceive they are very compatible perform better in the Java final exam than pairs that perceive they are satisfying compatible (Fig. 7).

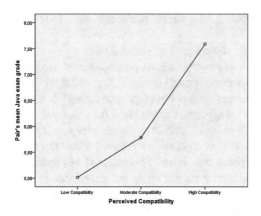

Fig. 7. Correlation between performance and perceived compatibility

5 Discussion—Conclusions

The benefits of DPP are numerous and have been extensively recorded in the literature. In this paper we attempted to find correlations between variables that might affect the performance of PP-students. Based on our findings we can draw some general conclusions and provide a practical contribution to the literature.

Teamwork in the form of DPP-assignments was once again evaluated very positively. The majority of students (79%) evaluated the overall experience in distributed and collaborative solution of assignments as a good or a very good experience. In previous studies DPP has always gained positive feedback from students. A similar evaluation result is also presented in the work of Muller and Padberg [7].

According to the results of the first research question pair's prior programming experience is associated with performance in an OOP course that is supported by DPP assignments. There is a statistically significant correlation between previous programming performance and overall performance in Java (exam and assignments grade). Since prior knowledge of each pair member is a determinant factor for his learning efficiency, students should have a deep knowledge on fundamental programming concepts or constructs before enrolling in an OOP course.

In our study we found that DPP pairs are more compatible and perform better when students with similar perceived compatibility are grouped together. Williams et al. [6] examined whether pairs are more compatible if students with similar perceived skill level are grouped together. As in our study they conclude that there's a significant positive correlation between compatibility and perceived skill level. Therefore, we can confirm the findings of previous studies indicating that pairing students with similar skill levels has positive results on students' performance.

In summary, the most important findings of our research are as follows:

- Pair's prior programming experience is associated with pair's performance in an OOP course (Q1).

- Pairs are more compatible if students with similar perceived skill level are grouped together (Q5).
- Also, pairs with higher confidence levels work faster (Q4) and perform better in the course and the programming assignments (Q2).
- Students seem to feel comfortable during DPP sessions regardless of their self-perception in programming competence (Q3).

In our future work we aim to extend the current research to study correlations at group and student level.

References

1. Williams, L., Kessler, R.R., Cunningham, W., Jeffries, R.: Strengthening the case for pair programming. IEEE Softw. **17**(4), 19–25 (2000)
2. Williams, L., McCrickard, D. S., Layman, L., Hussein, K.: Eleven guidelines for implementing pair programming in the classroom. In Proceedings of the Agile 2008 (AGILE '08), pp. 445–452 (2008)
3. Faja, S.: Pair programming as a team based learning activity: a review of research. Issues Inf. Syst. **XII**(2), 207–216 (2011)
4. Salleh, N., Mendes, E., Grundy, J.: Empirical studies of pair programming for CS/SE teaching in higher education: a systematic literature review. IEEE Trans. Softw. Eng. **37**(4), 509–525 (2011)
5. Braught, G., MacCormick, J., Wahls, T.: The benefits of pairing by ability. In Proceedings of the 41st ACM technical symposium on Computer science education, pp. 249–253 (2010)
6. Williams, L., Layman, L., Osborne, J., Katira, N.: Examining the compatibility of student pair programmers. In Agile Conference, 2006. IEEE (2006)
7. Muller, M.M., Padberg, F.: An empirical study about the feelgood factor in pair programming. In 10th International Symposium on Software Metrics, 2004 (Proceedings), pp. 151–158. IEEE (2004)
8. Thomas, L., Ratcliffe, M., Robertson, A.: Code warriors and code-a-phobes: a study in attitude and pair programming. In: ACM SIGCSE Bulletin, vol. 35(1), pp. 363–367. ACM (2003)
9. Hanks, B.: Student attitudes toward pair programming. In: ACM SIGCSE Bulletin, vol. 38 (3), pp. 113–117. ACM (2006)
10. Van Toll, T., Lee, R., Ahlswede, T.: Evaluating the usefulness of pair programming in a classroom setting. In 6th IEEE/ACIS International Conference on Computer and Information Science, 2007. ICIS 2007, pp. 302–308. IEEE (2007)
11. Tsompanoudi, D., Satratzemi, M., Xinogalos, S.: Distributed pair programming using collaboration scripts: an educational system and initial results. Inf. Educ. **14**(2), 291–314 (2015)

Assessment in Higher STEM Education: The Now and the Future from the Students' Perspective

Sofia Antera[1], Rita Costa[2], Vasiliki Kalfa[3], and Pedro Mendes[2(✉)]

[1] Stockholm University, Board of European Students of Technology,
Stockholm, Sweden
sofiante9@gmail.com

[2] University of Lisbon, Board of European Students of Technology, Lisbon,
Portugal
{ritaflscosta, pmendes1994}@gmail.com

[3] Aristotle University of Thessaloniki, Board of European Students of
Technology, Thessaloniki, Greece
vicky.kalfa@gmail.com

Abstract. The purpose of this paper is to provide input regarding the students' perspectives on the assessment methods used in Higher Science, Technology, Engineering, and Mathematics (STEM) Education. Are traditional methods still effective? What are the students' perspectives on the diverse evaluation methods in Higher Education? To answer these questions, the Educational Involvement Department of BEST (Board of European Students of Technology), a non-profit, non-governmental, non-political and non-representative student organization, organises BEST Symposia on Education, BSE (former Events on Education—EoEs), which aim to convene Higher Education stakeholders and raise the students' engagement in Higher STEM Education. By performing a secondary data analysis of the students' perspectives as they were expressed and recorded in EoE Gliwice (Manasova et al. in Be on the right track with SMART, learning - change the education of tomorrow!. Gliwice, 2016 [1]) and EoE Chania (Kloster Pedersen et al. in Refreshing education: update, rethink, grow. Chania, 2017 [2]) reports, the current study shows that laboratory settings are supportive for combining the three most preferred learning techniques: discussion groups, practicing by doing and teaching others/immediate use. Moreover, it was concluded that the assessment on every evaluation system should combine the students' attitude in class and feedback from professors. Final exams no longer appeal to students and cannot reflect the knowledge and skill set obtained. Professors, universities and particularly educational policymakers should consider the students' needs both when formulating a fair assessment system and creating/updating academic curricula.

Keywords: Assessment · Evaluation methods · Student perspective · BEST

© Springer Nature Switzerland AG 2019
M. E. Auer and T. Tsiatsos (Eds.): ICL 2018, AISC 917, pp. 772–781, 2019.
https://doi.org/10.1007/978-3-030-11935-5_73

1　Introduction

Learning assessment can significantly vary in the way it is perceived, the way it is addressed and surely in the way it is implemented, especially considering the different stakeholders involved in the process. What professors consider best practice students reject as invalid or non-reliable. The era in which students are awarded credit hours for attending classes and the transmission of knowledge is realised through lecturing is sadly not over [3]. With learning perceived as an act, it is "no longer seen as simply a matter of information transfer but rather as a process of dynamic participation, in which students cultivate new ways of thinking and doing, through active discovery, discussion, experimentation" [1].

Participating in this system, all students receive the same materials and teaching regardless of their cultural, social and knowledge background. Those requiring more time to learn fail, having no opportunity to follow at their own pace. This underestimation of the learning individuality leads many to believe that students' learning in tertiary education has not gained enough attention and has resulted in being one of the least sophisticated aspects of the Higher Education learning process [4].

In this paper, we briefly review the concept of validity of education assessment with a focus on fairness, in order to theoretically explain the students' views on STEM education assessment as expressed during the Events on Education of Gliwice and Chania.

2　Learning and Assessment

By approaching learning as a constructive act of the learner [5], we can assume that assessment is the process of evaluating this act. Despite the interactive and dynamic nature of learning and hence the difficulty to validly measure it, the academic community tends to evaluate learning, process, and outcomes for several purposes, such as ranking students or improving the learning experience. As Hattie [6] highlights, in Higher Education 'we implicitly trust our academics to know what they value in their subjects, to set examinations and assignments, to mark reliably and validly, and then to record these marks and the students move on, upwards, and/or out'.

In this line of thought, several assessment methods are implemented by tertiary education institutions representing different approaches to both learning and assessment and serving various purposes. Therefore, in education, assessment refers to a variety of methods and tools applied to measure, document and evaluate the academic readiness, the progress of learning, the competence acquisition and the students' educational needs.[1] Nowadays, the efficiency of these methods is highly doubted not only by the scientific community but also by the receivers of these practices, the learners.

Investigating students' general perceptions about assessment [7], aspects of perceptions emerging refer to the validity of assessment methods and to the concept of fairness, as perceived by students. Drew's study shows that learners value self-management and examinations are perceived as less supportive for the development of

---

[1] Edglossary.org.: https://www.edglossary.org/assessment/.

this competence, whereas deadlines per se are not considered as unhelpful. Deadlines increase self-discipline as well as the ability to study under pressure, and assessment can be seen as a powerful motivation to learn. In addition, clarity of professors' expectations and assessment criteria are highly valued by the learners. Effective communication between the learners and the teachers is also appreciated. Finally, students tend to link feedback with support, since the latter is critical to boost self-confidence [7].

2.1 Validity and Fairness

Validity is an integrated evaluative judgment of the degree to which empirical evidence and theoretical rationale support the adequacy and appropriateness of inferences and actions based on test scores or other modes of assessments [8]. According to this definition, the aim of assessment is to provide a comprehensive package of evidence that learning takes place, including the competences in practice and the degree they are developed [9].

Curricular validity ensures that the goals and objectives set by the curriculum agree with what students need to know. Curricular validity is a prerequisite for quality assessment since it examines the link between the learning objectives for the course and the desirable outcomes [10].

Construct validity refers to the relation between the assessment content and method and the learning objectives and is prone to dispute. Most assessments require general skills besides the subject domain (e.g. the ability to construct an essay). These competences can be assessed along with the subject-related ones or they can be considered implicit criteria in the learning objectives. Establishing construct validity, the professors need to ensure that the assessment content corresponds to the course learning objectives [10]. In the present study findings suggest that students doubt the construct validity of the assessment methods under discussion.

Finally, according to predictive validity, predictions made based on the assessment results are valid. This validity type is useful for selection purposes because it ensures that the current student performance is closely related to a future one [10].

Validity is related to the concept of fairness [5], considered a fundamental aspect of assessment, by students. Embracing more than cheating, the notion of fairness relates closely to the one of validity. Therefore, it is common that students express the view that a method may or may not be an accurate measure of learning.

Dimensions of fairness related to assessment methods included the control the students have over the examination process. Evaluation methods that require no active participation from the side of students are regarded as less fair. Moreover, if the assessment results are highly influenced by the factor of luck, the method is considered of low fairness. For example, when only one part of the course is examined, the factor of luck may lead to either good or bad results that do not reflect the actual learning [11].

Students tend to associate fairness with the time and effort invested in what they regarded as meaningful learning. Consequently, assessment methods which demonstrate this effort are perceived as fairer. Other dimensions of fairness include openness and clarity, meaning that methods which support better communication between the examiner and the examinee and allow feedback are seen as fairer [11].

3 Methodology

Reshaping education by offering students the opportunity to have a core role in its formation is the way towards quality education [12]. In this context, BEST Symposia on Education—BSEs (former Events on Education—EoEs) convene Higher Education stakeholders, with the purpose to strengthen students' involvement in several aspects of Higher STEM Education, through exchanging views and practices with academics, industry representatives, and other education experts. BSE's outcomes encompass interesting students' perceptions on education and, hence, their further exploration is considered helpful for making the students' voice heard.

To achieve a more objective diversity of students, more than 20 STEM students participated in each event, from different countries, educational and cultural backgrounds. The selection is based on gender, academic qualifications and origin to ensure diversity. In EoE Gliwice and EoE Chania, gender balanced is reported, while 17 countries are represented combining both events. Regarding the students' academic qualifications, more participants have already acquired a bachelor's degree, but some are still undergraduate, and others already have a master's degree (more details on Annex).

Based on the data from the EoE Gliwice [1] and EoE Chania [2] reports, a secondary data analysis has been done and its results are presented in the next chapter. This study adopts a qualitative research strategy, with aim to highlight the views of students regarding assessment methods. Qualitative research focuses on social processes assisting in demonstrating patterns emerging among participants' perceptions [13].

The data collected and analysed refers to the outcomes as they were reported by the facilitators of the EoEs. The facilitators consisted of both the event organisers, meaning BEST members—engineer students of various disciplines—and university professors that delivered sessions and workshops.[2] The group of authors performs qualitative content analysis, in an attempt to interpret the outcomes of the EoEs, identifying latent content [14]. Starting from the existing theory, the authors seek in the data same, similar or different issues in the assessment methods as expressed by students. The assessment methods discussed during both EoEs were collected and the most interesting ones were selected to be present in this study. The criterium was the amount of information available in the reports expressing the participants' viewpoints.

This study is limited to assessment methods discussed during the EoEs, as well as the data are analysed based on the reports and some theoretical approaches as mentioned above.

4 Results

While collecting students' views on evaluation methods, a distinction between traditional and alternative methods was made. The goal of this separation is to differentiate regular evaluation methods from the ones that are occasionally used in Higher STEM

---

[2] Find more details about facilitators in references [1] and [2].

Education. This perception is also shown through students' opinions. To facilitate the input gathering, students discussed the methods' advantages and disadvantages separately and proposed improvement points for evaluation systems in general.

4.1 Traditional Methods

Referring to traditional methods, oral, written and multiple-choice exams were discussed. Oral exams are found to enhance soft skills through practicing public speaking. This evaluation method promotes fairness since cheating gets harder. Oral exams are found to enhance soft skills through public speaking, which promotes fairness since it is harder to cheat, but at the same time may undermine it because of achieving a good grade which would correspond to their soft skills (for example presentation skills) and not actual knowledge on the subject. To increase objectivity in evaluation, students suggested the engagement of more professors as part of a jury, rather than a single examiner. On the other hand, written exams were mentioned to be objective and offer more chances to pass. However, students believe that they only cover a small part of the course syllabus and promote 'learning by heart', hence surface learning. To counteract these two points, students noticed the need for full-fledged course evaluation to encourage deep knowledge and raise fairness. As a third addition to the traditional methods, multiple choice tests were not considered to support actual learning either in knowledge, soft skills or creativity and only help surface learning [2].

4.2 Alternative Methods

Numerous alternative methods have been discussed, along with their advantages and disadvantages, as well as improving points.

4.2.1 Open book

Open book is a method that creates positive feelings to students, since they do not have to memorise trivial information. It was also perceived as less stressful. However, the examination difficulty level was discussed as very low, meaning that complete and proper evaluation could not take place [2].

4.2.2 Projects

Widely known to boost hard skills, projects were considered to promote the practical implementation of knowledge, according to students. Although projects were perceived as good evaluation methods, students raised some issues for further discussion. Project development may be time consuming, especially considering that projects cover only part of the course. Meanwhile, among team members, evaluating students' individual involvement is always a challenge. To counteract these issues, students proposed to divide bigger projects into smaller ones, to minimise the workload. At the end of the semester the smaller projects could lead up to a final presentation. Also, students acknowledge that there should be a way to evaluate individual equal contribution to a project. Competition between teams may be a way to promote participation [2].

4.2.3   Case Studies

Regarding alternative methods, case studies seem to be increasingly popular to students. Among the advantages discussed, case studies promote analytical thinking by asking students to work on a real case scenario, strengthening teamwork and team building and encouraging self-directed learning, since students perform their own independent research. Moreover, 3 out of 6 teams stated that case studies support soft skills development. Nonetheless, students claimed that there is a constant repetition of topics throughout the years and there is usually only one acceptable solution, the one favouring a professor's particular point of view. A yearly update of case studies is recommended to tackle the previously mentioned issue [2].

4.2.4   Presentations

Following the previous concepts, students discussed presentations. 4 out of 6 teams claimed that presentations help to develop soft skills and especially, gain knowledge on public speaking. Even though presentations were signified as important, the fear of evaluating presentation skills and not knowledge was expressed. Students identified presentations as stressful and mentioned the influence, importance and impact of the impression that they will leave on the professor. To overcome the obstacle of public speaking, students suggested that universities should offer courses on presentation skills. Thus, evaluation would be based on knowledge of the topic instead [2].

4.2.5   Practical Labs

Moving to a different type of evaluation method, practical labs were explored. They were perceived as a method to promote deep learning through the application of theoretical knowledge in real working situation. On that account, students expressed the view that practical labs lead to an increased possibility of passing the final exam, since they facilitate better understanding of the subject. However, students traced a loophole in this method by indicating that it cannot evaluate theoretical knowledge. The combination with other evaluation methods may help improve practical labs [2].

4.2.6   Online Exams

Following the advancements in technology, online exams were also discussed as a possible evaluation method. Lack of time restriction and speed in grading were among the advantages presented. Additionally, 3 out of 6 student groups emphasised the benefits of flexibility on the examination timing. Nevertheless, technical problems during the procedure seem to be common and as half of the students pointed out, it is relatively easy to cheat during online exams [2].

4.2.7   Homework

Even though some may consider it as a traditional method, homework at universities is ranked among the alternative methods. In this case, advantages balance the disadvantages. Students claimed that the main benefits are independence, flexibility in work scheduling and learning through exercising. Furthermore, during homework students can come up with more complex problem solutions, due to time flexibility. Contradicting the benefits, students mentioned that this method is hard to ensure objective evaluation of learning, since it is based on the individual's ability to do quality studying and it does not prevent cheating and copying the work of others. Having different homework with the same level of difficulty can be one of the solutions to this issue, as students proposed, as well as having a time restriction for completing homework assignments [2].

4.3   Improvement Points in Evaluation

During the discussions, students stated additional ideas on evaluation methods, with aim to better evaluate both theoretical and practical learning.

Students remarked that evaluation should be monitoring the progress of their studies rather than depend on the final exam. Furthermore, they claimed that feedback from professors is better than judging in an evaluation system, to promote deep learning. In a unanimous understanding, students expressed the necessity of combining more than one evaluation method to obtain a successful outcome in learning and deviate from the traditional approach of final exams. Considering a different vantage point, they proposed numerous possibilities to objectify the grading of students, by detaching evaluation from a single person, namely one professor, and distributing the role in various parties [2].

Recommended ideas such as self-evaluation and peer-evaluation, appointment of a 'jury' consisting of professors experienced in different fields and definition of diverse evaluation criteria per professor, show that students contemplate the importance of passing from traditional evaluation methods towards alternative ones [2].

During a Time in Teams (TiT) activity,[3] students were given a scenario of an assessment system to reflect and propose improvement points [2]. One of the teams, facing outdated lectures with no recording, low level of freedom and flexibility and an insufficient share of projects and topic diversification, pointed out the necessity for a continuous assessment of the evaluation methods to enhance their quality. For this purpose, both students and professors should be involved, sharing their feedback before and after the classes. In the same line of thought, continuous professional development for professors is crucial, especially with a focus on the pedagogical strategies they use. More specifically, students mentioned that professors should remain informed about trends in evaluation approaches and regularly work to develop communication skills.

Correlating different assessment methods with learning techniques, namely discussion groups, practice by doing (projects) and teaching others/immediate use, the only evaluation method that could evaluate all three simultaneously was lab work according to students, while case studies, real life case studies provided by companies and in engineering competitions, thought to assess two out of three. With only one, there were problem solving discussions, summer camps, mentoring and internships [1].

Regarding discussion groups, monitoring students' progress and evaluating discussion effectiveness through asking questions in an hourly basis were suggested. During projects, the importance of acknowledging the share of contribution by students

---

[3] For the TiT sessions the students were divided in four teams, each one representing a hypothetical institution with a given number of problems (six, in this case). The students were responsible to present improvement points, following brainstorming and debating practices.

was highlighted. Competition among teams was also considered of value. Finally, when it comes to teaching others, a system of points was mentioned, where points are awarded to mentors if their students successfully finish the tasks given. Evaluation can take place among team members as well [1].

5 Conclusions

Reflecting on findings, it is agreed that the evaluation process should not reside exclusively on examination results, since they promote surface learning and are not favoured by students due to their low construct validity. Students express a strong concern about the objectivity of evaluation in general and the validity of the given evaluation method. Their fear is explained by the importance of the grades in progressing with studies. In an effort to seek for more fair evaluation methods, students suggest the engagement of multiple examiner/professors and demand evaluation methods that examine the course as a whole and not partially. In the same line of thought, students tend to believe that the wider variety of methods, the better, with online exams, presentations and case studies being the favourite assessment types among students. Finally, the proposal of breaking bigger projects into smaller units, also serves the purpose of rising construct validity for students, ensuring that the content of the subject is thoroughly assessed.

Soft skills seem to have a quite controversial place when evaluation is under discussion. They are mentioned as competences practiced and developed in all assessment methods. However, although their practice and improvement are seen as positive, students perceived soft skills evaluation as negative, since they correlate it with poor evaluation of the course content.

Students' opinion aligns with Drew's conclusions with respect to the importance of feedback. Feedback is perceived a friendlier form of evaluation and refers in a big degree to the progress made while studying. Therefore, with feedback the focus moves from evaluating the outcome to assessing students' effort and progress. According to Drew [7], focusing on assessing progress and effort enhances the feeling of fairness.

Continuous assessment of the evaluation methods and their effectiveness is rendered as necessary by students, with feedback constantly shared between students and professors. This closer relationship between the stakeholders engaged in the process of evaluation can offer a multifaceted approach in selecting and applying evaluation methods corresponding to the needs of both the students and the institutions. This step is essential in moving towards assessment that is not solely translated into grades, but it actually monitors students' progress by supporting their learning.

Annex

See Tables (1 and 2).

Table 1. Number of participants per country of each of the Events on Education

Country	EoE Gliwice (9 male/9 female)	EoE Chania (10 male/13 female)
FYROM	1	0
Hungary	2	1
Italy	1	1
Belgium	1	1
Ukraine	1	1
Russia	4	3
France	1	0
Greece	1	2
Spain	4	2
Croatia	1	1
Bulgaria	1	0
Poland	0	5
Moldova	0	2
Portugal	0	1
Turkey	0	1
Estonia	0	1
Romania	0	1

Table 2. Qualifications of the participants of each of the Events on Education

Qualifications	EoE Gliwice	EoE Chania
Secondary education	4	9
Bachelor's degree	12	12
Master's degree	2	2

References

1. Manasova, D., Merlier, A., Guliaeva, A., Wippich, A., Trajkovikj, N.: Be on the right track with SMART, learning—change the education of tomorrow!. In: BEST Event on Education 2016, 21–31 July 2016. Gliwice, Poland (2016). https://issuu.com/bestorg/docs/eoe-gliwice
2. Kloster Pedersen, L., Sobrino Verde, C., Churyło, K., Pasovic, D.: Refreshing education: update, rethink, grow. In: BEST Event on Education 2017, 12–21 July 2017. Chania, Greece (2017). https://issuu.com/bestorg/docs/eoe_chania_2017
3. Johnston, H.: Proficiency-based education. Retrieved from ERIC database (ED538827) (2011)
4. James, R.: Academic standards and the assessment of student learning: some current issues in Australian higher education. Tert. Educ. Manag. **9**(3), 187–198 (2010)
5. Struyven, K., Dochy, F., Janssens, S.: Students' perceptions about assessment in higher education: a review. In: Joint Northumbria/Earli SIG Assessment and Evaluation Conference: Learning communities and assessment cultures, University of Northumbria at Newcastle, Longhirst Campus, 28–30 Aug 2002

6. Hattie, J.: The black box of tertiary assessment: An impending revolution. In: Meyer, L.H., Davidson, S., Anderson, H., Fletcher, R., Johnston, P.M., Rees, M.(eds.) Tertiary Assessment and Higher Education Student Outcomes: Policy, Practice and Research, pp. 259–275. Ako Aotearoa & Victoria University of Wellington, Wellington, NZ (2009)
7. Drew, S.: Perceptions of what helps learn and develop in education. Teach. High. Educ. **6**(3), 309–331 (2001)
8. Messick, S.: Validity. In: Linn, R.L. (ed.) (1993) Educational Measurement, 3, pp. 13–103. American Council on Education/Macmillan, New York, NY (1989)
9. Gyll, S., Ragland, S.: Improving the validity of objective assessment in higher education: steps for building a best-in-class competency-based assessment program. J. Compet.-Based Educ. **3**, e01058 (2018)
10. McAlpine, M.: Principles of Assessment. CAA Centre, University of Luton, Luton (2002)
11. Sambell, K., McDowell, L., Brown, S.: 'But is it fair?': an exploratory study of student perceptions of the consequential validity of assessment. Stud. Educ. Eval. **23**(4), 349–371 (1997)
12. Sambell, K., McDowell, L., Montgomery, C.: Assessment for Learning in Higher Education, pp. 10–11. Routledge, Milton Park (2012)
13. Bryman, A.: Social research methods, 4th edn. Oxford University Press, Oxford (2012)
14. Adamson, B., Morris, P.: Comparing curricula. In: Bray, M., Adamson, B., Mason, M. (eds.) Comparative Education Research: Approaches and Methods, pp. 263–283. The University of Hong Kong, Hong Kong (2007)

An Evaluation of Competency-Oriented Instructional Tasks for Internal Differentiation in Basics of Programming

Carolin Gold-Veerkamp$^{(\boxtimes)}$, Marco Klopp,
Patricia Stegmann, and Joerg Abke

University of Applied Sciences Aschaffenburg, Wuerzburger Strasse 45, 63743
Aschaffenburg, Germany
{carolin.gold-veerkamp,marco.klopp,patricia.stegmann,joerg.abke}@h-ab.de

Abstract. Internally differentiated instructional tasks are a strategy to handle the heterogeneity of learners' individual knowledge levels. The use of differentiated learning tasks can make an important contribution to teaching/learning management in order to efficiently promote the individual competency development of students. This paper covers the first formative evaluation results of implementing an inner differentiation approach covering instructional tasks in the basics of programming in an engineering bachelors' degree program.

Keywords: Instructional task · Scaffolding · Heterogeneity ·
Competencies · Taxonomy · Formative evaluation

1 Introduction

On the basis of evaluation results from an entry questionnaire and final evaluation of Informatics I (summer term 2017) and Informatics II (winter term 2017/18; i.e., the same cohort) in the Bachelor Degree Program Mechatronics (B.Eng.) several challenges were perceived, that have to be dealt with to improve teaching and learning. These problems are:

Heterogeneity in the level of knowledge. The students evaluate their level of knowledge in Informatics heterogeneously: 15 % describe their knowledge as much/very much, but the majority has little or no experience.

Complexity. The students rated the practical lessons to be more complex in comparison to the lecture. Same applies to the extent of the contents in lecture and practice.

Preparation and follow-up. The module manual [1] displays 4.5 hours/week in Informatics I (5 ECTS) and 3 hours/week in Informatics II (4 ECTS) for preparation and follow-up per semester (excl. exam preparation). This contrasts the students' median of "0–1 h" per week.

© Springer Nature Switzerland AG 2019
M. E. Auer and T. Tsiatsos (Eds.): ICL 2018, AISC 917, pp. 782–792, 2019.
https://doi.org/10.1007/978-3-030-11935-5_74

In order to meet these challenges written learning tasks are an important aspect. Thus, differentiated task sheets have been designed as a didactical approach [2]. These are used in "Informatics I" in the summer term 2018. The work of Zehetmeier et al. [3] was the inspiration for this approach, which provided a long version for programming novices and a shorter one for experienced students [2].

The evaluation concept and results are the core of this paper (Sect. 3); in advance the concept is shortly explained in Sect. 2. For more information about relevant theory regarding internal differentiation, instructional tasks, and competency oriented exercises, a previous publication shall be taken into account [2].

2 Concept: Design of Instructional Tasks

Since the module "Informatics I" consists of a weekly lecture and a practical lesson, a differentiation should take place with the help of task sheets. All students receive a homogeneous input in the lecture, while the exercises are constructed differentiated for preparation as well as during the practical lessons. The versions differ in quantity and quality/scaffolding, but cover the same scope. Therefore, the students have the opportunity to choose between the two different task sheets (short and long; see Fig. 1) based on their individual level of knowledge.

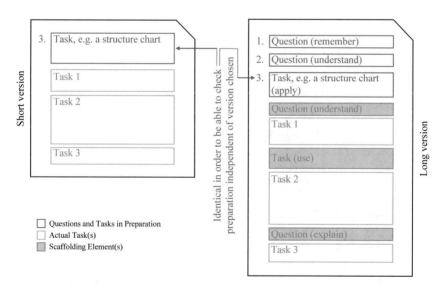

Fig. 1. Pattern for the two task sheet formats [2]

The Figure displays a pattern for the short and long version sheet formats, which cover: Preparation, the actual task, and scaffolding elements. Looking at the longer version, it is visible that scaffolding is used in order to support weaker learners by using additional questions, explanations, and/or small tasks.

The decision was made to only provide two versions, as (1) this is a first test whether the students benefit from two versions to chose from and (2) the students may be overburdened with too many alternatives. The actual task is summarized in the short version compactly, while in the long version intermediate steps or other helpful elements are incorporated. The solution results of the tasks are and therefore learning results should be identical in both cases.

Table 1 presents the eleven exercises with their content-related topics. In addition, each topic is assigned with the duration and it is indicated whether different versions are available.

Table 1. Outline of topics and versions for exercises in Informatics I

No.	Content	Duration [lessons]	Versions
Exercise 1	First Project (hands-on introduction to eclipse)	1	1 version
Exercise 2	Display output on EV3	1	Short/Long
Exercise 3	Structograms	1	1 version
Exercise 4	Variables and data types	2	Short/Long
Exercise 5	Arithmetic overflow	1	Short/Long
Exercise 6	Numeral systems (binary, octal, hexadecimal)	1	1 version
Exercise 7	Numeral representation (complements, IEEE 745)	1	1 version
Exercise 8	Control flow (if, else, switch case)	2	Short/Long
Exercise 9	Control flow (loops)	1	Short/Long
Exercise 10	Boolean Algebra	1	Short/Long
Exercise 11	Functions	1	Short/Long

3 Evaluation

3.1 Purpose

The objective is to use the concept and the developed task sheets and test them using a formative questionnaire (every lesson), which has already been designed. This shall contribute to gain insights into the manner and effort students prepare themselves for the practical lessons, whether experienced students use the short version and novices experience support when using the long version as well as how well the versions fit the students and their motivation.

3.2 Approach

To follow the research approach, the following research questions (RQ) shall be answered in order to resume the identified challenges (Sect. 1):

(a) How much experience do the students, who enter the course, have?
(b) Which version of the task sheets (short/long) do the students choose?
(c) Did they change versions during the assignments?
(d) How much time do they spend on the preparation of the task sheets?
(e) How satisfied are the students concerning the result of the practical lessons, which are based on the task sheets?
(f) How much fun[1] did they have during working on the task sheet?
(g) How good do they understand the content of the practical sessions/task sheets?
(h) Are they over- or under-challenged with regard to the task sheets provided?
(i) Which students (novice/expert) use which version of the task sheets (long/ short)?
(j) How good did the version chosen fit the students' knowledge levels?

The elicited results shall gain information about the (a) usefulness of the concept and the operationalized task sheets in general – i.e. whether this is a sensible approach to follow in the future – and (b) further optimization potentials to improve this method in the teaching and learning arrangement of Informatics.

3.3 Evaluation Results

In order to show the effect of the differentiated task sheets, a formative as well as summative questionnaire survey are carried out. The formative survey is embedded into each practical lesson by means of a short questionnaire.

The following results have been elicited through the initial questionnaire and the formative evaluations of all exercises (*Exercises 2* to *11*; see Table 1).

The evaluations took place in the summer term of 2018 in Informatics I ($N = 68$) in the Bachelor's degree program Mechatronics (B.Eng.).

---

[1] In reference to their motivation.

Level of Knowledge (a)

As shown in Fig. 2, according to their self-assessment, more than half of the students has no programming experiences (4 point Likert scale: 1 – none ... 4 – a lot), 35 % rated their knowledge in programming as low and only 9 % as relatively high at the beginning of the semester; furthermore, none is an expert (cf. RQ Sect. 3.2-a).

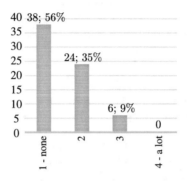

Fig. 2. Level of knowledge in programming at the beginning of the semester

This reflects the findings from the cohort before (see Sect. 1), where the majority (85 %) evaluated their knowledge as 1 – none ... 2.

Despite the fact that 68 students took part in the entry questionnaire, the mean value of participation in the practical sessions is 33.7 (see Fig. 3; response rate: 387 questionnaires in total). As we used the first practical lesson for a questionnaire pretest to adjust some questions, additionally, 17 individuals took part in *Exercise 2*, but their answers are not included in the following.

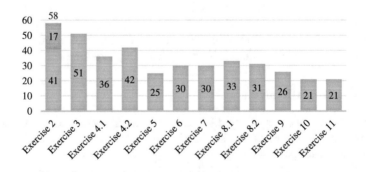

Fig. 3. Participation of exercises (see Footnote 2)

Version of the Task Sheet (b)

The answer to RQ Sect. 3.2-b is that most students prepared themselves using the long version of the exercises (61 % on average; see Fig. 4); averagely 26 %

used the short one. A percentage of 17 did not make themselves ready for the practical lessons. The lowest participation rate can be found in *Exercises 10* and *11* – at the semester end. *Exercise 3, 6* as well as *7* are not listed here as we did not construct differentiated task sheets for these two (see Table 1).

In sum, the participation diagram represents: 269 questionnaires could be used for identifying differences with regard to the chosen version of the task sheet (cf. Table 1).

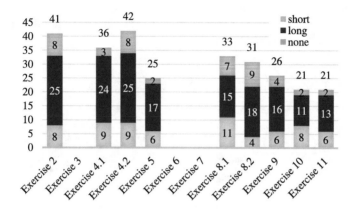

Fig. 4. Version the students' prepared themselves with (see Footnote 2)

Changes of Versions (c)

Concerning the question of changes during the preparation of the task sheets (RQ Sect. 3.2-c), Table 2 should be considered. The numbers shown are based on all task sheets having two versions; i.e. *Exercises 2, 4.1, 4.2*[2], *5, 8.1, 8.2*(see Footnote 2), *9, 10,* and *11*.

Out of the students that prepared themselves using – at least – one of the two versions ($N = 267$), most students stuck to their decisions (91 %).

Table 2. Percentages of changes during the assignments

Single choice	Total frequency (%)
No change	90.64
Short to long	3.00
Long to short	1.12
Several changes	5.24

---

[2] These numbers indicates that task sheets 4 and 8 are subject of two weeks each, whereby all lessons including preparation have been evaluated (see Table 1). Information on the version are based the one that was finally chosen.

Time Spend for Preparation (d)

Figure 5 shows differences in the time (see RQ Sect. 3.2-d) students needed to prepare the task sheets (here only data from students who did the preparation are compromised). The preparation for *Exercise 4.1* was the one that took the students averagely the longest (38.15 min), while *Exercise 11* was the quickest to do (18.21 min). Obvious from the graph, the numbers of the individuals greatly scatter (max: *Exercise 3*: 0...150 min). Also it can be recognized that the mean of the shorter versions is lower than the one of the long versions (again we do not use differentiated task sheets for *Exercises 3, 6,* and *7*). This seems to be a logical consequence due to the versions' number of pages that differ of about 2 on average (hereto see also i).

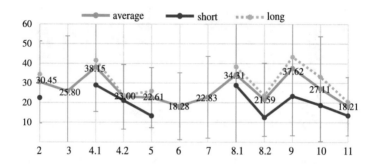

Fig. 5. Mean of minutes for preparation per exercise (divided into versions) (see Footnote 2)

In contrast to the evaluations of summer term 2017 (Sect. 1; median "0–1 h/week"), no enhancement in 2018 has taken place (median again is "0–1 h/week"). In detail, the average of $\bar{t} = 28.63\ min$ per week (see Fig. 5) is consistent with $\bar{t} = 25.91\ min$ per week the students estimated in the summative evaluation of 2018.

Therefore, we can definitely say that the outcome of 2018 are not inferior to the once of the year before.

Satisfaction (e), Fun (f), Understanding (g), and Challenge (h)

Students' assessment concerning their satisfaction, fun, understanding (see RQ Sect. 3.2-e, f, g), and challenge (RQ Sect. 3.2-h) during the whole exercise are shown in Fig. 6. Whereas the first three items have been measured on a 4 point Likert scale (1 – not at all ... 4 – very) visible on the left and the degree of challenge on a 5 point Likert scale (1 – over-challenged ... 3 – fitting ... 5 – under-challenged), which is shown on the right. The means of fun[1], understanding as well as satisfaction are rated positively; except for *Exercises 8.1* and *11*. The exercises are overall rated as a bit over-challenging when looking at the means. The most understandable part of the semester seems to be "Numeral Systems" – *Exercise 6*, whereas the most difficult content seem to be covered by *Exercise 11* – "Functions". Remarkably, the curves follow a similar course. Therefore, the

Spearman's rank correlation coefficient has been calculated for all combinations (see Table 3).

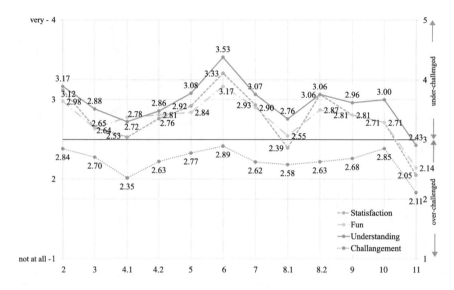

Fig. 6. Average of satisfaction, fun, understanding, and challenge (see Footnote 2)

Throughout all exercises, Table 3 shows strong positive correlations: *Understanding – Fun, Understanding – Satisfaction,* and *Fun – Satisfaction*, which seems to picture a consistent connection: The more one understands, the more fun he/she has, the more satisfied he/she is – or vice versa.

As well as medium negative effects regarding: *Challenge – Understanding, Challenge – Fun,* and *Challenge – Satisfaction*. This can be condensed to: The more challenging a task, the smaller the understanding, the smaller one is satisfied and has fun – or the other way round.

In a nutshell, this indicates that *Satisfaction* is not independent from *Fun, Understanding,* and *Challenge* – or vice versa. Noteworthy, only *Challenge, Understanding,* and *Satisfaction* weakly correlate to the *Level of Knowledge* the students themselves assessed at the beginning of the semester.

Version Chosen by Students (i)

Concerning RQ Sect. 3.2-i, the results of the Chi^2 test or Likelihood Ratio (LR)[3] in Table 4 give the information that there is a significant connection between the variable "*Version*" and the *Level of Knowledge, Challenge, Understanding, Satisfaction* as well as the level to which the tasks *Take up Knowledge Level*; i.e. the two groups "short version" and "long version" differ significantly.

---

[3] The decision between χ^2 and LR is based on [5].

Table 3. Spearman's rank correlation coefficient ρ

		Challenge	Understanding	Fun	Satisfaction	
Challenge	Correl. Coeff.	1.000		.486**	.385**	.468**
	Sig. (2-tailed)	.	.000	.000	.000
Understanding	Correl. Coeff.	.486**	1.000	.645**	.753**
	Sig. (2-tailed)	.000	.	.000	.000
Fun	Correl. Coeff.	.385**	.645**	1.000	.714**
	Sig. (2-tailed)	.000	.000	.	.000
Satisfaction	Correl. Coeff.	.468**	.753**	.714**	1.000
	Sig. (2-tailed)	.000	.000	.000	.
Level of Knowledge	Correl. Coeff.	.152**	.185**	.062	.176**
	Sig. (2-tailed)	.006	.000	.235	.001

According to Cohen [4, p. 82]:
$r = .10$ corresponds to a weak effect
$r = .30$ corresponds to a medium effect
$r = .50$ corresponds to a strong effect

Table 4. Results of χ^2 and LR in the contingency table: variable "*Version*"

Item	χ^2 or LR	p	Mean (short)	Mean(long)
Level of Knowledge	$\chi^2(4, n = 263) = 13.152^*$	.011	1.90	1.65
Challenge	$LR(8, n = 233) = 20.287^{**}$	.009	2.82	2.51
Understanding	$\chi^2(6, n = 274) = 19.649^{**}$	.003	3.22	2.81
Satisfaction	$\chi^2(6, n = 273) = 13.734^*$	.033	2.99	2.68
Take up Knowledge Level	$LR(4, n = 217) = 10.860^*$	.028	2.77	2.63
Fun	$\chi^2(6, n = 275) = 11.557$	.073	2.99	2.71
Minutes	$LR(20, n = 229) = 30.071$	.069	21.61	31.70

According to [6, p. 740]:
**. Significance at the .01 level (2-tailed).
*. Significance at the .05 level (2-tailed).

Looking at the means in Table 4, students choosing the short version:

- are **significantly more experienced** (self assessment; cf. Fig. 2).
- feel **highly significantly less challenged**.
- describe their **understanding as highly significantly better**.
- are **significantly more satisfied with their result** of the practical lesson.
- experience the **task sheet as significantly better connected to their state of knowledge**.

Same applies to students selecting the long version inversely (see "Mean (long)" in comparison to "Mean (short)", Table 4).

No significances are visible regarding *Fun* and the *Minutes* they spend on preparation; i.e. H_0 can not be rejected here.

Meeting Students' Knowledge Level (j)

To answer RQ Sect. 3.2-j, Fig. 7 gives insights. On average, the degree to which the task sheets took up the students' current skills is rated positively ($\overline{x} = 2.69$). Learners choosing the short version, evaluated *Exercises 4.1, 4.2, 9,* and *10* a lot better than the mean and higher than the other sheets.

All in all, the task sheets seem to meet the students' acquired competencies moderately, which should be further investigated using open-ended/qualitative survey methods.

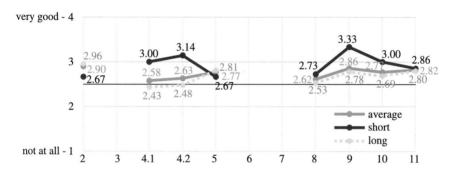

Fig. 7. Mean degree the exercises took up the students' current skills

4 Summary and Outlook

To address the identified challenges (heterogeneity, complexity, and lac of preparation; see Sect. 1) in higher education in Informatics, differentiated instructional tasks have been designed.

The evaluations show that the students' *Level of Knowledge* on average is low. As low experienced students choose the long version of the task sheet significantly more often, we should think about continuing to formulate these for the subsequent module "Informatics II". Furthermore, qualitative investigations could help to find out what could be improved in these exercises.

The results also indicate that the students rarely change the versions. Additional, the predictable expectation that it takes more time to prepare with the longer version could be proved. Concerning the final evaluation, it can be said that the students worked "0–1 h per week" (median) with an average of 25.91 min for the preparation of the task sheets. Hence, we could not manage to rise the weekly time for preparation through differentiated task sheets and therefore failed to achieve this goal.

Unfortunately, we noticed that the number of participants has steadily decreased (see Fig. 3); this must be addressed in the future.

Moreover, an interdependence between *Understanding, Fun, Challenge*, and *Satisfaction* has been detected with merely selective weak effects concerning the *Level of Knowledge*.

Concerning the *Level of Knowledge, Challenge, Understanding, Satisfaction*, and *Take up Knowledge Level* a statistically significant connection between these variables and the short vs. long version of the task sheets could be found; i.e. there is a statistically significant distinction between the two groups of students (short/long version).

The findings display that differentiated task sheets are one step into the right direction as a first trial of differentiated task sheets, since we could not find any destructive results. But they also reveal further potential for future investigations. Hence, we plan to integrate more data from the upcoming formative evaluations in this semester (cf. Table 1).

Acknowledgement. The present work as part of the EVELIN project was funded by the German Federal Ministry of Education and Research (Bundesministerium für Bildung und Forschung) under grant number 01PL17022B. The authors are responsible for the content of this publication.

References

1. University of Applied Sciences Aschaffenburg, "Mechatronik [German]", Module Manual (2017). https://www.h-ab.de/fileadmin/dokumente/fbiw/modulhandbuecher/module_m_spo3.pdf
2. Klopp, M., Gold-Veerkamp, C., Stegmann, P., Abke, J.: Using competency-oriented instructional tasks for internal differentiation in informatics. In: Proceedings of the 3rd ECSEE - European Conference of Software Engineering Education, pp. 26–33. ACM - Associated for Computing Machinery, New York (2018)
3. Zehetmeier, D., Böttcher, A., Thurner, V.: Differenzierte Übungsblätter – Für Experten und solche die es werden wollen [German]. In: MINT-Symposium, Nürnberg, Germany, pp. 94–98 (2017)
4. Cohen, J.: Statistical Power Analysis for the Behavioral Sciences, 2nd edn. Lawrence Erlbaum Associates, Hillsdale (1988). 0-8058-0283-5
5. McHugh, M.L.: The chi-square test of independence. Biochemia Medica, 143–149 (2013)
6. Bortz, J., Döring, N.: Forschungsmethoden und Evaluation für Human-und Sozialwissenschaftler [German], 4th edn. Springer, Heidelberg (2006)

Models of Educational Screenwriting: State of the Art and Perspectives

Jeanne Pia Malick Sene[1]([✉]), Ndeye Massata Ndiaye[2],
Gaoussou Camara[1], and Claude Lishou[3]

[1] Alioune DIOP University, Bambey (UADB), Bambey, Senegal
{jeannepia.sene, gaoussou.camara}@uadb.edu.sn
[2] Virtual University of Senegal (UVS), Dakar, Senegal
ndeyemassata.ndiaye@uvs.edu.sn
[3] Cheikh Anta Diop University (UCAD), Dakar, Senegal
claude.lishou@ucad.edu.sn

Abstract. In this article, we make a comparative study of models of educational scriptwriting. Learning scripting models such as SCORM (Shareable Content Object Reference Model), IMS LD (**IMS-Learning Design**) and LOM (Learning Object Metadata) place the relationships between learning activities and educational resources at the center of their respective models. These learning scripting models are implemented in distance learning platforms. In addition, they can be in the form of graphical tools, which help in the design and validation of various scenarios that can, then, be imported into the distance learning platform conceived following the same model. To ensure easy interoperability of online learning environments; we propose a learning scripting model based on the data from the educational information system. We review the advantages and shortcomings of teaching models through well-defined terms of reference, which will give us precisely the justification of our choice.

Keywords: Pedagogical scenario model · Meta model · Modeling language · Modeling tools · Scenario

1 Introduction

The Bachelor Master Ph.D. system (BMP) reform has led to the implementation of measures that modifies the Senegalese higher educational system and adapts it to international standards. Concerning the system, courses are organized in teaching units containing inherent elements (courses or subjects). To proceed to the next level, learners are required to pass all the teaching units or obtain 70% of the credits in order to be given a conditional passage.

As a result, students under conditional passage accumulate the modules of the current academic year and the lower-level uncommitted modules. The development of schedules does not take into account these cases of students. They are forced to make a choice on the different subject to follow in class (attendance), because schedules overlap. The Bachelor Master Ph.D. system also give the opportunity of the training

© Springer Nature Switzerland AG 2019
M. E. Auer and T. Tsiatsos (Eds.): ICL 2018, AISC 917, pp. 793–802, 2019.
https://doi.org/10.1007/978-3-030-11935-5_75

course. Indeed, based on the possibilities offered by higher educational institutions, students can build their careers from years after years according to their preferences and possibilities. In some cases, students are required to validate one or more lower-level specific teaching units if they wish to be specialized in a particular domain

Given this situation, we propose a new model of learning to the learner. This consists in setting up a distance learning platform that will offer online courses for students based on their academic performance. The online learning platform will have to implement an adaptive learning-scripting model. So, it will help learners to follow their courses according to the module to resume and necessary educational resources.

In this article, we propose a new pedagogical scenario model based on data from the educational information system. Several works on educational scriptwriting models have been published. Our scripting model is based on the study and the comparative analysis that we have carried out through the existing models.

At first, we will present a detailed analysis of different pedagogical scenario models. Then, the second part proposes a discussion on the main characteristics of these scripting models. Finally, we're going to present our new pedagogical scenario model.

2   Analysis of Educational Script Writing Models

2.1   What Is an Educational Scripting

Kindly for decades, models of pedagogical scenarios have been proposed in previous works. These models are often specific to particular or general learning situations. Michel Leonard and Gilbert Paquette in their article [1] conduct a comparative study of educational storytelling models. This study contextualizes the possibility of having a new representation of the pedagogical scenarios, especially as they highlight, in their synthesis, the advantages and shortcomings of all these educational writing models.

However, an educational scripting model is to answer a certain number of essential points that ensure a representation of the basic concepts of the pedagogical scenario. Several definitions of a learning scripting model were proposed in previous works. According to Michel Leonard and Gilbert Paquette [1], a pedagogical scripting model is a "system of representation that describes the scenarios or learning activities in a structure of its own".

Pernin [2], defines a model of educational scenario as the representation of handling result of modeling a precise learning situation. He adds that it is the "the description of the course (planned or observed) of a learning situation in terms of roles, activities and the environment necessary for its implementation, but also in terms of knowledge manipulated". Sadiq and Talbi 2010 [3], see a template of educational scripting as being "a typical representative of a scenario category, whether it corresponds to a pedagogical approach or to a referenced learning situation". All these definitions are intended to promote the representation of a learning situation that is broken down by the description of the links that exist between the entities of a pedagogical scripting model such as roles, actors, activities, scenarios etc.

In view of all these definitions we can retain that a model of pedagogical scripting must take into account a number of criteria. The existence of all these models brings a diversification of the representation of the following basic concept: teachers and learners' educational activities, pedagogical resources, roles and different actors involved in the activities. The normal course of these activities is based on the scripting of the concerned learning situation [4].

However, the pedagogical scenario aims to describe and build, from a synopsis, the unfolding of a situation in an exhaustive way. It is the organization of the activity in the form of several sequences based on objectives. These sequences must be linked by imported educational resources, off-product or reused ones, in an online learning platform. These resources are various depending on the activities, but also on the objectives.

We noticed that all of these models offer a manual require the basic resources (personal information of the players, the route information of the learner etc.) and similarly a default scenario, regardless of these imported resources. Models must mostly take into account generic scenario from specific resources, such as data from the systems of educational information.

2.2 Comparative Study of e-Learning Standards on Engineering Education

"The issue of modeling education is a question related to engineering", Georges-Louis Baron [5]. This modeling of education is based on the creation of standards, languages and modeling tools that give birth to pedagogical scenario models and LMS (Learning Management System). Gerard-Michel Cochard [6] defines a standard as a set of recommendations developed and proposed by a representative group of users (e.g. IETF RFC, W3C, IEEE recommendations).

There are several learning scripting models. The most popular is the SCORM model. It is a standard that proposes a model which, overall, performs the aggregation of the contents; however it is a mono actor. It is a simple sequencing", which translates the simple succession of content in its approach, and this limits the variation of the pedagogical approaches of the actors.

The SCORM model has a model of content aggregation and an environment of execution. The first component of the SCORM model is the aggregation scheme which is defined in three levels:

- The first, which is the basic level, consists in different types of educational resources (asset resource). It is image, file, text, video, etc.
- The second, intermediate level, SCO (Sharable Content Object) is the combination of several teaching resources, which gives a global teaching resource.
- The third, which is the higher level, is about the aggregation of content (courses).

The second component of SCORM is the online platform, which consists in the state of execution of an object.

The LAMS (Learning Activity Management System) model is concerned; it is graphic design that ensures online management of collaborative learning activities in sequence. At the level of this model, actors are not represented, which gives trend, to the creation of the sequences rather individual. But despite the non-representation of actors, it is a model that promotes the development of collaborative sequences.

The GMOT/TELOS model is also a graphical tool based on the language of "Object modeling by type". It defines the flow of activities while clarifying the links with the actors involved in each activity. On top of that, we have the consideration of some specific needs of the model's actors, particularly learners. For these latters, they have the opportunity to perform entrance tests, exercises, and updates of their knowledge. Concerning training actors, their needs in the model are limited to assessing learners' knowledge and validating their exercises.

There is the SCENARI (OPALE) model, which produces scripted educational documentation and ensures the division of the module into pedagogical blocks. However, it favors the scripting of the learning situation by reception. But it is important to mention the non-recursive aspect of these scenarios. So we talk about consultation and production.

There is the EML (Educational Modelling Language) model proposed by KOPER. In the model, KOPER proposes the description of the current learning situations using a pedagogical modeling language that puts the learning situations at the center of the process, and not the resources.

Thus, the study unit is considered as a composition of activities carried out by a set of actors in a given environment. So, there are several types of business including: learning activities, support activities, and instrumentation activities. Considering all these models, we have targeted two models of pedagogical teaching, which harbor a more general approach.

This is recognition for a representation in the broadest sense of the basic concepts of distance education. These concepts are: resources, actors, sequences, scenarios, activities, courses etc. As mentioned above in terms of the description of the models, the SCORM model and the IMS LD model constitute the two models whose approach is better suited for the realities of distance education. In addition to this advantage, the biggest platforms on distance learning have been implemented in these models.

For example, model which is currently considered the largest distance teaching platform from the francophone area implemented the SCORM model. Its latest version dates back to 2004. The IMS LD model is the generalization of the SCORM model, that is to say, it is the improved model of the SCORM model. At the level of its design, it has taken over all the basics of the SCORM model. With respect to this conception, we name them in various ways depending on the mode the. The IMS LD model is an improved version of the SCORM model.

Other models such as SCENARI, LAMS, EML, GMOT/TELOS are graphic tools that help to carry out educational scenarios that are especially focused on the resources. These scenarios will have to undergo a validation under the norms of the language which the graphical tool has implemented. Behind this validation, there lies a meta model of educational scenario [7]. All the scenarios that are performed through these tools are considered instances of the meta model. This description can be represented by the following concepts: meta model, modeling language, educational storyboard template, modeling tool, user requirements package. Figure 1 represents these concepts and the connections between them.

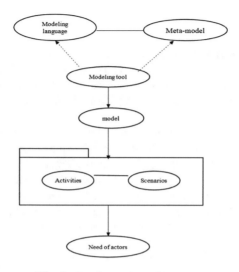

Fig. 1. Representation of concepts

This chart shows the operation of scripting tools. Starting from the definitions, a Meta model is like a model of a modeling language. It expresses the concepts common to all the models of the same domain. Each modeling language has a meta model as well as a graphic editor that respects the concepts of the meta-model of that very language. Each model of the modeling language inherits from the meta model, this to explain that the model becomes an instance of the meta model itself [8]. The modeling tool is in charge of graphically representing the objects of the models, but also ensuring their validity. Each model is important in relation to a specific situation. Thus, through the abovementioned terms of reference, our comparison will focus on the SCORM model and the IMS LD model. This comparison is represented by Table 1.

Table 1. Comparative table of models

Criteria	SCORM	ISM-LD
Reuse of resources	Yes	Yes
Resource production	Yes	Yes
Automatic transfer of resources	No	No
Diversification of scenarios	No	Yes
Generic scenarios	No	No
Diversification of activities	Yes	Yes
The representation of the actors	Yes	Yes
The representation of roles or function	Yes	Yes

3 Discussion

At the end of this study, it is important to report that all learning scripting model and learning scripting tools are equal. Certainly, we are at a time of changing pedagogical approaches, lessons, but also learning situations [9]. This evolution is also to mark a new approach to the representation of learning units the name of the model SCORM and IMS LD.

The revision of the modeling of the automatic transfer of teaching resources and taking into account student activity through dynamic scenarios is of paramount importance. Depending on their results, their performance but also the transferred resources will bear dynamic scripting. As of now, we are able to have dynamic scenarios depending on the needs of the players. These generic scenarios will no longer rely on a predefined structuration but on a model of more specific pedagogical scenario to scripting following exact transferred resources, but also on the learning situation of learners.

Work similar to ours is made. This, just in order to place the need to improve all these representations, and bet more on the learners. Oubahssi and Grandbastien [10] propose a model based on the IMS LD model. This one aims at representing the life cycle of e-learning. However they present the different phases of distance learning: a creation phase, an orientation phase, a learning phase, a monitoring and evaluation phase, a management phase [10]. Each phase of distance learning cycle goes by a process that consists of several activities, which are carried out in online platforms, but which vary according to the phases. They base their study on the representation of educational activities in general.

The proposed EML model by KOPER describes learning situations using a pedagogical modeling language that places learning situations and not resources at the center of the process. However, the study unit which corresponds to the name learning unit at SCORM and IMS LD is considered as a composition of activities carried out by a set of actors in a given environment. The types of activities taken into account in the Kopper model are: learning activities, support activities and instrumentation activities. These activities are located at the phase of learning of the complete cycle of the e-learning proposed by Oubahssi and Grandbastien [10]. This shows that the EML model does not address other activities in other phases of the e-learning cycle. It is based on the IMS LD model that is a model very close to EML model, which it is derived, with some differences. Instead of the unity of the study, it uses the concept of the learning unit, it also uses the concept resource instead of the object, and in the end, a business can use resources, but can also produce news.

The study of Lahcen Oubahssi and Monique Grandbastien states that the Kopper model does not cover all the cycle of ODL (Open and Distance Learning). For the extension of their model, IMS LD is more appropriate. The scripting learning activities is always putting forward research questions. Thus J. P. Pernin [2] proposes a conceptual model based on the notion of educational scenario. Their proposal is based on a set of precisely defined concepts and taxonomy of scenarios. It's a model which includes the activities and is particularly interested in the nature of the relationships linking the activity resources.

G. Paquette [11] offers a complete method of educational engineering ranging from the design of specifications to implementation within the distance-learning platform. In his model, the activities intervene in the description of learning units. All these models support the learning activity, but in a very different way. If we go back to the proposal of Lahcen Oubahssi and Monique Grandbastien, the model of Paquette [11] takes partially some e-learning or training activities, especially the links that exist between its concepts.

So, Oubahssi and Grandbastien [10] base their work on the IMS LD model, which represents best the concepts of e-learning in particular, the actors, the activities, the resources, considering the limits of the other models. In the end, an extension of the IMS LD model appears. The business model is offered in an open and distance learning environment. Each phase of the process of e-learning is associated with an environment in which the actors perform one or more activities. A represented e-learning environment is composed of a set of work units, links, rules and resources. Each working unit runs activities. The working unit is defined as a composition of activities carried out by a set of actors (author, counselor, tutor, learner, evaluator, administrator, general administrator, and pedagogical administrator) in a given e-learning environment.

The work units correspond to the different phases of the E-learning. Each activity is characterized by a set of prerequisites and objectives; it is defined by a state (for example in progress). The environment in which the activity takes place makes it possible to group together a set of resources of all types and the tools necessary for the accomplishment of the activity. The activity uses and produces resources. The new concept that stems from it is referred to as work unit which distinguishes the activities according to the five phases of the process.

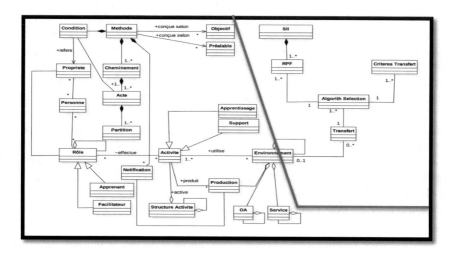

Fig. 2. Diagram of adaptive storyboarding

Previous studies on instructional scripting models show that research questions about scriptwriting are still a hot topic. Why so much model, proposal and the question is still relevant. From a global perspective some models are general, others specific to precise purpose. Hence, a generalist model is still not the most appropriate to answer all the problems of the users [12].

All these models put the emphasis on all activities and resources. But they bring variations at the representation of these concepts. Currently, the reuse of resources through other computerized system is an open door. The new concept of automatic transfer of resources without using an XML file generates the birth of generic model giving birth to our extension. This discussion has allowed us to understand that extent of our model based on IMS LD competes with all previous work. Thus Fig. 2 shows the structure of the proposed new model in conjunction with that of IMS LD.

4   A New Model of Educational Scriptwriting Based an Educational Information System

SII represents a computerized educational information system. FLR is stands for functional educational resources. Transfer criteria is means basic conditions for data transfer. A selection algorithm it is the selection of algorithm that will perform the election of the data. Transfers it is all automatic transfers to be made through the selection algorithm to the distance learning platforms.

A computerized educational information system is composed of several functionalities. Each of them implements a number of tables in the production database. This portion of the database manages data and once processed they generate actionable information through a graphical interface. Criteria for the transfer of these data will be specified. So at the level of our algorithm, we will have information about the functionality whose data is to be transferred. The annual deliberation module encompasses the overall learning outcomes of each student. Students in final passage have validated all teaching units and obtained 60 credits. Those with credits under conditional passage move to the upper class with at least 42 credits. But after, they take over the teaching units they did not validate. Students in repeating situation resume all the teaching units of the repeated class. In this context, distance education introduces face-to-face (attendance) and other modules in distance.

These students in conditional transition who are at the end of the cycle may opt to perform the resumption of the subjects to be taken back from a distance. So the data relating to his students will be transferred to the teaching platform that will accommodate these cases. This transfer will be done on the basis of an algorithm well defined and that will trigger the automation of the adaptive scenario to the learning situation of the student. Thus the descriptive table that is represented by Table 2 shows the different design levels of the new model and the steps of each level.

Table 2. Table descriptive levels and phases of the model

Level 1: Definition of the meta model	
Phase: P1	Analysis of the educational information system
Phase: P2	Representation of the attributes of each class
Phase: P3	Specification of Resource Transfer Criteria
Phase: P4	Definition and writing of the algorithm
Phase: P5	Generic scenario data transfer
Level 2: Model definition	
Phase: P1	Representation of concepts in class
Phase: P2	Representation of the attributes of each class
Phase: P3	Representation of relationships between classes
Level 3: Instantiation of the model defines	
Phase: P1	Definition of a case study
Phase: P2	Validation of the objects represented
Level 4: Implementing the model	
Phase: P1	Model implementation
Level 5: Test the implemented model	
Phase: P1	Practice with our targets
Phase: P2	Gathering impressions of our targets
Level 6: Evaluation of tests	
Phase: P1	Analysis of the impressions of our targets

5   Conclusion

In this article we present our research work on the state of the art of educational scripting models. This is the precision of our added value that sits on the problematic, generic script templates through educational resources transferred automatically from a computerized information system to a platform.

This study clarifies the need for an appropriate solution that meets a specific and usable need by any system of distance education. This model will be implemented in an LMS through algorithm of resource selection for the generation of scenarios related to the position of students in conditional passage. More detailed work on this extension is in project of production. The structure of our model will be better detailed and its implementation made in our next paper.

References

1. Billen, R., Laplanche, F., Zlatanova, S., Emgard, L.: Towards the creation of a generic meta-model of urban 3D spatial information. XYZ Rev. **114** (1 st quarter of 2008)
2. Perni, J.-P.: Which models and which tools for scriptwriting activities in new learning devices? In: Seminar: ICT, new jobs and new learning devices (Nov 2003)
3. Sadiq, M., Talbi, M.: Modeling of learning units on distance learning platforms Modeling project engineering & Module training and advice

4. Drira, R.: Assistance in modeling and contextualization of complex educational devices. Ph.D. Thesis, University of Science and Technology, Lille 1 (2010)
5. Baron, G.-L.: Education and Learning, EA 4071, Université Paris Descartes lab EDA. https://rechercheformation.revues.org/1565
6. Cochard, G.-M.: Standards and Standards in e-Learning (LOM, SCORM, etc.). University of Picardie Jules Verne, GMC/University Vivaldi Americas, 11–13 Jan 2010
7. Nodenot, T.: Contribution to the model-driven engineering in ILE: the case of cooperative problem situation, HDR. University of Pau and Pays de l'Adour (2005)
8. Lejeune, A.: IMS learning design: study of an educational modeling language, distances. Knowl. Rev. 2 (2004)
9. Villiot-Leclercq, E.: Advantages and challenges of pedagogical scenario approach. In: Presentation, University Joseph Fourier, Grenoble
10. Oubahssi, L., Grandbastien, M.: A generalization of the IMS LD business model for e-learning systems. J. e -RT, Electron. Inf. Technol. (July 2007)
11. Paquette, G.: Engineering distance learning: to build learning networks. University Press of Quebec, Sainte-Foy (2002)
12. Guéraud, V.: For engineering active learning situations. In: Desmoulins, C., Marquet, P., Bouhineau, D. (eds.), Proceedings of the Conference ILE 2003, Strasbourg (2003)

Primary School Student Ideas on Optics and Vision as Part of a Constructivist Teaching Involving Streaming Educational Video-Clips and ICT

Andreas K. Grigoropoulos and George S. Ioannidis[✉]

Department of Primary Education, University of Patras, Patras, Greece
{agrigorl,gsioanni}@upatras.gr

Abstract. An innovative teaching approach was designed and developed as a part of an already completed doctoral research, by incorporating various teaching and learning strategies suitable for 10-year-old students, combined with specially designed streaming video clips. These were integrated inside specific web pages following the course of the teaching sequence, and could be accessed through a browser at any point during teaching. This new teaching approach was extensively tested by measuring students' ideas before and after the intervention. The results concerning student's ideas on vision just prior to the experimental teaching are presented herein, with their total measurement errors computed. A full systemic network analysis of student ideas on vision is also presented.

Keywords: Student ideas · Light · Vision · Systemic network · Optics · Steaming video clips · Innovative teaching approach

1 Introduction

Constructivism has been hailed as the predominant theory for explaining human learning, for decades now. Despite being the accepted norm everybody agrees with, and tries to think accordingly, little has in reality changed in the way teaching is done on a practical level, even when ICT is used during teaching. The present trial is part of a serious and diligent attempt to adopt constructivism in practice, while teaching optics and vision to 10-year-olds.

2 Design and Implementation of a New Teaching Approach

Several researchers developed and tested teaching sequences trying to overcome obstacles in learning. An extensive study of this literature guided the design of a new teaching approach to geometrical optics and vision, incorporating various teaching and learning strategies combined with different types of streaming video clips. This teaching sequence was implemented in 8 different classes, while a total of 155, fifth-form 10 year-old pupils participated. To evaluate the effects of incorporating the streaming video clips in the new learning procedure, both an experimental group and a

© Springer Nature Switzerland AG 2019
M. E. Auer and T. Tsiatsos (Eds.): ICL 2018, AISC 917, pp. 803–814, 2019.
https://doi.org/10.1007/978-3-030-11935-5_76

reference-control group were utilized, and pre-tests were performed to establish (amongst other issues) the equivalence between the students in these two groups. The full dataset also involved an additional third group at a slightly different age, but this analysis will be dealt with in a future article. However, for the purposes of the present study, experimental and reference groups were consolidated into one single sample of 155 fifth-form students, on which prior ideas on optics and vision were examined.

The pre-test was designed to solicit free student ideas and mainly included multiple-choice questions and some "free text answer and associated drawing" type questions (totalling to 19 questions).

3   Data Collection and Analysis for Student Ideas on Vision

In this article we analyse one pre-test question in which students were asked to draw on a printed sketch provided to them, and in parallel to explain in words how a child sees a doll. The wording of the question was "How does the child see the doll?" (-the accompanying picture had a child looking towards a doll, while a light source was also depicted, as in Fig. 1). Written replies to the open question were combined with the associated student drawings on the image, and were subsequently analysed and compared with each other. This led to the painstaking construction of the corresponding (so-called) systemic network [1] as per Fig. 3, by the researchers. It was up to the researchers to decode and clarify what each student meant. This process has slowly filled a database with answers representing students' ideas from patterns emerged by grouping similar answers together, eliminating possible research biases that could be caused by the researchers preconceived ideas. The resulting categories and subcategories are presented in the systemic network at Fig. 3. The associated total measurement errors were also computed, for each of the categories and subcategories percentages.

Fig. 1. Image contained in the first question of the pre-test

From the total sample of **155** children, a $15.48 \pm 3.54\%$ gave no answer, or selected "I don't know". The percentage of children who gave vague, internally inconsistent, or irrelevant answers was very low ($3.23 \pm 2.46\%$), while the large majority of the children ($81.29 \pm 3.73\%$) gave useful answers, relevant to the question.

These answers were further categorised in 13 broad categories with their numerical distribution appearing (as percentages) in Fig. 2. These categories are the following (marked with letters from A to M):

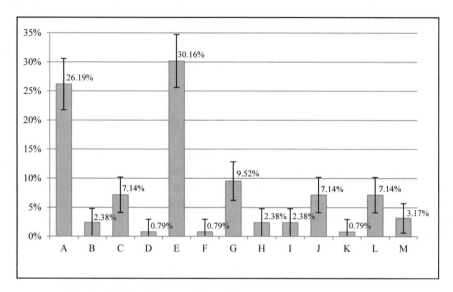

Fig. 2. Children ideas on vision before any educational intervention

A. We can just see, or We see using our eyes, or Our eyes can see, or We see using our vision (without any reference to light).
B. All the above plus "Light is not necessary".
C. Light is helping us see (- without direct or indirect reference that light is essential, and without reference on "where it is" or "where it goes").
D. Light is strong (uncertain if it's being considered necessary or not) and without reference whether it goes to the object or not.
E. Light is essential, or light allows us to see, or we see because there is light (- No mention as to if light moves towards object).
F. There is light everywhere ("bath of light" plus the concept that there is "something coming out of our eyes"—i.e. vision).
G. Light illuminates the object, or falls on the object.
H. We see because the light is reflected.
 I. We see because light goes to the eye (- *without any mention of the object*).
J. The light from the source goes to the eye *and* to the object.
K. Light from the source hits the object and is reflected off it, while colours and images are created *as it moves* towards the eye.
L. Light can inhibit our vision, or blur our vision, (as it falls on our eyes, or when it is too bright, or when it lights-up the object too much).
M. The object lies within our field of vision.

Figure 2 shows the percentages of the answers falling into each of the above categories. These have been computed with their associated total measurement errors (statistical and systematic) for each individual data-point. Specifically, the Bessel-corrected standard deviation was calculated to start with. Although extreme care was paid in avoiding large systematics, complete elimination is impossible [2]. Therefore, remaining systematic errors were evaluated and set at 2.0%, this being comparable with calculated statistical errors. By adding in quadrature the systematic with each statistical error (as these two are by definition independent), the total measurement error was computed.

4   The Systemic Network Analysing Student Ideas on Vision

The full systemic network produced for this question, together with its extended (and increasingly finer) subcategories to the broader ones is being presented in Fig. 3. Square brackets ([) indicate mutually exclusive subcategories (like a Boolean XOR) while angle brackets ({) signify that the subcategories are various independent dimensions within the broader category. An answer of the broader category appears in all the boxes constituting the "bra" as each one refers to a different (and independent) dimension of a matrix defining the answer of the individual student. The categories and subcategories of the systemic network are explained below.

Category A includes the answers of the children who in their attempt to interpret vision, they are simply referring to the *human ability of seeing* either involving the eyes or not (e.g. "we can see with the eyes" or "our eyes see"), *without any reference to light*. About one fourth of the students expressed this idea (26.19% ± 4.41%). Other researchers have reported similar findings, for various student ages. Piaget [3] was the first who reported that young children do not connect the object to the light nor the eye to the object. Osborne [4, 5] also reported that some children believe that "seeing with our eyes" is shelf evident and needs no further explanation.

In order to investigate the possible relation between the ideas of this category with those misconceptions collectively called "active vision" (or as Piaget [3] called it "eye centred vision" that could take the form of a model of "visual ray" [6], Jung [7] referred to the act of looking which sometimes is being accompanied by sight rays. Such ideas have been confirmed by [4, 5, 8–17] and many others), the answers of these students were further analysed into subcategories, the main criterion being the way the graphical representation depicts the relationship between eye and object.

Specifically, subcategory A1 (see Fig. 3) included the answers that contained no drawing connecting the eyes to the object. About half of the students belong to this category (51.52 ± 9.06%). Absence of drawing input means that these students may or may not adopt ideas of "active vision" in their interpretive scheme—it is just ambiguous.

In subcategory A2 the answers of the students contained a drawing where *the object is connected to the eyes* (usually with some lines, albeit not necessarily straight ones) *without any indication of directionality* from the eyes to the object or conversely. About one third of the students of category A were included in this subcategory (27.27 ± 8.12%).

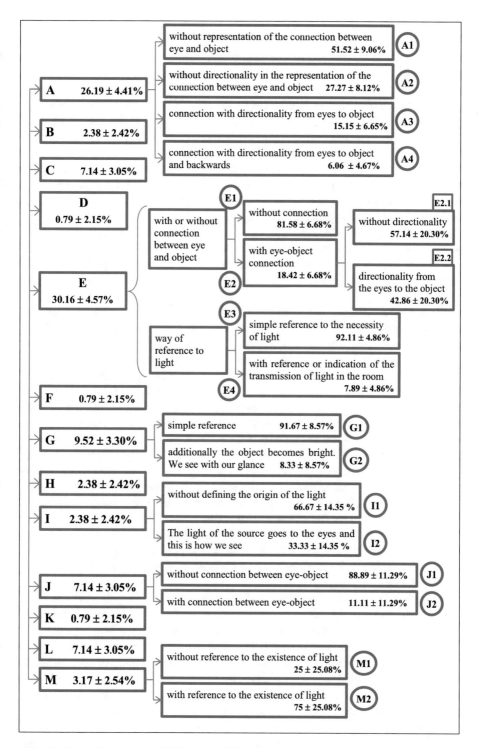

Fig. 3. Systemic network of fifth grade children ideas on vision (for explanation see text)

Subcategory A3 incorporated the answers of those students who drew a *connection between eyes and the object* (again usually using some lines albeit not necessarily straight) *with* some indication of *direction* (usually some arrows) *from the eyes to the object*. Even this directionality from the eyes to the object does not constitute definite proof of the existence of the idea of "active vision" (e.g. it could be an indication of where we turn our attention while looking)—however it does look likely. Although the percentage (15.15 ± 6.65%) of the students in this category is not very large, it has some importance precisely due to this possibility, i.e. the proximity of this representation to the idea of "active vision".

Subcategory A4 is more interesting since it includes the answers of the students who drew lines in the drawing indicating that eyes and object are connected bi-directionally, that is, with an indication of directionality from the eyes to the object and backwards. The percentage in this subcategory was very low (6.06 ± 4.67%).

The subcategories A2, A3 and A4 (especially the last two) indicate that it is possible that some of the children of this age have approached ideas of "active vision". It is a fear that representations like these may form the starting point (indeed the stepping-stone) for future establishment of active vision ideas, due to their conceptual proximity with it.

Category B includes the interpretive scheme of Category A (humans have the ability of seeing, "we see with our sight or with our eyes", "our eyes see") with the additional component that *the light is reported as not necessary for vision*. The percentage of the students who spontaneously expressed this idea was very small (2.38 ± 2.42%) and cannot be considered as being significant as the total measurement error exceeds the measurement. This fact does not mean that the idea that "the light is not necessary for vision" is not adopted by any students, it does show however that this idea is not necessarily (or spontaneously) expressed by the students along this type of questioning about how we see an object. This idea was not specifically expressed in writing, but it could still coexist lurking behind the ideas of categories A, C, D, L, and M. For this reason, the research included another question in the pre-test targeted on this (albeit not described herein). It can be reported, however, that the answers to this additional question clearly revealed that, without prior scientific teaching, the idea of light not being necessary for vision is very popular.

In the interpretations of **category C** a *supportive role is being ascribed to light* without stating or implying that it is necessary and without referring to where the light exists or where it goes. The percentage of students who expressed this idea was low (7.14 ± 3.05%).

Category D includes the answers of the students who interpret vision referring simply to the fact that *the light is intense* without any reference or connotation to as if it is necessary or to as if it goes to the object. Although this idea has been reported from other researchers, in our sample only one student expressed this idea (0.79 ± 2.15%).

Category E accommodates explanations which contain the *recognition of the necessity of light in vision* (e.g. "the light allows us to see", "we see because there is light"), *but with no reference as to whether the light goes to the object or not*. About one third of the student expressed this idea (30.16 ± 4.57%). Reports from other researchers (e.g. [12, 13, 18, 19] just to site a few), confirm the use of this type of reasoning, by children of various ages. In order to investigate the possible relation between the ideas in this category with "active vision" misconceptions, all answers were analysed to subcategories the criterion being the graphical representations used in students' drawings to support their views and describe the relationship between eye and object. The vast majority of the students (81.58 ± 6.68% did not draw any "line connection" between the eye and the object (subcategory E1). Although this does not specifically exclude the possibility of these students carrying some "active vision" ideas in their reasoning, it does preclude anything further coming out of this analysis. Subcategory E2 included the rest of the students who did make a graphical connection between eyes and object (18.42 ± 6.68%). For completeness reasons, subcategory E2 was further analysed with subcategory E2.1 referring to those students who didn't use any indication of directionality (arrow) in the connection between the eye and object, and subcategory E2.2 referring to those students who had drawn a connection from the eye towards the object. Results show that less than half of those who used the reasoning of category E and connected the eye to the object (18.42 ± 6.68%), did this with a direction from the eye towards the object. This leads us to the conclusion that *among the students who recognize the necessity of light in vision (without mentioning whether it goes to the object or not), very few connect the eye towards the object.*

In order to investigate the way of referencing to the light, all answers of category E were divided into subcategories E3 και E4. Subcategory E3 incorporated the answers of the students who only made a simple statement for the necessity of light in vision. The vast majority of the students of category E were classified under subcategory E1 (92.11 ± 4.86%). Subcategory E4 included the answers of the students who mentioned the transmission of light in space either in wording or with some kind of graphical representation. The percentage of students of this subcategory was low (7.89 ± 4.86%). Consequently, *among those students who recognise the necessity of light in vision (without mentioning whether it goes to the object or not) very few include in their reasoning some kind of light transmission in space.*

Category F was created to host the answers that incorporate or imply the idea of static existence-presence of light everywhere or around in space (*"bath of light"* or *"sea of light"* as described by [10, 12, 16, 20–22] and others) *combined with the idea that something comes out of the eyes (without any reference to the object).* This combination of ideas was by no means intentional, it was, however, forced upon the researchers by the fact that only an extremely low percentage expressed the idea of "bath of light" combined with "extramission" (the idea of something going out of the eye to reach the object reported by [5, 7–9, 12, 14, 16, 18, 23–25] and many others). This percentage (0.79 ± 2.15%) doesn't exceed the total measurement error and it is thus considered insignificant or ambiguous. It seems that the idea of "bath of light" was not clearly expressed by the students in this sample. Nevertheless, this doesn't mean that it may not coexist (or indeed to hide behind) the ideas expressed in categories C, D, E και M.

Category G includes the answers of students who in their explanation of vision expressed the idea that *the light "illuminates the object" or "falls on the object".* A small percentage of students (9.52 ± 3.30%) used this idea in their reasoning. The vast majority of these students (91.67 ± 8.57%) made a specific simple reference to this idea (subcategory G1), while a very small percentage of them (8.33 ± 8.57%) mentioned that the object becomes bright and that we see it by glancing on it. As this latter percentage doesn't exceed the total measurement error, it is considered insignificant.

Category H incorporated the answers of the students who, while explaining vision, they *mentioned that the light is being reflected (without any reference whether it goes to the eyes or not)*. The extremely low percentage of students who expressed this idea (2.38 ± 2.42%—the percentage is experimentally insignificant as being marginally lower that the measurement error) is in good agreement with the reports of other studies. The idea of light being reflected is rarely met in the thinking of students of similar age—before any relevant teaching (e.g. [3, 7, 10, 14, 15, 26] among others). It is worth mentioning that, although the idea of reflection of light is consistent with the scientific explanation of vision, the students who express it may accompany it with alternative (and not so scientifically correct) ideas. In our sample, this came from a single student who, although he mentioned light reflection, he also drew a line with direction from the eye to the object. Such is the predicament of teachers teaching science.

Category I concerns answers from students who in their interpretation of vision mentioned that *the light goes to the eye (without reference to the object)*. This idea was expressed by an extremely low percentage (2.38 ± 2.42%) which didn't exceed the measurement error and it is therefore marginal. The wording of this idea may, if examined superficially, seem to be in compliance with the scientific explanation of vision. Nonetheless, it may also carry alternative ideas. One of the students in the sample specified that the light of the source (lamp) goes to our eyes (implying that this makes us see).

Category J hosts the answers of those students who mentioned that the light of the source goes to the eye and to the object. This idea was detected in a small percentage of the sample (7.14 ± 3.05%). The vast majority of these students (88.89 ± 11.29%) didn't make any connection between the eye and the object (subcategory J1). Only a single student connected the eye to the object, but without any indication of directionality (subcategory J2).

Category K includes the interpretation of vision where the light of the source falls on the object, is reflected, and subsequently creates colours and images as it moves to the eyes. This category was created for completeness reasons since it was expressed by only one student. This idea, although it seems to be very close to the scientific model of vision, it still hides alternative ideas (e.g. "light carries images"). The insignificant percentage (0.79 ± 2.15%) in this category signifies that the idea of traveling images (even by reflected light) is extremely rare to children at this age.

Category L incorporates the idea that the light can inhibit vision and cause us blurry vision when it falls on our eyes or when it is too strong or when it lights up the object too much. A small percentage of students expressed this idea (7.14 ± 3.05%). Since this idea has been reported from other studies too (e.g. [7, 17, 27]), the present research confirms the presence of this way of reasoning in the population of students of this age.

Category M contains explanations of vision where the students state directly or indirectly that *we see because the object lies within the optical field*. A very small percentage of students expressed this idea (3.17 ± 2.54%). The majority of these students mentioned the existence of light ($75.0 \pm 25.08\%$—subcategory M2) while the rest did not (subcategory M1).

From a simple quantitative comparative examination of all categories presented above, one can easily draw the conclusion that categories A and E concentrate more than half of the students' answers. Consequently, they contain the dominant interpretation schemes of students of this age, about vision. The rest of the students prefer mainly the ideas in categories G, C, J, and L.

5   An Alternative Approach in Analysing Student Ideas on Vision

In order to get a better insight into the way students think about the eyes-object connection, the students' drawings were reclassified with these connections being the exclusive criterion. The new categories thereby created are shown in Fig. 4.

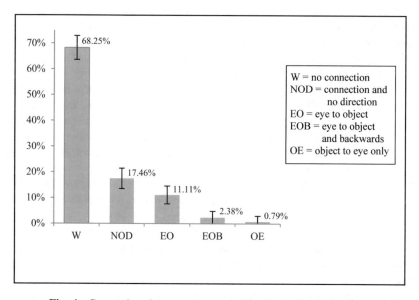

Fig. 4. Connections between eyes and object in students' drawings

As it can be easily seen, the majority of the students (category W) *did not draw any kind of connection between eye and object* in their drawings ($68.25 \pm 4.62\%$). About one fifth of the students ($17.46 \pm 3.94\%$) did provide a graphical connection between eye and object but they didn't provide any indication of directionality between these two ends (category NOD). Although students of category NOD represented the connection between eye and object (mainly using lines), they didn't provide any other indication that they may implicate "active vision" ideas in their reasoning.

The percentage of students who showed a connection from the eye towards the object (category EO) was relative low ($11.11 \pm 3.45\%$). Only an extremely small percentage of students ($2.38 \pm 2.42\%$) showed double directionality from the eye towards the object and backwards (category EOB). Since this percentage fails to exceed the total error, it is considered insignificant. Similarly, extremely small and insignificant was the percentage of students who showed a single connection from the object to the eye (category OE, i.e. $0.79 \pm 2.15\%$).

Although the students of categories EO and EOB give us a direction in the relationship between eye and object, there is no certainty that all of them have adopted ideas of "active vision", (e.g. the indication of directionality from the eye to the object could simply imply the direction of our attention).

6   Discussion and Conclusions

The analysis of these results leads us to the conclusion that for the majority of the students who took part in this research there was no evidence that they may use some form of the idea of "active vision" in their interpretive schema. This result is in agreement with other recent findings [21] with similarly aged students. Consequently, it is likely that previously reported ideas concerning active vision might arise from systematic error sources [2] like (formal or informal) previous teaching about optics and vision, and/or factors like social customs, everyday language influences, children TV programs, mobile apps, just to name a few. Alternatively, measurements/ observations may be the result of the specific wording used, terminology, and the method used in research, rather than the product of spontaneous reasoning at this age. However, one could only hypothesise that those students who indicated the connections of categories NOD, EO, EOB (of Fig. 4) could perhaps be more prone to adopting ideas of "active vision" due to the proximity-propinquity of such ideas to the representations they already use.

Naturally, nothing intricate or complete is expected from 10-year old children—it is mostly intuitive ways of thinking. However, taking a macroscopic view of Figs. 2 and 4 (with special attention to the measurement errors), we see two main categories of (simplistic) yet mutually exclusive ideas emerging, A & E. One centres on vision being a human ability, while the other revolves around the unspecified yet benevolent quality of light, as it allows us to see (i.e. something compatible with the "general illumination" idea). Both of these are rather superficial verbal creations (i.e. just for replying to the question), as opposed to revealing a deeper attempt to get to grips with what is asked of them. Even more notable is that the views expressed are *mostly static*, as opposed to indicating an attempt by the child to form a procedure—even of a non-causal variety— where something (anything) happens. The next 4 significant categories (C, G, J & L) all revolve around light, albeit in different ways. C & G (roughly equivalent in size) imply that some kind of procedure is seen as essential for explaining vision. For J & L light is either good and helpful or potentially harmful for our vision. Naturally, it is not expected for children at this age to be accustomed with "cause and effect" causal procedures as being vital to explain anything.

For a more detailed view of the confusion reigning in young students' minds, the full and elaborate systemic network of their ideas needs to be developed and examined in some detail, before designing and fine-tuning a constructivist teaching intervention.

Furthermore, as social customs, everyday language usage, TV programs, mobile apps and the like keep changing, student prior ideas change too—hence the need for up to date research before all major educational undertakings.

Observing the percentages (and associated measurement errors) for each of the sub-categories of the systemic network in Fig. 3, one can clearly deduce in detail the problems associated with student's prior ideas. These should be suitably adjusted during teaching, using explanation, rationalisation, everyday examples, school experiments, ICT, and various teaching techniques. A new teaching approach was therefore developed to specifically teach Optics and Vision, helping students test and explore various ideas. Special innovative experimental streaming videos were designed and produced, to attract student interest and either help them in their inquiry or induce constructive conflict of ideas in students' mind. With suitable use of ICT during teaching this novel teaching approach was subsequently tried out in class, data were taken and were compared with other two different samples, to measure where the increased knowledge was derived from. As all this forms the bulk of the research of an already completed doctoral thesis, its major findings are reserved to be presented in a forthcoming article. However, it should be stressed that the need for such up-to date investigations is great, especially since the present ubiquity of ICT based devices in children's hands is all but certain to keep altering their preconceptions.

References

1. Bliss, J., Monk, M., Ogborn, J.: Qualitative Data analysis for Educational Research: a Guide to uses of Systemic Networks, p. 215. Croom Helm, London (1983)
2. Ioannidis, G.S.: Data processing, systematic errors, and validity of conclusions in education research. In: International Conference on Interactive Collaborative Learning, ICL 2017, pp. 162–173. Springer International Publishing (2018)
3. Piaget, J.: Physical world of the child. Phys. Today **25**(6), 23 (1972)
4. Osborne, J.F. et al. (eds.): Light. Primary SPACE Project Research Report, p. 96. Liverpool University Press, Liverpool (1990)
5. Osborne, J.F., et al.: Young children's (7−11) ideas about light and their development. Int. J. Sci. Educ. **15**(1), 83–93 (1993)
6. Piaget, J., Garcia, R.: Understanding Causality. p. 192, Norton New York (1974)
7. Jung, W.: Understanding students' Understandings: the case of elementary optics. In: Novak, J.D. (ed.) Second International Seminar Misconceptions and Educational Strategies in Science and Mathematics, pp. 268–277. Cornell University, Department of Education, Ithaca, NY, USA (1987)
8. Andersson, B., Kärrqvist, C.: Light and its properties: EKNA Report 8. Institutionen for Praktisk Pedagogik. University of Gothenburg, Sweden (1981)
9. Andersson, B., Kärrqvist, C.: How Swedish pupils, aged 12−15 years, understand light and its properties. European J. Sci. Educ. **5**(4), 387–402 (1983)
10. Anderson, C.W., Smith, E.L.: Children's Conceptions of Light and Color: Understanding the Role of Unseen Rays. Research Series No. 166, p. 36. Michigan State University East Lansing Institute for Research on Teaching (1986)
11. Fetherstonhaugh, T., Treagust, D.F.: Students' understanding of light and its properties: teaching to engender conceptual change. Sci. Educ. **76**(6), 653–672 (1992)

12. Guesne, E.: Children's ideas about light. In: Wenham, E.J. (ed.) New trends in physics teaching, pp. 179–192. UNESCO, Paris (1984)
13. Guesne, E.: Light. In: Driver, R., Guesne, E., Tiberghien, A. (eds.) Children's Ideas in Science, pp. 10–32. Open University Press, Philadelphia (1985)
14. Eaton, J.F., Sheldon, T.H., Anderson, C.W., Light: A teaching module, Occasional Paper, p. 68. Institute for Research on Teaching, Michigan State University (1986)
15. Collis, K.F., et al.: Mapping development in students' understanding of vision using a cognitive structural model. Int. J. Sci. Educ. **20**(1), 45–66 (1998)
16. Selley, N.J.: Children's ideas on light and vision. Int. J. Sci. Educ. **18**(6), 713–723 (1996)
17. Ramadas, J., Driver, R.: Aspects of Secondary Students' Ideas About Light. University of Leeds Centre for Studies in Science & Mathematics Education, p. 138 (1989)
18. Boyes, E., Stanisstreet, M.: Development of Pupils' Ideas about Seeing and Hearing–the path of light and sound. Res. Sci. Technol. Educ. **9**(2), 223–244 (1991)
19. Anderson, C.W., Smith, E.L.: Student's conceptions of light, color and seeing. In: Annual Convention of the National Association for Research in Science Teaching. Fontana, Wisconsin USA (1982)
20. Jung W.: Conceptual frameworks in elementary optics. In: Jung, W., Pfundt, H., Rhoneck, C. (eds.) International Workshop on Problems Concerning Students' Representations of Physics and Chemistry Knowledge, Pädagogische Hochschule: Ludwigsburg, West Germany (1981)
21. Kokologiannaki, V., Ravanis, K.: Mental representations of sixth graders in greece for the mechanism of vision in conditions of day and night. Int. J. Res. Educ. Methodol. **2**(1), 78–82 (2012)
22. Dedes, C.: The mechanism of vision: conceptual similarities between historical models and children's representations. Sci. Educ. **14**(7–8), 699–712 (2005)
23. Crookes, I., Goldby, G.: How we see things: an introduction to light. The Science Curriculum review, Science Process Curriculum Group, Leicestershire (1984)
24. Cottrell, J.E., Winer, G.A.: Development in the understanding of perception: the decline of extramission perception beliefs. Dev. Psychol. **30**(2), 218–228 (1994)
25. de Hosson, C., Kaminski, W.: Historical controversy as an educational tool: evaluating elements of a teaching–learning sequence conducted with the text "Dialogue on the Ways that Vision Operates". Int. J. Sci. Educ. **29**(5), 617–642 (2007)
26. Eaton, J.F., Anderson, C.W., Smith, E.L.: Students' misconceptions interfere with learning: case studies of fifth-grade students. In: Research Series No.128, p. 32. Institute for Research on Teaching, Michigan State University (1983)
27. de Hosson, C.: Using historical reconstruction to implement inquiry-based teaching in primary school. In: Proceedings of the Third South-East European School for Hands-on Primary science Education, pp. 161–168. Belgrade, Serbia (2007)

Assessing Intercultural Competence of Engineering Students and Scholars for Promoting Academic Mobility

Raushan Valeeva[1], Julia Ziyatdinova[1(✉)], Petr Osipov[1],
Olga Oleynikova[2], and Nadezhda Kamynina[3]

[1] Kazan National Research Technological University, Kazan, Russian Federation
{firdausv,uliziat}@yandex.ru, posipov@rambler.ru
[2] Center for Vocational Education and Training Studies, Moscow, Russia
on-oleynikova@yandex.ru
[3] Moscow State University of Geodesy and Cartography, Moskva, Russia
kamyninan@gmail.com

Abstract. This article provides an assessment of the intercultural competence of engineering university students and professors as an essential tool for the development of academic mobility. EU states and the US are the leading countries in international academic mobility. In recent years, however, China also scores high results in this field and has developed a number of efficient mechanisms for promoting intercultural competence. The study of its experience can be of great importance for higher education in other countries. Thus the primary purpose of the research is to present and assess the mechanisms of intercultural competence development for academic mobility purposes of the engineering students and scholars in China and justify the possibilities of using them in higher education of other countries. To implement this an expert survey was conducted among university professors.

Keywords: Intercultural competence · Academic mobility · Expert opinion

1 Introduction

New trends of the modern society as worldwide globalization and internationalization have a strong influence over modern system of higher education and demand the growth of academic mobility since universities are an integral part of the global scientific community [1]. The issue of its promoting is very challenging. Intercultural competence of engineering students and university professors serves as one of the ways out as it implies the ability of each participant of communication to find the ways of interaction with the representatives of different cultures; the ability to analyse, accept and understand characteristics that distinguish one culture from another; the pursuit of self-improvement related to intercultural interaction. The paper aims at assessing the intercultural competence of engineering university students and professors and its role in promoting transnational academic mobility.

M. E. Auer and T. Tsiatsos (Eds.): ICL 2018, AISC 917, pp. 815–825, 2019.
https://doi.org/10.1007/978-3-030-11935-5_77

We consider intercultural competence of engineering students and scholars to be an essential tool for the development of academic mobility. These considerations are proven by recently published research of different scholars from different countries [2, 3]. The leading countries in inbound and outbound academic mobility are EU states and the US [2, 4]. In recent years, however, China also scores high results in this field and has developed a number of efficient mechanisms for promoting intercultural competence [5, 6]. The study of its experience can be of great importance for higher education in other countries. Thus the primary purpose of the research is to present and assess the mechanisms of intercultural competence development for academic mobility purposes of the engineering students and scholars in China and justify the possibilities of using them in higher education of other countries

2 Approach

The paper uses theoretical and empirical approaches. We studied the experience of China in intercultural competence development both through the research literature and personal experience of the authors. At the same time, we attracted distinguished experts in the field of intercultural communication and academic mobility, to analyse their expert opinions. In doing so, we went through the following procedures:

(1) proving the reliability of the expert opinion;
(2) defining the criteria of expert proficiency;
(3) selecting experts and identifying their proficiency level;
(4) developing a survey questionnaire for an expert opinion poll;
(5) analysing the survey results;
(6) determining the opportunities for the development of intercultural competence of engineering students and scholars in Russian universities based on the outcomes of the expert judgement.

We selected over 50 international university professors and administrators competent in higher education and intercultural communication, in particular. They represent Russian and foreign universities (Kazan National Research Technological University; Kazan National Research Technical University; Kazan State Medical University; Kazan State Conservatory; Kazan Institute of Economics, Management and Law; Confucius Institute at Kazan Federal University, Volga Region State Academy of Physical Culture, Sport and Tourism, Tomsk Polytechnic University, Ukrainian Engineering Pedagogics Academy, Association for Engineering Education of Russia, Liaoning University of Petroleum and Chemical Technology). The majority of the experts (89%) are university professors teaching foreign languages, because they introduce foreign culture, its customs, traditions and peculiarities to students. According to the opinion of one of the distinguished Russian scholars in the field of intercultural communication Doctor in philology S. Ter-Minasova "each English language class is a crossway of cultures, training intercultural communication, as each foreign word reflects foreign world and foreign culture: there is an idea of the world based on the national consciousness behind every word" [7].

6 experts (11%) are University administrators, involved in the international activities of the University with an extensive experience in the development of intercultural competence of students and professors.

The questions of the survey are relevant to each University professor, as intercultural competence becomes a priority area of higher education in many countries.

We were interested in the opinion of both experienced and young respondents, since senior university professors can provide in-depth and balanced assessment due to their wide experience, while young university professors being more sociable and mobile define the development of higher education in future. 44 experts have had work experience for more than 10 years, and 9 experts less than for 10 years.

To determine the competence of the experts in international communication we used the method of self-evaluation [8]. According to this method, we calculated the competence index, based on the evaluation of the experts' own experience, knowledge and abilities. In accordance with the self-evaluation, we revealed high and middle positions.

The general index was counted according to the formula:

$$K = \frac{k1 + k2 + k3}{3}$$

where k1—self-evaluation of theoretical knowledge, k2—self-evaluation of practical experience, k3—self-evaluation of the anticipation skills [8].

As a result, the average competence index is 0.7, which indicates that the level of competence of experts is above average.

We collected data asking their opinions on the most efficient mechanisms for promoting intercultural competence and possibilities of using them in higher education. The survey was conducted anonymously. We have developed a questionnaire in the form of an expert evaluation. The questionnaire was tested for validity and reliability and proved its efficiency.

The questionnaire consisted of two parts. The first part included a list of statements for the intercultural competence development mechanisms. The experts were asked to evaluate them according to a 5-point scale: (a) I agree with the statement; (b) I agree, but I would like to add some details; (c) I would like to share my teaching experience; (d) I have developed certain teaching materials for this mechanism; (e) I strongly disagree. In the second part of the questionnaire, the experts made supplementary remarks to the statements.

Thus, our approach was empirical based on theoretical data.

We will consider each of the statements, as well as the experts' supplementary remarks.

The first statement on the development of intercultural competence was the use of various approaches to the development of intercultural competence in major and minor degree programs via the following courses: (a) "Foreign Language"; (b) "Business Foreign Language"; (c) "Russian Language and Formal Discourse". 44 experts gave a positive assessment of this statement, 33 experts fully agreed, 6 would like to make supplements, 5 experts can share their teaching experience, 6 experts have developed teaching materials, 1 expert strongly disagreed, 2 experts abstained.

The experts believe that the development of intercultural competence is an integral part of the foreign language education. They draw attention to the significance of practical courses, to "Business Foreign Language" in particular, as it improves communication skills with foreigners. The experts offer to expand the list of courses; to strengthen the contextual approach in foreign language training of students; to study the history of the Russian language, starting with the ancient Greek language, which contains many root words and concepts of the modern Russian language.

The experts have teaching materials on such topics as gestures of native speakers, rules and regulations of behavior in foreign countries (comparative analysis), as well as the development of the course "Business Foreign Language", etc.

Within the framework of our research, this can be implemented via the course "Foreign Language". In order to diversify the approaches to intercultural competence development it is possible to introduce such courses as "Spoken Chinese Language", "Chinese Culture", "Business Chinese", "Hieroglyphics of Chinese Language", "Translation Practice" into a minor degree program "Philology of Chinese language".

The second statement concerning the intercultural competence development includes cultural and regional content while learning foreign languages. 49 experts gave a positive assessment of this statement: 33 experts fully agreed, 9 experts made supplementary remarks on this issue, 2 experts would like to share their teaching experience, 5 experts developed certain teaching materials, 2 experts strongly disagreed, 2 experts abstained.

As supplementary remarks on this statement the experts suggest:

- discussing topics as "National Traditions and Holidays", "National Character", "Speech Etiquette";
- studying courses as "Country Studies", "Anthropology" comparing several countries;
- language learning via culture studying;
- applying 4–6 h per semester into "Country Studies" program (depending on the studied language) using multimedia technologies;
- paying more attention to business communications in different countries.

The experts are ready to share their teaching experience during scientific conferences and meetings. Some experts developed their own teaching materials including manuals and presentations covering such topics as "Country Studies", "Gestures of Native Speakers", "Cultural Differences (comparative analysis)".

Thus, the results of the questionnaire proved the significance of the applying cultural and regional content when teaching foreign languages. Taking into account experts' opinions, we consider it is appropriate to study Chinese culture in the framework of both major and minor degree programs via such courses as "Foreign (Chinese) Language", "Spoken Chinese", "Chinese Culture", "Business Chinese Language", etc.

The third statement on the development of the intercultural competence was the introduction of a separate course "Intercultural Communication".

39 experts gave a positive assessment of this statement: 28 experts fully agreed, 10 experts made their supplements on this issue, only one expert has developed teaching materials, 13 experts strongly disagreed.

According to the experts' opinions this course helps students to integrate into a single economic space.

The experts offer to:

- introduce the course "Intercultural Communication" mainly for the students majoring in learning languages or as a special subject;
- introduce the course "Psychology of Foreign Language Teaching" as it helps students understand the importance of studying foreign languages;
- invite international professors involved in the study of both native and foreign cultures;
- introduce a new qualification as "intercultural communicator" with a one-year internship abroad for undergraduate and graduate students;
- apply various strategies for the intercultural competence development.

In addition, some experts believe that "Intercultural Communication" must be practically oriented, that implies a wide application of workshops and business games. All foreign language classes should be aimed at developing intercultural and professional communication.

The fourth statement of the questionnaire on the development of the intercultural competence was the development of training and methodological support of minor degree program "Philology (Foreign Language)". 45 experts gave a positive assessment, 29 experts fully agreed, 10 would like to make supplements, 2 experts can share their teaching experience, 4 experts developed teaching materials, 4 experts strongly disagreed, 4 experts abstained.

Experts believe that:

- intercultural competence can be developed by means of other courses, without getting too involved into linguistic aspects;
- a minor degree program "Philology (Foreign Language)" is not suitable for future engineers;
- this support will be useful for students taking a minor degree program "Translator in Professional Communication".

The experts are ready to share their experience in developing educational and methodological support of minor degree programs. While developing such programs they offer to take into account the particular characteristics of minor degree programs and adopt international experience.

We believe that the development of educational and methodological support of minor degree program "Philology (Foreign Language)" is very acute due to the lack of similar support. As part of our research it is required to develop minor degree programs for such courses as "Chinese Culture", "Chinese Grammar", "Hieroglyphics", "Business Chinese", "Spoken Chinese", "Practical Aspects of Translation", etc.

The fifth statement of the questionnaire on the intercultural competence development was the creation of language clubs. 51 experts gave a positive assessment, 34 experts fully agreed, 6 experts would like to make supplements, 7 experts would like to share their teaching experience, 4 experts developed teaching materials, 2 experts strongly disagreed, 2 experts attained.

There are several language clubs in Kazan. The experts in their supplementary remarks suggest:

- creating communication clubs in each institute;
- involving overseas students and professors into communication clubs;
- using multimedia technologies; case-study; brainstorming, didactic engineering.

We found out that the issue of organizing communication clubs is a very urgent and promising direction in studying cultures and languages. Therefore, we believe that the creation of Chinese communication club will contribute to the development of intercultural competence.

The next statement of the questionnaire on the intercultural competence development is involving students into various extracurricular activities. 50 experts gave a positive assessment, 28 experts fully agreed, 11 experts made supplementary remarks, 8 experts would like to share their teaching experience, 3 experts developed teaching materials, 1 expert strongly disagreed, 2 experts abstained.

According to the survey, extracurricular activities stimulate and develop an interest in learning foreign languages and cultures, as well as unite students and make their life more diversified. The experts proposed the following forms of extracurricular activities:

- radio broadcasts (dedicated to a particular country);
- language clubs;
- various events in foreign languages;
- student-led conferences;
- memory/mnemonics club—mnemonic techniques help to develop memory and imagination, which are important factors in the successful learning of foreign language;
- holidays, various competitions and quizzes.

The experts consider unconscious learning to be the best way of learning foreign languages. They suggest employing various extracurricular activities as the most effective forms of unconscious learning, such as:

- celebrating different holidays as Christmas, Halloween, New Year, March 8, Valentine's Day together with overseas students and professors;
- visiting cafes, museums, galleries around Kazan;
- arranging excursions in foreign languages;
- arranging competitions, quizzes in summer language camps.

Thus, the results of the survey, as well as the critical review of the educational resources, advanced domestic and overseas experience, proved that the experts pay great attention to extracurricular activities in the development of intercultural competence. Therefore, having studied the answers of the experts we consider it is necessary to employ various forms of extracurricular activities, such as contests, language quizzes, cultural events (festivals), because mastering language as well as cultural characteristics is achieved in a non-intrusive way. The development of students' creative thinking through art serves as a favorable source for the intercultural competence development. The activities carried out by language departments of the universities present the creative component of the intercultural competence and facilitate its development.

The impact of language and communication clubs as one of the most effective forms of extracurricular activities is undeniable. Various events held in communication clubs acquaint students and professors with the culture, traditions and reveal their creative talents.

46 experts gave a positive assessment, 34 experts fully agreed, 7 experts made their supplementary remarks, 5 experts would like to share their teaching experience, 2 experts strongly disagree, 3 experts abstained.

According to the survey, the experts called attention to attracting:

- overseas university professors with teaching experience for overseas students;
- international undergraduate and graduate students studying in Russia universities ready for communication with Russian students;
- overseas university professors for work in communication clubs, that enables integration of various strategies and adopting the experience of both Russian and overseas university professors;
- overseas university professors and scholars through certain programs;
- overseas scholars to collaboration in research.

10% of the experts consider this condition to be difficult to implement due to a number of challenges.

A number of Kazan universities (Kazan Institute of Economics, Management and Law, Kazan National Research Technological University, etc.) attract overseas professors within the framework of the Fulbright program. Overseas professors are engaged in conducting joint classes, for instance Jeffrey Grimm (USA) Fulbright scholar taught "Business English" at the Department of Foreign Languages of Kazan National Research Technological University in 2015–2016. Joint language classes help broaden outlook of students and university professors, contributing to the development of communicative competence. In addition, overseas professors are invited to take part in various events. An expert in French language and culture evaluated the knowledge of students at Kazan Institute of Economics, Management and Law and made a presentation "France is a country of the developed tourism".

Thus, taking into account experts' views, we believe that attracting Chinese university professors is of great interest to our work, since they promote intercultural competence development for students.

The next statement on the development of the intercultural competence is related to arranging communication between international and Russian students. 49 experts gave a positive assessment, 33 experts fully agreed, 12 experts made their supplementary remarks on this issue, 4 experts would like to share their teaching experience, 1 expert strongly disagreed, 2 experts abstained.

The experts strongly believe that interaction between international and Russian students leads to mutual understanding between them.

As a supplementary remark, the experts noted that the intercultural communication between students can be arranged directly in the classroom by employing multimedia equipment involving students from different countries and universities. Despite the widespread use of new strategies many experts emphasize the importance of real-life communication. In addition, some experts place emphasis on celebrating national cuisine days; watching films; short-term student exchanges; arranging communication clubs between Russian and international students, which help to identify the challenges and opportunities of education in different countries.

The experts shared their own teaching experience on this issue. University professors of Kazan National Research Technological University took part in the program SAL-Y (National Security Language Initiative for Youth) and in arranging communication between international and Russian students in the summer camp in 2009, 2010. Every year the camp attracts a large number of students and helps them to immerse into another culture. As a result, students gained deep knowledge about various cultures, made friends and improved their communication skills.

Thus, the results of the questionnaire showed that all experts put special emphasis on the significance of arranging communication between Russian and international students. There are many opportunities for arranging communication between Russian and Chinese students in Kazan, for instance through Confucius Institute, an academic unit focusing on Chinese culture and language which offers a wide range of courses, exams and contests [9].

Another way of attracting Chinese students, studying at Kazan National Research Technological University is through different extracurricular activities as competitions and a communication club.

A number of experts consider summer camps to be one of the most effective ways of arranging communication between international students, however they operate for a short period of time. Therefore, the creation of communication clubs can be an effective alternative to summer camps.

The ninth statement on the development of the intercultural competence implies preparing university professors ready for effective intercultural communication. 47 experts gave a positive assessment, 31 experts fully agreed, 10 experts made their supplementary remarks on this issue, 4 experts would like to share their teaching experience, 2 experts developed their own teaching materials, 1 expert strongly disagreed, 4 experts abstained.

The experts express quite opposite opinions on this issue. Some experts believe that this issue can be referred to the university professors teaching non-linguistic courses. Other experts suppose there is no need for special training, as any university professor can develop a work plan. Theoretically, all university professors who work with students and teach foreign languages should take such training, as it helps to look at language learning from the viewpoint of overseas professors and students. It is important to train university professors capable to teach intercultural communication. Such training can be implemented in the universities which arrange intercultural communication in a foreign language.

We fully agree with the experts emphasizing the importance of intercultural communication training for university professors. It is possible to implement this condition through collaboration with international colleagues, including:

- writing research papers;
- participating in scientific international conferences;
- collaborating with scientific communities directly involved in the intercultural competence development.

The tenth statement on the development of the intercultural competence relates to the arranging international scientific and language training for professors and students. 51 experts gave a positive assessment, 42 experts agreed completely, 7 experts made supplementary remarks on this issue, 1 expert is ready to share his teaching experience, 1 expert has developed some teaching materials, 1 expert strongly disagreed, 1 expert abstained. The experts highlight the significance of internships for foreign language university professors. It is possible to implement it by interactive cooperation through Skype with peer universities. The experts offer short-term internships at least once every 5 years.

Thus, according to the survey results, international scientific and language training of university professors and students is considered to be an effective mechanism for the intercultural competence development. It can be achieved by participating in research and training internships where they learn more about other countries and cultures. It can help to overcome stereotypical thinking in the perception of a particular culture.

In addition to the above statements the experts consider tolerance as one of the most important characteristics of the intercultural communication. "Currently, we are witnessing the interpenetration of cultures, states and their economies. The majority of countries are multicultural. One of the conditions for the development of the societies is a tolerant relationship" [10].

Teaching tolerance for children and university students is one of the ways of eliminating global threats.

In Russia, issues related to tolerance are a relatively new phenomenon, while in China it is cultivated since early childhood [11, 12]. Therefore, China's experience in teaching tolerance is invaluable.

Thus, the results of the expert evaluation allowed to determine the most effective mechanisms for the development of intercultural competence of students via:

(1) studying cross-cultural communication issues as part of major and minor degree programs through linguistic and non-linguistic courses;
(2) employing various extracurricular activities involving Russian and international students;
(3) involving international university professors and training Russian professors who are ready and able to develop intercultural communication skills;
(4) arranging scientific and language cooperation of Russian and international university professors including international internships.

These mechanisms practically coincide with the mechanisms of intercultural competence development of students and university professors revealed in the course of research in China.

2.1   Actual Outcomes

The results of the survey gave us a number of mechanisms for intercultural competence development. These results can be applied to university education in any country. At the same time, our investigation of the theoretical and practical research on the issues of academic mobility in China show that these mechanisms matched the approaches used in China.

The mechanisms included:

- infusing universal values education into the university degree programs;
- teaching foreign languages through different courses and inviting visiting professors from other counties;
- using the intercultural potential of different courses in the study plans;
- organizing different cultural extracurricular activities.

In general, the outcomes received proved the hypothesis that intercultural competence is a compulsory prerequisite for transnational academic mobility.

2.2   Conclusions

The study showed that intercultural competence development is an integral part of engineering education in the modern globalized world. Intercultural competence contributes to transnational academic mobility of students and professors [13]. This is of primary importance for engineering degree programs, where international teamwork is indispensible of engineering profession today.

The recent experience of Chinese universities reveals that it is possible to reach the leading positions in academic mobility development within a few years, and this can serve as a good example for other countries of the world. The Chinese mechanisms used for promoting academic mobility fully matched the mechanisms supported by the international experts, and this proves the universal value of intercultural competence and cooperation in education.

By applying the revealed mechanisms of intercultural competence development, an engineering university can enhance its competitiveness and visibility in the global market.

References

1. Valeeva, E.E., Kraysman N.V.: The impact of globalization on changing roles of university professors. In: International conference on interactive collaborative learning (ICL), Dubai, UAE (2014)
2. Egron-Polak, E.: Academic mobility in higher education worldwide. Where are we? Where might we go in the future? International Association of Universities. Retrieved 8 Feb 2017 from http://eacea.ec.europa.eu/erasmus_mundus/events/10_years_erasmus_mundus/1. Presentation%20Eva%20Egron%20Polak.pdf
3. Valeeva, R.S.: Academic mobility is the main tool of the intercultural competence development of engineering students and scholars in China and Russia. In: International Conference on Interactive Collaborative Learning (ICL), Kazan, Russia (2013)

4. De Wit, H.: International student mobility. European and US perspectives. Policy and Practice in Higher Education. Retrieved 10 Mar 2018 from https://doi.org/10.1080/13603108.2012.679752
5. Shibao, G., Yan, G.: Chinese education in the globalized world. Spotlight China Rotterdam **2**(8), 1–16 (2016)
6. Study in China. Retrieved 3 May 2018 from http://studyinchina.csc.edu.cn/#/home
7. Ter-Minasova, S. R.: Language and intercultural communication. Retrieved 16 Apr 2018 from http://www.gumer.info/bibliotek_Buks/Linguist/Ter/_04.php
8. Gorshkova, M.K., Sheregi, F.E.: How to Conduct Sociological Research. Politizdat, Moscow (1990)
9. Confucius Institute: promoting language, culture and friendliness. Retrieved 13 Feb 2018 from www.chinaview.cn 02 Oct 2006
10. Lopukhova, Y.V.: The Theoretical Basis of Tolerance Education of Students. Moscow Psychological and Social University, Moscow (2012)
11. Wang, G. A.: Comparative Study of Moral Education With Modern Information Resources in China and America's Universities (2011). Retrieved 12 Apr 2018 from https://link.springer.com/chapter/10.1007/978-3-642-23345-6_17
12. Wenzhong, H.: The Cultural Shift in Translation Studies (2014) Retrieved 15 May 2018 from http://www.ccpcc.com/jjxj/km1/020671.htm
13. Osipov, P.N., Ziyatdinova, J.N.: Faculty and students as participants of internationalization. Sotsiologicheskie Issledovaniya **2017**(3), 64–69 (2017)
14. Ruan, J., Leung, C.B.: Perspectives on Teaching and Learning English Literacy in China. Retrieved 5 Mar 2018 from https://docviewer.yandex.ru/view

Can Technology Make a Difference to the Level of Engagement Within Large Classes in Engineering Education? A Case Study in Sri Lanka

A. Peramunugamage[1(✉)], H. Usoof[1], and P. Dias[2]

[1] University of Colombo School of Computing, Colombo, Sri Lanka
anuradhask@uom.lk, hau@ucsc.cmb.ac.lk
[2] University of Moratuwa, Moratuwa, Sri Lanka
priyan@civil.mrt.ac.lk

Abstract. Engineering education is oriented towards problem based and project-based education. In Sri Lanka, teachers of first-year engineering undergraduate courses have the unique challenge of fostering student engagement especially in large classes. This study concerns a large classroom using the Moodle virtual learning environment. Data was collected qualitatively through focus-group discussions with undergraduates with respect to accessing the Moodle Log and quantitatively via the Moodle records. An important finding in this study is that when students are used to traditional teaching methods, they face difficulties in accepting TEL techniques, unless the latter is made compulsory for some aspects. Hence teachers must use a blended-learning approach and gradually increase the use of TEL techniques.

Keywords: Large classes · Engineering education · Engagement · Technology aided learning · Pedagogy

1 Introduction

The roots of engineering stem from science and mathematics; however education on engineering is oriented towards problem-based and project-based education [1–4]. Researchers have found that the said mixed-mode approach has become one of the most effective methods in imparting engineering edification to students. The combination of these two methods in engineering (together with theoretical ones) helps to meet industry requirements without posing a negative effect on the learning process. Successful learning is also attributed to student engagement [5].

Engineering education in Sri Lanka is largely confined to six state universities that enroll successful candidates of Advanced Level (A/L) examination; they provide free tertiary education for all. Within the state university platform, lecturers are especially challenged in fostering student engagement in large classrooms during the first undergraduate year. The large and crowded classrooms seemingly obstruct efficient student-teacher and student-student interactions, which is a key element of problem-based and project-based education. On the other hand, providing laboratory and other

© Springer Nature Switzerland AG 2019
M. E. Auer and T. Tsiatsos (Eds.): ICL 2018, AISC 917, pp. 826–833, 2019.
https://doi.org/10.1007/978-3-030-11935-5_78

essential facilities also creates a substantial financial burden. According to a report published by the Ministry of Finance Sri Lanka, the government investment on free education was $ 109 MN (3.5% of GDP) in 2016 which was mainly utilized for infrastructure development in universities related to lecture rooms, laboratories, hostels, sanitary and other facilities [6].

Developed countries use high-tech tools such as "clicker technology" to increase student engagement in large classes and support problem-solving. This technique enhances student participation through a unique signal that tracks responses from individual students[7–9]. In addition, hybridized teaching, a combination of conventional teaching and assessment methods, has also made headway in learning. Online discussion boards and blogs are also used in large classes for exchange of ideas. Research suggests that more studies are required on the effectiveness of online learning in engineering education [10, 11], since there is relatively little understanding of its impact on student engagement and learning.

2 Aims and Goal

Students in a learning environment get distracted in many ways. From chatting with colleagues, listening to music and watching videos to simply not focusing on lectures, they tend to disturb fellow students and teachers during a lecture and impair their own studying [12]. Self-directed learners must be motivated to take responsibility for their education through independent learning. During their first-year undergraduate courses, the majority is not familiar with learning techniques such as Problem Based Learning (PBL), online learning and student-centered learning. In Sri Lanka, the preliminary undergraduate year consists of a teacher centered theatre-based teaching system, which students have been made familiar with over their 13 years in school. As such, to promote student centric or problem based education with active learning, a progressive technique is required for undergraduates [13]. This research focuses mainly on the question "How can we improve Technology Enhanced Learning (TEL) in engineering education using a web-based system?"

This study addresses two vital gaps in engineering education research. Firstly, most TEL studies for engineering education examine class sizes of around 100 students or less. Secondly, the web-based technologies used to facilitate engineering education have not been adequately researched. In this study, TEL was applied to a large classroom of several hundred students using the Moodle forum, discussion board and chat for group discussions. The adoption of active learning is compared with the choice of passive learning by the same student population comprising of 829 first-year students.

3 Methods

The University of Moratuwa (UoM) has an annual intake of around 850 students for their undergraduate engineering degree programme. The starting semester of the first year comprises six compulsory modules that consist of lectures and practical sessions with tutorials spanning 12 weeks. We selected two modules for this study. One is the Statics component of the ME1032 Mechanics course; this component (duration of 6 weeks) involves a 2-hour face-to-face lecture session each week and a single 4-hour practical plus tutorial session where students work in small groups to complete a laboratory assignment and discuss tutorial problems. Due to the large numbers of students, the lectures were delivered to 3 separate groups in separate time slots, requiring the lecturer to deliver the same lecture thrice a week with around 275 students in each group. The challenge for the lecturer was to increase student engagement in the lecture and laboratory/tutorial sessions without incurring any material or resource costs (including additional hours, additional teaching assistants, or infrastructure). As Moodle was installed in the university and already utilised by the teacher, no technology cost was incurred.

Six lectures were video recorded and uploaded weekly on Moodle. The recording was done only for one of the 3 groups each week, with each group having 2 of their lectures recorded in the six week duration. The recording called for a one-time resource cost which was negligible and would not need to be sustained during following years. Hard copies of lecture material, tutorials and an assignment were distributed among students in the classroom while soft copies were uploaded on Moodle. Moreover, discussion forums on three open-ended questions were developed to promote a higher order of thinking and group discussion. Online assignment submissions, forum discussions and other activities were optional for students since this was implemented as a pilot project for 2018. The ME1032 module was conducted essentially as a traditional lecture series using PowerPoint presentations. Prior to the lecture, the students were not provided Power-point notes, recorded lectures and assignments. Thus, this demonstrated a teacher-centred approach similar to the students' previous school education system. In addition to the presentation, lecture time was also used to explain and answer questions raised by the students. The ME1032 Moodle page consisted of subject content, such as lecture notes, videos, tutorials and practice questions, and was used to send notices to students (with an immediate email to students) as shown in Table 1.

The second module selected, CS 1032 (Fundamentals of Programming) is also a compulsory module in the first year first semester but conducted via Moodle. Lectures were delivered physically and material was uploaded to Moodle whilst practical session instructions were given through the CS1032 course page on Moodle. Practical assignment submissions and quiz facilities were provided during the hands-on session where a desktop computer with Internet facilities aided the student for 3 h.

The 829 students were divided into 13 lab groups and all groups used the same lab for practicals per the semester 1 academic time table. Online activities of CS1032 module which consisted of take-home assignments, quizzes and reading practical instructions were compulsory. The online component was allocated 20% of the final grade contribution as showed in Table 1.

3.1 Data Collection

The study was based on a research design which is a non-experimental descriptive analysis. Focus group interviews were used to identify points of motivation, determine

Table 1. Components of ME1032 and CS1032 teaching in 2017 intake at UoM

Teaching learning activities	ME1032	CS1032
Resources	Recorded lecture, lecture power-point, lecture hand-outs, tutorial questions, practice questions	Lecture power-point, practice questions, additional reading materials
Activities during lecture	• Revise previous lecture • Lecture related questions • Interactive discussions	• Revise previous lecture • Lecture related questions • Interactive discussions
Activities during tutorials	• Illustrate complex problems • Summarize understanding and interactive discussion	
Activities during lab sessions	• Individual/group in Lab experimentation • Take-home experimentation • Interactive discussions • Reflections with Lectures	• Individual/group in Lab experimentation • Take-home experimentation • Interactive discussions
Activities for assessment	• Individual lab report 10% • In-class and Take-home assignment 10% • Final examination (closed book) 80%	• Individual programming labs exercises 6% • Individual programming assignments 14% • Final examination (closed book) 80%

attitudes towards Moodle and the ME1032 module (where the use of Moodle was not compulsory as in the CS 1032 module); interviews were conducted with 5–8 randomly selected students per group, 15 groups in total being tested with same questions. Secondly, Moodle Access logs were skilfully used to identify usage patterns; this second mode of data collection was used for the CS1032 module as well, for purposes of comparison. We intend subsequently to compare these outcomes with a second semester course offered only to 125 students from a single discipline (i.e. Civil engineering), which is different in subject matter (i.e. a conceptual design course) to the ME1032 one (i.e. a strongly mathematical mechanics course).

4 Analysis and Results

The thematic analysis using five categorised themes and two global themes is presented in Table 2 and used to identify the pattern of the students for the module ME1032. Accessing Resources/Moodle reflects how students access Moodle for their learning activities. The peak usage was when students were at home or during library hours, while some accessed Moodle while travelling. Five students in the focus groups accessed it during a computer lab session. The majority of students described Moodle/online support as being 24/7 accessible, and valued it especially for missed lectures.

Table 2. Themes identified by thematic analysis of the focus-group discussions

Temporal matters	Barriers & preferences
Convenience	*Barriers*
• Fast access to information and materials	• I'm not accustomed to online material
• Can Access lectures from anywhere, anytime	• Used to working with hard/printed copies
• Can access missed lectures easily	• We have no experience working online
• Ability to work online	• We have notes through hard copies given
• Helpful if implemented for other modules too	• Do not know how to use Moodle
• Guidance/references for further reading	• We were not aware on new updates
Accessing resources/moodle	*Assignments*
• At home	• Need more practical questions
• During library hours	• Need answers for tutorials
• While traveling the bus	• Implement for all subjects
• During computer lab session	Class size
	• We felt sleepy or tend to talk to neighbour
	• Cannot see the blackboard properly
	• Cannot hear properly with microphone
	• Too far to see lecturer movements

Certain *Barriers* for using Moodle existed due to daily routine and external factors. Some were reluctant to use Moodle due to the ready availability of printed notes and hard copies while others mentioned that they were unfamiliar with online learning. *Assignments* is a mode of testing student knowledge on the given course. The majority preferred more practice questions that encourage independent learning. *Class size* reflects students' attitude towards current class size and arrangement during lectures. It was evident that the current class set up created boredom and poor visibility of the blackboard for students at the rear and inaudible lecture delivery even through the microphone.

Table 3 depicts the Moodle access log records of the courses ME1032 and CS1032; the percentage of students who have viewed lecture notes, tutorials, forums and recorded lecture videos uploaded to Moodle during the first six weeks of ME1032 and CS1032 are presented in Table 3. During 1–2 weeks of CS1032 course there were no additional exercises but the third, fifth and sixth weeks' additional exercises were uploaded to Moodle.

Table 3. Distribution of student's access records on ME1032 and CS1032 course activities

Percentage of students accessing ME1032								Percentage of students accessing CS1032							
Week	1	2	3	4	5	6	Avg	Week	1	2	3	4	5	6	Avg
Lecture notes	32	19	18	20	14	12	20	Lecture notes	15	16	8	22	27	24	19
Tutorials			24		20		22	Assignments	99	99	99	99	99	98	99
Forums		19		17		13	16	Additional exercises			98	99	94	90	94

Throughout the semester, 20% of the students had viewed the ME1032 video recorded lectures; and 19% had accessed and downloaded materials and assignments.

On average 16% of them had viewed the forum discussions; but did not participate in them; also, 22% of them had accessed and downloaded tutorials on ME1032. The statistics for the CS1032 module were similar to the above for the lecture notes component; but close to 100% for the other two—this is discussed below.

5 Discussion and Conclusion

This study aims to identify means of improving TEL in engineering education; to identify patterns of TEL usage in large classes and determine experiential feedback of students' learning through Moodle using a web-based system. Data collected through focus groups yielded two global themes. The theme "Temporal Matters" represented students' perceptions towards Moodle, which were generally positive in relation to resources provided. The theme "Barriers and preferences" captured hindrances to use Moodle as against their expectations. These outcomes were very closely related to the four themes identified by Boz and Adnan [14]. The themes were Temporal Matters, Barriers to Online Learning, Preferences for Online Learning, and Teaching Process (Boz and Adnan, 2017). It was highly evident that class size impacted the learning activities in the theater based education system. Previous experience on learning technology also created willingness or vice versa to use TEL.

Students are also able to use lecture and tutorial times engaging in effective discussions with lecturers to learn through linking theories with real-world examples and recent developments. As large class size posed a problem for this, Moodle forums provided a potentially useful resource. But, Moodle log records indicated the opposite as students still prefer discussing issues face-to-face in private. Their reluctance for online discussions or response stemmed from the fact that they were unwilling to be in the spotlight as compared to students who did not actively participate in discussions. The large group size and the open-ended nature of online discussions did not allow for students to effectively collaborate as they would have in smaller face-to-face groups. As such, although technology allows convenience, it hinders collaboration due to unfamiliarity and the lack of feeling of community between individuals. Similar findings were reported by others [12, 15, 16]. Recorded lectures were viewed frequently (by over 20% of students). We presume that these numbers will increase when the end-semester examinations approach. Students appreciated the video recorded lectures; both to recall their own lecture (in 2 of the 6 cases) and to view lectures delivered to another group (in 4 of the 6 cases) in ME1032. It also provided them with the opportunity of catching up on missed lectures. Students had the opportunity to revise course content more effectively for difficult topics by replaying the recorded lectures.

This is evident in comments in their responses during focus groups. One said; "sometimes lecturer was very fast. I wasn't able to follow him. So I used Moodle video. They are great". Another constructively said; "The video recording and uploading of lecture to Moodle is highly useful and we recommend to do so for all lectures and also upload more questions". This was especially effective in overcoming certain barriers experienced in large classes as explained in Table 2. The Moodle page for ME1032 provided the opportunity of catering for students' needs as required, such as allowing less capable students a chance to meet the pace of the course, while also facilitating more competent students to enjoy the challenges in open-ended topics. The use of Moodle however was essentially optional.

However, the CS1032 module students did not have a method other than online for lab submissions since all had to be done via Moodle. Therefore, it compelled students to use it; and 99% of students used the system for assignment submission. On the other hand, the lab report writing task in ME1032 was the most difficult exercise for students, since it was required as a handwritten document during practical session. We understand from this that in situations when course materials are optional and participation is not rewarded, students use hard copies; but when the online system is made mandatory with associated rewards, students use it willingly. Therefore, as described by Nabushawo et al. (2018) online activities should direct students to work with online environments; this will help to promote TEL among students. Focus group participants expressed that all course modules should have a Moodle page with additional resources and more practice questions and/or assignments together with video recorded lectures. An important finding in this study is that when students are used to traditional teaching methods, they face difficulties in accepting TEL techniques. Therefore, teachers need to use a blended-learning approach and gradually increase the use of TEL techniques. Furthermore, when introducing online techniques for large classes, a proper mechanism to engage with all participants must be devised. Moreover, the nature of the course needs to be considered when introducing online activities.

Acknowledgements. The authors acknowledge the support received from the LK Domain Registry in publishing this paper. The conclusions and/or recommendations expressed in this paper are those of the author and may not necessarily reflect the views of the LK Domain Registry. This study was conducted at University of Moratuwa, Sri Lanka and we would like to express our gratitude to all the lecturers, non-academic staff and students those who have involved in this study. We express our special thanks and appreciation to Mr. Sanjaya Sooriyaarchchi, Mr. Shashi N Amarasinghe and all CIT studio staff at the university for recording lectures and uploading to the Moodle.

References

1. Mills, J.E., Treagust, D.: of Engineering based or project-based learning the. Australas. J. Eng. Educ. 3, (2003). ISSN 1324-5821
2. Newman, M.: A pilot systematic review and meta-analysis on the effectiveness of problem-based learning. Ltsn, pp. 1–74 (2003)
3. Clive, D. et al.: Engineering design thinking, teaching, and learning. J. Eng. Educ. (January), 103–120 (2005)
4. National, T.H.E., Press, A. : Educating the Engineer of 2020 (2005)
5. Neto, P., Williams, B.: More activity, less lectures: A technology stewardship approach applied to undergraduate engineering learning'. In: Proceedings—1st International Conference of the Portuguese Society for Engineering Education, CISPEE 2013 (2013)
6. Ministry of Finance Sri Lanka.: Annual Report, p. 25. (2015). doi: 1.1

7. Campbell, R., et al.: Clickers in the large classroom: current research and best practice tips. CBE—Life Sci. Educ. **6**, 1–15 (1993)
8. Czekanski, A.J., Roux, D.-M.P.: The use of clicker technology to evaluate short and long-term concept retention. 1–12 (2009)
9. Pritchard, D.: The Use of "Clicker" Technology to Enhance the Teaching/Learning Experience, ... /hedc/resources/Digital-Resources-for-your-Teaching ..., pp. 1–24 (2006)
10. Felder, R., Silverman, L.: Learning and teaching styles in engineering education. Eng. Educ. **78**(June), 674–681 (1988)
11. Fisher, K.D.: Technology-enabled active learning environments: an appraisal, CELE exchange. Centre Effective Learn. Environ. (**6–10**), 1–8 (2010)
12. Langan, D., et al.: Students' use of personal technologies in the university classroom: analysing the perceptions of the digital generation. Technol. Pedagogy Educ. **25**(1), 101–117 (2016)
13. Chokri, B.: Factors influencing the adoption of the e-Learning technology in teaching and learning by students of a university class. European Sci. J. **8**(28), 165–190 (2013)
14. Boz, B., Adnan, M.: How do freshman engineering students reflect an online calculus course? Int. J. Educ. Math. Sci. Technol. (IJEMST), **5**(4), 262–278 (2017). https://doi.org/10.18404/ijemst.83046
15. Juniu, S.: Use of technology for constructivist learning in a performance assessment class. Meas. Physical Educ. Exerc. Sci. **10**(1) (2006)
16. Kirkwood, A., Price, L.: Technology-enhanced learning and teaching in higher education: what is "enhanced" and how do we know? A critical literature review. Learn. Media Technol. **39**(1), 6–36 (2016)

Knowledge Structures as a Basis for Learning Strategies in Engineering Education

Charles Pezeshki and David A. Koch

Washington State University, Pullman, WA, USA
{pezeshki,david_koch}@wsu.edu

Abstract. Conway's Law, which states that 'the design of a system will map to the social structure that created it', is a well-established principle in social network theory, as well as design. But the concept has deeper implications as well. In order for a design to be executed, the knowledge for the design must first be constructed. That observation led to what has been called 'The Intermediate Corollary', which then implies that different types of social structures can be mapped to specific knowledge structures. What is then required is a system of social evolution that shows the different social structures in some evolutionary form. For the case of this paper, the social system evolution model is taken from Spiral Dynamics, an outgrowth of Clare Grave's seminal work, elaborated on by Don Beck and Integral Philosopher Ken Wilber. This system, when modified with the overarching principles contained in Conway's Law, yields a fundamental basis set of knowledge structures that maps well to fundamental concepts in engineering education. From knowledge fragments, to advanced synergistic heuristics, all have maps from the social structures that create them, to classroom situations where optimal social structures can be utilized for acceleration of student learning. This paper explores the nexus of these main concepts, as well as the notion of optimal classroom groupings for learning different engineering concepts.

Keywords: Bloom's taxonomy · Social structure · Knowledge structures · Spiral dynamics · Educational theory

1 Introduction

Various entities both in the United States and abroad have called for increased science and engineering literacy of students, as well as an increase in degreed individuals in all the STEM fields. For example, in the United States of America, in the following paper [1] published in the Proceedings of the National Academy of Sciences, says:

> The President's Council of Advisors on Science and Technology has called for a 33% increase in the number of science, technology, engineering, and mathematics (STEM) bachelor's degrees completed per year and recommended adoption of empirically validated teaching practices as critical to achieving that goal.

© Springer Nature Switzerland AG 2019
M. E. Auer and T. Tsiatsos (Eds.): ICL 2018, AISC 917, pp. 834–843, 2019.
https://doi.org/10.1007/978-3-030-11935-5_79

All this is well and good, especially when presented to a group of engineering faculty with a strong international complement, like the ICL. It signals potential increased support for our programs, as well as a strong message sent to the larger population to sign up for our various institutions. The intent of such efforts is to increase scientific, as well as engineering literacy across the larger society. The incumbent benefit, ostensibly, is an uptick of data-driven, rational decision making that can help the burgeoning global population adapt to challenges ranging from Average Global Warming, to potential crises in resource allocation.

With the rapid, technologically driven changes in modern society, there seems to be no counterargument toward increasing scientific and engineering literacy in the larger society. But at the same time, it behooves all of us to take a more profound look at what we are actually teaching—not just from the more surface-level perspective of the standard course material in a given engineering curriculum. Achieving a 33% increase in bachelor's degrees will require more than simply increasing enrollment; what will be needed is a deeper understanding into what prevents current students from graduating. To realize this, the more complex cognitive skills of how students actually arrive at engineering solutions need to be more deeply understood. The goal of understanding these is constructing what are called 'mental models' [2] that can enlarge the conversation and enable us to act explicitly in pursuit of particular goals in engineering education.

Various large governmental entities are also doing the asking. With partners, the National Research Council undertook the Transforming Undergraduate Engineering Education [3] process with the intent of triggering stakeholder discussion on the future needs of engineering education. There, a large group of industry and academic stakeholders created a list of 36 KSAs (Knowledge, Skills and Abilities), mapping roughly from Bloom's Taxonomy, that future engineers would need to master to compete in the global workforce. The idea was loosely based around the idea of increasing cognitive load, with Abilities requiring Skills, which then might include particular information, embodied as Knowledge. Yet when looking through the actual definitions of the KSAs, often Skills were defined in terms of abilities, and Knowledge was defined as a Skill, creating considerable ambiguity regarding what these individual items actually entail. [4]

These key problems with separability—establishing some level of cognitive clarity, much as engineers do when doing model order reduction on complex systems—were not addressed in these analyses. Clearly, the participants wanted to develop some checklist that engineering educators could use as a backstop for assessing larger curricular issues. But the labyrinthine logic, full of self-referential notions, left no clear path for implementation. The report contained no clear separable definition of knowledge, skills and abilities to start, and how those types of individual representations of specific human understanding might be (a) related to each other in any consistent manner, and (b) arranged in any larger meta-structural form.

The approach in this paper clears up these inconsistencies. Knowledge structures are inherently independent of content, effectively a meta-structure. Different types of knowledge structures can be used across disciplines—for example, a rule or algorithm structured such that it takes two inputs and delivers a decision can have as much viability in the field of engineering as in law. Working from knowledge structures allow an educator to consider not just particular classes of information, but to also reflect on ways that students might be required to execute that information in the course of future professional employment. Additionally, they establish a meaningful dialectic with other professionals, and allow either independent execution by an individual to complete a task, or recruitment of a team to more fully explore all engineering options and alternatives.

2   Bloom's Taxonomy

In 1948, a group of psychologists interested in education assessment and achievement testing met at the American Psychological association Convention in Boston. Their work was described in a series of texts, which were the consensus of expert opinion at the time, called Bloom's Taxonomy [5]. Bloom's Taxonomy has been written about extensively, and according to Google Scholar, as of this writing has had the original work cite over 10,000 times. Its use is widespread, and the context for the writing is well accepted. Yet at the same time, and this paper as an example, is not the first to argue [6] that Bloom's Taxonomy is not a true taxonomy. It does not follow systematically across education. Instead, what is created is a consensus of expert opinion that appeals strongly to intuition of educators about what intelligent subdivisions may exist, and therefore, be applied to seeing if there is appropriate covering of a given subject in a class or other, similar pedagogical endeavor. The idea of such a taxonomy as a reflective tool is also one that has strong merit. Once couched in the way educators think, instead of a more neurogenic framework, there is a natural tendency for instructors to present material as it was presented to them. This egocentricity does not necessarily help grow understanding in the field on how others might learn given topics.

In response, however, to this and other criticism, as well as evolution in educational thought, Bloom's Taxonomy was substantially revised. The revision was released in 2001, and systematized comparison of standards and objectives. It also included extensive material for adoption and use by teachers. [7] A diagram of Bloom's Taxonomy taken from open source material and checked for veracity is shown in Fig. 1.

As with many works done by experts, it is inclusive to the point of being exhaustive. By essentially including virtually every documented cognitive behavior, one can pick and choose as the content creator exactly what the content creator had in mind when they designed a particular course exercise. By working toward the idea of necessary

Fig. 1. Bloom's Taxonomy by K. Aainsqatsi—Own work, CC BY-SA 3.0. https://commons. wikimedia.org/w/index.php?curid=4000460

completeness, though, much is also lost as far as understanding the larger guiding principles that generate knowledge in the first place. Though the material in Bloom's larger rose is arranged from ostensible 'first principles' dimensions clustered in the center, such as 'synthesis', 'analysis', 'knowledge', etc., there is no clear divisibility between the afore-mentioned topics. How does one divide up 'synthesis' and 'analysis', for example? Doesn't synthesis demand analysis? And doesn't one need knowledge before one can do analysis? And so on. Bloom's Taxonomy may shine for a teacher establishing an assessment tool that others in a given field can recognize. Yet the deeper meaning of the application of the taxonomy is lost, and other than covering more particular, non-separable cognitive bases, evolution of the material—or the minds learning the material—is lost.

Bloom's Taxonomy succeeds, and has value, as tool of measuring what the author calls 'sophistication'—increasing fine detail that aids the complicated nature of a given piece of knowledge. But it gives precious little direction as far as evolution of knowledge. What knowledge might be more amenable to a less evolved thinker? And how does one know that one is moving students up on to higher, evolved, critical thought? There is reference to some of this in the taxonomy. But due to the lack of systematic underpinnings, it can be elusive.

3 Understanding Knowledge Structure as Social Structure

In the paper, It's All about Relationships—How Human Relational Development and Social Structures Dictate Product Design in the Sociosphere [8] those researchers mapped neurogenic function based on empathy to a given fundamental basis set of social structures taken from Beck's classic work in Spiral Dynamics [9], and then placed this in the context of design structure. These fundamental social structures are imbued with characteristics called Value Memes (v-Memes) that, in a global sense, show how information is processed by members of an organization. These information processes then create knowledge with particular structure that directly maps back to the social structure, creating a self-similar loop.

The principle undergirding this mapping process is called Conway's Law [10], which has been empirically proven in a number of studies. Figure 2a–c. show the basic relationships.

The Intermediate Corollary, discussed in [8], based on the inductive construct that one must have the assembled knowledge of what one is designing before one can build it, implies that given social structures will necessarily produce certain knowledge structures. Empathy and connection will inform synergy in the design. If information is not shared, nor generated with the context of combined coherence, synergy is also impossible.

Once this principle is understood, then a fundamental basis set of knowledge structures emerges that given social structures will create. This then generates a systemic taxonomy of knowledge structures built on first principles of socially constructed knowledge. When mapped with the social evolution, this also shows an evolution of knowledge, corresponding directly to societal and empathetic development.

Since knowledge and its aggregation is fundamentally a historical, emergent phenomena, the thesis can then be asserted that evolution of societies—and accessibility to students—correspond directly to the evolution of knowledge in a self-similar fashion. Once this consequential view of knowledge aggregation is accepted, the Intermediate Corollary yields directly a set of knowledge structures that map across the Spiral Dynamics social structures as shown in Table 1.

(a)

WHY Does this matter for Product Design?

Conway's Law

organizations which design systems ...
are constrained to produce designs which are copies
of the communication structures of these organizations

Hmmmm...... what's in that arrow in the middle?

(b)　　　　**The Intermediate Corollary**

How we socially interact directly creates the knowledge structure that creates our designs.

(c)　　　　**And...**

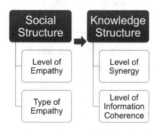

Fig. 2. Mapping social structures to knowledge structures, and thereby to design structures, using Conway's Law

Table 1. Knowledge structures from associated social structures

	Value Meme	Knowledge Structure	Example/Physical Analog	
Second Tier Value Memes	Bodhisattva (Coral)	Reflective and Selfless Feedback Loop-Based Heuristics	???/Evolving Mixed Turbulent Flow with Feedback Loops	Independently Generated, Trust-Based, Data-Driven Evolved Empathy Relationships
	Global Holistic (Turquoise)	Reflective/Feedback Loop-Based Heuristics Based on Connected Group Reflection and Integration	The Dynamics of the Internet/Fully Mixed Turbulent Flow with Large Scale Structures	
	Global Systemic (Yellow)	Reflective/Feedback Loop-Based Heuristics Based on Personal Reflection	Zen Buddhism/Feedback Loop Nonlinear System/Chaotic Motion	
	Communitarian (Green)	Multiple, Integrated Scaffolded Heuristics	OpenIDEO/Multi-constituent, mixing flow with entrained solids	
First Tier Value Memes	Performance/Goal-Based (Orange)	Fixed Scaffolded Heuristics	Waterfall Design Process/Turbulent to Laminar Transition Flow with entrained solids from lower modes	
	Legalistic/Absolutistic (Blue)	Rules, Determinate Processes, Created by a Group, Applied Across a Group	Professional Code of Ethics/Fully Developed, Crystalline Fractal Structure with transformed elements	Trust Boundary
	Authoritarian/Egocentric	Knowledge Fragments, Decided on by an Authority	Modulus of Elasticity/Beginning of Crystalline Structure with regular self-similar patterns	Externally Defined, Low Empathy Belief-Based Relationships
	Tribal/Magical (Purple)	Indeterminate Short Time Stories, based on Deep Long Time Narratives	Creation Myth/Amorphous Solid	
	Survival (Beige)	Knowledge Fragments, Short Time, Small Spatial Scale	"Where's the water cooler?"/Single Particle	

Every Value Meme above contains partial representations of all lower Value Memes. Colors from Don Beck SD.

4   Discussion

There are key elements in understanding knowledge structures that are listed in Table 1. The first is that lower Value Meme (v-Meme) levels are ensconced, much as a nested matryoshka doll, in the v-Memes above them. A Tribal v-Meme will include structures established in that v-Meme, plus the information in the Survival v-Meme below. This pattern continues up the Spiral, and into the more complex knowledge structures. A Performance-based v-Meme knowledge structure will therefore have content from five different social structures embedded in it, to varying degrees.

The second element necessary for understanding is to appreciate the emergent role of independent agency in the generation of knowledge. Knowledge structures associated with the lower Value Memes are largely separate from the individual and their perception of the world and would typically be labeled 'beliefs'. For example, an individual participating in a tribal society has little or no effect on modification of the creation myth for the culture. As one moves up the social evolution ladder, in an Authoritarian system, one must accept knowledge fragments given from above as fundamental truths. Legalistic systems will hand down rules of law and algorithms that apply to all, that may be decided through group consensus, that an individual may participate in—an example of slowly increasing agency. Academic institutions trade heavily in this kind of knowledge, which has high reliability due to exhaustive review, and slow aggregation. As an engineering example, there has been very little change in the content of teaching statics or thermodynamics for the last hundred years, as there is complete consensus on basic understandings of these engineering principles.

When making the jump to performance-based thinking, necessary for teaching synthesis and design, the primary difference is that individuals now have true, granted agency by the social system to affect establishment of shared goals. And that agency— the ability of an individual to individually actuate future actions and thoughts, will be reflected in the knowledge generated by this social structure. Final knowledge is now dependent on individual perspective and analysis. At this level, scaffolding of knowledge structures below still matters—an individual attempting to execute a design will still rely on algorithmic rule following and expert knowledge from past learning (one doesn't design a mechanical device without dynamics). But because varying trade-offs must now be considered, student agency is required in creating a unique personal opinion. Critical thinking appears here first. It involves the ability of the individual to assemble a unique perspective through various established heuristics that would be part of course content.

Communitarian v-Memes generate scaffolded, independent solutions that now are combined from a range of heuristic thinking. Organizations like OpenIDEO, with their various design challenges, accurately describe this v-Meme. Higher, second-tier v-Memes generate even more complex knowledge structures, with independent reflective practice built into the knowledge generation.

5   Application to Engineering Coursework

Breaking down knowledge into the Conway's Law social structure-derived forms helps instructors understand how students might comprehend content from a neurogenic perspective. A summary of the basic types is given in the Table 2. This suggests that the teaching methods used to educate students are more effective when tailored to the information being taught. While that statement will likely not be surprising to anyone, what is new here is the way knowledge complexity and educational methods are tied to the social complexity. Selecting educational methods that are of equal social complexity to the knowledge being imparted could then be beneficial to ensuring comprehension.

Table 2. Educational examples mapped to their associated knowledge structure

Spiral v-meme level	Associated knowledge structure	Educational example
Survival	Fragmented, ephemeral information	Classroom#, classroom time
Tribal/magical	Long Time stories, 'tribal knowledge'	Instructor reputation before class, material difficulty
Authoritarian	Knowledge fragments decided by experts	Modulus of elasticity, enthalpy tables
Legalistic/absolutistic	Algorithm and rule sets	Methodologies for free-body diagrams, structural analysis codes
Performance-based	Goal-based heuristics	Waterfall design processes
Communitarian	Merged design heuristics from different groups of stakeholders	OpenIDEO [11]
Global systemic	Reflective heuristic practice	Any number of design processes where awareness of designer bias is formally built into the process. A larger construct of design review built into a class

6   Contrast and Application

It is beyond the scope of a conference paper to compare and contrast completely the methods of Bloom's Taxonomy and a new, knowledge-formation template presented in this paper. Nonetheless, some important points can be made.

The proposed system of knowledge structures is a true, systematic and inclusive taxonomy. Lower level knowledge structures are included in the higher knowledge structures, and a course designer can consider whether enough appropriately structured information is included in the course to enable the more developed forms.

On a superficial level, it appears that Bloom's Taxonomy, by including emotional affective terms, has an advantage over a straight knowledge generation taxonomy. However, this discounts the emotional development required to facilitate empathetic exchanges central to the higher level knowledge structures. See [12] for more details. The minute that personal agency is elevated for more open-ended problems, as required by ABET in a capstone course, emotional development becomes important for intellectual exchange.

It is postulated that the different social structures that generate different types of knowledge structures also may be highly suggestive for the optimal social structures for teaching the material itself. Authority-based content, which primarily involves memorization, may be best taught through low-empathy lecture, with reflective practice in student pairs. Algorithmic processes, because of their demand for repetition of sequential steps, have been addressed with problem-solving sessions involving demonstration by the T.A., and copy of rote example, for decades.

7   Further Work

The knowledge structure set is easily recognizable to engineering educators. And while Bloom's Taxonomy may have some advantages with laying out educational outcomes in that they are constructed in the manner that Bloom's Taxonomy is constructed, by parsing a knowledge structures approach, professors can work on gaining deeper insight on the constituent levels that may be causing student incomprehension. The next step will be revision of course material and testing done with assessment experts to determine if the clarity provided by this approach helps students learn.

References

1. Freeman, S., Eddy, S.L., McDonough, M., Smith, M.K., Okoroafor, N., Jordt, H., Wenderoth, M.P.: Active learning increases student performance in science, engineering, and mathematics. Proc. Natl. Acad. Sci. **111**(23), 8410–8415 (2014)
2. Senge, P.M.: The Fifth Discipline: The Art and Practice of the Learning Organization. Crown Pub (2006)
3. National Research Council.: Transforming Undergraduate Education in Science, Mathematics, Engineering, and Technology. National Academies Press (1999)
4. Pezeshki, C.: Understanding the NSF Transforming Undergraduate Engineering Education Report–Why are Industry and Academic Pathways toward Knowledge Development at Odds?. 122nd ASEE Annual Conference & Exposition. pp. 26.1627.1–14 (2015)
5. Bloom, B.S.: Taxonomy of Educational Objectives. vol. 1: Cognitive domain, pp. 20–24, New York, McKay (1956)
6. Morshead, Richard W.: Taxonomy of Educational Objectives Handbook II: affective domain. Stud. Philos. Educ. **4**(1), 164–170 (1965)
7. Anderson, L.W., et al.: A Taxonomy for Learning, Teaching and Assessing: A Revision of Bloom's Taxonomy, Longman Publishing, New York (2001)
8. Pezeshki, C., Kelley, R.: It's all about relationships: how human relational development and social structures dictate product design in the socio-sphere. In: Product Development in the Socio-sphere, pp. 143–168. Springer International Publishing (2014)
9. Beck, D.E., Cowan, C.: Spiral Dynamics: Mastering Values, Leadership and Change. John Wiley & Sons (2014)
10. Conway, M.E.: How do Committees Invent. Datamation **14**(4), 28–31 (1968)
11. Fuge, M., Agogino, A.: How Online Design communities evolve over time: the birth and growth of OpenIDEO. In: ASME IDETC/CIE Vol. 7: 2nd Biennial International Conference on Dynamics for Design; 26th International Conference on Design Theory and Methodology. American Society of Mechanical Engineers (2014)
12. Pezeshki, C.: Influencing performance development in student design groups through relational development. Int. J. Eng. Educ. **30**(1), 91–100 (2014)

Providing Cognitive Scaffolding Within Computer-Supported Adaptive Learning Environment for Material Science Education

Fedor Dudyrev[1], Olga Maksimenkova[2(✉)], and Alexey Neznanov[2]

[1] Institute of Education, National Research University Higher School of Economics, Moscow, Russia
fdudyrev@hse.ru
[2] Faculty of Computer Science, National Research University Higher School of Economics, Moscow, Russia
{omaksimenkova, aneznanov}@hse.ru

Abstract. These days adaptivity is the cutting edge of modern education. Technologies are being developed rapidly and bringing new possibilities to educators. Thus, diverse types of adaptive learning environment have appeared during these last decades. Material Science and Engineering Education (MSEE) have a solid formalized foundation, which consists of standards, recommendations and clear rules. Moreover, investigators report on growing role of computer in teaching and learning in MSEE. These brings great perspectives to computer adaptive learning system based on a material science and engineering ontology. This paper aims to justify general pedagogical foundations of adaptivity and to collect requirements to a computer adaptive learning system. As an extra result we introduce the architecture of ontology-based adaptive learning system to MSEE.

Keywords: Material science education · Scaffolding · Adaptive learning system · MSEE · Computer adaptive learning

1 Introduction

These days materials science and engineering (MSE) remain a leading driver of technological progress as well as physics, chemistry, electrical, mechanical, aerospace, and civil engineering. It is well known that materials play a key role in the technology advances in a rapidly changing world [1]. Consequently, the part of STEM education which deals with materials-literate technicians, engineers and well-informed and innovative materials scientists is essential for future economic development and growth.

Fortunately, MSE education (MSEE) is a well-formalized domain. It can be explained by the fact that MSEE inherits most of the features from Engineering and Technology education (ETE) [2, 3]. In its turn ETE has been a topic of high interest among educators over the last two decades [4, 5].

© Springer Nature Switzerland AG 2019
M. E. Auer and T. Tsiatsos (Eds.): ICL 2018, AISC 917, pp. 844–853, 2019.
https://doi.org/10.1007/978-3-030-11935-5_80

A literature review demonstrates that MSE education (MSEE) calls concept and technological knowledge a significant learning outcome. Moreover, some studies report on growing role of computers in MSE [6] and in MSEE and give recommendations on integrating computer-based learning methods to educational processes [7, 8]. The scope of this paper lays in the field of MSEE and focuses on computer-based adaptive learning and its environment.

Despite the fact that modern computer-based educational techniques are claimed the topic of great interest in MSEE, a little has been done in their positioning and proving. Moreover, the idea that a computer can teach and help to shape some well-structured learning outcome is attractive but still is purely studied. For today, the most prospering area of the field is an adaptive assessment because achievements in intelligent systems allowed to use formal representation of domain knowledge [9].

This paper aims both to justify pedagogical applicability of computerized adaptive learning to MSEE and to speculate on an appropriate place for a computer adaptive learning system (CALS). It elaborates on ideas of student-centered learning and instructional scaffolding from the adaptive learning perspective and formulates a set of requirements for CALS design.

2 Foundations of Adaptivity

There is no common view on what foundations of adaptivity are. Investigators bring different meaning to them. Thus, McMullen et al. [10] deal with foundations of adaptivity to the class of problems solving in a specific domain. Pearson Corporate are wider and introduced the checklist for adaptivity self-assessment in [11]. This section gives an overall picture of the world of adaptive learning systems. It also delineates several educational methodology questions about adaptivity and collects general requirements to an adaptive system.

2.1 Adaptive Learning Systems Landscape

Undoubtedly, the field of computerized adaptive learning is the most formalized one in the modern education. Because of its high interdisciplinarity, the researches in this area touch aspects from educational design to artificial intelligence and emergency technology. As far as this paper focuses at CALS design and the justification of its applicability, this section aims to give a picture of adaptive learning systems and their pedagogical frameworks if such is presented.

A new era of adaptive learning systems started in the 2010s when cloud technologies had been shaped, accepted and begun rapidly widespread to all spheres of life. We may surely call reviews [12, 13] a milestone. These papers generalize nearly all we know about educational, mathematical, and technological foundations of adaptive learning systems and slightly open a door to the new generation of distributed and collaborative systems.

Since that time the researchers has been systematically reporting on different adaptive learning systems. The systems vary from relatively unsophisticated solutions with simple adaptivity to complicated cloud-based distributed systems which provide high-level interoperability by means of IEEE-LOM, QTI/APIP, SCORM, etc. Nevertheless, most of the mentioned adaptive systems contain a knowledge base. And the review by Magnisalis et al. [12] highlights the significance of domain knowledge representation for an adaptive learning system design and introduces so-called D-Type support (Domain Knowledge Support) for this purpose. In fact, the way of knowledge representation in an ALS has high impact on its design, implemented algorithms, process coordination (management) and even architectural style. Among the others ontology knowledge representation is known as advanced one because of its richness and usefulness for automatic processing. It is important to mention that technically there are several different methods for knowledge storage and retrieval from an ontology.

Moreover, some investigations in learning systems highlight a specific aspect of interoperability, which is called a semantic interoperability. According to classical paper by Heiler [14] semantic interoperability means that two interacting subsystems in a distributed system have a common understanding of the "meanings" of the requested services and data. Returning to learning systems we should mention that semantic interoperability is very difficult to achieve and requires special software engineering solutions (e.g. see [15, 16]). For today, using ontologies is quite popular to support semantic interoperability [17]. Moreover, using ontologies for this purpose in learning systems is also popular and comprehensively documented as at educational administration level as well at learning content management [18–21].

Summarizing, essential feature of every CALS is the way of its interaction with a learner during the educational process. So, a digital learning system is considered adaptive if it provides means to support this interaction and:

1. Is able to analyze the individual activities of the student during learning process.
2. Can adjust or adapt the content, methods and stylistics of training in accordance with the cognitive and psychological characteristics of individual students.
3. Includes explicit formally presented knowledge and learning outcomes description.

2.2   From Student-Centered Design to Computer Assisted Scaffolding

We will separate a special way of interaction and call it a pedagogical intervention. Pedagogical intervention is a process of giving a learner some useful hints or directions for successful solution of a task. This section carries out requirements to CALS and defines pedagogical interventions in it. For these purposes it reveals connection between constructivists pedagogy concepts and establishes pedagogical foundation of computer-based adaptive learning.

2.3   Scaffolding

Scaffolding is the most comprehensive concept that describes the whole variety of actions, which the teacher/instructor/coach uses to help the student in the situation of new types activity acquisition or solving new cognitive tasks within his ZPD. Even though the term "scaffolding" first came into use in an article written by Wood, Bruner, and Ross, it draws on the work of Vygotsky [26]. In education, scaffolding is a metaphor for the environment in which learners are provided with support they need to attain their goals, and which is removed bit by bit as it is no longer needed, much like a physical scaffold used in construction. It's important to distinguish traditional instructions which place the teacher at the center of the educational process and instructional scaffolding that firmly focuses at the student's current abilities and needs. The teacher's support can't be perceived as scaffolding if it is not fading and doesn't lead the student to independent problem solving [27].

The function of the teacher in the framework of scaffolding is to assist learning and development of individuals within their ZPD. By organizing collaborative dialogue with the student, the teacher firstly stimulates his cognitive interest and involves him in the problem-solving process. Offering complex, hitherto unbearable problems to the student, the teacher ensures the feasibility of the decision process, breaking the impossible task into easier subtasks. At the same time, he\she protects the student from frustration and loss of interest, encouraging and motivating him.

By pushing and accompanying the learner on the way of complex problem solving, the teacher ensures the internalization of the basic steps of this process and gradually delegates to the student additional degrees of freedom and independence. After that, scaffolding goes to the next round (a new cycle of training), when the student has to handle even more complex tasks, and the entire training cycle is repeated again.

The concept of more knowledgeable other (MKO) is widespread in constructivist pedagogy. "The MKO is anyone who has a better understanding or a higher ability level than the leaner particularly in relation to a specific task, concept or process. Traditionally the MKO is thought of as a teacher, an older adult or a peer" [28]. Scaffolding can be provided not only by the teacher, but by any peer who is better informed in the subject or particular activity than the student. It is important to emphasize that the MKO functions can also be performed by an information system that can provide and support the progress of the learner within the ZPD [28].

Ideas of scaffolding are successfully implemented in adaptive systems, lots of examples may be found in papers of Koedinger and the colleagues beginning from the signed work on the topic [29].

3 Adaptive Learning System Requirements

3.1 General Requirements

In this paper we suggest following foundations of adaptivity to a CALS:

- Explicit description of learning outcomes.
- Ontology controlled knowledge assessment.
- Availability of scaffolding including:
  - Automatic gap detection,
  - Instant self-assessment,
  - Formative feedback.
- Checking relevance of educational materials.

- Engaging active learning techniques.
- Reinforcing the training.
- Scalability.

Automatic pedagogical interventions can be used to achieve several goals in CALS.

3.2   Features of Adaptive Learning System for Material Science Education

In this section we enrich our general requirements with several specific features which make a CALS for MSEE different from any other system.

It is clear from Sect. 2.1, that any adaptive learning system in highly structured domain may be rather easily built on ontology base. As we know, MSE as a branch of engineering is well structured and tightly covered with standards and recommendations domain. And this is the reason why the essential part of MSEE connected with accurate knowledge and some skills which are directly related with them. So, using domain ontology to support adaptivity process is reasonable and is not very laborious.

Figure 1 shows main subsystems, ontologies and databases that play significant role in adaptivity of CALS. Classical *Learning Process Management Module* contains additional subsystem for selecting and activating pedagogical interventions. Any intervention is based on decision algorithm based on Bayes inference procedure that uses Learning Activity DB with current and historical data about student's actions and other modules for intervention's content: domain ontologies, learning materials and assessment materials. In complex 'Relevant Materials Searching Subsystem', 'Pedagogical Intervention Activation Subsystem', 'Automatic Item Generation Subsystem' and 'Ontology Querying Subsystem' present so-called 'scaffolding square' and are obligatory to be implemented.

For universality of ontology management tasks, we need common metaontology with basic semantic relations, units of measurements, scales, etc. The most of basic knowledge in MSE deal with one or another measurement system. Thus, it seems to be a good solution to use one units of measurement ontology in addition to domain ontology. Such a solution makes a CALS flexible enough and sustainable to measurement systems interchanges.

Learning materials are commonly represented in textual form with embedded multimedia objects. We can notice a fundamental problem with cross-references between domain ontology nodes and learning materials sections and paragraphs, especially definitions of entities and distinctions between entities. So, we need complex *References DB* with statuses of approval for each cross-reference. Such DB is very useful during automatic parsing of new materials by text mining techniques. Illustrative example of problem is two meaning of phrase "hard metal" in different Internet-sources: it is not only about material science—it is also about music genre.

An assessment materials bank should meet several requirements especially if it is used as a part of a CALS. The first requirement is data versioning to make bank suitable to diverse types of assessment. The second one is external references to domain and learning outcomes ontology. This last requirement is necessary to provide automatic assessment items generation, which is wished to represent high level of

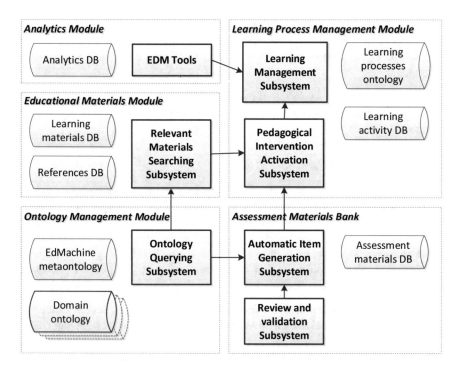

Fig. 1. Specific subsystems and data storages

adaptivity. Generated items should have similar characteristics to the item where student is mistaken. So, a generated item mush checks the same properties of a learning concept as the mistaken one. We present methodology of ontology-based generation based on Formal Concept Analysis (FCA).

4 Outcomes

We believe that E-learning environments have a number of functional characteristics that allow them effectively act as MKO. The greatest potential in this respect is provided by CALS. An information system that has the characteristics of adaptability can be extremely effective in scaffolding, because:

- It has the tools that provide opportunities for dynamic assessment of the student and record the errors or knowledge gaps of previous training. This function of the information system can be invaluable for determining the existing level of skills and competences, as well as determining the educational goals, which are potentially achievable for the student;
- The information system has a diversity of pedagogical interventions. Numerous hinting tools and clarification of difficult questions will provide the student with an easier transition to a new level of competence and internalization of the basic steps of problem solving;

- The information system is able to measure out the educational content and adjust its difficulty to the educational level of each student. This ensures the feasibility of the offered tasks and reduces the risks of student frustration and loss of motivation for learning.

Below we present a description of the technical requirements for the CALS, which provide scaffolding for students in the process of student-oriented training in MSEE disciplines.

4.1   Student-Centered Learning Model

As far as the paper discusses abilities of constructivist pedagogy for computer adaptive learning. This section gives an overview of constructivists' pedagogy foundation.

It is well-known that student-centered learning is one of the cornerstones of constructivist pedagogy. The theoretical basis for this model is provided by the construct zone of proximal development (ZPD), which was introduced by Vygotsky and widely disseminated in modern didactics and cognitive science [22, 23]. The basic characteristics of the model of the student-centered learning process are the following.

1. The inseparability of learning and development; the learning process is mediated by the process of individual mental development and cannot be successful and effective in isolation from it [24, 25].
2. The source of development is a situation arising spontaneously or formed by a teacher during the learning process. For example, a teacher suggests a learner some tasks which indicate a gap between the actual abilities and skills and the skills that are needed to handle a new complex problem that the student has not yet had to solve.
3. This gap is well known as ZPD can be overcome only in the process of interaction with a teacher or a more competent peer.
4. Actions of a teacher's in the ZPD are directed to:
   - Formation of the ZPD helps in understanding the student's own incompetence and his/her personal learning goals;
   - Provision of pedagogical actions facilitating the student to bridge the gap between actual and necessary competences (mentoring, coaching, scaffolding, etc.).

5. ZPD is dynamic and mobile. Rational and effective learning process with this concept is built on a constant transition to new, increasingly complex cognitive tasks [24]. Thus, students can incrementally develop their skills and acquire relevant competencies.

5 Conclusion

This paper followed two interconnected goals. First one was to study and justify pedagogical possibilities of computerized adaptive learning in MSE. Second one was to find a place for a computer adaptive learning system (CALS).

The review of student-centered learning and instructional scaffolding reviled powerful perspective and allowed to formulate a set of methodology and technical requirements for CALS design.

We put following principles as a foundation of adaptivity and scalability of a CALS to reinforce the training process.

- Ontology controlled knowledge assessment in the field of MSE with automatic generation of several types of assignments.
- Automatic knowledge gaps detection with permanent availability of instant self-assessment according to constructivist pedagogic approach.
- Educational content relevance estimation using Bayes inference on learning logs.
- Automatic pedagogical interventions after assessment actions with formative feedback.

The implementation of these principles allows to support effectively cognitive scaffolding in CALSs.

Acknowledgements. The article was prepared within the framework of the Basic Research Program at the National Research University Higher School of Economics (HSE) and supported within the framework of a subsidy by the Russian Academic Excellence Project "5–100".

References

1. Independent Administrative Institution National Institute for Materials Science: A Vision of Materials Science in the Year 2020. NIMS, Ibaraki (2007)
2. Hacker, M.: Engineering and technology concepts: key ideas that students should understand. In: de Vries, M. (ed.) Handbook of Technology Education, pp. 173–191. Springer International Publishing AG (2018)
3. Potter, P., France, B.: Influences of materials on design and problem solving learning about materials. In: de Vries, M. (ed.) Handbook of Technology Education, pp. 463–472. Springer International Publishing AG (2018)
4. Bassett, G., Blake, J., Carberry, A., Gravander, J., Grimson, W., Krupczak, J., Mina, M., Riley, D.: Philosophical Perspectives on Engineering and Technology Literacy, I, Electrical and Computer Engineering Books (2014)
5. Strimel, G., Grubbs, M.: Positioning technology and engineering education as a key force in STEM education. J. Technol. Educ. **27**(2), 21–36 (2016)
6. Thornton, K., Nola, S., Garcia, R., Asta, M., Olson, G.: Computational materials science and engineering education: a survey of trends and needs. Comput. Mater. Educ. **61**(10), 12–17 (2009)
7. Magana, A.J., Falk, M.L. Jr., Reese, M.J.: Introducing discipline-based computing in undergraduate engineering education. ACM Trans. Comput. Educ. **13**(4), 16–22 (2013)

8. NSF: The Future of Materials Science and Materials Engineering Education. Arlington (2008)

9. Litovkin, D., Zhukova, I., Kultsova, M., Sadovnikova, N., Dvoryankin, A.: Adaptive testing model and algorithms for learning management system. In: Joint Conference on Knowledge-Based Software Engineering (2014)

10. McMullen, J., Brezovszky, B., Rodríguez-Aflecht, G., Pongsakdi, N., Hannula-Sormunen, M., Lehtinen, E.: Adaptive number knowledge: exploring the foundations of adaptivity with whole-number arithmetic. Learn. Individ. Differ. **47**, 172–181 (2016)

11. The World's Learning Company [Online]. Available: https://www.pearson.com/content/dam/one-dot-com/one-dot-com/global/Files/efficacy-and-research/methods/learning-principles/Foundations_of_Adaptive_Learning.pdf. Accessed 20 May 2018

12. Magnisalis, I., Demetriadis, S., Karakostas, A.: Adaptive and intelligent systems for collaborative learning support: a review of the field. IEEE Trans. Learn. Technol. **4**(1), 5–20 (2011)

13. Mulwa, S. Lawless, M. Sharp, Arnedillo-Sanchez, I., Wade, V.: Adaptive educational hypermedia systems in technology enhanced learning: a literature review. In: Proceedings of the 2010 ACM Conference on Information Technology Education (2010)

14. Heiler, S.: Semantic interoperability. J. ACM Comput. Surv. (CSUR) **27**(2), 271–273 (1995)

15. Forte, E., Haenni, F., Warkentyne, K., Duval, E., Cardinaels, K., Vervaet, E., Hendrikx, K., Forte, W., Simillion, M.: Semantic and pedagogic interoperability mechanisms in the ARIADNE educational repository. ACM SIGMOD Rec. **28**(1), 20–25 (1999)

16. Rizzardini, R.H., Gütl, C., Amado-Salvatierra, H.: Tools, interoperability for cloud-based applications for education settings based on JSON-LD and hydra: ontology and a generic vocabulary for mind map. In: i-KNOW'14 Proceedings of the 14th International Conference on Knowledge Technologies and Data-Driven Business (2014)

17. Nascimento, V., Viamonte, M., Canito, A., Silva, N.: Improving semantic interoperability with ontology alignment negotiation and reutilization. In: IIWAS'13 Proceedings of International Conference on Information Integration and Web-based Applications and Services (2013)

18. Heiyanthuduwage, S., Schwitter, R., Orgun, M.: An adaptive learning system using plug and play ontologies. In: IEEE International Conference on Teaching, Assessment and Learning for Engineering (TALE) (2013)

19. Kotchakorn, B., Ngamnij, A., Somjit, A.: Ontology-based Metadata Integration Approach for Learning Resource Interoperability. In: 2010 Sixth International Conference on Semantics, Knowledge and Grids (2010)

20. Obrst: Ontologies for semantically interoperable systems. In: Proceeding CIKM'03 Proceedings of the Twelfth International Conference on Information and Knowledge Management (2003)

21. Mehedi: Collaborative e-learning systems using semantic data interoperability. Comput. Hum. Behav. **61**(C), 127–135 (2016)

22. Stanlaw, J.: Vygotsky, Lev Semenovich (1896–1934). In: Birx, H. (ed.) Encyclopedia of Anthropology, pp. 2292–2293. SAGE Publications Ltd., Thousand Oaks, CA (2006)

23. Schaffer, H.: Key concepts in development psycology. SAGE Publications, London (2006)

24. Vygotsky: Mind in Society: Development of Higher Psychological Processes, p. 175. Harvard University Press, Massachusetts (1980)

25. Berk, L., Winsler, A.: Scaffolding children's learning: Vygotsky and Early Childhood Education, p. 192. National Association for the Education of Young Children (1995)

26. Dennen, V.: Cognitive apprenticeships in educational practice: research on scaffolding, modeling, mentoring, and coaching as educational strategies. In: Jonassen, D. (ed.)

Handbook of Research on Educational Communications and Technology. Lawrence Erlbaum, Mahwah, NJ (2004)

27. Belland, B.: Instructional scaffolding: foundations and evolving definition. In: *Instructional Scaffolding in STEM Education*, pp. 17–53. Springer (2017)

28. Gould, J.: Learning Theory and Classroom Practice in the Lifelong Learning Sector. Learn. Matter. **152** (2012)

29. Ritter, S., Anderson, J., Koedinger, K., Corbett, A.: Cognitive tutor: applied research in mathematics education. Psychon. Bull. Rev. **14**(2), 249–255 (2007)

LMS Use in Primary School as an Internet-Accessible Notice Board

D. M. Garyfallidou$^{(\boxtimes)}$ and George S. Ioannidis

The Science Laboratory, University of Patras, Rio, Greece
d.m.garyfallidou@gmail.com, gsioanni@upatras.gr

Abstract. The educational use of LMSs in primary schools where all students entered the platform using their own e-mail account in order to enter in the platform and get access to homework or tasks assigned to them by the teacher has been presented and analysed before. However, there are practical problems involved in enrolling very young students into any type of platform, which resulted in the present trial to use it without such restrictions. The school use of the LMS as (effectively) an internet-accessible notice board came to provide solutions to the following educational needs: (a) Open communication between teacher and parents e.g. to inform them for upcoming events, assignments, and instructions on a day-to-day basis. (b) Help students who missed lessons to find reliable information as to what was taught, and download supporting material. (c) Offer extra teaching material (or supporting material) to students. (d) Attempt to build a sense of three-part educational community between students, parents, and teacher with obvious benefits to all. Specifically, the parents get the supervisory role to their child's education that they deserve. Details of the actual LMS deployment and its actual use at a primary school are given. The results from the analysis of a student and parent questionnaire are analysed and presented graphically herein, while conclusions are drawn.

Keywords: LMS · Primary school LMS · Parent-teacher communication · Educational games

1 Introduction

Traditionally, communication between teacher and parent was done with the parent (usually with the mother who in older times was not working) being present during the monthly school meeting, and generally acting as "an interface" between Schools and parents. This has gradually changed dramatically. As both parents are working nowadays, and their place of work is in many cases quite far away from the school, it is difficult for them to reach school at the arranged "meeting time". In many cases apart from the distance, parents face difficulties asking for a leave to visit their child's school. Moreover, in a traditional class, homework used to be given to students through photocopied pages, which often students used for "origami constructions", instead of paying attention and solving them during homework. LMS comes to help, if not to altogether solve the parent–teacher communication problem and alter dramatically the way extra homework is given to the students.

© Springer Nature Switzerland AG 2019
M. E. Auer and T. Tsiatsos (Eds.): ICL 2018, AISC 917, pp. 854–864, 2019.
https://doi.org/10.1007/978-3-030-11935-5_81

A further use of the LMS setup proposed herein is to alleviate possible school ICT equipment issues. Despite being generally accepted that ICT equipment, high-speed internet, and school projectors offer many educational advantages, not all schools are yet equipped accordingly, neither is their equipment necessarily up to date. In such schools, teachers have no choice but to adopt a conventional form of teaching, regardless of their better judgement. In case a classroom only has one computer, if the projector is unavailable, it is difficult for all students to manage to get even a glimpse of the computer screen.

After a suitable school LMS was set-up, in accordance to the quality characteristics exemplified by Horvat et al. [1], an educational trial followed investigating its use amongst all concerned. The results of this investigation are analysed herein.

2   Purpose or Goal

The educational use of LMSs in primary schools is faced with various special issues, mainly related with the age of the students. Previous analyses concerned older students, enrolling in the school LMS platform (usually Moodle-based). They used the LMS to get access to homework or tasks assigned to them by the teacher and all this process has been presented before [2]. However, as it is widely known, primary school students are not allowed to open e-mail accounts by themselves, and neither is it easy to have the consent of parents in order to open personal accounts for children (even for school use) or even use parent's accounts for it. This task is even more complicated when parents are divorced or separated. This study overcomes this problem by allowing students and parents alike to access the LMS's platform as "guests" (i.e. without prior enrolment hence bypassing the initial need for an external email account). At any rate, although the present LMS implementation by no means exhausts all facilities available, its scope is to be utilised in primary school and undergo educational trials. In parallel, this study also tries to investigate if students and parents (all of which could visit the platform in an anonymous basis) did in fact use it, and if they felt that it was helpful for them to do so.

This particular use of the LMS came to solve the following problems:

(a) Offer up-to-date communication between teacher and parents e.g. to inform parents about upcoming school events, ask them to participate, ask for their help, let them know of assignments to their children on a day-to-day basis.
(b) Help students who missed one (or even more) lessons to find reliable information as to what was taught, as well as to download educational material on whatever they missed.
(c) Offer extra teaching material (or supporting material) to students, in a convenient format.
(d) Attempt to build a sense of three-part educational community between students, parents, and teacher with obvious benefits to all. Specifically, the parents get the supervisory role to their child's education that they deserve. Teachers are endowed to offer students educational extra-curricular activities in selected subjects. Students are expected to benefit overall, resulting subtly from a higher quality education.

3 Approach

The LMS Moodle was installed onto a separate server especially for this trial, and both parents and students were given detailed printed instructions on how to enter the platform. There, they could select between 3 different so-called "lessons". The main one contains whatever students are asked to do for the following lesson—updated daily. They can also find announcements and information about events that would take place in the school e.g. forthcoming excursions, materials needed for some school handicrafts etc.

Events that were taken place outside school in which they have had free entrance (so as not to be seen as an advertisement) and they are educationally beneficial to the students e.g. a movie, lessons on how to ride a bike safely, free STEM lessons, free physics experiments, etc. are uploaded here.

In this, they can find extra material relevant to the lesson taught e.g. a video about how to use the triangle to draw right angles or parallel lines, videos simulating a historical battle or a naval battle etc.

They can also find extra exercises that students could decide if they want to solve, which at a later stage would be corrected by the teacher if they so wish (Fig. 1).

Fig. 1. Lesson 2: Additional material

The second "lesson" is divided in sub-subjects e.g. Physics, Maths, Geography etc. and each one contains extra-curricular material. All the material contained here is free of charge for unlimited use, free of "additional purchase offers", and without any other advertisements. It does not require prior enrolment to the LMS, and everything has been tried in real school. Extra material created by the teacher is also contained therein e.g. presentations with photos of the places of interest around Greece. Also linked-in or included is free-to use material from the Internet, such as games designed on behalf of the European commission with the aim to teach young students about Europe and its history.

Some other material is books offered for a free download. Students can also find here material not necessarily closely connected to school curriculum, yet eminently suitable for the students' age. For example, the ancient Greek history lesson contains nothing about the Tunnel of Eupalinos, or the Antikythera mechanism, however such interesting information was offered to the students to excite their interest by explaining the brilliant thinking involved in their design and construction. Similarly, online dictionaries are offered as add-on material for the language lesson, as well as videos with easy to do physics experiments, and representations of historical battles, on-line books etc.

4 Data Analysis

The analysis of all relevant experimental data was performed using specially constructed software, while all results formed the input to a plotting package. The Bessel-corrected standard deviation was computed for all data points. All measurements are subject to systematic errors. In the present study, special care was taken to avoid large systematics from all sources. However, such sources were reviewed, any possible biases were considered independently, while any remaining systematic errors evaluated. The nominal systematic error was set at 2.5%, this being considered to be fair (albeit on the low side) and consistently comparable with all current statistical errors. This results to the total experimental error computed being neither statistics-dominated nor systematics-dominated, and this holds true for every single data-point presented. The total measurement error was then computed by adding in quadrature systematic with each statistical error, these two being independent, by definition.

In this trial, a total of 24 primary school students, aged 10, participated.

In the first half of the questions presented herein, more than one answers could be selected. Therefore, each point in such diagrams should be considered as an independent yes or no question. For this reason, the sum of these percentages is something much higher than 100%.

4.1 Question 1. *For What Purpose* Did You Use the LMS?

We can observe that the majority of the users 87.5 ($\pm$7.3%) used the system to find out what next day's teaching program involved, or to be suitably prepared for a history chapter, or exercises in mathematics and language). The rest of the answers are not so great, yet four of them were comparable with each other. A 37.5% ($\pm$10.4%) used the history videos suggested, and a similar percentage used the platform in order to learn about other school activities. Only one third of the students 33.3% ($\pm$10.1%) used the physics experiments offered. Surprisingly low is the percentage 16.7 ($\pm$8.2%) of the students that used the educational games. Is it perhaps because the appearance of these games is "too simplistic" for them, perhaps? Could it be that parents show their displeasure when any games are played? (—i.e. irrespective as to if these are the educational or not). This warrants further investigation.

The 45.8% ($\pm$10.7%) of the student s that claimed that used the platform for the Greek language lesson exercises was something of a surprise. These grammar exercises offered limited interactivity. They can only be printed, and after completion, someone

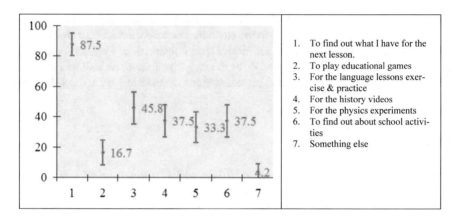

1. To find out what I have for the next lesson.
2. To play educational games
3. For the language lessons exercise & practice
4. For the history videos
5. For the physics experiments
6. To find out about school activities
7. Something else

Fig. 2. Answers to question 1

has to check them. An explanation of these results might be that interested parents had done precisely this, as language is considered by many parents to be very important and parents are accustomed to do just that supervising grammar exercises

In general, the above results were somehow expected for three important reasons. (A) It is the first time students and parents alike faced an LMS platform, and therefore some more time is needed to familiarise themselves with it and appreciate the advantages of its use. However, research has shown that students in particular adapt themselves quickly to this new way of learning [3] (B) Not all students have the same abilities or interests therefore they may have used only those parts that were within their interests. (C) Self-directed learning addresses mostly the most capable of the students, or those ones most motivated to learn new things. (D) Not all parents allowed their children to access the platform (e.g. by prohibiting access onto the net).

4.2   Question 2. What Did You *like Most* About the LMS?

The majority of the users 66.7% (±10.1%) reported that what they liked most about the LMS was the ability to be informed about the content of the next lesson. Also, a 62.5% (±10.4%) liked most that they could be informed about forthcoming school activities. Despite the fact that (as it is deduced from Fig. 2) not all supplementary educational material was visited by all students, most of them must have seen at least something interesting in it. This, surely, must be the explanation for the 58.3% (±10.6%) of the students that liked the fact that some extra material was on offer, something encouraging indeed (Fig. 3).

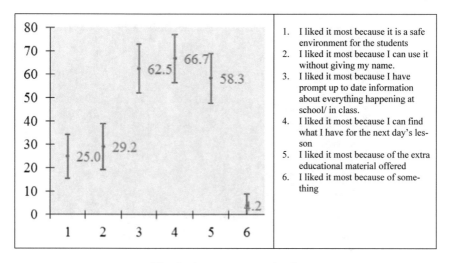

Fig. 3. Answers to question 2

4.3 Question 3: *What Else* You Would like to Have on the LMS Platform

It can be observed that the overwhelming majority of the students 75.0% (±9.4%) valued the ability to communicate with their classmates. This was expected as it came to meet a social need. The school where the educational trial took place covers a wide area of a large city, with children living all over the place, thereby having few opportunities to meet each other outside school. How would young students use this feature will be investigated in a future study. Will they use it in order to enhance their learning performance as the research has shown [4], or will they use it just for chat?

Half of the users, 50.0% (±10.7%), answer that they would like to have some on-line exercises. Not all children are familiar with that type of task. However, there are reasons to believe that what they liked was the idea that the computer would correct their mistakes, and hence none would know how many mistakes they made or how many times they tried, in order to finish with the exercise.

A 41.7% (±10.6%) would have liked to be able to communicate with their teacher, to ask for help when facing difficulties with exercises perhaps, while a 45.8% (±10.7%) believe that if they had the chance they could offer material appropriate for their age. This is a challenge for the next year to investigate as to what they could, in reality, offer.

Another 45.8% (±10.7%) believes that the extra material already offered is useful and therefore wished to be able to access it in the following years, no matter if they would have the same teacher or not.

It seems that all students that claimed that they visited the material about physics experiments wished to have more of them. The percentage is an exact match of the students 33.3% (±10.1%) that wished for more physics experiments.

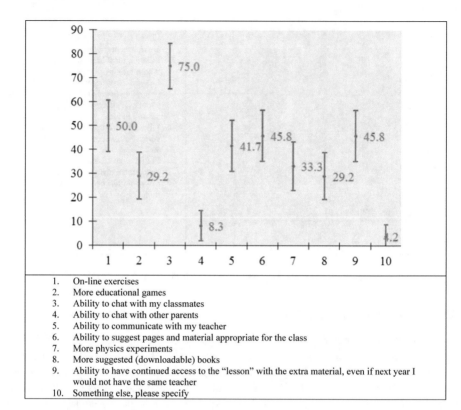

1. On-line exercises
2. More educational games
3. Ability to chat with my classmates
4. Ability to chat with other parents
5. Ability to communicate with my teacher
6. Ability to suggest pages and material appropriate for the class
7. More physics experiments
8. More suggested (downloadable) books
9. Ability to have continued access to the "lesson" with the extra material, even if next year I would not have the same teacher
10. Something else, please specify

Fig. 4. Answers to question 3

4.4 Question 4: Which of the Following Did You *Use*? You Can Choose More Than One Answer

It can be observed that the vast majority 87.5% (±7.3%) of the students used the platform mostly in order to find out next day's lessons and the forthcoming school activities Another 41.7% (±10.6%) claims they accessed the physics experiments. This is compatible with previous answers (as in Figs. 2 and 4) concerning physics experiments. The 37.5% (±10.4%) used the suggested grammar books and the novels (offered for free, as their copyright expired). As we can observe the percentages of the students that used the suggested educational games at home are very low. This has some explanations: (A) Parents did not allowed them to connect to the internet in order to play either due to fear of the net or due to fear of the games. (B) Because children are not familiar with this way of learning. (C) Because the suggested games were not as elaborate or captivating as the (commercial) ones students were used to. Most educational games were developed by individual teachers in order to teach a very specific subject, and they therefore have neither the graphics a commercial game has, nor do they have a large amount of user-levels to choose from (i.e. they are not so adaptable).

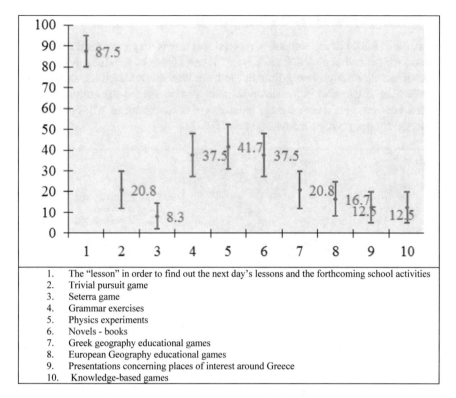

1. The "lesson" in order to find out the next day's lessons and the forthcoming school activities
2. Trivial pursuit game
3. Seterra game
4. Grammar exercises
5. Physics experiments
6. Novels - books
7. Greek geography educational games
8. European Geography educational games
9. Presentations concerning places of interest around Greece
10. Knowledge-based games

Fig. 5. Answers to question 4

4.5 The Use of the LMS Was a Pleasant Experience for Me

The 79.2% (±8.8%) of the students responded that they found the use of the LMS platform a pleasant experience. A 4.2% (±4.9%) of the students disagrees with that statement while none of the students "disagreed totally" with this (Fig. 6).

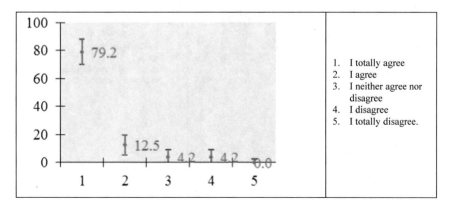

1. I totally agree
2. I agree
3. I neither agree nor disagree
4. I disagree
5. I totally disagree.

Fig. 6. Answers to question 5

4.6 It Was *Easy* for Me to Use the LMS Platform

The 20.8% (±8.8%) of the students answered that it was very easy for them to use the platform, while another 41.7% (±10.65) find it easy. None of the students claimed that the use of the platform was very difficult for them. Previous research has shown that the system's ease of use also influences e-learning system use [5] therefore these results obtained here are very encouraging, indicating that the platform will continue to be used in the future, perhaps increasingly so (Fig. 7).

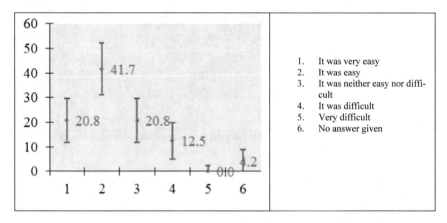

Fig. 7. Answers to question 6

4.7 I Believe that the Educational Games Uploaded in the LMS Platform Offered Me Extra Knowledge in a Pleasant Way

The 58.3% (±10.6%) responded "I totally agree" and therefore considered that the educational games offered him/her extra knowledge, while another 20.8% (±8.8%) agreed with this. As it is observed in Fig. 5, the students tried only some of the educational games. However, they generally believe that they learned something out of whatever they did try (Fig. 8).

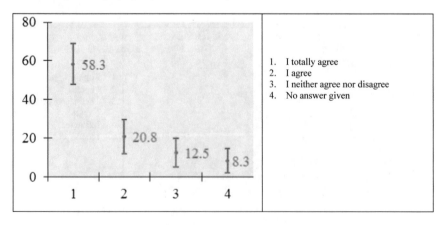

Fig. 8. Answers to question 7

4.8 I Believe that the Virtual Tour Around Greece Offered Me Extra Knowledge in a Pleasant Way

The 54.2% (±10.7%) "totally agree" that the PowerPoint presentations with the highlights of the "points of interest" of different places of Greece offered them extra knowledge, while another 20.8 (±8.8%) also "agree" with this (Fig. 9).

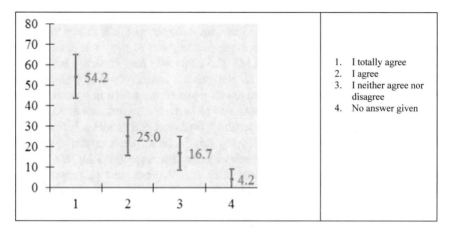

Fig. 9. Answers to question 8

5 Conclusions

Using an LMS this way in a primary school is distinct and particular, as the teacher has no individual one-to-one communication with his/her students through the platform. Besides, he/she has no way of monitoring which student visited which pages, this being left for the parents to do, as part of their job of supervising their offspring. The use of LMS is, therefore, akin to that of an "internet accessible notice board".

With the use of LMS, students that missed the lessons know what to prepare for the next day. The system also offers quick one-way communication with the parents. The speed with which parents responded whenever their help was asked was impressive.

Furthermore, this approach can (reportedly) be used just to reduce photocopying costs [6], in addition to reducing (potentially) the environmental impact through reduced paper wastage. Exercises can be uploaded and let parents decide to either print them or allow online use of the LMS. As almost all uploaded material can be characterised as "optional", this can be interpreted as: (a) only there for those children that wished to use it, and (b) being here and available at any-time children might feel like wanting to use it (i.e. anytime during the whole week), thereby alleviating complaints of temporary student overload, at one particular day.

A further use for the LMS can be to upload mainly material already taught in class, so as students can use it for revisions. Such might be: (a) presentations that were already shown in class (so as students could see them again through the LMS or, as they have expressly said. to share it with their parents). Or (b) videos already presented in class or just having some affinity to the curriculum, while also being suitable for the age of the students, and also suitable for sharing. Also, (c) educational games, (d) Suitable novels (offered freely, as their copyright expired), (e) exercises, (f) information addressed to the parents.

The majority of children believe that they gained something by using the platform. A pleasant surprise: the two parents that answered the questionnaire have also admitted that they have indeed learned something from the material offered in the platform.

There are, therefore, reasons to conclude that the first trial to use the school LMS as an internet accessible notice board was met with some success. It is certain that there is a need for it: teachers often have to improvise in order to distribute more learning resources to their students, and the LMS plays this role magnificently. It also offers "utility value" in communicating with the parents (albeit one-way), all the while utilising a more attractive teaching and learning procedure, with a minimum of hassle. In addition, this distinct LMS use does seem to be well received, and to arouse good will from the parents, as it creates a sense of three-part collaboration between themselves, the teacher, and the child. Having this in mind, the use above-mentioned approach is well balanced for primary school usage. However, this LMS use (i.e. as an internet accessible notice board) needs further development and refinement. It may eventually form the open-access part of a more comprehensive primary school LMS, whereas a second part where students shall have only private access will be used for other LMS functions. This would naturally require further testing under real school conditions.

References

1. Horvat, A., Dobrota, M., Krsmanovic, M., Cudanov, M.: Student perception of Moodle learning management system: a satisfaction and significance analysis. Interact. Learn. Environ. **23**(4), 515–527 (2015)
2. Skellas, A.I., Garyfallidou, D.M., Ioannidis, G.S.: Suitably adapted LMS used to teach science to primary school students using blended learning—utilising a novel educational design to tech heat and thermal phenomena. In: Auer, M.E. (ed.) Proceedings of 2014 International Conference on Interactive Collaborative Learning, ICL2014, pp. 473–478. Dubai, 3–6 Dec 2014. ISBN: 978-1-4799-4438-5/14, © 2014. IEEE
3. De Smet, C., Valcke, M., Schellens, T., De Wever, B., Vanderlinde, R.: A qualitative study on learning and teaching with learning paths in a learning management system. JSSE-J. Soc. Sci. Educ. **15**(1), 27–37 (2016)
4. Damnjanovic, V., Jednak, S., Mijatovic, I.: Factors affecting the effectiveness and use of Moodle: students' perception. Interact. Learn. Environ. **23**(4), 496–514 (2015)
5. Islam, N.: Investigating e-learning system usage outcomes in the university context. Comput. Educ. **69**, 387–399 (2013)
6. Nurakun Kyzy, Z., et al.: Learning management system implementation: a case study in the Kyrgyz Republic Interactive Learning Environments. ISSN: 1744-5191, Jan 2018

Poster: Improvement of Professional Education Quality by Means of Mathematics Integration with General Education and Vocation-Related Subjects

Svetlana V. Barabanova[1], Natalia V. Kraysman[1(✉)],
Galina A. Nikonova[1], Natalia V. Nikonova[1],
and Rozalina V. Shagieva[2]

[1] Kazan National Research Technological University, Kazan, Russia
n_kraysman@mail.ru
[2] Russian Presidential Academy of National Economy and Public
Administration, Moscow, Russia

Abstract. The paper focuses on the improvement of professional educational quality. From the beginning of training it is necessary to teach students an applied mathematical thinking as a combination of intellectual fundamental structures, oriented on the study and construction of mathematical, ecological and technological objects. The development of a thinking logic, an ability to put into practice axiomatic, multiple-theoretical, inductive and deductive methods, statistics and probability methods of discrete mathematics and methods of optimization permits to resolve any issues of any disciplines and any field successfully. Therefore, we can note an increasing part of the mathematical foundation as a mean of professional tasks resolution through mathematical modeling.

Keywords: Vocational education · Fundamental training on mathematics · Mathematical methods · Integration of mathematics with general education subjects

The basis of society socio-economic development is the scientific and technical progress, which motor are the scientific and technical engineering human resources.

One of the main issues in the preparation of high-skilled engineers for the present economic context is the warranty of a high quality education. The achievement of this goal is a possible way that permits to protect the founding principles and satisfaction of the current and potential demands of science and industry fields. Besides, the modern specialist has to have a creative personality allowing him to make unusual but good decisions. He has to be continuously working on the development of his self-education, creativity skills, oriented thinking and self-development. New technical and technological discoveries are possible nowadays in a special cultural environment of interdisciplinarity across the various sciences that makes the special demands to engineering education.

The technical and technological development of universities offering numerous multilevel and multidisciplinary education in Russia permitted to have a new look at the question of the preparation of specialized engineers. A new kind of university has to

© Springer Nature Switzerland AG 2019
M. E. Auer and T. Tsiatsos (Eds.): ICL 2018, AISC 917, pp. 865–870, 2019.
https://doi.org/10.1007/978-3-030-11935-5_82

dedicate itself to ensuring the foundation principles; depth and amplitude of the trainings provided and at the same time reinforce the professional orientation of education. The final and main goal, characterizing the quality of the course, is the professional competence of the specialist. Therefore, arises the problem of an optimal correlation between fundamental and professional components of education, which balances the content of textbook programs.

The labour market dictates its conditions. The exigency towards young professionals is getting toughened. The majority of high-scale production companies' executives, working on new strategies, hire students who just graduated if they justify a positive stable technical and professional knowledge and sufficient abilities allowing them to create new technical products, as well as juridical nature knowledge and skills permitting to prove the viability of the implementation of the new product into practice. Also the future engineer has to master at least one, but several would be better, languages to become competitive and achieve his goals in accordance with the current conditions of the market. Obviously, the knowledge of foreign languages is a real necessity for the present-day engineer, expanding his possibilities and, consequently, opening doors to a new world in which he can integrate himself easier into the global market.

The goals and tasks of the university coincide. The growth of competition level of the labour market set serious challenges before students: in addition of their deep knowledge of their specialty they have to put it into practice resolving actual issues, also, they should be skilled of a good system-analytical reflection, to be able to prognosticate the consequences of the decisions given, or even other decisions, and to possess juridical and managing skills.

Accordingly, the problem of the future engineer mathematical culture development arises. The future engineer has to have a fundamental training in mathematics to be able to assimilate fast and successfully the most advanced technologies and work them out to present a final product.

The multidisciplinary approach of the course training is a source of motivation for students. From the beginning of training it is necessary to teach students an applied mathematical thinking as a combination of intellectual fundamental structures, oriented on the study and construction of mathematical, ecological and technological objects. The development of a thinking logic, an ability to put into practice axiomatic, multiple-theoretical, inductive and deductive methods, statistics and probability methods of discrete mathematics and methods of optimization permits to resolve any issues of any disciplines and any field successfully. Therefore, we can note an increasing part of the mathematical foundation as a mean of professional tasks resolution through mathematical modeling. Using mathematical modeling methods gives the possibility to define properly the resources required, to analyze the activity of organizations from an all-intechnic-economic point of view, the improvement of organizational managing structures, the prognosis of the most efficient parts of their development. It is precisely from the interaction of mathematics, physics, chemistry, biology, but also of law, philosophy, astronomy, geography, that were born concrete forms of rational reflection and rational life. The integration of mathematics lessons to history, economics, biology, physics, geometry and other classes permits to study important phenomena on all levels, and link the classes with the real life, show the richness and the complexity of the surrounding world, recharge the student's curiosity batteries and creative energy. The necessary professional competences of future engineer which are especially high-demanded by practice in modern conditions are formed during such interdisciplinary training.

Consequently, the knowledge of mathematics is a pillar of the main disciplines majority of general and profiled education. The construction of mathematical models is carried out through mathematics without making use of substantial reflection. The elaboration of the theoretical part of the mathematical modeling is based on studies, carried out by the specialized departments. The main directions of research and tasks are highlighted and some fragments have been added to the textbook "Mathematics" [2]. The prerequisite for a proper and justified use of mathematics is the defined compatibility between the means required and the actual state of the apparatus provided in the study and research premises. The mathematical apparatus, especially designed to study the sociological scientific description of human behavior, has not been created yet, but some of its elements are in the process of being made. The existence of a socio-juridical reality of laws of statistics in some phenomena is one the impartial prerequisite to choose the methods of mathematics that best correspond, from which we can give qualitative descriptions of realizations of law sociology. Statistics approach and methods can be used effectively in theoretical and empirical study of a certain number of legal governmental issues. In that case should be taken into account the fact that, the subjective nature of the processes studied doesn't represent numerous changes and modeling. For example, in criminology we successfully use a branch of mathematics "Differential equations".

The most important role is up to the theories of probability and statistics. If any action of statistics' laws appears in the empirical material collected (forms, outcomes of experiment), then the use of probability methods and statistics' theories in a socio-legal study for the analyze and interpretation of the results obtained is not only appreciated but also needed. A significant interest for the sociology of law is such scientific methods like the identification of different modes, the measurement's theory, the theory of information and others. Consequently, the sociology of law is linked to the amplitude and complexity of the mathematical apparatus. During the learning process of socio-legal phenomena, the mathematical methods fulfill diverse functions:

- the precision and improvement of the linguistic knowledge. The mathematics are seen as a language of science permitting to describe processes of the objective reality.
- the merger of juridical science with other social and natural sciences. Some methods and means of modern mathematics, conceptions, means of informatics start actively to infiltrate the juridical scientific field.
- the enhancement of precision of researches and results.
- the development of the vision quality of the law sociology studied object. The description of phenomena in mathematical language supposes a selection of the points, which are by their definition and uniformity easier to analyze.

Accordingly, in the current conditions the mathematical demonstration and juridical proof are cultural algorithms, by which we can develop and go into the inter-determination of the rationality of ideas and humans actions in depth. This inter-determination widen the borders of rationality, permits to exclude the unilateral approach, fixes the multidimensionality and flexibility of demonstrative constructions, merges science and law as part of socio-cultural practices. The value of the demonstrative activity as a condition of rationalization of culture makes it an autonomous object of study. The necessity of such a union can be explained by the fact that, in the fields of mathematics and law, it is important to save the precision and succession of the demonstrative thinking, and that way avoiding monotony. Any mathematical demonstration, as a theoretical reasoning, induces that the initial premises are proper means to make conclusions and, finally, the own form of the expected result. Speaking in a more formal descriptive way, it is the basic axioms and the mathematical language corresponding to the theory, that allows the rule of logical conclusion and the own conclusion in the form of a series of deductive reasoning and affirmations, which ends with the formula of the affirmation demonstrated. For instance, during a judicial practice, the judges are to be careful when it comes to what can be produced as a proof and what are the questions that can be asked to the witnesses. The formal procedure normalizes the means of judicial demonstrations, including declarations and evidences, as initial presumptions, and the eloquence and credibility of layers as a mean of reasoning and exoneration. It is obvious that mathematical methods, used in the juridical scientific field, is not disconnected to real life, juridical practice and theory. On the contrary, they result from actual scientific practical necessities of jurisprudence. For example, the question of the formalization of branches of law, of the mathematical description coming from tasks of legal regulation, the measuring of socio-legal information are studied in the literature of law through the prism of government and law sectorial juridical sciences theory. The use of modern means of mathematics in the criminology is based on theoretical concept applications into the scientific criminology.

It should be pointed out the use of mathematics in the juridical medicinal expertise.

Definitive attributes of the crime place permits to reveal information about the case. Thanks to mathematical models it is possible to estimate the time of death, identify digital printings of the suspect and analyze the structure of the blood founded:

(1) Newton's law of cooling.
    To solve this issue, judicial experts use data of temperature to determine and establish an approximate time of death. The estimation is based on the velocity of body warmth loss directly proportional to the difference between temperatures of the body and the surrounding environment.
(2) Analyze of the blood structure.
(3) Geographic profiling includes the analyze of spatial models of connected crimes in order to determine the probability of every suspect, starting from the fact that criminals inclined to act in the area they well know. Kim Rossmo, Canadian criminologist, is a pioneer of the geographic profiling branch, proposing to determine the serial crimes sites through a mathematical formula.

Mathematical methods are used for the resolution of applied cases in the criminology and judicial expertise.

(1) There is a caliber of the bullet under the form of an ellipse. They can determine the angle of the given rifle according to the obstruction of the weapon.

(2) In case of fatal rifle into the area of the heart, it is possible to establish the type of melee weapon.

(3) It is possible to determine the volume of the stolen sand from a warehouse.

(4) They manage to calculate the distance and the height from which the shoot has been executed.

(5) It is successfully the investigation of road accidents connected to the changes of means of transport.

(6) It is possible to determine the number of versions according to a series of study of several reconstruction crime scenes.

(7) Choice in the data basis of the possible criminals according to the characteristics indicated is carried out.

Unfortunately, giving a bigger freedom in the formation of the main educational programs to the educational organizations by the federal law "About Education in the Russian Federation" not always contributes to the development of interdisciplinarity. So, in the training program "Legal studies" we do not observe the interest in inclusion of the mathematical subjects in educational program.

It opens the new opportunities for demand of universities graduates, as a rule having a profound mathematical knowledge in the boundary spheres and professions, on the one hand, and provides the formation of special requirements to their preparation during the training, on the other hand.

In particular, to reach a successful integration of mathematics into the general and professional education disciplines, it is essential to put at the disposal of students textbooks with a differential approach, presenting two-three courses' levels, in which we assigned a main role to significant professional themes. Furthermore, it is necessary to put into these textbooks a big part of theoretical mathematics and its practical realizations, avoiding voluminous theoretical display and an abundance of useless formulas.

One of the possible solutions of a problem of the new type textbook creation it is realized by authors in a didactic set [1, 2].

Generalization of the didactical information is based on the idea of putting at the disposal of students an all-round material of study and an optimal visualization of the goals of intensification and improvement of the ergonomic quality of the textbook. With regard to this purpose, a system of basic notes is designed for every submodule in combination with the use of modern mathematical symbolic, conception of modules through fundamental mathematical methods, allowing the knowledge and the use in practice leading to one meaning: the use of mathematical methods in modeling and of digital methods in differential models. The construction of informational basis is being regularized by the next principles:

- entirety—by the way of a training of conception using fundamental mathematical methods which permit to knowledge and practice to form a whole;
- regularity and consistency—unifying inductive and deductive means of exposure, as well as unifying the abstract and the concrete under the schema of "concrete-abstract-concrete";

– the accessibility via the pedagogic rule of "from the easier to the most difficult" corresponding to scientificity, by the determination of the degree of abstract, required to expose mathematical knowledge.

The textbooks of oriented practical training contain a pedagogic method, tasks with correct answers and their explanation, exercises with correction, among which we can find some unusual exercises, also versions of exercises to get ready for evaluations, which permit to master a practical knowledge according to the following schema:

(1) understanding of the supportive notes, analyze of the problem along with its solution
(2) autonomous solution of problems, exercises of routine calculation
(3) in case of difficulties, go back to the point (1)
(4) solution of different type of evaluation versions.

The electronic versions of these textbooks are designed to distance education and also to a much deeper study of the subject autonomously. The intensification of the learning process permits to add the study of mathematics using knowledge of programs and tools nature pragmatic possibilities.

References

1. Danilov, Y.M., Jurbenko, L.N., Nikonova, G.A., Nikonova, N.V., Nurieva, S.N.: Mathematika. Infra-M, 3-e izdanie, 2016.—496 c. Danilov, Y.M., Jurbenko, L.N., Nikonova, G.A., Nikonova, N.V., Nurieva, S.N.: Mathematics. In: Infra-M, 3rd edn., p. 496 (2016)
2. Jurbenko, L.N., Nikonova, G.A., Nikonova, N.V., Nurieva, S.N., Degtyareva, O.M.: Mathematika v primerakh i zadatchakh.—Infra-M, 3-e izdanie, 2017.—372c. Jurbenko, L.N., Nikonova, G.A., Nikonova, N.V., Nurieva, S.N., Degtyareva, O.M.: Mathematics in exercises and problems. In: Infra-M, 3rd edn., p. 372 (2017)

Two Revolutions in the Engineering Settings, Many Learning Opportunities

José Figueiredo[(✉)]

CEG-IST, Instituto Superior Técnico, Universidade de Lisboa, Lisbon, Portugal
jdf@tecnico.ulisboa.pt

Abstract. Engineering practices evolved from prehistory till now in a pace that sometimes went fast, others slowed down, according to the gradient of environmental change. Simplifying we can say that the evolution of engineering practices shows basically two revolutions. The first, centuries ago, was commanded by the absorption of exact sciences in the way engineering is practiced. The second, timidly in motion, is pulled by the absorption and internalization of social sciences. We repute that learning/teaching strategies should pace accordingly to the evolution of practices and we also stress that learning in general should be driven by practice. Collaborative practices and group work with "local" autonomy are the feasible support for learning strategies, and should be explored in a wide range of different settings. Internalizing the need for a new education paradigm in engineering is fundamental. Our research method is a mix of qualitative narratives and abductive reasoning in cultural contexts of engineering practice, through history, exploring conjecture. The results point out the lack of consensus on some concepts, definitions, and objectives, thus, based on reflective reasoning, some future directions are proposed. The novelty about this article is only the tentative internalization of new engineering practices, some of them already known, others less recognised, but all underrated, seldom assumed, and even less presumed.

Keywords: Active learning · Collaborative learning · Cooperation and teamwork · Translation · Complexity

1 Introduction

Engineering is about practice, about doing, designing and developing and using technological artefacts. Engineering practices were born with the primitive man creating instruments of war, hunting and sheltering. That phase of engineering practice was by far the longest one. After a certain time, man begun creating abstract models and symbolic reasoning and concepts of what we now call mathematics and physics. Although these concepts were born in antiquity they evolved, creating solid bodies of knowledge that we now use to address as exact sciences. These concepts are available as instruments and support, and engineering practice begun to absorb them and use them to sustain practice. This is what we call the *first engineering revolution*. Examples of this revolution are the creations of the Babylon's, Egyptians, Greek and Romans [1], but also all the kind of engineering produced till now, or at least till the last quarter of

M. E. Auer and T. Tsiatsos (Eds.): ICL 2018, AISC 917, pp. 871–883, 2019.
https://doi.org/10.1007/978-3-030-11935-5_83

the twentieth century [2, 3]. From the last quarter of the twentieth century on, we have been influenced by what we call the *second engineering revolution*. In fact, engineering nowadays has to absorb and integrate concepts from the social sciences in order to respond at his best to the challenges of our modern life [4]. Our modern life has many surprising characteristics, mainly related to the increasing number of relations between things, volatility of contexts of work, (both indicators of a higher complexity) and also due to the fact that decision about what to do and how to do it tend to displace from an appointed leader to the team, from a stable plan to the emergence of action. We now live with plans in progress, a continually in motion draft. Complexity and the need for negotiating the course of action with stakeholders, as well as definition of requirements and other stuff lead to the need of socializing, bargaining and negotiate [5, 6] to develop new knowledge and understand the situational social relations and tensions that are always emerging. The engineer needs to understand the social, meaning the complex socio-economic environment of engineering design. Knowledge is both situational and predictive, as it is cumulative and the main support for the ongoing learning processes. So, there is a need to develop and endure social skills [7] in order to be able to practice in the desired way, in an effective manner, today an towards the future. Engineers, as actors in situated environments, in situated circumstances, need to develop different knowledge abilities, such as capacity to develop new knowledge, creativity, critical thinking and resilience. That is what we point out to be prepared to participate in the *second engineering revolution*. Engineering in full is our purpose, an engineering that practices over all kind of different concepts, soft and hard, qualitative and quantitative, using deduction, induction and abduction, and able to create solutions in a universe in motion, using flexible decision. We talk about an engineering able to explore the system and the changing environment, including the technology in its contexts of design and use, and also, whenever feasible, able to anticipate the consequences of technology use. We need an ethical mind able to produce sustainable design or, better, an effective design that generates sustainable values [8–10]. Just to be clear we think that soft sciences are often much more difficult in action (practice) than the hard ones [11, 12].

With this paper, we intend to construct a narrative proving that social sciences represent a primal influence in engineering design, in engineering practices, and consequently [2], in designing the learning models and school's approaches that should align with this practice reality. So, we pretend to appoint some directions for improvement in engineering education and engineering practices. We use different supports glued with narrative. This tentative approach looks adequate as theory results from observation, comparing incidents and different phenomena [13, 14]. We also explored conjecture as a "methodological" approach [15].

Concerning research approaches in an ontological and epistemological perspective, we can say that the gap between the view we explore in this paper and the usual way of seeing these things can be paralleled by the gap between the way Positivists (Comte) structure their actions and thoughts, their concern on measuring and quantifying, and the way Interpretivists (Husserl & Schutz) try to understand systems (systemic reality) and things, usually within situational qualitative reasoning. Interpretivists are generally grounded in phenomenology, particularly aligned with permanent changing environments and complexity.

Our intended results are assumedly discursive, we think that if we stress some evidences they can perhaps be understood and eventually internalized and, as a result, important attitudes towards engineering and engineering teaching/learning can be moulded. Our paper doesn't intend to be an ultra-robust final proof, but just an argument on the way.

This paper is structured in sections, this introduction, followed by the roots of the problem (the divide), then a description of things around what we call the *first engineering revolution*, then we move into what is necessary to prepare and consolidate what we refer as *second engineering revolution*, then we concentrate on adopting and stimulating the effects of this second revolution, after we describe two simple case studies, and finally we appoint some reflexions and state some conclusions.

2   The Divide and Expression of Needs

What was a philosopher in Hellenic times? He was an integrated system with different capabilities and competences, always ready to learn, he was knowledge in progress. The Greek philosopher reflected about the reason of things, constructed reasons and explanations for objective and subjective things, interpreted reality, but he was also many other things. He was a doctor, diagnosing and healing people, he was an engineer, developing technological artefacts, he was a jurist, preparing, commenting and applying laws. He was also a poet as he expressed and analysed emotions, he was above all a man in full, a man of great knowledge, knowledge constructed through deep reflection, and experiment, knowledge always in progress. His maturity was earned by observing, experimenting and testing, and he was always developing well-organized mental models.

But let us jump some centuries, from the splendour of the Hellenic world into the XVII century. The XVII century was the time where René Descartes attained his huge success. Philosopher and "modern" scientist, Descartes wrote some important books and established some very strong points about reason, and reasoning. Descartes pearls dominated occidental culture, namely scientific culture [6]. One of these pearls was the Discourse of Method. Descartes was someone very influent in occidental education, on the way how we think about things and how we learn. He built an architecture of knowledge that marked our (occidental) way of thinking. For example, the dualism between spirit and body, and the mechanist reasoning, that is, the need to divide and decompose in order to understand [16], these were pillars of Descartes philosophy, now a little outdated.

This Cartesianism divide, always present in the evolution of knowledge, and very present even today in science, is the basis for our reasoning in the two revolutions.

Cartesianism expressed a global philosophy, a philosophy of everything. Its starting point was to reject all authority and question everything. So far a very good working principle in science. Descartes claimed cogito, ergo sum, or I only exist if I'm able to think. Encouraged by his own success at the time, Descartes developed a rationalist philosophy that was the ground for part of the "modern" occidental thinking. A central idea in this philosophical approach was the dualism between mind and body. Mind and body being the two basic substances of reality. In Descartes view, each of these substances can exist separately, body as realized in inanimate objects and lower forms of life, mind as realized in abstract concepts and mathematical algorithms.

This divide, however, was not invented by Descartes. It is mainstream since the origin of times. Aristoteles and Plato already discussed these matters. Aristoteles, Plato's disciple, stated that it was an error, a Platonic error, to consider a duality between idea and form. Modern times and the development of neurosciences (mechanisms of mind) give reason to Aristoteles. Form cannot live independently of physical content, and it is not independent of perception, a translation from sensations [6]. And Aristotle was right, Aristotle's Holism stated that knowledge is derived from the understanding of the whole and not of the single parts [17].

Based on the work of neuroscientists [18, 19] and modern theory of systems [20–22], this divide, as strong as it was, became to blur and now we know that there are no pure rational approaches as well as there is no pure emotional reasoning. What we experience and use in our professional and individual life, the substrate in which we base our actions, is a blend of both approaches. At any activity of mind and body we use reason and emotion.

Apart from that, things are naturally complex and intricate. Boolean algebra, based on Aristoteles logic, was the basis for modern computers. Boolean algebra was born of philosophy (with some mathematics). All things blend, all things integrate and interact. In our divide, there is a need for conciliation [6]. We need to break the divide that separates things in parts.

In a survey in Australia, employers were concerned with the lack of social literacy and human competences in engineering graduates [23]. This skills and competences are important by themselves, but they also provide significant improvements in critical thinking and capacity to formulate important questions. In a compared study between engineering [24] and humanities students, altogether with Hudson [25] the conclusion was that engineering students were more convergent, focused in solutions, but less conceptual and weaker in conceptualization. Enduring engineering students with humanities and social sciences skills will provide stronger and more effective actors in the entire system, being able to conceptualize as well as deep down in focused objectives.

In a similar line of reasoning [26] recommended that, to explore a wider view and mature conscience, engineering students should explore subjects as diverse as Ethics, and industrial and technology history in their engineering curricula.

At that time, we could verify a certain consistency about these subjects. For example, [27] stated that in the engineering curricula 30% should be social sciences and humanities; The Accreditation Board for Engineering and Technology [28], a body responsible for the accreditation the different USA engineering courses, stated for a minimum of 12.5%; the similar body in Australia, Institution of Engineers Australia, set for 24%. As we see at these times there was an alignment on this subjects that pacifically promised a better future.

3 Engineering Basic Grounds and the First Engineering Revolution

Engineering practices were born with the primitive man, creating instruments of war, hunting and sheltering. First as hunter-gatherers and tribe lieders, after with company professionals and men of action, inventing and designing artefacts inscribed in different tools, embodying patterns of use, translated in a variety of technologies.

Considering the ill-defined structure of most engineering problems, we must recognize that the positivist paradigm ruling in engineering schools and some of the strategies adopted to teach, as the problem-solving approach, doesn't prove to be the most fruitful. It becomes more problematic when we observe the same limitative approach in research. Part of the research production is based on the deployment of hypothesis that needs to be validated, usually applying statistics. That kind of research is focused on validating the hypotheses. But, as [29] says, this isn't a sound prove, in fact, knowledge creation and discover of new realities cannot be attained this way, we can prove a hypothesis is wrong, we can never prove it is right.

But returning to the problem solving, why are engineers being given well defined problems with well-defined data sets and why are they trained to apply basically the fundaments (physics and mathematics) to solve this class of (closed) problems? Two errors in the same take! Physics and mathematics are crucial in the development of engineering models but they need to be paced by the needs, by the social value they need to provide.

In fact, the most important approach for engineering problems is much more the problematization, the way we define the problem and the negotiations and coalitions we need to arrange and construct in dealing with the problem itself and all the technical and social settings that envelops the problem. We need to study the problem within its network of stakeholders. To align complex sets of different factors and technologies within a same goal is only one of the topics in this concern [30, 31]. In a similar perspective Donald Schön [32] referred to problem setting and Michel Callon [30] referred to actor networks, detours, translations, alignments, and obligatory passage points.

4 Preparing for the Second Revolution

Engineering practice is becoming more and more the domain of integrated teams, carefully built and motivated, well managed and of course well prepared [33]. FET advisory Group [34] stated some of the strong needs to 2020 "technological innovation need to pay close attention to the social contexts in which they are to be placed".

The engineer as an individual used to be a lonely runner, a hero. Not anymore, nowadays the engineer is impelled to work within a team, not only because of the increasing specialization of action but also because action is taken in complex systems and complex environments, where change occurs and new problems emerge constantly, and that calls for different expertise's. So, due to this specialization-systemic facet of technologic artefacts, teamwork is essential [35]. The challenge is to act in teams in situated problems and to be able to formulate problems and resolve them using experience, knowledge, creativity and innovation.

Some decades ago, in the fifties in the United States, in a meeting of the best engineering schools, social science was considered to be mandatory in the engineering curricula. "In professional engineering practice the "new situation" often involves social and economic as well as technical elements, and these are not entirely separable". Grinter report [27] (page 79). Engineers alone are no longer viable, they need collaborative teams [36], and engineer's expertise is no longer enough, technology and social are intertwined [27] and the engineer needs to, on top of his technological background, develop a social mind [37]. We need engineering education oriented to this goal [38].

The formulation of the problem, the problematization is one of the most significant and noble moments of the engineering design act. Solutions are much more the result of negotiations then the strict result of a plan. Negotiating amid all stakeholders, producing good consequences to all of them [39] in a win-win basis, is the only basis that can ensure sustainability and quality. And we need to understand that we are always acting in a context of scarce resources, meaning work availability, financial availability, infrastructure reliability, and knowledge readiness [5]. Scope management has to do with all these variables in a systemic proportion [39] and is inhabited with technical and social factors and constrains.

Considering what we just stated why are engineering schools insisting in training future engineers in problem solving only? We risk having good solvers of wrong problems! This traditional and wrong approach is limiting the value of engineering practice [40]. To confirm what is said we can address the work of [41–43] as we could refer many others, but we can also consult statistics of project success. Worldwide these statistics [44–46] wherever they come from, they show an embarrassing level of failure, with high rates of projects cancelled and running under specifications (either in cost, time, or scope, usually with cumulative problems in all the referred dimensions) [47].

In fact, engineering, these days, has to absorb and integrate concepts from the social sciences in order to respond at its best to the challenges of our modern life [48]. Distributed power and boundary issues [49] that change quicker and quicker drives engineers to new ways of acting, demanding for new ways of preparing themselves. Engineers have a strong scientific background, so the phrase written by Star and Griesemer applies also to them: "Scientific work is heterogeneous, requiring many different actors and viewpoints. It also requires cooperation. The two create tension between different viewpoints and the need for generalizable findings" [50].

5 Adopting the Second Revolution, Two Simple Case Studies

The case of "Engineering Project Management", in an Electronics Master in Instituto Superior Tecnico, Universidade de Lisboa, Portugal

To develop social skills and a social awareness in engineering settings where technology tends to play an important role [51] we will need project oriented education practices, defining goal oriented tasks [7], cultivate self-sufficiency of the student (they should have a "space" to try and experiment by themselves) [52], if possible we should design an experimental educated competition just to feel that things are for real. We should also try to push some limits and constrains in order to feel the shape of ambiguity and extrapolate the potentiality of eventual risks. Risk analysis [53] and estimating risk impact, duration and costs (educated guessing) should be pushed as far possible, as they do represent major constrains/opportunities in the planning and development of projects. So, these aspects should be covered in educational initiatives, (and they rarely are).

In our course of engineering project management, that takes all a semester and is exercised in the first and second semester in IST, Universidade de Lisboa, we have students from electronics (mandatory), engineering and management of energy (optional), mechanics (optional), and bio medics (optional). This project management subject is a 4th or 5th year subject in a five-year engineering master in Electronics. It is good that the students are last year students, but sometimes it is not enough. Very often students don't have the maturity to appreciate subjects that, for them, are out of scope in their speciality, they are there to study technology!

We designed two training strategies that are particularly oriented to the described goal.

The first strategy is materialized organizing presenting/discussion sessions that involves groups and takes all the semester long. We distribute project management papers from top journals, like International Journal of Project Management, or Project Management Journal, to the groups of students (normally four or five students per group). Each group has a paper to present in an appointed and scheduled PowerPoint session. In these presentation sessions, there are other two groups that study the same paper and should question the presentation either in formal terms and in terms of content (was the presentation aligned with the paper? were there wrong translations?). This discussion should also extend to the value of the paper, and eventual critics to the authors. Iroups are invited (and it is a component of the evaluation) to explore a critical view of the paper presented. These debates, moderated by the professor, which can raise new questions, take about one hour and a half. During the semester, all groups present once, and discuss twice, covering altogether three different papers. So, every student in the course, which is typically a 150 students course, will study in deep three top articles, and less deeply all the others. This component helps to develop many important skills [54], and assures a kind of literature review in project management.

The second training component is the planning of a project, with real data and constrains that just happen (we force them to happen), or emerge [55]. Given the description of a project, groups assume reasonable settings and exerts their project management capabilities deciding on course of action, priorities, and estimations. All the exercise is focused in project management so the specific details of the engineering problem are only addressed in very general and light terms. To be evaluated the students present a report (succinct narrative of assumptions, decisions and actions taken), the correspondent MS Project file, and a PowerPoint file used to further explain their project options to a jury. The group has a fixed time to present the project and roughly the double to discuss it with the jury. The jury evaluates taking information from the discussion, from the presentation (what is said), from the MS file, and from the report. Students learn how to use MS Project by themselves, no classes on this topic. They all can download a copy of Microsoft Project to their own computer due to a IST partnership with Microsoft. This component exerts the ability to work autonomously [56] and learn with experiments, and in the course of the semester all students master MS Project just by using it within the project. To validate this assumption during the project discussion the jury asks any element of the group (or all of them) to explain options on the MS Project file. Learning MS Project by themselves and developing a solution for the (open) case are activities that demand the need for autonomy, a basic condition for learning unbounded.

The Case of "Project" in an Energy and Management Master (IST, UL, PT)

The extended importance of soft skills is so noticeable that some companies are bypassing academic qualifications altogether and using their own methods to assess a candidate. In 2015 the global accountancy firm Ernst & Young said that they were going to use their in-house assessment programme and numeracy tests. The CHRO of Ernst & Young explicitly said that "At EY we are modernising the workplace, challenging traditional thinking and ways of doing things. Transforming our recruitment process will open up opportunities for talented individuals regardless of their background and provide greater access to the profession" [57].

Our project, project-based learning and workout ways of formulating the problems to be solved, is intended to exercise relational skills and problem formulation skills, along with collaborative/competitive abilities and developing of an ethical mind. Beyond our main mindset is our experience on ethics, opening minds and tuning critical thinking.

In our classes, a company, with whom we negotiate a case and a company strategy, launches a need (always real), a general problem with technological and business choices, to which they are looking for an interesting solution. Our students, divided in heterogenic groups (preferably with students of different countries in a same group) will follow a step by step approach to a final solution that needs to be aligned with the company strategy. This step by step evolution can be defined as control-gates of a project. This control-gates usually pass by a technology/business review, then a business model, then a technological solution, and finally a systemic solution, mingling technology and business. Each step has a PowerPoint presentation by each group and an overall discussion in the class with questions and answers, as well as advices and coaching approaches. Each of these steps is evaluated by the company element (or elements) and the teacher. These discussions sometimes propose redoing, thinking things in completely different terms, enduring resilience, a skill that all engineers should also work on.

We stress that, with this activity, we facilitate and promote group working, soft-skills development, improve problem formulation, explore autonomous work (decisions) and the development of an ethical mind.

This project takes all a semester and is exercised in the first and second semester at IST, Universidade de Lisboa. At the beginning of the semester we meet with the students, we explain what is the purpose of the project, we arrange groups of students and we state a scheduling for the control-gates for all the semester, using about five sessions overall. In the first session, we play a kick-off meeting, the element of the company presents the company and the intended problem, and range of expected solutions. Then in each session groups present their work using Power Point and we all discuss options and quality of the work in progress.

In either case, we explore participatory methods [58], we coach students exploring co-opetition [59], meaning cooperation in competition (for the grade), we facilitate the reading of exogenous variables (their potential treats and/or opportunities) [2, 7]. We always encourage the pursuit of eventual solutions within their operational contexts [4], see the solution see the consequences, and whenever possible integrate stakeholders in the process [8]. We also explore the formulation of problems [31] and [32] as in dissociated part of the engineering act and part of a path to a solution [32]. We also explore the sense of risk and associated responsibility to any choice. Stakeholders are not passive actors in our plot, they change with time, they change entities, and positions, we need to be always negotiating with them. An all these factors are more or less simulated in our concerns, our problems always incentive an alert to these "disturbances" because they are not side-effects, they are part of the problem. Apart from these considerations we have a very high percentage of students that enjoyed the experience and considered it very challenging and very formative. In either case, we add no complaints and always felt students were enjoying course of action.

6 Conclusions

Our main conclusions imply some new roles on engineering education in order to create more flexible practitioners and more educated engineers. Our work looks relevant for the academic and practitioner community not because it is sound innovative, but because it can contribute to alert relevant actors in this community, define adequate contexts and situational settings, and above all align goals.

In preparing and through the second revolution the engineer grows wider, generates and accumulates more knowledge, expands its capabilities in transversal axis. Abilities to collaborate, work together, listen to the others, develop a sense of the world and social values, are now crucial topics to work on and internalize. Engineering education should address strategies to assure the necessary transitions. And internalization is not the same of "having heard", "I already saw it once", "we once talk about it", internalization is a multi-step approach that drives from your knowledge to your capacity to act.

It is mandatory to improve and enrich our knowledge base. Using a flexible roadbook comprehending different domains and forms (technical, social, law, norms, culture, common sense, ethics, and sustainability) always with a content on engineering, we explore how to make better decisions, how to better undertake problem formulation and how to solve under a project management approach, adjusting quality and ethics, and perceiving a sense of value (social, economic, sociotechnical, organizational, project), as well as analysing best practices in management. Acting in this controlled settings, engineers gain a different conscience of the engineering act, and the act itself transforms and expands. As action depends on negotiations among the different actors, and problems and constrains emerge continually, engineers gain a different sense of mission and readiness to action that positively affects the way they behave, design and decide their action. This approach to the engineering practice tends to strongly link the situated problem (framed) with the context, to understand technology as a sociotechnical thing, defining a network of different interests, where values are permanently negotiated and renegotiated, eventually aligned.

For example, from being guided by a stable plan of action, engineers now need to be attentive and ready for the emergence of action, anticipating and reacting to what drops from action in its core concern and also in its limits. That means engineers need to adapt to plans in progress, based on a continual and evolving draft, an activity like craft, as [60] would surely say. Engineers need being attentive to emergent details, reactions and opportunities. Complexity and the need for negotiating courses of action, as well as definition of requirements and other designing [61] stuff lead to the need of socializing [5, 6]. Socializing, by debating reflecting, internalizing and experimenting that's how knowledge is created. We believe that to internalize this knowledge we need a second (and third, and fourth) round of action-reflection, because learning is an elaborate process that happens in recursive cycles.

The knowledge produced is cumulative and contributes to the maturity of minds and quality of decisions and actions. This cumulative knowledge leads us to understand situational social relations and tensions that always exists, and emerges. The production of knowledge is both situational and predictive, as knowledge is cumulative and a support for the learning process.

We talk about an engineering able to explore the system and the changing environment, technology in its context of use and also, whenever feasible, able to anticipate the consequences of its use. We look for effective designs that generates sustainable value [8, 9].

And we intend to do it from the whole to the parts. From the whole to the parts, versus from the parts to the whole. These two paradigmatic approaches are so strong that they can facilitate or difficult things, namely the way we see and feel reality, roles and goals. From the parts to the whole is related with our atavist feelings, deeply founded in mechanistic reasoning. From the whole to the parts is more holistic, based on a systemic approach and has an advantage on the way links and connections among parts are percept.

In our view, from the whole to the parts is recommended to a significant number of situations. We look at the overall system and understand how parts are linked, we explore this links and connections, and only then we invest on the details, understanding or designing the parts. That is the way a system should be designed (or understood) as a whole, and parts are designed (understood) in their interrelationship with this whole.

In fact, as neuroscientists [18, 62] clearly described, the mechanistic approach stated by the Descartes legacy [19] is no longer the best approach to deal with complex systems. Engineering needs to enlarge its duties, the engineer needs to become more like a man/woman in full, more versatile and with and extended conscience of things.

Acknowledgements. This paper was financially supported by InnoEnergy as a part of Techers' Benefit project 2018 (Case 2).

References

1. Figueiredo, A.D.: On the historical nature of engineering practice. In: Williams, Bill, Figueiredo, José (eds.) Engineering Practice in a Global Context: Understanding the Technical and the Social. CRC Press, James Trevelyan (2013)
2. Kendra, J., Nigg, J.: Engineering and the social sciences: historical evolution of interdisciplinary approaches to hazard and disaster. J. Eng. Stud. Eng. Risk Disaster **6**(3), 134–158 (2014)
3. Rojter, J: The Role of Humanities and Social Sciences in Engineering Practice and Engineering Education. In: International Conference on Engineering Education and Research "Progress Through Partnership", pp. 1562–3580. Ostrava, ISSN (2004)
4. Picon, A.: Engineers and engineering history: problems and perspectives. Hist. Technol. **20** (4), 421–436 (2004)
5. Nonaka, I.: The Knowledge Creating Company, Harvard Business Review, pp. 2–9 (1991)
6. Nonaka, I., Takeuchi, H.: The Knowledge Creating Company, How Japanese Companies Create the Dynamics of Innovation. Oxford University Press, Oxford (1995)
7. Lappalainen, P.: Can and Should Social Competence be Taught to Engineers? Int. J. Eng. Pedagogy **1**(3), 13–19 (2011)
8. Dutta, E.A., Sengupta, I.: Engineering and sustainable environment. Int. J. Eng. Res. Gen. Sci. **2**(6) (2014)
9. Bocken, N.M.P., Short, S., Rana, P., Evans, S.: A value mapping tool for sustainable business modelling. Corp. Governance. **13**(5), 482–497 (2013)
10. Ihde, D.: The designer fallacy and technological imagination. In: Vermaas, P.E., Kroes, P., Light, A., Moore, S.A. (eds.) Philosophy and Design, pp. 51–59. Springer, New York (2008)
11. Berliner, David: Educational research: The hardest science of all. Educ. Res. **31**, 18–20 (2002)
12. Strobel, J., Morris, C.W., Weber, N., Dyehouse, M., Klingler, L., Pan, R.: Engineering as a caring and empathetic discipline: Conceptualizations and comparisons. Paper presented at the Research in Engineering Education Symposium, Madrid, Spain (2011)
13. Gilgun, J.F.: Grounded theory and other inductive research methods. In: Thyer, B. (eds) The Handbook of Social Work Research Methods, pp. 344–364. SAGE Publications, Inc. City: Thousand Oaks   Print (2001)
14. Berliner, D.C.: Educational research: the hardest science of all. Educ. Res. **31**(8), 18–20 (2002)
15. Fila, N.D., Hess, J., Hira, A., Joslyn, C.H., Tolbert, D., Hynes, M.M.: The people part of engineering: Engineering for, with, and as people. In: IEEE Frontiers in Education Conference (2014)
16. Giudice, F.: Eco-packaging development integrated design approaches. In: Handbook of Sustainable Engineering, Springer (2013)
17. Mele, Cristina, Pels, Jacqueline, Polese, Francesco: A brief review of systems theories and their managerial applications. Serv. Sci. **2**(1–2), 126–135 (2010). https://doi.org/10.1287/serv.2.1_2.126
18. Bechara, A., Damasio, H., Tranel, D., Damasio, A.R.: The Iowa Gambling Task and the somatic marker hypothesis: some questions and answers. Cogn. Sci. **9**(4), 159–162 (2005)
19. Damasio, A.: Descartes Error, Penguin Books (2005)
20. Checkland, P.: Systems theory and management thinking. Am. Behav. Sci. **38**(1), 75–91 (1994)
21. Capra, F.: The web of Life: A New Scientific Understanding of Living Systems, Anchor edition (1997)

22. Meadows, D.H.: Thinking in Systems, Chelsea Green Publishing (2008)
23. Beswick, D., Julian, J., Macmillan, C.: A national survey of engineering students and graduates. In: Williams, B. (Chairman) Review of the Discipline of Engineering, vol. 3, pp. 39–105. Australian Government Publishing Service, Canberra (1988)
24. Pan, R.: Engineering as a caring and empathetic discipline: Conceptualizations and comparisons. Paper presented at the Research in Engineering Education Symposium, Madrid, Spain, 3–7 Oct 2011
25. Hudson, L.: Human Beings: An Introduction to the Psychology of Human Experience. Cape, London (1975)
26. Ashby, E.: Technology and the Academics—An essay on Universities and the Scientific Revolution. Macmillan, London (1966)
27. Grinter, S.: Final Report of the Committee on Evaluation of Engineering Education, Journal of Engineering Education, September, p. 25–60 (1955)
28. ABET (Accreditation Board for Engineering and Technology): Criteria for accrediting engineering programs: effective for evaluations during the 2002–2003 Accreditation Cycle (2002)
29. Popper, K., Conjectures and Refutations: The Growth of Scientific Knowledge. (2nd Revised edn), Taylor & Francis Ltd, 2002 (1963)
30. Callon, M.: Some elements of a sociology of translation: domestication of the scallops and the fishermen of St Brieuc Bay. In: J. Law, Power, action and belief: a new sociology of knowledge? pp. 196–223, London, Routledge, (1986)
31. Argyris, C., Schon, D.: Organisational Learning: A theory of Action Perspective. Addison Wesley, Reading, Mass (1978)
32. Schön, D.A.: The Reflective Practitioner: How Professionals Think in Action. Basic Books, New York (1983)
33. Schmidt, J., Schmidt, L., Schmidt, P.: Besteams: A Curriculum for Engineering Student Team Training by Engineering Faculty, 2006, American Society for Engineering Education (2006)
34. FET: Future & Emerging Technologies Advisory Group (FETAG), European Comission (2018)
35. Riemer, M.J.: English and Communication Skills for the Global Engineer. Global J. of Eng. Educ. 6(1) (© 2002 UICEE Published in Australia) (2002)
36. Vincenti, W.G.: What Engineers Know and How They Know It: Analytical Studies from Aeronautical History. The Johns Hopkins University Press, Baltimore (1990)
37. Ohland, M.W., Sheppard, S.D, Lichtenstein, G., Eris, O., Chachra, D., Layton, R.A.: Persistence, engagement, and migration in engineering programs. J. Eng. Educ. (2008)
38. Hynes, M., Swenson, J.: The humanistic side of engineering: considering social science and humanities dimensions of engineering in education. J. Pre-Coll. Eng. Educ. Res. (J-PEER). 3 (2) (2013) (Article 4)
39. Cipolla, C.: Allegro ma non troppo, Il Mulino, Italy (1988)
40. Bransford, J., Preparing people for rapidly changing environments. J. Eng. Educ. (2007) (Editorial)
41. Ryan N.R.: Project Retrospectives: Evaluating Project Success, Failure, and Everything. MIS Quarterly Executive. 4(3) (2005)
42. Burger, S.: Industry 4.0 requires factories of learning, Creamer Media Engineering News. (2015) [Online]. Available at: http://www.engineeringnews.co.za/article/industry-40-requires-factories-of-learning–festo-2015-04-24/rep_id:4136. Accessed 12 Jan 2016
43. Prabhakar, G.P.: What is project success: a literature review. Int. J. Bus. Manage. 3 (9), 3–10. ISSN 1833-3850 Available from: http://eprints.uwe.ac.uk/14460 (2008)
44. PMI: A Guide to Project Management Body of Knowledge, 6th edn. (2017)

45. Standish Group: Chaos Report (2015)
46. KPMG: Driving business performance Project Management Survey (2017)
47. Williams, B., Figueiredo, J., Trevelyan, J. (eds.): Engineering Practice in a Global Context: Understanding the Technical and the Social Hardcover (2013)
48. Bromley, A.: Engineering Education: Designing an Adaptive System, Board on Engineering Education, National Research Council, ISBN: 0-309-52051-7 (1995)
49. Philipose, L., Geramany: Hard facts about soft skills in universities, University World News, 03 October 2010 **142** (2010)
50. Star, S.L.: In: Bowker, G.C., Timmermans, S., Clarke, A.E., Ellen, B (eds.) Boundary Objects and Beyond, Working with Leigh Star, Chapter 7, p. 171, MIT Press (2015)
51. Pertegal-Felices, M.L., Costa, J.L.C., Morenilla, A.J.: Personal and emotional skill profiles in the professional development of the computer engineer. Int. J. Eng. Educ. **26**(1), 218–226 (2010)
52. Kasher, A.: Ethical perspectives. J. Eur. Netw. **11**(1), 67–98. European Centre for Ethics, K. U. Leuven (2005)
53. de Carvalho, M.M., Junior, R.R.: Impact of risk management on project performance: the importance of soft skills. Int. J. Prod. Res. **53**(2), 321–340. https://doi.org/10.1080/00207543.2014.919423
54. Nguyen, D.Q.: The essential skills and attributes of an engineer: a comparative study of academics, industry personnel and engineering students. Glob. J. Eng. Educ. **2**(1) (1998)
55. Liebenberg, L., Mathews, E.H.: Integrating innovation skills in an introductory engineering design-build course. Int. J. Technol. Des. Educ. **22**(1), 93–113 (2012)
56. Christensen, S.H., Didier, C., Jamison, A., Meganck, M., Mitcham, C., Newberry, B.: Engineering Identities, Epistemologies and Values: Engineering Education and Practice in Context, 2015, Philosophy of Engineering and Technology 21, vol. 2. Springer (2015)
57. EY transforms its recruitment selection process for graduates, undergraduates and school leavers. Building a Better Working World, 03 Aug 2015
58. Rodriguez, A., Cavieres, E., Negrete, C.: Effect of participatory methodologies in academic performance. In: 4th International Conference on Education, Research and Innovation IPEDR vol. 81. IACSIT Press, Singapore (2014)
59. Nalebuff, Barry, Brandenburger, Adam: Co-opetition. Harper Collins Publishers, New York (1996)
60. Mintzberg, H., Joseph L.: Reflecting on the strategy process. Sloan Manage. Rev. Spring 1999. **40**(3), 21 (1987) (ABI/INFORM Global)
61. Bucciarelli, L. L., 1994, Designing Engineers, Inside Technology. W.E. Bijker, W.B. Carlson, and T.J. Pinch (eds.), MIT Press, Cambridge, MA
62. Goleman, D.: Vital Lies, Simple Truths: The Psychology of Self Deception, Simon & Schuster (1985)

Formation of Hybrid Educational Structures in Russian Engineering Education

Senashenko Vasiliy and Makarova Amina[✉]

Institute of Comparative Educational Policy, RUDN University, Moscow,
Russian Federation
vsenashenko@mail.ru, amina.somnium@gmail.com

Abstract. In this paper, a substantiation of the hybrid nature of the modernization of the Russian engineering education is given. The influence of integration preferences in the higher education reform on the formation of hybrid educational structures in engineering education of Russia is revealed. The definition of a hybrid system is given and the processes of hybridization of the education sphere are considered. Various aspects of educational hybridization are discussed. The peculiarities of hybrid transformations and their connection with educational reality are considered. The hybrid nature of the competence approach in Russian engineering education is emphasized.

Keywords: Hybrid educational systems · Hybridization of higher education ·
Hybrid transformations in education · Engineering education · Hybrid nature of
modernization of engineering education in russia · Knowledge approach in
engineering education · Competence approach in engineering education

1 Introduction

The particularities of the modern international life and various challenges of time generate new educational trends and initiate the renewal of educational traditions. Higher education is becoming increasingly internationalized. One of the key trends in enhanced international educational contacts is the hybridization of higher education systems [1]. The hybridization of education, which in the international context is a consequence of the globalization processes, represents the mutual penetration of educational ideas, concepts, theories and representations of various countries into the systems of traditional national education [2]. In essence, educational hybridization is one of the directions of education sector's globalization.

Increasingly, the question arises—how will the engineering education develop? Most believe that the current difficulties experienced by the higher education are not the errors in choosing a route, but only temporary difficulties. In addition, many have a misunderstanding of what is happening. There is still no clear answer to what constitutes advantages and disadvantages in a traditional education system. Which traditions should be continued, and from which we should refuse. Maybe it is time to remember that the system of engineering education should be an integral part of each country economical complex. In this context, it is necessary to develop the capabilities of future engineers for self-realization in international cooperation conditions.

M. E. Auer and T. Tsiatsos (Eds.): ICL 2018, AISC 917, pp. 884–894, 2019.
https://doi.org/10.1007/978-3-030-11935-5_84

Engineering education programs should be aimed at solving key technical problems, some of which were presented by the National Academy of Engineering earlier [3].

One of the basic reference points of pedagogical activity is innovation, which is aimed at increasing the effectiveness of educational activities. The formation of various models of the educational process, new mechanisms and forms of its organization does not cease. However, innovations do not always promote development they can also destroy the education system. That is why special attention should be paid to the directions and mechanisms for improving and further developing of engineering education.

A special place among the transformations of educational systems takes the hybridization of educational sphere and its educational institutions, which has not been given due attention for a long time and which represents one of the key tools for the educational systems modernization. Hybridization of educational sphere has a twofold focus. On one hand it is planned to develop education in the desired direction in order to ensure its further improvement and on the other hand there is a lack of hybrid transformations' uniqueness which leads to undesirable results. A difficult question arises: how to transform the educational system in such a way that the expected positive changes take place in it. How to make a monitoring of the hybrid elements penetration in traditional directions and types of educational activities in order to avoid unpleasant consequences?

Since the end of 1980-s, the ideas about new trends in engineering education were actively realized. They were considered as an essential condition for overcoming crisis phenomena. During the subsequent period, attempts were made to solve them, and create conditions for a constructive analysis of positive transformations in the educational space, as well as methods for their practical realization. However, by the end the first decade of the 21st century it became clear that the implementation of the hybrid educational model without proper analysis of its specific features is clearly insufficient to ensure the successful development of the key educational institutions.

The study and realization of effective ways which ensure the transition of hybrid engineering education system to sustainable functioning and further development should be aimed at identifying the level of commitment of the transforming educational system to its mission. Obviously, a hybrid system of education will be more sustainable if it directs its integrative efforts to overcome national isolation, while at the same time retaining the connection to national educational traditions.

At the same time, the educational process acquires a distinctly expressed hybrid orientation. A transformation of educational stereotypes takes place, there are significant changes in the organization of the educational process, the structure of subject field of engineering education is changing, but the possibility of taking into account social and cultural features of national educational landscape remains.

The purpose of this work is to describe the hybrid nature of the modernization of the engineering educational system in Russia, and to identify the influence of integration preferences in the reform of higher education on the hybrid educational structures formation in engineering education.

The mixing of educational systems, as the initial stage of the hybrid system formation, in conditions of value asymmetry and structural heterogeneity, generally presents the particular interest, since in the framework of the concept of social hybridization the processes of the mixing complex of social objects, one of which are educational systems, have not previously been investigated. Perhaps, this is the reason why there is still a lack of recognition of the influence of the above factors on the nature of hybrid educational systems formation.

At present, we can talk about the European educational hybrid: "The creation of the European higher education sphere is a fascinating project, likely to result in something far different from and in many respects more interesting than the "Anglo-Saxon" system with which it is supposed, erroneously, to be compatible" [4, p. 262].

2 Hybrid Educational Systems

The term "hybridization" is actively used in sociology, political science, needless to say in biology and zoology, as well as other natural sciences. Perhaps, therefore, this concept does not have a single definition and is perceived differently in various fields of science. The process of cultural hybridization was described by Jan Nederveen Pieterse as the adaptation by local cultures of global trends to "local" specifics, combining the homogeneity and heterogeneity of the cultural environment [5]. He also drew attention to the fact that cultural hybridization can be interpreted negatively as an inability to rethink creatively foreign culture and as a consequence new cultural elements take possession, resulting in the mechanical transfer of alien forms of organization.

The devastating consequences of educational reforms are usually caused by the structural and value differences of mixed educational systems. Their hybrid transformations, losing their creative origin, can lead to undesirable results. At the same time, it is necessary to distinguish the different types and methods of mixing education structures, and especially their different assessments in specific cultural and educational environments.

If it is a question of mixing the complex social systems, it is necessary to envisage the possibility to minimize the destructive consequences of hybrid transformations, based on the results of the best social and educational practices analysis. It is important to note that blending is only the initial stage of the educational hybrid formation. Hybrid training is accompanied by the formation of hybrid educational technologies. Therefore, educational hybridization means not only a mixture of the educational programs realization forms, but also the updating of didactic principles of the educational process organization. Furthermore the character of professor—student interactions changes. Due to the hybrid nature of educational reforms there are contradictions between new organizational forms of learning process and much deeper attitudes and values. There are discrepancies between the ways of connecting individual structural elements in the educational system, which is closely tied to the national educational context.

Any changes in the educational landscape are accompanied by the appearance of converted forms in educational reality. They do not arise instantaneously, but pass different stages of formation and characterized by varying level of completeness. Thus, the hybrid educational system generates predetermined conditions for converted forms emergence in educational reality. This is an immanent property of hybrid systems regardless to which surrounding reality they belong to a natural phenomenon or one of the social activities. In this case we can say that reforming a complex system by a hybrid algorithm is inevitably accompanied by the appearance of converted forms. They have the function of replacing the inherent properties of the system. A system which is characterized by a converted form constitutes its "true face". If the converted form generated by the environment objectively reflects the character of interrelations in it, then there is an opportunity to realize the creative potential which is embedded in it. In other cases the distortions of the converted form may occur—as a reflection of its inconsistency to the properties of the immersion environment.

At the same time, there are artificially created mechanisms for the emergence of converted forms in educational reality, among which are [6, 7]:

- the lack of necessary tools to achieve the set goals, solve emerging problems;
- the gap between educational reforms and the educational sphere realities and society needs;
- the lag of theoretical development from the pace of "implementation";
- the concept which has been implanted for years, "education—outside politics and it should be in the service domain";
- the haste of changing the educational model of a specialist;
- the coercive nature of the implementations of the Bologna process provisions;
- the complexity of the management object is not adequately taken into account, also disproportionate goals and methods of management;

It should be emphasized that the mechanisms listed above of the emergence of converted forms in educational reality are due to the hybrid nature of educational reform. Attention is drawn to the rather strong correlation between the causes of the converted forms appearance and the causes of the failures of educational reform, among which are [6]:

- the ill-considered use of badly studied foreign experience;
- the lack of information on national educational systems, about their advantages and disadvantages;
- the forced transition to the three-level structure of the basic educational program of the higher education;
- the accelerated adaptation of the education system to the market laws;
- the educational reform was not prepared and for a long time was considered as an intra-departmental activity not as a national social problem;
- the priorities of the educational reform were determined in isolation from the most important directions of the country development and renovation;

As for the sphere of higher engineering education the educational systems are called hybrid or mixed, because the elements of that mix are borrowed from different educational systems. These educational systems were formed in various cultural contexts and not only cultural, but also economic, political, and lifestyle differences in particular countries. Hybrid educational systems are systems that draw their origins from two or more sources, often initially very different from each other.

It is possible to construct hybrid educational systems continuum, since the hybrid nature of educational structures can be divided in terms of the specific components' weight of that mixture. On one side of the continuum there is an assimilation hybridism, where the original educational system dominates. On the opposite side there is a destabilizing hybridity with a large specific gravity of the mixture, which erodes (passively) or destabilizes (activates) the initial educational system.

The special interest is given to the problem of academic freedoms which is solved differently in each educational system. Academic freedoms can be considered at the level of separated disciplines and at the level of individual educational paths formation, which is related to the Anglo-Saxon educational model. For the Russian educational system, academic freedoms are considered at the level of profession choice. It is obvious that a certain approach to the academic freedoms problem is regulated by the goals and structural features of the educational system. This approach to the problem of academic freedoms in higher education makes it possible to preserve the integrity of engineering education.

3   Hybridization of Educational Sphere

The phenomenon of educational hybridization is defined as the process of mixing educational systems of different "social nature", as a result of which new properties are formed that allow formation under more favorable circumstances to become more stable to new conditions. On detailed examination it can be seen that hybridization is manifested in the borrowing of ideas, social institutions, management systems, etc. As a result, structures are formed which have new properties, different from their predecessors.

Hybridization is defined as the ways by which forms are separated from existing practices and recombined with new forms in new practices [8]. The processes of hybridization are characterized by spontaneity, unpredictability of development and growth, "fluidity" of the educational environment, which objectify the actions of subjects of the educational space. This interdisciplinary phenomenon is at the intersection of didactics, engineering pedagogy and sociology of education, but for its description, as a rule, natural science categories are used.

The formation of educational hybrids is associated with the modernization of the education system, represented by three vectors. The first vector of modernization is structural, connected with the updating of the structure of educational institutions with the aim of their rational functioning. The second vector is technological, which implies improving the technological equipment of the educational process. The third vector of modernization is informative, connected with the content of educational programs.

Educational hybridization is becoming an instrument for the reorganization of educational systems. In particular, structural hybridization is accompanied by the emergence of new educational practices, activates the competitive potential of educational systems. At the same time, the diversity of forms of international cooperation is accompanied by the emergence of new educational structures, projects, etc. Hybridity of educational systems is an instrument for interpreting and understanding the current changes in the system of engineering education. The principled position of hybridization is the problem of asymmetry and inequality of educational systems in the process of mixing, which requires a special study.

Educational hybridization means not only a mixture of forms, but also beliefs—educational values, the result of which is the change in the properties of all interacting educational systems. For example, as a result of complex structural transformations by mixing Russian and "Bologna" educational systems, a new educational system with inherent properties is formed.

The hybrid approach is of particular interest to the modeling of complex technical systems, which, in order to obtain the expected result, requires a high level of formalization of the problems under consideration, the development of a mathematical algorithm for their description. Both structural and functional models are used. Structural models consider the organization and mechanisms of functioning of the considered system, investigating the changes in its internal state under the influence of external factors, whereas functional models consider the whole system under investigation, abstracting from the features of its internal structure [9].

Formation of a hybrid educational space occurs with the participation of the surrounding (not only educational) environment in which the education system is immersed. A certain influence on the properties of hybrid educational structures is provided by the mentality of participants in educational activities. Therefore, few hybrids adapt to the new conditions of functioning. It is connected with the fact that the newly formed hybrid educational structures for their successful "acclimatization" are forced to make efforts to adapt the environment "for themselves" that is, as they see it fit. Such a difficult task often turns out to be unsolvable.

4 Technological and Structural Aspects of Hybridization

In education the phenomenon of hybridization takes place both at the structural and technological levels, at the level of using the educational technologies when it comes to transferring educational technologies from one educational environment to another, methodical equipping the formation of skills in a hybrid educational environment. The complexity of the practical aspects of social hybridization, including educational sphere, was noted especially: "The hybrid or the meeting of two media is a moment of truth and revelation from which new form is born" [10]. This statement can be fully attributed to the educational systems' mix.

In education, hybridization is also manifested at the level of structural changes in curricula. The changes that are introduced affect the educational process organization, but in fact they do not concern the content of the educational programs. However, structural changes in curricula without proper methodological and organizational support are one of the reasons for the sharp drop in the quality of education.

Attempts to switch to new principles of constructing engineering education, using a "hybrid context", to "force into" new curricula that correspond to the renewed structure of the basic educational programs of higher education, methodologically, methodically and organizationally reconciled for decades the content of higher education demonstrates its insolvency. Attention is drawn to a noticeable change in the vision of not only traditional, but also on-line learning, in which it is about creating courses that do not require a face-to-face meeting with the teacher. At the same time, the mixed form of the educational process' organization is becoming more popular. This is especially important in the implementation of engineering education programs. The scientific literature actively discusses the possibility of creating the "hybrid university campuses" instead of traditional ones, implementing mixed educational programs based on hybrid educational technologies. There is a variety of webinars and online forums which replace some students' sessions with teachers to the virtual forms of study, when students have more time to think about the subject before answering or making comments on the studying material.

In essence, this is a blended form of education equipped with modern information technologies. One of the problems is library and information facilities of the university in case of transition to hybrid educational technologies in the organization of educational process. The advantage of hybrid courses is that, unlike conventional online programs, their mastering is possible without complete loss of contact with the teacher. Hybrid models of the educational process can be less controversial among teachers than fully online courses. Inclusion of hybrid courses in educational programs will make it possible to use educational areas (the study room fund) more effectively. For example, the University of Central Florida offers about 100 hybrid courses that take half of the time in the classroom and half in online mode.

Supporters of hybrid courses believe that their main feature is improving the quality of student training. Training with a teacher is not always the best form because not all students well perceive traditional forms of study. Also, face-to-face training has long been not considered as a "gold standard" which must be followed to. However, engineering education has its own characteristics and therefore the "real" contacts between students and teacher do not lose their relevance.

By mastering hybrid courses students have more opportunities to turn in assignments in a variety of forms, which confirms the diversity of their engagement in the learning process.

Generally, the hybrid educational programs can consist of "pure" online courses that are studied in contact with the teacher and mixed courses, some of which are studied online, and the other part is studied in classroom with the direct teacher's participation. It is very important to think through and organize the educational material well in order to determine in advance which part is best offered for study online and which one is best done in the classroom with the teacher's participation.

What is the right proportion in which the combination of these possibilities should vary from course to course? The experience of many universities shows that the most attractive ratio can be 50/50. This proportion makes it possible to control the student's work and makes the form of mastering learning material more free, and also opens up additional opportunities for organizing students' independent work.

The question of which course type is best for achieving the educational program goals is decided separately in each specific case. At the same time, students should have the opportunity to study online courses in other universities.

Designing and teaching the hybrid courses can take place in various formats. Using the Internet, it is possible to introduce to students the main issues that will be addressed in the lesson in order to make the classroom work more effective; it is also possible to discuss the material that was previously presented on the website, which will allow for a more meaningful and in-depth discussion of the training material; it is possible to use the different kind of online classes, etc. The practice of many universities shows that courses that are partially studied with the teacher and partly using the various information technologies' capabilities can be most effective.

It should be also noted that online courses offer certain advantages in setting the teachers' workload. Compared to traditional courses the less labor-intensive possibility of their actualization opens up by gradually adding new material using new methods and tools of e-learning, adapting them to the available on-line teacher's knowledge.

The advantages of hybrid learning technologies are seen in increasing the quantity and quality of participants' interaction in the learning process. Hybrid training includes both traditional and new technologies. The learning process becomes less expensive, more effective and should not lead to a decrease in the level of students' knowledge.

Hybrid learning provides additional opportunities for active communication in the learning process, helps students prepare for traditional classes, participates in discussions at seminars, and adds new types of interactive learning activities. There is an opportunity to use the classroom time for active learning, pushing the learning content of the course into the online environment, allows you to freely access (at any convenient time) the course materials and develop them to each student at a convenient time in a convenient place and the most importantly—at their own pace.

5   Examples of Hybrid Transformations

The processes of globalization and the internationalization of education have a significant impact on the transformation of doctorate level education all over the world, and Russia is not an exception here. A particularly striking example of a hybrid educational construction is modern postgraduate study as a crossbreed between "Bologna prescriptions" and the traditional Russian model of training highly qualified scientific and pedagogical personnel. Due to the inherent hybrid transformations of uncertainty the final result, the question still remains open: "Is postgraduate study an educational program with a research component or a research program with an educational component"? It should be noted that the allocation of specialties corresponding to the priority areas of modernization and technological development of the Russian economy, an increase in financial support for graduate students of these specialties through the system of grants and scholarships of the President and the Government attest to the importance of advanced training of highly qualified personnel precisely in the field of engineering education.

The hybrid nature of the competence-based approach in engineering education requires a separate consideration. The key characteristic of competences, as is known, is their nature grounded in activity. Competences are the "living" constructions, which reflect the situational nature of the subject's activities. Moreover they are individualized and reflect the personality characteristics.

At present, the competence-based approach in Russian education is nothing more than a hybrid of "knowledge" and "activity" approaches. In practical implementation, this approach blurs the traditional forms of the organization of the educational process, offering nothing in exchange, except perhaps for the idea of modular construction of educational programs. But can such a substitute become a full-fledged equivalent of traditional educational relations?

There is the most striking example of the Bologna reforms in engineering education, which entailed the emergence of educational hybrids. They are the structural changes in the educational programs of engineering universities. The new structure of higher education, the emergence of educational programs for undergraduate and graduate programs, could not provide the level of professional training of graduators which is required by the labor market. The reason was a significant reduction in the terms of mastering the traditional educational program of higher education: instead of 5-6 years there were established 4 years of bachelor's studies with a simultaneous decrease in the number of "study in classrooms" and an increase in the amount of "independent work of students". Most of the universities were not prepared for this. And the master program for a long time could not compensate for the loss of engineering education's quality. The liberal algorithm of its formation deprive its effectiveness as an educational program of higher education's second level.

The increased diversity of educational programs implemented by universities in the specialties and areas of training at the same time is accompanied by a considerable scatter in the names and labor intensities of related educational disciplines in various specialties and areas of training, which adversely affects the quality of engineering education.

The phenomenon of hybridization in engineering education can be considered at the level of structural changes in curricula and educational programs. Considering the simultaneous calculation of the labor intensity of educational programs in credits—bachelor level—240 credits, master level—120 credits and the duration of their development in years of study (bachelor's qualification—4 years, master's qualification —2 years), the number of weeks and academic hours—the student's working week consists in 54 academic hours (40.5 astronomical hours) it turns the using of credits in some formality.

A similar situation with the simultaneous assessment of student's academic achievements on a 5-point scale and the ECTS system with alphabetic characters, currently accepted by Russian universities complicate the professor's work and, ultimately, become redundant. At the same time, maintaining the 5-point scale of assessments prevents the full-scale implementation of the ECTS system, its rating component.

Therefore, the system of Russian engineering education is still in a state of searching for hybrid constructions of educational programs that can optimize the organization of the educational process, which would ensure the required professional quality of engineering universities' graduates.

6 Features of Hybrid Transformations and Educational Reality

At the initial stage of the mixing of educational systems, surface elements hybridize, while deeper attitudes and values, the structural ensemble of the educational sphere, are closely tied to a specific national context. There is an effect of non-acceptance of hybrid transformations. Overcoming it requires colossal efforts, distributed over time.

Currently, the educational sphere is so deeply rooted in the notion of higher education in the framework of traditional stereotypes that a lot of professors in the universities are not able to take seriously the entire range of issues related to the hybridization of the educational structures of various educational systems; to those uncertainties that arise as a result of the "blind" use of hybrid algorithms for building an updated education system.

In analyzing educational systems of mixed types it is necessary to keep in mind the existence of objective prerequisites for the risks' emergence as a consequence of the unpredictability of educational hybridity. Destructive consequences are caused in many respects by the presence of structural and value differences of mixed educational systems. Their hybrid transformations can lead to unexpected results.

7   Conclusion

International experience in education does not exist in isolation from specific economic, political and social entities. International standard cannot be set here. It is possible to talk only about a balanced approach to the use of its hybrid nature. But this requires a clear formulation of the goals and directions of the educational strategy for transforming the system of engineering education, on the basis of which a sequence of organizational and tactical steps is determined. To achieve the desired results, it is necessary to develop an action plan that provides various options for the development of the education system and determine responses in each specific situation.

At the same time a scientific and theoretical basis of the educational hybridity is required, as one of the most complex social phenomena. Ultimately, it is necessary to answer the key challenges of the hybrid educational systems' formation. The Russian system of engineering education is planned to correspond to the level of academic freedoms accepted in the academic community and the norms of university's autonomy sufficient for its successful practical implementation. The new Russian system of engineering education should be more focused on solving national problems in the context of international projects and interactions.

Hybrid educational systems should be perceived as an objective reality and educational hybridity as a mechanism for improving the system of engineering education. And the main thing is that the hybrid educational systems are "grown" and not implanted—this is their characteristic feature.

References

1. Senashenko V.S., Makarova A.A.: Educational hybridization as a result of the globalization of educational sphere. In: International Congress Globalistics-2017. Section Globalisation and education. Electronic edition https://lomonosov-msu.ru/archive/Globalistics_2017/data/10146/uid162424_report.pdf (2017). Образовательная Гибридизация как Результат Глобализации Сферы Образования

2. Senashenko V.S., Makarova A.A.: Educational hybridization as an instrument of higher educational system's modernization. Alma Mater. №1, 11–15 (2017) Образовательная Гибридизация как Инструмент Модернизации Системы Высшего Образования
3. National Academy of Engineering. http://www.engineeringchallenges.org
4. Ash, M.: Bachelor of what, master of whom? The Humboldt Myth and historical transformations of higher education in German-speaking Europe and the US. Eur. J. Educ. **41**(2), 245–267 (2006)
5. Pieterse, J.N.: Globalization as hybridization. Int. Sociol. **9**(2), 49–51 (1994)
6. Sukharev, O.S.: Dysfunction of education and science of russia: trajectory of overcoming. Priorities of Russia. №1 (238), 2–17 (2014). Дисфункция Образования и Науки России: Траектория Преодоления. Приоритеты России
7. Grebnev L.S. The current round of the bologna process: russia and not only...(according to the works of V.I. Bidenko and N.A. Selezneva) Higher Education in Russia. №1, 5–18 (2018). Нынешний Раунд Болонского процесса: Россия, и не только ... (по Материалам В. И. Биденко и Н. А. Селезневой)
8. Rowe, W., Schelling, V.: Memory and Modernity: Popular Culture in Latin America. p. 231, Verso, London (1991)
9. Korolkova A.V.: Modern approaches to simulation of difficult systems. control systems, engineering systems: stability, stabilisation, ways and methods of research. In: Scientific-Practical Seminar of Young Scientists and Students. Yelets State Ivan Bunin University (2016). Современные Подходы к Моделированию Сложных Систем
10. McLuhan, M.: Understanding Media: The Extensions of Man. p. 67, McGraw Hill. London and New York, (1964)

Development of Critical Thinking and Reflection

Tiia Rüütmann(⊠)

Tallinn University of Technology, Tallinn, Estonia
tiia.ruutmann@ttu.ee

Abstract. The present paper gives the overview of the basic principles and strategies for development of students' critical thinking and reflection. Ideas of contemporary teaching and methodology for development of critical thinking and reflection presented in the article are implemented in teacher training programs at the Estonian Centre for Engineering Pedagogy, Tallinn University of Technology. Analysis of the qualitative research carried out among the students is introduced. Students evaluated strategies implemented in teaching with the aim of developing and supporting students' critical thinking, reflection and metacognition.

Keywords: Critical thinking · Active learning · Reflection · Methodology · Strategies · Metacognition

1 Introduction

Engineering has never mattered more. What engineers know, how they think and what they can do are critical resources in our society today.

According to the statement of Partnership for 21st Century Learning [1], international education leaders have committed to fusing the 4C's making up the learning and innovation skills in engineering education: Creativity and Innovation, Critical Thinking and Problem Solving, Communication, and Collaboration.

For critical thinking and problem solving, the framework of Partnership for 21st Century Learning emphasizes the need to [1]:

- Reason effectively, using various types of reasoning (inductive, deductive, etc.) as appropriate to the situation;
- Use systems thinking, analyzing how parts of a whole interact with each other to produce overall outcomes in complex systems;
- Make judgments and decisions, effectively analyzing and evaluating evidence, arguments, claims and beliefs; analyzing and evaluating major alternative points of view; synthesizing and making connections between information and arguments; interpreting information and drawing conclusions based on the best analysis; reflecting critically on learning experiences and processes;
- Solve problems, solving different kinds of non-familiar problems in both conventional and innovative ways; identifying and asking significant questions that clarify various points of view and lead to better solutions.

© Springer Nature Switzerland AG 2019
M. E. Auer and T. Tsiatsos (Eds.): ICL 2018, AISC 917, pp. 895–906, 2019.
https://doi.org/10.1007/978-3-030-11935-5_85

The main purpose of engineering education is to provide the learning required by students to become successful engineers or knowledge workers – technical expertise, critical thinking, social awareness and knowledge of innovation. The combined set of knowledge, skills and attitudes is essential for strengthening productivity, entrepreneurship and excellence in an environment being based on technologically complex and sustainable products, processes and systems. Thus the goal of engineering education should be educating students who are ready to engineer, in the real world, not only in classrooms, making them deeply knowledgeable of technical fundamentals.

These are important signposts, as the education world has long been averse to change—in many ways we are still preparing students for a world that no longer exists.

Contemporary engineering students must be able to set goals for their learning and choose the most effective ways for achieving those goals, independently, with intrinsic motivation regulating their cognitive activities from goal setting up to evaluation of results, choosing criteria for evaluation and classification. The main aims of education today are mastering competencies and reaching expertise, developing intellectual skills, learning to learn, and thinking reflectively about thinking. Reflection helps teachers to reach the level of students, to choose suitable teaching strategies and methodology, construct knowledge and evaluate learning activities, thus reaching metacognition.

The present paper gives the overview of the basic principles and methodology for development of critical thinking and reflection implemented in the teacher training programs at the Estonian Centre for Engineering Pedagogy, Tallinn University of Technology.

2   Critical Thinking and Reflection—Basis of Constructivism

According to Halpern [2] critical thinking is the use of those skills or strategies that increase the probability of a desirable outcome, it is used to describe thinking that is purposeful, reasoned, and goal directed. Halpern elaborates that "critical" part of critical thinking denotes an evaluation component [2]. When we think critically, we evaluate the outcomes for our thought processes—how good a decision is or how well a problem has to be solved [2]. Critical thinking is also called directed thinking as it focuses on a desired outcome [2]. Critical thinking also involves evaluation of the thinking process—the reasoning that went into the conclusion we've arrived at or the kinds of factors considered in making a final decision [3].

Different definitions for critical thinking are used: sometimes it is called analytical thinking, logical thinking, higher level thinking, creative thinking etc. Greek word "*kritike*" means evaluate, analyze, review, judge, debate. Thus critical thinking should be analytical, creative, constructive and reflexive.

Critical thinking is open, reflexive, evaluative higher level thinking helping students actively process new information and thus construct their own knowledge. Development of critical thinking skills will help students to acquire creative, self-evaluative and communication skills for building their own learning journey in lifelong learning. From pedagogical point of view, critical thinking is active, interactive, creative and reflexive cognition.

Critical thinking helps to find personal priorities, teaches the skills of individual responsibility, problem solving, self-control, openness to multiple ideas and solutions, communication, active listening, and helps to see connections.

Teaching reflection may be divided into three stages:

- Goal setting, creating learning environment and positive atmosphere;
- Didactical methods for supporting reflection, workshops, active learning, portfolio, case studies, PBL, creative and critical thinking, debates;
- Technological support, educational technology, innovation, communication, collaboration.

Construction of knowledge is based on reflection. Self-regulation, self-orientation and self-evaluation as interior processes are also based on reflection, and development of these processes is always very individual. Reflexive learning could help students to practice reflection in the study process. When our students control their cognitive learning process and understand the function of a teacher, they develop their metacognitive skills by taking responsibility for their own learning, by understanding what they already know and what they need to learn; by setting goals for their learning process and finding the most suitable ways for reaching these goals; by carrying out the designed action plan, compiling their own learning materials; regulating their study process and evaluating it; by analyzing the results and comparing these to their goals; by setting plans for further development—thus they develop abilities of self-regulation, self-evaluation and self-motivation.

3   Development of Reflection and Metacognition

Technical teachers and engineering educators should act as examples and demonstrators of a thinking process in the process of cognition and learning.

The reflection may be divided into primary reflection (reflection of a student) and secondary reflection (reflection of a teacher). Teacher should create conditions for development of student reflection, analyzing and understanding the development of students' metacognitive skills, including their self-development.

Criteria for determination of the level of students' reflective thinking may be divided as follows:

- *Activity approach*—according to the development of reflexive skills a cognitive experience is shaped for managing the cognitive learning process. The following metacognitive skills are needed on this level: evaluating one's knowledge (what I know and what I need to learn), setting learning goals; designing the plan for reaching the goals; organizing the research activities; evaluation critically the results; analyzing individual activities in the learning process (success and failure, what I have gained and learned, what to do differently next time etc.);
- *Dialogue approach* (generalized)—discussions, educational games, multiply possibilities of group work and decision making, experience of teamwork; questioning techniques—asking and answering questions; valuing and taking account of

different viewpoints; determining common ideas, active listening, reaching consensus; acquiring skills of open discussions, respecting each other;

- *Individual approach*—individual reflection, self-determination, finding the meaning of life, value appraisal, ethical norms; evaluation of intellectual possibilities and strengths; self-esteem, self-actualization, self-regulation.

Reflection is a universal mechanism for self-development and self-changing as a part of students' learning process and development.

Metacognition is developed on the basis of reflection supporting following abilities of teachers and students:

- Planning and designing the learning process, assignments and activities;
- Analysis and self-analysis;
- Evaluation and self-evaluation;
- Selection of learning strategies.

Though most teachers strive for making critical thinking one of the primary goal of their instruction, most of them do not realize that students must pass through the following 6 stages of development of learning critical thinking [4]: the Unreflective Thinker; the Challenged Thinker; the Beginning Thinker; the Practicing Thinker; the Advanced Thinker; the Accomplished Thinker. And of course students' intellectual development is of essential importance.

4 Development of Critical Thinking

Didactical strategies for development of critical thinking have been designed based on the three main stages: evocation, realization and reflection, supported by activities of students during the learning process and multiply of different ideas.

The Basic model for developing critical thinking and effective teaching (including planning and design) used at the Estonian Centre for Engineering Pedagogy is designed taking account of above listed didactical strategies, and consists of following phases:

- *Evocation and involvement*—goal setting, evoking interest, motivation, refreshing prerequisite knowledge, creating associations of real world implementation, systematization of knowledge, activating of students, demonstrations, videos etc.;
- *Learning and research*—teaching and learning new information, rethinking and creating connections, design, research, practical lessons and labs, testing, multiple possible solutions, implementing active learning structures (including reading and writing for development of critical thinking, group work, writing keywords, active reading, questioning, etc.) supported by e-learning materials (depending on the content the flipped classroom may be used);
- *Analysis and explanation*—analyzing the results, deep understanding, collaborative teamwork, discussions, organizing the material, graphical presentation of material, mind maps etc., questioning (using *what if?* questions), learning from mistakes, finding solutions, giving feedback, communication;

- *Development*—modelling and building prototypes, testing, improvement, imple-mentation the results of previous phases in new real life situations, creation, building connections between STEM and real life situations;
- *Evaluation and reflection*—processing, evaluation, analyzing, integrating, and reflecting, formulating goals for further independent learning, selection of learning strategies and methods for independent work, self-analysis, self-regulation, and analysis of acquisition of learning outcomes.

The aim of these didactical strategies of the described Basic model is to develop students' metacognitive skills and intellectual knowledge needed not only in the process of learning but also in everyday life; to teach students how to use specialty literature and knowledge information bases, find and implement new information, use it in real world situations, integrate it with the knowledge already acquired and analyzed, reproduce the knowledge and solve problems. It is important to know facts, concepts, theory—as all skills are based on acquired specialty knowledge, and competencies in turn are based on knowledge, skills and values (students have to know what, know how, and know why).

The Basic model assists teachers to design a logical and effective teaching process and support the development of critical thinking skills within contemporary teaching and learning, using strategies for development of critical thinking and reflection as essential tools.

5 Strategies for Development of Critical Thinking and Reflection

Inherent in any inductive teaching method and active learning method is that students actually have to do something, but the central aim of active learning is to give students possibilities to explore new concepts, solve problems, and reflect on their experiences, in order to improve their performance in the iterative cycle [9]. Thus a learning activity is appropriate if it supports students' analysis and reflective practice, the key component should be feedback and self-analysis.

Strategies for development of critical thinking from the pedagogical point of view teach students the following skills:

- Active reading skills, collection of information and analysis the quality of data;
- Analysis of learning situation in general, seeing the "big picture";
- Identifying and defining of a problem, clarifying the cause and effect, drawing logical conclusions;
- Designing individual opinion, finding alternatives and connections;
- Self-analysis, self-evaluation, feedback, and reflection.

Strategies for teaching critical thinking and reflection used at the ECEP are:

- *Collaborative group work*—for learning with deep understanding students must critically analyze their own views, make presentations, participate in brainstorming, analyze other students' views, work out the best solutions [3];

- *Muddy cards* [5]—finding the "muddiest" point of the lecture—students give feedback to analyze, increase learning retention and determine gaps in their comprehension—students are asked to reflect what they have learned—they write down the most unclear points of the lecture. The instructor may correct misconceptions by the next class or use the cards in seminars;
- *Peer instruction*—interactive teaching strategy promoted by Harvard Professor Eric Mazur [6, 7] it is a form of a flipped classroom and provides a structured way to guide student preparation, in-class active learning and rich feedback/analysis opportunities. The peer instruction cycle is implemented in following five steps:
  - A problem with multiple choice answers is displayed, students are given time to think over;
  - Students record their answers, using flash cards, rising hands, or electronic response system;
  - When there is a disagreement in the group, students are given a few minutes to convince their neighbors (peer instruction) of their choice. Students think individually and practice to express their views;
  - Students record their revised answers;
  - The instructor leads a discussion on ways of reasoning and arguing in relation to the problem.
- *Quizzes for reflection*—may be used in the beginning of a lecture to help students to recapitulate a previous lecture. This method was developed at Chalmers University of Technology for enhancing student active reflection [8]. It is related to peer-instruction and uses six multiple-choice questions that students discuss with peers at the beginning of the lecture. The instructor concludes by explaining why each choice is correct or incorrect [9];
- *Visual diagrams—Mind map, Fishbone diagram, Venn's diagram,* etc. These methods help to develop students' analytical and critical thinking abilities and reflection;
- *Student-led recitations (Ticking)*—appropriate for using in teaching problem solving, first used at Royal Institute of Technology (KTH) in Stockholm [9]. For weekly recitation sessions students are asked to work through a set of problems. At session students tick on the list which problems they are prepared to present, solve, explain, and lead the classroom discussion. Students need to tick for at least 75% of problems in order to past the course. Students spend time on task, are active, reflect on how to explain the methods and argue on problem-solving strategies;
- *Project-based learning*—experiential learning method, solving real world problems or situations ending with some kind of a real result—documentation, drawings, video report, model, prototype, written report etc.;
- *Simulations*—instructor's role is to explain rules, the situation and the role of students, to monitor, to help students to reflect on the experience, to lead a debriefing session;
- *Case studies*—giving real engineering experiences, case discussions develop independent thinking, decision-making skills, analytic and problem-solving skills through practice and reflection;

- *Journals and portfolios*—perfect methods for self-evaluation, self-analysis, reflection, develop critical thinking skills, reasoning skills, recording and analyzing the steps of an engineering design process;
- *Problem-based learning*—different methods may be used, for example IDEAL method (I-Identify the problem, D-Define and represent the problem, E-Explore possible strategies or solutions, A-Act on a selected strategy or solution, L-Look back and evaluate) or CDIO method (Conceive—Design—Implement—Operate);
- *Compilation of test questions* (Blooming questions)—students are asked to compile questions for a test/assignment/seminar based on the new material according to Bloom taxonomy levels (questions for the levels of remembering, understanding, implementation, analysis, evaluation and creation). Questions may be used for home assignments, tests, group works at seminars etc.;
- *INSERT method* (Interactive Noting System Effective Reading and Thinking)— founded by Vogan and Estes [10] developing critical thinking ability in reading scientific texts. This method visualizes the process of knowledge accumulation from known information to the new one. Student can mark text or information as following: "v"—already know, "+"—new information, "–"- had another idea, "?"— unclear, and needs to be clarified.

All active learning strategies for developing of critical thinking and reflection presented above have been implemented at the Estonian Centre for Engineering Pedagogy (ECEP) within the Basic model presented hereinbefore.

6   Research

A questionnaire consisting of 15 questions was designed for the research of the students' feedback. The questionnaire was derived from previous research studies according to the suggestions of TUT academic feedback program. The aim of the qualitative research was to evaluate the used methodology for developing students' critical thinking, reflection, and the quality of teaching in order to develop the curriculum.

In the research 36 participants (of total 45 students) answered to the questionnaire, 62% of them were male and 38% female students. Average age of the participants was 24 years.

For responses the Likert-type forced choice 6-point scale was chosen as the most widely used scale in survey research. The questions were presented as statements. A response scale used had scale of 6 points (from "0—absolutely do not agree" up to "5—fully agree").

All the students experienced all of the above described strategies. Students (N = 36) evaluated the methodology used in the teaching process, including a set of problems like: whether the used methodology supported the learning of the new material; whether the technology used developed student's skills of critical thinking; whether the used methodology supported the development of student's reflection, metacognition, analytical thinking, problem-solving skills, learning with deep understanding and comprehension, communication, finding connections, self-evaluation, self-regulation, feedback, self-control, responsibility, taking account of other opinions etc.

The strategies were evaluated relatively highly; the average score given by students to the used strategies being 4.52 (of the possible maximum 5) (see Fig. 1). Teaching strategies received the following scores:

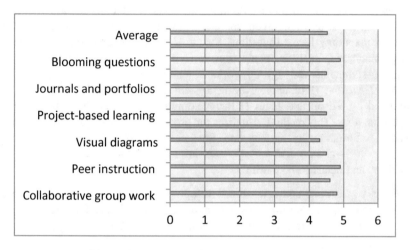

Fig. 1. Student evaluation of teaching strategies

- Collaborative group work (4.8);
- Muddy cards (4.6);
- Peer instruction (4.9);
- Quizzes for reflection (4.5);
- Visual diagrams (4.3);
- Ticking (5.0);
- Project-based learning (4.5);
- Case studies (4.4);
- Journals and portfolios (4.0);
- Problem-based learning (4.5);
- Blooming questions (4.9);
- INSERT (4.0).

Students wrote multiple comments on the implemented and analyzed teaching strategies (see Table 1).

The positive feedback of students motivates the ECEP to continue with the process of development of students' critical thinking, metacognition and reflection using strategies introduced above. Critical thinking is of high importance in the everyday work of engineers, it assists in decision making, reading scientific literature, logical thinking, solving of real life problems and in argumentation. But critical thinking is also needed while analyzing one's experiences and mistakes, in evaluating and reflecting.

Table 1. Students'comments on used teaching strategies

No	Students' comments
1	Collaborative group work helped more easily to understand difficult concepts and the process of decision making
2	I liked Muddy cards the most—sometimes it is not easy to ask questions in the large classroom, but when i wrote down the problems i did not understand, professor started with these problems on the next lecture and i understood everything
3	Please use more Peer instruction, i worked with high responsibility, i like flipped classroom
4	Quizzes for reflection was a good strategy, thus material passed in the last lecture was revised at the beginning of the next lecture and it helped to understand better and see connections
5	Visual diagrams helped to realize and analyze the problem, it became more clearer, i liked Venn's diagram the most, it was fun
6	Ticking is the best—i received bonus for my comprehension and it was good to solve problems in teams, holding discussions and arguing, it was fun and motivating, but still a hard work
7	Project-based learning has been used for a long time in engineering, but it still works, especially with real life tasks, i liked the video report presentations at the final seminar and the discussion
8	I liked the case studies most, at first we tried to solve the case, and then professor showed us how the problem was solved in a real life situation
9	At first i did not like the portfolio compilation, but afterwards i understood how it helped me to cope with self-analysis, reflection and metacognition—i started to think how to learn and work better in the future and learned from my mistakes
10	Problem-based learning—the problems were so interesting that we worked together for several hours, professor helped as a professional mentor and engineer
11	Using Blooming questions was motivating, especially because the professor used our questions at the tests and discussions
12	I have never understood the text so well before—using INSERT helped with supporting deeper understanding
13	I liked CDIO the best, it was like a real engineering process and we worked as a real team

7   Discussion and Further Developments

According to Robert Ennis [11] Critical thinking is a reasonable and reflective thinking focused on deciding what to believe or do.

Critical thinking assumes the understanding that everyone could make mistakes. Critical thinking assist in asking higher level questions (*what if? why?*), finding the pros and cons, in argumentation, self-evaluation, and self-critique.

The basic elements of critical thinking are: identification of a topic or a problem, finding connections, collecting information and critical analysis, situation analysis, logical conclusions, cause and effect analysis, combination, integration, modelling and defending one's views.

For teaching and developing critical thinking abilities, the following principles outlined by Ennis [11] have been taken account of at the ECEP in teaching engineering:

- Care that beliefs are true, and that decisions are justified; that is, care to "get it right" to the extent possible; including to [11]:
  - Seek alternative hypotheses, explanations, conclusions, plans, sources, etc.; and be open to them;
  - Consider seriously other points of view than their own;
  - Try to be well informed;
  - Endorse a position to the extent that, but only to the extent that, it is justified by the information that is available;
  - Use critical thinking abilities;
- Care to understand and present a position honestly and clearly, theirs as well as others' including to [11]:
  - Discover and listen to others' view and reasons;
  - Be clear about the intended meaning of what is said, written, or otherwise communicated, seeking as much precision as the situation requires;
  - Determine, and maintain focus on, the conclusion or question;
  - Seek and offer reasons;
  - Take into account the total situation;
  - Be reflectively aware of their own basic beliefs;
- Care about every person. Caring critical thinkers [11]:
  - Avoid intimidating or confusing others with their critical thinking prowess, taking into account others' feelings and level of understanding;
  - Are concerned about others' welfare.

In order to learn how to think critically, students have to focus on a problem or a question, analyze arguments, ask multiple clarifying questions, judge the credibility of a source analyzing every step, asking questions (*what if? why?*), being prepared for argumentation, feedback and other opinions, paying attention to self-control and respect.

How to help students to develop their critical thinking skills? The following set of recommended principles has been compiled for the teaching staff of the ECEP: the teaching material should be based on main concepts and generalizations; think about how to make students to think along; take account of students' preliminary knowledge; ask higher level questions (Why you think so? What may be alternatives? What if? What conclusions could be made? Why?); visualize students' argumentation; let students explain what they understood and what needs clarification; let them describe how the system and its elements function, are connected and influence each other; let them describe how they reached the solution; ask them to explain the concepts or problems in their own words; use peer instruction; analyze scientific articles; let them design a product according to given specification; ask them to check the solution and find mistakes; find multiple solutions; ask to integrate new details to a product; analyze the results of a research; analyze conclusions and find relevant data moving in reverse direction of a research; ask them to argue on superstitions related to your subject/topic;

write the list of criteria according to which the quality of a product may be evaluated; think aloud and ask your students also to think aloud; use 1- sentence summary method at the end of a lesson; be consistent and patient—teaching and supporting critical thinking is a time consuming process and you will see the results step by step.

The main strategies for teaching critical thinking are:

- *Affective strategies*—developing of independent thinking—What do you think? How you reached this decision? What are the other opinions?
- *Cognitive strategies*—processes related to knowledge and thinking—Why? What? What is the problem? What we have to solve?
- *Analytical strategies*—analysis, higher level questioning, logical thinking and discussions—What if? What will happen next? What may be the reason? What may be the consequences?

According to Brent and Felder [12] in order to help the talented and motivated students become much better at creative and critical thinking than they were at the beginning of the course teachers should show them examples of the kind of thinking teachers have in mind themselves; ask students to complete tasks that require that kind of thinking in class and in assignments; give them feedback; and repeat.

Thinking is an active process. Thus inductive teaching and active learning structures may be used with the aim of teaching critical thinking. Critical thinking must always be reflective and evaluative. This is the reason why all active learning strategies will have to be concluded with conclusion, reflective analysis and self-evaluation. Reflection assists learning from experiences.

Development and teaching of critical thinking is more effective in theoretical than in practical classes. It is recommended to develop critical thinking through a social experiences, collaborative learning and communication, using scaffolding, supporting and encouraging thinking.

Further developments include the research of the instructors' opinions on the implemented strategies but also providing support for continuing to incorporate analyzed strategies in the teacher training courses. After the research at TTU the results will be analyzed and published informing about how these strategies could be implemented elsewhere or how they might otherwise be applied at other universities.

8 Conclusions

The aim of contemporary engineering education is to educate future engineers. They will have to solve future engineering problems. What to teach them so they could cope with their job in the future? It is a difficult question and we do not have the answers today—we know nothing about the future problems. We should teach our students how to solve today's problems, how to think critically, find correct and relevant information, evaluate their decisions and to continue learning through their whole life. Critical thinking is the central pile of the development and success.

The key person in teaching critical thinking is a teacher who should plan assignments that make students think critically, taking account of strategies of critical thinking and students' preliminary knowledge. Let them learn from their own mistakes, let them analyze these mistakes, and let them find a better way.

"There is always a way to do it better – find it!"—Thomas A. Edison.

References

1. P 21—Partnership for 21 Century Learning. http://www.p21.org/our-work/p21-framework. Accessed on 29 May 2018
2. Halpern, D.F.: Critical Thinking Across the Curriculum: A Brief E, Edition of Thought and Knowledge, 3rd edn. Lawrence Erlbaum, Mahwah, NJ (1997)
3. Kalman, C.S.: Successful science and engineering teaching. theoretical and learning perspectives. Innovation and Change in Professional Education. vol. 3, Springer (2008)
4. Elder, L., Paul. R.: Critical thinking development: a stage theory. The Foundation of Critical Thinking (2017). http://www.criticalthinking.org/pages/critical-thinking-development-a-stage-theory/483. Accessed on 30 May 2018
5. Mosteller, F.: The muddiest point in the lecture as a feedback device on teaching and learning. **3**, 10–21 (1989). Available at: http://bokcenter.harvard.edu/fs/html/icb.topic771890/mosteller.html. Accessed 28 May 2018
6. Mazur, E.: Peer Instruction: A user's manual. Prentice Hall, Upper Saddle River (1997)
7. Crouch, C.H., Mazur, E.: Peer instruction: ten years of experience and results. Am. J. Phys. **69**(9), 970–977 (2001)
8. Knutson Wedel M.: Activating deep approach to learning in large class through quizzes. In: Proceedings of the 7th International CDIO Conference, Copenhagen, Denmark (2011)
9. Crawley, E.F., Malmqvist, J., Östlund, S., Brodeur, D.R., Edström, K.: Rethinking Engineering Education. The CDIO Approach. 2 edn, Springer (2014)
10. Шахакимова М. Т. Influential factors of increasing the quality in ESP teaching // Молодой ученый.—2017.—№18.—С. 350–352.—URL. https://moluch.ru/archive/152/43068/. Accessed on 31 May 2018
11. Ennis, R.H.: The Nature of Critical Thinking: An Outline of Critical Thinking Dispositions and Abilities. (2011). http://faculty.education.illinois.edu/rhennis/documents/TheNatureof CriticalThinking_51711_000.pdf. Accessed on 25 May 2018
12. Brent, R., Felder, R.M.: Want your students to think creatively and critically? How about teaching them? Chem. Eng. Educ. **48** (2), 113–114 (2014). http://www4.ncsu.edu/unity/lockers/users/f/felder/public/Columns/CreativeCritical.pdf. Accessed on 31 May 2018)

The Entrepreneurial Student's Journey Through Engineering Education—A Customer Centric View

Mario Fallast[1]([⊠]) and Stefan Vorbach[2]

[1] Research and Technology House, Graz University of Technology, Graz, Austria
mario.fallast@tugraz.at
[2] Institute of General Management and Organisation, Graz University of Technology, Graz, Austria
stefan.vorbach@tugraz.at

Abstract. The goal of this paper is to present a new perspective on how to design entrepreneurial education in a wider sense. By applying the method of "customer journey mapping", which is widely used in the field of marketing, the perspective of the student ("customer") is put in the centre of attention. It intends to raise awareness for a more holistic and customer-centric view of the entrepreneurship-related "customer experience". The paper will support engineering educational institutions (colleges, universities) in actively designing curricula as well as surrounding activities and initiatives to foster entrepreneurial intention.

Keywords: Entrepreneurship education · Entrepreneurial intention · Experience design · Customer journey

1 Introduction

Educating future entrepreneurs is increasingly requested from Universities and higher education institutions in general by policy makers and society at large. The most successful ways to educate future entrepreneurs have not yet been clearly identified, nor have the best approaches been established for releasing the entrepreneurial potential of students who are free to choose the career path of an entrepreneur.

1.1 Situation Analysis, Context

Designing engineering education curricula based on the necessary theoretical knowledge and practical skills is a widely accepted approach. When talking about "entrepreneurship", mindset and attitudes play important roles. The overall influencing factors for an individual person's decision to start entrepreneurial activities or even becoming an entrepreneur have to be taken into account as well. Curricular activities and providing access to knowledge seem to be only a small share of the influencing factors towards (or against) this decision. The possible measures to foster

© Springer Nature Switzerland AG 2019
M. E. Auer and T. Tsiatsos (Eds.): ICL 2018, AISC 917, pp. 907–916, 2019.
https://doi.org/10.1007/978-3-030-11935-5_86

entrepreneurial intention and entrepreneurial self-efficacy are far more wide-ranging in scope than merely offering a set of university courses.

1.2 Goals of the Paper

The mechanisms leading to the decisions by students to start up in business themselves and become entrepreneurs have not yet been clarified—the goal of the paper is to contribute towards a better understanding of what really influences students during their time spent under the influence of an educational institution. In this context we present a new perspective on how to design entrepreneurial education in a wider sense. It aims to support engineering educational institutions (colleges, universities) in observing and understanding what happens in this process and consequently to enable them in actively designing curricula together with the surrounding activities and initiatives to foster entrepreneurial intention.

By applying the method of "customer journey mapping", which is widely used in the field of marketing, the perspective of the student ("customer") is put in the centre of attention. This raises awareness for a more holistic and customer-centric view on the entrepreneurship-related "customer experience". The importance of every single touchpoint, any trigger-event throughout the journey will be highlighted. The paper encourages critical questioning of the existing elements in education and invites practitioners to take a closer look at the effect of events and incidents together with basically any contact students have with a higher education institution.

2   Theoretical Background

Two areas of research, "customer journeys" and "entrepreneurial education" historically evolved from the two very different fields of marketing and education shall be summarized in the following to set the foundation for combining them:

2.1   Entrepreneurial Education

In the course a report produced for the European Commission [1] 20 case studies about innovative entrepreneurship education (EE) practices at European universities were collected and analysed: "When looking at the design, the objectives of entrepreneurship education are usually a combination of the development of theoretical knowledge and practical skills for entrepreneurial thinking and acting." It makes the further point that the academic literature summarizes the impacts of EE as being still unclear and not well understood. In another meta study with a total sample size of more than 35.000 individuals, after controlling for pre-education entrepreneurial intentions, the relationship between entrepreneurship education and post-education entrepreneurial intentions was found to be not significant [2]. Something more seems to be happening besides classical entrepreneurship education.

In order for people to believe in themselves it is necessary for them to experience sufficient success when using what they have learned [3]. One of the mechanisms that occupies a central role in the regulatory process between motivation and performance attainments works through the beliefs people have in their personal efficacy [3]. Entrepreneurial self-efficacy specifically refers to a belief in one's ability to successfully perform the various roles and tasks of entrepreneurship [4].

The results of a study [5] underscore the importance of entrepreneurial self-efficacy to be a key component in understanding entrepreneurship interest and actual career choice. Perceived self-efficacy influences the beliefs people have in their capabilities to activate the motivation, cognitive resources and the actions needed to exercise control over events in their lives [3].

The exact mechanisms involved in this are not clear yet:

Results from empirical studies on the effect of entrepreneurship education on the entrepreneurial self-efficacy and entrepreneurial intention of students show that the effects are mixed [2]. The impact of entrepreneurship education on entrepreneurial self-efficacy is still unclear [6]. Souitaris et al. [7] tested the effect of entrepreneurship programs on entrepreneurial attitudes in order to find out whether entrepreneurship education stimulates and reinforces the intention to start a business. They also summarize psychological literature, where intention proved to be the best predictor of planned behavior. Entrepreneurship is a typical example of such planned, intentional behavior which is described as being particularly the case when that behavior is rare, hard to observe, or involves unpredictable time lags. In summary, the study illustrated that entrepreneurship programs are a source of trigger-events, which inspire students (arouse emotions and change mindsets). Inspiration is the benefit that raises entrepreneurial attitudes and intentions. The study additionally suggests investigating what kinds of emotions are experienced after each of the various trigger-events during an entrepreneurship program [7].

We wish to put a focus in our research on these trigger-events. We assume that placing a certain importance on the detailed design of such trigger events will affect the results of entrepreneurial education. The effects of certain trigger-events and their accumulation are already known to be important in other fields of research, e.g. marketing or service design. These effect the overall impression on a customer—essential for the survival of organizations e.g. depending on the customer's decision in a purchase process.

Customer journey mapping (CJM) is seen as a useful and increasingly popular strategic management tool to understand an organization's customer experience. It is valued by both academics and practitioners for its usefulness in understanding organization customer experience [8].

2.2 The Customer (Experience) Journey/Touchpoints

Rosenbaum et al. [8] describe the approach of customer journey mapping as follows: It is a visual depiction of the sequence of events through which customers interact with a service organization during an entire [purchase] process. These "events" are typically named touchpoints and are often accompanied by emotional indicators. Touchpoints occur whenever a customer "touches" an organization, across multiple channels and at various points in time [9].

The literature, e.g. [9] extends the range of a customer journey to all activities and events related to the delivery of a service and highlights the perspective from which the process is seen in a customer journey: the customer's perspective. It is of importance and value to skillfully manage the entire journey and not only single touchpoints [10].

The customer journey is also described as one of the frameworks in the service design process, used to understand how customers behave across a journey, what they are feeling, and what their motivation and attitude are across that journey [9].

We see a direct analogy to e.g. institutions of higher education.

Students can be seen as the "customers". The institution of higher education can be seen as a kind of "service organization". It has the ability to consciously design curricular activities. Single touchpoints are designed e.g. by the staff, the responsibility for an entire journey made by the students lies at the university management level.

When the literature [11] describes a growing tendency for outsourcing elements in the service delivery process, thus leaving the customer engaged with several complementary service providers we can extend the analogy to extracurricular activities. The literature [12] states in summary that customer experience is often considered predominantly as an overall evaluation, which restricts the understanding of the key "moments of truth" between the customer and an organisation.

We see an analogy with curriculum design, where the lowest level of detail in designing is usually a description of the taught content and objective of a course. Often, no further consideration of the intended effect of a course and its specific elements is undertaken.

The set of contacts between customers and organizations—the customer journeys—are generally becoming more and more complex, which raises the need to focus on the overall customer experience.

Customers now interact with organisations through an immense number of touch points in multiple channels and media, resulting in more complex customer journeys: Omni-channel management has become the new norm and firms are confronted with accelerating media and channel fragmentation [13].

Meanwhile, creating strong customer experiences is now a leading management objective and is the one that received the highest number one rankings when executives were asked about their top priorities for the next 12 months [13] quoting a study by Accenture (2015; in cooperation with Forrester).

2.2.1 Touchpoints as Elements of a Journey

Touchpoints are the key elements of every customer journey. Their description differs according to the actual journey to be described, some examples from the literature shall be given. Although this is a complex and difficult endeavor, it is important to identify critical touch points ("moments of truth") [13].

According to [11], the following criteria are essential for a touchpoint:

- *it must be visible to the customer, meaning that if the customer does not encounter it in any way, it is not a touchpoint;*
- *it must be a discrete event that can be appointed in time; and*
- *it must involve communication or interaction between the customer and a service provider.*

Other literature [13] partly confirms these categories but also extends them by the addition of "social/external/independent" touchpoints, which are affecting and including the customer but cannot be managed by the organization directly. We include a description for the application at a university:

- brand-owned (designed and managed by the university itself or under its control)
- partner-owned (jointly designed, managed, or controlled by the university and one or more of its partners. e.g. academic incubator, partner university, …)
- customer-owned (owned by students; the university, its partners, or others do not influence or control)
- social/external/independent (recognizes the important roles of e.g. peers in the customer experience).

In an analysis of customer narratives of experiences with retailers, seven distinct elements of customer experience touch points were identified by Stein and Ramaseshan [12]:

- atmospheric (surrounding observed by the customer)
- technological (direct interaction with any form of technology)
- communicative, One-way (communication from the retailer to the customer, e.g. promotional and informative messages)
- process, (actions or steps customers need to take in order to achieve a particular outcome with a retailer)
- employee–customer interaction (direct and indirect interactions customers have with employees of the retailer)
- customer–customer interaction (direct and indirect interactions customers have with other customers)
- product interaction (direct or indirect interactions with the tangible or intangible product)

Besides the content of a touchpoint, the moment of time when it occurs in the overall journey might also be important [13].

2.2.2 Planned Versus Actual Journey

Even if a journey and its included touchpoints are planned very well, during the execution of the service the actual journey can differ significantly from the planned journey. When analyzing planned and actual journeys, this deviation plays an important role.

The planning of customer journeys is already known in the literature as "service blueprinting", which describes what "an organization plans for a customer.

Touchpoints and journeys represent what actually happens from the customer's point of view" [9].

Broken down on the level of touchpoints, some literature sources [11] suggest touchpoints as also to be denoted as "expected" when reflecting the corresponding touchpoint in the planned journey. By contrast, an "ad hoc" touchpoint denotes an unexpected touchpoint or deviation. A touchpoint with an unwanted outcome is termed a "failing touchpoint." Another type of deviation is when a touchpoint is absent in the journey; these are referred to as "missing touchpoints."

3   The Customer Journey

Customer journey mapping or customer experience mapping are valuable tools for understanding and designing the interactions between single persons and an organization.

As a result of our research we suggest the use of the customer journeys and customer experience mapping to analyze and map—and further eventually design—the journey of students during their time of being directly and indirectly influenced by an institution of higher education.

The literature research showed, that the necessary tools are not described very well in the academic literature. The analysis of customer experiences is widely used in marketing and support processes, and often performed within firms or by consultancy firms focusing on that field. There is very little literature available on how experiences for customers of higher education in general and entrepreneurial education specifically (could) look alike. Educational programs and their impact are described, but mostly focusing on the view from outside and not seeking to understand the actual "touch-points" themselves.

3.1   Step-Wise Approach to Customer Journeys

The literature, e.g. [11] generally suggests a five-phase approach to analyse customer journeys. Figure 1 describes these five phases as follows:

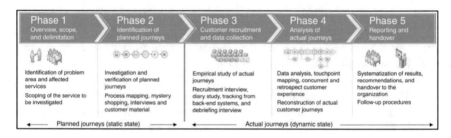

Fig. 1.   The five phases of the customer journey analysis according to [11]

In the case of Graz University of Technology (TUG)—and this seems to be the case in most institutions of higher education—there is no "planned journey" within entrepreneurship education existing.

As a first step we therefore performed phase 3 (empirical study of actual journeys) at the very beginning. We conducted a series of qualitative interviews at TUG to find out which experiences were the most influential ones. We used semi-structured interviews to find out which of these were remembered and recognized by entrepreneurs as being influential for their entrepreneurial intention.

It quickly became noticeable that some touchpoints, which had been thought to be crucial and supportive for students, in fact had very little impact ("failed touchpoints"), whereas others ("ad hoc" touchpoints), which were never intended by university management (but happened at university) had a major influence for the entrepreneurial intentions of students.

We also found that entrepreneurs often did not judge university courses in a traditional sense to be supportive towards their entrepreneurial intention—in some cases the courses were judged by university management to be of a high positive impact.

Other examples showed seemingly randomly occurring events—such as a supportive interaction with a professor—to be very influential.

One of the few examples targeting specifically the entrepreneurial intention is the course "Gruendungsgarage" (GG, "Founders' garage"), offered at Graz University of Technology (TUG). The GG course aims to offer an authentic entrepreneurship experience, as it guides start-ups from their first idea to a professional business that can stand on its own legs. The course exposes students to real world startup problems and opportunities and offers a model of how to effectively encourage entrepreneurship experience [14].

Another example offered at TUG is the two-semester course "product innovation project". The aim of the "product innovation project" is to expand the high level of the students' academic education towards a more practical and "market-like" view.

Interdisciplinary and international teams of students work on a task given by an industrial sponsor [15].

3.2 Sample Journey Through a Student's Entrepreneurial Education in a Wider Sense

We wanted to find out in which university-related activities in the field of entrepreneurship (curricular, extracurricular) students get in touch with and how they perceive these touchpoints and specific elements of them.

An empirical approach with qualitative interviews has thus been chosen to identify characteristics of entrepreneurship related touchpoints. We set up a list of descriptive elements which were mentioned in the literature to prepare the semi-structured interviews. These are intended on the one hand, to act as a guideline for the interviewer, and on the other hand as a "checklist" to cover all fields of interest. In the first interviews we carried out in the preparation phase we found out that interview partners tended to not mention certain elements by themselves. For example, there was usually no description given about the emotions involved, when describing a specific situation.

On the other hand, in the very first interviews the interviewed persons mentioned descriptive elements which were not prepared or mentioned in our interview guideline.

Figure 2 shows a first concept of a sample student's journey through entrepreneurial education in a wider sense. It includes university courses such as "product innovation project", but also touchpoints which are not managed by the university. Touchpoints perceived as supportive to entrepreneurial intention were placed above the horizontal centerline, touchpoints decreasing entrepreneurial intention placed below.

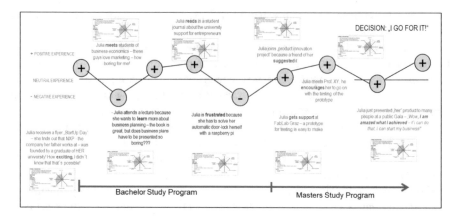

Fig. 2. A sample student's journey through entrepreneurial education

3.3   Touchpoint Elements and Description

A detailed description should finally be available for every single touchpoint encountered in the course of a journey finally.

We propose to start the description of a touchpoint with a verbal summary, e.g. using the answer given by the "customer" (in our case the student) during the interview.

The following descriptive elements in table form shall also be included

- Initiator (who initiated the touchpoint, who is in control)
- Time (in the curriculum, timing in the semester, day of the week, time of the day)
- Elements mentioned by Stein and Ramaseshan:
  - Emotions, Atmospheric, technological, communicative, One-way, process, employee–customer interaction customer–customer interaction, product interaction

To give a quick overview on the effects that emerged through the touchpoint, the following categories were summarized visually using a radar chart:

- theoretical knowledge
- motivation
- inspiration
- learning of new methods
- strengthening self-confidence
- application of methods

Figure 3 shows an example of a touchpoint description template, including a visualization.

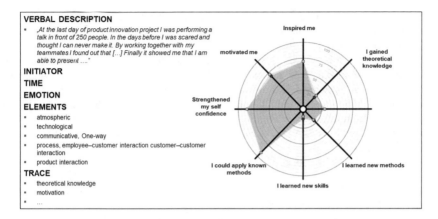

Fig. 3. Example of a touchpoint description template

4 Discussion and Outlook

Literature research shows that the tools to map and design an entrepreneurial student's journey are not described yet in the literature. Methodology to analyse and describe a customer journey exists in the field of marketing, where—compared to entrepreneurship education—relatively simple processes and clear goals (e.g. the purchase of a product) are in place. Multichannel-customer-journeys and ways to plan, analyse and describe them also already exist and can be used as a basis for our field of research.

There are still limitations to the capability of existing tools, however, because the journey of students through the educational system is highly complex, even when it is filtered down to entrepreneurship-relevant topics.

During the research and the development of sample touchpoint descriptions and journeys the need for further development of the tools became evident. Further research in this field is highly suggested.

References

1. Lilischkis, S., Volkmann, C., Gruenhagen, M., Bischoff, K., Halbfas, B.: Supporting the entrepreneurial potential of higher education, Final report (2015)
2. Bae, T.J., Qian, S., Miao, C., Fiet, J.O.: The relationship between entrepreneurship education and entrepreneurial intentions: a meta-analytic review. Entrep. Theory Pract. **38**(2), 217–254 (2014)
3. Wood, R., Bandura, A.: Social Cognitive Theory of Organizational Management University of New South Wales. Acad. Manag. Rev. **14**(3), 361–384 (1989)
4. Chen, S.C., Hsiao, H.C., Chang, J.C., Chou, C.M., Chen, C.P., Shen, C.H.: Can the entrepreneurship course improve the entrepreneurial intentions of students? Int. Entrep. Manag. J. **11**(3) (2015)
5. Wilson, F., Kickul, J., Marlino, D., Barbosa, S.D., Griffiths, M.D.: An analysis of the role of gender and self-efficacy in developing female entrepreneurial interest and behavior. J. Dev. Entrep. **14**(02), 105–119 (2009)

6. Shinnar, R.S., Hsu, D.K., Powell, B.C.: Self-efficacy, entrepreneurial intentions, and gender: assessing the impact of entrepreneurship education longitudinally. Int. J. Manag. Educ. **12** (3), 561–570 (2014)
7. Souitaris, V., Zerbinati, S., Al-Laham, A.: Do entrepreneurship programmes raise entrepreneurial intention of science and engineering students? The effect of learning, inspiration and resources. J. Bus. Ventur. **22**(4), 566–591 (2007)
8. Rosenbaum, M.S., Otalora, M.L., Ramírez, G.C.: How to create a realistic customer journey map. Bus. Horiz. **60**(1), 143–150 (2017)
9. Zomerdijk, L.G., Voss, C.A.: Service design for experience-centric services. J. Serv. Res. **13** (1), 67–82 (2010)
10. Rawson, A., Duncan, E., Jones, C.: The truth about customer experience. Harv. Bus. Rev. **91** (9) (2013)
11. Halvorsrud, R., Kvale, K., Følstad, A.: Improving service quality through customer journey analysis. J. Serv. Theory Pract. **26**(6), 840–867 (2016)
12. Stein, A., Ramaseshan, B.: Towards the identification of customer experience touch point elements. J. Retail. Consum. Serv. **30**, 8–19 (2016)
13. Lemon, K.N., Verhoef, P.C.: Understanding customer experience throughout the customer journey. J. Mark. **80**(6), 69–96 (2016)
14. Vorbach, S.: Lecturing Entrepreneurship at Graz University of Technology, pp. 54–61, September 2017
15. Reinikainen, M.T.T., Fallast, M.: A platform for innovative entrepreneurship and European innovation education, experiences from PDP, PIP and FLPD. In: Proceedings 36th European Society for Engineering Education SEFI Conference Quality Assessment, Employability and Innovation, November 2008

Work-in-Progress: Global Experiences for Engineering Programs

Claudio R. Brito[1], Melany M. Ciampi[2], Victor A. Barros[3,4(✉)],
Henrique D. Santos[5], Rosa M. Vasconcelos[6], and Luis A. Amaral[7]

[1] IEEE Education Society, COPEC – Science and Education Research
Organization, Santos, Brazil
drbrito@copec.eu

[2] IEEE Education Society, WCSEIT – World Council on Systems Engineering
and Information Technology, Braga, Portugal
drciampi@copec.eu

[3] COPEC – Science and Education Research Organization, Jataí, Brazil
victor@copec.eu

[4] Researcher of Information Systems Department,
University of Minho, Guimarães, Portugal

[5] IEEE Education Society, Information Systems Department,
University of Minho, Guimarães, Portugal
hsantos@dsi.uminho.pt

[6] IEEE Education Society, Council of Engineering College,
University of Minho, Guimarães, Portugal
rosa@det.uminho.pt

[7] CCG – Computer Graphics Centre, Information Systems Department,
University of Minho, Guimarães, Portugal
amaral@dsi.uminho.pt

Abstract. This paper has the goal to present and discusses the internationalization process of a university starting by the engineering programs that the engineering school offers. The objective is to make it more attractive and promote the double diploma in order to higher the quality of the programs.

Keywords: International cooperation · Double diploma · Strategic plans · Multiculturality

1 Introduction

When people have the chance to dive into a second culture, some interesting things may occur. It can increases overall openness to new experiences, that often leads to more creative ideas; it helps people to recognize that everything in the world can be viewed in many different ways, for instance, some cultures see mealtimes as opportunities to fuel the body, others see mealtimes as a time for social interaction [1].

The practical advantages are the diplomatic skills and sensitivity to different management styles as well as the capability to operate successfully in different settings and to adapt to a wide range of situations. Besides professionals with cross-cultural

© Springer Nature Switzerland AG 2019
M. E. Auer and T. Tsiatsos (Eds.): ICL 2018, AISC 917, pp. 917–922, 2019.
https://doi.org/10.1007/978-3-030-11935-5_87

experience can think critically and act logically to evaluate situations, solve problems, make decisions and are able to respond quickly to changing circumstances.

The key aspect of learning to adapt to a new culture is not that people have to decide which approach is better rather than recognize that everything in the world can be looked at in many different ways [2].

Seen by the current perspective this inter institutional international cooperation, in its essence, implies some fundamental conditions such as:

- to recognize the existence of people, protagonists of cooperation;
- all participants in the process should be involved and committed to the forms of cooperation, in accordance with the availability of available human and financial resources;
- the goals of cooperation have to be clearly defined and consistent with implementation strategies;
- projects should be included in the strategic development plans of the institutions optimizing the benefits and improving the levels of development of the partners; - the activities must be specifically established, respecting previously defined schedules and budgets;
- and finally, it is important the establishment of mechanisms directed to the development and evaluation of cooperation actions.

The study demonstrates that establishing internationalization strategies in an academic environment requires the observation of some characteristics that may limit the area of activity, such as: restrictions in terms of geographical location of partner universities; existence of language barriers; areas of excellence in teaching or research, as well as the level of development of the country where the institution is located [3].

2 Contemporary Education Demands

Many newly trained engineers do not always have the skills they need to succeed upon completion of the course. Several studies show that, on average, 40% of young graduates worldwide do not have the skills to make appropriate decisions outside of the university environment in which they have spent 4 or 5 years. The problem may arise from different environments:—On one hand, the highly structured environment of educational programs, and on the other, the changing, dynamic and challenging modern working environment. One way to improve this transition from the academic environment to the work environment is to modernize the students' curriculum. Nowadays there are harsh criticisms of the teaching methods that lead the professional to a long period of adaptation. However, there has to be a starting point that can start a new academic journey that allows students to acquire the necessary skills needed for a short adaptation to the job market [4]. Still there is no better place than universities to offer these opportunities, pushed by the enterprises collaboration offering internships opportunities.

3 The Project Boldness

To understand the relevance between school disciplines and how a career building journey takes place, it is helpful for young people to understand how work dynamics work. Engineering skills are internationally portable, leading to international mobility, which makes the profession extremely interesting. Intercultural skills, knowledge of languages and cultural prejudice management are very important also, because opportunities are extra boundaries and it is important to be able to adapt to any different cultural environment [5]. This is one of the proposals of COPEC's engineering education research team:—o embed a course with a more interesting activity for students, sooner, in the first year. It is a short-term workshop in order to show students the possibilities of performing as engineers in a global environment—a project developed for a private university in order to reduce retention rate among students of engineering courses. The main idea is during the first 3 months' period, in the second semester of the 1st. year, provide the students with different classes, which are more dynamic, due to the mix of site visits, lectures, project proposals, travel period and project presentation. It is a way to reduce the evasion of engineering courses, showing a glimpse of what it is to be an engineer and to see that there are wide varieties of opportunities worldwide [6].

4 Science and Education Research Organization (COPEC)

This is an organization of about 18 years of existence a multi-disciplinary organization that is a leader on advance science and its application to the development of technology serving society. The objectives of COPEC are to promote professionalism, integrity, competency, and education; foster research, improve practice and encourage collaboration in different fields of sciences. Contents, tools and services provided by COPEC, through courses, publications and consultations, with national and international experts, contribute to the promotion of the professional who wants to be privy of new achievements and service of men to technology. COPEC enjoys respect and recognition internationally characterized by the open discussion, the free exchange of ideas, respectful debate, and a commitment to rigorous inquiry. Its IIE—International Institute of Education—is a bold and resilient source of innovation in higher education [7].

5 The Project Design and Implementation

The starting point of this project development is the establishment of some actions in phases, in steps that leads to the 1 goal of internationalize the Engineering College Programs in a first moment, followed by other programs that the University offers.

The Phases are:

Phases	Description
1 - Conception	At this stage, it is established the estimates of achievement, dates, allocated resources, establishment of guidelines and elaboration of project outlines are discussed. The product of this stage is the project design
2 - Structuring	This stage involves all the technical part and activities necessary for the accomplishment of the project as: Hiring the operational team people; Basic schedule; Services to be contracted; Establishment of organization chart; among others. The product of this step is the entire action plan and timeline of final activities
3 – Execution	This stage consists of executing the project according to the action plan and timeline established between the stakeholders. The product of this step is the execution of all activities programmed in the project action plan
4 - Closure	This stage consists of the closing of the project and the preparation of the reports with information about the project, with the necessary documents and documents to render accounts with the institutions involved in the project. The product of this step is the final report with the detailed description of all the actions of the project

The Developed Activities are:

Terms	Activities
Short	- Establishment of an International Relations Office in Europe - Identification of partners to establish agreements in Europe and USA - Scheduling visits to partner universities
Mid	- Short-term courses outside the country for undergraduate, graduate and high school students - Agreements with universities in Europe, USA, and Canada - Implementation of top pioneer courses with unique characteristics[a]
Long	Courses offered with double diploma International Courses Exchange of students and teachers Receiving students from outside[b] Opening of the advanced campus in Europe

[a]The creation and offer Cyber Security Course on 3 different levels:
High school; University under graduation; Graduation; Short international courses for:
1. Specialization; 2. University Graduation; 3. Executive MBA and Specialization Course for Engineering Teachers - International Engineering Educator
[b]Focus on receiving students from CPLP—Community of Portuguese Language Countries component countries

About the agreements with other universities, they are the umbrella type with several possibilities such as student and teacher exchange, project development, double enrollment and others stipulated by the colleges involved. It is important to start in Europe for several reasons, as follows, being the main one the visa exemption for citizens, the second one Europe has a very high level of cultural life, depending on time of the year, depending on the season, the flying tickets have a more reasonable price.

It is important to go to Canadian universities and US as well. The perspective is to offer: Short courses—Workshops—Joint projects and courses. The last accomplishment will be the establishment of an advanced Campus of the University in Europe once it is a very important economic block besides the fact that there is the Portuguese Speaking Country—Portugal considered one of the most important colonizer countries in the world history [8].

6 Implications of Internationalization for Engineering Schools

Different circumstances will be prevalent in different countries and for different cultures. In all cases, COPEC's engineering education research team strongly recommends serious reflection and much preparation. By taking the courses abroad, universities are replicating their educational models to another set of circumstances, a different place and culture, with a different job market, demands, needs, and expectations. Becoming international or even global can be one of the most intelligent and profitable moves to improve the quality of the organization, attract new talent and increase its resources whether human or financial. Otherwise, institutions would be neglecting increasing quality of education, acquiring new knowledge and experience and, more importantly, more opportunities for research development [9].

7 Students Personal Achievements

Although for many universities the needs of international students are for monetary reasons too, however universities will bring longer-term ethical and intellectual considerations to bear on the profit motive. Internationalization is a project with an agenda with intellectual implications. It gives students and scholars the opportunity—and indeed the pressure—to view themselves and their cultures in new ways. Universities must invest in the economic agenda of internationalization being prepared to embrace its intellectual consequences; otherwise they can embark in a doomed project. And so for universities to fully internationalize themselves, it is clear that they must look to internationalize their fellow professionals, the teachers and staff first [10].

8 Final Considerations

For professional careers many employers states that no skill is more valuable than a foreign language. Many consider English as the "lingua franca". English as a medium of instruction is a prevalent way to offer international higher education as it continues to attract international students. English-taught degree programs continue to develop in many countries worldwide. With the multilingual environment come new challenges and responsibilities. Although the continued prominence of English is uncertain English continues to play an important role in international higher education as well as in work places [11]. People are considered literate when s/he speaks the mother tongue,

English and another foreign language. So the first positive aspect of an experience abroad is the learning of another language besides English. The internationalization project comes as an answer for the university that currently feels the need fostered by the incentives of governmental agencies as well as a way to attract more bright students and be at the top ranking. So far the organized actions that have been put in place are the agreements with universities abroad and a workshop in Europe for a group of students of the Civil Engineering Program.

References

1. Silva, A.F.C.: The diplomacy of professorships: German cultural foreign policy and higher education in São Paulo: the cases of the University of São Paulo and Escola Paulista de Medicina (1934–1942). História **32**(1), 401–431 (2013)
2. Perna, L.W., Orosz, K., Gopaul, B., Jumakulov, Z., Ashirbekov, A., Kishkentayeva, M.: Promoting human capital development: a typology of international scholarship programs in higher education. Educ. Res. **43**, 63–73 (2014)
3. Cabrera, Á., Le Renard, C.: Internationalization, higher education, and competitiveness. In: Ullberg, E. (ed.) New Perspectives on Internationalization and Competitiveness. Springer, Cham (2015)
4. BRITISH COUNCIL. Capacity Building & Internationalization for Higher Education. https://www.britishcouncil.org.br/atividades/educacao/internacionalizacao/capacitacao-ensino-superior
5. CAPES. Bolsas e Auxílios Internacionais. http://www.capes.gov.br/bolsas-e-auxilios-internacionais
6. Moskal, M.: International students path ways between open and closed borders: towards a multiscalar approach to educational mobility and labor market outcomes. Int. Migr. **55**(3), 126–138 (2017)
7. COPEC. Mission. http://copec.eu/menu/mission/
8. Daneva, M., Chong, W., Patrick, H.: What the job market wants from requirements engineers?: an empirical analysis of online job ads from then ether lands. In: Proceedings of the 11th ACM/IEEE International Symposium on Empirical Software Engineering and Measurement. IEEE Press (2017)
9. Brito, C.R., Ciampi, M.M., Vasconcelos, R.M., Amaral, L.A., Santos, H.D., Barros, V.A.: Engineering course specially designed to face retention issue. In: Proceedings of the 45th SEFI Annual Conference (2017)
10. Hudson, P.F., Hinman, S.E.: The integration of geography in a curriculum focused to internationalization: an interdisciplinary liberal arts perspective from the Netherlands. J. Geogr. High. Educ. **41**(4), 549–561 (2017)
11. Gould, R.R.: Why Internationalization matters in universities. http://theconversation.com/why-internationalisation-matters-in-universities-72533

Technical Teacher Training

Teachers' Professional Development for International Engineering Education in English

Tatiana Polyakova[✉]

Moscow Automobile and Road Construction
State Technical University, Moscow, Russia
kafedra101@mail.ru

Abstract. The global world determines internationalization as one of the trends of higher education development, including higher engineering education. In the sphere of international education there is a strong competition among universities. One of the approaches to attracting more students from other countries is offering study programs in English to help them to overcome the language barrier. Before introducing programs in English for international students, Moscow Automobile and Road Construction State Technical University (MADI) decided to design a special program aimed at training educators for teaching technical disciplines in English. The goal of the project is to design a new teachers' professional development program intended to improve their communicative competence in English. The paper describes the technology of the curriculum design, the educational objectives and the structure of the program, universal, general professional and professional competencies to be developed, the disciplines included in the curriculum.

Keywords: International education · Teachers' professional development · English training curriculum design

1 Context

The global world determines internationalization as one of the trends of higher education development, including higher engineering education. It transcends national borders and the number of engineering students studying at oversees campuses as part of study abroad or exchange programs is increasing. Academic mobility of students was stimulated by the Bologna process that stated it as one of its main action lines. Traditionally the biggest recruiters of international students are the USA, Great Britain, France, Australia, and Germany [1]. But engineering universities from other countries are also doing their best to attract more and more international students. The ability of a university to attract undergraduates and postgraduates is becoming the key to its success on the world stage and it is taken into account as one of the criteria to evaluate the global greatest universities. As a result, in the international education there is a strong competition among higher education institutions in the sphere of international education.

M. E. Auer and T. Tsiatsos (Eds.): ICL 2018, AISC 917, pp. 925–935, 2019.
https://doi.org/10.1007/978-3-030-11935-5_88

Universities pay special attention to the factors that influence decision-making of international students when they are choosing the country and an educational institution for studying oversees. As a rule they take into consideration the quality of study programs, tuition fee, access to learning environment, the language of instruction, the number of teachers and students, compatibility of diplomas, the opportunity to work part-time while studying, the accommodation, etc. But one of the main barriers to studying successfully in a foreign engineering university is the language [2].

Universities practice two approaches to solving this problem. The first one is connected with the introduction of various courses aimed at helping international students to master the native language of the country. The other one is offering study programs in English.

2   Goal

Moscow Automobile and Road Construction State Technical University (MADI) has significant experience in the sphere of international engineering education. The Preparatory Faculty for International Students was founded in 1960. Since that time it has trained more than 15000 students for secondary and higher professional education in Russian in the fields of engineering, technology, medicine, biology, and economics. Besides the main courses of the Russian language students study mathematics, physics, chemistry, geography, Russian culture, Russian literature and other subjects. After graduation from the Faculty international students start tuition in various universities all over Russia. Some of them enter MADI taking undergraduate and postgraduate programs. Being MADI students they have compulsory classes to improve their fluency in Russian. As one of the most popular field of studies is construction, the Department of Foreign languages compiled three-languages learners' terminology glossaries on Transportation Tunnels, Automobile Roads and Automobile Bridges for international students from Vietnam and China [3, 4]. In 2017 five dictionaries were compiled for Kirgizia students. Their publication was financed by the foundation "Russian World". But nevertheless students have difficulties while note-taking lectures, writing thesis, and taking exams in Russian.

In order to attract more international students and overcome language difficulties, like some other Russian Universities, MADI took the decision to offer short-long programs instructed in English. That means that MADI teaching staff should be ready for instruction in a foreign language.

The survey to reveal the English language competence level of the University teachers was undertaken in 2006. It showed that the teachers had different levels of foreign language competence, but 75% of them did not feel that they are competent enough to deliver lectures in English. These results have been confirmed recently (2017), when only 5% of the faculty members gave positive answers to be instructors of international students in English. At the same time, the survey of 2006 showed that the majority of the respondents are eager to improve their English.

The ability to use English as a means of international communication is also significant for teachers with the promotion of European dimension. It involves international cooperation in teaching, research and management. In the sphere of teaching this activity includes curriculum design, teachers exchange and inviting professors from other countries. The success of the cooperation depends on many factors but the ability to speak English is not the least.

So the first step in offering study programs in English is to train educators for teaching technical disciplines in English. The goal of the project is to design a new teachers' professional development program intended to improve their communicative competence of English.

3 Approach

The needs of technical teachers in English refer to the variety of foreign communicative needs of engineers. The needs of engineers in using foreign languages vary depending on many factors: the sphere of the language application (engineering activity, educational activity, translation, etc.), types of engineering activities implemented and the types of organizations they are employed at. This requires from the engineers professional communicative competence in a foreign language of various content and of different levels. Diversity of foreign language communicative needs with engineers is one of the principles explaining the necessity of foreign language life-long training diversification for engineers which can be viewed as one of the directions for its development and a condition for ensuring its continuity [5].

The Curriculum design can be carried out by means of the technology worked out within the concept of foreign language long-life training diversification [6]. The technology ensures the design and implementation of diversified compulsory and additional curricula and syllabi. The technology requires the following stages: determination of the sphere of the language application (1); determination of activity type in this sphere (2); defining the organization type where the language will be applied (3); communicative needs' analysis of learners (4); specification of the educational aims and objectives, selection of content and description of learning outcomes (5); concretization, modification or minimization of the aims and content of teaching (6); the description of the curriculum (7); selection and development of teaching materials (8); development of the whole documentation package (9); monitoring the learners' needs and correction of the curriculum (10).

According to Stage 1 educational process was defined as the sphere of the language application focusing on teaching. Stage 2 allowed identifying instruction, research, teaching materials development, communication with colleagues and international students in academic environment as the main types of activities in this sphere. A technical university with the Russian language environment was chosen at Stage 3. Stage 4 did not suppose any survey for identification of learners' needs. For that purpose the needs were defined on the basis of National Professional Standard "Pedagogue of professional training, professional education and additional professional education (2015) [7]. According to this document instruction includes lecturing, conducting seminars, organization of students self-study work, assessment of students outcomes (formative, progress and summative assessment), advising international students on study matters. Research implies advising and tutoring of international students research, assessment of projects, reviewing final qualification projects and research works of students, search for educational, scientific information for own research, and information necessary for teaching materials development, presenting the results of research in the form of papers and oral presentations. Teaching materials development in English requires curriculum design for three cycles of higher education and additional professional programs, teaching materials compiling, analysis and evaluation of teaching materials in English, including electronic materials in order to choose the most suitable for the teaching process, the development of all the documents necessary for realization of graduate and post graduate study programs in English, including additional professional programs. Communication with colleagues and international students in academic environment sets the objectives of participation in formal communication in the teaching/learning process and participation in informal communication with international students on the campus.

At Stage 5 the aim was formulated as the development of the English language communicative competence up to the level sufficient for instruction at the undergraduate and postgraduate levels. The objectives were described as communicative competences necessary to satisfy the needs selected at Stage 3. Learning outcomes were described as a projection of the teaching objectives. By the end of the program learners should acquire universal, general professional and professional competences necessary for realization of the technical program in English for international students. Universal competences do not depend on the content of the program to be realized in English. General professional competences are determined by the field of studies of learners. Professional competences are the abilities necessary for a special profile of the program in English within the field of studies.

The learners of the program designed should have the following universal competences (UC):

- The ability to participate in the oral form of intercultural interaction in order to provide cooperation in the teaching/learning process, to be tolerant to social, ethnic, confessional, cultural and other differences of international students from various countries (UC-1).
- The ability for written academic intercultural interaction (UC-2).
- The ability to produce correct, logical oral and written English texts with necessary argumentation (UC-3).

Besides the above mentioned universal competencies the learners of the program designed should have the following general professional competences (GPC):

- The ability to search and find paper and electronic sources of scientific, educational and methodic information in English, including textbooks and manuals (GPC-1).
- The ability to work simultaneously with several sources of information in English, to search for information, to store, to process and analyze it, and present it in the appropriate format (GPC-2).
- The ability to evaluate the quality and contents of information, to find most relevant facts and concepts, to interpret them and to express own opinion (GPC-3).
- The ability to compress written scientific, educational and methodic texts in English, including abstracting and annotating (GPC-4).

- The ability to present in public research results, to participate in the discussion on the own speciality (GPC-5).
- The ability to understand, analyze and evaluate oral texts (GPC-6).

The learners of the program designed should have the professional competences (PC) according to the objectives of the activities defined at Stage 2. They are instruction, research, teaching materials development, communication with colleagues and international students in academic environment.

Instruction in English requires the following professional competences:

- The ability to deliver a lecture on a particular technical subject in English (PC-1).
- The ability to conduct a seminar on a particular technical subject in English (PC-2).
- The ability to examine students in order to assess actual outcomes of the particular technical subject (PC-3).

For research in English the following professional competences are necessary:

- The ability to read, analyze students research work, including a final project, to review it and present the results of reviewing in writing (PC-4).
- The ability to write a scientific paper and its abstract in English (PC-5).
- The ability to make an oral presentation of research results (PC-6).
- The ability to organize and moderate a discussion among international students (PC-7).

Teaching materials development in English can be provided by the following professional competences:

- The ability to write an abstract of lectures (PC-8).
- The ability to design a syllabus of a particular course (PC-9).
- The ability to work out and keep the documents necessary for realization of the program in English (PC-10).
- The ability to work out tests in English for assessment of students actual outcomes (PC-11).

For communication with colleagues and international students in academic environment the learners should have the following professional competences:

- The ability to use formal academic English to maintain appropriate pedagogic relations with international learners (PC-12).
- The ability to use informal academic English according to the pedagogic ethics (PC-13).

Later, at Stage 6, the aims, objectives and content of teaching were modified where necessary. Stage 7 is very time-consuming as it implies defining the range of subjects, compiling the competence matrix, etc. At Stage 8 the set of materials "English for Academic Mobility" [8] was selected as the main teaching aid for a number of reasons. Firstly, teachers of technical universities are one of the target groups of this teaching set as it is addressed to students, lecturers and administrators of technical universities. Two components of the set, the main textbook and one of the practice books, are intended especially for them. Secondly, the teaching set is focused on teaching Global English and this is the language international students are expected to speak. Thirdly, the teaching set includes CD that contains audio- and video-materials and various tasks for self-study work. Audio-materials provide samples of pronunciation, monologues and dialogues for listening skills development. Video-materials give valuable cultural information and demonstrate a great variety of Global English accents of international students who are representatives of different countries. The tasks for self-study work give the opportunity of interaction. The lexical units and terminology refer to the teaching/learning process and engineering education.

4 Actual Outcomes

The Curriculum has been designed according to the technology described. It is titled "Methods of teaching technical disciplines in English". On successful completion of the course the learners will be awarded with the certificate with a qualification of higher engineering education teacher instructing in English. As a result, the program (26 credit points) comprises three blocks: Block 1—Compulsory Subjects, Block 2—Elective Subjects, Block 3—Final Achievement Assessment. The first Block (19 credit points) contains the disciplines "Oral academic communication", "Professional English", "Oral professional communication", "Methods and techniques of instruction in English", "Academic writing". The second Block (5 credit points) includes two disciplines "Introduction to technical literature translation" and "Introduction to sequential oral translation". The third Block (4 credit points) supposes a final project on the basis of learners' electronic portfolio and its defense (Table 1).

Table 1. The program "Methods of teaching technical disciplines in English"

Program components	Workload
Block 1. Compulsory disciplines/modules	
Oral academic communication	144 h /4 C.P.
Professional English	144 h /4 C.P.
Oral professional communication	72 h /2 C.P.
Methods and techniques of instruction in English	144 h /4 C.P.
Academic writing in English	180 h /5 C.P.
Scientific paper writing	72 h /2 C.P.
Teaching materials development	108 h /3 C.P.
Block 2. Elective disciplines/modules	
Introduction to technical literature translation	72 h /2 C.P.
Introduction to sequential oral translation	36 h /1 C.P.
Block 3. Final achievement assessment	
Writing a final project	108 h /3 C.P.
Preparing and defensing a final project	36 h /1 C.P.
Total workload of the program	936 h /26 C.P.

The module "Oral academic communication" is aimed at further development of foreign language communicative competence in formal and informal situations of educational activity (Table 2). It requires the skills of speaking in the forms of a monologue and dialogue and the skills of listening. The skills of listening should provide comprehension of oral utterances of representatives of various countries mainly those for whom English is not their mother-tongue. The content of this module is the basis for learning the modules "Oral professional communication", "Methods and techniques of instruction in English" and elective module "Introduction to sequential oral translation". Examination in the form of role play gives the opportunity to assess integrated habits and skills of oral communication in the situations simulating real academic communication.

The module "Professional English" trains learners to acquire technical English. It includes technical terminology, lexical and grammatical features of special scientific and technical discourse (Table 2). The module is based on the actual outcomes of the module "Oral academic communication". The content of this module is the foundation for learning the modules "Oral professional communication", "Methods and techniques of instruction in English", "Academic writing" and the elective module of Block 2 "Introduction to technical literature translation". At the examination the skills to use technical terminology, the ability to comprehend oral and written technical texts are assessed. At this stage learners begin to compile an electronic portfolio. As a result of this module they select sources of information necessary for the development of teaching materials for future course of studies they are planning to teach.

The module "Oral professional communication" is aimed at the development of communicative competence necessary for oral presentation and participation in the discussion (Table 2). The skills of speaking in the form of monologue, dialogue and polilogue as well as the skills of listening are improved. The content of this module is based on studying the modules "Oral academic communication" and "Professional English". In its turn the module is the foundation for the modules "Methods and techniques of instruction in English" and the elective module of Block 2 "Introduction to sequential oral translation". The role play "Conference" as an examination gives learners the opportunity to make presentations on technical issues and the English teacher can assess the learners' skills of moderating the conference, participation in a discussion, the ability to enquire information. The results of self-study work in the form of written text for oral presentation and visual aids in Power Point are included in the electronic portfolio of learners. Later they may be used as samples.

The module "Methods and techniques of instruction in English" is supposed to prepare learners for delivering lectures and conducting seminars (Table 2). It is based on the outcomes of learning the modules "Oral academic communication", "Professional English" and "Oral professional communication". The contents of the module is the foundation for the module "Academic writing in English" and preparing a final project and its defense. At the examination learners present a fragment of a lecture with visual aids in Power Point and answer the questions connected with the information of the lecture. They include the fragment of a lecture, a list of literature recommended to their future students, a list of examination questions and tests for international students' assessment.

Table 2. Matrix of competencies

Discipline	UC-1	UC-2	UC-3	GPC-1	GPC-2	GPC-3	GPC-4	GPC-5	GPC-6	PC-1	PC-2	PC-3	PC-4	PC-5	PC-6	PC-7	PC-8	PC-9	PC-10	PC-11	PC-12	PC-13
Oral academic communication	+		+																		+	+
Professional English				+	+	+	+															
Oral professional communication								+	+						+	+						
Academic writing in English																						
Methods and techniques of instruction in English										+	+	+								+		
Scientific paper writing														+								
Teaching materials development		+											+				+	+	+			
Introduction to technical literature translation		+			+		+			+				+			+	+	+	+		
Introduction to sequential oral translation	+																				+	+

The module "Academic writing in English" is aimed at training learners for written communication in the sphere of educational activity. It consists of two modules "Scientific paper writing" and "Teaching materials development" (Table 2).

The aim of the module "Scientific paper writing" is to develop the ability to write papers and abstracts to them in English for publishing them both in Russian and English journals. The module is based on the outcomes of studying the module "Professional English". The content of the module is the foundation for "Teaching materials development". For the assessment learners present their own paper and its abstract that later they also include in their electronic portfolio as a sample.

The module "Teaching materials development" is intended to develop foreign language habits and skills necessary for compiling all sorts of documents that provide realization of a program in English for international students (Table 2). It is based on the modules "Professional English", "Oral professional communication" and "Methods and techniques of instruction in English". The content of the module is essential for writing a final paper. For assessment of this module learners prepare a syllabus of the future subject and the text of lectures for international students.

There are two elective modules "Introduction to technical literature translation" and "Introduction to sequential oral translation". The first one gives learners the opportunity to develop basic translation competence of written scientific and technical texts that is desirable in the process of teaching, research and methodic work, when a teacher has to translate written sources of information. The module is based on the outcomes of the module "Professional English". It is a foundation for further improvement of translation competence and may be useful in writing a final project.

The elective module "Introduction to sequential oral translation" helps to develop initial habits and skills of oral translation. These skills may be necessary in the process of everyday communication between international students and University employees who do not speak English and in this case the teacher has to take the functions of an interpreter. The discipline is based on the modules "Oral academic communication", "Professional English", "Oral professional communication". In its turn it may be the foundation for further improvement of the competence in professional activities.

The final project is the main component of the assessment of learners' actual outcomes on the program "Methods of teaching technical disciplines in English". It is supposed to be written during the third and fourth terms. The work on the final project makes learners use and improve all the knowledge and competences acquired. It is recommended to apply electronic portfolio. The final project should contain the following components in English:

- annotation of the project work;
- syllabus of the discipline a learner is planning to teach;
- the text of a lecture;
- the text of oral presentation of this lecture fragment;
- visual aids in Power Point;
- a list of questions for this discipline examination and tests;

- one examination card as a sample;
- the review of a specialist in Russian and its translation into English prepared by the learner;
- references.

The main adviser for the final project is a teacher of English but at the same time there is a possibility to receive valuable information from the educator specializing in this discipline who reviews the final project. The text of the final project and the results of its review are the conditions of the admission to the final examination. The examination board consisting of three members assesses the universal, general professional and professional competences developed during the study program.

According to the procedure of the final examination learners make a presentation of a lecture fragment and answer the questions of the examiners concerning the contents of their presentation and the final project.

5 Conclusions

Instruction of international students at the undergraduate and postgraduate levels of engineering education in English is a way of helping them in overcoming the language barrier. The university technical discipline teachers who are not ready for instructing their courses in English need a special language program of teachers' professional development. The aim of the program is further development of teachers' English communicative competence necessary for instruction, research, teaching materials development, communication with colleagues and international students in academic environment. The Curriculum design of the program has been carried out by means of the technology worked out within the concept of foreign language long-life training diversification. In comparison with other programs the curriculum "Methods of teaching technical disciplines in English" of MADI defines leaners' needs on the basis of National Professional Standard "Pedagogue of professional training, professional education and additional professional education" and differentiates universal, general professional and professional communicative competencies. Besides, the Curriculum pays special attention to training learners for teaching materials development.

References

1. The Times higher education world university ranking 2012–13. Supplement 2012–13 in association with IDP, p. 66. Thomson Reuters
2. The formation of European Higher Education Area: the objectives of Russian higher school. State University of Higher School of Economics, Moscow, p. 524 (2004). Формирование общеевропейского пространства высшего образования: задачи для российской высшей школы. Москва, ГУВШЭ, с. 524 (2004)
3. Learners' Russian-Vietnamese-English terminology concise dictionary "Automobile roads". MADI, Moscow, p. 311 (2014). Учебный русско-вьетнамско-английский терминологический словарь-минимум "Автомобильные дороги". Москва, МАДИ, с. 311 (2014)

4. Learners' Russian-Chinese-English terminology concise dictionary Automobile roads. MADI, Moscow, p. 292 (2015). Учебный русско-китайско-английский терминологический словарь-минимум "Автомобильные дороги". Москва, МАДИ, с. 292 (2015)
5. Polyakova, T.Y.: Variety of engineers' needs in the foreign language usage as a basis for their training diversification. Soc. Behav. Sci. Procedia **214**, 86–94 (2015)
6. Polyakova, T.: Designing methods-and-curriculum basic for foreign language training courses in modern non-linguistics universities. Methods-and-curriculum basis of foreign language vocational training at a non-linguistics institution for higher professional education, Issue 14 (725), Moscow, FSFEI HPE MSLU, pp. 64–76 (2015). Полякова, Т.: Технология разработки программно-методического обеспечения курсов иностранного языка в условиях современного неязыкового вуза. Программно-методическое обеспечение профессионально ориентированной подготовки по иностранному языку в нелингвис-тическом вузе. Выпуск 14 (725), Москва, ФГБОУ ВПО МГЛУ, с. 64–76 (2015)
7. Degree of the Ministry of Labor and Social Protection.: On approving professional standard Pedagogue of professional training, professional education and additional professional education, № 608, 08 September 2015. Приказ Министерства труда и социальной защиты РФ от 08.09.2015, N 608 "Об утверждении профессионального стандарта "Педагог профессионального обучения, профессионального образования и дополнительного профессионального образования"
8. Polyakova, T.: English for academic mobility, p. 256. Publishing house "Academia", Moscow (2013). Английский язык для академической мобильности. Под ред. Т.Ю. Поляковой. Учебник. М.: Издательский центр "Академия", с. 256 (2013)

Some Problems of Advancement of Modern Information Technology in the Russian Engineering Education

Maxim Morshchilov, Larisa Petrova[✉], Vjatcheslav Prikhodko, and Sergey Abrakov

Moscow Automobile and Road Construction State Technical University (MADI), Moscow, Russia
mvmorshchilov@gmail.com, {petrova_madi, sergey_abrakov}@mail.ru, prikhodko@madi.ru

Abstract. New technologies in engineering education are becoming more and more relevant every year. Under the term "new technologies" in the educational process we understand primarily and mainly various kinds of on-line courses, electronic educational resources and other types of information resources. The application of all of the above mentioned tools works to improve the process of students learning and helps to increase their professional relevance on the labor market. In the paper the features of information technology application for the purposes of engineering education in Russia are considered. The review has been conducted on the basis of analysis of the main factors affecting the use of information technology in engineering education. These factors run a whole gamut of the student personalities of the Z-generation, to computer literacy of the teaching staff and the structure of the educational process in engineering education. The factors considered are comprehensively analyzed and the ways of interaction between teachers and students are proposed and on their basis. Google Classroom usage is considered, as a variant of the system of interaction between teachers and students.

Keywords: Engineering education · Educational resources · Online courses · Z-generation · Computer literacy

1 Introduction

Modern engineering education in Russia is facing the necessity to speed up IT technology introduction [1–3]. To meet the challenge teachers should be prepared to apply advanced methods of teaching process organization. And at that socio-psychological features of the new generation of students must be taken into account. Representatives of the so-called Z-generation cannot be imagined without gadgets whereas the teachers restrict their usage. Instead the gadgets could be used for information search and perception. The contradiction is that teachers use traditional methods of working with a student audience of Z-generation.

M. E. Auer and T. Tsiatsos (Eds.): ICL 2018, AISC 917, pp. 936–944, 2019.
https://doi.org/10.1007/978-3-030-11935-5_89

The approach suggests an identification of a contradictions between the requirements of the new generation of students and educational techniques in Russian technical universities. The purpose of the study is to analyze factors preventing active IT introduction in Russian technical universities and to find possible solutions of the problems caused by application of information technology. Practical goal is to test possibilities of advanced educational environment development using the Google Classroom service.

2 Background

2.1 Urgency of Creation of Universities' "Information-Educational Environment"

Contemporary educational and mass-media activities are characterized by increasing application of e-learning and the development of all kinds of on-line courses. There was even the suggestion by Y.I. Kuzminov, the rector of the National Research University "Higher School of Economics", to replace lecturers having no scientific publications by on-line courses [4]. Today there is a lot of available resources for mastering various on-line courses of different quality both in Russia and abroad. There is a wide choice of really high quality courses with perfect treatment of materials by professional lecturers with proper IT skills. But often the so-called DIYs develop their own courses to make their skills and experience publicly accessible.

On-line courses may be considered as independent educational units of Universities' "Information-Educational Environment" (IEE). Modern IT-technologies contribute not only to the development of IEE but also provide an access to these courses, university owned libraries, and to the lesson plans on disciplines. Besides, administrative activities may be organized via the IEE including workflow, students' progress reporting, and other forms of communication between participants of educational process. New forms of teacher-student communication by access to personal profiles will forger mutual understanding and professional competence of the teachers.

All of the above looks nice in black and white and rhetorically when discussed at different fora. The question is a rationality of various innovations, but truth is stranger than fiction. That is why our paper is an attempt to discover and analyze the cause of the contrast. Here are the questions to be considered—What are the modern students? Are teachers ready to teach them according to their needs? Does the existing model of educational process in technical university of the so called "mass higher education" [5] meet modern challenges?

2.2 Students of Z-Generation—Who Are They and How to Teach Them?

The term Z-generation is applied to children born at the end of 90th of the last century and later [5–7]. Therefore modern students who are about 18 just belong to this specific generation. Some experts [6] argue that in Russia the evolvement of Z-generation started at 2001–2003, and now they are 15–17 years old. Anyway the children of Z-generation are now in our universities or are going to enter any time soon. There is an expert view that Z-generation has the following features in terms of the learning process [6, 7]: they mature earlier, have a good command of information processing and assimilation, but suffer the so called "mosaic thinking". They habitually commence too many things but finish a few. But on the other side, they are focused on quick results; time-consuming and painstaking work brings them to loose dynamics. Often they are not quite prepared to absorb the knowledge in general and abstract concepts in particular. They are a common sight of our classrooms, indeed.

The following recommendations on methods and forms of such students teaching are given by the experts [6–8]. They can be introduced both by the teachers and university administration:

- To equip classrooms with modern gadgets and IT facilities, to provide free internet access;
- Do not restrict the usage of personal devices, but encourage their application in teaching process;
- To replace wherever possible printed learning materials by electronic books and manuals or at least by digital copies of textbooks;
- To teach difficult topics as presentations with vivid images and minimum text fragments, equations, etc.; for example, to present teaching materials as comics, sketches and iconographic elements;
- To focus more on specifics than abstraction, i.e. to give more examples from the real practice on a subject (case study);
- To organize tests with the usage of special software; ideally to exclude entirely "manual" work especially wherein similar operations are computerized in a business environment;
- To work out an individual approach to a student instead of a commonplace "leveling"; ideally to suggest personal tasks corresponding to individual preparedness to speed up the process of acquisition of specific knowledge;
- To create motivation and to promote competition among students;
- To provide for classes aimed at forging the team building spirit and shared decision-making skills.

3 Discussion I: Objective and Subjective Obstacles for IT Progress in Educational Process

3.1 The Problem of Computer Literacy of University Teachers

Not every teacher of technical university is able to follow the above mentioned recommendations or at least an essential part of them. Objective and subjective reasons of the fact are quite evident—age imbalance of teaching staff. Subjectively the teachers of the older generation are not quite prepared to use IT innovations due to their inadequate computer literacy.

According to the statistics [9] not more than 1/3rd of university teachers are under 40, and only 1/5th of them—under 30 according to the most optimistic estimation. Due to the human physiology synoptic contacts slow down with age [10]: the older a person becomes, the more difficult it is for him to master new skills. Of course, there are exceptions to this rule: this refers to the teachers who are motivated to the life-long learning. However, it is difficult to resist the physiology; degradation of computer literacy with age is confirmed by the results of surveys [11]. In the paper the level of computer literacy of teachers is evaluated as 7.2 according to the ten-point scale. But in the study [11] only day-to-day computer skill were assessed; such skills are not enough for modern IT content development. Microsoft Office or some other well-known standard software cannot provide for the implementation of the above mentioned tasks. If we analyze the teachers' skills in special software, evaluation indicator of computer literacy will go down sharply. As far as teachers' basic programming skills for educational soft writing are concerned, the situation seems to be close to the total "computer illiteracy".

It is to argue that teachers do nothing to improve computer literacy. But the approach to the problem is not consistent. Occasionally teachers can develop and introduce something they need to perform their professional functions but usually without any encouragement and recognition.

In our paper we tried to suggest a tool to solve the problem of better navigation of teachers in IT environment by application of simple and readily available IT recourses.

3.2   Technical Universities: Education or Socialization?

The other problem is the absence of mechanisms motivating the teacher to bring their disciplines up to date. It is not a secret that during the last 20–30 years student teaching process remains unchanged; on some subjects no changes were made for the last 40 years and more.

For example, to perform the tasks on "Engineering Graphics" students have to prepare paper drawings, and you can see them carrying special drawing tubes. And this is a fact in our high-tech age with a lot of software products to perform such tasks. How long ago did you watch a designer using drawing board and a ruler or an engineer making "hand" calculations?

Teachers who are committed to a "manual" way of doing exercise argue this reduces the chances of cheating. But we are sure that such problem may be minimized by suggesting of a number of task options.

Some of the Russian authoritative representatives of educational community refer to the national higher education as "mass higher education" and say that our universities do not give qualification being just institutions of social adjustment. The rector of Skolkovo Institute of Science and Technology "SkolTech" academician of the Russian Academy of Science A. Kuleshov has pointed out that our technical universities train engineers in nothing but name who do not even understand the basic principles of modern engineering. "They are still being taught The Performance of Materials – it is as good as to be taught to use a sliding ruler. As a result, they can do only primal operations assigned to the machines long ago" [5]. And he concludes that 90% of Russian technical universities perform the function of socialization rather than of education.

We are not going to interpret this comment as a call to remove The Performance of Materials from engineering educational programs. The point is: theory should always be backed with practice. And computer technologies make this practice a reality.

3.3   Conditions for Computer Literacy Breakthrough

The analysis of the problem gives the basis for recommendations to make up for IT technologies introduction in engineering education. These recommendations have been made both for teachers and universities' administration.

First of all it is necessary to have not just a separate IT operational body of a university itself, but set up the IT-teams at the departments and chairs. This will contribute to creation of qualitative educational content of specific subjects. Such teams may include young teachers mastering advanced computer knowledge skills and experienced teachers with deep knowledge in methodology of disciplines.

Secondly it is necessary to develop competitive environment among teachers by carrying out different contests to select the best teaching IT-materials. This should be accompanied by effective and transparent incentive system to encourage the development of new educational products. For example, university's financial assistance for teachers' qualification improvement may become an incentive taking into account the high quality programs cost.

Self-studying should become an essential requirement for the teachers to introduce IT-technologies into educational process. Such skills should be not only a professional advantage but a vital necessity. IT skills acquisition should be proven in action. Computer literacy in basic office software packages should be expanded to the study of specialized educational software and to the acquisition of programming skills.

The problem of better navigation of teachers in IT environment may be solved by application of simple and readily available IT recourses. Google Classroom service was selected as possible intermediary between gadget-dependent students and less advanced teachers. Testing of the service in educational process was carried out on "Technology of Structural Materials" subject during one semester with the following feedback analysis by teachers and students.

4   Discussion II: Google Classroom: Solution or Compromise?

Google Classroom is one of the numerous IT web-resources for educational purposes. It was developed by Google to make a process of task setting, distribution and arrangement straight forward and without any paperwork; it is free and accessible for the teachers. The main purpose of Google Classroom consists in streamlining of data exchange process between teachers and students. The service was created in May 2014 on the basis of G Suite platform, and it was intended specially for education. Public access was open on 12 August 2014. Till March 2017 the service was accessible only for educational institutions registered in the system; it was opened for wide range of users from April 2017 [12].

Google Classroom unites a lot of Google products focused on work with documents: text processors, tables, equations editors, presentations, etc. G-mail accounts are used for communication purposes and for users' identification. Google calendar can be applied for classes and tasks planning. Students are able to join the class via private code or to be imported automatically from an institutional domain if it has its own account in the system. Each group of students has a separate folder on the disc; a teacher places tasks, exercises and accompanying materials there; students place their completed work for evaluation. Figure 1 shows an example of attached teaching materials on the topic "Machining Parameters" of the discipline "Technology of Structural Materials". It contains tasks, pdf-files with text materials, pictures, videos, etc.

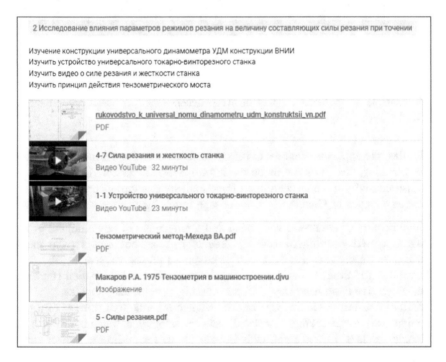

Fig. 1. Example of training materials on the topic "Machining parameters" of the lecture on "Technology of structural materials"

There is a function of different access modes to materials placed by teachers: view, individual downloading, and free access for teamwork. After the job evaluation students receive comments, questions, remarks, etc.

Mobile applications are available for devices running iOS and Android that makes possible for users to take and to attach pictures, to exchange files from other applications, and to receive information offline. Teachers can follow the progress of each student and analyze the statistics (Fig. 2).

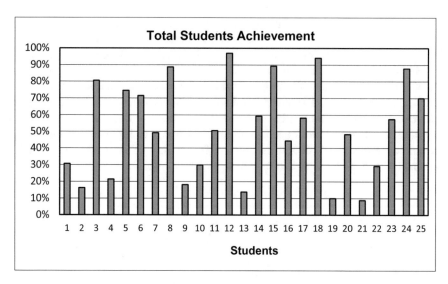

Fig. 2. Example of academic progress of every student (data per semester for the subject studied)

Besides, the application allows to inform students on some changes in a schedule, classes planning, etc. In online mode the information is accessible immediately. Students also can inform the teacher about force majeure circumstances.

The advantages of Google Classroom usage in teaching process are as follows:

- Simplification of courses development and arrangement of large amount of materials structured according to sections; students' tasks are not limited to the search of materials, but come to master it with an advanced purposeful search;
- Possibility to introduce the students to a large number of information sources and gradually give them necessary skill for working with primary sources;
- Possibility to test the knowledge of each student regardless of the size of a group;
- Saving time for checking completed tasks; possibility of integration with other Google services, for example, with Google Forms for on-line testing of students (Fig. 3);
- Possibility to differentiate educational process depending on students' levels: i.e. advanced students could be suggested more difficult tasks;
- The students have no way to pretend that information resources are unavailable: "The library is closed...", "There is no book in the library...", etc.;
- Significant costs saving on paper handouts;
- No need to equip and use a specialized computer class.

When the teacher uses the whole range of available tools (questions, setting time frames for an answer, evaluation system defining how prompt and correct is an answers) students do realize it as a lack of bias and this is a solid educational impact.

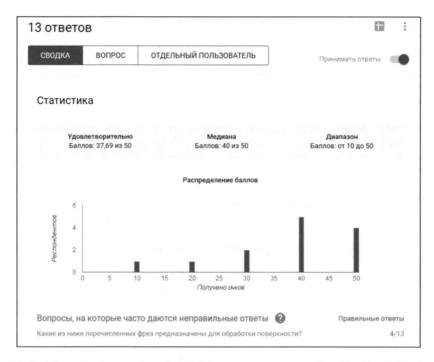

Fig. 3. Example of processing of statistics on answers to a test placed in Google Forms

5 Conclusions

Our study is an attempt to clarify if Google Classroom could become an effective tool of communication between teachers and students. Feedback analysis has shown positive response of students to Google Classroom application: they are ready to use it as information source, and they highly appreciate an unbiased assessment of their knowledge. This service gives them opportunity to use their favorite gadgets on a legal basis, and saves a lot of teacher's effort to interact with the audience in a proper way.

Recommendations have been made considering the feedback data from the teachers. IT teams at each chair may be useful to unite young teachers with advanced computer skills and senior teachers with high methodological experience. Teachers' advanced training on IT skills improvement should include special software and basic programming.

Google Classroom service may be successfully used to teach technical disciplines tailored to the needs of modern students. The service gives the opportunity to structure learning materials for easy search and provides an access to a number of information sources. Individual approach and differentiated on-line interaction with students, and minimization of printed materials usage are provided. The key advantage is that a teacher can communicate with students by means of a clear, familiar, and friendly tool.

References

1. Prikhodko, V., Soloviev, A.: What is modern engineering education? (Reflections of forum participants). High. Educ. Russ. **3**, 45–56 (2015)
2. Solovyev, A., Petrova, L., Prikhodko, V., Makarenko, E.: Quality of study programmes or quality of education. In: Interactive Collaborative Learning. ICL 2016. Advances in Intelligent Systems and Computing, vol. 544, pp. 362–366 (2017)
3. Prikhodko, V., Petrova, L., Soloviev, A.: A new format for implementing the tasks of the international integration of engineering education. High. Educ. Russ. **8–9**, 18–24 (2013)
4. Kalyukov, E. (2018). https://www.rbc.ru/society/26/02/2018/5a93f2889a79475483ae233e?from=main
5. Konstantinov, A.: Digital-born. Shroedinger's Cat **1-2**, 18–25 (2018)
6. Tretyakova, G.: Psychological features of Z-generation (2016). http://mansa-uroki.blogspot.ru/2016/04/z_12.html
7. http://kak-bog.ru/pokolenie-z-chto-eto-takoe-i-kakie-ih-harakternye-cherty
8. Tsulaya, O., Abrakov, S.: Foreign language loan words as the element of linguistic communication of students of the technical universities. Automobil. Road. Infrastruct. **1**, 19 (2017)
9. Pugach, V.: Age of teachers in Russian universities: what's the problem? High. Educ. Russ. **1**, 47–55 (2017)
10. Lesnjak, A.: The brain in which …. Shroedinger's Cat **12**(3940), 84–91 (2018)
11. Glushko, T.: Intrafirm training as a tool of increasing the level of computer training of a university teacher. Psychol. Pedagogical J. Gaudeamus **1**(15), 96–105 (2010)
12. Wikipedia - Classroom. https://en.wikipedia.org/wiki/Classroom

Training of New Formation Engineering Pedagogical Personnel to Implement the Industrial and Innovation Policy of Kazakhstan

Gulnara Zhetessova[1], Marat Ibatov[2], Galina Smirnova[3],
Svetlana Udartseva[4], Damira Jantassova[5(✉)], and Olga Shebalina[6]

[1] Vice-Rector on Strategic Development of Karaganda State Technical
University, Karaganda, Kazakhstan
zhetesova@mail.ru
[2] Rector of Karaganda State Technical University, Karaganda, Kazakhstan
imaratk@mail.ru
[3] Head of the Center of Engineering Pedagogy of Karaganda State Technical
University, Karaganda, Kazakhstan
smirnova_gm@mail.ru
[4] Head of the Educational-Methodical Department of Karaganda State Technical
University, Karaganda, Kazakhstan
s.udartseva@mail.ru
[5] Head of Department of Foreign Languages of Karaganda State Technical
University, Karaganda, Kazakhstan
damira.jantassova@gmail.com
[6] Center of Engineering Pedagogy, Karaganda State Technical University,
Karaganda, Kazakhstan
cep.kstu@mail.ru

Abstract. This article looks at an innovative approach to engineering and pedagogical personnel training based on the continuity of professional and pedagogical education, and its implementation, at Karaganda State Technical University (college—higher education institution—post-graduate education—professional development). Presented are the results of a project entitled "Innovative approach development to engineering and pedagogical personnel training for industrial and innovative development of Kazakhstan"; functional components of pedagogical system of engineering and pedagogical personnel continuous professional education based on competence approach by means of dual model implementation; engineer-teacher matrix of competencies. The latter is an auxiliary tool that combines the results of training with the disciplines in which these competencies will be formed.

Keywords: Engineer-teacher · Vocational training · Continuity and modularity principle

© Springer Nature Switzerland AG 2019
M. E. Auer and T. Tsiatsos (Eds.): ICL 2018, AISC 917, pp. 945–956, 2019.
https://doi.org/10.1007/978-3-030-11935-5_90

1   Context

Breakthrough directives of innovative development, determined by N. A. Nazarbayev, President of the Republic of Kazakhstan, in his Addresses to the People of Kazakhstan, focused on the socio-economic modernization of the nation, and are based on the priority development of leading industries and creation of economic clusters with the potential to increase the country's competitiveness. In the near future, realization of these directives will demand thousands of highly qualified workers and engineering—pedagogical personnel—specialists in new fields with a wide range of competencies and competitive skills that the vocational education system should train. The level of training of these specialists directly depends on the competence of vocational education teachers (Master of Industrial Training and teachers of special disciplines) including engineering teaching personnel. In Kazakhstan the training of such specialists is carried out within the framework of a "vocational training" specialty and also as part of postgraduate education. After completing a discipline-based master's degree program, the graduate can master an additional educational program in pedagogy providing the right to teach in a higher educational institution in the field in which he is trained.

2   Purpose or Goal

Pedagogical staff training in engineering does not have analogues in the practice of higher education, and its uniqueness is in the integration of such components as: sectoral, psycho-pedagogical training and training in the work profession. The sites of engineer-teacher activity are technical and vocational education colleges specializing in training of working staff and middle-level specialists for relevant branches of economy. The role performed by the engineer-teacher is a complex phenomenon in its structure and direction, which differs from the specialists of other professions activities since the components of technical, labor and pedagogical work are fully integrated.

It is professional and pedagogical personnel who provide an expanded reproduction of the main social capital, specifically, who are capable of creative self-determination and self-realization in their professional activities.

Under current new social, economic realities, when there is an avalanche-like growth of information—which is rapidly out of date—increased development of electronics and rapid change of production and information technologies, widespread computerization of production and education, there is a problem for introducing new approaches to teachers' training within vocational training, retraining and upgrading.

At Karaganda State Technical University, from 2007 to the present time, bachelors' training in vocations of educational programs (profiles) corresponding to prior areas of country's industrial and innovative development: machinery production, building; Information Technology; operation and repair of motor vehicles has been the norm. The main direction of training of future specialists' in both theory and practice is vocational training for teaching. The vocational training teacher should be competent, not only in pedagogical strategies, but also well-informed regarding the sectors of the national economy for which personnel are trained in a professional educational institution. The teacher should provide high-quality vocational training for students in a particular profession based on educational programs, vocational qualifications and requirements of modern production as it develops. Moreover, the teacher must not only be aware of the content of the profession, but also be experienced in its practice, and be prepared to transfer professional knowledge and skills to others; confidently use active methods and training facilities, be able to follow innovations in the field and incorporate new developments into the pedagogical activity.

In order to improve the level of engineering and pedagogical personnel training on the basis of "Vocational training and Pedagogy" department" and Center of Engineering Pedagogy, a study was conducted for the Ministry of Education and Science of the Republic of Kazakhstan on the subject: "Innovative approach Development to engineering and teaching personnel training for Kazakhstan industrial and innovative development".

Within the framework of the research, we hypothesized that improvement of engineering and training of teaching personnel using an innovative approach based on the continuity principle and on competency-oriented modular educational programs developed in cooperation with employers, would produce a practical-oriented curriculum that meets the requirements of the current labor market as well as international standards.

The background for these studies was the fact that in the vocational education system there is an urgent need for competent engineering and pedagogical personnel related to standards for international best practices in professional education in order to provide the economy with professionally mobile and competitive specialists.

3 Approach

In order to confirm the project hypothesis and success of implementation we carried out a review of leading domestic and foreign experience in competence approach vocational training and its use for personnel training in the Republic of Kazakhstan; a degree analysis of modular and dual training technologies in foreign countries, post-Soviet countries and Kazakhstan; an analysis of social partnership as an effective means of increasing the skills of engineering and pedagogical personnel, and identified optimal ways of interaction between employers and vocational schools. In addition, we performed analysis of continuous vocational training principles of implementation in Kazakhstan from the competence approach perspective; target vectors for engineering and pedagogical personnel training based on the competence approach in conditions of innovative and industrial development of Kazakhstan were explained.

Carried out analysis of working training programs of "Vocational training" specialty, requirements study of modern production in the conditions of industrial and innovative development of the Republic of Kazakhstan to the engineering and pedagogical personnel training for the system of vocational education allows us to talk about the need to form continuous professional education model of engineering and pedagogical personnel based on the competence approach. Therefore, the process of vocational training teacher's professional activity was modeled.

System modeling of continuous professional education for engineering and pedagogical personnel based on competent approach is inextricably linked with production and technological areas as a component of professional and pedagogical activity. In turn, bachelor's training in technological fields is accomplished in a professional and pedagogical environment, which must be modeled taking into account Kazakhstan industrial and innovative developments.

The following functional components of the pedagogical system were defined: *target (planning), comprehensive, activity (organizational and managerial), analytical and effective.*

1. *Target component* includes all the diversity of pedagogical activity goals and objectives beginning with the general goal—comprehensive and harmonious personality development of the student's individual qualities aligned to specific tasks. The pedagogical system is an activity system, and any activity is aimed at achievement of a specific goal. Aims, tasks are the first system-forming factor.

   The target component of engineering and pedagogical personnel is continuous professional education toward the aim of developing a vocationally trained competent teacher.

2. *Comprehensive component* reflects the overall goals of the pedagogical project and its specific tasks. It is implemented on the basis of defined principles and methods. The comprehensive component is determined by the availability of training professional programs at various levels (from bachelor's to PhD programs, and various professional development programs).

3. *The third system-forming factor is an activity component,* which is implemented as an interaction between educators and learners and their cooperation, which are inseparable constituents in achieving the final result.

   The most productive element of the whole pedagogical arsenal is interactive learning, based on the principles of interaction, learners' participation, and reliance on group experience. A vivid example of interactive learning is the "learning in cooperation" technology, which perfectly meets the requirements of the labor market. It allows one to form a specialist who is able to: work in a team; make independent decisions; easily improve his methods of work; set and solve new professional tasks; self-study and introduce professional innovations.

4. *Analytical and effective component* of pedagogical process reflects its effectiveness and its progress; it characterizes the progress achieved in accordance with the stated aims. It takes into account that the final result depends on the fact that solution of the educational tasks subordinated to the achievement of overall desired results by the entire educational system can be high-, medium-, or low-productive. In this case, it is necessary to decide what particular elements of the system contributes to the result. Thus, we can say that the result depends more on the second and the third components of the system.

Considering the specifics of designated vocational training concepts, we designed a hypothetical model for the professional and pedagogical personnel continuous training in Kazakhstan.

The levels of vocational education are characterized as follows:

1. Technical and vocational education:

- award for graduates: (1) qualification Master of production; (2) qualification in working profession (3, 4 rank).
- period of study: 4 years on the basis of compulsory secondary education; 3 years on the basis of general secondary education; 2 years on basis of technical and professional education obtained after 9-th grade of secondary school; 1 year on basis of technical and professional education obtained after 11-th grade of secondary school.

Taking into account the professional and pedagogical education duality and sectoral training specifics implying the implementation of production training, it is advisable to involve college graduates with working qualification and clear motivation in the pedagogical activity.

In order to ensure the prestige and attractiveness of professional and pedagogical education, preferential admission to the college in a "Vocational training" specialty in the relevant industry is advised to attract the best graduates of vocational lyceums. This should consist of a shortened duration of studies due to their working qualifications obtained at college (2 levels). Foundation: The Law of the Republic of Kazakhstan "On Education", Article 20, p. 1.

Educational institutions of VET are the field of industrial training masters' activity.

2. Higher education:

- award for graduates: (1) academic degree "Bachelor of Vocational Training" in the appropriate branch of training, focused on the performance of special and general technical disciplines as a teacher and Master of Industrial Training; (2) qualification in working profession (3 rank);
- period of study 4 years on the basis of compulsory secondary education, 3 years—on basis of technical and professional, post-secondary education.

For college graduates (graduated with honors) to provide for educational grants based on interview results. Foundation: in Part 5, article 25, p. 6 The Law of the Republic of Kazakhstan "On Education" to provide quota for the target state order (in the form of grants number determined by VET Department of the MES RK).

The field of bachelor's professional activity—VET educational establishments where a Bachelor of Vocational Training can work as a teacher of general technical and special disciplines and a Master of Industrial Training within the appropriate field.

In the system of multi-level continuous education, it is necessary to ensure the level of content has continuity with professional and pedagogical education (university program conjugation with colleges profile program). It allows implementation of reduced education programs at a level of high quality.

Vocational education reform is aimed at attracting experienced and mature vocational training teachers. The young age of current vocational school graduates, and the limited period of pedagogical practice do not provide such an opportunity. Under such conditions, it is advisable to consider internships, postgraduate training for Bachelors of Vocational Training (for 1 year) with subsequent assignment of vocational training specialist—teacher qualification.

3. Post-graduate education

Scientific and pedagogical personnel training is carried out in master's and PhD programs.

Training in master's program is carried out in two directions:

- scientific and pedagogical (period of study 2 years);
- specialized (period of study 1–1.5 year).

Educational programs of specialized master's programs are aimed at professional managers training (managers in the system of professional education in all aspects of management activity).

Educational programs of the scientific and pedagogical master's programs are focused on the professional and pedagogical personnel training for the system of higher and postgraduate education and research scope.

Graduates of the master's program are awarded the academic degree "Master of Vocational Training".

Scientific personnel Training in PhD program is carried out in two directions with the award of a higher academic degree:

- Philosophy Doctor (PhD)—(3 years of study);
- Education Doctor (Ed.D)—(no less than 2 years of study).

Prior stage of education is master's degree.

Vocational educational programs of post-graduate education are aimed at scientific and teaching personnel training of higher qualification and graduate improvement of their professional training level.

Under modern conditions, economic growth is identified with scientific and technological progress, and its provision with highly qualified labor resources. The quality of personnel is determined by its professional competence and viability for improvement. This statement is especially relevant for the sphere of vocational education, since the level of personnel training and competitiveness of technical and service labor directly depends on the professional competence of pedagogical personnel: masters of industrial training and teachers of special disciplines.

Under the conditions of the Republic industrial-innovative development, the urgency of additional professional and pedagogical education for higher technical or higher pedagogical (teacher) education increases.

With such specialists training for the vocational education system, it is possible to organize the training process in the following educational directions:

- additional training of industry specialists on the programs of psychological and pedagogical disciplines cycle (second higher education, professional development);
- additional preparation of subject teachers who received traditional pedagogical education, within the programs of sectoral training;

– training of Vocational training teachers at the relevant faculties in technical and profile universities without deep integration of psychological, pedagogical, special and sectoral components of education with training for the work profession.

Professional and pedagogical personnel high level of professionalism is one of the effective conditions for quality of specialists' training, which is based on the use of the latest achievements in pedagogical science and production. This leads to the systematic professional development of vocational training teachers.

A qualitative breakthrough in the professional and pedagogical personnel competitiveness requires the modernization of the system for their professional development taking into account the priorities of the State Program for the Development of Education in the Republic of Kazakhstan for 2010–2020.

Today in Kazakhstan in the system of technical and vocational education there are six interregional centers for the VET pedagogical personnel professional development, which were established by order of the Ministry of Education and Science of the Republic of Kazakhstan within the framework of the State Program for the VET Development in Kazakhstan for 2008–2012. Training in the centers should be carried out taking into account foreign experience, tested in professional educational institutions of the Ministry of Education and Science of the Republic of Kazakhstan within the framework of international projects of the European Union "Social partnership in the system of VET RK", the German Society for International Cooperation (GIZ) "Regional Program "Vocational education and training in Central Asia I. Regional Teacher Training Network".

In the system of professional development, it is necessary to shift to the professional development model that most fully corresponds to the top professional qualities and competencies for modern specialists in the system of education.

Specific features of the professional development model are: publicity; strengthening the importance of activity, personal (focus on satisfying students' needs and requirements) and practical oriented aspects in the learning process for the acquisition of personal experience in solving a variety of professional tasks; increasing the role of self-education to ensure the possibility of lifelong learning; as well as providing an opportunity for different categories of learners' active interaction in the process of training, counseling and experience exchange.

In this regard, the distinctive features of the professional development programs, which should be guided, are:

– reliance on competence approach;
– modularity principle allowing to develop and build up necessary competencies for the professional tasks solution in pedagogical activity.

In the system of professional development teaching for personnel, it is advisable to introduce training formats with target audiences within which framework are included the urgent areas of professional development:

General pedagogical format (increase of level of pedagogical competence, regardless the specialists' training profile).

Profile-pedagogical format (increase the level of professional-pedagogical competence in accordance with the profile of specialists' training).

Technological format (qualification improvement of VET master in accordance to training profile).

Information technology format (increase of information and technological competence).

Projective format (increase of competence in the field of normative educational documents design, research works organization);

Organizational and management format (increase of organizational and managerial competence).

Our analysis served as the basis for an engineer-teacher's matrix of competence development, which is an auxiliary tool allowing to combine the results of training with disciplines within which these competences will be formed. The matrix of competencies can be considered as quality monitoring for determining learning outcomes composition, disciplines list and contents for training modules.

The advantage of modular programs based on competencies is that, as the requirements of the labor sphere change, the necessary changes can be promptly made in modules, and based on various modules combinations, it is possible to form a variety of training courses depending on the students' needs, their initial level of education and professional experience.

4 Actual or Anticipated Outcomes

The practical result of the research: a model for continuous professional development for engineering and pedagogical personnel based on a competence approach; definition of employers' requirements for engineering and pedagogical personnel professional activity; design of a matrix of engineering and pedagogical personnel competencies based on the group of required competencies and developing conditions.

The model for engineer-teachers training at university is carried out on the basis of the proposed model which is formulated on integration of education, science and production within the framework of the Corporate University having the status of Innovation-Educational Consortium, comprising 86 enterprises of mining and metallurgy. The quality of engineering and pedagogical personnel training is ensured by implementation of a dual model of training, i.e. preparation and combining students' practical training with their part-time employment in production and traditional training at university. Such training is carried out on the basis of:

- Working professions centers (Mechanical engineering, Mining equipment, Road construction machines, Transport, Telecommunications);
- "Corporate University" innovation-educational consortium enterprises;
- Colleges of corresponding profile.

The centers of working professions and "Corporate University" enterprises have modern technological equipment in their arsenals, possessing advanced technologies and highly trained specialists, so it is possible to use them as the foundation for educational and industrial practices.

At industrial enterprises, students in the process of industrial practice get acquainted with modern industrial technological processes, machine tools and materials, receive a holistic view of materials processing, finishing and quality control, as well as production automation in general and modern methods of services providing.

Students develop their psychological and pedagogical competencies at the college level within the framework of pedagogical practice. During the practice period, future Vocational Training teachers:

- get acquainted with general technical, and special disciplines, teacher's work, industrial training master's work, and class teacher work;
- consolidate theoretical knowledge obtained in the study of psychological and pedagogical cycle of the discipline;
- develop skills of general professional and special disciplines teaching, of educational work organization in the framework of the integral system of pedagogical processes.

In the dual system training mechanism, future teachers' new psychology creation is implied—high motivation of knowledge and skills acquisition in their activity. Interconnection, interpenetration and mutual influence of various systems (science and education, science and production, etc.) are ensured, thus it results in qualitative changes in vocational education. Educational institution, working in close contact with the enterprise, takes into account the production requirements for future specialist in the course of training. The newly trained employees can immediately be involved in production as soon as their training is completed: there is no need for mentors and professional adaptation. Companies benefit from investing in education, because "at the end of the process" they have a ready-made specialist who is thoroughly familiar with the specifics of work at a particular enterprise (organization). Employers are confident that after receiving a diploma the graduate will work for them, moreover, on the employer's terms. This is an effective model for providing VET Educational Institutions with young qualified teachers of a new modern type.

Close interaction with professional educational institutions and industrial enterprises makes it possible to identify the group of engineer-teacher's professional competences presented in Table 1.

Interactive teaching methods are widely, used as a tool employs the means of training in cooperation, best suited to the demands of labor market and enabling the specialist formation who is capable: to work in team, to make independent decisions, to rebuild themselves, to set and solve new professional tasks, and to study and implement professional innovations independently. Engineer-teacher gets systemic education consisting of three integrated components: psychological-pedagogical and industrial training, as well as training in working specialty.

Improvement of training methods for engineering and pedagogical personnel and implementation of the proposed training model based on a competence approach led to a 4-year leading position in Kazakhstan (1-st place) according to the rating of the Independent Accreditation and Rating Agency for the "Vocational Training and Pedagogy" Department for training future teachers in technical and vocational education. In 2017 "Vocational training" specialty passed international accreditation at the Institute of Accreditation, Certification and Quality Assurance (ACQUIN).

Table 1. The group of engineer-teacher's professional competences

Profile of training	Professional competences (PC)	Units of PC
1	2	3
Mechanical engineering	Conduct lessons based on methodological basis of professional pedagogy in the field of mechanical engineering	Design lesson and learning process for discipline mastering using intersubject links and innovative learning technologies
		Transfer learning information using modern learning technologies
		Perform practice-oriented training
		Teach independent learning skills
		Organize and manage students' learning activity
		Perform career guidance
	To educate students to the system of social values	Conduct educational and cultural events based on universal and national values
		Develop learner's personality professional orientation
		Develop civic position among students on the basis of moral and spiritual values
		Ensure psychological and pedagogical assistance
		Develop personal qualities
		Assist students' adaptation to new educational environment
	To provide methodological support for educational process	Plan the lesson in accordance with educational program and regulations requirements
		Develop educational and methodical support based on optimally selected training technologies
		Use IT technologies to optimize the learning process and effectively master learning information
		Execute report documentation
	To carry out industrial and technological activity at mechanical engineering	Develop designs and technological processes
		Carry out the production, machinery and equipment maintenance
	To study mastering level of educational content, to study educational and production environments	Know research methodology
		Plan and organize research and experimental work at college
		To plan and organize research and experimental work at enterprise in the field of mechanical engineering and metal working
		Carry out reflection of pedagogical activity
		Register the results of research and experimental work

(*continued*)

Table 1. (*continued*)

Profile of training	Professional competences (PC)	Units of PC
	To interact with professional community and parties involved in education	Have skills of pedagogical management
		Communicate with students, colleagues and representatives of interested parties following the business communication etiquette
		Apply group and collective technologies in students' educational activities
		Develop students' skills of interaction in socio-cultural and professional environments

KSTU established a Center of Engineering Pedagogy (CEP) for the formation of competent and competitive specialists among engineering teaching staff. Its work is focused on university teachers' professional development based on innovative educational technologies and best practices of modern engineering pedagogy.

The main activities of the Center are: organization of training seminars on innovative pedagogical technologies of teaching for the faculty; organization of training seminars based on the curriculum of the International Society for Engineering Education (IGIP) on the basis of KSTU; training in the additional educational program of pedagogical profile for people who graduated with a master's degree; projects development for participation in competitions on grant financing of fundamental, applied and other scientific research on issued problems of professional technical education; establishment and development of international scientific contacts, etc.

Participation in scientific and international projects allows teaching staff of the "Vocational Training and Pedagogy" Department to be on the crest of the wave of pedagogical innovations and introduce them to professional development courses in the university's teaching process.

The work of The Center of Engineering Pedagogy and "Vocational Training and Pedagogy" Department in the area of engineering and pedagogical personnel training was highly appreciated by the European Training Foundation (ETF). In the framework of ETF project at the meeting of the Central Asian Academy, the project "Development of technological training for vocational training bachelor based on the competence approach" was recognized as an example of a best practice and was the winner of the competition among educational institutions.

5 Conclusions/Recommendations/Summary

Improved system of engineering and pedagogical personnel training taking into account international experience (including participation in the programs of the European Foundation for Education "Development of Professional Institutions of Central Asia for the purpose of lifelong learning", German Society for International Cooperation "Support of Vocational and Technical Education in the Republic of

Kazakhstan") allowed to: create the necessary didactic conditions for engineering and pedagogical education, science and production integration in order to achieve the conformity of training quality to society and changing socio-economic conditions; improve the graduates' training quality by introducing practical-oriented training based on modular competence-oriented educational programs developed to ensure continuity of education levels and labor market requirements.

This system allowed creating necessary didactic conditions for engineering and pedagogical education, science and production integration, in order to achieve the conformity of training quality to society and changing socio-economic conditions; to provide vocational training with competent teaching personnel having competitive skills and potential for further professional development.

The success indicator for engineers-teachers training quality is their demand in the labor market and a high percentage of employment in technical and vocational education colleges, as well as positive feedback from employers about graduates of Karaganda State Technical University "Vocational Training" specialty.

Transfer of Information by Mentor Teachers

Istvan Simonics[(✉)]

Trefort Ágoston Centre for Engineering Education, Óbuda University, Budapest,
Hungary
simonics.istvan@tmpk.uni-obuda.hu

Abstract. Since 2011, we have been educating mentor teachers in four
semester postgraduate courses. Mentor teachers can support the preparation
process of our engineering teacher students in secondary vocational schools by
coaching their teaching practice. Therefore updated educational technology is
important in their studies. It was important and interesting to get acquainted with
mentor teachers' presentation skills, their knowledge about, preparedness and
success on that field. In 2016 we organized a survey to measure mentor teachers'
presentation skills and how they can transfer information in teaching learning
process and their professional life. For the survey, we had elaborated a questionnaire. The purpose of the questionnaire was to survey the mentor teachers'
presentation skills and their transfer of information in four parts. The goal of the
survey was to define how frequently they use presentations in their work, what
kind of lack they have in editing process and how we can support their application of presentations in mentor and teaching work. We elaborated a hypothesis: We have to emphasize the development of mentor teachers' presentation
skills in our training and support them to learn the effective information process.
In this study we describe, more than the basic elements of the survey was shown
previously. More detailed evaluation of data will be shown especially concerning the practice and expectation of mentor teachers. The survey raised our
attention to the fact that the further development of the presentation skills needs
methodological upgrading of our teaching process.

Keywords: Presentation skills · Mentor teaching · Theoretical training
Learning pedagogy

1 Introduction

Since 2011, we have been educating mentor teachers. Mentor teachers can support the
preparation process of our engineering teacher students in secondary vocational schools
by coaching their teaching practice, and they also help their trainee colleagues in the
first two years at the beginning of their career. This postgraduate course was new for
our colleagues as well. It was a challenge to prepare ourselves as well as possible, to
give lectures to our mentor teacher students, who are practicing teachers at schools and
have good experience in teaching their subjects. They have high standard expectations
to learn the newest knowledge based on the most modern educational technology.

© Springer Nature Switzerland AG 2019
M. E. Auer and T. Tsiatsos (Eds.): ICL 2018, AISC 917, pp. 957–967, 2019.
https://doi.org/10.1007/978-3-030-11935-5_91

The best mentor teacher can accelerate changes in teaching-learning process, which needs creativity.

In the past 5 years, the content and methodology of mentor teacher training courses have changed. In the preparation process of course design, we had to take into account the changing learning styles as well [1, 2].

Our mentor teacher students have several practical lessons. They have to work together in groups, find, collect and share information with their colleagues [3–5]. In this way they have to use Information and Communications Technology – ICT, edit presentations and show them to their colleagues. Presentation skills for teachers were surveyed nationally and internationally as well. Cranton and Cohen emphasized "To explore how educators learn through teaching based on a transformative learning in theoretical framework" [6]. "Competence in using ICT in teaching is required for Qualified Teacher Status and considerable attention" [7]. There are several books and online courses about presentation techniques, but only a few researches about the effectiveness of presentations.

2 Background, Goal and Methodology of Research

Mentor teachers have high standard expectations to learn the newest knowledge based on the most modern Educational Technology. It was important and interesting to get acquainted with mentor teachers' presentation skills, their knowledge about, preparedness and success on that field. In 2016 we organized a survey to measure mentor teachers' presentation skills, and and how they can transfer information in teaching learning process and their professional life.

For the survey we had elaborated a questionnaire. The purpose of the questionnaire was to survey the mentor teachers' presentation skills in four parts. In the first part, we studied their personal data: age, sex, area of teaching, type of school of work. In the second part, we involved questions about preparation of presentation skills, if they learnt Educational Technology at university during the teacher training period, how they learnt making presentations, if they knew the basic rules of editing presentations. In the third part, we were interested in managing resources for editing presentations. In the fourth part, we measured the application of presentations in their education process.

The goal of the survey was to define how frequently they use presentations in their work, what kind of lack they have in editing process and how we can support their application of presentations in mentor and teaching work.

We elaborated a hypothesis: *We have to emphasize the development of mentor teachers' presentation skills in our training and support them to learn the effective information process.*

The survey was conducted in November 2016. We elaborated a questionnaire for mentor teachers. These target groups involved mentor teacher students from three different semesters. The questionnaire consisted of four different parts and contained 32 questions.

We could ask 16 mentor teachers from the first semester, 18 from the second semester and 75 from the third semester, altogether 109 mentor teacher students. Results were analyzed on the levels of significance, correlation, factors tests and used

SPSS software. In the previous study we could present the basic data of the first and the second part. In this paper we focus on the results of part 3 and part 4 of the questionnaire.

In mentor work, teachers need several competencies and skills which help them with the preparation and development of their students or colleagues. These competences are various e.g. knowledge of pedagogy, professional subjects, methodology, teacher training, psychology etc. They had to accept the requirement of self-development. Application of presentations in pedagogical work is essential for mentors. Cooperation, communication and transfer of information are also very important factors in mentors' work, so we support activities which can accelerate them.

3   Planning and Preparing Presentations

3.1   Resources of Presentations

In the third part of the questionnaire, we dealt with planning and preparation of presentations. We tried to measure how the mentor teachers use resources for planning and preparing presentations.

The possible answers were as follows: *printed material, data from the internet, Wikipedia, articles from the internet, professional pages from the internet, information broadcasted by radio or TV, information from colleagues, relatives, information from pupils, other.* They could select more than one item Fig. 1.

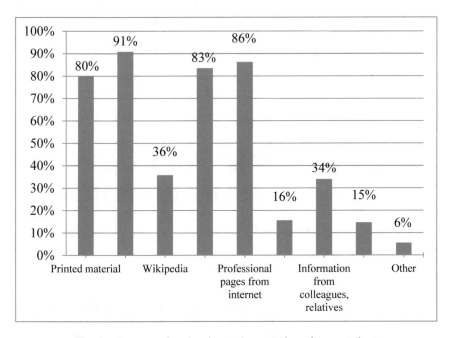

Fig. 1.   Resources for planning and preparation of presentations

More than 80% of the answerers used printed materials, data, articles and professional pages from the internet. Only 36% used Wikipedia and 34% collected information from colleagues, relatives. All the other resources were below 20%. These results mean, they generally believe the information from printed material and the internet, reliability of information from Wikipedia and colleagues, relatives is more or less acceptable and all the others not or not useful.

Printed materials in most cases are checked by publisher's readers. All the other information is not so reliable. When our students were giving presentations, we had to recognize they believe in data, and they never check their reliability.

We have to teach them to manage information carefully and wisely! It happened several times that they used 5–10-year-old statistical data to explain problems and crisis, but at that time they weren't reliable nor true, because economical circumstances have changed. If the basis of the system? is false in our presentation, our professional prestige can be destroyed with data which are not valid or valuable.

3.2 Editing Pictures and Diagrams for Presentation

We asked the mentor users who edited pictures and figures for presentation. We elaborated separate questions asking about pictures, videos, diagrams and charts. There wasn't any significant difference among the questions, trends were the same regardless of the type of pictures, diagrams or videos. Half of the answerers edited the pictures or diagrams him or herself. 49% of mentor teachers downloaded the necessary pictures from the internet. 18% of users asked friends to edit diagrams. 8% of mentor teachers never edited diagrams. It can be explained with the following reasons: Being a mentor teacher needs minimum three years of experience in teaching, so below 29 nobody can be a mentor teacher. So the majority of the target group was between 35–50 years. Most of them are not so well prepared on ICT applications Fig. 2.

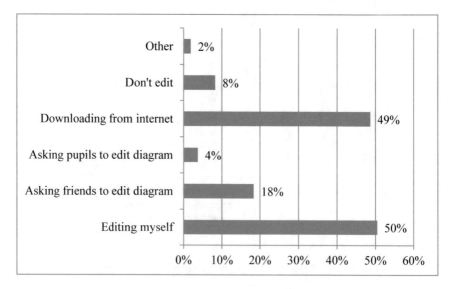

Fig. 2. Editing pictures and diagrams for presentation

3.3   Problems of Citation

In the next part of the third chapter, we examined how the mentor teachers think about citation: Whether they know the rules of citation, if it is important to them, or what they use in their presentation that is their own, if it is not so important to indicate the name of authors of used texts or pictures. It was important for us, educators to clarify and emphasize to our students that they should learn and apply the right rules of citation. The problem is twofold. Generally teachers are regarded by students as an example, a good practice. If they take no notice of citing, it will be the rule in the schools they teach, supporting the students' general idea that everything can be use freely from the internet. The other task of educators at the university is preparing mentor teachers to edit their thesis well to finish their studies. When they finish the research work and edit their thoughts in a thesis work, they have to upload their text to a plagiarism software application. If they don't use the rules of citation well and correctly, they can fail this plagiarism test. In that case, the thesis cannot be assessed and the mentor teacher can't finish his or her studies. This is why it was important to ask their opinion. The educators of the university raise their attention to making essays, home work to learn and keep to the correct rules of citation. We support this process with information brochure about how to elaborate thesis that contains some examples of citation as well. The result of the answers can be seen on Fig. 3. Fortunately more than 50% of the mentor teachers answered that they give the resources of texts, videos, diagrams in every case using the citation rules in presentations. 46% of the answerers gave negative answer and 2% didn't give an answer.

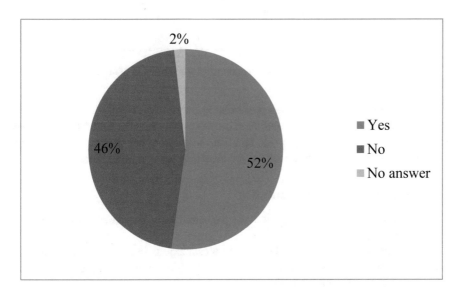

Fig. 3. Every case indicating resources of texts and diagrams using the citation rules in presentations

4   Application of Presentations in Professional Work and Education

4.1   Frequency of Application of Presentations in Education

In the fourth part, we measured the application of presentations in their professional work and education. The first question dealt with the frequency of application of presentation Fig. 4.

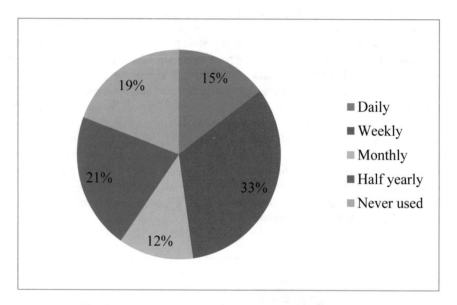

Fig. 4. Frequency of application of presentations in education

15% of the mentor teachers used presentations every day. 1/3 of the answerers used them weekly, 12% monthly, 1/5 used them only every six months, and 19% never used presentations in their work. But no student can finish his or her studies without editing and presenting several presentations in our mentor teacher training lessons.

4.2   Using Presentations in Education

In the second question of this section we measured how they used presentations in education Fig. 5.

39% of the mentor teacher students used presentations in theoretical lessons, 10% used them in practical lessons, 19% used them in both theoretical and practical lessons. 22% never used them and 10% did not answer.

Fortunately two third of the mentor teachers used presentations, in their educational work.

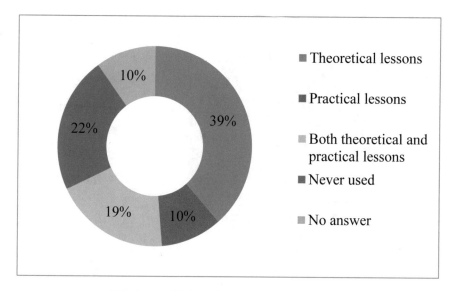

Fig. 5. Application of presentations in education

4.3 Feedback from Our Mentor Teacher Students About the Contribution of Their Studies to the Development of Competencies of Presentation Skills

In the third part of this section, we asked some feedback from our mentor teacher students. In some questions, we asked them to give their opinion about how much our university could contribute to the development of their competencies on the area of the elaboration of presentations, the application of presentations in professional work and the performance of presentations.

They could select from the following answers:

- Not a bit
- Some degree
- Mostly
- Substantially.

It is important to mention, they could select only one possible answer. The answers for elaboration of presentations can be seen on Fig. 6.

When we analyzed the results we had to recognize that our students gave definite results according to their feeling. Only 10% thought he/she was highly satisfied. 36% were mostly satisfied with the contribution of their study on presentation preparation skills.

We asked them how much the university could contribute to the development of their competencies on the area of application of presentations in professional work Fig. 7.

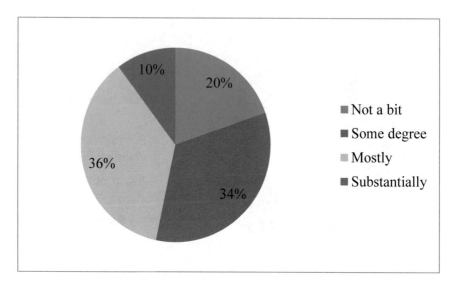

Fig. 6. Our studies contributed to the elaboration of presentations

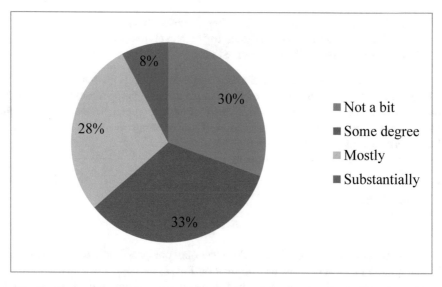

Fig. 7. Our studies contributed to the application of presentations in professional work

The results were similar to those of the previous question. According to the figures, the majority of the mentor teachers, almost 2/3 of them were not satisfied with this area either.

With the last question, we analyzed how the university could contribute to the development of their competencies on the area of the performance of presentations Fig. 8.

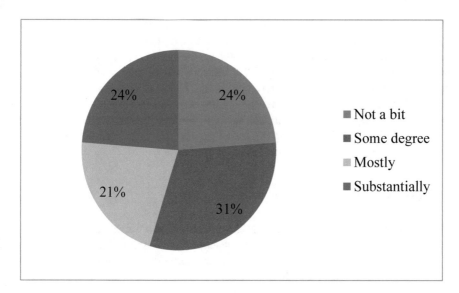

Fig. 8. Our studies contributed to the development of the competencies on the area of the performance of presentations

This was the area where we had received a really positive feedback: 24% of the answerers were substantially satisfied. For the author personally, it was a really good result because he involves mentor teacher students to work in groups, collect information and give presentations. Finally, we try to prepare them to improve their presentation techniques in the right way.

That result can explain why our mentor teachers have less and less problems with elaborating new presentations, and can provide more and more effective presentations at the final exam.

4.4 Feedback from Our Mentor Teacher Students About Needs to Develop Competencies of Presentation Skills

In the last part of this section, we asked some feedback from our mentor teacher students. In some questions, we asked them to write about their needs to develop competencies on the area of the elaboration of presentations, the application of presentations in professional work and the performance of presentations.

To express their needs they could select from the following answers:

- I have no needs
- Less needs
- Average needs
- Significant needs.

We analyzed these answers finding cross connections with other questions where we could recognize correlations. This question was "How frequently do you use presentations in education?" 9,5% of those answerers who use presentations weekly have no needs, 11,4% required less needs, 10,5% needed an average and only 2,9% indicated significant needs to develop the elaboration of presentations. 9,8% of those answerers who use presentations weekly have no needs, also 9,8% required less needs, 12,7% needed an average and only 2,9% indicated significant needs to develop the application of presentations in professional work. 11,9% of those answerers who use presentations weekly have no needs, also 11,9% required less needs, 4,8% needed an average and 9,5% indicated significant needs to develop the performance of presentations. *This last result proves that several mentor teachers recognized they need further improvement in the performance of presentations.*

5   Results and Conclusions

The survey was organized for mentor teacher students at our university. The purpose of the questionnaire was to study the mentor teachers' presentation skills in four parts.

The goal of the survey was to define how frequently they use presentations in their work, what kind of lack they have in editing process and how we can support their application of presentations in mentor and teaching work.

In this study, we described the second part of the survey. However, these analyses have proved our hypothesis, we have to emphasize the development of the mentor teachers' presentation skills in our training and support them to learn the effective information process.

The survey raised our attention to the fact that the further development of the presentation skills needs methodological upgrading of our teaching process.

References

1. Toth, P.: New possibilities for adaptive online learning in engineering education. In: Ali, A., Abab, M., Plaha, R., Rahim, A. (eds.) International Conference on Industrial Engineering and Operations Management. Curran Associates Inc., United Arab Emirates, Dubai, pp. 2242–2246 (2015). (ISBN:978-0-9855497-2-5)
2. Holik, I.: Achievement motivation of engineering students. In: Borsos, E., Námesztovszki, Z., Németh, F. (eds.): The Challenges of Contemporary Education 6th International Methodological Conference; 11th Scientific Conference; 4th ICT in Education Conference. University of Novi Sad Subotica, Serbia, p. 47 (2017). (ISBN:978-86-87095-75-5)
3. Simonics, I.: Presentation skills of mentor teachers. In: Auer, M.E., Guralnick, D., Simonics, I. (eds.): Teaching and Learning in a Digital World: Proceedings of the 20th International Conference on Interactive Collaborative Learning, vol. 1, 968 p, pp. 450–459. Springer, Cham (2018). (ISBN:978-3-319-73209-1)
4. Holik, I.: Experience and possibilities of information processing in training of mentor teachers. In: Szakál, A. (ed.) IEEE 13th International Symposium on Applied Machine Intelligence and Informatics SAMI 2015, pp. 219–222. IEEE Hungary, Herlany, Slovakia (2015). (ISBN:978-1-4799-8220-2; 978-1-4799-8221-9)

5. Molnár, G.: Open content development in ICT environment. In: Gómez Chova, L., López Martínez, A., Candel Torres, I. (eds.): INTED2017 Proceedings: 11th International Technology, Education and Development Conference. International Association of Technology, Education and Development (IATED), Valencia, Spain, pp. 1883–1891 (2017). (ISBN:978-84-617-8491-2)

6. Cranton, P., Cohen, L.R.: Learning through teaching: a narrative analysis. In: Wang, V.X. (ed.) Handbook of Research on Teaching and Learning in K-20 Education. Atlantic University, Florida, pp. 17–32 (2013). (ISBN13: 9781466642492)

7. Darling-Hammond, L.: Evaluating Teacher Effectiveness – How Teacher Performance Assessments Can Measure and Improve Teaching, p. p36. Center for American Progress, Washington (2010)

E-Learning in Electropneumatics Training

Gabriel Bánesz[1], Alena Hašková[1], Danka Lukáčová[1],
and Ján Záhorec[2($\boxtimes$)]

[1] Constantine the Philosopher University, Nitra, Slovak Republic
{gbanesz, ahaskova, dlukacova}@ukf.sk
[2] Comenius University, Bratislava, Slovak Republic
zahorec@fedu.uniba.sk

Abstract. The issue presented in the paper is connected with a project aimed at implementation and use of remote laboratories in distance education. One of the project tasks has been to create an e-learning course designed for higher education students enrolled in technical study programs. The main topic of the created course are electropneumatic system principles. The goal of the course is to familiarize students with these principles based on the use of an electropneumatic panel, which is a commonly used teaching aid in the subject *Machines and parts of machines*. The electropneumatic panel enables students to train basics of machine system management by means of electrical control elements, what is the essence of automation systems applications in the technical practice. Practical activities (training) precede just the e-learning course, which serves as a mean of preparation of the students for their practical activities (training). The course consists of theoretical knowledge on air properties and principles of pneumatic valves and pneumatic cylinder activity, and there are exercise tutorials, which the students use to prepare themselves to be able to work with the electropneumatic systems. The paper presents and analyses results achieved by the students in the electronic test of the given e-learning course, relevant to the students´ theoretical preparation and concrete electropneumatic circuits, which were recorded in working sheets in frame of the students' face-to-face teaching. To assess the test results and to assess the practical circuit connections, descriptive statistics was used. As there was indicated a mutual causal relationship between the results of the tests and elaboration of the practical assignments, also the coefficient of correlation was determined. Further, a content analysis was used to evaluate the working sheets to assess students' practical skills to connect and work with electropneumatic circuits.

Keywords: E-learning · Electropneumatics · Tertiary students education and training

1 Context of the Research

The use of remote real experiments in teaching process is not a new phenomenon either abroad or in Slovakia. However, in Slovakia this kind of experiments has not still become a common teaching tool for science and engineering education and has not been ranked among the conventional ways of teaching [1, 2]. To promote their

© Springer Nature Switzerland AG 2019
M. E. Auer and T. Tsiatsos (Eds.): ICL 2018, AISC 917, pp. 968–976, 2019.
https://doi.org/10.1007/978-3-030-11935-5_92

implementation in practice a project entitled *Remote laboratories in distance forms of education* has been solved at Faculty of Education, Constantine the Philosopher University in Nitra (SK). One of the project goals has been to create an e-learning course, which could be used by students of the Department of Technology and Information Technologies in frame of the face-to-face education as well as in frame of the distance education form. The topic of the course are pneumatic systems and their use in technical practice.

In the traditional way of teaching, operation principles of pneumatic systems are explained and training in connecting these circuits is carried out in the working environment of a common classroom equipped with relevant teaching facility (teaching aids) produced by Lucas-Nulle [3]. The facility enables students to create both basic and complex pneumatic circuits so as they are used in practice. As the students coming to study to the university are graduates of different types of secondary vocational schools and grammar schools, not all of them have acquired necessary math and physics knowledge to understand by electrical circuits controlled pneumatic systems operation. For this reason, a course for the students was created to support them to broaden their theoretical knowledge and to prepare them for practical exercises (activities) in frame of face-to-face lessons.

The above-mentioned means that students obtain a necessary information relevant to the taught topic through a lecture given by the teacher, and to enable students to broaden their understanding of the lectured subject matter an e-learning course for them was created [4–10], which they can use in frame of their self-study (home-preparation to next lessons).

In the first part of the course physical characteristics of air and their utilization in pneumatic systems are explained. In the next parts the students are familiarized with basic pneumatic components and their graphic presentations as well as with pneumatic schemes designing. In particular, the course consists of following parts:

- Pneumatics—theory
- Electropneumatics—measurement instructions
- Electropneumatics—measurements

Purpose of the course is to prepare students in both theoretical as well as practical side.

In the theory of pneumatic systems air physical attributes are presented. Main attention is paid to air composition and characteristics reflected in Boyle—Mariotte and Gay—Lussac's laws. Consequently, advantages and disadvantages of pneumatic systems are presented. A separate part (sub-chapter) is devoted to pneumatic basic components used in particular circuits and schemes. These are mainly pneumatic engines (single-acting and double-acting cylinders), and valves and switchgears as control pneumatic elements. Each of the presented basic components is described and explained from the point of view of their technical configuration and operation principles. Although each of these basic pneumatic components is accompanied by its schematic symbols, an overview of the most widely used pneumatic schematic symbols is presented in the further separate chapter.

In the next part of the course the students are familiarized with the own work with the concerned pneumatic systems facility. Here they obtain detail information on all its components. Special attention is paid to its electrical part (Fig. 1), through which the pneumatic parts are controlled (power source, functions of switching elements such as cut off switches, change-over switches, electric relay, time-adjustable switching relay etc.). A very important part of the electrical systems are also electrically actuated control pneumatics valves, actuating of which is in the electrical part of the panelboard.

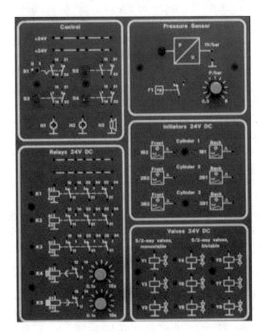

Fig. 1. Electrical part of the pneumatic systems facility

The second part of the pneumatic system facility comprises pneumatic components of various mechanisms, which may be connected according to given assignments (Fig. 2). It relates mainly to a valve block, which comprises 5/2 monostable valves and 5/2 bistable (flip flop) valves, one single-acting cylinder and three double-acting cylinders. Connecting of these pneumatic components is carried out by means of tubes, through which compressed air is supplied.

The last part of the course is focused on basic ways how the mentioned valves and cylinders can be connected. This is done through three assignments which the students are expected to fulfil during face-to-face lessons when they work with the own panelboard. The assignments are the following ones:

- Assignment 1
  Actuation of the single-acting cylinder through the monostable 5/2 valve.
- Assignment 2
  Actuation of the double-acting cylinder through the monostable 5/2 valve.
- Assignment 3
  Actuation of the double-acting cylinder through the bistable 5/2 valve with the airflow regulation.

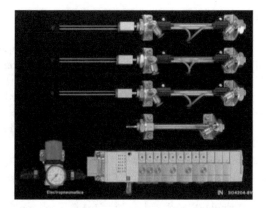

Fig. 2. Pneumatic part of the pneumatic systems facility

At the end of the course (following the theoretical part) there is included an electronic test to check students' theoretical knowledge before their own work within the practical training (practical exercises realization). The test consists of ten items, which are related to the following issues (students' knowledge):

1. properties of gases, Gay—Lussac and Boyle—Mariotte's law,
2. air composition,
3. gas pressure units,
4. 2/3 electro-pneumatic valve scheme,
5. electro-pneumatic scheme of monostable 5/2 valve, electro-pneumatic scheme of bistable 5/2 valve,
6. schemes of the compressed air supply,
7. possibilities to control capability of compressed air in pneumatic systems,
8. identifiers of pneumatic valves inputs,
9. use of the 5/2 valve for controlling single acting linear pneumatic motor,
10. use of the 5/2 double solenoid valve for controlling double-acting linear pneumatic motor.

Philosophy of the use of the created e-learning course [11] by students follows next three steps:

1. Passing the lectures, within their home preparation for the next lessons, students study the theoretical bases of the presented subject matter in the electronic course and confirm their knowledge. As the course is complemented by relevant pictures, they can obtain also a needed information about the pneumatic systems facility.
2. From the electronic course students should print working sheets to the particular assignments, which they will do at the forthcoming lessons.
3. Before their own work on the assignments at the concerned lesson students' knowledge is checked by means of an electronic test and consequently students work on the assignments using the electropneumatic panelboard and they record the circuit connections of the particular assignments in their working sheets.

2 Goal and Methodology of the Research

Following implementation of the created electronic course into the teaching process of students, a research survey was carried out to assess its influence on students' learning achievements.

Assessment of the students' learning achievements was based on recording and comparing their results in a knowledge test and fulfilment of the given assignments stated in the working sheets. Research data processing was done by descriptive statistics, Pearson correlation coefficient and paired t-test.

The knowledge test measured students' level of proficiency at the acquired knowledge on practical aspects of measurement of pneumatic systems, i.e. it was a cognitive, output, NR-distinctive, non-standardized knowledge test. Its content reflected the subject curricula (theoretical basis of the subject, see the above-mentioned ten test items). Evaluation of the test was done automatically by Moodle [12]. For each correct answer one point was given, in case of incorrect, incomplete or missing response the system gave zero points. At an association response in case of some mismatched answers, the system counted out a proportion of one point. So the maximal score, which each respondent could achieve in the test, was 10 points. Time to take the test was set to 12 min.

Maximal number of points, which the students could achieve at each of the given assignments, was 4 points. A student obtained four points if the circuit presented in his/her working sheet was correct. In case of some mistakes/incorrectness the number of the given points was smaller, zero was given for completely incorrect solution. So the total number of points, which a student could obtain was 12, as each student had to fulfil all 3 assignments.

Possible dependency between the achievements of students obtained in knowledge tests (x) and working sheets elaboration (y) was tested through Pearson's correlation coefficient, according the formula

$$r_{xy} = \frac{\sum_{i=1}^{n} (x_i - \bar{x})(y_i - \bar{y})}{\sqrt{\sum_{i=1}^{n} (x_i - \bar{x})^2 \sum_{i=1}^{n} (y_i - \bar{y})^2}}$$

Further a null hypothesis.

H0: *There is no significant difference between the means of the input test (i.e. the electronic test) and output tests (i.e. the results of the particular assignments).*
was tested.

The stated hypothesis represents de facto 3 null hypotheses (one for each of the assignments). To verify the hypotheses a paired test was used. A precondition to use a paired test is more than 30 statistical items in the statistical set or its normality [13]. As the research survey was done with a research sample consisted of 34 students enrolled for the subject *Machines and their parts* (during the period of academic years 2016–2018), so this precondition was fulfilled. The paired test was carried out in Excel software means.

3   Research Results

Descriptive statistics to the students' results achieved in the knowledge test and total results achieved in the three given assignments are presented in Tables 1 and 2, respectively.

Table 1. Descriptive statistics of the knowledge test results

Knowledge test results	
Mean	5.0
Standard error	0.3
Median	5.5
Mode	6.0
Standard deviation	1.8
Sample variance	3.5
Minimum	2.0
Maximum	9.0
Count	34.0

Table 2. Descriptive statistics of the assignments 1–3

Assignments results	total	Assign. 1	Assign. 2	Assign. 3
Mean	6.8	2.2	2.8	1.8
Standard error	0.6	0.3	0.2	0.1
Median	8.0	4.0	4.0	2.0
Mode	10.0	4.0	4.0	2.0
Standard deviation	3.6	2.0	1.6	1.0
Sample variance	13.1	4.0	2.6	1.0
Minimum	0.0	0.0	0.0	0.0
Maximum	12.0	4.0	4.0	4.0
Count	34.0	34.0	34.0	34.0

As Table 1 shows, average number of points achieved by the students in the knowledge test was 5.0 with the sample variance 3.5 points. Maximal number of the achieved points was 9 and minimal number was 2 points.

As to the results of the assignments (Table 2), average number of points achieved by the students for all three assignments elaboration was 6.8 with the sample variance even 13.1 points. This is no mistake, it follows the fact that one of the students fulfilled all assignments excellently and gained the whole sum of the points, i.e. 12 and one of them completely did not manage the assignments and finished with the final score of 0 points.

Testing the dependency between the achievements of students obtained in knowledge tests and working sheets elaboration (assignments results), the value of the Pearson correlation coefficient was $r = -0.1$, what does not prove the tested dependency. This means that there is no relationship between the results achieved by the students in the knowledge test and the working sheets elaboration (practical training).

Possible relationship between the particular test items and assignments results was tested at first for the test item 9 and assignment 1 (actuation of the single-acting cylinder through the 5/2 monostable valve). There was identified a weak negative relationship ($r = -0.2$), i.e. the lower result was achieved by a student in the test item 9, the better was result s/he got in assignment 1 working sheet assessment.

Finally we tested possible relationship between the test item 10 and assignment 2 (actuation of the double-acting cylinder through the 5/2 monostable valve). The achieved value of the Pearson's correlation coefficient $r = 0.01$ has showed that there is no relationship.

As to the verification of the null hypothesis (or de facto the hypotheses), its results are presented in Table 3.

Table 3. Verification of the null hypothesis: results of the paired test

	Variable 1	Variable 2
Mean	5	6.882352941
Variance	3.515151515	13.1372549
Observations	34	34
Pearson correlation	−0.098103478	
Hypothesized mean difference	0	
df	33	
t Stat	−2.588072052	
P(T <= t) one-tail	0.007120547	
t Critical one-tail	1.692360309	
P(T <= t) two-tail	0.014241095	
t Critical two-tail	2.034515297	

The null hypothesis

H0: *There is no significant difference between the means of the input test (i.e. the electronic test) and output tests (i.e. the results of the particular assignments).*

was rejected, based on the value $p = 0.0142$, as well as $t = 2.588 > 2.034$ at the significance level $\alpha = 0.05$. This means that despite of the previous presented partial findings, there is a significant difference between the results achieved by the students in the electronic test and in the working sheets elaboration, i.e. results which the students achieved in the working sheets elaboration are statistically significantly better.

4   Discussion of the Research Results

The presented research results point at several very interesting findings.

On the one hand it was found out that the electronic course focused only on theoretical knowledge does not support students' practical performance, it does not help them at their practical activities.

On the other hand students' results in the working sheets elaboration were significantly better than their results in the knowledge test. A question is what could be the reason of this, what could had a positive impact on students' performance in their practical activities. Here four facts should be mentioned and considered. One of them is the fact that working with the electropneumatic panelboard the students worked in groups and could consulted together. So a higher measure of their success can be linked with the given possibility of their mutual co-operation. A next one is the fact that the students manipulated directly with the real panelboard, what could be for them a very motivating factor leading them to logical thinking about the situation they saw on it, and consequently about the ways how the circuit could be connected and recorded, too. A factor, in our opinion of a minority influence on the survey results, could be the time aspect. The time for the knowledge test fulfilment was limited, while during the training the students could verify freely their solutions of the given assignments, e.g. based on the "experiment—error" procedure, and consequently they could revise the solutions they proposed. For this reason more correct answers appeared in the working sheets than in the knowledge test. Contrary to that, in our opinion a factor with a very high impact on the results is the fact that quality of the students' home preparation was increased by the possibility to try the given connections of the circuits through a remote access to the panelboard. In a matter of fact just this is the main goal of the solved project—to enable students a remote access to the electropneumatic panelboard in a form of a remote real experiment.

5 Conclusion

The use of the training based on practical activities and implementation and use of remote laboratories is not important only in tertiary education. It should be a dominant aspect on which also technical education at lower levels of education is based, not excluding neither the lower level of secondary education (ISCED 2). In this way the presented issue corresponds also with the project entitled *Development of teaching materials supporting pupils' orientation for technical study programs* which has been currently also solved at Faculty of Education, Constantine the Philosopher University in Nitra (SK).

Acknowledgment. This work has been supported by the Cultural and Educational Grant Agency of the Ministry of Education, Science, Research and Sport of the Slovak Republic under the projects No. KEGA 011UKF-4/2017 and No. 021UKF-4/2018.

References

1. Bánesz, G., Hašková, A., Lukáčová, D.: Elimination of barriers for a broader use of remote experiments in Slovakia. In: ASEE International Forum, p. 8. Ohio, Columbus (2017). https://peer.asee.org/29282
2. Kozík, T., Šimon, M., Kuna, P., Arras, P., Tabunschyk, G.: Remote Experiment at Universities. In: IDAACS'2015: The crossing point of Intelligent Data Acquisition &

Advanced Computing Systems and East & West Scientists. In: Proceedings of the 2015 IEEE 8th International Conference on Intelligent Data Acquisition and Advanced Computing Systems: Technology and Applications, pp. 929–934. Warsaw, Piscataway IEEE (2015)
3. Lucas—Nule, Pneumatics with UniTrain (2016). http://www.lucas - nuelle.com/316/apg/3751/Pneumatics-with-UniTrain.htm
4. Bánesz, G.: LMS Moodle vo vzdelávaní bezpečnostných technikov. In: Modernizace vysokoškolské výuky technických předmětů, Hradec Králové: UHK, pp. 11–14 (2013)
5. Tomková, V.: Requirement Communication Skills Teacher in the Application of ICT in Education. In: New Technologies in Education, pp. 1–4. Masaryk University, Brno (2010)
6. Klement, M., Dostál, J.: Learning styles according to vark classification and their possible uses in tertiary education carried out in the form of e-learning. J. Technol. Inf. Educ. 6(2), 58 (2014)
7. Serafin, Č.: Reflection of Technical Education in a Globalizing World. In: Trends in Education, pp. 5–12. UP, Olomouc (2014)
8. Salata, E.: Motywy dokształcania i doskonalenia oraz samokształcenia nauczycieli. Institut Technologii Eksploatacji, pp. 208–215. Wspólczesne problemy pedeutologii i edukacji, Radom (2007)
9. Neuschlová, M., Nováková, E., Kompaníková, J.: Application of e-learning in education of immunology in Jessenius Medical Faculty. In: 7th International Conference on Education and New Learning Technologies (Edulearn 2015), pp. 7739–7744. IATED Academy, Barcelona (2015)
10. Paulsen, M.F.: Online Education and Learning Management Systems—Global Elearning in a Scandinavian Perspective. NKI Forlaget, Oslo (2003)
11. Constantine the Philosopher University in Nitra, Education portal (2016). https://edu.ukf.sk/
12. Neradová, S., Horálek, J.: Use of the statistics at evaluation in the e-learning systems. J. Technol. Inf. Educ. 4(1), 10–14 (2012)
13. Markechová, D., Tirpáková, A., Stehlíková, B.: Štatistika v praxi. UKF, Nitra (2011)

Poster: Teachers-Researchers Training at Technological University

Farida T. Shageeva, Roza Z. Bogoudinova,
and Natalia V. Kraysman$^{(\boxtimes)}$

Kazan National Research Technological University, Kazan, Russia
n_kraysman@mail.ru

Abstract. The article focuses on a currently important issue: teachers-researchers training at Technological University. It presents the educational goals, the structure of the content, educational and control methods and means that provide formation and development of a new professional competence: ability to organize the process of professional training basing on the methodology, theory and technologies of the contemporary educational science. The training addresses all the post-graduate student of the university.

Keywords: Post-graduate studies · Professional competences · Methodology and technologies of professional training

The modern situation in engineering education is defined by a number of social tendencies among which: a dynamic development of information society with indications of knowledge-based economy; a special role and value attributed to educational system in the modernization of all spheres of life demanding advancing educational content and new pedagogical technologies; evolutions of educational process among the participants in the process [1, 2]. The importance of top academic qualification and teaching skills of scientific-pedagogical personal has been increasing lately.

The postgraduate study is the third level of the higher education in the Russian Federation and is carried out according to the educational programs independently developed by each higher education institution. The analysis of corresponding Federal State Educational Standards has shown that the higher education training in the corresponding direction is specified in the section "Characteristic of the postgraduate study graduates professional activity". Teaching activities for educational programs in higher education are among the types of training. It is supposed that after a successful passing of the State Final Certification procedure the graduates will obtain diplomas and will be designated as: "Researcher. Teacher-Researcher". Therefore, it is in no doubt that the postgraduate study programs for all departments have to include a pedagogical component.

The best conditions for the realization of such training are created at the universities of a new type—namely the national research universities—which are not only capable to organize the effective process of training but also to carry out its integration with the scientific research conducted in the university laboratories and scientific centers [3]. The postgraduate study programs are implemented at Kazan National Research Technological University in nineteen departments. The discipline "Methodology,

© Springer Nature Switzerland AG 2019
M. E. Auer and T. Tsiatsos (Eds.): ICL 2018, AISC 917, pp. 977–980, 2019.
https://doi.org/10.1007/978-3-030-11935-5_93

theory and technologies of vocational education", totaling two credits, 72 h is introduced into curricula of each postgraduate study program. The primary objective of the discipline is to help the future teacher develop his ability to elaborate independently a professional and pedagogical activity implying the solution of difficult professional tasks at a research university.

As the main result of the discipline mastering we have formulated a new professional competence: "Ability to the organization of the vocational education process from the perspective of the modern pedagogical science methodology, theory and technologies development".

During the development of the indicators and criteria evaluation competence we have marked its mastering levels:

Entry level: readiness for the analysis and evaluation of the professional education development tendencies, methodologies and design methods of the educational systems, scientific research, characteristics of innovative educational technologies, quality of professional education, active training methods and technologies;

Advanced level: readiness for understanding and development of training courses in the professional activity fields, generalization of the theoretical and empirical research results, preparation of the resource materials, development of the training active methods, design of the training technologies elements, quality management and measuring means of the educational process quality;

Excellent level: readiness for teaching activity on the basis of the developed courses in the professional activity fields, application of the educational technologies in education activities for the higher education programs, guiding of students' research work, elaboration of the teaching materials, textbook and training manuals, use of the quality diagnostics results in the improvement of teaching activity.

A content is structured on three sections, there are several subjects in each section. For example, the first section "Theoretical Bases of Vocational Education" consists of the following subjects: vocational education in the context of economic globalization; integration of the Russian Federation professional education modernization into the world educational space; scientific and pedagogical innovations in the educational activity; integration of the natural-science and humanities education; social and personal focused at the base of the Russian Federation educational system. The second section "Methodology of Vocational Education" includes the following subjects: specificities of formation and development, structure and functions of pedagogical methodology; development of pedagogy as a scientific system; research priority areas; methodological fundamentals of higher school pedagogy; competence-based focused training of specialists; teacher as a scientific and pedagogical unit; professional competence of the teacher; concept of professional education quality. The third section "Technologies of Vocational Education" consists of the following subjects: innovative technology as a basis of the multilevel educational process organization; essence and substantial characteristics of innovative educational technologies; principles, design algorithms and use of educational technologies in the university teaching and learning process.

Lecture and practical lessons are given with the use of modular, contextual training technologies elements and teaching units consolidation [4, 5]. Except the use of various multimedia training tools we practice supportive notes and didactic hand-out during lectures. Seminar classes are given in an active form.

So, for example, practical classes on the subject "Technologies of the Concentrated Training" are given in the "discussion in a low voice" method. After the lecture (with the use of the supportive notes) students receive a task: make the schedule of a concentrated educational process of one semester of bachelor training curriculum at the technological university. The group divides into three subgroups each of which, according to one of the concentrated training models, "discusses in a low voice" the version of the lesson schedule, and then reports and defends the project. Under the guidance of the teacher there is an analysis from the point of view of two positions: "Would these working methods fit me if I was a student?" and "Would these working methods fit me if I was a teacher?" Students have a real opportunity to see and feel advantages and disadvantages of the concentrated technology.

Independent work on discipline assumes the writing of reference paper, mini-report, essay, article and preparation for seminar [6, 7]. We will give several examples of learning activities according to the results of independent work.

"Pluses and minuses of the educational process intensification". The performance result takes the form of a speech on a round table. Students prepare the performances on a subject, carry out the analysis of intensified educational process in terms of efficiency, advantages and disadvantages. They draw conclusions about the efficiency of the teaching and educational process intensification in the system of vocational education. The performance is carried out in a classroom with the use of the Internet resources, observations, studying of the educational institutions experience.

"Industrial and educational technologies in general and particularly". Students have to prepare the essay with the justification of their problem vision, understanding of the concept "technology" transfer adequacy from the technics sphere in the education sphere.

"Examples of the innovative educational technologies used in the system of vocational education". The task is performed with use of a case-study method: the teacher offers a number of concrete educational technologies developed by the university teachers. Students have to examine them, evaluate the correspondence of some technologic components with design rules, analyse their implementation expediency.

The credit test assignment consists in development and public defence of a lecture presentation on a certain subject at the choice of the teacher and the student, for example:

Problems of the vocational education information technology.
Competences as a result of training quality in the system of vocational education.
Modernization of the higher education structure and content in the context of the Bologna agreement.
Integration of traditional and innovative technologies as a factor of multilevel educational process optimization of the higher education institution.
Distance learning as means of an educational process intensification.
Realization of the training active methods during the professional competences formation of future engineers.
Design of the main high school educational programs training.
Consistency of the high school teacher activity during the preparation for the lessons.

In our opinion, the organization of all types of lessons has to rely on the pedagogical stimulation as deliberate, purposeful influence on the requirements and the motivational sphere of graduate students: development of identity, involvement in activity, promotion of the fascinating perspective purposes, encouragement of real-life communication, consideration of real psycho-physiological opportunities, etc. Thus, the formation of the competence necessary for the future professional activity of the teacher-researcher is provided in the context of the research technological university.

References

1. Prikhodko, V.M.: What will be modern engineering education? Vysshee obrazovanie v Rossii [Higher Education in Russia]. No. 3, pp. 45–56 (2015)
2. Ivanov, V.G.., Sazonova, Z.S., Sapunov, M.B.: Engineering Pedagogy: Facing Typology Challenges. Vysshee obrazovanie v Rossii = Higher Education in Russia. No. 8/9 (215), pp. 32–42 (2017). (In Russ., abstract in Eng.)
3. Zhurakovskij, V.M., Vorov, A.B.: System innovations in model of education of engineers. Professional'noe obrazovanie. Stolica. [Prifesional Education. The capital]. No. 8. pp. 17–24 (2016)
4. Shageeva, F.T., Ivanov,V.G.: Contemporary technologies for training future chemical engineers. In: 16th International Conference on Interactive Collaborative Learning, ICL, Article number 6644546 (2013)
5. Shageeva, F.T., Gorodetskaya, I.M.: EdTech competence of engineering university professors: Research on pshychology and education. In: Proceedings of 2015 International Conference on Interactive Collaborative Learning, ICL 2015. Taly, pp. 177–180 (2015). https://doi.org/10.1109/icl.2015.7318021
6. Valeeva, E.E., Kraysman, N.V.: The impact of globalization on changing roles of university professors. In: 43 IGIP International Conference on Engineering Pedagogy. IGIP 2014, ICL 2014 (2014). https://doi.org/10.1109/icl.2014.7017901
7. Kraysman, N.V., Ziyatdinova, Yu.N., Valeeva, E.E: Advanced training in French with practical application in professional and scientific activities at KNRTU. In: 44 IGIP International Conference on Engineering Pedagogy. IGIP 2015, ICL 2015, pp. 1091-1092 (2015). https://doi.org/10.1109/icl.2015.7318183

Poster: Improving Skills for Teaching at an Engineering University

Farida T. Shageeva$^{(\boxtimes)}$ and Maria Suntsova

Kazan National Research Technological University, Kazan, Russia
faridash@bk.ru, emci2008@gmail.com

Abstract. This paper was inspired with discussions of the status, definition, and typology of engineering education. The ambiguous status of a professor at the engineering university as a teacher, researcher, and methodologist generates the following questions: How can all the above functions be balanced? How to secure the efficiency of educating activities without breaking their integrity? Considering these questions, we inevitably come to the issue of a professor's pedagogical skills and the necessity of finding the ways to develop and improve them. This paper deals with the essence and structure of pedagogical skills, which is really relevant for those teaching at an engineering university and, as a rule, having no basic educational background in pedagogy and psychology; as well as with the paths in developing and improving such skills. An advanced training program consisting of several modules has been proposed as one of potential options. We have also considered the contents of such modules and the features of their implementation at an engineering university.

Keywords: Engineering university professor · Pedagogical skills · Advanced training program

Article titled *Engineering Pedagogy: Facing Typology Challenges* [1] and published in one of the latest issues of the *Higher Education in Russia* touches on the matters of the unity and polarity of positions of a professor at an engineering university as a teacher, researcher, and methodologist. The authors invite the audience to discuss on the status of engineering pedagogy, as well as its definition and typology. However, we can see another aspect of the problem here. In his speech held at the Round Table within the Tatarstan Oil, Gas and Petrochemical Forum on September 7, 2017, on the problems of staffing enterprises, M.B Sapunov, one of the authors of the above article, emphasized that it is very hard for a professor, not to say 'impossible', to perform all his or her pedagogical activities with the equally high quality: Teaching as such, including the tasks of training, educating, personal development, and structuring the activities of both the professor and his or her students, scientific research, and the development of teaching aids. Questions: How can all the above functions be balanced? How to secure the efficiency of pedagogical activities without breaking their integrity? Hence, we inevitably come to the issue of a professor's pedagogical skills and the necessity of finding the ways to develop and improve them. In one of his lectures, Dr. A.A. Verbitsky asked: "Where would pedagogical skills come from without any training in pedagogy? Where can a potential be found for its development?"

© Springer Nature Switzerland AG 2019
M. E. Auer and T. Tsiatsos (Eds.): ICL 2018, AISC 917, pp. 981–986, 2019.
https://doi.org/10.1007/978-3-030-11935-5_94

The problem of pedagogical skills is very relevant for those teaching at an engineering university and usually having no basic educational background in pedagogy and psychology. However, they have wide experience in teaching and master pedagogical skills empirically [2–4]. Pedagogical skills are developed and become evident in activities, so the activity approach is applicable to considering their essence [5].

1. In the current psychological and pedagogical literature, the term of pedagogical skills is characterized by multiple and ambiguous interpretations, each of them being supported by a certain model of pedagogical activities, which is, in a way, justified and efficient in one socio-cultural situation or another [6–8]. A consistent concept of pedagogical skills as a required component within the professional competence of a teacher and as an independent academic course in a teachers training college was formulated and put into teachers training practice by Zyazyun [9]. Many of his ideas are quite applicable to the pedagogical skills of those teaching in higher education, since they involve professional approach and ability to solve a wide variety of pedagogical problems. We adhere to his definition of the term: "Pedagogical skills are a complex of the teacher's personal properties, which provides the high-level self-organizing of his or her professional activities and includes the following components: Humanistic personality orientation, professional knowledge, pedagogical ability, and technical educational skills".

2. *The humanistic personality orientation* of a professor means his or her aims, values, and interests inspired with the main idea—love to his or her students, taking their needs and interests into account, and selecting reasonable means and methods, as well as what can be called "good self-assertiveness" (it is important for a professor that his or her students peg him or her as a knowledgeable, experienced professional, an interesting person, a mentor, etc.). *Professional knowledge* is a combination of three knowledge components: The subject to be taught, the methods of teaching it, and the basics of pedagogy and psychology; this is the core component of pedagogical skills, characterized by its integrated and personalized nature. The stability of pedagogical skills is determined by both of these two components.

3. The possibility and speed of developing a professor's pedagogical skills depend on his or her *pedagogical abilities*. Researchers usually name a wide variety of them: Communication skills, perceptual abilities, creativity, emotional strengths (ability to manage one's own emotional state), personal dynamism (ability to manage a pedagogical situation), optimistic forecasting (the teacher's belief in his or her students' intellectual abilities), etc. [10, 11]. A component that is impossible to be just "known", but needs to be mastered or possessed is *technical educational skills*. They include both the competence in interacting within the educational process (it is commonly known that teaching comprises communication and interaction between a professor and his or her students in various forms) and ability to manage one's own behavior and emotional state, and elocution, as well as appearance and body language.

4. Students are trained for their pedagogical activities and skills during their studies in master's and doctor's degree courses [12, 13]. Further, in the course of their professional activities, educational experience, as well as their lecturing and methodological skills are developed. In combination with their research and

methodological activities, based on their abilities of self-development and self-actualization, the professional competence is developed in university professors.

5. There are different ways of improving the pedagogical skills in professors of an engineering university: Research activities and defending theses; attending specialized psychological and pedagogical workshops; participating in research-to-practice conferences at various levels and in tutorial workshops at their respective departments; giving demonstration lessons, including with the involvement of expert teachers; advanced training in higher education pedagogy, and training under specific programs aimed at improving pedagogical skills [14, 15].

From our point of view, a specific training program should be introduced that would directly involve the matters of developing and improving pedagogical skills, containing the following areas: Psychological and Pedagogical Basics of Teaching/Learning Process; A System of Improving Professors' Pedagogical Skills; Recommended Practices in Enhancing the Efficiency of a Lecture Course; Technologies of a Professor's Educational Activities; Didactic Grounding for Using Teaching Equipments; Pedagogic Bases of and Ways to Improve the Students' Individual Work Management, etc.

At the Department of Engineering Pedagogy and Psychology in Kazan National Research Technological University, the Professors' Pedagogical Skills advanced training program was developed, the main purpose of which is mastering professional competences securing the readiness of engineering university professors for efficient educational activities on the basis of systematized psychological and pedagogical knowledge and skills, as well as forming the image of a professor, developing the components of pedagogical skills, and mastering the elements of educational technical skills and acting skills in teaching.

The program includes a number of modules, the first of which considers the problems, such as higher education in the current context: Competence-based approach and key trends in the new paradigm of higher education; teacher as a subject of educational activities; pedagogical skills as a system: Definition, structure, and interrelation of its components; professionalism, skills, and competence of a professor: Interrelations and polarity; and the composition of professor's competences. Theory is reinforced at a practical lesson titled "Testing Pedagogical Abilities" with some discussion and developing a self-development program.

The next module is devoted to the psychological basics of professor's activities: Learning and cognitive activities of students and a student as the subject of learning activities; factors and laws of a student's personal development and age peculiarities of students; and upbringing activities of a professor: Techniques and innovations. The module supposes the independent acquisition of the theory using instructional materials distributed in advance and a number of class activities, such as a training session aimed at elaborating a generalized map of aims, motives, emotions, and states in learning activities; an active workshop titled "Interrelation of Teaching and Development"; and business game named "Vzglyad" (Look), in the course of which the experiences in and problems of education in different universities are identified at individual, microgroup, and group levels.

The third module is one of the most difficult ones—professional knowledge as a basic component of pedagogical skills: Didactics of higher education, methodology as specific didactics, didactics of laboratory classes, and modern educational technology. The materials are studied and reinforced at practical lessons: Basics of Methodology for Developing and Giving Class-Room Lessons of Various Types in Multilevel Education —with training elements, and Developing a Laboratory/Practical Session Using Educational Information Technology—with the elements of discussion. The modules culminates in the Active Teaching/Learning Forms and Methods As a Part of Educational Technology workshop in form of questions and answers, comments, and sharing trainees' experiences.

The next module is of an enduring interest for trainees, since it covers the problems of skills in the area of pedagogical interaction and considers advanced communication and interaction management psychotechnology, communication and interaction management mechanisms, techniques of establishing contact to the audience, communicative practices and techniques; cooperative pedagogy, subject-subject relationship between a teacher and a student, cooperative pedagogical communication style and communication techniques of a professor; professional ethics as an integral part of a professor's moral culture, business etiquette, and codes of conduct in relationships, such as professor—student, professor—colleagues, or professor—university management. The module cannot be mastered, of course, without proactive cognitive activities at the lessons, such as the Conflict Management in Pedagogic Interaction active training workshop and the Moral and Legal Culture of a Professor workshop in form of questions and answers, comments, and sharing experiences.

The Pedagogical Technical Skills module is perhaps the most interesting and popular with trainees, since it covers matters that are usually beyond the traditional beliefs about the importance of methodological and technological components of professor's preparations for his or her lessons. The trainees start understanding the importance of details, such as appearance, correct pronunciation, intonations, and body language of a professor.

The module includes topics, such as pedagogical technical skills as a form of organizing professor's behavior; social and perceptual qualities, such as attention, observation skills, and imagination; emotional state management and self-direction of a teacher; professor's image as a generalized character, and its typical features, such as competence, culture, lifestyle, and behavior; elocution; interrelation between acting and pedagogical skills; Stanislavsky Acting Method in pedagogic situations; and pedagogical artistry. These topics certainly need careful consideration at several workshops and trainings planned, such as the Emotional State Management practical class with training elements and developing affirmation programs; a workshop aimed at analyzing situations and solving problems of professor-students communication culture; the Basics of Body-Language Schematonics practical class with training elements; and trainings using the Art of Communication and Audience Impact case technology.

The program culminates in a round table titled "Developmental Paths of Pedagogical Skills." The trainees prepare their projects in advance. Then the professors of engineering universities come to certain conclusions while discussing the topic proactively and share their thoughts and plans for the future. The two-year experience in implementing the program has proven its being much in demand by and popular with the audience.

The authors are planning to develop the discussion and prepare a more detailed paper, probably Full Paper, in future, having supported the information on the outcomes with statistical data and more information on the process of improving teaching skills of engineering professors at technical universities.

References

1. Ivanov, V.G.., Sazonova, Z.S.., Sapunov, M.B.: Engineering Pedagogy: Facing Typology Challenges. Vysshee obrazovanie v Rossii=Higher Education in Russia. No. 8/9 (215), pp. 32–42 (2017). (In Russ., abstract in Eng.)
2. Medvedskaya, T.M., Soboleva, E.L.: Improvement of Educational Skills of the Technological University Teachers. Aktual'nye voprosy obrazovaniya [Topical Issues of Education]. No. 1, pp. 136–139 (2015). (In Russ., abstract in Eng.)
3. Minin, M.G., Benson, G.F., Belomestnova, E.N., Pakanova, V.S.: Lecturers Pedagogic Training at Engineering University. Vysshee obrazovanie v Rossii=Higher Education in Russia. No. 4, pp. 20–29 (2014). (In Russ., abstract in Eng.)
4. Lyubayeva, G.V.: Formation of Pedagogical Skills of Higher Engineering School Teachers. Al'manach sovremennoy nauki I obrazovaniya (Almanac of Modern Science and Education). No. 4, pp. 132–136 (2008). (In Russ.)
5. Khidoyatova, M.A.: Features of Pedagogical Activity and Pedagogical Skill of University Teachers. Aktual'nye problemy gumanitarnych i socialno-ekonomicheskih nauk (Actual Problems of Humanitarian and Social - Economic Sciences). No. 6, pp. 78–79 (2012). (In Russ.)
6. Sergievich, A.A., Batcevich, A.E.: The Problem of Formation of University Teacher Pedagogical Skills]. Vestnik Omskogo universiteta (Herald of Omsk University). No. 2, pp. 191–193 (2014). (In Russ., abstract in Eng.)
7. Dremova, N.B.: Raising Skill Level of University Pedagogical Staff. Vysshee obrazovanie v Rossii=Higher Education in Russia. No. 1, pp. 117–120 (2010). (In Russ., abstract in Eng.)
8. Dremova, N.B., Konoplya, A.I.: Formation of Pedagogical Skills of Higher School Teachers. Vysshee obrazovanie v Rossii=Higher Education in Russia. No. 1, pp. 127–132 (2015). (In Russ., abstract in Eng.)
9. Zyazuyn, I.A.: Osnovy pedagogicheskogo masterstva (Basics of Pedagogical Excellence), p. 358. Prosveshenie, Moscow (1989)
10. Nechayeva, M.S.: Personal and Professional Qualities as an Indicator of Pedagogical Mastery of a University Teacher. Vestnik Voronegskogo gosudarstvennogo universiteta. Seriya: Problemy vysshego obrazovaniya (Proceedings of Voronezh State University. Series: Issues of Hugher Education). No. 4, pp. 65–69 (2015). (In Russ., abstract in Eng.)
11. Khidoyatova, M.A.: The Nature and Content of the Pedagogical Skills Instructor. Aktual'nye problemy gumanitarnych i socialno-ekonomicheskih nauk (Actual problems of humanitarian and social and economic sciences). No. 7–2, pp. 157–158 (2013). (In Russ.)

12. Shageeva, F.T., Bogoudinova, R.Z.: Professional-Pedagogical Training of Post-graduate Students at Research University. Kazanskaya nauka (Kazan Science). No. 10, pp. 181–183 (2016). (In Russ., abstract in Eng.)
13. Muratova, E.I., Popov, A.I., Rakitina, E.A.: Technology of Formation of Readiness of Post-graduates for Lecturing Activity. Alma mater (Vestnik vysshey shkoly) (Alma mater (High school Herald)). No. 1, pp. 52–59 (2017). (In Russ., abstract in Eng.)
14. Ivanova, K.E., Chemerilova, I.A.: About Self-Development and Formation of Professional Competence of Higher School Teachers. Vestnik Kazanskogo technologicheskogo universiteta (Herald of Kazan Technological University). No. 17, pp. 225–228 (2014). (In Russ., abstract in Eng.)
15. Sucovatitsin, N.A.: Pedagogical Skills and the Ways to Formation. Voenniy nauchno-prakticheskiy vestnik (Herald of Military Scientific and Practical). No. 2, pp. 116–121 (2015). (In Russ.)

Author Index

Printed in the United States
By Bookmasters